Fundamentals of Linear Electronics
Integrated and Discrete

2nd edition

James Cox
Daytona Beach Community College

DELMAR

THOMSON LEARNING™

Australia Canada Mexico Singapore Spain United Kingdom United States

DELMAR

THOMSON LEARNING

Fundamentals of Linear Electronics: Integrated and Discrete

by

James Cox

Business Unit Director:
Alar Elken

Executive Editor:
Sandy Clark

Senior Acquisitions Editor:
Gregory L. Clayton

Developmental Editor:
Michelle Ruelos Cannistraci

Editorial Assistant:
Jennifer A. Thompson

Executive Marketing Manager:
Maura Theriault

Channel Manager:
Mona Caron

Marketing Coordinator:
Karen Smith

Executive Production Manager:
Mary Ellen Black

Production Manager:
Larry Main

Senior Project Editor:
Christopher Chien

Art & Design Coordinator:
David Arsenault

Library of Congress Cataloging-in-
Publication Data

Cox, James F., 1945–
   Fundamentals of linear electronics :
   integrated and discrete / James Cox.—
   2nd ed.
     p. cm.
   Includes index
   ISBN 0-7668-3018-7 (alk. paper)
   1. Linear integrated circuits. I. Title.
TK7874.C744   2001
621.3815—dc21   2001028356

NOTICE TO THE READER

Publisher does not warrant or guarantee any of the products described herein or perform any independent analysis in connection with any of the product information contained herein. Publisher does not assume, and expressly disclaims, any obligation to obtain and include information other than that provided to it by the manufacturer.

The reader is expressly warned to consider and adopt all safety precautions that might be indicated by the activities herein and to avoid all potential hazards. By following the instructions contained herein, the reader willingly assumes all risks in connection with such instructions.

The publisher makes no representation or warranties of any kind, including but not limited to, the warranties of fitness for particular purpose or merchantability, nor are any such representations implied with respect to the material set forth herein, and the publisher takes no responsibility with respect to such material. The publisher shall not be liable for any special, consequential, or exemplary damages resulting, in whole or part, from the readers' use of, or reliance upon, this material.

To my wife, Betty Lynn,
whose love, patience and help made this book possible.

Contents

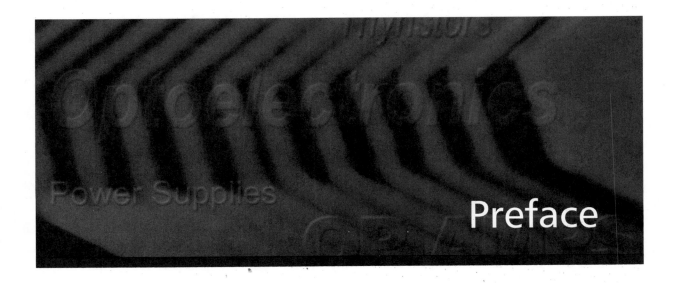

Preface

INTENDED AUDIENCE

Fundamentals of Linear Electronics is designed for Solid State Devices and Linear Electronics courses within two- and four-year Electronics Technology and Electronics Engineering Technology programs.

TEXTBOOK ORGANIZATION

The text is divided into two sections. Section I covers components that are used to build electronics circuitry, while Section II covers electronics circuitry that is used to build electronics systems.

Section I—Devices and Op-Amps includes Chapters 1 through 10 and covers the building blocks of discrete devices. The information concerning discrete devices has been reduced when compared to many texts. This reduced coverage of discrete devices is by design and mirrors the needs of the electronics industry. Today's technician needs to understand the fundamentals of discrete devices but does not need to spend hours of mathematical analysis of discrete circuits that are no longer used in the electronics industry. Chapters 9 and 10 provide detailed coverage of the op-amp that is the building block of most modern circuitry. Chapter 9 analyzes the ideal op-amp, Chapter 10 goes into limitations the technician should understand about op-amps.

Section II—Subsystems includes Chapters 11 through 21 and focuses on systems and electronics circuitry. Chapters 11 through 14 cover a variety of circuitry including active filters and oscillators that use the standard op-amp as the basic building block. Starting with Chapter 15, special ICs and components are covered that are used in up-to-date electronic systems. Topics in Chapters 15 through 19 cover differential amplifiers, voltage regulators, analog-to-digital converters, digital-to-analog converters, power amplifiers, and phase-control circuits using SCRs and triacs. Chapter 20 is an introduction to optoelectronics, including laser diodes and fiber optics. The last chapter, Chapter 21, covers transducers and actuators, including temperature sensors, pressure sensors, relays, and motors.

FEATURES

- Integrated pre-labs provide opportunities to solve circuit problems similar to hardware lab experiments. The students are expected to solve these pre-lab exercises using a calculator and, if possible, verify their results using MultiSIM™ or other electronic analysis software. Note that all pre-lab circuits have been verified using MultiSIM.

- Integrated hardware labs demonstrate practical use of the circuit.

- The pre-labs, lab experiments, questions, and problems at the end of each chapter are perforated pages so they can be easily removed for evaluation. A list of all components used in the lab experiments is contained in Appendix A along with the address of several mail-order companies that can supply the components. Data sheets for all components used in the experiments are contained in Appendix B. The last appendix, Appendix C, contains answers to the odd-numbered questions and problems found at the end of each chapter.

- Chapters on optoelectronics, transducers, data conversion, and filters illustrate circuits showing practical applications.

- The use of numerous illustrations and examples helps the user analyze schematics and predict voltage readings and signal waveforms.

- There is a balance between a systems and a components approach; the first half of the text covers building blocks and devices, and the second half focuses on subsystems and how inputs and outputs are related.

- The text provides detailed coverage of op-amps, differential amplifiers, thyristors, sensors, power supplies, and special ICs.

- Computer simulation using MultiSIM is an optional section covered in Chapter 1.

- The CD includes pre-created MultiSIM files to help students solve circuit problems.

NEW TO THIS EDITION

- All pre-labs have been modified so they can be easily verified using MultiSIM to help prepare students for hardware labs. An optional section introducing MultiSIM is added to Chapter 1.

- Reduced coverage of discrete circuitry permits more study time for integrated circuitry.

- Addition of second color allows for better readability.

USING THE CD-ROM

The accompanying CD includes MultiSIM circuit files and the textbook version of MultiSIM. Students can use these pre-created files for troubleshooting and simulation. MultiSIM files were created from the pre-labs. The Student Edition of MultiSIM may be purchased through your college bookstore or through www.electronictech.com.

SUPPLEMENTS

The complete ancillary package was developed to achieve two goals:

1. To assist students in learning the essential information needed to prepare for the exciting field of electronics.
2. To assist instructors in planning and implementing their instructional programs for the most efficient use of time and other resources.

Instructors' Guide

The Instructor's Guide contains solutions to end-of-chapter textbook questions, problems, pre-labs, and labs. (ISBN: 0-7668-3019-5)

e.resource

The *e.resource* is an educational resource that creates a truly electronic classroom. It is a CD-ROM containing tools and instructional resources that enrich your classroom and make your preparation time shorter. The elements of *e.resource* link directly to the text and tie together to provide a unified instructional system. With the *e.resource,* you can spend your time teaching, not preparing to teach. (ISBN: 0-7668-3020-9)

Features contained in the *e.resource* include:

- **PowerPoint Presentation.** These slides provide the basis for a lecture outline that helps you to present concepts and material. Key points and concepts can be graphically highlighted for student retention.

- **Computerized Testbank.** Over 900 questions in multiple-choice, true/false, short answer, and completion format so you can assess student comprehension.

- **Image Library.** Over 200 images from the textbook to create your own transparency masters or to customize your own PowerPoint slides. The Image Library comes with the ability to browse and search images with key words and allows quick and easy use.

Online Companion

The text has a companion website at www.electronictech.com, which will have high appeal to both educators and students. The Online Companion provides access to text updates, online quizzes, RealAudio broadcasts, and more.

ACKNOWLEDGMENTS

The author and Delmar would like to thank the following reviewers for their feedback:

- Russell Bowker, Northeast Community College, Norfolk, NE
- Robert Diffenderfer, DeVRY Institute of Technology, Kansas City, MO
- Phil Golden, DeVRY Institute of Technology, Irving, TX
- A. Kent Johnson, Brigham Young University, Provo, UT
- Kenneth Lowell, Brown Institute, Mendota Heights, MN
- Leei Mao, Greenville Tech, Greenville, SC

- Daniel Metzger, Monroe College, Monroe, MI
- Robert Peeler, Lamor State College, Port Arthur, TX
- Michael Miller, DeVRY Institute of Technology, Phoenix, AZ
- Jim Pannell, DeVRY Institute of Technology, Irving, TX
- Previn Raghuwanshi, DeVRY Institute of Technology, N. Brunswick, NJ
- Mike Sanderson, DeVRY Institute of Technology, Columbus, OH
- Jim Stewart, DeVRY Institute of Technology, N. Brunswick, NJ
- Jay Templin, DeVRY Institute of Technology, Kansas City, MO
- Chia-chi Tsui, DeVRY Institute of Technology, Long Island City, NY
- Manfred Ueberschaer, DeVRY Institute of Technology, Decatur, GA

SECTION I

Devices and Op-Amps

Fundamentals of Linear Electronics is comprised of two major sections. Section I includes Chapters 1 through 10 and discusses the devices that are used to build electronic circuits. Section II includes Chapters 11 through 21 and examines electronic circuits and subsystems used to build electronic systems.

Resistors, capacitors, and inductors are basic components studied in electrical fundamentals. Section I will cover diodes, transistors, and op-amps. All three of these are important devices used to construct electronic circuits. The diode junction is fundamental to all electronic devices; therefore, the text starts with a detailed study of diodes, diode applications, and special diodes. Next, the text covers bipolar and field effect transistors. Basic circuits using transistors as switches and as amplifiers are discussed. Op-amps are introduced at the end of Section I. The op-amp, an integrated circuit containing numerous transistors, has become a standard building block for many electronic circuits. Basic op-amp switching and amplifying circuits are covered. The limitations of op-amps are also addressed.

Engineers use the devices studied in Section I to design electronic subsystems. In Section II, electronic subsystems will be examined. All electronic systems are comprised of subsystems, and most of the circuitry that comprises these subsystems will be covered in Section II.

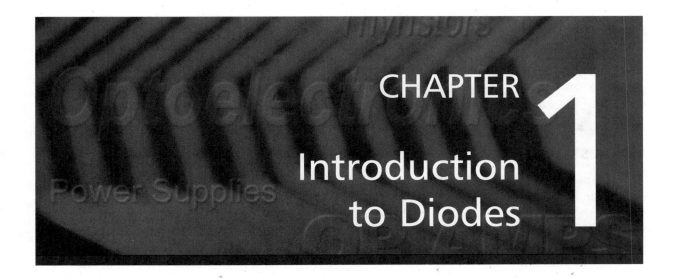

CHAPTER 1

Introduction to Diodes

OBJECTIVES

After studying the material in this chapter, you will be able to describe and/or analyze:

○ the function of diodes,

○ the internal operation of the diode,

○ three diode models,

○ computer circuit analysis, and

○ MultiSIM software.

CONVENTIONAL VERSUS ELECTRON CURRENT FLOW

Which way does current flow? If you are an engineering student, you will answer positive to negative, because you are using conventional current flow terminology. On the other hand, if you are a technology student, you may answer negative to positive because you are using electron current flow terminology.

In Figure 1–1, the current flows through resistor R2, causing the top of R2 to be positive. Notice, I have just defined the current flowing through R2 and everyone is happy. Those who think in terms of electron current flow believe the current is flowing up, while those who abide by the conventional current flow terminology believe the current is flowing down. In most cases, there is no need to show current arrows, and without arrows, the reader can use electron or conventional current flow. If there is a need to show current arrows, the word electron or conventional will precede the word current. Since there are only a few places in this text that contain current arrows, this approach adds little confusion. The majority of technicians (60% to 80%) use the electron flow terminology. Almost all engineers, however, use the conventional flow terminology. Since you will be working with engineers and reading technical material written by engineers,

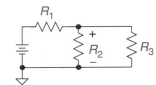

FIGURE 1–1   Serial/parallel circuit

you should know both options. It is hoped the approach used in this text will make it easy for you to read technical literature using either type of current flow terminology.

1.1   INTRODUCTION TO DIODES

DIODE SYMBOL AND TWO RULES OF OPERATION

This chapter is an introduction to the function and operation of diodes in circuits. The diode is one of the building blocks of electronic circuits, and the understanding gained by studying this chapter will permit you to analyze and troubleshoot numerous circuits using diodes. In addition, your understanding of diodes will serve as a foundation for understanding other devices covered in later chapters.

Before we examine the internal workings of the diode, let us look at the function of the diode and analyze a simple diode circuit. The diode is a two-terminal device that permits current flow in only one direction. One terminal is called the *anode*, and the other terminal is called the *cathode*. Figure 1–2 shows the diode symbol. The anode is the terminal connected to the base of the arrow. The cathode is the terminal connected to the bar.

FIGURE 1–2   Schematic symbol of a diode

Remember these two important rules about diodes:

1. **A diode permits current flow when the voltage on the anode is positive with respect to the voltage on the cathode.**
2. **A diode inhibits current flow when the voltage on the anode is negative with respect to the voltage on the cathode.**

A diode is called *forward biased* when the anode is positive with respect to the cathode. A forward-biased diode functions like a closed switch permitting current flow. A diode is called *reverse biased* when the anode is negative with respect to the cathode. A reverse-biased diode functions like an open switch inhibiting current flow.

THE DIODE IN A BASIC CIRCUIT

Figure 1–3(a) shows a battery, forward-biased diode, and bulb. Let us calculate the meter readings in the circuit. Since the diode is forward biased, it can be replaced with a closed switch as shown in Figure 1–3(b). Solving for the voltage and current meter readings is now a simple Ohm's law problem as shown in Example 1.1.

EXAMPLE 1.1

> **Step 1.** Determine the biased condition of the diode. Examine the circuit in Figure 1–3(a). Note that the anode of the diode is positive with respect to the cathode. The diode is, therefore, forward biased.

EXAMPLE 1.1 continued

> **Step 2.** Draw the equivalent circuit. Figure 1–3(b) shows the equivalent circuit. The forward-biased diode has been replaced with a closed switch.
>
>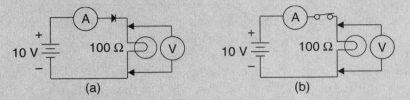
>
> (a) (b)
>
> **FIGURE 1–3** The forward-biased diode in a simple circuit
>
> **Step 3.** Use Ohm's law to calculate the current through the bulb in the equivalent circuit.
>
> $$I = V/R = 10 \text{ V}/100 \text{ }\Omega = 100 \text{ mA}$$

Figure 1–4(a) shows a battery, reverse-biased diode, and bulb. Example 1.2 shows how to calculate the meter readings in the circuit.

EXAMPLE 1.2

> **Step 1.** Determine the biased condition of the diode. Examine the circuit in Figure 1–4(a). Note that the anode of the diode is negative with respect to the cathode. The diode is, therefore, reverse biased.
>
>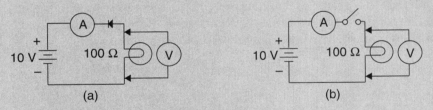
>
> (a) (b)
>
> **FIGURE 1–4** The reverse-biased diode in a simple circuit
>
> **Step 2.** Draw the equivalent circuit. Figure 1–4(b) shows the equivalent circuit. The reverse biased diode has been replaced with an open switch.
>
> **Step 3.** Use Ohm's law to calculate the current through the bulb in the equivalent circuit. Since the reverse-biased diode acts like an open circuit, it has an infinite resistance.
>
> $$I = V/R = 10 \text{ V}/\infty\Omega = 0 \text{ mA}$$
>
> **Step 4.** Use Ohm's law to calculate the voltage across the bulb in the equivalent circuit. The circuit is open, and the current flow is zero ($I = 0$ mA); therefore, the voltage across the bulb would also equal zero.
>
> $$V = I \times R = 0 \text{ mA} \times 100 \text{ }\Omega = 0 \text{ V}$$

It should be noted that the voltage across the diode will equal the value of the power source (10 V in Example 1.2).

1.2 INSIDE THE DIODE

SEMICONDUCTOR MATERIAL

In basic electronics, you learned that the ability of a material to support current flow is dependent upon its atomic makeup. Good conductors have one electron in the outer ring called the *valence ring*. This electron can be easily influenced by an energy force to become free to support current flow. Most metals are good conductors of electricity. Good insulators, on the other hand, have a full outer valence ring. Because of this, it is difficult to free these electrons from the atom. Glass, plastic, and rubber are examples of materials that are good insulators.

As one might expect, the name *semiconductor* is given to materials that lie between these two extremes: *conductors* and *insulators*. Silicon, germanium, and carbon are examples of materials that are semiconductors and contain four electrons in the valence ring. While valence electrons in semiconductor materials are more tightly bound to the atom than the valence electrons in conductor material, they are not nearly as tightly bound as the valence electrons in insulating material. Figure 1–5 shows the atom and the valence ring for the three types of materials. Diodes are made from semiconductor materials and are called *semiconductor devices*.

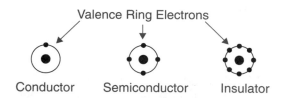

FIGURE 1–5   Valence ring of atoms

Figure 1–6 shows the effect of connecting a battery across a conductor, insulator, and semiconductor. The conductor has a large electron current flow, the semiconductor has a smaller electron current flow, and the insulator has no electron current flow.

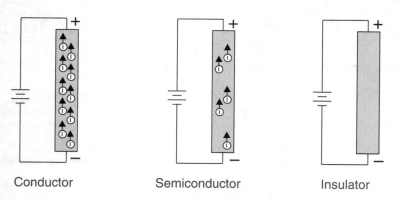

FIGURE 1–6   Current flow through materials

MAKING THE DIODE

The first step in manufacturing semiconductor diodes is to refine and process silicon into a pure crystalline form as shown in Figure 1–7. In this crystalline form, all the silicon atoms are arranged symmetrically, and a sharing action takes place among the valence electrons. Each atom has four of its own valence electrons, but in addition, the atom shares one electron from

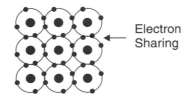

FIGURE 1–7 Covalent bonding

each of its four neighbors. The net effect is that each atom in the crystalline structure acts as if it has eight electrons in its valence ring. In other words, the valence ring appears to be full. This sharing of valence electrons is called *covalent bonding*. Because of this covalent bonding, the electrons are tightly bound in the silicon crystal, which causes the material to exhibit the characteristics of an insulator. The word *intrinsic* is used to describe semiconductor material in its pure crystalline form.

Now imagine what would happen if, during the process of forming the pure crystalline structure, a small number of atoms with five electrons in the valence ring (pentavalent) were incorporated into the crystalline structure. The result would be that four of the five valence electrons would combine with neighboring atoms in covalent bonding. But the fifth electron would be left out of the covalent bonding structure as shown in Figure 1–8(a). This extra electron is not tightly bound in the crystalline structure and is free to support current flow.

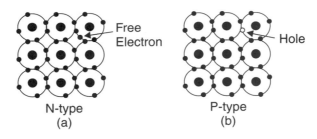

FIGURE 1–8 Doped semiconductor material

The process of adding a pentavalent element to the crystalline structure is called *doping with donor atoms*. The resulting material is called *N-type semiconductor material*. Arsenic (As) is an example of a donor element that is often used for doping N-type material. An important fact to remember is that the name *N-type material* is derived from the fact that current supporting particles are negatively charged electrons, but the N-type material itself has an overall neutral polarity. This is because for every extra electron added during the doping process, there was also a proton added.

Next imagine what would happen if, during the process of forming the pure crystalline structure, a small number of atoms with three electrons in the valence ring (trivalent) were incorporated into the crystalline form. This is called *doping with acceptor atoms*. The three valence electrons combine into a covalent bond with neighboring atoms as shown in Figure 1–8(b). The resulting material is called P-type semiconductor material. Gallium (Ga) is an example of an acceptor element that is often used for doping P-type material. The crystalline structure is missing one electron to complete the covalent bonding. This missing electron forms a hole in the covalent bonding structure. The hole acts like a positive current carrier. The name *P-type material* is derived from the fact that current supporting particles are positively charged holes, but the material itself has an overall neutral polarity. The overall charge of the P-type material is neutral because each atom added with one less electron also had one less proton. The word *extrinsic* is used to describe silicon crystalline material that has been doped.

Figure 1–9 shows the current flow through the three types of semiconductor materials when a battery is connected across the materials. Figure 1–9(a) shows there are no current carriers available to support current flow in the pure undoped crystalline silicon. Figure 1–9(b) shows that electrons support current flow in the N-type silicon. Since the electrons are negatively charged particles, the electrons flow from negative to positive. Figure 1–9(c) shows that holes support current flow in the P-type silicon. Since holes are positively charged particles, the holes flow from positive to negative.

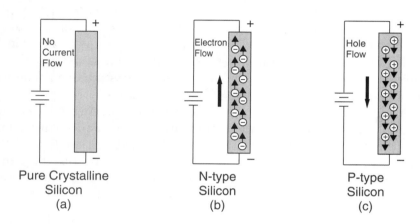

FIGURE 1–9   Current flow through different types of semiconductors

Both N-type and P-type materials are needed to make a diode. A diode is made from a single bar of semiconductor material containing a P-type material at one end and an N-type material at the other end. The place where the two types of silicon meet is called a *PN junction*. As shown in Figure 1–10, the terminal connected to the P-type material is called the *anode*, and the terminal connected to the N-type material is called the *cathode*.

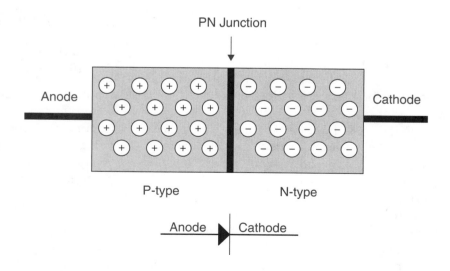

FIGURE 1–10   The PN junction

REVERSE-BIASED PN JUNCTION

Figure 1–11 shows the effect of connecting the positive terminal of a battery to a cathode (N-type material) and connecting the negative terminal of the battery to the anode (P-type material). Note that the positive current carriers in the P-type material are attracted away from the junction by the negative charge. Likewise, the negative current carriers in the N-type material are attracted away from the junction by the positive charge. This leaves an area near the junction depleted of current carriers. This area is called the *depletion region*. The depletion region behaves like an insulator, which causes the reverse-biased diode to act like an open switch. When the anode is negative with respect to the cathode, the diode is reverse biased, and no current flows through the diode.

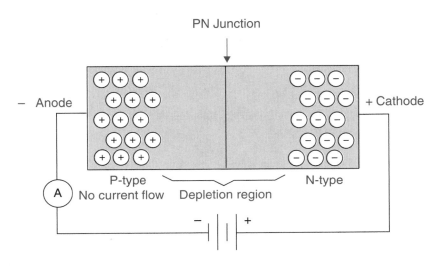

FIGURE 1–11   Reverse-biased PN junction

Actually, the reverse-biased diode will have a small leakage current. The majority current carriers in N-type material are electrons, but the N-type material will contain a few holes that are minority current carriers. The majority current carriers in P-type material are holes, but the P-type material will contain a few electrons that are minority current carriers. Minority carriers are mainly caused by the thermal breakdown of covalent bonding. When a junction is reverse biased for majority carriers, it is forward biased for minority carriers, and there will be minority current flow that is called *leakage current*. The leakage current is so small for silicon diodes that it is normally considered negligible.

FORWARD-BIASED PN JUNCTION

Figure 1–12 shows the effect of connecting the positive terminal of a battery to the anode (P-type material) and the negative terminal of the battery to the cathode (N-type material). Note that the positive current carriers in the P-type material are pushed toward the junction by the positive charge. Likewise, the negative current carriers in the N-type material are pushed toward the junction by the negative charge.

At the junction, the negative current carriers (electrons) cross over the junction and combine with the positive current carriers (holes). When a negative and positive current carrier combine, their charges cancel; they become electrically neutral and, therefore, disappear as current carriers.

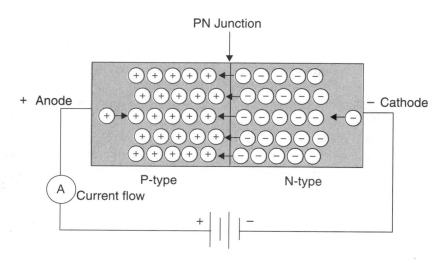

FIGURE 1–12   Forward-biased PN junction

The combining of an electron and hole causes the N-type material to lose a negative charge. The lost negative current carrier is quickly replaced by an electron from the negative terminal of the battery. The positive current carrier lost in the P-type material is quickly replaced by the positive terminal of the battery. The battery attracts an electron from the P-type material, leaving a hole that is pushed toward the junction. This action causes continuous electron current to flow into the cathode and continuous electron current to flow out of the anode.

The diode acts like a conductor or a closed switch. When the anode is positive with respect to the cathode, the diode is forward biased, and current flows through the diode. Actually, a diode is not a perfect conductor when forward biased. In order to cause electrons to cross the PN junction, a small voltage drop called the *barrier voltage* must be overcome by an external voltage source. For silicon diodes, the barrier voltage is approximately 0.7 V. Diodes made of other semiconductor materials have different forward voltage drops; for example, germanium and Schottky diodes have a barrier voltage of approximately 0.3 V, and light-emitting diodes (LEDs) have a barrier voltage drop in the range of 1.5 V to 3 V. In this text, unless noted differently, all diodes will be assumed silicon and, therefore, have a forward voltage drop of 0.7 V.

1.3   THREE DIODE MODELS

It is often useful to simplify diodes when analyzing circuits. For this reason, three models can be used to represent a diode in a circuit: the ideal diode model, the practical diode model, and the detailed diode model.

IDEAL DIODE MODEL

Figure 1–13 shows a voltage versus current graph of the ideal diode model. The anode voltage is plotted on the horizontal axis, and the current through the diode is plotted on the vertical axis. By examining the graph, it can be seen that when the diode is forward biased, current flows through the diode, and the voltage across the diode is zero. Therefore, the ideal forward-biased diode acts as a closed switch with zero resistance. In the reverse-biased area of the graph, it can be seen that the current through the diode is zero. In other words, the ideal reverse-biased diode has an infinite resistance and acts as an open switch.

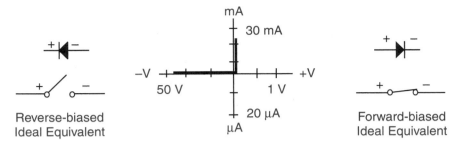

FIGURE 1–13 Graph of ideal diode model

PRACTICAL DIODE MODEL

Up to this point, we have assumed the forward-biased diode acts like a perfect switch. Actually, the forward-biased diode has a small voltage across its terminals. The voltage drop is approximately 0.7 V for silicon diodes. Figure 1–14 shows a forward-biased diode equivalent circuit for a practical diode model. The equivalent circuit consists of an ideal diode and a 0.7 V source that opposes the external supply. The voltage versus current graph in Figure 1–14 shows that, as the current increases, the voltage across the diode stays constant at 0.7 V. In many applications, the practical diode model is best suited for analyzing circuits. In reverse bias, the practical diode model assumes the diode to be equivalent to an open switch.

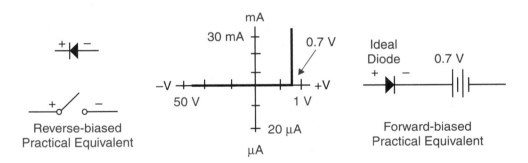

FIGURE 1–14 Graph of practical diode model

DETAILED DIODE MODEL

The graph in Figure 1–15 shows that in reverse bias, the diode does have a small current flow. This small current is called *leakage current*. The amount of leakage current flow is a function of the number of minority carriers present. Minority carriers are mainly caused by thermal breakdown of the covalent bonding. Therefore, the amount of leakage current is directly related to the temperature of the PN junction. At normal operating temperatures, the leakage current for a silicon diode is normally in the nanoamps and, therefore, the reverse-biased resistance is in megohms. The equivalent circuit for a detailed diode model in reverse bias is shown in Figure 1–15 as a large resistor in parallel with an open switch. It is important to remember that in most cases at normal operating temperatures, the resistance of a reverse-biased diode can be considered to be an open. However, at high operating temperatures, the resistance of the reverse-biased diode drops and could affect the circuit operation.

Figure 1–15 shows the voltage versus current graph for the detailed diode model. Note that in the forward-biased area, current gradually increases between zero volts and 0.7 V. When the voltage reaches approximately 0.7 V, the slope of the curve increases greatly but it does not go totally vertical. The equivalent circuit of the forward-biased detailed diode model, shown in

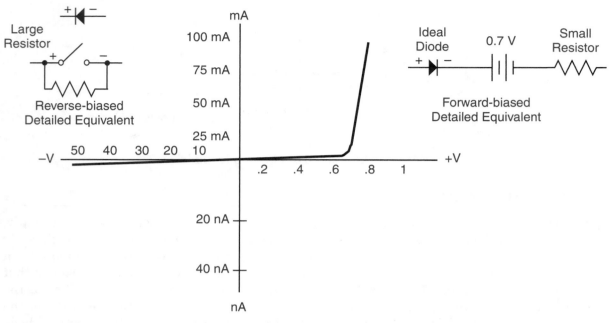

FIGURE 1–15   Graph of detailed diode model

Figure 1–15, consists of an ideal diode, a 0.7 V battery, and a small resistor all connected in series. As the current through the diode increases, the voltage drop across the small resistor increases. This explains why the voltage increases slightly when current increases through the diode.

The detailed diode model is useful in examining the resistance of a forward-biased diode junction. There are two methods of calculating the resistance of a device. One method is used when calculating the static resistance (DC) and the other method is used when calculating the signal resistance (AC). The static resistance of a diode junction is simply equal to the voltage across the junction divided by the current through the junction. In forward bias, the current increases greatly with small changes in voltage. In fact, the voltage across the forward-biased junction stays relatively constant at 0.7 V. Hence, as the forward-biased current goes up, the static resistance of the diode goes down.

The dynamic resistance is the opposition a diode junction will present to an AC signal riding on a DC forward-biased current. The dynamic resistance of the junction is dependent upon the DC bias current. Therefore, the dynamic resistance can be adjusted by controlling the bias current. Figure 1–16(a) shows a 2.7 V battery driving a resistor (R) in series with a diode. The DC bias current flowing through the diode is a function of the battery voltage and the size of the resistor. By plotting a DC load line for a resistor on the same graph of the diode curve, the bias current can be determined.

To find the end points for the DC load line, first assume the diode is an open and calculate the current through the diode and the voltage across the open diode. If the diode is an open, the current through the diode will equal zero amps and the voltage across the diode will equal the power supply voltage (2.7 V in this case). Therefore, one end of the DC load line is located at zero amps and the power supply voltage of 2.7 V. The second point on the DC load line is calculated by assuming the diode is a short and then calculating the voltage across the diode and the current through the diode. The voltage across a shorted diode is zero volts. The current through the shorted diode is a function of the resistor and power supply in series with the diode. Dividing power supply voltage by the resistance of the series resistor results in the current that will flow through the shorted diode. Figure 1–16(b) shows the results of plotting two DC load lines: one for a 2 kΩ resistor and the second for a 4 kΩ resistor. Notice the starting point is the

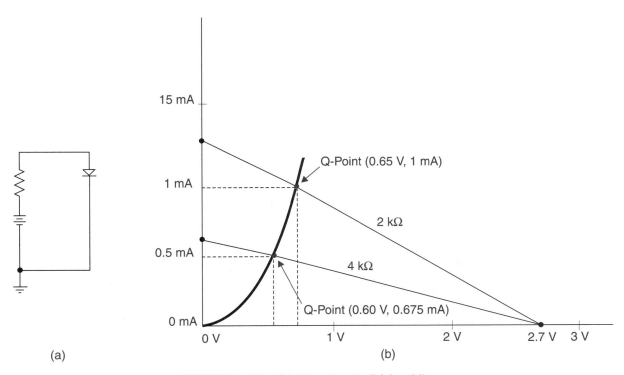

FIGURE 1–16 (a) Bias circuit; (b) load lines

same for both load lines on the horizontal axis; however, the slope of the load lines depends on the resistor value. The end point on the vertical axis for the 2 kΩ resistor is 1.35 mA (2.7 V/2 kΩ) and the end point on the vertical axis for the 4 kΩ resistor is 0.675 mA (2.7 V/4 kΩ).

Where the load line intersects the diode curve is the Q-point (Quiescent Operating Point). The Q-point is the DC bias point of a circuit; the dynamic signal (AC signal) will be superimposed on this bias point. Changing the series resistor or changing the power supply voltage can control the location of the Q-point on the diode curve. In Figure 1–16(b) there are two load lines and two Q-points.

The dynamic resistance of a diode junction at a Q-point can be calculated by taking a small change in voltage and dividing it by the change in current caused by the change in voltage ($r_d = \Delta v/\Delta i$). The delta symbol (Δ) is used to denote change, and lowercase letters are used to designate dynamic signal values. The steeper the slope of the diode curve at the Q-point, the lower the dynamic resistance. Notice the diode curve has a much steeper slope at the Q-point on the 2 kΩ load line than at the Q-point on the 4 kΩ load line. While the dynamic resistance could be calculated by using a graph, this is seldom done because accurate graphs are not normally available. Because of this, the noted semiconductor physicist, Dr. William Shockley, developed an approximation formula that states that dynamic junction resistance is equal to 25 mV divided by the forward-biased current through the diode junction ($r_d = 25$ mV/I).

The technician does not normally need to be concerned with the dynamic resistance of diodes. However, as we expand our study to transistors, the dynamic resistance of the PN junction inside the transistor will be important. Example 1.3 shows how controlling the Q-point can change the dynamic resistance. It also demonstrates how an AC signal can be superimposed on a DC bias.

EXAMPLE 1.3

> **Step 1.** Calculate the signal at TP_1 for the circuit in Figure 1–17(a) if (a) $R_2 = 2$ kΩ and
> (b) $R_2 = 4$ kΩ.

EXAMPLE 1.3 continued

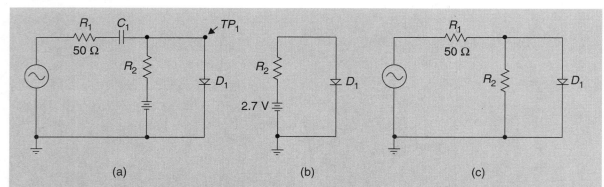

FIGURE 1–17   (a) Example 1.3; (b) DC equivalent; (c) Signal equivalent

(a) $R_2 = 2$ kΩ

Step 2. Calculate the DC forward-biased current (I_F). Remember the capacitor is an open for DC current, so Figure 1–17(b) is the DC equivalent circuit.

$$V_{R2} = 2.7 \text{ V} - 0.7 = 2 \text{ V} \quad \text{(Assume the diode has a 0.7 V drop.)}$$
$$I_F = V_{R2}/R_2 = 2 \text{ V}/2 \text{ k}\Omega = 1 \text{ mA}$$

Step 3. Calculate the dynamic resistance using Shockley's equation.

$$r_d = 25 \text{ mV}/I_F = 25 \text{ mV}/1 \text{ mA} = 25 \text{ }\Omega$$

Step 4. Use the voltage divider formula to calculate the signal voltage at TP_1. The capacitor and the DC power supply are shorts for the signal current; the signal equivalent circuit is, therefore, equal to the circuit shown in Figure 1–17(c). The parallel resistance of 2 kΩ and 25 Ω is approximately equal to 25 Ω.

$$V_{TP1} = V_{in} \times r_d/(R_1 + r_d) = 50 \text{ mV}_{\text{p-p}} \times 25/(25 + 50) = 16.7 \text{ mV}_{\text{p-p}}$$

Step 5. Draw the waveform at TP_1. (Read the bias voltage from the graph in Figure 1–16(b).)

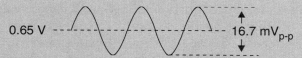

(b) $R_2 = 4$ kΩ

Step 1. Calculate the DC forward-biased current (I_F). Remember the capacitor is an open for DC current, so Figure 1–17(b) is the DC equivalent circuit.

$$V_{R2} = 2.7 \text{ V} - 0.7 = 2 \text{ V} \quad \text{(Assume the diode has a 0.7 V drop.)}$$
$$I_F = V_{R2}/R_2 = 2 \text{ V}/4 \text{ k}\Omega = 0.5 \text{ mA}$$

Step 2. Calculate the dynamic resistance using Shockley's equation.

$$r_d = 25 \text{ mV}/I_F = 25 \text{ mV}/0.5 \text{ mA} = 50 \text{ }\Omega$$

Step 3. Use the voltage divider formula to calculate the signal voltage at TP_1. The capacitor and the DC power supply are shorts for the signal current; the signal equivalent circuit is, therefore, equal to the circuit shown in Figure 1–17(c). The parallel resistance of 4 kΩ and 50 Ω is approximately equal to 50 Ω.

$$V_{TP1} = V_{in} \times r_d/(R_1 + r_d) = 50 \text{ mV}_{\text{p-p}} \times 50/(50 + 50) = 25 \text{ mV}_{\text{p-p}}$$

EXAMPLE 1.3 continued

> **Step 4.** Draw the waveform at TP_1 (Read the bias voltage from the graph in Figure 1–16(b).)
>
> 0.60 V --- 25 mV$_{p-p}$

CHOOSING THE CORRECT DIODE MODEL

When analyzing diode circuits, the first step is to decide which of the diode models should be used. Using the ideal model simplifies circuit analysis, and therefore, it is the logical choice whenever possible. If the 0.7 V dropped across the diode is negligible, the ideal model should be used.

The practical diode model is a good choice when the junction voltage of 0.7 V is large enough to be considered. This is often the case when analyzing circuits with low voltages.

The detailed diode model is hardly ever used in analyzing and troubleshooting circuits, although there are always exceptions. For example, we used the detailed model when discussing the dynamic resistance of a diode junction.

The two examples that follow will help you understand when the ideal diode model can be used and when the practical diode model should be used.

EXAMPLE 1.4

> Figure 1–18 shows a 3 V supply connected to a 100 Ω load through a forward-biased diode. The current for the circuit in Figure 1–18 is calculated twice below. In the first calculation, the 0.7 V dropped across the forward-biased diode is considered. In the second calculation, the 0.7 V is disregarded.
>
>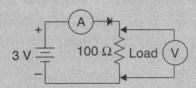
>
> **FIGURE 1–18** Diode circuit with low voltage supply
>
> Diode voltage drop considered:
> $$V_L = E - V_D = 3\ V - 0.7\ V = 2.3\ V$$
> $$I = V_L/R = 2.3\ V/100\ \Omega = 23\ mA$$
>
> Diode voltage drop *not* considered:
> $$V_L = E = 3\ V$$
> $$I = V_L/R = 3\ V/100\ \Omega = 30\ mA$$
>
> Error calculation:
> $$(\text{difference/actual}) \times 100\% = (7\ mA/23\ mA) \times 100\% = 30\%$$

As the calculations show, the actual circuit current is 23 mA. If the voltage drop across the diode is not considered, the circuit current is calculated to be 30 mA. Not considering the diode voltage drop in Example 1.4 resulted in a 30% error. The practical diode model is the correct choice for Example 1.4.

EXAMPLE 1.5

Figure 1–19 shows a 200 V supply connected to a 100 Ω load through a forward-biased diode. The current for the circuit in Figure 1–19 is calculated twice. In the first calculation, the 0.7 V dropped across the forward-biased diode is considered. In the second calculation, the 0.7 V is disregarded.

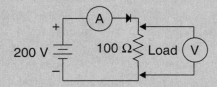

FIGURE 1–19   Diode circuit with high voltage supply

Diode voltage drop considered:
$$V_L = E - V_D = 200 \text{ V} - 0.7 \text{ V} = 199.3 \text{ V}$$
$$I = V_L/R = 199.3 \text{ V}/100 \ \Omega = 1.993 \text{ A}$$

Diode voltage drop *not* considered:
$$V_L = E = 200 \text{ V}$$
$$I = V_L/R = 200 \text{ V}/100 \ \Omega = 2 \text{ A}$$

Error calculation:
$$(\text{difference/actual}) \times 100\% = (.007 \text{ A}/1.993 \text{ mA}) \times 100\% = 0.35\%$$

The actual circuit current is 1.993 A. If the voltage drop across the diode is not considered in the calculations, the result is 2 A. Not considering the diode voltage drop has resulted in an error of only 0.35%. The ideal diode model would be the correct choice in Example 1.5.

After examining the two examples, it should be apparent that the voltage drop of the diode must be considered in the first example. In the second example, however, not considering the diode voltage drop resulted in less than a 1% error. There is no specific rule stating under what conditions the diode voltage drop can be disregarded. A rule of thumb for troubleshooting purposes is to disregard the diode voltage drop if its omission causes less than a 10% error.

1.4 COMPUTER CIRCUIT ANALYSIS

INTRODUCTION

This text is not tied to computer analysis. In fact, all circuit problems should be analyzed using the knowledge you gained in your study of electronics circuits before using computer analysis to verify your results. However, computer analysis is a powerful tool for learning electronic circuits; therefore, the author feels it should be part of a technician's education when available.

In this text, it is hoped that you will use computer circuit analysis to verify and stimulate your thought processes. If you are analyzing a circuit, wouldn't it be great to get immediate feedback on whether your thought processes are correct? You are trying to understand the function of a component in a circuit; wouldn't it be nice to be able to play the "what if game" to see what happens when the component is changed in value? The fact is, with an investment of a few hours and access to a PC running MultiSIM, you have the tools available.

Excellent opportunities to make use of electronic circuit analysis are the pre-labs that precede each hardware lab experiment in this text. It is an excellent idea to perform the pre-lab before attempting the hardware lab session. The pre-lab circuits are similar to the circuits that will be assembled and tested in hardware labs. The best procedure for completing the pre-lab is to use the "wetware" between your ears to analyze the circuit and then verify your results using computer analysis.

There are numerous electronic analysis software programs available and all are based on the same core program: SPICE (Simulation Program with Integrated Circuit Emphasis). SPICE was developed at the University of California and was first used by engineers as a design tool. Several companies, however, have enhanced SPICE, making it more user-friendly, and these modified versions of SPICE are now useful tools for anyone in electronics. The human interface developed by these companies is straightforward; you simply draw the schematic and watch it come alive.

The product dominating the educational market is MultiSIM™ from Electronics Workbench. MultiSIM is an excellent product that permits the user not only to draw the circuit but also, after simulation, to read icon instruments connected to the circuit to check the performance of the circuit.

Analyzing a circuit using a circuit analysis program is a three-step process: (1) drawing a schematic of the circuit, (2) having the computer analyze the circuit, and (3) examining the results. It sounds easy, and it is easy. In order to become comfortable with an analysis program you must use it continuously over a period of time. By doing so, it is hoped you will enhance your learning of analog circuits. Remember, the reason for using computer circuit analysis is to learn electronics, so always analyze the circuit using the information you learned, and then verify your results using computer analysis.

1.5 MultiSIM LAB EXERCISE

INTRODUCTION

In the following exercise we will analyze three circuits. The first is an RC circuit, the next circuit has two diodes controlling a lamp, and the third circuit examines the dynamic resistance of a diode. As stated previously, the first step in using computer analysis as a learning tool is to calculate the circuit values. The calculations for the RC circuit are shown below. Following the calculations is a step-by-step procedure that enters the RC circuit into MultiSIM, and then shows how the results can be examined using MultiSIM analog instruments. After you have worked through the first circuit, you will be ready to analyze diode circuits, enter the circuits in MultiSIM, and execute computer analysis.

ANALYZING THE RC CIRCUIT

Figure 1–20 shows an RC circuit driven by a signal source of 1 V_p at 1 kHz. Let us determine and draw waveforms for the generator voltage, generator current, and output voltage.

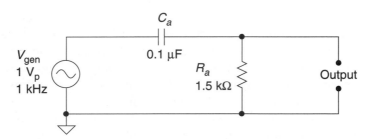

FIGURE 1–20   Circuit to be analyzed

Step 1. Calculate the reactance of C_a at a frequency of 1 kHz.

$X_c = 1/(2\pi fc) = 1/(2\pi \times 1 \text{ kHz} \times 0.1 \text{ μF}) = 1592 \text{ Ω}$

Step 2. Calculate the total circuit impedance.

$$Z = \sqrt{R^2 + X_C^2} = \sqrt{1.5\,\text{k}^2 + 1.592\,\text{k}^2} = 2.19\,\text{k}$$

Step 3. Calculate the phase angle.

Phase angle = $\text{Arctan}(-X_c/R) = \text{Arctan}(-1.592 \text{ k}/1.5 \text{ k}) = -46.7^\circ$

Step 4. Calculate the generator current.

$I_{gen} = V_{gen}/Z = 1 \text{ V}_p < 0^\circ/2.19 \text{ k} < -46.7^\circ = 0.457 \text{ mA}_p < 46.7^\circ$ (The generator voltage is assumed to have a phase angle of zero.)

Step 5. Calculate the output voltage.

$V_o = R \times I_{gen} = 1.5 \text{ k} < 0^\circ \times 0.457 \text{ mA} < 46.7^\circ = 0.686 \text{ V}_p < 46.7^\circ$

Step 6. Using the information calculated, the waveforms shown in Figure 1–21 can be drawn.

The drawing created in Figure 1–21 is an amplitude versus time graph. The amplitude is plotted on the vertical axis, and the time is plotted on the horizontal axis. This is the type of display the technician would expect to see on an oscilloscope. Using computer circuit analysis,

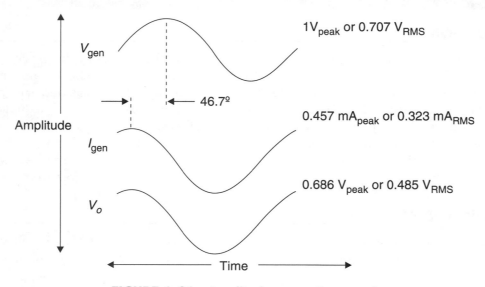

FIGURE 1–21   Amplitude versus time graph

this type of display can be created. The technician will find the amplitude versus time graph most useful. By clicking the mouse, voltage waveforms can be displayed at any point in the circuit.

If the frequency of the generator in Figure 1–20 is changed, the output voltage will change. Table 1–1 shows the data obtained by running through the calculations for eight different frequencies. If the data in Table 1–1 were plotted with the frequency on the horizontal axis and the voltage amplitude on the vertical axis, the graph shown in Figure 1–22 would be obtained.

TABLE 1–1 Calculated Data

Frequency	V_{out}
10 Hz	0.009 V
100 Hz	0.093 V
500 Hz	0.425 V
1 kHz	0.686 V
2 kHz	0.884 V
5 kHz	0.978 V
10 kHz	0.994 V
50 kHz	0.999 V
100 kHz	1.000 V

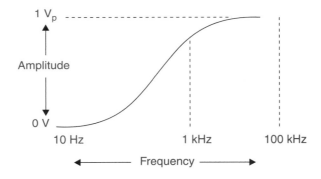

FIGURE 1–22   Frequency response graph

This curve shows how the circuit responds to changes in frequency. The generator voltage is held constant at 1 V_p as the frequency sweeps from 10 Hz to 100 kHz. Using computer circuit analysis, this type of display is created when an AC sweep analysis is executed. The technician will find AC sweep analysis useful when working with circuits that are frequency sensitive.

EXAMINE THE MultiSIM WINDOW

Step 1. To open MultiSIM, click on the **Start** button, highlight **Programs,** find and highlight **MultiSIM** folder, and then click the **MultiSIM** icon. The schematics window shown in Figure 1–23 will open. This window will be your interface to MultiSIM. Your screen should look the same as that in Figure 1–23 except it should be maximized to full screen.

Step 2. Examine the MultiSIM window. There are eight items on the menu bar. If all the commands under each menu are added, there are approximately 70 commands.

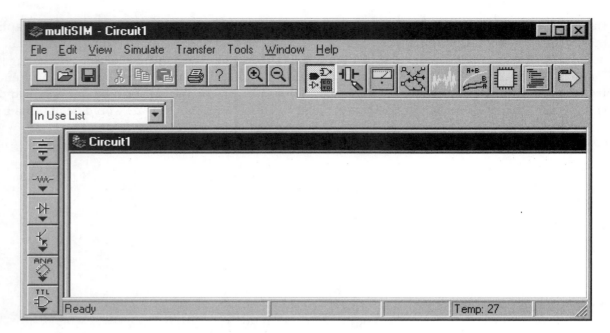

FIGURE 1–23   MultiSIM window

The popular commands can also be executed using the icons on the Toolbar and Design Bar located under the menu items. We are going to use a tutorial approach. This means we will only consider the commands we need to complete our task. Using built-in Help, you can obtain the functions of all commands. To use Help, select the Help menu or press the F1 key.

The Component toolbar is displayed on the left side of the work area (Circuit window). The Component toolbar can be minimized using a button on the Design Bar. The electronics components are stored in toolbars based on the type of part. A specific toolbar is selected by placing the cursor on an icon on the Component toolbar. The parts on the toolbar are shown in the vertical window displayed on the left side of the screen. The circuit is drawn in the Circuit window shown in Figure 1–23.

ENTERING THE RC CIRCUIT

Step 3. Select the **Basic** toolbar by placing the cursor on the **resistor** icon located on the **Component** toolbar. The parts in the Basic toolbar will appear in a window.

Step 4. To move a resistor onto the workspace, click the **resistor** in the **Basic** toolbar window. A browser window will appear. Select the desired resistor value (1.5 kΩ) from the **Component List** and then click **OK.** Your cursor will now appear in the Circuit window as a rectangle with an arrow on one end. Point the arrow to the location where the resistor is to be placed and click. (See Figure 1–24.) The resistor will now appear on the screen.

Step 5. The resistor needs to be rotated. To rotate the resistor, first select it by clicking and then use the **90 clockwise** command from the pull-down **Edit** menu. Control keys can be used to perform most commands. The control keys are listed on the pull-down menu.

Step 6. Repeat steps 3 and 4 to place the 100 nF capacitor on the screen.

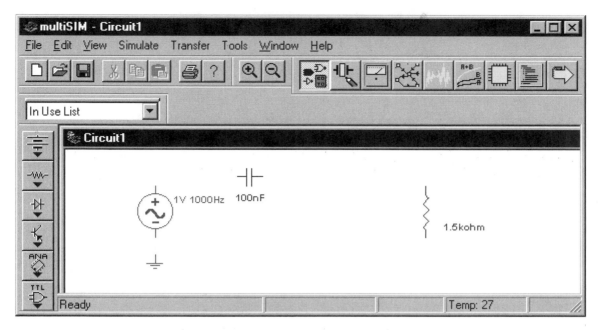

FIGURE 1–24 Parts moved to workspace

Step 7. Select the **Sources** toolbar by placing the cursor on the **battery** icon located on the **Component** toolbar. The parts in the **Sources** toolbar will appear in a window.

Step 8. To move an AC source onto the workspace, click the **AC_VOLTAGE_SOURCE** in the **Sources** toolbar window. Your cursor will now appear in the Circuit window as a rectangle with an arrow on one end. Point the arrow to the location where the AC source is to be placed and click. (See Figure 1–24.) The AC source will now appear on the screen.

Step 9. Repeat step 8 to place the circuit ground on the screen.

Step 10. Wire the circuit. Point to the top of the ground symbol and the cursor will change from a pointer to a crosshair. When the crosshair cursor is over the ground symbol lead, click and then move the crosshair cursor to the bottom terminal of the signal source and a dash line will follow the crosshair cursor as it is moved. When the crosshair cursor is over the terminal of the signal source, click and the dash line will change to a solid line indicating the connection has been completed. Connect all the components as shown in Figure 1–25. If you want to move components, point to the component and press the mouse button and drag. Clicking and dragging can also move the text on the screen.

CONNECTING INSTRUMENT

Step 11. Click the **Instruments** icon on the **Design** toolbar. This will cause the Instruments toolbar to appear at the bottom of the Circuit window. Click the multimeter on the **Instruments** toolbar. Move the cursor to the location where the multimeter is to be placed and click. The multimeter will now appear on the screen as shown in Figure 1–26.

Step 12. Repeat step 11 to place the oscilloscope on the screen as shown in Figure 1–26.

Step 13. Select the **Indicators** icon (Seven-segment Display) on the **Component** toolbar. The Indicators toolbar window will appear; click on the **ammeter.** A browser

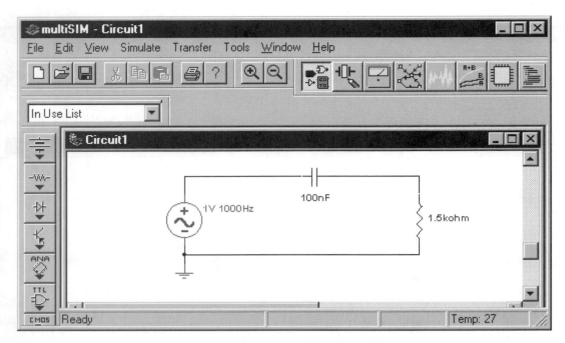

FIGURE 1–25   Wired circuit

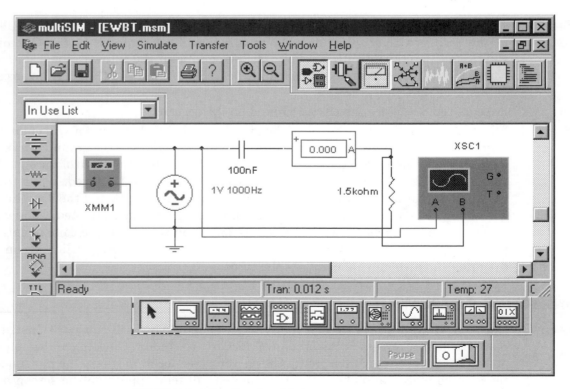

FIGURE 1–26   Circuit with instruments connected

window will appear. Select the generic ammeter from the **Component List** and then click **OK.** If an ammeter is placed over a wire segment, the ammeter will automatically be inserted in the wire segment. Move the cursor to the location of the ammeter shown in Figure 1–26 and click. The ammeter should automatically be inserted into the wire segment.

Step 14. Using the cursor, wire the multimeter and oscilloscope as shown in Figure 1–26. An oscilloscope ground is not needed as long as the circuit is grounded.

Step 15. Right clicking a wire and selecting **Color** from the shortcut menu can change the color of the wire. A Color window appears where wire color can be selected. The color of the oscilloscope trace is a function of the color of the wire connected to each channel of the oscilloscope. Change the wire connected to Channel A to red and the wire connected to Channel B to blue.

Step 16. Select **Show Simulate Switch** from the **View** menu. This will cause the simulate switch to appear below the Instruments toolbar as shown in Figure 1–26.

ADJUSTING INSTRUMENT

Step 17. Double click the **Multimeter** icon. The icon will expand so the meter settings can be read as shown in Figure 1–27. Set the multimeter to read AC volts by clicking the **V** and the **Sinewave** buttons. Close the window and the multimeter will return to a small icon.

FIGURE 1–27 Multimeter

Step 18. Double click the **Oscilloscope** icon. The icon will expand so the oscilloscope settings can be read as shown in Figure 1–28. Set the oscilloscope Time Base to **500 us/div,** Channel A to **1 V/Div,** and Channel B to **1 V/Div** as shown in Figure 1–28. Close the window and the oscilloscope will return to a small icon.

Step 19. Double click the **ammeter** and a dialog box will open. Set the Mode to **AC** and click **OK.** The system will return to the schematics window.

ACTIVATE THE RC CIRCUIT AND EXAMINE RESULTS

Step 20. Click the **Simulate Switch** on. The circuit is activated and the computer performs an analysis of the circuit. Notice the ammeter now displays 0.325 mA. This is the RMS current flowing through the circuit. This value compares nicely to the value we calculated. (Check Figure 1–21.)

Step 21. To read the multimeter, double click the **Multimeter** icon and the meter will expand. The meter should read 707 mV, which is the RMS voltage set on the signal source. Close the window and the multimeter will return to a small icon.

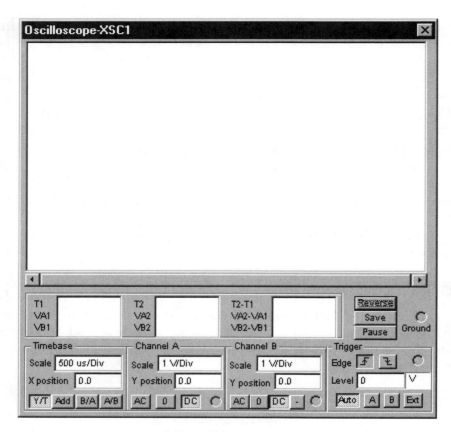

FIGURE 1–28   Oscilloscope

Step 22.   To read the oscilloscope, double click the **Oscilloscope** icon and the oscilloscope will expand as shown in Figure 1–29. The screen of the scope displays traces for Channel A and Channel B. The traces continually update. Click **Sing.** on the oscilloscope. This will cause the oscilloscope to display a constant waveform.

Step 23.   The controls of the oscilloscope can be adjusted to change the display. For example, select the **0** (AC-0-DC) for Channel B coupling. This will cause trace B on the display to be a straight line since the Channel B input is grounded. Change Channel A V/Div from 1 V/Div to **500 mV/Div.** Trace A will double in size. Change the Timebase setting from 0.500 us/div to **1.00 ms/div.** If the instrument does not respond as expected to a change in setting, always try cycling the simulate switch. Return the oscilloscope to the settings shown in Figure 1–29.

Step 24.   There are two cursors: one located on the left side of the screen and the second located on the right side of the screen. Point to the triangle at the top of the left cursor, press the mouse button, and drag the cursor to the right until the cursor is at the negative peak of the larger sinewave. Point to the triangle at the top of the right cursor, press the mouse button, and drag the cursor to the left until the cursor is at the positive peak of the larger sinewave. The display below the XY graph gives readings that correspond to the cursor locations as shown in Figure 1–29. The reading labeled $VA1$ is the voltage amplitude of the signal on Channel A (approximately $-1V_p$) at the location of cursor 1. The reading labeled $VA2$ is the voltage amplitude of the signal on Channel A (approximately $1\ V_p$) at the location of cursor 2. The reading labeled $VA2\text{-}VA1$ is the difference voltage

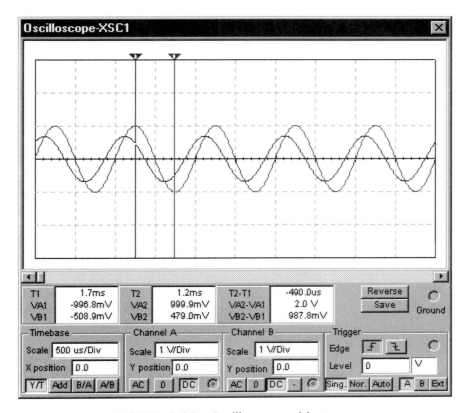

FIGURE 1–29   Oscilloscope with traces

between the two cursor locations (approximately 2 V_p). With the readings labeled T_1, T_2 and T_2-T_1, you can use the cursors to measure timing.

Step 25. If you want to save the circuit, select **Save** from the **File** menu and give the circuit a name.

Step 26. Remove the oscilloscope, multimeter, and ammeter. **Click** each instrument and, when the instrument is selected, press the **Delete** key to remove the instrument. Removing the ammeter will cause an open circuit; reconnect the circuit.

MEASURE THE FREQUENCY RESPONSE

Step 27. Select the **Bode plotter** from the **Instrument** toolbar and connect the Bode plotter as shown in Figure 1–30.

Step 28. Double click the **Bode plotter** icon and it will expand as shown in Figure 1–31. The Bode plotter is used to measure the frequency response or the phase shift of a circuit. The Bode plotter is not an instrument found in the usual electronics lab. In most labs the technician will have to use a combination of instruments to perform the function of the Bode plotter. We will use the Bode plotter in this exercise to measure the frequency response of the output signal.

Step 29. Set the Bode plotter to measure the **Magnitude** of the output signal as shown in Figure 1–31. The vertical range is set to **Log** starting with an initial (I) value of **−40 dB** and ending with a final (F) value of **0 dB**. The horizontal range is set to **Log** starting with an initial (I) frequency of **10 Hz** and ending with a final (F) frequency of **100 kHz.**

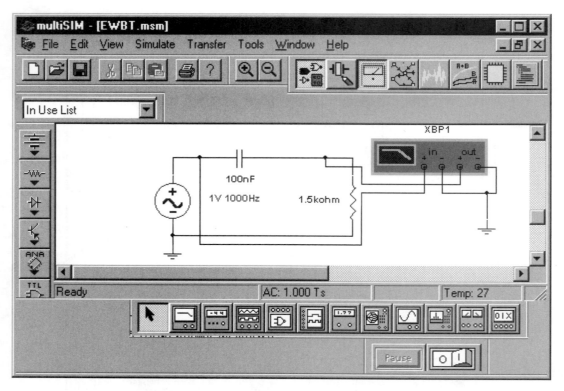

FIGURE 1–30   Circuit with Bode plotter connected

FIGURE 1–31   Bode plotter

Step 30.   Click the **Simulate Switch** to activate the circuit. A band response curve will be displayed on the Bode plotter as shown in Figure 1–32. The cursor is located to the far left, and the values at the cursor location (–40.5 dB and 10 Hz) are displayed.

Step 31.   Use the right pointing arrow to move the cursor to the right until the magnitude reads approximately **–3 dB** as shown in Figure 1–32. With the cursor set for –3 dB, the cutoff frequency can be read (approximately 1 KHz). Signals that pass through the circuit with less than –3 dB attenuation are considered in the bandpass of the filter circuit.

Step 32.   Select **Print** from the **File** menu. A dialog box will appear. Note you can select to print a variety of documents such as schematic, parts list, instruments, and so

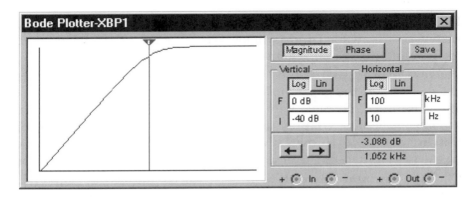

FIGURE 1–32   Band response curve

on. Since we will not generate a hard copy in this exercise, you can close the dialog box by clicking **Cancel.**

Step 33. If you want to save the circuit, select **Save As** from the **File** menu and give the circuit a name.

DIODE CONTROLLED LAMP CIRCUIT

Step 34. Use the knowledge gained about diodes in this chapter to determine which condition in Figure 1–33 will cause the lamp to light. Indicate lamp off or lamp on below.

Switch Up _____   Switch Down _____

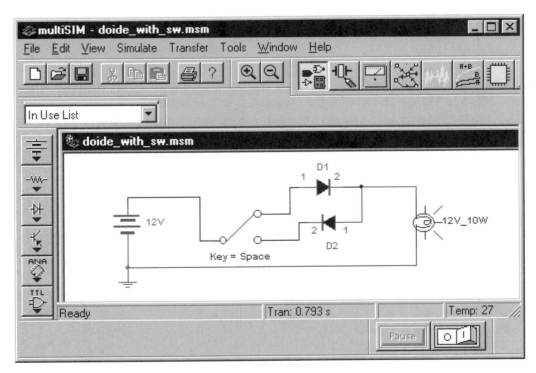

FIGURE 1–33   Diode control lamp circuit

Step 35. Draw the circuit shown in Figure 1–33. Select the ideal diode from the Component List on the **Browser Diode** window. The 12 V lamp is found on the **Indicators** toolbar, and the switch is found on the **Basic** toolbar.

Step 36. The text "IDEAL DIODE" will show on the screen. To change the text to D1 and D2, double click the upper diode and a **Diode** window will open. The window has four tabs: Label, Display, Value, and Fault. Click the **Label** tab and type **D1** for the label. Click the **Display** tab and unselect **Use Schematic Option Global Setting** and select **Show Labels.** Click **OK** and the label D1 will be displayed on the screen. Repeat the procedure for the lower diode. Give it a label of D2.

Step 37. Click the **Simulate Switch** to activate the circuit. Use the **Space bar** to open and close the switch and note how the circuit responds below. Were you correct in step 34? _____

Switch Up _____   Switch Down _____

Step 38. If you want to save the circuit, select **Save As** from the **File** menu and give the circuit a name.

DYNAMIC RESISTANCE CIRCUIT

Step 39. The circuit in Figure 1–34 can be used to demonstrate the dynamic resistance of a diode. Calculate the dynamic resistance of the diode for a potentiometer setting of 9.5 kΩ (95%) and a setting of 500 Ω(5%).

Diode resistance (pot = 9.5 kΩ) _____   Diode resistance (pot = 500 Ω) _____

Step 40. Use the calculated dynamic resistances to calculate the signal across the diode for each setting.

Signal (pot = 9.5 kΩ) _____   Signal (pot = 500 Ω) _____

Step 41. Construct the circuit in Figure 1–34 using MultiSIM and measure the signal voltage across the diode. Do measured values confirm your calculations? Note that the lowercase *a* changes the potentiometer in one direction and the uppercase *A* changes it in the opposite direction.

Signal (pot = 9.5 kΩ) _____   Signal (pot = 500 Ω) _____

Step 42. If you want to save the circuit, select **Save As** from the **File** menu and give the circuit a name.

Step. 43. Exit MultiSIM.

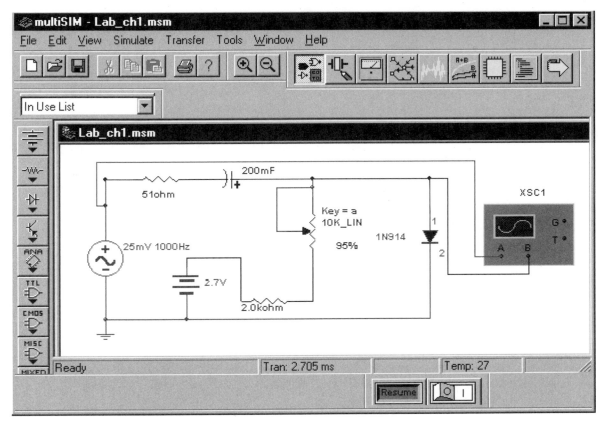

FIGURE 1–34   Analyzing the dynamic resistance of a diode

QUESTIONS

1.1 Introduction to Diodes

____ 1. The diode schematic symbol is ____.

   a.   b.   c.   d.   e.

____ 2. When forward biased, the diode is essentially a/an _____ .
   a. high resistor
   b. open switch
   c. closed switch
   d. low capacitor

1.2 Inside the Diode

____ 3. Materials that are good conductors normally have ____ valence electrons.
   a. 1
   b. 4
   c. 8
   d. 16

____ 4. Covalent bonding is the _____.
   a. glue used to connect the leads to the PN junction
   b. sharing of electrons in the crystalline structure
   c. action that occurs when intrinsic and extrinsic material come together
   d. none of the above

____ 5. In P-type materials, the majority carriers are _____.
   a. neutrons
   b. electrons
   c. holes
   d. ions

____ 6. When a PN junction is reverse biased, a/an _____ is formed that prohibits current flow.
   a. blocking voltage
   b. depletion region
   c. extrinsic area
   d. reverse current

____ 7. When a PN junction is forward biased, the electrons will combine with _____ at the junction to permit current flow.
   a. electrons
   b. ions
   c. neutrons
   d. holes

____ 8. The barrier voltage for a silicon PN junction is approximately _____.
      a. 0.1 V
      b. 0.3 V
      c. 0.7 V
      d. 1.5 V

1.3 Three Diode Models

____ 9. Which of the following is not one of the diode models discussed in the text?
      a. The zener model
      b. The detail model
      c. The ideal model
      d. The practical model

____ 10. A circuit is operating with 3 V; therefore, the junction voltage drop is significant. The
          circuit should be analyzed using the _____ model.
      a. zener
      b. detail
      c. ideal
      d. practical

1.4 Computer Circuit Analysis

____ 11. Computer circuit analysis is a powerful tool for _____.
      a. learning electronic circuits
      b. designing electronic circuits
      c. replacing hardware so real circuits are no longer needed
      d. both a and b

____ 12. There are numerous electronic analysis software programs available and all are based
          on the same core program called _____.
      a. Windows 418
      b. CIRCUITS
      c. SPICE
      d. none of the above

1.5 MultiSIM Lab Exercise

____ 13. Figure 1–23 shows the _____ toolbar displayed on the left side of the work
          area (Circuit window).
      a. Design
      b. Instrument
      c. Help
      d. Component

_____ 14. To rotate a component, first _____ and then use the **90 clockwise** command from the pull-down **Edit** menu.

    a.   select it by clicking

    b.   select it by right clicking

    c.   type rotate in the command line

    d.   type 90 in the command line

_____ 15. The Bode plotter is a MultiSIM instrument that is used to measure the _____.

    a.   frequency response

    b.   voltage response

    c.   current

    d.   power

CHAPTER

Diode Circuits

2

OBJECTIVES

After studying the material in this chapter, you will be able to describe and/or analyze:

- ○ the operation of rectifier circuits,
- ○ the operation of voltage multiplier circuits,
- ○ the operation of clipper circuits,
- ○ the operation of clamper circuits,
- ○ the switching circuits, and
- ○ diode troubleshooting procedures.

INTRODUCTION

Chapter 1 covered the basic operation of a diode. In this chapter we will examine several circuits using diodes. Diodes, like all electronic components, have many different applications. It is important to understand how the diode functions and then apply this knowledge to the circuit being analyzed. Remember that when the diode is forward biased, it acts like a closed switch with 0.7 V dropped across its terminals. When the diode is reverse biased, it acts like an open switch.

2.1 RECTIFIER CIRCUITS

HALF-WAVE RECTIFIER

The function of a rectifier circuit is to change alternating current (AC) to direct current (DC). Figure 2–1(a) shows an AC source connected to a 1 kΩ resistor.

Notice that the lower side of the AC source is referenced to ground and the upper side produces a voltage that swings sinusoidal between 100 V positive and 100 V negative as shown in Figure 2–1(b). The current in the circuit is shown in Figure 2–1(c). The current through the load will be a sinusoidal waveform between 100 mA positive and 100 mA negative.

Figure 2–2(a) shows a circuit with a diode in series with the load. Using the ideal diode model, let us analyze the output voltage and current. When the source swings positive with respect to ground, a positive voltage is applied directly to the anode of the diode. The cathode of the diode

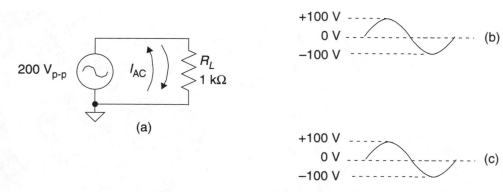

FIGURE 2–1   AC source with load

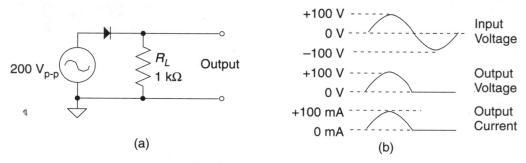

FIGURE 2–2   Half-wave rectifier

is tied to ground through the load. This forward biases the diode, and it acts like a closed switch. The output voltage and current are shown in Figure 2–2(b).

When the source swings negative, a negative voltage is applied to the anode. The cathode is still tied through the load back to ground. The diode is now reverse biased, since the anode is negative with respect to the cathode. The diode acts like an open switch, inhibiting current flow through the load during the negative alternation. Since no current flows in the circuit, no voltage will be produced across the load during the negative alternation.

Figure 2–2(b) shows that the input is a sinusoidal AC waveform varying between 100 V positive and 100 V negative. But the output voltage is a pulsating DC that varies between 0 and 100 V during the positive alternation of the input. During the negative alternation, the output voltage remains at 0 V. The output current varies between 0 and 100 mA during the positive alternation of the input and stays at zero during the negative alternation of the input.

Because the current flows only in one direction, it is defined as a pulsating DC current. The circuit in Figure 2–2(a) allows an output only during the positive alternation and is referred to as a positive half-wave rectifier. A negative half-wave rectifier can be constructed by turning the diode around. In this case, there would be an output only during the negative alternation of the input.

CENTER-TAPPED TRANSFORMER FULL-WAVE RECTIFIER

Figure 2–3(a) shows a transformer used to produce two sine waves that are out of phase by 180º with respect to each other if the center tap is used as the reference. If diodes are connected to the transformer as in Figure 2–3(b), a full-wave rectifier is formed.

On one alternation, TP_1 will go positive with respect to the center tap, which will cause D_1 to be forward-biased. Current will flow through D_1 and the load, causing the top of the load to be positive. During this alternation, D_2 is cut off because the voltage at TP_2 is negative with respect to the center tap. On the next alternation of the input, TP_2 will go positive with respect to the

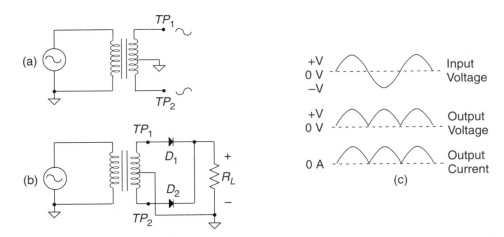

FIGURE 2–3   Center-tapped transformer full-wave rectifier

center tap, and this will forward bias D_2. Current will flow through D_2 and the load, causing the top of the load to be positive. During this alternation, D_1 is cut off because the voltage at TP_1 is negative.

Figure 2–3(c) shows input and output waveforms. The current flowing through the load during the total input cycle is always in one direction, causing the top of the load to swing only positive. Because of this, the output is a positive pulsating DC voltage during the positive and negative swings of the input. Output voltage and current are present during the complete cycle.

FULL-WAVE BRIDGE RECTIFIER

Figure 2–4(a) is another type of full-wave rectifier called a bridge. The full-wave bridge rectifier does not require a center-tapped transformer but does require four diodes.

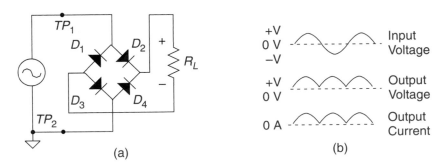

FIGURE 2–4   Full-wave bridge rectifier

When the input signal at TP_1 goes positive, D_3 and D_2 are forward biased and are in series with the load. During this positive alternation, current will flow through the load and through D_3 and D_2 in such a manner as to cause the top of the load to be positive. D_1 and D_4 are both reverse biased so they are effectively open during this alternation. Figure 2–5 shows the effective circuit when TP_1 is positive. The darken diodes (D_2 and D_3) are conducting.

When the input signal at TP_1 is on its negative alternation, D_1 and D_4 become forward biased and are in series with the load. During the negative alternation, current will flow through the load and D_1 and D_4 in such a manner as to cause the top of the load to be positive. D_2 and D_3 are

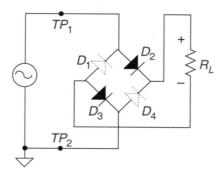

FIGURE 2–5   The effective circuit with TP_1 positive

both reverse biased, so they are effectively open during this alternation. Figure 2–6 shows the effective circuit when TP_1 is negative. The darkened diodes (D_1 and D_4) are conducting. In Figure 2–4(b), it can be seen that the current through the load and the voltage across the load are pulsating DC.

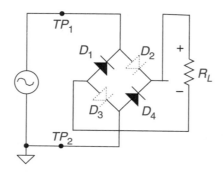

FIGURE 2–6   The effective circuit with TP_1 negative

The full-wave bridge rectifier has become extremely popular, and there are commercial packages available that contain all four diodes internally connected in the full-wave bridge configuration. Figure 2–7 shows a full-wave bridge package. It has four leads—two for AC inputs and two for DC outputs—that are marked plus and minus.

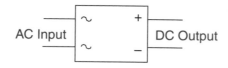

FIGURE 2–7   Bridge rectifier package

A transformer is not necessary at the input of a bridge rectifier, but it is often used for two reasons. It permits the AC input voltage to be stepped up or down, which is a good way to control the magnitude of the DC output. It also provides an AC input to the bridge that is isolated from the AC line. This isolation permits the DC output of the bridge to be referenced to ground without shorting a diode.

Figure 2–8(a) shows how diode D_3 would be shorted if the negative side of the load were referenced to ground. This short would cause excessive current through D_1 when forward biased. In Figure 2–8(b), the AC input to the bridge rectifier has been isolated with a transformer permitting a ground reference to be connected to the DC side of the circuit without shorting any circuit components.

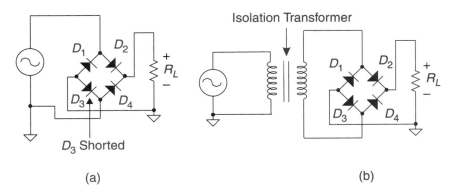

(a)

(b)

FIGURE 2–8 Bridge circuit with isolation transformer

If a bridge rectifier is connected to the secondary of a transformer, the rectified peak output will equal the secondary peak voltage minus two diode junction drops (1.4 V). The center tap of the transformer is not used. If a center-tapped transformer full-wave rectifier circuit uses the same transformer, its output would equal one-half of the secondary peak voltage minus one diode junction drop. Example 2.1 demonstrates how to determine the peak output voltage for each circuit.

EXAMPLE 2.1

Calculate the peak output voltage for each circuit in Figure 2–9.

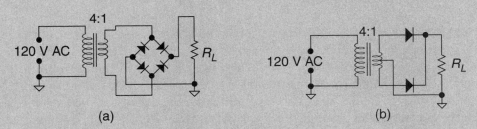

(a)

(b)

FIGURE 2–9

Step 1. Calculate the secondary voltage (a equals the turns ratio).
$$a = 4$$
$$V_s = V_p/a = 120 \text{ V}/4 = 30 \text{ V}$$

Step 2. Convert the secondary RMS voltage to peak voltage.
$$V_{peak} = V_{RMS}/0.707 = 30 \text{ V}/0.707 = 42.4 \text{ V}_{peak}$$

Step 3. Calculate the peak output voltage for the bridge rectifier circuit.
$$V_{Opeak} = V_{peak} - 1.4 \text{ V} = 42.4 \text{ V} - 1.4 \text{ V} = 41 \text{ V}$$

Step 4. Calculate the peak output voltage for the center-tapped transformer full-wave rectifier circuit. Since only half of the secondary voltage is rectified during each alternation, the peak secondary voltage must be divided by two.
$$V_{Opeak} = (V_{peak}/2) - 0.7 \text{ V} = (42.4 \text{ V}/2) - 0.7 \text{ V} = 20.5 \text{ V}$$

2.2 FILTERING PULSATING DC

All rectifiers produce pulsating DC. The frequency of pulsating signals is called the *ripple frequency*. To obtain a useful steady-state DC voltage, the ripple frequency must be filtered out. Figure 2–10 shows the relationship between input AC signal frequency and the output ripple frequency of rectifiers. The half-wave rectifier has a ripple frequency output equal to the frequency of the input AC signal. The full-wave rectifier has a ripple frequency output equal to twice the frequency of the input AC signal. Figure 2–10 shows that a 60 Hz AC input signal causes an output ripple frequency of 60 Hz for the half-wave rectifier but an output ripple frequency of 120 Hz for the full-wave rectifier.

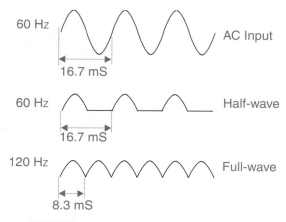

FIGURE 2–10   Ripple output signals

A filter circuit consisting of a single capacitor connected across the output of the rectifier can remove most of the ripple signal. Figure 2–11 shows a full-wave bridge rectifier circuit. The AC signal present at the secondary of the transformer is 17.8 V_{peak}. If the output of the rectifier is connected to the load without the filter capacitor, a DC signal pulsating between 0 V and 16.4 V_{peak} is present across the load. If a 200 µF capacitor is connected across the output of the rectifier, most of the ripple component is removed from the output. Figure 2–11 shows that the ripple component is reduced to a 1.4 V_{p-p}.

For a short period each cycle, the rectifier diodes are forward biased (on) and charge the filter capacitor. During the rest of the time, the diodes are reverse biased (off). During the off time of the diode, the filter capacitor supplies current to the load. Figure 2–12 shows the off/on cycle of a diode rectifier driving a capacitor filter. The diodes conduct approximately 15% of the time and are off 85%. The capacitor starts to discharge during the off time but is quickly recharged each cycle when the diodes turn on. If a larger capacitor is used, the magnitude of the ripple voltage can be decreased. However, this means the "on time" of the diodes will decrease to less than 15% of the period. The diodes must pass the same coulombs of charge to support the load current, so reducing their "on time" causes the diodes surge current to become excessive. Engineers normally design the filter circuit so the diodes conduct 15% to 20% of the period.

The amplitude of the output ripple is a function of the size of the capacitor, the load current, and the period of the ripple signal. The formula below shows how these variables are related.

$$Ripple\ voltage_{p-p} = \frac{Load\ current \times Diode\ off\ time}{Filter\ capacitor}$$

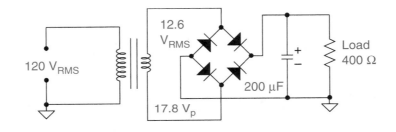

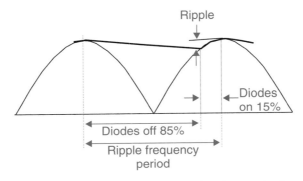

FIGURE 2–11 Outputs with and without filter

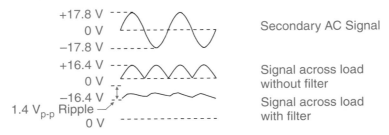

FIGURE 2–12 Rectifier conduction period

EXAMPLE 2.2

Calculate the peak-to-peak ripple voltage for the circuit in Figure 2–11.

Step 1. Calculate load current. First calculate the approximate DC output voltage. The output voltage is equal to the peak secondary voltage minus the voltage drops of the diodes. This will only give an approximate value of DC output voltage; it does not account for transformer drops or for the fact that the average DC voltage will be reduced by ripple variations.

$$V_p = V_{RMS}/0.707 = 12.6\ V_{RMS}/0.707 = 17.8\ V_p$$
$$V_{DC} = V_p - diodes\ voltage\ drops = 17.8\ V_p - 1.4\ V = 16.4\ V_{DC}$$
$$I_{load} = V_{DC}/R_{load} = 16.4\ V/400\ \Omega = 41\ mA$$

Step 2. Calculate diode off time. The diode off time is assumed to be 85% of the period of the ripple.

$$P = 1/F = 1/120\ Hz = 8.3\ mS$$
$$T_{diode\ off} = p \times 85\% = 8.3\ mS \times 85\% = 7\ mS$$

Step 3. Calculate ripple voltage. Use the values calculated and the size of the filter capacitor to calculate the peak-to-peak ripple voltage.

$$V_{Ripple\ p\text{-}p} = (I_{load} \times T_{diode\ off})/C = (41\ mA \times 7\ mS)/200\ \mu F = 1.4\ V_{p\text{-}p}$$

Since the ripple frequency of a half-wave rectifier is one-half the ripple frequency of a full-wave rectifier, a half-wave rectifier circuit requires a filter capacitor twice as large as a full-wave rectifier to obtain the same ripple magnitude. Consequently, most power supplies use full-wave rectifiers.

2.3 BLOCK DIAGRAM OF A COMPLETE POWER SUPPLY SYSTEM

Normally, the filter circuit is not designed to remove all of the ripple components of the output signal. Instead, the filter circuit is followed by a regulator circuit that will remove the remaining ripple in the DC. Figure 2–13(a) shows a block diagram of a complete power supply system. The power supply system contains a rectifier, filter, and regulator. Figure 2–13(b) shows the waveform at the input and output of each section of the power supply. The output of the filter is called raw DC, and the output of the regulator is called regulated DC. Regulator circuits will be covered in a later chapter.

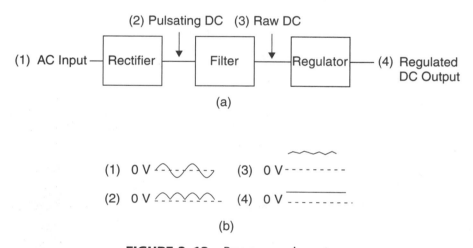

FIGURE 2–13   Power supply system

2.4 VOLTAGE MULTIPLIERS

HALF-WAVE VOLTAGE DOUBLER

Figure 2–14 shows a half-wave voltage doubler. On the negative alternation of the input, C_1 charges to 100 V through the forward-biased diode, D_1. On the positive alternation of the input, the 100 V across C_1 is in series aiding with the input voltage. The 100 V positive peak of the input adds with the 100 V across C_1 causing 200 V that forward biases diode D_2 and charges capacitor C_2 to 200 V. Capacitor C_2 will maintain 200 V across the load during the time D_2 is off. There will be some discharge of C_2, causing a ripple voltage on the output signal. The magnitude of the ripple voltage will depend on the size of the capacitors and the load current. The ripple frequency at the output of the half-wave voltage doubler will be the same as the input frequency. Recall that the ripple frequency of a half-wave rectifier is equal to the input frequency.

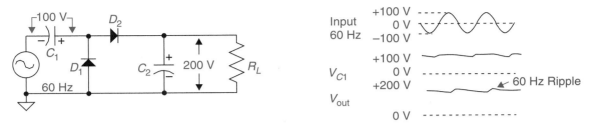

FIGURE 2–14   Half-wave voltage doubler

FULL-WAVE VOLTAGE DOUBLER

Figure 2–15 shows a full-wave voltage doubler. On the positive alternation, capacitor C_1 charges through diode D_1 to positive 100 V. On the negative alternation, capacitor C_2 charges through diode D_2 to negative 100 V. The voltage across the load is equal to the sum of the voltage across the capacitors, or 200 V in this case. The ripple frequency of the full-wave voltage doubler will be twice the frequency of the input frequency. With an input frequency of 60 Hz, the ripple frequency is 120 Hz.

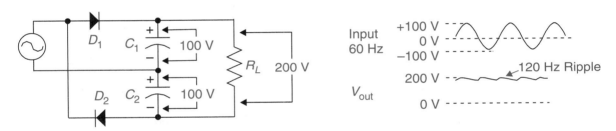

FIGURE 2–15   Full-wave voltage doubler

2.5 CLIPPER CIRCUITS

Clipper circuits, also called limiters, are designed to limit the signal swing. They are often used at the input of a system to protect the system from overvoltage. When clipping occurs, the shape of the waveform is changed. Circuits designed to respond to frequency variations often use limiters to clip the signal and remove unwanted amplitude variations. For example, FM radio systems use frequency variations to transfer information. Therefore, one of the first circuits in an FM radio receiver is a limiter circuit that removes the unwanted amplitude variations but keeps the frequency variations. Clipper circuits use diodes to perform the clipping. If the diode is in series with the load, the circuit is called a series clipper. If the diode is in parallel with the load, the circuit is called a parallel clipper.

SERIES CLIPPER

The half-wave rectifiers shown in Figure 2–16 is a series clipper circuit. During the negative half-cycle, the diode is in reverse bias and no voltage is developed at the output. The positive half-cycle is passed to the output through the forward-biased diode. Figure 2–16 shows the output waveform reaching 19.3 V as the input signal peaks at 20 V. The 0.7 V difference is dropped across the forward-biased diode.

FIGURE 2–16   Series clipper circuit

Since the diode is in series with the load resistor, there will be an output signal across the load only when the diode is forward biased. Therefore, the clipping level can be determined by calculating the minimum signal voltage necessary to forward bias the diode. The clipping level of this circuit is 0.7 V. Signal levels above 0.7 V will pass to the load undistorted; however, signal levels below 0.7 V will be clipped off. The input and output waveforms are shown in Figure 2–16.

If a bias supply is added, as shown in Figure 2–17, the amount of clipping performed on the output waveform can be controlled by the bias supply voltage. By changing the magnitude of the bias supply, any portion of the waveform can be clipped. The voltage level at which clipping occurs can be calculated by determining the minimum input signal voltage necessary to forward bias the diode. Example 2.3 shows how the output waveform can be calculated for the circuit in Figure 2–17.

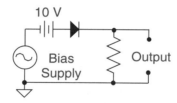

FIGURE 2–17   Series clipper circuit with bias supply

EXAMPLE 2.3

Calculate the waveform for the circuit shown in Figure 2–17.

Step 1. Assume the signal input voltage is zero, and determine if the diode is forward biased or reverse biased. Figure 2–18 shows an equivalent circuit with the signal input adjusted to zero. In this example, the diode is forward biased by the bias supply.

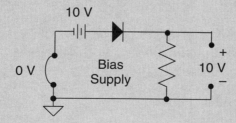

FIGURE 2–18   Equivalent circuit with 0 V signal input

Step 2. Determine the magnitude and polarity of the input signal necessary to change from forward bias to reverse bias. In this example, a –9.3 V input signal is needed to reverse bias the diode. Figure 2–19 is an equivalent circuit showing –9.3 V of signal voltage.

EXAMPLE 2.3 continued

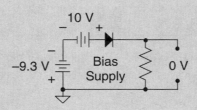

FIGURE 2–19   Equivalent circuit with –9.3 V signal input

Step 3. Draw the clipping level on the input waveform as shown in Figure 2–20. Then determine which part of the waveform will be clipped. The output waveform will be clipped during time periods when the signal input causes the diode to be reverse biased. In this example, all voltages more negative than –9.3 V will cause the diode to be reverse biased. The wave shape of the output signal will be the same as the unclipped portion of the input shown below.

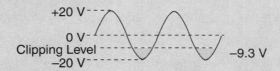

FIGURE 2–20   Input showing clipping area

Step 4. Determine the DC reference of the output waveform. Since current only flows in one direction through the diode, the output signal will be unidirectional with respect to ground. The output signal will vary between 0 V DC and +29.3 V DC as shown in Figure 2–21.

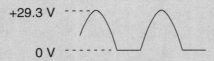

FIGURE 2–21   Output signal

PARALLEL CLIPPER CIRCUITS

Figure 2–22 shows a parallel clipper circuit. In this circuit, the diode is in parallel with the load (R_L) and the output signal is developed across the diode. When the diode is forward biased, clipping occurs and the output signal is limited to 0.7 V. When the input signal voltage is less than the clipping level, the diode is off and resistors R_S and R_L form a simple series circuit. With the diode off, if the resistance of R_S is small compared to the resistance of the load ($R_S < R_L/10$), then the output signal can be considered equal to the input signal. If the resistance of R_S is not small compared to the resistance of the load, then the voltage divider rule should be used to calculate the output signal. Figure 2–22 shows that during the negative alternation, the output

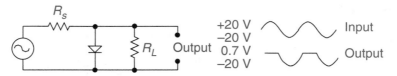

FIGURE 2–22   Parallel clipper circuit

signal tracks the input signal. However, during the positive alternation, the output is clipped at 0.7 V.

Figure 2–23 shows a parallel clipper circuit with a bias supply. Any portion of the output waveform can be clipped by a combination of adjusting the bias supply and/or reversing the diode. The procedure used in Example 2.4 can be used to calculate the output waveform.

FIGURE 2–23   Parallel clipper circuit with a bias supply

EXAMPLE 2.4

Calculate the output waveform for the circuit shown in Figure 2–23.

Step 1. Mentally replace the signal source with an adjustable supply as shown in Figure 2–24. Determine the magnitude and polarity of the voltage needed to forward bias the diode. The voltage needed to forward bias the diode is the voltage at which clipping will occur. In this example, clipping will occur at 10.7 V.

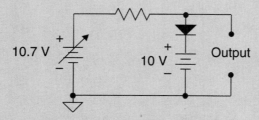

FIGURE 2–24   Replacing signal source with supply

Step 2. Draw the clipping level on the input waveform as shown in Figure 2–25, and determine which portion of the waveform is clipped off.

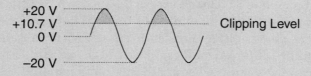

FIGURE 2–25   Input signal with clipping level shown

EXAMPLE 2.4 continued

Step 3. Draw the output waveform as shown in Figure 2–26.

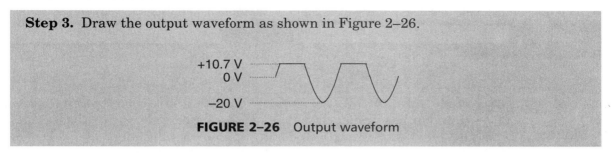

FIGURE 2–26 Output waveform

Figure 2–27(a) shows a bi-directional parallel clipper circuit that is designed to limit both positive and negative alternations of the output waveform. If the positive voltage across the load (R_L) exceeds 0.7 V, diode D_1 turns on and clips the positive alternation. If the negative voltage across the load (R_L) exceeds –0.7 V, diode D_2 turns on and clips the negative alternation. Figure 2–27(b) shows the output waveform with a 10 $V_{\text{p-p}}$ signal applied to the input. If the diodes are stacked, the clipping voltage can be controlled. The circuit in Figure 2–27(c) clips the output signal at 1.4 V in each direction since two silicon diodes are in series.

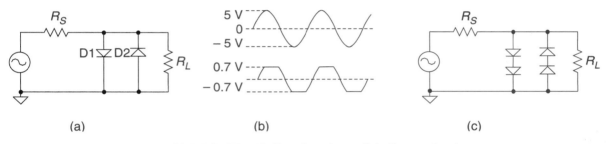

FIGURE 2–27 Bi-directional parallel clipper circuits

2.6 CLAMPER CIRCUITS

Clamper circuits are designed to change the DC reference of the waveform, but not its shape. By using a capacitor in conjunction with a diode, a clamper circuit can be designed.

Figure 2–28(a) shows a negative clamper circuit that clamps the output signal so it swings only negative with respect to ground. The input is a 20 $V_{\text{p-p}}$ square-wave signal riding on a ground reference. The output and input waveforms are shown in Figure 2–28(b). The output has a magnitude

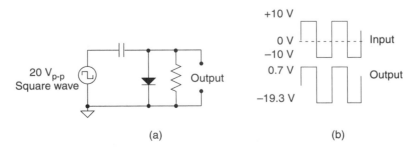

FIGURE 2–28 Clamper circuit

of 20 V$_{p-p}$, but it swings from 0.7 V to negative 19.3 V. The magnitude and the shape of the output signal remain unchanged, but the DC reference does change. Example 2.5 analyzes the circuit operation.

EXAMPLE 2.5

Calculate the output waveform for the circuit shown in Figure 2–28.

Step 1. Draw an equivalent circuit, as shown in Figure 2–29, for the positive alternation of the input for the clamper circuit. The load resistor is shorted by the forward-biased diode, which is shown as a closed switch in the equivalent circuit. During the positive alternation, the output is 0.7 V, and the capacitor is charged to 9.3 V with the polarity shown.

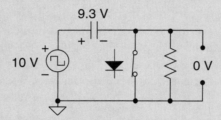

FIGURE 2–29   Clamper equivalent circuit on positive alternation

Step 2. Draw the equivalent circuit, as shown in Figure 2–30, for the negative alternation of the input. The voltage across the capacitor and the signal generator voltage are in series aiding and add together to drive the output to negative 19.3 V. The output will remain at negative 19.3 V during the negative alternation (assuming that the capacitor does not discharge). When the input switches back to the positive alternation, the capacitor recharges to the value of the positive input and the cycle begins again.

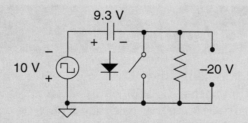

FIGURE 2–30   Clamper equivalent circuit on negative alternation

Step 3. Using the information obtained in steps 1 and 2, draw the output waveform as shown in Figure 2–31.

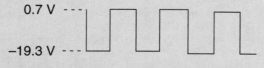

FIGURE 2–31   Clamper output signal

In summary, when the input is positive, the capacitor charges to the value of the input because the load is bypassed by the forward-biased diode. When the input swings negative, the input voltage and the capacitor voltage are in series and add to produce the output voltage. In order for the discharge across the capacitor to be negligible during the negative swing of the input, the RC time constant of the circuit must be long with respect to the period of the input signal.

The clamper circuit in Figure 2–28(a) can be modified to perform positive clamping by reversing the diode. If this is done, the output signal would swing from −0.7 to +19.3 V.

Clamper circuits can be used to clamp sine wave signals as shown in Figure 2–32. The capacitor charges to the peak value of the input waveform. The polarity of the charge will depend on the direction of the diode. Once charged, the capacitor is a DC source in series with the sine wave input. On one alternation, the input signal will add to the capacitor voltage, and on the next alternation, the input signal will subtract from the capacitor voltage. Notice that the output signal in Figure 2–32 is riding on a −9.3 V DC reference and swings between 0.7 V and −19.3 V.

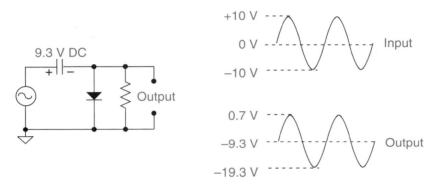

FIGURE 2–32 Clamper circuit with sine wave input

If a DC source is added, a clamper circuit can be designed to clamp the output to any desired DC reference. Figure 2–33(a) shows a clamper circuit designed to clamp the output to positive 3.3 V DC. When the input swings negative, the diode is forward biased, and the capacitor is charged to 13.3 V as shown in Figure 2–33(b).

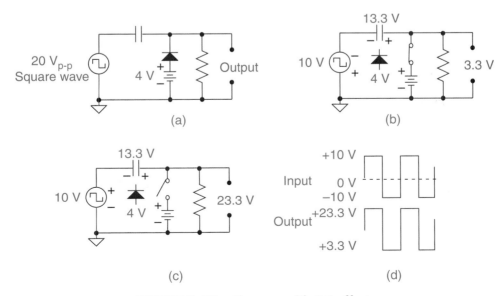

FIGURE 2–33 Clamper with DC offset

The output voltage is equal to positive 3.3 V on the negative swing of the input, since the diode is forward biased and the 3.3 V DC is connected directly across the output. When the input swings positive, the diode becomes reverse biased and is shown as an open switch, as in Figure 2–33(c). The positive 10 V across the input is in series aiding with the 13.3 V across the capacitor, and the output voltage goes to positive 23.3 V. There will, of course, be some discharge of the capacitor, but in a well-designed circuit, it should be negligible and the output should stay at 23.3 V until the input swings negative. When the input swings negative, the output will drop back to 3.3 V and the cycle will begin again. Figure 2–33(d) shows the input and the output waveforms.

In summary, a signal can be referenced to any DC value desired by using a clamper circuit that consists of a diode, capacitor, and DC source. In most cases with clamper circuits, the capacitor can be thought of as charging immediately when the diode is forward biased. When the diode is reverse biased, the capacitor maintains a constant voltage charge. In order for the clamper circuit to operate properly, the RC time constant of the capacitor and the load should be several times the period of the signal being clamped.

2.7 DIODE SWITCHING CIRCUITS

Figure 2–34(a) shows how diodes can be switched with bias voltage to control the flow of signal current. With switch S_1 in the up position, diode D_1 is forward biased and has approximately 2 mA [(12 V – 0.7 V)/($R_1 + R_2$)] of bias current flowing. This causes D_1 to have a dynamic resistance of 12.5 Ω (25 mV/I_D). Figure 2–34(b) shows the signal equivalent circuit. Diode D_1 is shown as a 12.5 Ω resistor that can be considered a short since all the other resistors are 3 kΩ. Therefore, R_1, R_2 and R_L can be considered in parallel having an equivalence of 1 kΩ. The 10 kHz signal from V_1 will be present across the load with an amplitude of 50 mV. With the switch in the up position, there is no bias current through D_2; therefore, it has a high dynamic resistance and the 20 kHz from V_2 will not be present at the load.

If S_1 is switched to the down position, the bias current through D_1 will be turned off and the 10 kHz signal from V_1 will not pass to the load. However, with the switch in the down position, D_2 is now forward biased. This permits the 20 kHz signal from V_2 to be present across the load. The signal voltage across R_2 is riding on a DC reference of approximately 5.7 V. Capacitor C_3

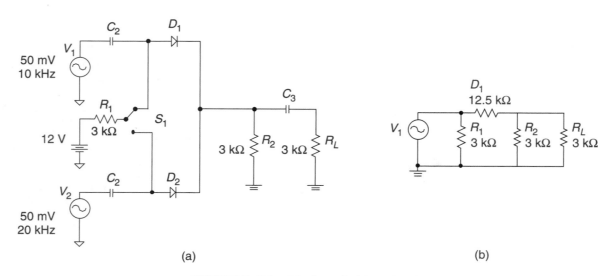

(a)   (b)

FIGURE 2–34 Diode switching circuit

blocks the DC voltage from the load; therefore, the signal voltage across R_L is riding on ground reference.

The diodes are turned "on" and "off" for the signal by controlling the bias voltage. In Figure 2–34(a), a mechanical switch controls the bias voltage; however, in many electronic circuits using diode switches, the bias voltage is controlled by electronic circuitry. This means signal flow can be controlled without the need for moving parts.

2.8 DIODE CHARACTERISTICS AND DATA SHEETS

There are hundreds of diodes on the market, and they all permit current flow when forward biased and inhibit current flow when reverse biased. The differences in diodes can be found by examining the electrical characteristics listed on data sheets. Peak repetitive reverse voltage, sometimes referred to as peak inverse voltage (PIV), and average forward current are the two characteristics most important to the technician. The following six characteristics are the most common characteristics found on data sheets.

1. **Peak repetitive reverse voltage** (V_{RRM}) is the maximum voltage that can be repeatedly applied to the diode in reverse bias. The reverse voltage rating could be as low as 5 V or as high as 2000 V.
2. **Average forward current** (I_o) is the maximum continuous current that can flow through a forward-biased diode. The maximum forward current could range from a few milliamps to over 1000 A.
3. **Peak surge forward current** (I_{FSM}) is the maximum amplitude of forward current pulse the diode can withstand. The value is typically ten to thirty times higher than the average forward current rating. For example, a diode with an average forward current rating of 12 A could easily have a peak surge forward current rating of 250 A.
4. **Forward voltage drop** (V_F) is the maximum voltage drop across the forward-biased diode. The typical voltage drop across a forward-biased silicon diode is 0.7 V; however, under high current conditions, the diode voltage may be considerably higher. Normally, the maximum forward voltage drop is given for maximum average forward current (I_o).
5. **Reverse current** (I_R) is the maximum leakage current that will flow through the diode when reverse biased. Reverse current is affected greatly by the operating temperature of the diode.
6. **Reverse recovery time** (t_{rr}) is the time required for the diode to stop conduction once it has been reverse biased. This is a particularly important parameter for high-speed switching diodes.

2.9 TROUBLESHOOTING DIODES

Diodes can be checked when out of the circuit by using a digital multimeter (DMM). When using a DMM, it is important to use only the scale designed to check diode junctions. The normal ohms scale on DMMs do not have a high enough output voltage to forward bias the diode junction; therefore, if a diode is read on these scales, it will read open in both directions. The scale used for checking diodes is normally indicated on a DMM with a diode symbol. Figure 2–35(a) shows the use of a DMM for checking a diode in the forward-biased direction. The diode scale is selected, then the "V-Ω-A" lead is connected to the anode of the diode, and the "Com" lead is connected to the cathode of the diode. The meter will read approximately 0.7 V, indicating the diode is conducting in forward bias. The leads are then reversed, and "Com" is connected to the anode and "V-Ω-A" is connected to the cathode. The meter will now give an open circuit reading, indicating that the diode is functioning as an open in reverse bias.

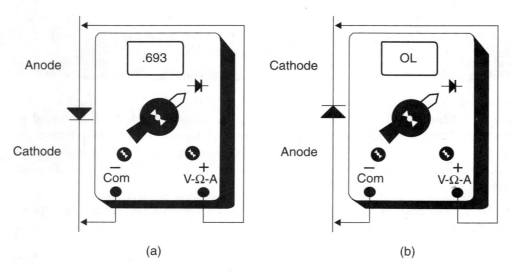

FIGURE 2–35   Diode checked with a DMM

Diodes can also be checked with an analog meter (VOM). Figure 2–36 shows a VOM being used to check a diode. Recall from your study of basic electronics that a VOM on the ohms scale is basically a battery in series with a current meter. If the battery in the meter forward biases the diode, there will be current flow, and the meter will read lower on the ohms scale as shown in Figure 2–36(a). On the other hand, if the battery reverse biases the diode, there will be no current flow (except for leakage current), and the meter will show a high ohms reading or infinite ohms as shown in Figure 2–36(b).

In most meters, the "V-Ω-A" lead is the more positive terminal with respect to the common lead. If this is the case, the meter readings will correspond to those in Figure 2–36. If the battery

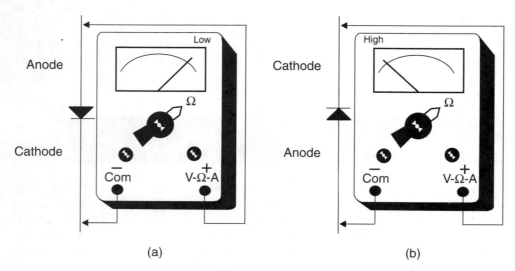

FIGURE 2–36   Diode checked with a VOM

in your meter is reversed, the readings will be reversed. The actual ohms reading is not important, but the ratio of forward-to-reverse resistance is important. The reverse-biased reading should be a minimum of ten times more (usually much more) than the forward-biased reading on the same ohms scale. Meters normally have several ohms scales; the lowest scale ($R \times 1$) causes the largest current to flow through the forward-biased diode. A small diode with a low forward current rating could be damaged using the $R \times 1$ ohms scale. It is, therefore, better to use the $R \times 10$ or $R \times 100$ scale for checking diodes. If a forward-biased diode is measured on two resistance scales, the ohm values will be different. The forward voltage dropped across the diode remains constant at approximately 0.7 V. As resistance scales are changed, the current through the diode changes, causing different meter readings.

If the diode is shorted, it will read a low resistance in both directions, and if the diode is an open, it will read infinite in both directions. In either case, the diode is bad and needs to be replaced. If a diode passes the meter test by reading a relatively low resistance when forward biased and a high resistance when reverse biased, the diode is probably good. The ohmmeter test is not perfect. Even if the diode passes the meter test, it still could break down under circuit voltages, have a slow switching speed, or experience other dynamic problems.

Precautions must be taken when checking diodes in circuits. First, make sure the power is removed from the circuit. Second, realize that diodes in circuits may have parallel paths connecting their anodes and cathodes together through other circuit components. Readings can be made, but the parallel circuitry must be considered. One way to check a diode in a circuit is to disconnect either lead from the circuit. Then the out-of-circuit check can be performed.

Diode Rectifier Circuits

1. Does the meter in Figure 2–37(a) read high or low? _____

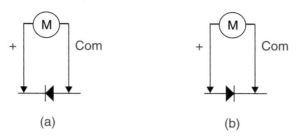

FIGURE 2–37

2. Does the meter in Figure 2–37(b) read high or low? _____
3. The circuit in Figure 2–38 has an input of 1 V$_p$. Calculate, draw, and label the output signal at TP_1 in the graph.

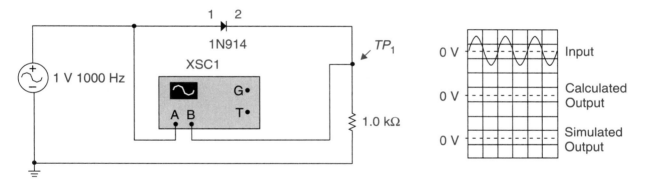

FIGURE 2–38

4. Simulate the circuit in Figure 2–38. Draw and label the output signal at TP_1 in the graph.
5. The circuit in Figure 2–39 has an input of 20 V$_p$. Calculate, draw, and label the output signal at TP_1 in the graph.

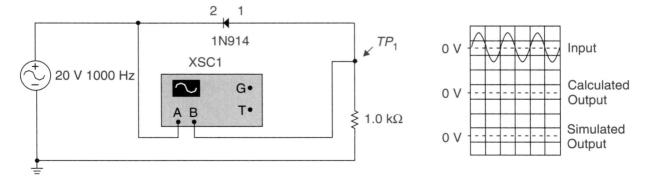

FIGURE 2–39

6.   Simulate the circuit in Figure 2–39. Draw and label the output signal at TP_1 in the graph.

7.   The circuit in Figure 2–40 has an input of 120 V_{RMS}. Calculate, draw, and label the output signal at TP_1 in the graph.

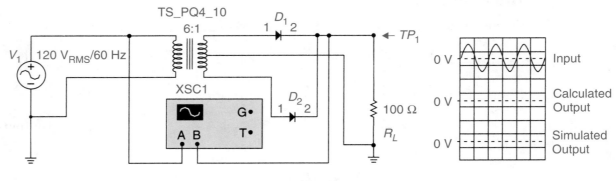

FIGURE 2–40

8.   Simulate, draw, and label the output signal at TP_1 for the circuit in Figure 2–40.

9.   The circuit in Figure 2–41 has an input of 120 V_{RMS}. Calculate, draw, and label the output signal at TP_1 in the graph.

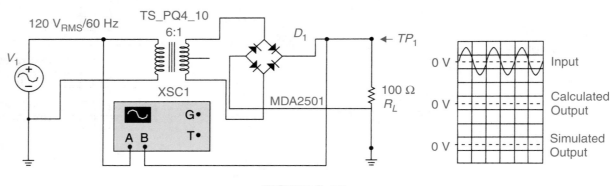

FIGURE 2–41

10.   Simulate, draw, and label the output signal at TP_1 for the circuit in Figure 2–41.

LAB 2.1

Diode Rectifier Circuits

Caution! This experiment uses line voltage that can cause severe electrical shock.

I. Objectives

1. To test diodes for shorts or opens using a multimeter (VOM) and a digital multimeter (DMM).
2. To examine the outputs of three types of rectifiers and observe how the measured outputs compare to the calculated values.

II. Test Equipment

(1) 0 to 10 V DC power supply   (1) 1 kHz sine wave generator

(1) VOM   (1) DMM

(1) dual trace oscilloscope

III. Components

(1) 2N4148

(4) 1N4004 diodes or equivalent

(1) 120 V primary, 12.6 V center-tapped secondary

(1) 1 kΩ resistor (R_L in all circuits)

IV. Procedure

1. Set the VOM to the $R \times 100$ range and connect the meter to forward bias the diode as shown in Figure 2–42(a). Record the ohms reading in Table 2–1.
2. Leaving the meter set on the $R \times 100$ range, connect the meter to reverse bias the diode as shown in Figure 2–42(b). Record the ohms reading in Table 2–1.

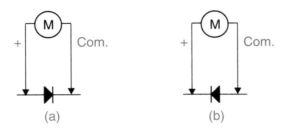

FIGURE 2–42

3. Change the VOM to $R \times 1$ k and repeat steps 1 and 2.
4. Use a DMM set to the diode range that is designed for reading diodes. Repeat steps 1 and 2.

TABLE 2–1

VOM $R \times 100$		VOM $R \times 1$ kΩ		DMM	
Forward bias	Reverse bias	Forward bias	Reverse bias	Forward bias	Reverse bias

5. Using a VOM or DMM, check the remaining three diodes, repeating steps 1 and 2.
6. For an input of 10 V DC, calculate and record the DC voltage at TP_1 for the circuit in Figure 2–43. Calculated voltage at $TP_1 =$ _____.

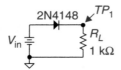

FIGURE 2–43

7. Construct the circuit in Figure 2–43. Measure and record the output voltage at TP_1. Measured voltage at $TP_1 =$ _____.
8. For an input signal of 10 $V_{p\text{-}p}$, draw and label the signal at TP_1 for the circuit in Figure 2–44.

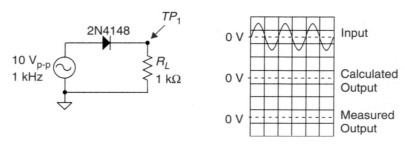

FIGURE 2–44

9. Use an oscilloscope to measure the signal at TP_1. Draw and label the output signal in Figure 2–44.
10. For an input of 120 V AC, draw and label the output signal at TP_1 for the circuit in Figure 2–45.

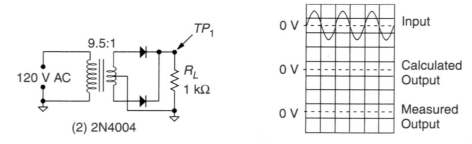

FIGURE 2–45

11. Use an oscilloscope to measure the signal at TP_1. Draw and label the output signal in Figure 2–45. Be sure to measure the input under load conditions.

  Caution! This step uses line voltage that can cause severe electrical shock.

12. For an input of 120 V AC, draw and label the output signal at TP_1 for the circuit in Figure 2–46.

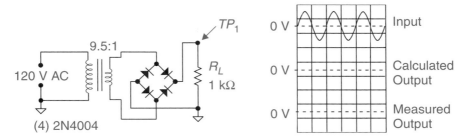

FIGURE 2–46

13. Use an oscilloscope to measure the signal at TP_1. Draw and label the output signal in Figure 2–46. Be sure to measure the input under load conditions.

 Caution! This step uses line voltage that can cause severe electrical shock.

V. Points to Discuss

1. Explain why the forward-biased ohms reading on the $R \times 100$ scale was different from the forward-biased ohms reading on the $R \times 1$ k scale.

2. If a DMM is used on one of the ohms scales not designed for diode checking, what type of reading would you expect and why?

3. If the diode passed the ohmmeter out-of-circuit test, is it guaranteed to work properly in the circuit? Explain.

4. Why must the forward voltage drop across diodes be considered when the input voltage is small?

5.   Why is the circuit in Figure 2–44 not as efficient as the circuits in Figures 2–45 and 2–46?

6.   With 120 V AC input, which circuit (Figures 2–45 or 2–46) delivers the largest peak signal to the load and why?

7.   Why can't DC voltages be applied to the circuits in Figures 2–45 and 2–46?

Clipper and Clamper Circuits

1. With a sine wave input set to 2 kHz and 25 V_{p-p}, calculate the signal at TP_1 for the circuit in Figure 2–47. Draw and label the calculated waveform.

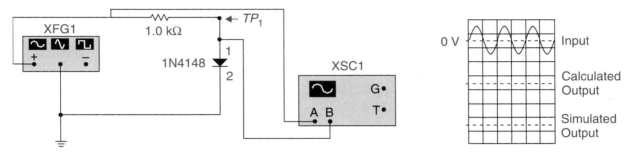

FIGURE 2–47

2. Simulate the circuit in Figure 2–47. Draw and label the waveform at TP_1.

3. With a sine wave input set to 2 kHz and 20 V_{p-p}, calculate the signal at TP_1 for the circuit in Figure 2–48. Draw and label the calculated waveform.

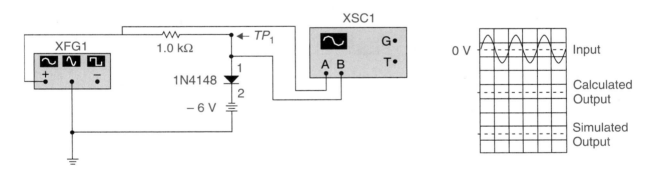

FIGURE 2–48

4. Simulate the circuit in Figure 2–48. Draw and label the waveform at TP_1.

5. With a square-wave input set to 2 kHz and 30 V_{p-p}, calculate the signal at TP_1 for the circuit in Figure 2–49. Draw and label the calculated waveform.

6. With a square-wave input, simulate the circuit in Figure 2–49. Draw and label the waveform at TP_1.

7. With a sine wave input set to 1 kHz and 20 V_{p-p}, calculate the output at TP_1 for the circuit in Figure 2–49. Draw and label the calculated waveform.

8. With a sine-wave input, simulate the circuit in Figure 2–49. Draw and label the waveform at TP_1.

9. With a square-wave input set to 1 kHz and 25 V_{p-p}, calculate the output signal at TP_1 for the circuit in Figure 2–50. Draw and label the calculated waveform.

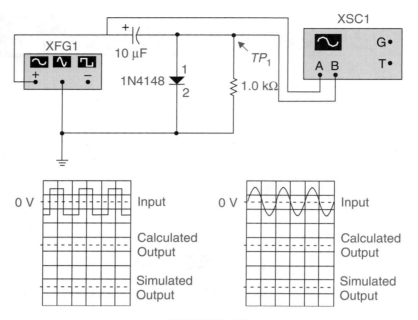

FIGURE 2–49

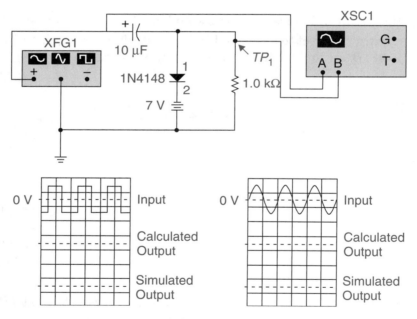

FIGURE 2–50

10. With a square-wave input, simulate the circuit in Figure 2–50. Draw and label the waveform at TP_1.

11. With a sine wave input set to 1 kHz and 16 $V_{p\text{-}p}$, calculate the output signal at TP_1 for the circuit in Figure 2–50. Draw and label the calculated waveform.

12. With a sine-wave input, simulate the circuit in Figure 2–50. Draw and label the waveform at TP_1.

LAB 2.2

Clipper and Clamper Circuits

I. Objective

To analyze the outputs of clipper and clamper circuits and compare the calculated outputs to the values measured experimentally.

II. Test Equipment

(1) function generator   (1) dual-trace oscilloscope

(1) 5 V DC power supply

III. Components

(1) 1 kΩ resistor   (1) 10 μF capacitor

(1) 1N4148 diode or equivalent

IV. Procedure

1. With a sine wave input set to 2 kHz and 10 V_{p-p}, calculate the signal at TP_1 for the circuit in Figure 2–51. Draw and label the calculated waveform.

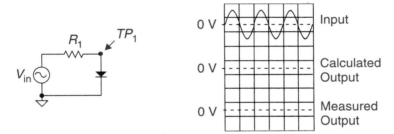

FIGURE 2–51

2. Construct the circuit in Figure 2–51 using a dual-trace oscilloscope to measure the waveform at TP_1. Draw and label the measured waveform.

3. With a sine wave input set to 2 kHz and 10 V_{p-p}, calculate the signal at TP_1 for the circuit in Figure 2–52 ($V_1 = 5$ V DC, $R = 1$ kΩ). Draw and label the calculated waveform.

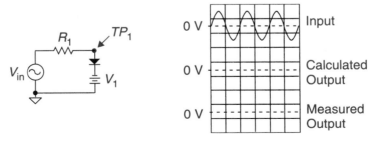

FIGURE 2–52

4. Construct the circuit in Figure 2–52 using a dual-trace oscilloscope to measure the waveform at TP_1. Draw and label the measured waveform.

5. With a square-wave input set to 2 kHz and 10 $V_{p\text{-}p}$, calculate the signal at TP_1 for the circuit in Figure 2–53 ($R = 1\ k\Omega$ and $C = 10\ \mu F$). Draw and label the calculated waveform.

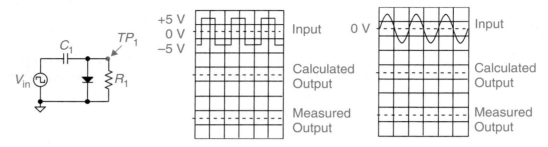

FIGURE 2–53

6. Construct the circuit in Figure 2–53 using a dual-trace oscilloscope to measure the waveform at TP_1. Draw and label the measured waveform.

7. With a sine wave input set to 2 kHz and 10 $V_{p\text{-}p}$, calculate the output at TP_1 for the circuit in Figure 2–53. Draw and label the calculated waveform.

8. With a 2 kHz and 10 $V_{p\text{-}p}$ sine wave connected to the circuit in Figure 2–53, use a dual-trace oscilloscope to measure the waveform at TP_1. Draw and label the measured waveform.

9. With a square-wave input set to 2 kHz and 10 $V_{p\text{-}p}$, calculate the signal at TP_1 for the circuit in Figure 2–54 ($V_1 = 5\ V\ DC, R = 1\ k\Omega$, and $C = 10\ \mu F$). Draw and label the calculated waveform.

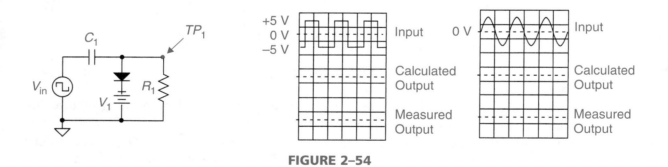

FIGURE 2–54

10. Construct the circuit in Figure 2–54 using a dual-trace oscilloscope to measure the waveform at TP_1. Draw and label the measured waveform.

11. With a sine wave input set to 2 kHz and 10 $V_{p\text{-}p}$, calculate the output at TP_1 for the circuit in Figure 2–54. Draw and label the calculated waveform.

12. With a 2 kHz and 10 $V_{p\text{-}p}$ sine wave connected to the circuit in Figure 2–54, use a dual-trace oscilloscope to measure the waveform at TP_1. Draw and label the measured waveform.

V. Points to Discuss

1. Explain any differences between calculated outputs and measured outputs for the clipping circuits shown in Figures 2–51 and 2–52.

2. How could the clipping level be changed in the circuit in Figure 2–52?

3. Discuss why the circuits in Figures 2–53 and 2–54 change the DC reference of the signal but do not change the wave shape.

4. How could the DC reference of the output of the circuit in Figure 2–54 be changed?

5. When a sine wave was connected to Figures 2–53 and 2–54, the positive peak of the sine wave was clipped. Explain why. (Hint: consider loading.)

QUESTIONS

2.1 Rectifier Circuits

_C_ 1. Which of the following is not a rectifier circuit?
  a. bridge
  b. half-wave
  c. clamper
  d. center-tapped transformer

_A_ 2. A _____ rectifier circuit uses four diodes.
  a. bridge
  b. half-wave
  c. clamper
  d. center-tapped transformer

_A_ 3. A _____ rectifier will have two diodes forward biased at the same time.
  a. bridge
  b. half-wave
  c. clamper
  d. center-tapped transformer

2.2 Filtering Pulsating DC

_C_ 4. An increase in load current will cause the ripple voltage to _____.
  a. decrease
  b. stay the same
  c. increase
  d. drop to zero

2.3 Block Diagram of a Complete Power Supply System

_A_ 5. The order of subsystems of a complete power supply from input to output is _____.
  a. rectifier, filter, and regulator
  b. filter, rectifier, and regulator
  c. rectifier, regulator, and filter
  d. none of the above

2.4 Voltage Multipliers

_A_ 6. Voltage multiplier circuits are constructed with diodes and _____.
  a. capacitors
  b. inductors
  c. only diodes and resistors
  d. clippers

2.5 Clipper Circuits

B 7. The clipper circuit changes the _____ of the output waveform.
    a. DC offset
    b. shape
    c. frequency
    d. bandwidth

2.6 Clamper Circuits

____ 8. The clamper circuit changes the _____ of the output waveform.
    a. DC offset
    b. shape
    c. frequency
    d. bandwidth

2.7 Diode Switching Circuits

A 9. A diode can be switched on for signal current by _____.
    a. forward biasing the diode with DC current
    b. forward biasing the diode with AC current
    c. reverse biasing the diode with DC voltage
    d. reverse biasing the diode with AC voltage

2.8 Diode Characteristics and Data Sheets

A 10. Normally the two most important diode ratings for the technician are _____ .
    a. peak repetitive reverse voltage and average forward current
    b. peak surge forward current and forward voltage drop
    c. reverse current and reverse recovery time
    d. average forward current and reverse current

2.9 Troubleshooting Diodes

____ 11. You remove a diode from the circuit and check it with a DMM. It reads over scale when forward biased and over scale when reverse biased. What is wrong with the diode?
    a. The diode is open.
    b. The diode is shorted.
    c. The diode is operating normally.

____ 12. You remove a diode from the circuit and check it with a DMM. It reads 0.636 when forward biased and over scale when reverse biased. What is wrong with the diode?
    a. The diode is open.
    b. The diode is shorted.
    c. The diode is operating normally.

_____ 13. You remove a diode from the circuit and check it with a DMM. It reads 0.636 when forward biased and 0.752 when reverse biased. What is wrong with the diode?

    a. The diode is open.

    b. The diode is shorted.

    c. The diode is operating normally.

_____ 14. The circuit in Figure 2–55 has a sine wave input. Diode D_1 is open. The output waveform is shown in _____ below.

    a. 0 V ——   b. 0 V ⌄⌄   c. 0 V ⌃⌃   d. 0 V ⌃⌄⌃

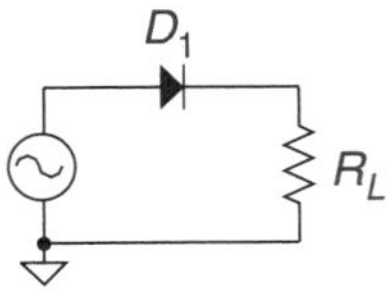

FIGURE 2–55

_____ 15. The circuit in Figure 2–55 has a sine wave input. Diode D_1 is shorted. The output waveform is shown in _____ below.

    a. 0 V ——   b. 0 V ⌄⌄   c. 0 V ⌃⌃   d. 0 V ⌃⌄⌃

_____ 16. The circuit in Figure 2–56 has a sine wave input. Diode D_1 is open. The output waveform is shown in _____ below.

    a. 0 V ——   b. 0 V ⌄⌄   c. 0 V ⌃⌃   d. 0 V ⌃⌄⌃

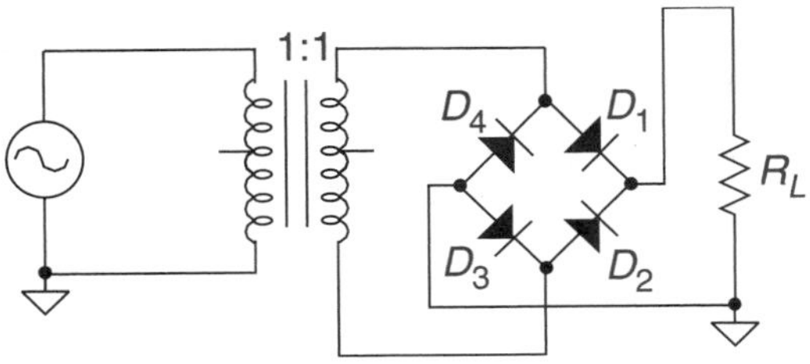

FIGURE 2–56

_____ 17. The circuit in Figure 2–56 has a sine wave input. Diode D_2 is open. The output waveform is shown in _____ below.

    a. 0 V ——   b. 0 V ⌄⌄   c. 0 V ⌃⌃   d. 0 V ⌃⌄⌃

_____ 18. The circuit in Figure 2–57 has a 50 V_p sine wave input. The output waveform is shown in _____ below.

    a. 0 V ——   b. 25 V ——   c. 50 V ——   d. 100 V ——   e. 50 V_{p-p} ⌃⌄⌃

_____ 19. The circuit in Figure 2–57 has a 35 V_{RMS} input. The output waveform is shown in _____ below.

    a. 0 V ——   b. 25 V ——   c. 50 V ——   d. 100 V ——   e. 50 V_{p-p} ⌃⌄⌃

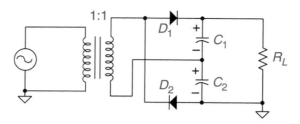

FIGURE 2-57

____ 20. The circuit in Figure 2–58 has a 20 $V_{p\text{-}p}$ square-wave input (+10 V to –10 V). Capacitor
C_1 is shorted. The output waveform is shown in ____ below.

    a. 0 V ____   b. 10 V ⎍⎍⎍   c. 6.7 V ⎍⎍⎍   d. 6.7 V ⎍⎍⎍
                                  –10 V   –10 V   –13.7 V

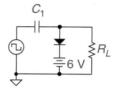

FIGURE 2-58

PROBLEMS

1. If 18 V is applied to the circuits, what voltage will the meter read in Figure 2–59(a)?
         _____ In Figure 2–59(b)? _____

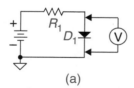
                                                     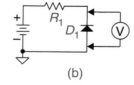

            (a)   (b)

FIGURE 2-59

2. If diode D_1 in Figure 2–60 has a maximum forward current rating of 300 mA, what is the
minimum ohms value of R_1 that can be used in the circuit? _____

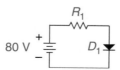

FIGURE 2-60

3. What will be the power dissipated in diode D_1 in the circuit in Figure 2–61(a)? _____
    In Figure 2–61(b)? _____

(a)   (b)

FIGURE 2–61

4. If D_1 in Figure 2–62 has a PIV rating of 400 V, what is the maximum voltage that can be
    applied to the circuit in Figure 2–62(a)? _____ In Figure 2–62(b)? _____

(a)   (b)

FIGURE 2–62

5. Draw and label the output waveform at TP_1 for the circuit in Figure 2–63.

FIGURE 2–63

6. Draw and label the output waveform at TP_1 for the circuit in Figure 2–64.

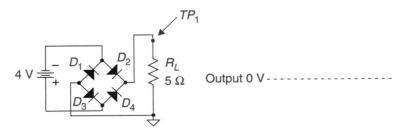

FIGURE 2–64

7. Draw and label the output waveform for the circuit in Figure 2–65.

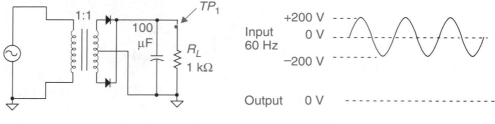

FIGURE 2–65

8. The circuit in Figure 2–66(a) has an input voltage of 2 V DC (V_{AB}). Calculate the signal at TP_1.

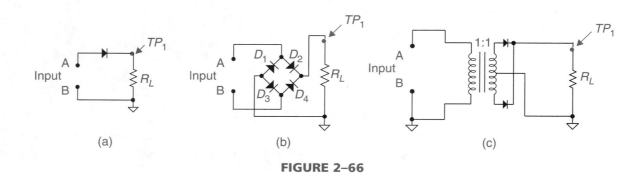

FIGURE 2–66

9. The circuit in Figure 2–66(b) has an input voltage of 2 V DC (V_{AB}). Calculate the signal at TP_1.

10. For an input voltage of 100 V_{p-p}, calculate and draw the signal at TP_1 for the circuit in Figure 2–66(a).

11. For an input voltage of 100 V_{p-p}, calculate and draw the signal at TP_1 for the circuit in Figure 2–66(b).

12. For an input voltage of 100 V_{p-p}, calculate and draw the signal at TP_1 for the circuit in Figure 2–66(c) (60 Hz input with a load equal to 1 kΩ).

13. Estimate the power dissipated by a bridge rectifier when it supplies 5 A of current.

14. Estimate the power dissipated by the diodes in a full-wave rectifier using a center-tapped transformer when it supplies 5 A of current.

15. Calculate the percentage of error in the output caused by ignoring the forward drop of the diodes in a full wave-bridge rectifier when (a) the input is 200 V_{p-p}, and (b) the input is 10 V_{p-p}.

16. Draw and label the output waveform for the circuit in Figure 2–67. The input signal is a 50 V_p sine wave at a frequency of 2 kHz.

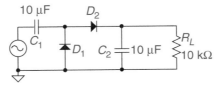

FIGURE 2–67

17. Draw and label the output waveform for the circuit in Figure 2–68. The input signal is a 50 V_p sine wave at a frequency of 2 kHz.

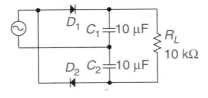

FIGURE 2–68

18. Draw and label the output waveform for the circuit in Figure 2–69.

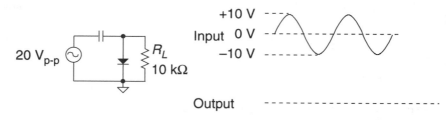

FIGURE 2–69

19. Draw and label the output waveform for the circuit in Figure 2–70.

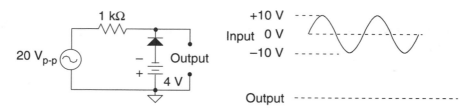

FIGURE 2–70

20. For a sine wave input voltage of 22 $V_{p\text{-}p}$, draw the signal at TP_1 for the circuit in Figure 2–71(a) (R_1 = 1 kΩ).

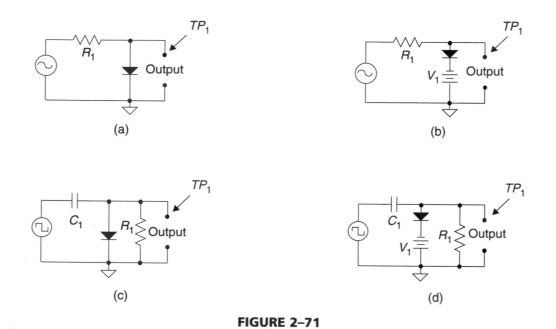

FIGURE 2–71

21. For a sine wave input voltage of 8 $V_{p\text{-}p}$, draw the signal at TP_1 for the circuit in Figure 2–71(b) (R_1 = 1 kΩ, V_1 = –3 V).

22. For a square-wave input voltage of 8 V_{p-p}, draw the signal at TP_1 for the circuit in Figure 2–71(c) ($R_1 = 1$ kΩ).

23. For a square-wave input voltage of 28 V_{p-p}, draw the signal at TP_1 for the circuit in Figure 2–71(d) ($R_1 = 1$ kΩ, $V_1 = -3$ V).

CHAPTER 3

Special Diodes

OBJECTIVES

After studying the material in this chapter, you will be able to describe and/or analyze:

○ the function of zener diodes,

○ zener diode applications,

○ the function of voltage surge protectors,

○ the function of varactor diodes,

○ the function of high frequency switching diodes,

○ the function of light-emitting diodes,

○ the function of photodiodes, and

○ troubleshooting techniques for special diodes.

3.1 ZENER DIODES

Figure 3–1 shows the current versus voltage curve for a normal diode and a zener diode. In the forward-biased direction, the normal diode and the zener diode both permit current flow if 0.7 V is exceeded. When in reverse bias, the normal diode will not permit current flow unless the peak reverse voltage rating is exceeded (Figure 3–1[a]). Once the normal diode breaks down and starts conducting in reverse bias, the power rating of the diode will be exceeded, causing the diode to be destroyed. When in reverse bias, the zener diode will permit current flow once the zener breakdown voltage is reached (Figure 3–1[b]). The zener diode is designed to operate in the zener region without causing damage to the zener. The zener breakdown voltage (V_z) is determined during the manufacturing process. Zener diodes are available with zener voltages in the range of 1.8 V to 400 V.

An important fact to note from the graph in Figure 3–1(b) is that once the zener voltage is exceeded and the diode is operating in the zener region, the voltage across the diode stays almost constant even with changes in current. The fact that the zener diode is capable of maintaining a near constant voltage with changes in current makes it a very useful device. In the forward-biased direction, the zener diode functions like an ordinary diode. In most applications,

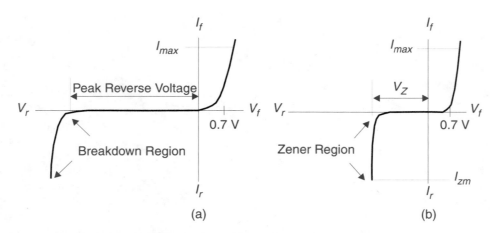

FIGURE 3–1   Voltage versus current curves

zener diodes are used in the reverse-biased mode. Figure 3–2 shows the symbol used for zener diodes. Notice that the ends of the cathode bar are slanted.

FIGURE 3–2   Zener diode symbol

ZENER DIODE LIMITATIONS

Zener diodes, like all other electronic devices, can be destroyed by excessive heat. The major limiting factor is the power that can be dissipated by the device. Zener diodes are available with power ratings from 1/4 W to 50 W. The difference between a zener diode and an ordinary diode is the ability of the zener to handle current in the zener breakdown region, but even the zener diode has limitations. The maximum current that the zener diode can handle in the zener breakdown region is called *maximum regulator current* (I_{zm}). I_{zm} is a function of the power and voltage rating of the particular diode. Example 3.1 demonstrates how I_{zm}, V_z, and the power rating of a zener diode are related.

EXAMPLE 3.1

A 5 W zener diode is rated for a zener voltage (V_z) of 15 V. What is the maximum regulator current (I_{zm}) for this zener?

Using the maximum power and voltage rating of zener diodes, calculate the maximum regulator current (I_{zm}) by applying the power formula:

$$I = P/V = 5 \text{ W}/15 \text{ V} = 0.333 \text{ A or } 333 \text{ mA}$$

3.2 ZENER DIODE APPLICATIONS

VOLTAGE REFERENCE

Zener diodes are often used to provide reference voltages in circuits. If a zener diode is placed in series with a resistor connected to a voltage source as shown in Figure 3–3, the voltage at the

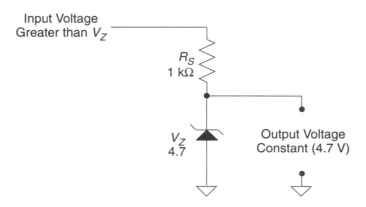

FIGURE 3–3   Voltage reference

junction of the resistor and zener will remain constant. The input voltage applied to the circuit must be larger than the zener breakdown voltage (V_z). Example 3.2 calculates the voltage and power dissipated for different values of input voltages.

EXAMPLE 3.2

Calculate the output voltage, power dissipated by R_s, and the power dissipated by the zener for the circuit in Figure 3–3 for an input voltage of (a) 3 V, (b) 6 V, and (c) 12 V.

Step 1. Calculate the output voltage of the circuit. The voltage drop across the zener is equal to the output voltage of the circuit. If a zener is reverse biased in a circuit, the voltage across the zener will be equal to or less than the rated voltage of the zener. To calculate the voltage drop, assume the reverse-biased zener to be an open, and calculate the voltage across the open. If the calculated voltage is less than the rated voltage of the zener, the zener will not conduct, and the voltage calculated is the voltage across the zener. If the voltage is greater than the rated voltage of the zener, the zener will conduct, and the voltage across the zener will equal the rated voltage of the zener (V_z).

(a) 3 V input voltage would cause 3 V to appear across an open zener, so the voltage across the 4.7 V zener would equal 3 V.

(b) 6 V input voltage would cause 6 V to appear across an open zener, so the voltage across the 4.7 V zener would equal 4.7 V.

(c) 12 V input voltage would cause 12 V to appear across an open zener, so the voltage across the 4.7 V zener would equal 4.7 V.

Step 2. Calculate the power dissipated by R_s.

(a) With an input of 3 V, the zener is not conducting; therefore, no current is flowing in the circuit. With zero current flowing, resistor R_s will dissipate zero watts.

$$P_{Rs} = 0 \text{ W}$$

(b) With an input of 6 V, the zener is conducting and drops 4.7 V, and the remaining voltage is dropped across R_s.

$$V_{Rs} = V_{\text{input}} - V_z = 6 \text{ V} - 4.7 \text{ V} = 1.3 \text{ V}$$
$$I_{Rs} = V_{Rs}/R_s = 1.3 \text{ V}/1 \text{ k}\Omega = 1.3 \text{ mA}$$
$$P_{Rs} = V_{Rs} \times I_{Rs} = 1.3 \text{ V} \times 1.3 \text{ mA} = 1.69 \text{ mW}$$

EXAMPLE 3.2 continued

(c) With an input of 12 V, the zener is conducting and drops 4.7 V, and the remaining voltage is dropped across R_s.

$$V_{Rs} = V_{input} - V_z = 12\ V - 4.7\ V = 7.3\ V$$
$$I_{Rs} = V_{Rs}/R_s = 7.3\ V/1\ k\Omega = 7.3\ mA$$
$$P_{Rs} = V_{Rs} \times I_{Rs} = 7.3\ V \times 7.3\ mA = 53.3\ mW$$

Step 3. Calculate the power dissipated by the zener diode.

(a) With an input of 3 V, the zener is not conducting; therefore, no current is flowing in the circuit. With zero current, the zener will dissipate zero watts.

$$P_Z = 0\ W$$

(b) With an input of 6 V, the zener is conducting, and the current flowing through the zener is the same as the current flowing through resistor R_s.

$$I_{Rs} = 1.3\ mA\ \text{(calculated in step 2)}$$
$$I_z = I_{Rs} = 1.3\ mA$$
$$P_Z = V_Z \times I_Z = 4.7\ V \times 1.3\ mA = 6.11\ mW$$

(c) With an input of 12 V, the zener is conducting, and the current flowing through the zener is the same as the current flowing through resistor R_s.

$$I_{Rs} = 7.3\ mA\ \text{(calculated in step 2)}$$
$$I_z = I_{Rs} = 7.3\ mA$$
$$P_Z = V_Z \times I_Z = 4.7\ V \times 7.3\ mA = 34.3\ mW$$

Once the input voltage exceeds the zener voltage, the output voltage remains constant. The current flow does increase with increases in input voltage; consequently, circuit components dissipate more power as the input voltage increases.

VOLTAGE REGULATION

A voltage regulation circuit is a circuit that delivers a constant DC voltage to an electronic load. Good voltage regulator circuits maintain a constant voltage across the load with changes in input voltage or changes in the current demands of the load. Zener diodes can be used in regulator circuits. Figure 3–4 shows a voltage regulator circuit consisting of a series dropping resistor (R_s) and a zener diode. The zener is in parallel with the load, and this parallel combination is in series with R_s.

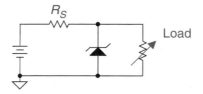

FIGURE 3–4 Zener diode voltage regulator

The current through R_s divides: some of it flows through the load, and the remainder flows through the zener. If the load resistance goes down, the load current will go up, and less current will flow through the zener. If the load resistance goes up, the load current will go down, and more current will flow through the zener. The maximum current will flow through the zener when the load is open.

The input voltage to the regulator can change, but as long as current flows through the zener, the circuit will provide a constant output voltage across the load. Example 3.3 shows a design of a voltage regulator circuit that maintains a constant voltage across a load.

EXAMPLE 3.3

Design a voltage regulator to maintain a constant 8 V across the load for load currents ranging from 0 mA to 20 mA. The circuit should be designed to operate from an input voltage of 12 V to 15 V DC.

Step 1. Draw the schematic of the voltage regulator circuit showing the range of input voltages, the desired range of output current, and the voltage rating of the zener. (See Figure 3–5.)

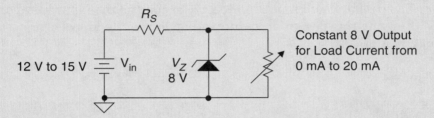

FIGURE 3–5 Voltage regulator

Step 2. A zener diode with a V_z rating of desired output is selected.
$$V_z = 8 \text{ V}$$

Step 3. Determine the current flow through R_s. The zener diode should draw some current even when the load current is at maximum. Setting the current through R_s to 110% of the maximum load current will ensure zener current at maximum load current.
$$I_{Rs} = 20 \text{ mA} \times 110\% = 22 \text{ mA}$$

Step 4. Calculate the value of the voltage drop across R_s. Use Kirchhoff's voltage law to calculate V_{Rs}. By examining the circuit in Figure 3–5, you should see a Kirchoff's voltage loop containing V_{in}, V_{Rs}, and V_z. To ensure that adequate current flows through R_s under all conditions, the lowest value of V_{in} should be used for this calculation.
$$V_{Rs} = V_{in(min)} - V_z$$
$$V_{Rs} = 12 \text{ V} - 8 \text{ V} = 4 \text{ V}$$

Step 5. Use Ohm's law and the values calculated in steps 3 and 4 to calculate the value of R_s.
$R_s = V_{Rs}/I_{Rs}$ 4 V/22 mA = 182 Ω (Next lower standard value is 180 Ω.)

Step 6. Calculate the power dissipated in R_s under the worst case conditions. The worst case power dissipation will occur when the input voltage is at maximum. First, calculate voltage across R_s when the input is maximum, then calculate the power dissipated.
$$V_R = V_{in} - V_z \qquad\qquad V_{Rs} = 15 \text{ V} - 8 \text{ V} = 7 \text{ V}$$
$P = V^2/R = 7 \text{ V}^2/180 \Omega = 0.27 \text{ W}$ (Normal practice is to select a resistor with a power rating approximately twice the power dissipated. In this case, a 0.5 W resistor will be adequate.)

EXAMPLE 3.3 continued

> **Step 7.** Calculate the power dissipated in the zener diode under worst case conditions. The zener will dissipate maximum power when it has maximum current flow. Maximum zener current flow will occur when the load is drawing zero current and the input voltage is maximum. First, calculate the maximum zener current, then calculate the power dissipated by the zener under this condition.
>
> $I_{Rs} = 7\text{ V}/180\ \Omega = 38.9\text{ mA}$ (If the load current is 0 mA, then 38.9 mA flows through the zener.)
>
> $P = IV = 38.9\text{ mA} \times 8\text{ V} = 311.2\text{ mW}$ (0.5 W zener or larger should be used.)

The voltage regulator in Figure 3–5 should use a 180 Ω/0.5 W series resistor and an 8 V/0.5 W zener diode. This will permit the voltage to stay constant at 8 V over a load current range from 0 mA to 20 mA. The circuit would function properly as long as the input voltage stays between 12 V and 15 V. Studying this design procedure will give you the knowledge needed to analyze zener regulator circuits.

EXAMPLE 3.4

> For the circuit shown in Figure 3–6, find the values of I_{Rs}, I_L, I_z, P_{Rs}, and P_z.
>
>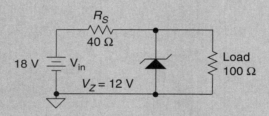
>
> **FIGURE 3–6**   Zener voltage regulator
>
> $$I_{Rs} = (V_{in} - V_z)/R_s = (18\text{ V} - 12\text{ V})/40\ \Omega = 150\text{ mA}$$
> $$I_L = V_z/R_L = 12\text{ V}/100\ \Omega = 120\text{ mA}$$
> $$I_z = I_{Rs} - I_L = 150\text{ mA} - 120\text{ mA} = 30\text{ mA}$$
> $$P_{Rs} = I_{Rs}^2 \times R_s = 150\text{ mA}^2 \times 40\ \Omega = 0.9\text{ W}$$
> $$P_z = V_z \times I_z = 12\text{ V} \times 30\text{ mA} = 0.36\text{ W}$$

I_{Rs} must be larger than I_L or the circuit is not regulating the output voltage. If I_L calculates larger than I_{Rs}, the circuit is functioning as a simple series circuit consisting of R_s and R_L. In this case, disregard the zener diode and calculate the current for the series circuit consisting of R_s and R_L. (See Example 3.5.)

EXAMPLE 3.5

> For the circuit shown in Figure 3–7, find the values of I_{Rs}, I_L, I_z, V_{Rs}, V_{RL}, P_{Rs}, and P_z.
> $$I_{Rs} = (V_{in} - V_z)/R_s = (18\text{ V} - 12\text{ V})/40\ \Omega = 150\text{ mA}$$
> $I_L = V_z/R_L = 12\text{ V}/50\ \Omega = 240\text{ mA}$ (I_L is larger than I_{Rs}. The circuit is not regulating and is operating as a simple series circuit consisting of R_L and R_s.)

EXAMPLE 3.5 continued

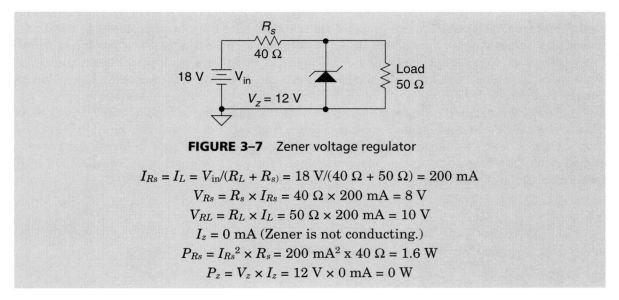

FIGURE 3–7   Zener voltage regulator

$$I_{Rs} = I_L = V_{in}/(R_L + R_s) = 18 \text{ V}/(40 \text{ }\Omega + 50 \text{ }\Omega) = 200 \text{ mA}$$
$$V_{Rs} = R_s \times I_{Rs} = 40 \text{ }\Omega \times 200 \text{ mA} = 8 \text{ V}$$
$$V_{RL} = R_L \times I_L = 50 \text{ }\Omega \times 200 \text{ mA} = 10 \text{ V}$$
$$I_z = 0 \text{ mA (Zener is not conducting.)}$$
$$P_{Rs} = I_{Rs}^2 \times R_s = 200 \text{ mA}^2 \text{ x } 40 \text{ }\Omega = 1.6 \text{ W}$$
$$P_z = V_z \times I_z = 12 \text{ V} \times 0 \text{ mA} = 0 \text{ W}$$

When we consider that the zener in Figure 3–7 is in parallel with the load, it is clear that the voltage across the zener is 10 V ($V_z = V_{RL}$). This voltage is below the zener voltage ($V_z = 12$ V), so the zener will act as an open and will not provide regulation. In order for a zener diode to provide regulation, the voltage across the load set up by the voltage divider action of R_s and R_L must be equal to or exceed the zener voltage rating of the diode being used.

3.3 VOLTAGE SURGE PROTECTORS

Figure 3–8 shows back-to-back zener diodes connected across the power line input to protect electronic equipment from power surges. The peak voltage of a 120 V AC line is 170 V_p. Electronic equipment has power supply circuitry that is designed to input AC line voltage and convert it to the needed DC voltages. Occasional power surges will occur on the power line, causing the peak voltage to reach thousands of volts. Lightning is one of the main causes for power surges, but other factors such as large inductive loads cutting on or off can also cause surges on the line.

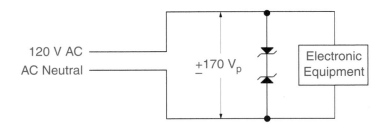

FIGURE 3–8   Zener surge protection

If the zener diodes in Figure 3–8 are rated for a breakdown voltage of 180 V, then under normal conditions, zener diodes will not conduct, and the equipment will receive the needed line voltage. If a power surge occurs and the line exceeds 180.7 V peak, the zener diodes will conduct, clamping the input voltage to 180.7 V.

A high voltage surge will cause a large current to flow through the zener diodes, causing a large power dissipation. Fortunately, the duration of surges are normally less than a millisecond. Manufacturers have developed a special group of zener diodes that have been designed to handle large power dissipation for short periods of time. These devices are called *transient voltage suppressors* (TVS). Motorola, for example, has a line of TVS devices rated for a peak power dissipation (P_{pk}) of 1500 watts. The data book states that these devices can dissipate 1500 watts for 1 mS. Transient voltage suppressor devices are available as single zener devices. Transient voltage suppressors are available with breakdown voltages starting as low as 6 V and going as high as 250 V. Unidirectional TVS devices use a schematic symbol similar to the normal zener diode symbol shown in Figure 3–10. Two unidirectional TVS devices must be connected back-to-back to give bidirectional protection. Bidirectional TVS devices are available containing two internally connected TVS devices. The symbol for a bidirectional TVS device is shown in Figure 3–9.

FIGURE 3–9   Unidirectional and bidirectional TVS devices

A metal-oxide varistor (MOV), also called a varistor, performs the same function as a zener transient voltage suppressor. The difference is that varistors are made from zinc oxide granules rather than from normal semiconductor materials. The construction of varistors permits them to dissipate more power, so varistors can handle larger surge voltages and current. Varistors are available with breakdown voltages from 8 V to 1000 V. One disadvantage of varistors is their high capacitance; consequently, they are seldom used across telecommunication lines. The symbol for the varistor is shown in Figure 3–10.

FIGURE 3–10   Varistor symbol

3.4 VARACTOR DIODES

The varactor diode is a device that functions as a variable capacitor. Recall that a capacitor is a device that consists of two conductive plates separated by a dielectric (insulating material). The value of capacitance is a function of three factors: the area of the plates, the distance between the plates, and the type of dielectric material separating the plates. The following formula shows the relationship between these factors and the magnitude of capacitance. The capacitance is directly proportional to the area of the plates (A) and the dielectric constant (K). The capacitance is indirectly proportional to the distance (d) between the plates.

$$C = KA/d$$

Figure 3–11(a) shows the inside of a reversed-biased diode. Notice that it consists of two regions that have current carriers (the P region and the N region) separated by a depletion region without current carriers. A reverse-biased diode acts like a capacitor; the P and N regions function as conductive plates, and the depletion region functions as a dielectric. Figure 3–11(b) shows that as the reverse-biased voltage is increased, the depletion area widens. The device still has capacitance, but since the conductive regions are farther apart, the capacitance has been reduced.

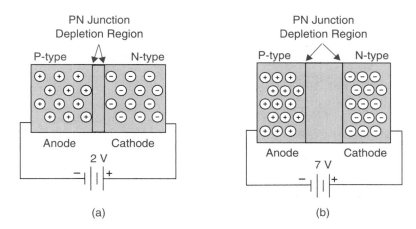

FIGURE 3–11 The PN junction as a capacitor

A varactor is a diode designed to have high junction capacitance. The capacitance of the varactor is controlled by the magnitude of the reverse-biased voltage applied. The greater the reverse-biased voltage, the less the capacitance. Figure 3–12 shows the symbol for the varactor and a graph of reverse-biased voltage versus capacitance for a MVAM108 varactor diode. The relationship between the reversed-bias voltage and the capacitance of a varactor is a nonlinear function. The graph in Figure 3–12 uses a linear scale for the voltage and logarithmic scale for the capacitance. The graph shows that a small voltage change causes a large capacitance change.

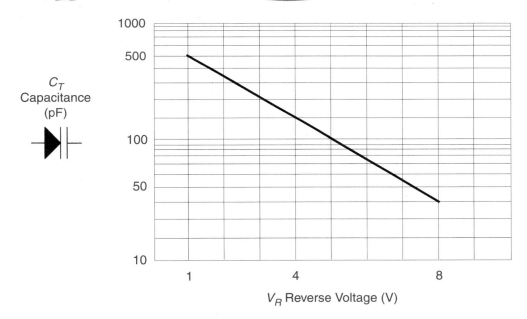

FIGURE 3–12 MVAM108 capacitance versus reverse voltage

The circuit in Figure 3–13 shows how a varactor can be used to control the selection of a signal from an antenna. At resonance, a parallel tuned circuit has a high impedance. A signal present on the antenna at the resonant frequency of the tuned circuit will develop a voltage drop across the high impedance of the tuned circuit, and the signal will be amplified. Signals at other frequencies will see the tuned circuit as a low impedance to ground and will not be amplified. The equivalent capacitance of the tuned circuit is equal to 500 pF in parallel with the series combination 0.1 µF capacitor and the capacitance of the varactor. Example 3.6 shows how the MVAM108 varactor can be used to adjust the resonate frequency of the tuned circuit.

EXAMPLE 3.6

Calculate the resonant frequency of the tuned circuit in Figure 3–13 for a voltage setting of (a) 1 V and (b) 7 V.

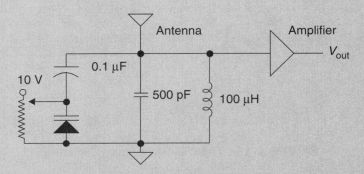

FIGURE 3–13   Varactor tuner

Step 1. Read the graph in Figure 3–12 and determine the capacitance of the MVAM108 varactor at 1 V reverse bias and 7 V reverse bias.

(a) 500 pF @ 1 V

(b) 55 pF @ 7 V

Step 2. Calculate the equivalent capacitance of the tuned circuit. Since the varactor capacitance is much smaller than 0.1 μF, the capacitance of the series combination will equal the capacitance of the varactor. The total equivalent capacitance of the tuned circuit is equal to the value of the varactor in parallel with 500 pF.

(a) C_{eq} @ 1 V = 500 pF + 500 pF = 1000 pF

(b) C_{eq} @ 1 V = 55 pF + 500 pF = 555 pF

Step 3. Calculate the resonate frequency for both settings of the varactor.

$$F_R = \frac{1}{2\pi\sqrt{LC}}$$

$$(a)\, F_R = \frac{1}{2\pi\sqrt{100\ \mu H \times 1000\ \rho F}} = 504\ kHz$$

$$(b)\, F_R = \frac{1}{2\pi\sqrt{100\ \mu H \times 555\ \rho F}} = 676\ kHz$$

3.5 HIGH FREQUENCY SWITCHING DIODES

INTRODUCTION

The varactor diode is an example of utilizing the capacitance present in a reverse-biased PN junction diode. All PN junction diodes have some capacitance. The PN junction capacitance is not a problem at low frequencies, but at high frequencies, the capacitive reactance (X_c) of the PN junction may drop to the point where the diode no longer inhibits current flow in reverse bias.

Reverse recovery time (t_{rr}) is the time required for the diode to stop conduction once it has been reverse biased. The reverse recovery time becomes an important factor at high switching speeds. Low frequency rectifier diodes will have reverse recovery time rated in microseconds, whereas high speed diodes will have reverse recovery time rated in nanoseconds. Manufacturers have developed high frequency switching diodes that can operate at frequencies above 3000 MHz.

STEP-RECOVERY DIODES

The step-recovery diode is a PN junction diode. The P and N material near the junction is lightly doped. The doping of the semiconductor material gets gradually heavier as the distance from the junction increases. The design of the step-recovery diode reduces the junction capacitance and permits high frequency operation. This device uses the normal diode symbol.

PIN DIODES

Figure 3–14 shows the inside of the PIN diode. The PIN diode is constructed of heavily doped P and N semiconductor material separated by a region of undoped or intrinsic semiconductor material. The name *PIN diode* is derived from the type of semiconductor materials used to construct the diode. The intrinsic material acts as a dielectric separating the two conductive regions. The separation of the two conductive regions reduces the junction capacitance of the diode and makes the PIN diode usable at high frequencies. This device uses the normal diode symbol.

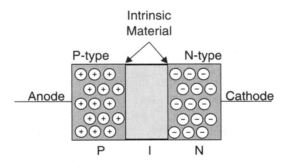

FIGURE 3–14 PIN diode

SCHOTTKY DIODES

The Schottky diode, also called the *hot-carrier diode*, does not contain a PN junction. The Schottky diode junction is formed between a barrier metal (gold, platinum, silver) and N-type semiconductor material as shown in Figure 3–15. The device acts like a normal diode except it drops approximately 0.3 V when forward biased, and the reverse breakdown voltages are normally less than 50 V. The advantages of the Schottky diode are its low forward voltage drop

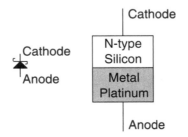

FIGURE 3–15 Schottky diode

and its switching speed. Schottky diodes can be designed to go from on to off in 10 nS. Because of their fast switching speed and the low forward voltage drop, Schottky diodes are often used in switching power supply applications. The symbol for the Schottky diode is shown in Figure 3–15.

3.6 LIGHT-EMITTING DIODES (LEDs)

INTRODUCTION

When a diode is forward biased, current flows, causing a junction voltage drop. This voltage drop times the magnitude of current flow results in the amount of power dissipated in the junction. In a normal diode, this power is dissipated in the form of heat energy. However, special diodes have been developed that dissipate this energy as light. These diodes are called *light-emitting diodes* (LEDs). There are two differences between normal diodes and LEDs. One difference is that LEDs are made from elements like phosphor, arsenic, or gallium. Diodes manufactured from these elements will radiate visible light energy when forward biased. When the normal diode is forward biased, electrons combine with holes at the junction and energy is given off in the form of heat. When the LED is forward biased, electrons combine with holes and energy is given off in the form of light. The second difference is that LEDs are designed so that they have a window over the junction so the light energy can be seen.

ANALYZING LEDs

Light-emitting diodes are available in various sizes and colors. When analyzing an LED in a circuit, the general rules used for normal diodes apply. However, LEDs have a higher forward voltage drop than normal silicon diodes. The drop across the forward-biased junction of an LED will range from 1.5 V to 3 V. The actual voltage drop across an LED will depend on the size, color, and type of LED. The current needed to properly illuminate an LED will range from 10 to 50 mA. Generally, the larger the LED, the greater the current requirement. In reverse bias, an LED functions the same as a normal diode and does not permit current flow. However, the PIV is extremely low, normally less than 6 V, so care must be taken to avoid damaging LEDs when reverse biased.

Figure 3–16(a) shows the symbol for an LED, and Figure 3–16(b) shows an LED connected to a power supply through a current limiting resistor (R_s). The resistor is necessary to limit current when forward biased to prevent the LED from being damaged. The value of the resistor can be calculated by subtracting the voltage drop of the LED from the supply voltage and dividing by the desired forward current $[(V_{cc} - V_{LED})/I_{LED}]$. The voltage drop across the LED and current requirements can be obtained from the manufacturer's specification sheet. For troubleshooting

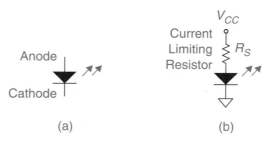

FIGURE 3–16   LEDs

purposes, a good nominal value to use for the forward voltage drop is 2 V. Example 3.7 demonstrates how the series resistor can be calculated.

EXAMPLE 3.7

An LED is to operate from 15 V. From the data sheet, it is determined that 10 mA of current is required. Calculate the value of the current limiting resistor.

Step 1. Calculate the voltage across the current series limiting resistor.
$$V_{Rs} = V_s - V_{LED} = 15 \text{ V} - 2 \text{ V} = 13 \text{ V}$$

Step 2. Calculate the resistance of the series limiting resistor. Once the resistance value is calculated, normally the next lower standard value resistor is selected for use in the circuit. This ensures sufficient current flow through the LED.
$$R_s = V_{Rs}/I = 13 \text{ V}/10 \text{ mA} = 1.3 \text{ k}\Omega \text{ (Next lower standard value} = 1.2 \text{ k}\Omega.)$$

SEVEN-SEGMENT DISPLAY

A popular application of LEDs is in display units. One of the most widely used is the seven-segment display. Seven LEDs are arranged in a pattern that forms an eight. Each LED is referred to as a segment and is identified with a letter as shown in Figure 3–17. By lighting different combinations of segments, all digits in the base ten number system can be formed (0 through 9). To simplify connecting the display, one terminal of all the LEDs is connected together and brought out to a common pin. If all the cathodes are connected, the display is referred to as a *common cathode display*, and if all the anodes are connected, the display is referred to as a *common anode display* (shown in Figure 3–17). Often an eighth segment is added to function as a decimal point.

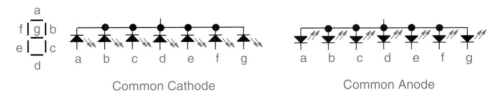

FIGURE 3–17   Seven-segment display

OTHER LED DISPLAYS

Light-emitting diode displays are also available in dot-matrix and alphanumeric displays. Figure 3–18(a) is a dot-matrix display. There are thirty-five LEDs making up a 5 × 7 matrix. By forward biasing selected diodes, it is possible to form all the letters of the alphabet, numbers, and numerous graphical characters. Light-emitting diodes are normally connected in a matrix as shown in Figure 3–18(b). There are five columns of inputs and seven rows of inputs. By applying a positive voltage to the selected row and a ground to the selected column, the diode at the junction of the row and column will be illuminated. In order to use all the LEDs in the matrix, it is necessary to use them one column at a time. If a positive voltage is applied to the selected row inputs and the first column is supplied with a ground, the desired LEDs in that column will be illuminated. If this process is repeated, one column at a time, all selected LEDs can be illuminated. It is true that only one column of LEDs can be illuminated at any given moment, but if this process is repeated fast enough (50 Hz minimum), the human eye cannot detect it, and all selected LEDs will appear illuminated simultaneously. This time-multiplexing process is called scanning. It greatly reduces the number of connections needed for the display. By using the scanning technique, only twelve connections are needed. If scanning is not used, thirty-six connections are required.

Dot-matrix Display

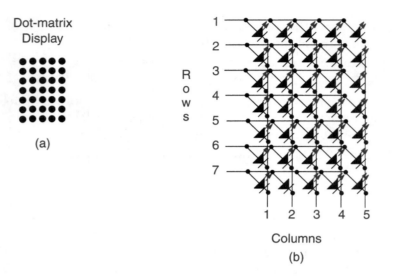

(a)

(b)

Fourteen-segment Alphanumeric Display

(c)

FIGURE 3–18   Types of LED displays

Alphanumeric displays have the ability to display letters as well as numbers. The dot-matrix is in this category, and there are other displays available with fourteen or sixteen segments that also can be used. Figure 3–18(c) shows the arrangement of a fourteen-segment display. These alphanumeric displays are available in both common anode and common cathode configurations.

Light-emitting diode displays are normally driven by digital circuitry. One logic level causes the LED segment to light, and the other logic level turns the LED segment off. Figure 3–19 shows how a common cathode seven-segment display could be connected to seven switches to control the digital display. There is one current limiting resistor required in series with each segment. If all the resistors are of equal value, the current flowing through each segment will be approximately equal, and all segments will emit the same amount of light. Example 3.8 shows how the digital display can be determined.

1. a-Anode   14. Common Cathode
2. f-Anode   13. b-Anode
3. Common Cathode   12. No Pin
4. No Pin   11. g-Anode
5. No Pin   10. c-Anode
6. Dec Pt-Anode   9. No Connection
7. e-Anode   8. d-Anode

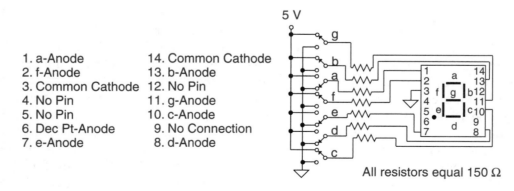

All resistors equal 150 Ω

FIGURE 3–19   Common cathode LED display

EXAMPLE 3.8

Calculate current flowing through an active segment and determine what digit is illuminated on the display shown in Figure 3–19.

Step 1. Calculate the voltage across an active current limiting resistor.
$$V_R = V_s - V_D = 5\text{ V} - 2\text{ V} = 3\text{ V}$$

Step 2. Calculate the current through each active segment. The current through the diode will equal the current through the current limiting resistor.
$$I_D = I_R = V_R/R = 3\text{ V}/150\ \Omega = 20\text{ mA}$$

Step 3. Determine the digit displayed in Figure 3–19. By examining the circuit in Figure 3–19, it is determined a high voltage will activate the LED segment. Further examination reveals the following segments are active: b, c, f, and g. With these segments illuminated, the digit 4 will be displayed.

3.7 PHOTODIODES

Figure 3–20(a) shows the construction of a photodiode, and Figure 3–20(b) shows the symbol for the photodiode. A P-type pocket is fused into an N-type substrate to form a PN junction. The P-type material is exposed to light through a window. Photodiodes are designed to operate with reverse bias as shown in Figure 3–20(c).

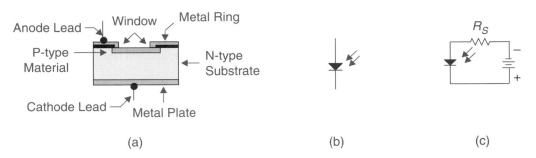

(a)   (b)   (c)

FIGURE 3–20   Photodiode

In reverse bias, the junction is actually forward biased for minority carriers, and the amount of current flow through the device is a function of the number of minority carriers present. In a photodiode, the amount of minority carriers is directly related to the amount of light coming through the window. A photodiode is essentially a light sensitive resistor; the greater the light intensity, the lower the reverse-biased resistance. The photodiode has the fastest response to changes in light intensity of all photo devices.

If, in the construction of a photodiode, an intrinsic layer is added between the P-type and N-type layers, the performance of the device can be improved. This added undoped region has the effect of widening the depletion area. Because of the wider depletion area, the photons entering the diode window have a greater chance of producing an electron/hole pair, which makes the diode more efficient over a wider range of light frequencies. In addition, a wider depletion area lowers the capacitance of the diode and allows faster response time. Photodiodes with the added intrinsic layer are often referred to as *PIN photodiodes*.

3.8 TROUBLESHOOTING SPECIAL DIODES

Being able to read a circuit schematic is one of the most important steps in troubleshooting. In order to read schematics, one needs to know the symbols of the components used in the circuit. In this chapter, we have studied special diodes. Some of these devices have their own symbols while some use the normal diode symbol. Figure 3–21 shows the symbols of the devices discussed in this chapter. The technician should realize some devices may use more than one symbol. For example, TVS devices often use the normal zener symbol. It is also not unusual to find schematics using the incorrect symbol, so the technician must take care when identifying components on a schematic diagram.

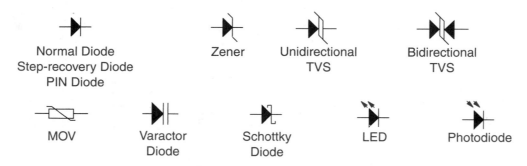

Normal Diode
Step-recovery Diode
PIN Diode

Zener

Unidirectional
TVS

Bidirectional
TVS

MOV

Varactor
Diode

Schottky
Diode

LED

Photodiode

FIGURE 3–21   Schematic symbols

Zener diodes can be checked out of circuit for shorts or opens in the same manner as normal diodes. This test does not, however, check the zener voltage of the diode. This can be checked by placing a resistor in series with the zener, connecting it to a voltage source greater than the zener voltage, and measuring the voltage across the zener as shown in Figure 3–22. The series resistor should be large enough to limit the current flow below the current rating of the zener.

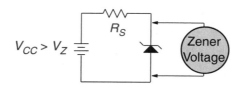

$$V_{CC} > V_Z$$

R_S

Zener Voltage

FIGURE 3–22   Checking a zener diode

In a circuit, a forward-biased zener should read approximately 0.7 V. A higher reading indicates an open. In reverse bias, the voltage across the zener should be the rated voltage of the zener, assuming the circuit is supplying a voltage higher than the rated zener value. At voltages below the rated zener value, it should act like an open.

Transient voltage suppressor and varistor devices are designed to protect electronic circuitry by consuming the energy caused by over voltage conditions. If the energy exceeds the rating of the TVS or varistor device, it will fail. The shorting of a TVS or varistor device is the preferred type of failure, since the rest of the circuit will be protected. The surge protector can be replaced, and the circuit can be returned to operation. If a line surge causes a TVS or varistor device to open, there is a good possibility the surge protector did not do its job and the system circuitry is damaged. The system should never be returned to service without a known good surge protector. The circuit can operate without a functioning surge protector; however, the next line surge may damage expensive circuitry. Out-of-circuit testing using an ohmmeter can detect a shorted surge protector, but the open surge protector cannot be detected using a simple ohmmeter test.

An out-of-circuit test can be executed on LEDs in the same manner as on normal diodes. However, the meter (VOM or DMM) must deliver enough voltage to its lead to forward bias the diode: approximately 2 V for an LED. Some meters do not deliver enough voltage and cannot be used. It is normally easy to make a test circuit by using a power supply and a current limiting resistor. Do not forget the current limiting resistor—connecting the LED directly across the power supply will quickly destroy the device.

Zener Diode Voltage Regulator Circuit

1. Calculate and record the output voltage for the circuit shown in Figure 3–23 for the load currents given in Table 3–1.

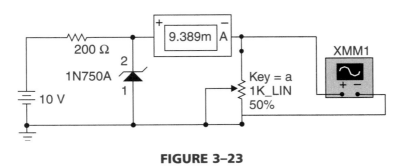

FIGURE 3–23

TABLE 3–1

Load Current	V_{out} (Calculated)	P_z	V_{out} (Simulated)
0 mA			
5 mA			
10 mA			
15 mA			
20 mA			
25 mA			
30 mA			
35 mA			

2. Calculate and record in Table 3–1 the power dissipated by the zener at each load current.
3. Complete a graph of output voltage versus output current in Figure 3–24.

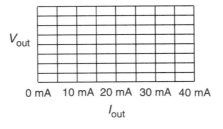

FIGURE 3–24

4. Simulate the current in Figure 3–23 and record the output voltage for the load currents given it in Table 3–1.

Zener Diode Voltage Regulator Circuit

I. Objective

To analyze the output of a zener diode voltage regulator circuit under varying loads.

II. Test Equipment

(1) multimeter
(1) 12 V power supply
(1) dual-trace oscilloscope

III. Components

(1) 270 Ω resistor
(1) 1 kΩ potentiometer
(1) 1N753 zener diode or equivalent

IV. Procedure

1. Calculate and record the output voltage for the circuit shown in Figure 3–25 for the output currents given in Table 3–2.

TABLE 3–2

Load Current	V_{out} (Calculated)	V_{out} (Measured)
0 mA		
10 mA		
15 mA		
20 mA		
25 mA		
30 mA		
35 mA		

2. Construct the circuit in Figure 3–25.

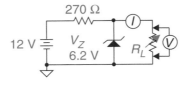

FIGURE 3–25

3. Adjust the input voltage to 12 V. Maintain the input at 12 V throughout the experiment.
4. Adjust the load resistance so zero load current flows (load open). Record the value of the output voltage in Table 3–2.
5. Adjust the load resistance to obtain each of the load currents listed in Table 3–2. Record the value of the output voltage in Table 3–2 at each current setting.

6. Complete a graph of output voltage versus output current in Figure 3–26.

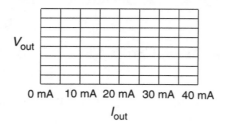

V_{out}

0 mA   10 mA 20 mA   30 mA 40 mA

I_{out}

FIGURE 3–26

V. Points to Discuss

1. Compare your measured data with the calculated data. What is the average percent of error? What caused this difference?

2. At what load current did the regulator drop out of regulation? Explain why.

3. At what value of load current is the current through the zener maximum? Explain.

4. What could be done to allow the circuit to regulate the voltage under a larger load current condition?

LED Circuits

1. Calculate the value for the current limiting resistor needed in the circuit in Figure 3–27 to limit the diode current to 10 mA. _____

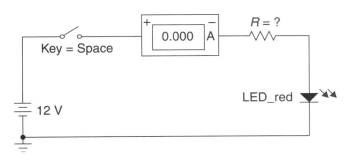

FIGURE 3–27

2. Simulate the circuit in Figure 3–27 and record the current flow. _____
3. What voltage will be dropped across the LED in Figure 3–27? _____
4. What power is dissipated by the resistor in Figure 3–27? _____
5. What power is dissipated by the LED in Figure 3–27? _____
6. Design a circuit to display a "4" using the MAN72A seven-segment display shown in Figure 3–28. The circuit will use a 12 V power supply. Draw your schematic below. Use current limiting resistors of the same value calculated in step 1.

1. a-Cathode
2. f-Cathode
3. Common Anode
4. No Pin
5. No Pin
6. Dec Pt-Cathode
7. e-Cathode

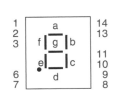

14. Common Anode
13. b-Cathode
12. No Pin
11. g-Cathode
10. c-Cathode
 9. No Connection
 8. d-Cathode

FIGURE 3–28

7. When the digit "4" is being displayed, what is the power dissipated by the seven-segment LED display in the circuit you designed in step 6? _____

LED Circuits

I. Objective

To design and analyze a circuit for driving LEDs.

II. Test Equipment

(1) multimeter
(1) power supply

III. Components

(7) resistors valued as calculated
(1) LED
(1) MAN72A seven-segment LED display or equivalent

IV. Procedure

1. Calculate the value for the current limiting resistor needed to limit the diode current to 15 mA in Figure 3–29.

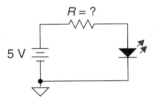

FIGURE 3–29

2. Construct the circuit in Figure 3–29 using a standard value resistor with the next lower value to the value calculated in step 1.

3. Apply power to the circuit—the LED should light. Estimate and record the voltage across the LED in Table 3–3.

TABLE 3–3

V_D (Estimated)	V_D (Measured)

4. Measure and record the voltage across the LED.

1. a-Cathode
2. f-Cathode
3. Common Anode
4. No Pin
5. No Pin
6. Dec Pt-Cathode
7. e-Cathode

14. Common Anode
13. b-Cathode
12. No Pin
11. g-Cathode
10. c-Cathode
 9. No Connection
 8. d-Cathode

FIGURE 3–30

5. Design a circuit to display a "7" using the MAN72A seven-segment display shown in Figure 3–30. Draw your schematic below. Use current limiting resistors of the same value used in the circuit in Figure 3–29.

6. Construct and test the circuit designed in step 5.

V. Points to Discuss

1. Compare your estimated diode voltage drop to the measured diode voltage drop. Were the differences within reason? Explain.

2. What is the power dissipated by the LED and the total circuit in Figure 3–29?

3. What is the power dissipated by the LEDs and the total circuit designed in step 5?

4. Would a seven-segment LED display circuit always dissipate the same power regardless of what number is being displayed? Explain.

QUESTIONS

3.1 Zener Diodes

____ 1. A zener diode normally operates in _____.

    a. forward bias

    b. reverse bias

    c. no bias

    d. resistance bias

____ 2. How does the zener diode compare to the normal diode in the forward-biased region?

    a. The junction drop is much less.

    b. The junction drop is much larger.

    c. The junction drop is approximately the same.

    d. The junction drop is always approximately 2 V.

____ 3. When a zener diode is operating in its zener region, a large change in zener current will produce _____.

    a. a large change in zener voltage

    b. a small change in zener voltage

    c. no change in zener voltage

    d. a large change in forward current flow

3.2 Zener Diode Applications

____ 4. A properly designed zener voltage regulator circuit is capable of maintaining a relatively constant output voltage _____.

    a. with changes in load current

    b. with changes in load resistance

    c. with changes in input voltage

    d. all the above

____ 5. The input voltage to a zener voltage regulator should be _____.

    a. equal to the zener voltage

    b. less than the zener voltage

    c. greater than the zener voltage

    d. equal to zero

____ 6. A zener diode regulator circuit uses a dropping resistor _____.

    a. in series with the load resistor

    b. in parallel with the zener diode

    c. in series with a parallel combination of the load and the zener diode

    d. in series with the zener diode

3.3 Voltage Surge Protectors

_____ 7. Voltage surges can be generated on power lines by _____.
   a. changes in light
   b. loud sounds
   c. lightning
   d. all the above

_____ 8. A TVS (transient voltage suppressor) is essentially a _____.
   a. silicon diode
   b. carbon resistor
   c. electrolytic capacitor
   d. zener diode

_____ 9. The letter "V" in the acronym MOV stands for _____.
   a. voltage
   b. varistor
   c. varactor
   d. variable

3.4 Varactor Diodes

_____ 10. A varactor is a _____-terminal device that utilizes the voltage-variable capacitance of a PN junction.
   a. one
   b. two
   c. three
   d. four

_____ 11. The internal capacitance of a varactor diode decreases as _____.
   a. the forward current flow increases
   b. the forward voltage increases
   c. the reverse-biased voltage decreases
   d. the reverse-biased voltage increases

3.5 High Frequency Switching Diodes

_____ 12. _____ is the main factor that limits the switching speed of a diode.
   a. Junction resistance
   b. Junction inductance
   c. Junction capacitance
   d. Junction temperature

_____ 13. Which of the following is not considered a high frequency switching diode?
   a. PIN
   b. step-recovery
   c. Schottky
   d. varactor

____ 14. The PIN diode contains an N section, a P section, and ____.

    a. an inverted layer

    b. a heavily doped IP section

    c. a doped IN section

    d. an intrinsic layer

____ 15. Which diode is also referred to as a hot-carrier diode?

    a. PIN

    b. step-recovery

    c. Schottky

    d. varactor

3.6 Light-Emitting Diodes (LEDs)

____ 16. The approximate voltage drop across a forward-biased LED is ____.

    a. 0.0 V

    b. 0.3 V

    c. 0.7 V

    d. 2 V

____ 17. LEDs emit light only when they are ____.

    a. reverse biased

    b. forward biased

    c. hot

    d. cold

____ 18. Seven-segment LED displays are available in ____ packages.

    a. common anode

    b. common cathode

    c. common-emitter

    d. both a and b

3.7 Photodiodes

____ 19. A photodiode normally operates with ____.

    a. forward bias

    b. reverse bias

    c. no bias

    d. resistance bias

____ 20. As the light hitting the junction increases, the resistance of the device ____.

    a. decreases

    b. increases

    c. is not affected

    d. switches to an open circuit

3.8 Troubleshooting Special Diodes

____ 21. The schematic symbol of a varactor is _____.

       a.   b.   c.   d.   e.

____ 22. The schematic symbol of a photodiode is _____.

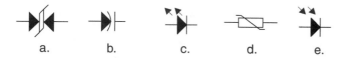

       a.   b.   c.   d.   e.

____ 23. The voltage drop across a forward-biased zener diode is approximately _____.
   a. 0.7 V
   b. V_z
   c. $V_z + 0.7$
   d. 2 V

____ 24. The voltage drop across a conducting reverse-biased zener diode is approximately _____.
   a. 0.7 V
   b. V_z
   c. $V_z + 0.7$ V
   d. 2 V

Select the most probable cause of each situation listed below for the circuit shown in Figure 3–31.

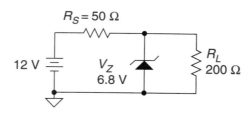

FIGURE 3–31

____ 25. The output reads 9.6 V.
   a. R_s is open.
   b. R_s is shorted.
   c. R_L is open.
   d. R_L is shorted.
   e. Zener is open.
   f. Zener is shorted.
   g. The circuit is functioning properly.

____ 26. The output is zero volts and all components are at room temperature.
   a. R_s is open.
   b. R_s is shorted.
   c. R_L is open.
   d. R_L is shorted.
   e. Zener is open.
   f. Zener is shorted.
   g. The circuit is functioning properly.

____ 27. The output is 6.8 V.
   a. R_s is open.
   b. R_s is shorted.
   c. R_L is open.
   d. R_L is shorted.
   e. Zener is open.
   f. Zener is shorted.
   g. The circuit is functioning properly.

PROBLEMS

1. Draw and label the output waveform (TP_1) for the circuit in Figure 3–32.

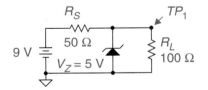

FIGURE 3–32

2. What is the power dissipated by the zener diode in Figure 3–32?

3.   If R_L in Figure 3–32 is decreased, the load current will increase. Determine the maximum load current that can be supplied without the circuit dropping out of regulation.

4.   For an input voltage of 28 V DC, calculate the voltage at TP_1 for the circuit in Figure 3–33 ($R_1 = 1$ kΩ, $R_L = 1$ kΩ, $V_z = 9.6$ V).

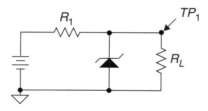

FIGURE 3–33

5.   For an input voltage of 15 V DC, calculate the voltage at TP_1 for the circuit in Figure 3–33 ($R_1 = 1$ kΩ, $R_L = 1$ kΩ, $V_z = 9.6$ V).

6.   In Figure 3–33, if $V_{in} = 28$ V, $R_1 = 1$ kΩ, $V_z = 12$ V, and $R_L = 1$ kΩ, (a) find the zener current, (b) find the load current, and (c) find V_{out}.

7.   For Figure 3–33, if $V_{in} = 25$ V, $R_1 = 1$ kΩ, and $V_z = 9$ V, find the power dissipated by the zener when (a) R_L is 1 kΩ, and (b) R_L is 10 kΩ.

8. Design and draw a zener regulator circuit similar to the one in Figure 3–33 with an output voltage of 9.1 V. The input voltage range is between 12.3 V and 14.6 V, and output load current ranges from 0 to 100 mA.

9. How much power will be dissipated by a 9.1 V zener when 100 mA flows through it in the forward-biased direction?

10. How much power will be dissipated by a 9.1 V zener when 100 mA flows through it in the reverse-biased direction?

11. The bidirectional TVS in Figure 3–34 is rated for a breakdown voltage of +/– 28 V. The RS232 line is working within limits (switching –15 V to +15 V). What is the current flow through the TVS?

Logic Levels
0 = 15 V (25 V Max)
1 = –15 V (–25 Max)

RS232 Input — Computer System

FIGURE 3–34

12. The varactor diode in Figure 3–35 has a capacitance of 200 pF with 4 V of reverse bias applied. At what frequency setting of V_1 will the circuit in Figure 3–35 have a maximum output signal?

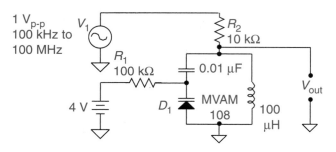

FIGURE 3–35

13. If the 4 V DC supply in Figure 3–35 is adjusted to 8 V DC, then at what frequency setting of V_1 will the circuit have a maximum output signal? (Refer to the graph in Figure 3–12.)

14. Calculate the voltage at TP_1 and TP_2 for the circuit shown in Figure 3–36.

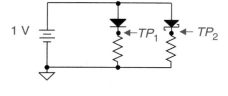

FIGURE 3–36

15. An LED is used to indicate when a 20 V circuit is energized. Calculate the series resistance needed to limit the forward current of the LED to 15 mA.

16. What number will be displayed on the seven-segment display in Figure 3–37?

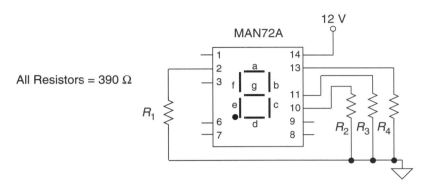

FIGURE 3–37

17. Calculate the current flow through the conducting diode segments in the circuit shown in Figure 3–37.

18. Calculate the total power dissipated by the circuit shown in Figure 3–37.

19. Show a diagram, with circuit values, of an LED connected to 12 V so it will illuminate. The current through the LED should be limited to 25 mA.

20. Diagram the connections to a common anode seven-segment LED display (MAN72) showing the inputs, necessary resistors, and the power connections to display the numeral 5.

CHAPTER 4

The Bipolar Transistor

OBJECTIVES

After studying the material in this chapter, you will be able to describe and/or analyze:

○ transistor architecture,

○ transistors functioning as switches,

○ transistor characteristics,

○ base biasing of amplifier circuits,

○ signal parameters of the amplifier circuit,

○ the procedure for measuring input and output impedance,

○ transistor output characteristic curves, and

○ troubleshooting techniques for transistor circuits.

4.1 INTRODUCTION TO TRANSISTORS

Bipolar transistors, also called BJTs (bipolar junction transistors), are three-terminal devices that can function as electronic switches or as signal amplifiers. In this chapter, you will learn how transistors perform both of these important functions. Figure 4–1 shows the schematic symbols for the two types of bipolar transistors.

The schematic symbols show the three terminals of the transistor. The leg with the arrow is called the emitter, the collector is the other slanted leg, and the base is the leg connected perpendicular to the line connecting the emitter and the collector. If the arrow of the schematic

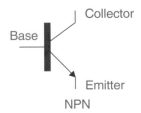

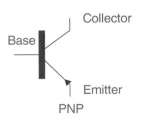

FIGURE 4–1 Transistor symbols

symbol points outward, the transistor is an NPN transistor. If the arrow points inward, the transistor is a PNP transistor.

4.2   INSIDE THE TRANSISTOR

A transistor is a three-layer device, whereas a diode has only two layers. Figure 4–2(a) shows how two layers of N-type material sandwich one layer of P-type material to make an NPN transistor. Figure 4–2(b) shows how two layers of P-type material sandwich one layer of N-type material to form a PNP transistor. In both cases, the layer in the center is the base, and the two outer layers are the emitter and collector. The arrow on the schematic symbol identifies the emitter and always points to the N-type material. By remembering this, you should always be able to determine whether a transistor is an NPN or PNP type.

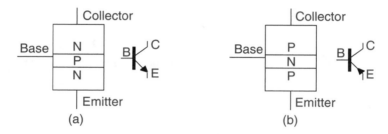

FIGURE 4–2   Transistor architecture

Let us examine the operation of an NPN transistor. At first glance, you might assume the NPN transistor would function like two diodes where the anodes are connected to the base terminal. Indeed, if each junction of the transistor is used independently, it does function as two diodes. However, when the base/emitter junction is forward biased and the base/collector junction is reverse biased, a phenomenon happens that makes the transistor possible.

Figure 4–3 shows the bias voltages needed to turn on an NPN transistor (current limiting resistors have been left out for simplification). Supply V_{BB} is connected in such a manner as to forward bias the base/emitter junction. The voltage on the base is positive 0.7 V with respect to the emitter, which is at ground potential. Supply V_{cc} sets the collector to +10 V potential with

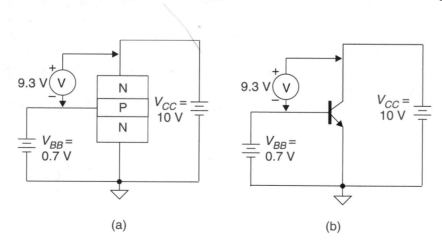

FIGURE 4–3   Bias voltages

respect to the emitter. The collector is 9.3 V more positive than the base; therefore, the base/collector junction is reverse biased. With the base/emitter junction forward biased and the base/collector junction reverse biased, the transistor is turned on.

Using electron current flow, let us discuss the current flow in the circuit in Figure 4–4. Begin by assuming switch S_1 open, thus disconnecting V_{cc} from the collector. If this is the case, we only need to be concerned with V_{BB}. Supply V_{BB} is forward biasing the base/emitter junction, causing current to flow.

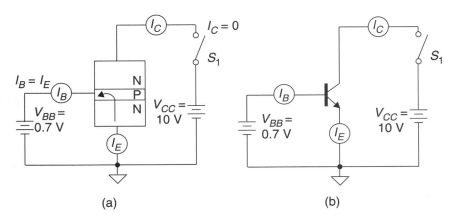

FIGURE 4–4   Base/emitter current with collector open

Meters I_E and I_B have the same current flowing through them. Meter I_c has zero current flow because switch S_1 is open. If the bias supply (V_{BB}) is adjusted more positive, the current through meters I_E and I_B will increase. If the bias supply (V_{BB}) is reduced, less current will flow through meters I_E and I_B. In other words, with only supply V_{BB} connected, the base/emitter junction functions like a normal forward-biased diode.

If we close switch S_1, 10 V will be applied to the collector, reverse biasing the base/collector junction. The base/emitter junction is still forward biased with supply V_{BB}. At first glance, one would think the current flow would remain the same ($I_E = I_B$ and $I_C = 0$ A), but this is not the case.

What actually takes place in the circuit when S_1 is closed (see Figure 4–5) is majority current carriers from the emitter (electrons) cross the forward-biased base/emitter junction, and some combine with the majority current carriers in the base (holes) and cause base current. However,

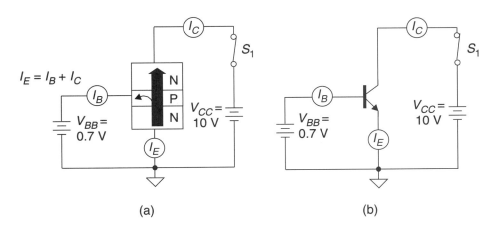

FIGURE 4–5   Bias currents

the base is physically thin and lightly doped when compared to the emitter. The emitter injects a large number of electrons into the base region relative to the limited number of holes available in the base region. The result is a large number of electrons present in the base region that do not immediately combine with holes.

Once in the base region, if the electron does not combine with a hole, it will fall under the influence of the collector supply V_{cc}. Remember that the base/collector junction is reverse biased, and the majority current carriers are prohibited from crossing the junction. However, when a junction is reverse biased for majority current carriers, it is forward biased for minority current carriers. The electrons in the base, being minority carriers, find it easy to pass through the base/collector junction. Once in the collector, the electrons are attracted by the positive terminal of the V_{cc} supply and flow through meter I_c, causing the meter to indicate current. As Figure 4–5(a) shows, the current leaving the emitter divides, and a small amount flows out as base current (I_B), but the majority of the current flows out as collector current (I_C). The current flow can be expressed mathematically by the formula $I_E = I_B + I_C$.

BETA OF THE TRANSISTOR

The ratio of how the emitter current divides into base and collector current is a function of the particular transistor being used. Typically only about 1% of the emitter current will exit the base, and 99% will exit the collector. The ratio of collector current to base current is a parameter of the individual transistor and is called beta (β). Beta can be stated mathematically as $\beta = I_C/I_B$. Beta is sometimes referred to as h_{FE}, which stands for forward current gain in the common emitter configuration.

Beta is the current gain of the transistor where I_C is output current and I_B is input current (beta has no units, since it is a ratio of two currents). The important thing to remember about beta is that it is a fixed ratio between collector current and base current. Therefore, a small change in base current will cause a large change in collector current. The base current is the control current.

Let us assume values for the circuit in Figure 4–6. Let $\beta = 100$ and $I_B = 10$ μA. Then I_C equals 1 mA ($I_C = 100 \times 10$ μA). If the base current increases to 15 μA, the collector current will go to 1.5 mA. This is because beta, the ratio between I_C and I_B, is constant, and an increase in input current will result in an increase in output current. Likewise, if the base current decreases to 5 μA, the collector current will decrease to 0.5 mA. The following two formulas should be remembered:

$$I_E = I_B + I_C \qquad \text{Transistor current divider}$$
$$\beta = I_C/I_B \qquad \text{Transistor current gain (beta)}$$

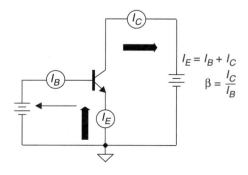

FIGURE 4–6   Beta of the transistor

EXAMPLE 4.1

A transistor is connected as shown in Figure 4–7. It has a base current of 8 μA and a collector current of 1.2 mA. What is the emitter current and the beta of the transistor?

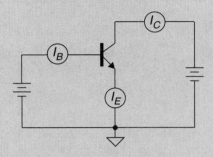

FIGURE 4–7 Transistor currents and beta

Step 1. Calculate emitter current.

$$I_E = I_B + I_C$$
$$I_E = 8 \text{ μA} + 1.2 \text{ mA}$$
$$I_E = 1208 \text{ μA}$$

Step 2. Calculate beta.

$$\beta = I_C/I_B$$
$$\beta = 1200 \text{ mA/8 mA}$$
$$\beta = 150$$

EXAMPLE 4.2

A transistor is connected as shown in Figure 4–7. It has an emitter current of 2.42 mA and a collector current of 2.4 mA. What is the base current and beta of the transistor?

Step 1. Calculate base current.

$$I_E = I_B + I_C$$
$$I_B = I_E - I_C$$
$$I_B = 2.42 \text{ mA} - 2.40 \text{ mA}$$
$$I_B = 0.02 \text{ mA or } 20 \text{ μA}$$

Step 2. Calculate beta.

$$\beta = I_C/I_B$$
$$\beta = 2400 \text{ μA} / 20 \text{ μA}$$
$$\beta = 120$$

EXAMPLE 4.3

A transistor is connected as in Figure 4–7 and has a base current of 16 μA and a beta of 80. What is the collector current and emitter current of the transistor?

Step 1. Calculate the collector current.

$$\beta = I_C/I_B$$

EXAMPLE 4.3 continued

$$I_C = \beta \times I_B$$
$$I_C = 80 \times 16 \ \mu\text{A}$$
$$I_C = 1280 \ \mu\text{A or } 1.28 \ \text{mA}$$

Step 2. Calculate the emitter current.

$$I_E = I_B + I_C$$
$$I_E = 16 \ \mu\text{A} + 1280 \ \mu\text{A}$$
$$I_E = 1296 \ \mu\text{A or } 1.296 \ \text{mA}$$

Transistors of the same type will have large variations in beta. However, in this chapter, we will consider beta to be a constant. In the next chapter, we discuss circuits that correct for variations in beta.

4.3 TRANSISTOR SWITCHES

INTRODUCTION

A transistor can function as an SPST (single-pole single-throw) switch, but rather than being mechanically controlled, it is controlled by an electronic signal driving the base terminal. Figure 4–8 shows a comparison between an open SPST switch and an NPN transistor.

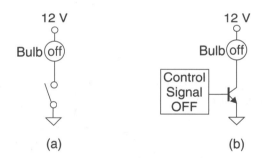

FIGURE 4–8   Open switch equals off transistor

When the switch is open, as shown in Figure 4–8(a), there is no current flowing in the circuit and the bulb is off. When the control signal on the base terminal of the transistor turns the transistor off, as shown in Figure 4–8(b), the transistor acts like an open switch. The resistance between the collector and the emitter terminals rises infinitely high and prevents current flow in the circuit. The bulb in series with the transistor is off.

Figure 4–9 shows a comparison between the SPST switch and the transistor when turned on. When the switch is closed, current flows in the circuit and lights the bulb. Likewise, when the control signal on the base terminal turns the transistor on, the resistance between the collector and emitter drops to zero, and the current flow lights the bulb.

Actually, the transistor is not a perfect switch. When it is off, the resistance between the collector and emitter does not go to infinity, and when it is on, the resistance between the collector and emitter does not drop to zero. Even though the transistor is not a perfect switch, it is close enough to function well in most circuits.

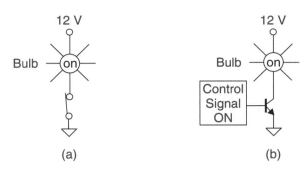

FIGURE 4–9   Closed switch equals on transistor

DESIGNING A TRANSISTOR SWITCHING CIRCUIT _____

Figure 4–10 shows a transistor switching circuit designed to control the on/off condition of a 12 V bulb. When the control signal equals 5 V, the bulb is on, and when the control signal equals 0 V, the bulb is off. The current flowing through the bulb is the collector current. The collector current is zero when the bulb is off and is limited by the bulb resistance when the bulb is on. Example 4.4 shows how to design the switching circuitry.

EXAMPLE 4.4

Design the transistor switching circuit shown in Figure 4–10.

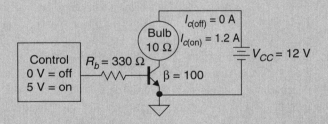

FIGURE 4–10   Transistor switching circuit

Step 1. Calculate the collector current when the bulb is in the on state. The supply voltage is divided by the resistance of the load (bulb).
$$I_C = I_L = V_{CC}/R_L = 12 \text{ V}/10 \text{ }\Omega = 1.2 \text{ A}$$
Step 2. Using beta, calculate the needed base current.
$$I_B = I_C/\beta = 1.2 \text{ A}/100 = 12 \text{ mA}$$
Step 3. Calculate the value of V_{RB}.
$$V_{RB} = V_{\text{control}} - V_{BE} = 5 \text{ V} - 0.7 \text{ V} = 4.3 \text{ V}$$
Step 4. Calculate the value of R_b.
$$R_b = V_{RB}/I_B = 4.3 \text{ V}/12 \text{ mA} = 358 \text{ }\Omega \text{ (Use the next lower standard value, 330 }\Omega\text{.)}$$
Step 5. Draw the switching circuit. (The switching circuit is shown in Figure 4–10.)

If the control voltage in Figure 4–10 is zero, the base/emitter junction will be reverse biased, and there will be zero base current. With zero base current, the collector current would also be zero ($I_C = \beta \times I_B$). With the collector current being zero, there would be no current through the bulb, and the light would be off. The transistor is functioning as an open switch.

If the control voltage in Figure 4–10 is 5 V, the base/emitter junction is forward biased. The junction drops 0.7 V, leaving a 4.3 V drop across R_b. The current through R_b is 13 mA (4.3 V/330 Ω = 13 mA). Since the transistor has a minimum beta of 100, the minimum expected collector current (I_C) would equal 1.3 A (100 x 13 mA). However, the resistance of the load limits the collector current to 1.2 A (12 V/10 Ω). The collector current flows through the bulb, causing the bulb to light.

When a transistor is fully on and its current is limited by the load resistance, it is in saturation. The collector/emitter voltage (V_{CE}) across a saturated transistor is approximately 0.2 V to 0.3 V. For calculations, assume 0.2 V is dropped across a saturated transistor. In Example 4.4, we assume an ideal transistor with zero voltage drop from collector to emitter in the on state. If we assume a 0.2 V drop across the saturated transistor, the voltage across the bulb will be 11.8 V, and the current through the bulb will be 1.18 A.

HIGH POWER LOAD BEING CONTROLLED WITH LOW POWER SIGNAL

At this point, you may think transistor switching circuits do nothing more than allow one voltage source to turn off and on another voltage source, but let us examine the circuit more closely. In the off state, the current through the load is equal to zero. Therefore, the power dissipated by the load is equal to zero watts ($P = IV$). In the off state, zero volts are applied to the base circuit, so the base current equals zero, and the control signal power is also equal to zero.

When the transistor switch in Example 4.4 is on, the load current is equal to 1.2 A, and the voltage drop across the load is equal to 12 V. Therefore, the power dissipated in the load is equal to 14.4 W (1.2 A × 12 V). The input control signal is 5 V at 12 mA, and the input power dissipated is equal to 60 mW (12 mA × 5 V).

Notice that a low power input signal (60 mW) is able to control a relatively high power output (14.4 W). It is important to realize that the transistor is a control device, and not a power generator. The transistor provides the control function of opening or closing a current path, but the power dissipated in the load comes from the power supply (V_{cc}).

The wattage rating of the switching transistor used in the circuit is determined by calculating the power dissipated by the transistor. When the transistor is off, the current flowing through the transistor is zero. Consequently, zero power is dissipated by the transistor. When the transistor is on, the current through the transistor is determined by the resistance of the load and the supply voltage. The voltage across the transistor is approximately 0.2 V. The transistor in Example 4.4 dissipates approximately 240 mW (1.2 A × 0.2 V) when the transistor is conducting.

IMPROVING TRANSISTOR SWITCHING SPEED

Figure 4–11 shows two transistors being switched by the same control signal. At low switching speed, the signal present on the collectors of both transistors would look like the signal shown

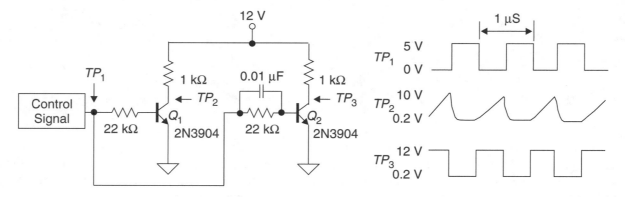

FIGURE 4–11 Improving transistor switching speed

at TP_3. However, as the switching speed increases, transistor Q_1 will not switch sharply and will have a collector voltage that ramps up and down as shown in Figure 4–11. Adding a capacitor in parallel with the base resistor, as shown Figure 4–11, will permit transistor Q_2 to switch at a higher speed. The capacitor passes a current spike that clears the current carriers out of the junction area and permits the transistor to cut off quickly. When the transistor is switched on, the capacitor passes the needed voltage directly to the base and drives the transistor quickly into saturation.

SUMMARY OF THE TRANSISTOR IN THE SWITCHING MODE _____

When a transistor is used for switching, it is in one of two states: on or off. In the off state, the base bias current (I_B) is zero, and the transistor is cut off. (Actually there is a small amount of leakage current, but it can be ignored in most cases.) In the on state, the base bias current (I_B) is set large enough to drive the transistor into saturation. In saturation, the voltage drop across the transistor (V_{CE}) decreases to zero, and the voltage across the load goes to V_{cc}. (Actually the voltage drop across the transistor decreases to approximately 0.2 V, and the voltage drop across the load rises to V_{cc} minus 0.2 V). Transistors find many uses in the switching mode. They are used to switch off and on electrical loads of all types and are used extensively in digital electronics where the circuitry requires an on or off state. Example 4.5 will review transistor switching circuit concepts.

EXAMPLE 4.5

Design a circuit to control the on/off conditions of a 50 Ω load connected to 30 V. The control signal is a voltage switch between 0 V and 4 V. The load will be on when the control voltage is 4 V and off when the control voltage is 0 V. The transistor used in the circuit will have a beta of 50.

Step 1. Draw a diagram of the switching circuit. (Figure 4–12 shows a diagram of the circuit.)

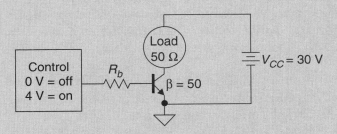

FIGURE 4–12   Transistor switching circuit

Step 2. Calculate the collector current when the load is in the on state. The supply voltage is divided by the load resistance. The saturation voltage (0.2 V) of the transistor could be subtracted from the supply voltage, but it is not significant in this case.
$$I_C = I_L = 30 \text{ V}/50 \text{ } \Omega = 600 \text{ mA}$$
Step 3. Calculate the base current needed.
$$I_B = I_C/\beta = 600 \text{ mA}/50 = 12 \text{ mA}$$
Step 4. Calculate the value of V_{RB}.
$$V_{RB} = V_{\text{control}} - V_{BE} = 4 \text{ V} - 0.7 \text{ V} = 3.3 \text{ V}$$
Step 5. Calculate the value of R_b.
$$R_b = V_{RB}/I_B = 3.3 \text{ V}/12 \text{ mA} = 275 \text{ } \Omega \text{ (Use the next lower standard value, 270 } \Omega \text{.)}$$

EXAMPLE 4.5 continued

> **Step 6.** Calculate the wattage rating of R_b.
> $$P = IV = 12 \text{ mA} \times 3.3 \text{ V} = 39.6 \text{ mW}$$
> The resistor selected should be able to dissipate approximately twice the calculated wattage or more. A 1/8 W or larger resistor is needed in this case.
>
> **Step 7.** Calculate the wattage rating of the transistor.
> $$P = IV = 600 \text{ mA} \times 0.2 \text{ V} = 120 \text{ mW}$$
> The transistor selected should be able to dissipate 120 mW or more.

4.4 TRANSISTOR CHARACTERISTICS AND DATA SHEETS

There are hundreds of transistors on the market. The differences in these transistors can be found by examining the electrical characteristics listed on the data sheets. The three most important characteristics for the technician to know are the maximum collector current (I_C), maximum power dissipation (P_D), and small signal beta (β). Certain technical literature uses H-parameters to identify transistor characteristics. We will not use H-parameters in this text, although in the following list of transistor characteristics, we will give the H-parameter names when relevant. The following paragraphs state the most common transistor characteristics.

Maximum collector current (I_C) is the maximum continuous current that can flow in the collector leg of the transistor without damage to the transistor. Bipolar transistors are available with maximum collector current ratings from 50 mA to 50 A.

Maximum power dissipation (P_D) is the maximum power the transistor can dissipate without being damaged. An approximation of the power being dissipated by a transistor can be calculated by multiplying the voltage across the transistor from collector to emitter (V_{CE}) times the collector current (I_C). Bipolar transistors are available with maximum power dissipation ratings from 0.2 W to 250 W.

Small signal beta (β) or (h_{fe}) is the signal current gain of the transistor in the common-emitter configuration. The input terminal is the base, the output terminal is the collector, and the emitter is the reference terminal. The formula for small signal beta is $\beta = i_c/i_b$.

DC beta (β) or (h_{FE}) is the DC current gain of the transistor in the common-emitter configuration. In most applications, the small signal beta and the DC beta are interchangeable. In this text, we will assume they are interchangeable unless otherwise noted. The formula for DC beta is $\beta = I_C/I_B$. Transistors are available with beta ratings from 10 to 1000.

Maximum base current (I_B) is the maximum current that can flow in the base leg of the transistor without damage to the transistor.

Collector to base breakdown voltage (V_{CBO}) is the maximum reverse-biased voltage that can be applied across the collector to base junction. Bipolar transistors are available with collector to base breakdown voltage ratings from 20 V to 1500 V.

Collector to emitter breakdown voltage (V_{CEO}) is the maximum voltage that can be applied across the transistor from collector to emitter. Bipolar transistors are available with collector-to-emitter breakdown voltage ratings from 20 V to 800 V.

Emitter to base breakdown voltage (V_{EBO}) is the maximum reverse-biased voltage that can be applied across the emitter to base junction. Bipolar transistors are available with emitter to base breakdown voltage ratings from 4 V to 20 V.

Gain-bandwidth product (f_T) is the frequency at which the gain of the transistor drops to unity. The gain-bandwidth product is normally given in megahertz and can range from less than 1 MHz to 5000 MHz.

4.5 THE TRANSISTOR AMPLIFIER

GAIN AND AMPLIFICATION

Transistor amplifiers are circuits that provide signal gain. Gain is an important concept in electronics. Let us take a moment to make sure we have a common understanding of the word. *Gain* is the ratio of output to input. The general formula is *Gain = Output / Input*. Gain has no units because the output and the input must be in the same units, and the units cancel.

Figure 4–13 shows the symbol used for an amplifier. An amplifier is an electronic circuit used to obtain gain. In Figure 4–13, the amplifier has an input of 0.6 V$_\text{p-p}$ and an output of 6 V$_\text{p-p}$. By using the gain formula, we can calculate the voltage gain to be 10 (6 V$_\text{p-p}$/0.6 V$_\text{p-p}$ = 10). The word *voltage* preceding *gain* indicates that the gain ratio is comprised of the output voltage and the input voltage. It is also possible to have current and power gains.

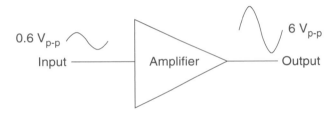

FIGURE 4–13   Amplifier

THE TRANSISTOR AS AN ADJUSTABLE CURRENT SOURCE

Figure 4–14 shows a conventional current source in series with a resistor. The variable current source is adjustable from zero to 2 mA. If the current source is adjusted to 1 mA (I_C = 1 mA), the current flow through the resistor will be 1 mA. The 5 kΩ resistor will drop 5 V, causing the output voltage to be 5 V. The calculation follows:

If I_C = 1 mA, then

$V_{RC} = I_C \times R_c = 1 \text{ mA} \times 5 \text{ k}\Omega = 5 \text{ V}$

$V_\text{out} = V_{CC} - V_{RC} = 10 \text{ V} - 5 \text{ V} = 5 \text{ V}$

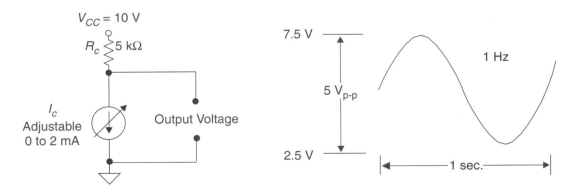

FIGURE 4–14   Adjustable current source

If the current source is adjusted to 1.5 mA, the output voltage will drop to 2.5 V. Adjusting the current source to 0.5 mA will cause the voltage to rise to 7.5 V. The calculation of each output voltage follows:

If $I_C = 1.5$ mA, then

$V_{RC} = I_C \times R_c = 1.5$ mA $\times$ 5 kΩ = 7.5 V

$V_{out} = V_{CC} - V_{RC} = 10$ V $- 7.5$ V = 2.5 V

If $I_C = 0.5$ mA, then

$V_{RC} = I_C \times R_c = 0.5$ mA $\times$ 5 kΩ = 2.5 V

$V_{out} = V_{CC} - V_{RC} = 10$ V $- 2.5$ V = 7.5 V

Any voltage between 2.5 V and 7.5 V can be obtained at the output by adjusting the current source between 0.5 mA and 1.5 mA. If the current source is properly adjusted over a one-second period, it is possible to obtain the sine wave shown in Figure 4–14. The output signal voltage is a function of the changing current flowing through the 5 kΩ resistor.

Figure 4–15 shows a transistor in series with a 5 kΩ resistor. It is possible to control the collector current between 0 mA and 2 mA by controlling the base signal. The 2 mA limit is set by the 10 V supply and the 5 kΩ resistor (10 V/5 kΩ). If the control signal on the base is set to cause 1 mA of collector current (I_C), the output voltage will be 5 V. By controlling the signal on the base of the transistor, the collector current can be adjusted to 1.5 mA, which will cause the output voltage to drop to 2.5 V. If the signal on the base is adjusted so the collector current decreases to 0.5 mA, the output voltage will rise to 7.5 V. If the proper control signal is applied to the base of the transistor over a one-second period, it is possible to obtain the waveform in Figure 4–15.

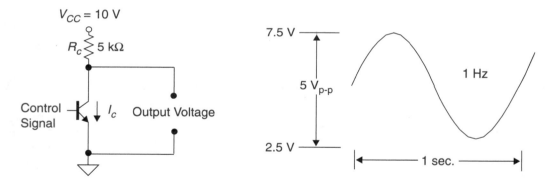

FIGURE 4–15   The transistor as an adjustable current source

The signal voltage output in Figure 4–15 is a function of the changing collector current flowing through the 5 kΩ resistor. The transistor functions as a current source in series with the 5 kΩ resistor. The collector current through the transistor is adjustable by applying a control signal to the base terminal of the transistor. In the coming sections, it will be shown how a small input signal on the base of a transistor amplifier can control a large output signal on the collector.

BIASING THE TRANSISTOR AMPLIFIER

Biasing a transistor amplifier circuit is the process of setting the DC voltage and current to the proper level so a signal can be accepted and amplified by the circuit. A comparison can be made between an automobile idle and the transistor amplifier bias current. With the idle set properly, the automobile is ready to receive an input from the accelerator. With the bias current set properly, the transistor amplifier is ready to receive an input from the signal source. In the case of the automobile, we are trying to set the engine to a constant rpm with no input applied. In the case of the transistor amplifier, we are trying to set the collector current (I_C) to a constant value with no input signal applied. Theoretically, the idle of an automobile could be set anywhere from zero rpms (engine off) to the maximum rpms (engine wide open). The proper idle point is somewhere between the two extremes. The same is true of a transistor amplifier; theoretically, the bias could be set anywhere from $I_C = 0$ amps (cutoff) to maximum I_C (saturation). The proper bias point is somewhere between the two extremes and is called the *quiescent operating point*, or Q-point.

When an amplifier is biased so the Q-point is near the middle of its operating range (halfway between cutoff and saturation), it is said to be operating class-A. This operation permits the output current to increase and decrease from the Q-point as the input swings through a complete cycle. In other words, output current flows for the full 360º of the input cycle. Figure 4–16(a) shows a class-A amplifier that permits the output current to flow for the full 360º of the input cycle.

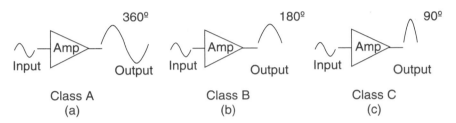

FIGURE 4–16 Classes of transistor operation

When an amplifier is biased at cutoff, it is operating class-B. In class-B operation, the output current will flow only for one alternation, or 180º, of the input cycle. Figure 4–16(b) shows a class-B amplifier.

In class-C operation, the amplifier is biased below cutoff. This means the input signal must cross zero and move partway through an alternation before output current starts to flow. In other words, output current flows less than 180º. Figure 4–16(c) shows a class-C amplifier.

Class-B and class-C amplifiers do not provide a true reproduction of the input signal; however, they are both much more efficient that class-A amplifiers. Chapter 16 describes how the higher efficiencies of class-B and class-C amplifiers can be put to use in power amplifiers.

In this chapter, we will limit our study to class-A amplifier circuits. Since class-A amplifier circuits have output current flow for the complete cycle of the input signal, the output wave shape will be the same as the input. However, because the circuit has gain, the magnitude of the output signal is larger.

BASE BIAS AMPLIFIER

Figure 4–17(a) shows a transistor amplifier circuit using base bias, also called *fixed bias*. The function of bias circuitry is to set the Q-point by setting the collector current (I_C) to a constant value without a signal applied to the circuit. Since bias currents are steady-state DC, the coupling capacitors block bias current flow. For bias analysis, the circuit can be simplified to the circuit shown in Figure 4–17(b). Example 4.6 shows how the bias voltages and currents can be calculated.

EXAMPLE 4.6

Calculate V_E, V_B, V_C, I_B, I_E, and I_C for the circuit in Figure 4–17(a).

Step 1. Simplify the circuit as shown in Figure 4–17(b) for analyzing bias voltages and currents.

EXAMPLE 4.6 continued

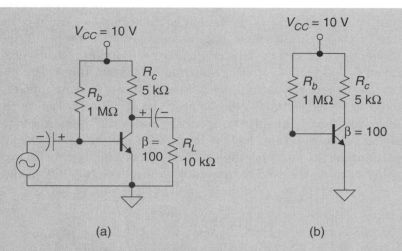

FIGURE 4–17   Base bias circuits

Step 2. Calculate V_E. The emitter is connected to ground; therefore, $V_E = 0$ V.

Step 3. Calculate V_B. The base/emitter junction is forward biased; therefore, $V_B = 0.7$ V.

Step 4. Calculate I_B. First, the voltage drop across R_b is calculated, and then Ohm's law is used to calculate I_B.

$$V_{RB} = V_{CC} - V_B = 10 \text{ V} - 0.7 \text{ V} = 9.3 \text{ V}$$
$$I_B = V_{RB}/R_b = 9.3 \text{ V}/1 \text{ m}\Omega = 9.3 \text{ μA}$$

Step 5. Calculate I_C. The formula for beta is used to calculate I_C.

$$I_C = \beta \times I_B = 100 \times 9.3 \text{ μA} = 0.93 \text{ mA}$$

Step 6. Calculate V_c. First, the voltage drop across R_c is calculated, then this voltage is subtracted from the supply voltage (V_{CC}) to obtain the voltage on the collector.

$$V_{RC} = R_c \times I_C = 5 \text{ k}\Omega \times 0.93 \text{ mA} = 4.65 \text{ V}$$
$$V_c = V_{cc} - V_{Rc} = 10 \text{ V} - 4.65 \text{ V} = 5.35 \text{ V}$$

Step 7. Calculate I_E.

$$I_E = I_C + I_B = 0.93 \text{ mA} + 0.0093 \text{ mA} = 0.9393 \text{ mA}$$
$$I_E \approx I_C \approx 0.93 \text{ mA}$$

In most cases, the emitter and collector currents are assumed to be equal.

Figure 4–18 shows the transistor amplifier with the important voltage and current values calculated in Example 4.6. The technician can measure voltage values and compare them to the calculated values to determine if a circuit is operating properly. If the circuit in Example 4.6 is constructed and the collector voltage reads approximately 5.3 V, the bias circuit is functioning properly. If the collector voltage reads low, the collector current of the transistor is too high. If the collector voltage reads high, the collector current of the transistor is too low.

Engineers normally design class-A amplifiers for midpoint bias; thus, the transistor collector voltage will read approximately one-half of V_{cc}. There are other factors the engineer must consider when designing the class-A amplifier. For this reason, the collector voltage may not read exactly 50% of V_{cc} but will normally read between 30% to 70% of V_{cc}. Hence, the technician can quickly get an indication of proper circuit operation by measuring the collector voltage of a class-A amplifier.

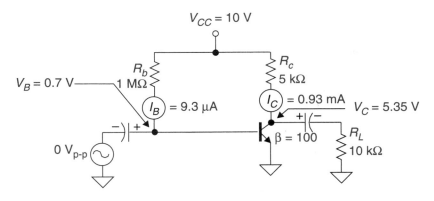

$V_{CC} = 10$ V

R_b
1 MΩ

$V_B = 0.7$ V

$I_B = 9.3$ μA

R_c
5 kΩ

$I_C = 0.93$ mA $V_C = 5.35$ V

β = 100

R_L
10 kΩ

0 V$_{p-p}$

FIGURE 4–18 Base bias circuit

4.6 SIGNAL ANALYSIS OF THE BASE-BIASED AMPLIFIER

Class-A transistor amplifiers are set to a Q-point by DC bias currents. The first step in troubleshooting these circuits is to verify that the bias circuit is functioning properly. You cannot, however, completely troubleshoot or analyze the amplifier circuit without knowing the signal parameters.

In this section, you will learn how to calculate signal parameters. When referring to signal values, we will use lowercase letters. Beta and transistor current divider formulas, as shown below for bias parameters, are also valid for signal parameters.

$I_E = I_B + I_C$ also $i_e = i_b + i_c$ Transistor current divider

$β = I_C / I_B$ also $β = Δi_c / Δi_b$ Beta

INPUT IMPEDANCE (z_{in})

Input impedance (z_{in}) is the impedance the signal source sees when looking into the amplifier. Let us examine the amplifier in Figure 4–19(a) and determine its equivalent input impedance. The input coupling capacitor (C_1) has a very low reactance (X_c) for the frequency of the signal and looks like a short. The DC power source has a very low internal impedance for signal frequency; therefore, the DC power bus is always considered to be signal ground. This means, from the perspective of the signal source, resistor R_b connects directly to ground. Figure 4–19(b)

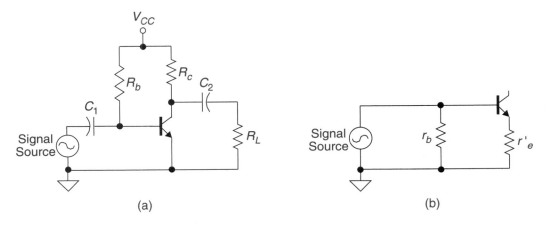

V_{CC}

R_b R_c C_2

C_1

Signal Source R_L

(a)

Signal Source r_b r'_e

(b)

FIGURE 4–19 Input impedance

shows an equivalent circuit with C_1 removed and r_b connected directly to ground. Lowercase r_b is used for the signal resistance from the base terminal to ground. The resistance r_b will include all signal paths to ground with the exception of the path through the transistor. In this case, there is only one path through R_b, so r_b will equal R_b.

The transistor base/collector junction is reverse biased. The signal source sees this reverse-biased junction as a open circuit, so the signal path through the collector is considered an open circuit as shown in Figure 4–19(b). The signal source sees a path to ground through the forward-biased base/emitter junction. The dynamic resistance of the base/emitter junction will limit signal current flow through the junction. The dynamic resistance of the junction is shown in the equivalent circuit as resistor r'_e.

Let us review the concept of dynamic resistance. In Chapter 1, we discussed the detailed diode model that shows how forward-biased current increases steeply with changes in forward-biased voltage. We stated that the dynamic resistance of a diode junction can be calculated by taking a small change in voltage and dividing it by the change in current caused by the change in voltage ($r = \Delta v/\Delta i$). However, these changes in voltage and current are difficult to obtain. Dr. Shockley developed an approximation formula that simplifies solving for dynamic resistance. Shockley's formula states that the dynamic resistance of a diode junction is equal to 25 mV divided by the forward-biased current through the diode junction ($r = 25$ mV$/I$). So, once the bias current is known, the dynamic resistance can be easily calculated.

In a simple diode, there is only one current flowing; however, in a transistor, there are three currents: emitter current, base current, and collector current. The formula for calculating the dynamic resistance of the forward-biased base/emitter junction is $r'_e = 25$ mV$/I_E$. The emitter bias current (I_E) is used to calculate the dynamic resistance (r'_e) of the base/emitter junction because the emitter current is the current flowing through the base/emitter junction. The fact that only a small percentage of emitter current exits the base has no effect on the dynamic resistance of the junction.

To calculate the input impedance, we now have two parallel paths for signal current flow: one through the base resistance (r_b) and another through the forward-biased base/emitter junction, which has dynamic resistance in the emitter leg. The resistance in the emitter leg will be seen as beta times larger from the perspective of the base. The explanation that follows will show why the resistance in the emitter leg must be multiplied by beta.

Figure 4–20(a) shows an equivalent circuit with the three possible Kirchhoff's voltage loop equations. We will use the Kirchhoff's voltage loop equation containing the signal voltage source and base/emitter junction ($V_s - V_{r'e} = 0$) to start our explanation.

$$V_s - V_{r'e} = 0$$

Using Ohm's law to expand our equation gives us:

$$V_s - (i_e \times r'_e) = 0$$

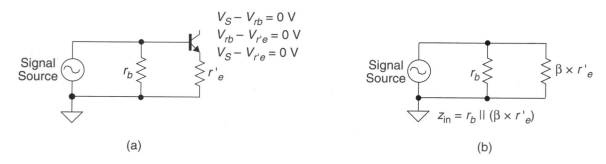

FIGURE 4–20   Input impedance equivalent circuits

From the perspective of the signal source, the current flowing into the transistor is base current. We can rewrite the voltage loop equation using base current.

$$i_e = (\beta + 1) \times i_b$$

Since beta is much larger than one:

$$i_e \approx \beta \times i_b$$

Substituting $\beta \times i_b$ for i_e:

$$V_s - (i_b \times \beta \times r'_e) = 0$$

The previous equation could be written as:

$$V_s = i_b \times (\beta \times r'_e).$$

Since the signal source drives the base circuitry, the value of r'_e looks beta times larger when viewed from the perspective of the signal source. Figure 4–20(b) is an equivalent circuit showing the input impedance from the perspective of the signal source. Note that the base/emitter junction and r'_e are replaced with a resistor that is equal to $\beta \times r'_e$. This branch is in parallel with all other signal paths to ground. The general formula for input impedance is $z_{in} = r_b \parallel (\beta \times r_e)$, where r_b is the signal resistance of the base circuitry and r_e is the signal resistance of the emitter circuitry. Example 4.7 shows how the impedance of an amplifier circuit can be calculated.

EXAMPLE 4.7

Calculate the input impedance for the circuit in Figure 4–21.

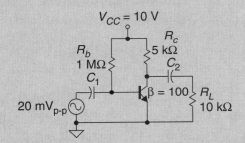

FIGURE 4–21 Base bias amplifier circuit

Step 1. Calculate the bias emitter current (I_E).

$$V_B = 0.7 \text{ V}$$
$$V_{RB} = V_{CC} - V_B = 10 \text{ V} - 0.7 \text{ V} = 9.3 \text{ V}$$
$$I_B = V_{RB}/R_b = 9.3 \text{ V}/1 \text{ M}\Omega = 9.3 \text{ μA}$$
$$I_C = \beta \times I_B = 100 \times 9.3 \text{ μA} = 0.93 \text{ mA}$$
$$I_E \approx I_C = 0.93 \text{ mA}$$

Step 2. The signal input impedance can now be calculated.

$$z_{in} = r_b \parallel \beta \times r'_e$$
$$\beta = 100 \text{ from schematic}$$
$$r'_e = 25 \text{ mV}/I_E = 25 \text{ mV}/0.93 \text{ mA} = 27 \text{ } \Omega$$
$$r_b = 1 \text{ M}\Omega$$
$$z_{in} = 1 \text{ M}\Omega \parallel (100 \times 27 \text{ } \Omega) = 2.63 \text{ k}\Omega$$

An analogy that is helpful when calculating input impedance is to picture the forward-biased base/emitter as a telescope that magnifies the resistance in the emitter leg by a factor of beta when viewed from the base terminal. Figure 4–22 shows this analogy.

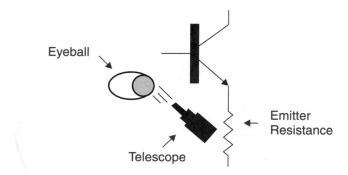

FIGURE 4–22   Emitter resistance when viewed from base

VOLTAGE GAIN

The voltage gain of an amplifier is defined as signal voltage out divided by signal voltage in $(A_v = V_{\text{out}}/V_{\text{in}})$ and can be easily calculated if both the input and output signal voltages are known. In troubleshooting, however, it is helpful to be able to estimate the voltage gain from circuit values. In this section, we will develop a formula that makes this possible.

The input signal voltage (V_{in}) is equal to the terminal voltage of the signal source (v_s) connected to the circuit. The Kirchhoff's voltage equation for the loop containing the signal voltage source (v_s) and base/emitter junction is $v_s - (i_e \times r_e) = 0$, where r_e is the signal impedance in the emitter leg. Therefore, the signal voltage (v_s) is equal to the signal emitter current (i_e) times the signal impedance in the emitter. Expressed in mathematical terms, the input signal voltage is $V_{\text{in}} = v_s = i_e \times r_e$.

The output signal voltage (V_{out}) is equal to the output signal current (i_c) times the signal impedance (r_c) in the collector leg. Figure 4–23(a) shows that the collector signal current has two paths to ground: one through R_c and the other through R_L. The power source (V_{cc}) and the coupling capacitor are shorts for signal current. Therefore, the signal equivalent circuit in Figure 4–23(b) shows R_L and R_c connected in parallel. Expressed in mathematical terms, the signal impedance in the collector leg is $r_c = R_c \parallel R_L$. The signal output voltage equals $V_{\text{out}} = i_c \times r_c$. Now we have sufficient information to determine a formula for voltage gain using circuit components.

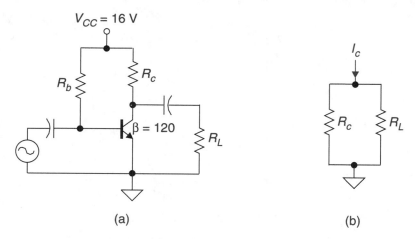

(a)   (b)

FIGURE 4–23   Signal load impedance

$A_v = V_{out}/V_{in}$

$V_{out} = i_c \times r_c$ (where r_c is the signal impedance in the collector leg)

$V_{in} = i_e \times r_e$ (where r_e is the signal impedance in the emitter leg)

$A_v = i_c \times r_c / i_e \times r_e$ (since $i_e \approx i_c$, the currents cancel)

$A_v = r_c / r_e$

EXAMPLE 4.8

Calculate the voltage gain for the circuit in Figure 4–24.

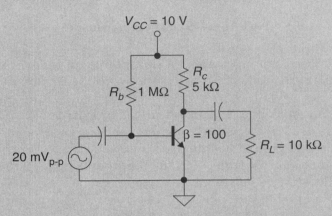

FIGURE 4–24 Amplifier circuit

Step 1. Determine the signal resistance in the emitter leg. Examining the circuit in Figure 4–24 reveals that r'_e is the only resistance in the emitter leg. The bias emitter current must be calculated in order to calculate r'_e.

$$V_B = 0.7 \text{ V}$$
$$V_{RB} = V_{CC} - V_B = 10 \text{ V} - 0.7 \text{ V} = 9.3 \text{ V}$$
$$I_B = V_{RB}/R_b = 9.3 \text{ V}/1 \text{ M}\Omega = 9.3 \text{ }\mu\text{A}$$
$$I_C = \beta \times I_B = 100 \times 9.3 \text{ }\mu\text{A} = 0.93 \text{ mA}$$
$$I_E \approx I_c \approx 0.93 \text{ mA}$$
$$r'_e = 25 \text{ mV}/I_E = 25 \text{ mV}/0.93 \text{ mA} = 27 \text{ }\Omega$$
$$r_e = r'_e$$

Step 2. Determine the signal resistance in the collector leg.

$$r_c = R_c \parallel R_L$$
$$r_c = 5 \text{ k}\Omega \parallel 10 \text{ k}\Omega = 3.3 \text{ k}\Omega$$

Step 3. Calculate the voltage gain.

$$A_v = r_c/r_e$$
$$A_v = 3.3 \text{ k}\Omega/27 \text{ }\Omega = 123$$

OUTPUT IMPEDANCE (z_{out})

The output of an amplifier is usually connected to some type of load. It may be a resistor (R_L), a transducer, or the input impedance of the next stage. In any event, the amplifier becomes the signal source for the load. All signal sources have some internal impedance, and so does an

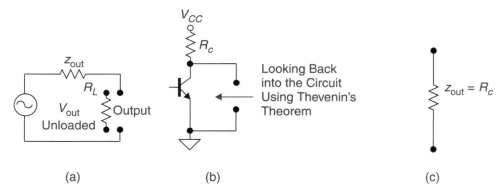

FIGURE 4–25   Output impedance

amplifier. From the perspective of the amplifier, this internal impedance is referred to as *output impedance*. Figure 4–25(a) shows that z_{out} is in series with the load resistance connected to the amplifier.

The value of the output impedance (z_{out}) is calculated by looking back into the circuit from the load as shown in Figure 4–25(b) and calculating the Thevenin equivalent impedance. There are two parallel signal paths to ground looking back into the circuit. One is through the reverse-biased collector/base junction of the transistor, and the other is through R_c and the power source. The reverse-biased collector/base junction has a signal impedance in the megohms. The impedance of the other path is equal to the value of R_c in series with the signal impedance of V_{cc}. The signal impedance of V_{cc} is close to zero, so the impedance of this path is equal to R_c. Since R_c is in kilohms and the impedance through the transistor is in megohms, the equivalent output impedance (z_{out}) is simply equal to R_c ($z_{out} = R_c$) as shown in Figure 4–25(c).

SIGNAL PHASE INVERSION

A positive voltage swing of the input signal increases base current (i_b), which causes the collector current to increase. The increased collector current causes a decrease (negative swing) in the output signal voltage. The inverse is also true—a negative swing of the input signal causes a positive swing in output signal. We describe an amplifier as having phase inversion when the output is 180º out of phase with respect to the input.

SUMMARY OF TRANSISTOR PARAMETERS

The concepts developed in this section permit you to analyze circuit signal parameters by using circuit component values. This is demonstrated in Example 4.9.

EXAMPLE 4.9

What is the (a) input impedance, (b) voltage gain, (c) output impedance, (d) signal output voltage, and (e) signal output phase for the circuit in Figure 4–26?

Step 1. Calculate input impedance.

$$z_{in} = r_b \parallel \beta \times r_e$$
$$r'_e = 25 \text{ mV}/I_E$$
$$I_B = V_{cc} - V_B/R_b = 23.3 \text{ V}/1.165 \text{ M}\Omega = 20 \text{ μA}$$
$$I_C = \beta \times I_B = 100 \times 20 \text{ μA} = 2 \text{ mA}$$
$$I_E \approx I_C = 2 \text{ mA}$$

EXAMPLE 4.9 continued

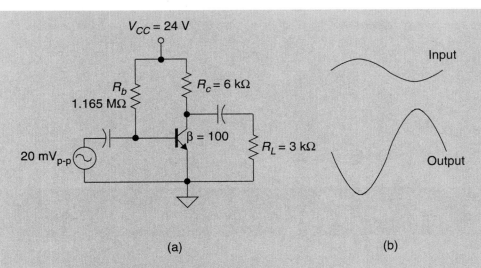

FIGURE 4–26 Amplifier with input and output signals

$$r'_e = 25 \text{ mV}/I_E = 25 \text{ mV}/2 \text{ mA} = 12.5 \ \Omega$$
$$r_e = r'_e$$
$$z_{\text{in}} = r_b \parallel \beta r_e = 1.165 \text{ M}\Omega \parallel (100 \times 12.5 \ \Omega) = 1250 \ \Omega$$

Step 2. Calculate voltage gain.
$$A_v = r_c/r_e$$
$$r_c = R_c \parallel R_L = 6 \text{ k}\Omega \parallel 3 \text{ k}\Omega = 2 \text{ k}\Omega$$
$$r_e = r'_e$$
$$A_v = r_c/r_e = 2 \text{ k}\Omega/12.5 \ \Omega = 160$$

Step 3. Calculate output impedance.
$$z_{\text{out}} = R_c$$
$$z_{\text{out}} = 6 \text{ k}\Omega$$

Step 4. Calculate V_{out}.
$$A_v = V_{\text{out}}/V_{\text{in}}, \text{ therefore } V_{\text{out}} = A_v \times V_{\text{in}}$$
$$V_{\text{out}} = 160 \times 20 \text{ mV}_{\text{p-p}} = 3.2 \text{ V}_{\text{p-p}}$$

Step 5. Determine the phase of the output signal with respect to the input signal. A positive swing in base voltage causes an increase in collector current. The increase in collector current causes an increase in the voltage dropped across R_c, which reduces the collector voltage. Hence, the output signal is 180º out of phase with respect to the input signal.

The base-biased amplifier can be analyzed using a schematic diagram of the circuit containing component values and knowing the beta of the transistor. The first step is to calculate the DC bias currents and voltages. The second step is to calculate the signal circuit parameters of input impedance, output impedance, and voltage gain. With these values known, you have the needed information to troubleshoot the circuit. In actual troubleshooting, you might not use paper and pencil to do exact calculations, but you should have approximate calculations in your mind. Having a list of formulas to plug values into and then cranking out the answers is not the way to understand electronic circuits. However, there are a few relationships one needs to know in order to work with bipolar transistor circuitry. The list of relationships below are important and should be memorized.

$I_E = I_B + I_C$	Transistor current divider
$\beta = I_C/I_B$	Current gain (beta)
$r'_e = 25\ \text{mV}/I_E$	Signal junction resistance
$z_{\text{in}} = r_b \parallel (\beta \times r_e)$	Input impedance
$z_{\text{out}} = R_c$	Output impedance
$A_v = V_{\text{out}}/V_{\text{in}}$	General voltage gain
$A_v = r_c/r_e$	Voltage gain

4.7 MEASURING INPUT AND OUTPUT IMPEDANCE

The input and output impedance of a transistor amplifier circuit cannot be measured with an ohmmeter. Using our knowledge of series circuits and a little common sense, however, we will develop a method that can be used in the lab to determine the input and output impedance of any circuit.

The box in Figure 4–27(a) is an electronics circuit that has an input lead, an output lead, and a common lead. The generator signal is applied across the input and common leads, and the load is connected across the output and common leads. We do not know what is in the box, but we do know that all circuits look like a single impedance to the generator driving the circuit. Figure 4–27(b) shows the circuit with the signal generator adjusted to 1 V and a 1 k resistor connected in series between the generator and the input lead. A voltage of 0.8 V is read from the input lead to the common lead. Example 4.10 demonstrates how the information shown in Figure 4–27(b) can be used to calculate the input impedance of the unknown circuit.

EXAMPLE 4.10

Calculate the input impedance of the unknown circuit shown in the box in Figure 4–27(b).

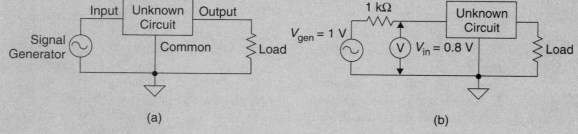

FIGURE 4–27 Input impedance test circuit

Step 1. Calculate the voltage across the 1 kΩ resistor.
$$V_R = V_{\text{gen}} = V_{\text{in}}$$
$$V_R = 1\ \text{V} - 0.8\ \text{V} = 0.2\ \text{V}$$

Step 2. Calculate the generator current. Since the 1 kΩ resistor is in series with the generator, it has the same current as the generator.

EXAMPLE 4.10 continued

$$i_{gen} = V_R/R$$

$$i_{gen} = 0.2 \text{ V}/1 \text{ k}\Omega = 0.2 \text{ mA}$$

Step 3. Calculate the input impedance of the circuit. The current flowing into the circuit equals the current through the 1 kΩ resistor, since the resistor is in series with the circuit. The voltage across the circuit is known to be 0.8 V.

$$i_{in} = i_R = i_{gen}$$

$$z_{in} = V_{in}/i_{in}$$

$$z_{in} = 0.8 \text{ V}/0.2 \text{ mA} = 4 \text{ k}\Omega$$

Thevenin's theorem states that from the perspective of the load, any circuit can be replaced with a source (V_{Th}) and a series resistance (R_{Th}). Figure 4–28(a) shows the original circuit driving the load, and Figure 4–28(b) shows the Thevenin's equivalent circuit driving the load. The Thevenin's resistance is equal to the output impedance of the original circuit. Therefore, if Thevenin's resistance of the circuit can be determined, we will know the output impedance (z_{out}).

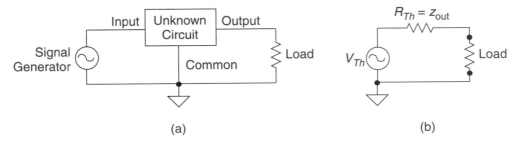

(a)   (b)

FIGURE 4–28 Thevenin's equivalent

Figure 4–29 shows the procedure for determining the output impedance in the lab. First, the load is removed from the circuit. The load will remain out of the circuit for the rest of the procedure. Second, the open circuit voltage is read and recorded [shown in Figure 4–29(a)]. The open circuit voltage is equal to the Thevenin voltage (V_{Th}). Third, a resistor with a known ohms value (1 kΩ in this case) is connected across the circuit as shown in Figure 4–29(b), and the voltage across the resistor is read and recorded. Fourth, the Thevenin equivalent circuit is drawn showing the known voltage values [shown in Figure 4–29(c)]. Fifth, the circuit is solved for the value of the Thevenin resistance, which equals the output impedance.

Step 1. Calculate the current in the Thevenin equivalent circuit.

$$i_{Th} = V_R/R = 3 \text{ V}/1 \text{ k}\Omega = 3 \text{ mA}$$

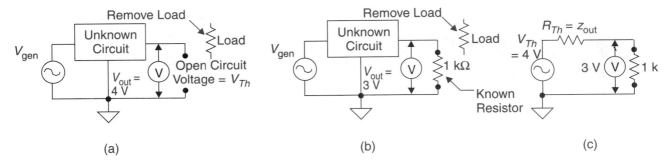

(a)   (b)   (c)

FIGURE 4–29 Output impedance test circuit

Step 2. Calculate the voltage drop across z_{out}.

$V_{zout} = V_{Th} - V_R = 4 \text{ V} - 3 \text{ V} = 1 \text{ V}$

Step 3. Calculate z_{out}.

$z_{out} = V_{zout}/i_{Th} = 1 \text{ V}/3 \text{ mA} = 333 \text{ }\Omega$

In the lab, using the procedure described, you will be able to measure the input and output impedance of all types of circuits. If the impedances are not pure resistance, the phase angle needs to be considered. In this text, we will consider the input and output impedance to be pure resistance unless otherwise noted.

In Figure 4–30(a), the unknown circuit is a bipolar amplifier. Figure 4–30(b) shows how a variable resistor can be connected in the circuit to determine input impedance. The resistor can be adjusted to any known value, and the input impedance can be calculated. However, if the resistance were adjusted until the signal at the input of the circuit equaled 50% of the generator signal, the resistance of the variable resistor would equal the input impedance of the circuit. After the variable resistor is adjusted to the correct value, it can be removed from the circuit and measured with an ohmmeter.

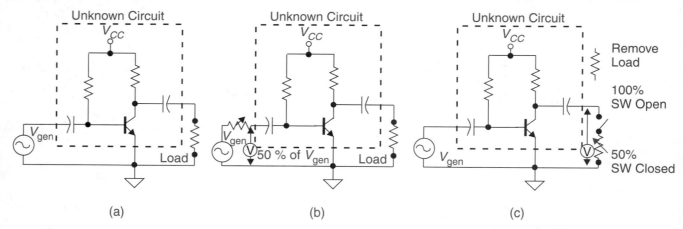

FIGURE 4–30   Experimentally determining input and output impedance

Figure 4–30(c) shows how a variable resistor can be connected in the circuit to determine the output impedance. Again, the resistance can be adjusted to any known value, and the needed data can be obtained to calculate the output impedance. However, if the variable resistor were adjusted until the closed switch signal equaled 50% of the open switch signal, the value of the variable resistor would equal the output impedance.

The output impedance of most function generators is 50 Ω, and this value is normally displayed on the face plate of the generator. Next time you are in the lab, verify the impedance of your function generator using the procedure discussed.

4.8 TRANSISTOR OUTPUT CHARACTERISTIC CURVES

A graph showing collector current (I_C) versus collector/emitter voltage (V_{CE}) with base current (I_B) held constant is called an output characteristic curve. Figure 4–31(a) shows an experimental circuit used to obtain the needed data to draw an output characteristic curve. Figure 4–31(b) shows the output characteristic curve obtained by holding the base current constant at 5 µA and monitoring the collector current as the collector/emitter voltage is varied from 0 V to 20 V.

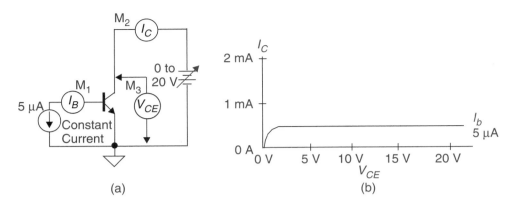

FIGURE 4–31   Circuit for developing characteristic curves

If the base current is adjusted to a second value and the process is repeated, data can be obtained for a second curve. This process can be repeated several times, and a family of output characteristic curves can be obtained as shown in Figure 4–32.

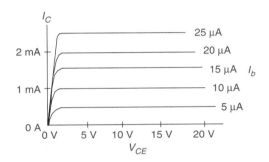

FIGURE 4–32   Family of output characteristic curves

Output characteristic curves can be used to examine how a transistor will function over a dynamic range. Figure 4–33(a) shows a fixed bias amplifier with a 10 kΩ resistor in the collector leg (R_c). Figure 4–33(b) shows the output characteristic curves for the transistor with the DC load line for the 10 kΩ resistor plotted. The end point of the DC load line on the horizontal axis is calculated by assuming the transistor is cut off and then calculating the voltage across the transistor. This would equal the power supply voltage (20 V in this case). The end point of the DC load line on the vertical axis is calculated by assuming the transistor is in saturation. With the transistor in saturation, the only resistance limiting current flow is resistor R_c, so the collector current at saturation is found by dividing the power supply voltage by the value of R_c. In this case, the collector current at saturation is equal to 2 mA (20 V/10 kΩ). The DC load line can be used to select the Q-point. Figure 4–33(b) shows that for class-B operation the Q-point would be selected at the cutoff point on the DC load line. For class-A operation the Q-point should be selected near the center of the DC load.

Based on the DC load line, it would appear that the circuit in Figure 4–33(a) could have an undistorted output signal of 20 V_{p-p} as shown in Figure 4–33(b). However, the signal flowing in the collector leg not only sees R_c but also sees R_L. Therefore, in order to determine maximum undistorted output signal, an AC load line must be plotted. The AC and DC load lines share the same Q-point; hence, we already have one point on the AC load line. The second point on the AC load line is determined by calculating the signal saturation current (i_c). The signal saturation current is equal to quiescent collector current (I_c) plus the change in collector voltage (V_{CE})

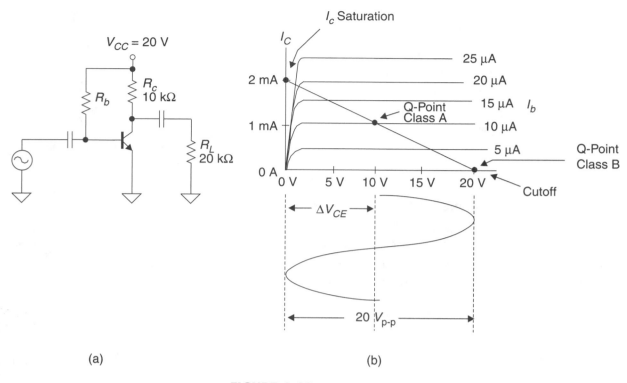

FIGURE 4–33   DC load line

divided by the signal resistor in the collector leg (r_c). Example 4.11 shows that the AC load line can be used to determine the maximum undistorted output signal.

EXAMPLE 4.11

For the circuit in Figure 4–33(a) draw the AC load line and determine the maximum output swing without distortion.

Step 1. Determine the signal resistance in the collector leg.

$$r_c = R_c \parallel R_L = 10 \text{ k} \parallel 20 \text{ k} = 6.7 \text{ k}$$

Step 2. Determine the quiescent collector current from the graph in Figure 4–33(b).

$$I_c = 1 \text{ mA}$$

Step 3. From the graph in Figure 4–33(b) determine the change in V_{CE} between the Q-point and saturation.

$$\Delta V_{CE} = 10 \text{ V}$$

Step 4. Calculate the signal saturation current.

$$i_c = I_c + \Delta V_{CE}/r_c = 1 \text{ mA} + 10 \text{ V}/6.7 \text{ k} = 2.5 \text{ mA}$$

Step 5. Draw the AC load line on the graph in Figure 4–34 from the signal saturation current point on the vertical axis through the Q-point to the horizontal axis.

Step 6. Draw the maximum symmetrical output signal without distortion below the horizontal axis. The signal cannot exceed saturation or cutoff on the AC load line. Notice the signal can swing only 6.5 V from the Q-point before cutoff is reached on the horizontal axis. Therefore, the maximum symmetrical signal the circuit could handle without distortion is 13 $V_{\text{p-p}}$ as shown in Figure 4–34.

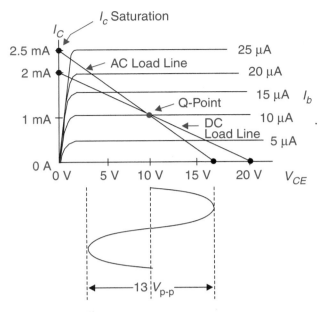

FIGURE 4–34 AC load line

Engineers make use of characteristic curves for designing devices and circuits. The technician does not normally utilize characteristic curves in circuit troubleshooting; however, understanding characteristic curves is useful for technicians when using a curve tracer. A curve tracer is an instrument that permits the technician to check the dynamic conditions of an electronic device. By using a curve tracer, the technician can display a family of output characteristic curves to check the dynamic performance of transistors. For example, if a circuit needs matching transistors, a curve tracer can compare the performance of transistors over a dynamic range. If the transistors are matched, their characteristic curves will be matched. If they are not matched, the characteristic curves will be different.

4.9 TROUBLESHOOTING TRANSISTORS

With the transistor out of the circuit, the two diode junctions of the transistor can be checked with an ohmmeter. Figure 4–35 shows that the transistor can be thought of as two diodes connected back-to-back. This test also can be used to identify the transistor as an NPN or a PNP.

For an NPN transistor, as shown in Figure 4–35, the test is as follows: connect the positive lead of the ohmmeter to the base, and connect the negative (common) lead to the emitter. This should forward bias the base/emitter junction, and the ohmmeter will read low. Now connect the negative lead to the collector. This will forward bias the base/collector junction, and the ohmmeter will again read low. These readings are shown in Figure 4–35(b). Now, if the ohmmeter is reversed, the two junctions can be read in reverse bias, and both junctions should read high as shown in Figure 4–35(c). If one of the junctions reads low in both directions, the junction is shorted and the transistor is bad. If one of the junctions reads high in both directions, the junction is open and the transistor is bad. Next, make a reading between the collector and emitter, then reverse the meter leads and make a second reading. Both of these readings should be high. If they are not, the transistor has a collector-to-emitter short and should be replaced.

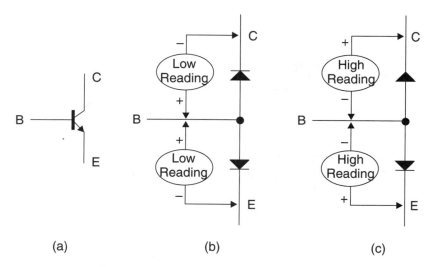

(a)   (b)   (c)

FIGURE 4–35   Transistor out-of-circuit test

If the terminals are unknown, you can find the base terminal by identifying which terminal has a low reading in relation to the other two terminals. This terminal is the base. If the low readings are caused by the base being positive, the transistor is an NPN. If the low readings are caused by the base being negative, the transistor is a PNP. If the checks are being performed by a DMM, the emitter and collector terminals can be determined. Most DMMs give a millivolts reading on the diode scale. Carefully read and record the millivolts obtained when checking the forward-biased base/emitter and base/collector junctions. The junction that has the higher reading is the base/emitter junction.

If the transistor passes the ohmmeter test, it is probably good. However, the transistor could be breaking down under circuit conditions not present with the meter test. In addition, this test does not check the signal characteristics, for example current gain (β) of the transistor. There are many transistor testers on the market that do check the signal characteristics, and if one is available, it would be preferable to using a meter. The procedure for using the transistor tester will be provided in the operator's manual of the tester. The technician also can use a curve tracer to check the dynamic response of a transistor. No form of out-of-circuit testing totally simulates the circuit in which the device is operating. For this reason, a device may pass all out-of-circuit tests and still cause the circuit to fail.

Figure 4–36 shows a transistor functioning as a switch. Reading the collector voltage will quickly tell the technician the condition of the switch. If the switch is off, the voltage on the

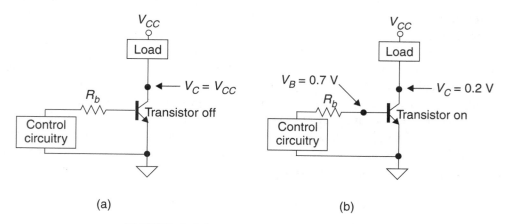

(a)   (b)

FIGURE 4–36   Transistor switching circuits

collector should measure approximately the supply voltage (V_{cc}). The technician can turn the transistor off by connecting a jumper between the base and emitter. If this procedure does not turn the transistor off, it is bad and needs to be replaced.

If the switch is on, the voltage on the collector will measure approximately 0.2 V, and the base will measure approximately 0.7 V. The technician can connect the base to V_{cc} through a pull-up resistor, and this should cause the transistor to turn on. If the transistor does not turn on, the transistor is bad or the control circuitry is holding the base at ground.

Always remember that a transistor functioning as a switch should be in cutoff ($V_c = V_{cc}$) or in saturation ($Vc \approx 0.2$ V). If the collector voltage is measuring some other voltage, the transistor is not operating as a switch and the transistor is bad or the control circuitry is not functioning properly.

Figure 4–37 shows a transistor functioning as a class-A amplifier. The technician should make a quick check of the collector voltage (V_c), which should be approximately one-half of the supply voltage (V_{cc}). If the collector voltage reads under 30% or over 70%, the operating conditions on the amplifier should be questioned. If the collector voltage is in question, then measure the voltage between the base and emitter (V_{BE}), which should measure approximately 0.7 V. If the DC readings are within range, the technician can then proceed with signal insertion to check the gain of the amplifier.

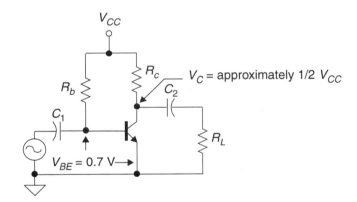

FIGURE 4–37 Transistor class-A amplifier circuit

Transistor Switch and Transistor Amplifier

1. Assuming the transistor to have a minimum beta of 100, calculate and record the value of R_b for the circuit in Figure 4–38.

   $R_b =$ _____

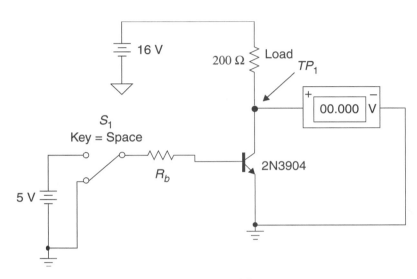

FIGURE 4–38

2. Calculate and record the voltage at TP_1 with S_1 down and with S_1 up for the circuit in Figure 4–38.

   S_1 down = _____ S_1 up = _____

3. Simulate and record the voltage at TP_1 with S_1 down and with S_1 up for the circuit in Figure 4–38.

   S_1 down = _____ S_1 up = _____

4. Calculate and record the control power in and the power delivered to the load when the transistor is on.

   $P_{in} =$ _____ $P_{out} =$ _____

5. Calculate and record V_B, V_C, z_{in}, z_{out}, A_v, and V_{out} for the circuit in Figure 4–39. Assume a beta of 150.

   $V_B =$ _____ $V_C =$ _____ $z_{in} =$ _____

   $z_{out} =$ _____ $A_v =$ _____ $V_{out} =$ _____

6. Simulate and record V_B, V_C, A_V, and V_{out} for the circuit in Figure 4–39.

   $V_B =$ _____ $V_C =$ _____ $A_V =$ _____ $V_{out} =$ _____

7. Calculate the beta of the transistor based on the simulated readings.

   Beta = _____

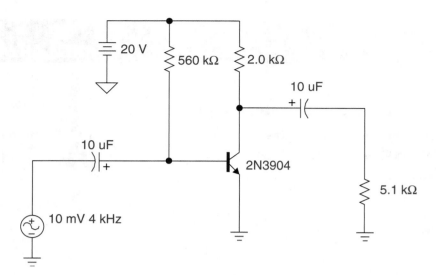

FIGURE 4–39

LAB **4.1**

Transistor Switch and Transistor Amplifier

I. Objectives

1. To construct and analyze a transistor switching circuit.
2. To construct and analyze a base-biased transistor amplifier.

II. Test Equipment

(1) sine wave generator   (1) 0 to 15 V power supply

(1) dual-trace oscilloscope   (1) multimeter

III. Components

Bulb: (1) 12 V @ 80 mA

Capacitors: (2) 10 µF

Resistors: (1) 150 Ω/2 W, (2) 1 kΩ, and (1) 220 kΩ

Potentiometers: (1) 5 kΩ and (1) 100 kΩ

Transistors: (1) 2N3904

IV. Procedure

1. Assuming the transistor to have a minimum beta of 50, calculate and record the value of R_b for the circuit in Figure 4–40.

   R_b = _____   Next smaller standard value = _____

2. Construct the circuit in Figure 4–40 using the standard value resistor determined in step 1.

3. Measure the voltage across the load with the transistor on and off.

   $V_{R_L(\text{off})}$ = _____ $V_{R_L(\text{on})}$ = _____

4. Calculate and record the control power in and the load power out when the transistor is on.

   P_{in} = _____ P_{out} = _____

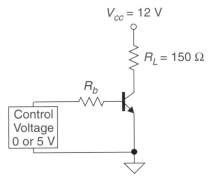

V_{cc} = 12 V

R_L = 150 Ω

R_b

Control Voltage 0 or 5 V

FIGURE 4–40

5. Replace the load resistor in Figure 4–40 with a light bulb. Adjust the supply voltage (V_{cc}) to the voltage rating of the bulb. With the control voltage on, make the needed adjustment

in R_b to ensure the transistor is functioning as a closed switch. Note the collector voltage should measure approximately 0.2 V when the transistor is on.

6. Measure the current through the bulb.

7. Calculate and record the control power in and the bulb power out when the transistor is on.

    P_{in} = _____ P_{out} = _____

8. Calculate and record V_B, V_C, z_{in}, z_{out}, A_v, and V_{out} for the circuit in Figure 4–41. Assume a beta of 150.

    V_B = _____ V_C = _____ z_{in} = _____

    z_{out} = _____ A_v = _____ V_{out} = _____

9. Construct the circuit in Figure 4–41. Use a transistor tester to find a transistor with a beta near 150. Adjust the 100 kΩ potentiometer to 66 kΩ. This will give a starting value of R_b equal to 286 kΩ: the correct value if the transistor has a beta of 150.

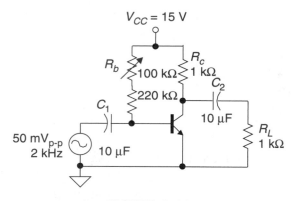

FIGURE 4–41

10. Apply DC power (V_{cc}) to the circuit. Adjust the 100 kΩ potentiometer until the collector voltage measures 7.5 V.

11. Measure V_B and V_c.

    V_B = _____ V_C = _____

12. Turn on the signal generator and adjust the input signal to 50 mV$_{p-p}$ at 2 kHz. Measure the output signal (V_{out}). Note the output signal of this amplifier will have some distortion, but do not worry about it at this point. The next chapter will examine amplifiers with less distorted output.

    V_{out} = _____

13. Calculate the voltage gain using measured values for V_{in} and V_{out}.

    A_v = _____

14. Using the procedure discussed in this chapter, measure the input impedance of the amplifier circuit.

    z_{in} = _____

15. Using the procedure discussed in this chapter, measure the output impedance of the amplifier circuit.

    z_{out} = _____

V. Points to Discuss

1. Why should the next smaller standard resistor be used for R_b in the transistor switch rather than the next larger standard value? *Because if you use a smaller resistor there is less of a risk of losing the forward bias in the emitter.*

2. What would be the power delivered to the 150 Ω load if the control voltage were reduced to 2 V?

928mW B/c you're still in saturation

3. With only 2 V of control voltage, would the transistor be operating in saturation? Explain.

yes b/c 2v is enough to be put in saturation.

4. Could the transistor switch circuit be redesigned to operate properly from a control of (a) 2 V and (b) 0.5 V? Explain.

2v is enough to forward bias emitter, but .5v is not because you have to have at least .7v to forward bias the emitter.

5. Would you expect the values to change greatly if a second 2N3904 transistor were used to replace the one in the amplifier circuit? Explain.

yes the betas are different in every transistor

6. Could the input and output impedances of this circuit be measured using an ohmmeter? Explain.

Potentiometer.

QUESTIONS

4.1 Introduction to Transistors

____ 1. What are the three terminals of a bipolar transistor?
   a. emitter, base, collector
   b. T_1, T_2, T_3
   c. source, gate, drain
   d. emitter, gate, collector

____ 2. The arrow on the transistor symbol always points to what type of material?
   a. P-type
   b. N-type
   c. Base-type
   d. PN-type

____ 3. Bipolar transistors are classified as _____.
   a. PPN and PIN devices
   b. NPN and PNP
   c. NNP and PPN devices
   d. N-type and P-type

____ 4. The schematic symbol of a bipolar PNP transistor is shown below as _____.

   a.   b.   c.   d.   e.

4.2 Inside the Transistor

____ 5. How many PN junctions are there in the bipolar transistor?
   a. 0
   b. 1
   c. 2
   d. 3
   e. 4

____ 6. What type of material is the base section of a PNP transistor?
   a. P-type
   b. N-type
   c. Base-type
   d. PN-type

____ 7. Compared to the collector and emitter, the base section of a transistor is _____.
   a. very thick
   b. very thin
   c. very soft
   d. very hard

____ 8. In a transistor, the base current is _____ when compared to the collector and emitter currents.
   a. small
   b. large
   c. fast
   d. slow

____ 9. A bipolar transistor is constructed so that its base region is very thin and _____.
   a. heavily doped
   b. doped the same as the collector
   c. lightly doped
   d. doped the same as the emitter

____ 10. The collector current of a transistor is always _____.
   a. much lower than its emitter current
   b. lower than its base current
   c. equal to the emitter current
   d. equal to the emitter current minus base current

____ 11. During normal operation of an NPN transistor, most electrons entering the emitter terminal _____.
   a. leave the transistor through the collector terminal
   b. leave the transistor through the base terminal
   c. are absorbed by the transistor
   d. none of the above

____ 12. Which equation expresses the correct relationship between the base, emitter, and collector currents?
   a. $I_E = I_B + \beta$
   b. $I_C = I_B + I_E$
   c. $I_E = I_B + I_C$
   d. $I_B = I_E + I_C$

____ 13. The ratio of collector current to base current is called _____.
   a. rho
   b. pi
   c. omega
   d. beta
   e. alpha

4.3 Transistor Switches

____ 14. When a transistor switch is in saturation, V_{CE} is approximately equal to _____.
   a. V_{cc}
   b. V_B
   c. 0.2 V
   d. 0.7 V

____ 15. When a transistor switching circuit is on, the collector current is limited by _____.
   a. the base current

   b. the load resistance
   c. the base voltage
   d. the base resistance
____ 16. When a transistor switch is cut off, V_{CE} is approximately equal to _____.
   a. V_{cc}
   b. V_B
   c. 0.2 V
   d. 0.7 V

4.4   Transistor Characteristics and Data Sheets

____ 17. The three important transistor characteristics are beta, maximum power dissipation, and _____.
   a. minimum rho
   b. maximum pi
   c. maximum collector current
   d. minimum hold current
____ 18. The H-parameter h_{fe} is equal to the _____ of the transistor.
   a. alpha
   b. beta
   c. maximum collector current
   d. minimum hold current

4.5   The Transistor Amplifier

____ 19. When the transistor amplifier is biased properly for class-A operation, the _____.
   a. base/emitter junction is forward biased and the base/collector junction is reverse biased
   b. base/emitter junction is reverse biased and the base/collector junction is reverse biased
   c. base/emitter junction is forward biased and the base/collector junction is forward biased
   d. base/emitter junction is reverse biased and the base/collector junction is forward biased
____ 20. For class-A amplifier operation, the collector/base junction of the transistor has to be _____.
   a. an open circuit
   b. a closed circuit
   c. forward biased
   d. reverse biased
____ 21. The voltage gain of a transistor amplifier is equal to _____.
   a. V_B/V_E
   b. V_{in}/V_{out}

c. V_{out}/V_{in}
d. V_{cc}/V_c

____ 22. The bias voltage at the collector (V_c) of a class-A amplifier is approximately equal to
          _____.
          a. V_{cc}
          b.   one-half of V_{cc}
          c.   0 V
          d.   0.2 V

4.6   Signal Analysis of the Base-Biased Amplifier

____ 23. The input impedance of the amplifier covered in this chapter is ____.
          a.   equal to 1 kΩ
          b.   inversely related to beta
          c.   directly related to beta
          d.   none of the above
____ 24. The output impedance of the amplifier covered in this chapter is _____.
          a. R_c
          b.   inversely related to beta
          c.   directly related to beta
          d.   1 kΩ
____ 25. The phase difference between the input and output signals of the amplifier covered in
          this chapter is _____.
          a.   0º
          b.   90º
          c.   180º
          d.   270º
____ 26. The general formula for calculating voltage gain for the amplifiers covered in this
          chapter is _____.
          a. $A_v = V_{cc}/V_c$
          b. $A_v = V_B/V_E$
          c. $A_v = r_c/r_e$
          d. $A_v = R_L \times \beta$

4.7   Measuring Input and Output Impedance

____ 27. The input impedance of a transistor amplifier can be measured using a/an _____.
          a.   ohmmeter
          b.   impedance meter
          c.   curve tracer
          d.   potentiometer in series with the function generator
____ 28. The output impedance of a transistor amplifier can be measured using a/an _____.
          a.   ohmmeter
          b.   impedance meter

    c.   curve tracer

    (d.) potentiometer in place of the load resistance

4.8 Transistor Output Characteristic Curves

____ 29. A curve tracer can be used to display output characteristic curves, which is a graph of the _____.

    a.   base current versus collector/emitter voltage

    b.   collector current versus base/emitter voltage

    c.   collector current versus collector/emitter voltage

    d.   emitter current versus base/emitter voltage

4.9 Troubleshooting Transistors

____ 30. When checking a good transistor with an ohmmeter, the bipolar transistor _____.

    a.   should exhibit a high ratio of forward-to-reverse resistance across both junctions

    b.   should exhibit a high ratio of forward-to-reverse resistance across the collector/base junction

    c.   should exhibit a high ratio of forward-to-reverse resistance across the emitter/base junction

    d.   none of the above

____ 31. When the positive probe of an ohmmeter is connected to the base and the negative probe is connected to the collector of an NPN transistor, what resistance will be measured?

    a.   0 Ω

    b.   low resistance

    c.   5 kΩ

    d.   high resistance

____ 32. When the negative probe of an ohmmeter is connected to the base and the positive probe is connected to the emitter of an NPN transistor, what resistance will be measured?

    a.   0 Ω

    b.   low resistance

    c.   5 kΩ

    d.   high resistance

____ 33. What resistance is read between the collector and the emitter of a good transistor?

    a.   0 Ω

    b.   low resistance

    c.   5 kΩ

    d.   high resistance

____ 34. What is the voltage on the collector of the transistor in Figure 4–42(a)?

    a.   0.2 V

    b.   0.7 V

    c.   7.5 V

    d.   15 V

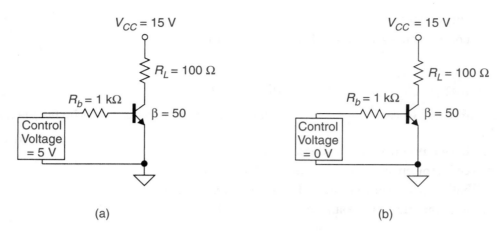

(a)   (b)

FIGURE 4–42

_____ 35. What is the voltage on the collector of the transistor in Figure 4–42(b)?

    a.   0.2 V

    b.   0.7 V

    c.   7.5 V

    d.   15 V

_____ 36. What is the DC voltage on the collector of the transistor in Figure 4–43?

    a.   0.2 V

    b.   0.7 V

    c.   7.5 V

    d.   15 V

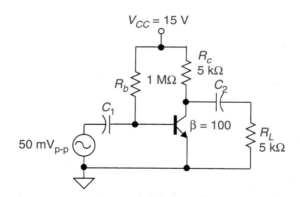

FIGURE 4–43

_____ 37. What is the DC voltage on the base of the transistor in Figure 4–43?

    a.   0.2 V

    b.   0.7 V

    c.   7.5 V

    d.   15 V

____.38. What is the signal voltage on the collector of the transistor in Figure 4–43?
  a. 50 mV$_{p-p}$
  b. 0.2 V$_{p-p}$
  c. 7.5 V$_{p-p}$
  d. 15 V$_{p-p}$

____ 39. If the output capacitor opens (C_2) in Figure 4–43, what would be the signal voltage on the collector of the transistor?
  a. 50 mV$_{p-p}$
  b. 0.2 V$_{p-p}$
  c. 7.5 V$_{p-p}$
  d. 15 V$_{p-p}$

PROBLEMS

1. If β = 80 in the circuit in Figure 4–44, what are the readings on meters I_B, I_E, I_C, and V_{CE}?

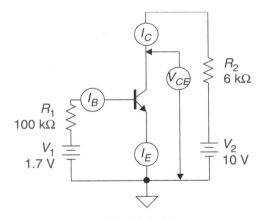

FIGURE 4–44

2. If the base current is 30 μA and the collector current is 4 mA, what is the value of the emitter current?

3. If the base current is 20 μA and the collector current is 4 mA, what is the value of beta?

4. The transistor in Figure 4–45 is used as a switch. Find the voltage at the collector when the transistor is conducting.

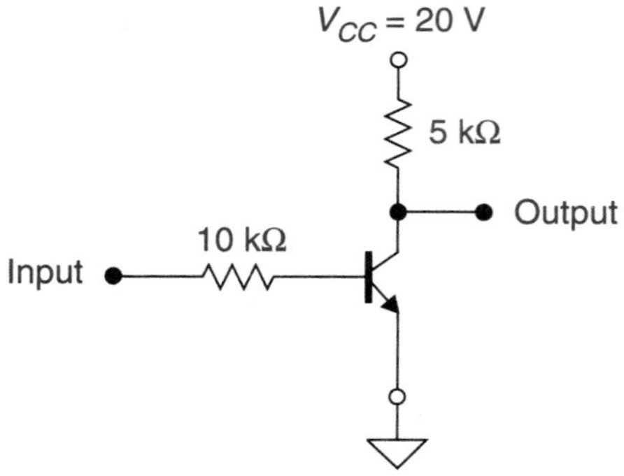

$V_{CC} = 20$ V

5 kΩ

Output

10 kΩ

Input

FIGURE 4–45

5. Find the collector voltage when the transistor in Figure 4–45 is turned off.

6. If the beta of the transistor in Figure 4–45 is 50, what minimum voltage is necessary at the input to saturate the transistor?

7. When the input to the circuit in Figure 4–45 is 5 V, what beta is required to saturate the transistor?

8. In the circuit in Figure 4–45, suppose the collector current is 4 mA, and the input current is 0.5 mA, if the input current is increased to 1 mA, what will the collector current equal?

9. If the "on" voltage at the input is known to be a minimum of 3.7 V, what is the maximum value of the input resistor that can be used to achieve saturation in Figure 4–45 (β_{min} = 50)?

10. Find the values requested below for the circuit in Figure 4–46 (β = 80).

    V_B = _____ V_C = _____

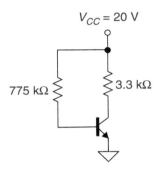

FIGURE 4–46

11. What resistance value would R_b (775 kΩ) in Figure 4–46 need to be changed to in order to cause the collector voltage to equal 10 V (β = 80)?

12. In the circuit of Figure 4–47, if R_c is 3.3 kΩ and beta is 120, find the value for R_b, which will provide midpoint bias (V_c = 7.5 V).

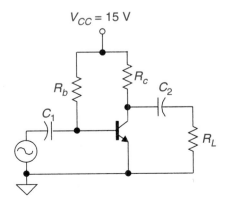

FIGURE 4–47

13. What will the collector voltage equal in the circuit in Figure 4–47 if $R_b = 755$ kΩ, $R_c = 3.3$ kΩ, and $\beta = 240$?

14. In Figure 4–47, if R_b is 600 kΩ and beta is 200, what value of R_c would be appropriate for midpoint bias ($V_c = 7.5$ V)?

15. In Figure 4–47, if R_b is found to be 800 kΩ and R_c is found to be 4.7 kΩ, what beta is required to provide midpoint bias ($V_c = 7.5$ V)?

16. What is the value of r'_e, z_{in}, and z_{out} for the circuit in Figure 4–48?
     $r'_e =$ _____ $z_{in} =$ _____ $z_{out} =$ _____

17. What is the voltage gain for the circuit in Figure 4–48?
     $A_v =$ _____

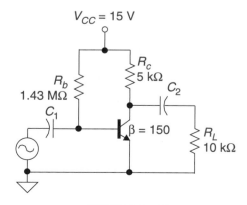

FIGURE 4–48

18. If capacitor C_2 is open, what would be the voltage gain for the circuit in Figure 4–48?

$A_v =$ _____

19. For the circuit in Figure 4–49, find:

$V_c =$ _____ $V_B =$ _____ $V_E =$ _____ $V_{CE} =$ _____

$z_{in} =$ _____ $z_{out} =$ _____ $V_{out} =$ _____ $A_v =$ _____

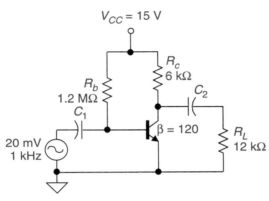

FIGURE 4–49

20. For the circuit in Figure 4–50, find:

$V_c =$ _____ $V_B =$ _____ $V_E =$ _____ $V_{CE} =$ _____

$z_{in} =$ _____ $z_{out} =$ _____ $V_{out} =$ _____ $A_v =$ _____

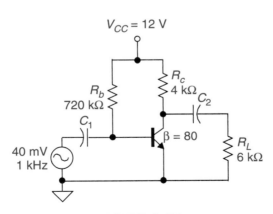

FIGURE 4–50

CHAPTER 5

Transistor Circuits

OBJECTIVES

After studying the material in this chapter, you will be able to describe and/or analyze:

○ the need for biased and signal stabilizing circuitry,

○ common-emitter amplifier circuits using voltage divider biasing,

○ common-emitter amplifier circuits using emitter biasing,

○ common-emitter amplifier circuits using voltage feedback biasing,

○ RC coupled multistage amplifiers,

○ direct coupled multistage amplifiers, and

○ troubleshooting procedures for transistor circuits.

5.1 INTRODUCTION

In the previous chapter you studied the bipolar transistor and learned how to make a simple amplifier circuit. Our study was limited, however, in that we assumed an ideal transistor and considered only single-stage amplifiers. In this chapter, we will consider transistor limitations and multistage amplifier circuits.

Q-POINT INSTABILITY CAUSED BY BETA

Because manufacturers cannot produce transistors with a precise beta value and because beta also changes with environmental conditions, the exact value for beta is unknown. The typical beta, given on data sheets, is an approximation. The actual beta for the individual transistor can range from minus 50% to plus 100% of the typical beta given by the manufacturer. Example 5.1 shows how the acceptable range of beta is calculated.

EXAMPLE 5.1

If a transistor has a typical beta of 150, what is the acceptable range of beta for the transistor?

EXAMPLE 5.1 continued

> **Step 1.** Calculate the minimum beta:
> $$\beta_{min} = \beta_{typical} - (\beta_{typical} \times 50\%) = 150 - (150 \times 50\%) = 75$$
> **Step 2.** Calculate the maximum beta:
> $$\beta_{max} = \beta_{typical} + (\beta_{typical} \times 100\%) = 150 + (150 \times 100\%) = 300$$

A transistor with a typical beta of 150 is considered good if its actual beta is within the range of 75 to 300.

The beta of a transistor also varies with changes in temperature. Figure 5–1 shows a graph of beta versus temperature. For an operational temperature of 0ºC, the beta of the 2N3904 is 140, and at a temperature of 100ºC, the beta is 220.

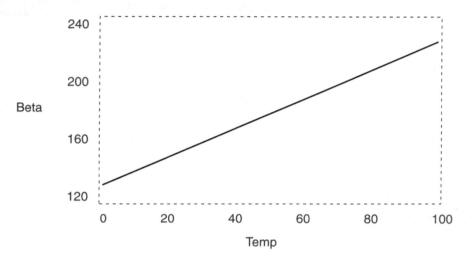

FIGURE 5–1   Beta versus temperature

In order for a transistor amplifier circuit to be useful, the circuit must be stable and predictable. With no signal applied to the amplifier circuit, the amplifier is in the quiescent state. At quiescence, the voltage across the transistor (V_{CE}) and the current through the transistor (I_C) should be constant. When the signal is applied to a class-A amplifier, the signal will cause variations around the quiescent point, also referred to as the Q-point. It is the function of the bias circuit to hold the circuit stable at the designed Q-point. Example 5.2 shows how the base-biased amplifier is not effective at holding the Q-point constant with changes in beta.

EXAMPLE 5.2

> Calculate the values of I_c and V_{CE} for the transistor amplifier in Figure 5–2, assuming the transistor has a beta of (a) 100 and (b) 180.
> **Step 1.** Calculate I_C and V_c for a beta equal to 100.
> $$I_B = V_{Rb}/R_b = 9.3 \text{ V}/930 \text{ k}\Omega = 10 \text{ }\mu\text{A}$$
> $$I_c = \beta \times I_b = 100 \times 10 \text{ }\mu\text{A} = 1 \text{ mA}$$
> $$V_{Rc} = I_c \times R_c = 1 \text{ mA} \times 5 \text{ k}\Omega = 5 \text{ V}$$
> $$V_c = V_{cc} - V_{Rc} = 10 \text{ V} - 5 \text{ V} = 5 \text{ V}$$

EXAMPLE 5.2 continued

Step 2. Calculate I_C and V_c for a beta equal to 180.

$$I_B = V_{Rb}/R_b = 9.3 \text{ V}/930 \text{ k}\Omega = 10 \text{ }\mu\text{A}$$
$$I_c = \beta \times I_b = 180 \times 10 \text{ }\mu\text{A} = 1.8 \text{ mA}$$
$$V_{Rc} = I_c \times R_c = 1.8 \text{ mA} \times 5 \text{ k}\Omega = 9 \text{ V}$$
$$V_c = V_{cc} - V_{Rc} = 10 \text{ V} - 9 \text{ V} = 1 \text{ V}$$

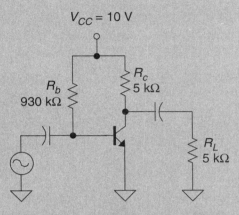

FIGURE 5–2   Base-biased amplifier

THE VOLTAGE DIVIDER

A simple voltage divider circuit can be used to provide reference voltage. Two resistors connected in series can provide any voltage level below the supply voltage at the junction of the two resistors by adjusting the ratio of the resistors. If the junction point is connected to other circuitry, the voltage at the junction may drop because of the loading effect of the added circuitry. However, if the resistance of the added circuitry is high compared to the resistors in the voltage divider, the voltage loading may be negligible. Example 5.3 illustrates how a lightly loaded voltage divider circuit can maintain a near constant output voltage.

EXAMPLE 5.3

Calculate the voltage at the junctions of R_1 and R_2 for both circuits shown in Figure 5–3 and determine the percentage of change in voltage caused by adding the load.

Step 1. Calculate the voltage at the test point for the circuit in Figure 5–3(a). The circuit is an unloaded voltage divider, and the voltage at the test point can be calculated by using the voltage divider formula.

$$V_{\text{out}} = V_{\text{in}} \times R_2/(R_1 + R_2) = 12 \text{ V} \times 2 \text{ k}\Omega/\text{x}\ddot{\text{u}}0 \text{ k}\Omega + 2 \text{ k}\Omega) = 2 \text{ V}$$

Step 2. Calculate the voltage at the test point for the circuit in Figure 5–3(b). The parallel equivalent resistance (R_{EQ}) of R_2 and the load resistor must first be calculated before the voltage divider formula can be used.

$$R_{EQ} = R_2 \parallel R_L = 2 \text{ k}\Omega \parallel 20 \text{ k}\Omega = 1.82 \text{ k}\Omega$$
$$V_{\text{out}} = V_{\text{in}} \times R_{EQ}/(R_1 + R_{EQ})$$
$$V_{\text{out}} = 12 \text{ V} \times 1.82 \text{ k}\Omega/(10 \text{ k}\Omega + 1.82 \text{ k}\Omega) = 1.85 \text{ V}$$

EXAMPLE 5.3 continued

Step 3. Calculate the percent of change in voltage caused by loading the voltage divider circuit.

$$\%\Delta = \Delta V / V \times 100\%$$
$$\Delta V = 2\ V - 1.85\ V = 0.15\ V$$
$$\%\Delta = 0.15\ V / 2\ V \times 100\% = 7.5\%$$

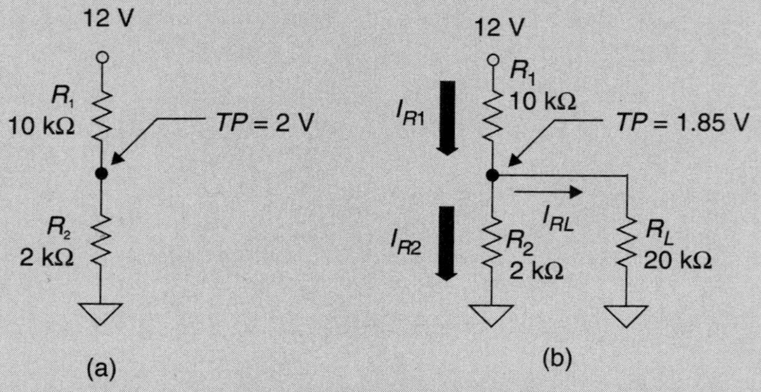

FIGURE 5–3   Voltage divider circuit

The voltage divider circuit in Figure 5–3(b) can be considered a lightly loaded voltage divider because the load resistor is ten times the resistance of R_2. The load causes the voltage at the junction of the voltage divider to drop by 7.5%. Loads with higher resistance will cause a smaller drop in voltage, and loads with lower resistance will cause a greater drop. A good rule of thumb is to consider the voltage divider lightly loaded if the load is ten times or greater the value of the parallel resistor (R_2 in this case). For lightly loaded voltage dividers, the voltage at the junction of the divider can be approximated using the simple voltage divider formula.

Figure 5–3(b) shows that the conventional current through R_1 divides, and a portion of the current flows through R_2 while the remainder flows through the load resistor. If the voltage divider is lightly loaded, most of the current will flow through R_2, and less than 10% of the current will flow through the load.

5.2 VOLTAGE DIVIDER BIASING

The amplifier circuit shown in Figure 5–4(a) uses voltage divider biasing. This method of biasing has proven to be a popular way to design bipolar transistor amplifiers. The circuit counteracts the effects of the uncertainty of beta. For DC bias analysis, the circuit can be simplified as shown in Figure 5–4(b) because the capacitors are considered open to the DC bias currents.

The name *voltage divider* comes from the fact that the base voltage is provided by a voltage divider formed by R_{b1} and R_{b2}. The voltage at the junction of R_{b1} and R_{b2} holds the base voltage constant. The circuit is always designed so the base current (I_B) is less than 10% of the current flowing through R_{b2}. Therefore, the voltage divider is essentially not loaded by the base current flow.

Since the voltage divider formed by R_{b1} and R_{b2} is lightly loaded, the base voltage (V_B) can be easily calculated by using the simple voltage divider formula. The base/emitter junction is forward-biased. Therefore, the emitter voltage (V_E) will be one junction drop different than the

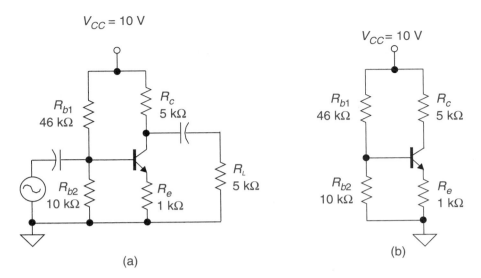

FIGURE 5–4   Voltage divider biasing

base voltage. Once the voltage on the emitter is known, the emitter current (I_E) can be easily calculated using Ohm's law, since R_e is connected between the emitter and ground. The collector current (I_C) can be approximated, since it is almost the same as the emitter current. With the collector current known, the collector voltage (V_C) can be calculated.

The preceding paragraph gave the procedures for calculating all of the bias voltages and currents for a voltage divider biased amplifier. Note that the calculations were completely independent of the beta of the transistor. This means the voltage divider biased amplifier will maintain a constant Q-point (I_C will be constant) even with large changes in beta. The circuit design does depend on a minimum beta, but we can be sure the engineer designed the circuit for the minimum beta of the transistor. Example 5.4 will show you how easy it is to calculate the bias voltages and currents for a voltage divider biased amplifier circuit.

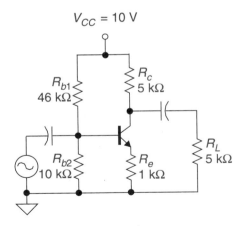

FIGURE 5–5   Voltage divider biased amplifier circuit

EXAMPLE 5.4

> Calculate V_B, V_E, I_E, I_C, V_C, and V_{CE} for the amplifier circuit in Figure 5–5.
>
> **Step 1.** Calculate the base voltage (V_B). Since the base voltage divider is lightly loaded, we can use the simple voltage divider formula.

EXAMPLE 5.4 continued

$$V_B = V_{cc} \times R_{b2}/(R_{b1} + R_{b2})$$
$$V_B = 10\text{ V} \times 10\text{ k}\Omega/(46\text{ k}\Omega + 10\text{ k}\Omega) = 1.8\text{ V}$$

Step 2. Calculate the emitter voltage (V_E). The base/emitter junction is forward-biased. The circuit is using a silicon NPN transistor, so the base must be 0.7 V more positive than the emitter.

$$V_E = V_B - V_{BE}$$
$$V_E = 1.8\text{ V} - 0.7\text{ V} = 1.1\text{ V}$$

Step 3. Calculate the emitter current (I_E) and collector current (I_C). Ohm's law can be used to calculate the emitter current, since $R_{E(total)}$ is connected between the emitter and ground.

$$I_E = V_E / R_{E(total)} \; (R_{E(total)} = R_e \text{ in this case.})$$
$$I_E = 1.1\text{ V}/1\text{ k}\Omega = 1.1\text{ mA}$$

$I_E \approx I_C$ (Base current is small; therefore, I_E is approximately equal to I_C.)

$$I_C \approx 1.1\text{ mA}$$

Step 4. Calculate the collector voltage. With the collector current known, the voltage drop across R_c can be calculated. If the voltage across R_c is subtracted from the supply voltage (V_{CC}), the voltage on the collector can be found.

$$V_{Rc} = I_c \times R_c = 1.1\text{ mA} \times 5\text{ k}\Omega = 5.5\text{ V}$$
$$V_c = V_{cc} - V_{Rc} = 10\text{ V} - 5.5\text{ V} = 4.5\text{ V}$$

Step 5. Calculate the collector to emitter voltage (V_{CE}). The voltage across the transistor can be calculated by finding the difference in voltage between the collector and the emitter.

$$V_{CE} = V_C - V_E$$
$$V_{CE} = 4.5\text{ V} - 1.1\text{ V} = 3.4\text{ V}$$

Voltage divider biasing uses current-mode feedback to stabilize the circuit. The current flowing through the emitter resistor causes negative feedback and stabilizes the circuit. Voltage divider biasing is a practical way to make the bias circuitry independent of beta. This approach is not the only means of stabilization, but it is one of the most popular.

5.3 SIGNAL PARAMETERS IN VOLTAGE DIVIDER CIRCUITS

The price paid for independence from beta is a decrease in signal voltage gain. Fortunately, these lower voltage gains are predictable. The input impedance is also affected, but it increases, and in most cases this is a plus. We will discuss each of these parameters and then look at some examples.

The size and the type of resistance in the emitter leg has a great effect on the operation of the voltage divider amplifier. There are five types of resistances in the emitter leg, each of which is defined in the following list. As you continue with the chapter, use this list to make sure you understand the type of emitter resistance being discussed.

r'_e is the internal signal resistance of the base/emitter junction ($r'_e = 25\text{ mV}/I_E$).

R_e is an external emitter resistor that opposes signal current (AC) and bias current (DC).

R_E is an external emitter resistor that only opposes bias current (DC).

r_e is the total signal (AC) resistance in the emitter leg and is equal to r'_e plus R_e.

$R_{E(\text{total})}$ is the total bias (DC) resistance in the emitter leg and is equal to R_E plus R_e.

INPUT IMPEDANCE

The general formula for input impedance is $z_{\text{in}} = r_b \parallel (\beta \times r_e)$. Resistance r_b is the equivalent signal resistance to ground on the base leg, and r_e is the equivalent signal resistance to ground on the emitter leg. Figure 5–6(a) shows a voltage divider biased amplifier circuit. From the perspective of the signal generator, resistors R_{B1} and R_{B2} appear in parallel as shown in Figure 5–6(b). Remember the power supply bus is at signal ground. This parallel combination is shown as r_b in Figure 5–6(c).

The path through the reverse-biased base/collector junction is considered an open. Figure 5–6(b) shows that the values of r'_e and R_e are seen through the forward-biased base/emitter junction. The resistance r_e is the series equivalent of r'_e and R_e as shown in Figure 5–6(c). The signal resistance (r_e) in the emitter leg must be multiplied by the value of beta as shown in Figure 5–6(c). Example 5.5 shows how input impedance of a voltage divider biased amplifier can be calculated using circuit values.

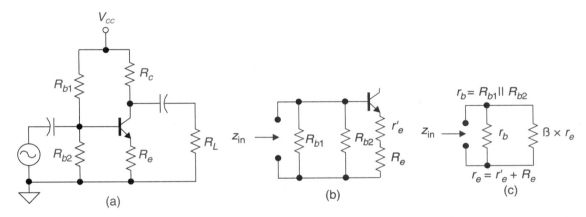

FIGURE 5–6   Input impedance of the voltage divider circuit

EXAMPLE 5.5

Assume a beta of 100 and calculate the input impedance of the circuit in Figure 5–7.

$V_{CC} = 10\text{ V}$

R_{b1} 46 kΩ

R_c 5 kΩ

R_{b2} 10 kΩ

R_e 1 kΩ

R_L 5 kΩ

FIGURE 5–7   Voltage divider biased amplifier circuit

EXAMPLE 5.5 continued

Step 1. Calculate the biased emitter current. The biased emitter current is needed to calculate the value of r'_e.

$$V_B = V_{cc} \times R_{b2}/(R_{b1} + R_{b2})$$

$$V_B = 10 \text{ V} \times 10 \text{ k}\Omega/(46 \text{ k}\Omega + 10 \text{ k}\Omega) = 1.8 \text{ V}$$

$$V_E = V_B - V_{BE}$$

$$V_E = 1.8 \text{ V} - 0.7 \text{ V} = 1.1 \text{ V}$$

$$I_E = V_E/R_{E(\text{total})} \ (R_{E(\text{total})} = R_e \text{ in this case.})$$

$$I_E = 1.1 \text{ V}/1 \text{ k}\Omega = 1.1 \text{ mA}$$

Step 2. Calculate the signal resistance of the base/emitter junction.

$$r'_e = 25 \text{ mV}/I_E = 25 \text{ mV}/1.1 \text{ mA} = 23 \ \Omega$$

Step 3. Draw an equivalent circuit showing how R_{b1}, R_{b2}, R_e, and r'_e appear from the perspective of the signal generator. The equivalent circuit is shown in Figure 5–8.

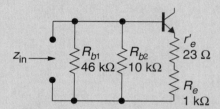

FIGURE 5–8   Input impedance equivalent circuit

Step 4. Calculate the signal resistance in the base leg (r_b).

$$r_b = R_{b1} \parallel R_{b2} = 46 \text{ k} \parallel 10 \text{ k}\Omega = 8.2 \text{ k}\Omega$$

Step 5. Calculate the signal resistance in the emitter leg (r_e).

$$r_e = r'_e + R_e = 23 + 1 \text{ k}\Omega = 1023 \ \Omega$$

Step 6. Multiply the emitter leg signal resistance times beta. (Assume beta equals 100.)

$$\beta \times r_e = 100 \times 1023 = 102.3 \text{ k}\Omega$$

Step 7. Draw the simplified equivalent circuit showing the base signal resistance in parallel with the emitter path signal resistance. The equivalent circuit is shown in Figure 5–9.

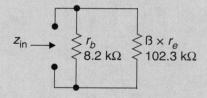

FIGURE 5–9   Simplified equivalent circuit

Step 8. Calculate the circuit input impedance by finding the parallel equivalent of the signal base resistance and the signal emitter path resistance.

$$z_{in} = r_b \parallel \beta r_e = 8.2 \text{ k}\Omega \parallel 102.3 \text{ k}\Omega = 7.6 \text{ k}\Omega$$

Two points should be noted. First, if the value of R_e is large in comparison to the value of r'_e, then r'_e can be dropped from the calculation. Second, note that beta is still in the equation, but by examining Example 5.5, you can see that the signal base resistance (r_b) is the main influence on input impedance. Because of this, changes in beta do not have a dramatic effect.

VOLTAGE GAIN OF THE VOLTAGE DIVIDER BIASED AMPLIFIER

The general formula for the voltage gain of the common emitter amplifier is $A_v = r_c/r_e$, where r_e is the signal resistance in the emitter leg and r_c is the signal resistance in the collector leg. In the base biased circuit, the only signal resistance in the emitter leg is r'_e. In the voltage divider circuit, the signal resistance in the emitter leg (r_e) is equal to r'_e plus R_e, where R_e is equal to the unbypassed resistance in the emitter leg. The signal resistance in the collector leg r_c is equal to R_c in parallel with R_L. Example 5.6 shows how the voltage gain of a circuit can be calculated from circuit component values.

EXAMPLE 5.6

Calculate the voltage gain of the circuit in Figure 5–10.

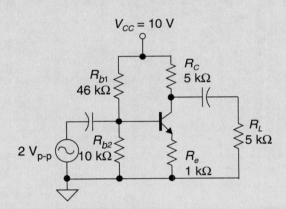

FIGURE 5–10 Voltage divider biased amplifier

Step 1. Calculate the signal resistance in the collector leg.
$$r_c = R_c \parallel R_L = 5 \text{ k}\Omega \parallel 5 \text{ k}\Omega = 2.5 \text{ k}\Omega$$

Step 2. Calculate the signal resistance of the base/emitter junction.
$$V_B = V_{cc} \times R_{b2}/(R_{b1} + R_{b2})$$
$$V_B = 10 \text{ V} \times 10 \text{ k}\Omega/(46 \text{ k}\Omega + 10 \text{ k}\Omega) = 1.8 \text{ V}$$
$$V_E = V_B - V_{BE}$$
$$V_E = 1.8 \text{ V} - 0.7 \text{ V} = 1.1 \text{ V}$$
$$I_E = V_E/R_{E(\text{total})} \quad (R_{E(\text{total})} = R_e \text{ in this case.})$$
$$I_E = 1.1 \text{ V}/1 \text{ k}\Omega = 1.1 \text{ mA}$$
$$r'_e = 25 \text{ mV}/I_E = 25 \text{ mV}/1.1 \text{ mA} = 23 \text{ }\Omega$$

Step 3. Calculate the signal resistance in the emitter leg.
$$r_e = r'_e + R_e = 23 \text{ }\Omega + 1000 \text{ }\Omega = 1023 \text{ }\Omega \text{ or } 1.023 \text{ k}\Omega$$

Step 4. Calculate the voltage gain (A_v).
$$A_v = r_c/r_e = 2.5 \text{ k}\Omega/1.023 \text{ k}\Omega = 2.44$$

The value of r'_e has little affect on the voltage gain of the circuit as long as R_e is much larger than r'_e. The voltage gain is greatly reduced, but the gain is predictable because it is independent of r'_e. In the following sections you will learn techniques to regain some of the lost gain and still have circuit stability.

Once the voltage gain of an amplifier is known, the output signal can be easily calculated by multiplying the input signal by the voltage gain. It should be noted that the output signal swing is limited by the power supply voltage and the Q-point. A good rule of thumb to follow for estimating the maximum peak-to-peak signal swing is to calculate the voltage across the transistor at quiescence (V_{CE}) and multiply this value by two. If the calculated output signal exceeds this value, clipping will occur on the peaks of the output signal. Example 5.7 shows how to calculate the output signal and the maximum output swing.

EXAMPLE 5.7

Calculate the output signal and the maximum output swing for the circuit in Figure 5–10.
Step 1. Calculate the output signal.
$$V_{\text{out}} = V_{\text{in}} \times A_v = 2\ V_{\text{p-p}} \times 2.44 = 4.88\ V_{\text{p-p}}$$
Step 2. Calculate V_{CE}. The value of V_E was calculated in Example 5.5 and equals 1.1 V. The value of V_c needs to be calculated.
$$V_{Rc} = I_c \times R_c = 1.1\ \text{mA} \times 5\ \text{k}\Omega = 5.5\ \text{V}\ (I_E \approx I_c\ \text{was calculated in Example 5.5.})$$
$$V_c = V_{cc} - V_{Rc} = 10\ \text{V} - 5.5\ \text{V} = 4.5\ \text{V}$$
$$V_{CE} = V_c - V_E = 4.5\ \text{V} - 1.1\ \text{V} = 3.4\ \text{V}$$
Step 3. Calculate the maximum output swing.
$$\text{maximum } V_{\text{out p-p}} = 2 \times V_{CE} = 2 \times 3.4\ \text{V} = 6.8\ V_{\text{p-p}}$$

Since the output signal (4.88 $V_{\text{p-p}}$) is less than the calculated maximum output swing (6.8 $V_{\text{p-p}}$), the output signal will not be clipped.

5.4 VARIATIONS OF VOLTAGE DIVIDER BIASED AMPLIFIERS

There are three variations of the voltage divider biased amplifier: the unbypassed, the fully bypassed, and the split emitter. The bias circuitry on all three variations functions the same. The difference is in the signal parameters. We will examine each variation and work an example of each.

UNBYPASSED VOLTAGE DIVIDER BIASED AMPLIFIER

The circuit in Figure 5–11 is the unbypassed voltage divider biased amplifier. This is the same circuit we have studied in the previous sections. Making two assumptions will simplify the circuit calculations. First, no value is given for the beta of the transistor in the circuit. Rather than wasting time looking it up, let us assume beta equals 100. One hundred is a reasonable guess of beta for low and medium power transistors. Second, the unbypassed resistor in the emitter (R_e) is large compared to r'_e, so r'_e will be negligible. These assumptions permit us to use a practical approach to calculate the bias and signal parameters of the voltage divider amplifier circuit using component values. The answers may not be precise because of the

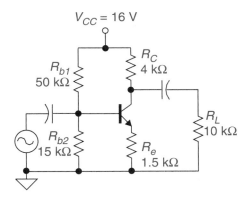

FIGURE 5–11 Unbypassed voltage divider biased amplifier

assumptions made, but they will be close enough to use in troubleshooting. Example 5.8 demonstrates this approach.

EXAMPLE 5.8

Analyze the circuit in Figure 5–11. Calculate (a) the DC bias voltages for V_B, V_E, V_c, and V_{CE}, and (b) the signal voltage gain (A_v), input impedance (z_{in}), output impedance (z_{out}), and the approximate maximum signal output voltage swing.

Step 1. Calculate the DC biased voltages and currents.

$$V_B = V_{cc} \times R_{b2}/(R_{b1} + R_{b2})$$
$$V_B = 16 \text{ V} \times 15 \text{ k}\Omega/(50 \text{ k}\Omega + 15 \text{ k}\Omega) = 3.7 \text{ V}$$
$$V_E = V_B - V_{BE} = 3.7 \text{ V} - 0.7 \text{ V} = 3 \text{ V}$$
$$I_E = V_E/R_{E(total)} = 3 \text{ V}/1.5 \text{ k}\Omega = 2 \text{ mA} \quad (R_{E(total)} = R_e \text{ in this case.})$$
$$I_C \approx I_E = 2 \text{ mA}$$
$$V_{RC} = I_C \times R_c = 2 \text{ mA} \times 4 \text{ k}\Omega = 8 \text{ V}$$
$$V_c = V_{cc} - V_{RC} = 16 \text{ V} - 8 \text{ V} = 8 \text{ V}$$
$$V_{CE} = V_c - V_E = 8 \text{ V} - 3 \text{ V} = 5 \text{ V}$$

Step 2. Calculate the signal input impedance.

$$R_b = R_{b1} \parallel R_{b2} = 15 \text{ k}\Omega \parallel 50 \text{ k}\Omega = 11.5 \text{ k}\Omega$$
$$r_e \approx R_e = 1.5 \text{ k}\Omega \quad (R_e \text{ is large compared to } r'_e.)$$
$$z_{in} = r_b \parallel (\beta \times r_e)$$
$$z_{in} = 11.5 \text{ k}\Omega \parallel 100 \times 1.5 \text{ k}\Omega$$
$$z_{in} = 10.7 \text{ k}\Omega$$

Step 3. Calculate the output impedance.

$$z_{out} = R_c = 4 \text{ k}\Omega$$

Step 4. Calculate the voltage gain.

$$r_c = R_c \parallel R_L = 4 \text{ k}\Omega \parallel 10 \text{ k}\Omega = 2.86 \text{ k}\Omega$$
$$A_v = r_c/r_e$$
$$A_v = 2.86 \text{ k}\Omega/1.5 \text{ k}\Omega = 1.9$$

Step 5. Calculate the maximum output signal swing. An approximation of the output peak-to-peak signal swing equals $2 \times V_{CE}$.

$$V_{CE} = 5 \text{ V}$$

Therefore:

$$\text{Maximum output swing} = 2 \times 5 \text{ V} = 10 \text{ V}_{p-p}$$

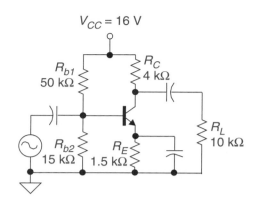

FIGURE 5–12   Bypassed voltage divider amplifier

THE FULLY BYPASSED AMPLIFIER

The voltage divider biased amplifier using an unbypassed emitter resistor (R_e) is stable and predictable but has a greatly reduced voltage gain when compared to the base biased amplifier. The voltage gain of the base biased amplifier is in the range of 50 to 500, but the voltage gain of the unbypassed voltage divider biased circuit is in the range of 2 to 10. In this section we will examine a circuit that retains the bias stability but has high voltage gain.

The circuit in Figure 5–12 is identical to the circuit in Figure 5–11, except that a capacitor has been added that bypasses the emitter resistor (R_E). The capacitor is an open for DC bias, hence, the biased currents and voltages are unchanged by the addition of the bypass capacitor. The circuit retains good Q-point stability, but the voltage gain is greatly increased.

Note the bypassed emitter resistor is called R_E with a capital "E" subscript. This denotes that the resistor will exist for the DC biased current but will appear as a short for the signal current. An unbypassed emitter resistor will be denoted as R_e and will exist for both biased currents (DC) and signal currents (AC).

The voltage gain is equal to the ratio of signal resistance in the collector leg to signal resistance in the emitter leg ($A_v = r_c/r_e$). In the unbypassed circuit, the signal resistance in the emitter leg is equal to r'_e plus the signal emitter resistor. The bypass capacitor shorts out the emitter resistor for the signal, leaving the signal resistance in the emitter leg equal to r'_e. With only the value of r'_e in the emitter leg, the voltage gain returns to the high voltage gain obtained with the base biased circuit.

The addition of the bypass capacitor will cause a decrease in input impedance. The output impedance will not be affected. Example 5.9 illustrates how signal values can be calculated for the fully bypassed amplifier.

EXAMPLE 5.9

Calculate the input impedance, output impedance, and the voltage gain for the circuit in Figure 5–12.

Step 1. Calculate r'_e.

$$V_B = V_{cc} \times R_{b2}/(R_{b1} + R_{b2})$$
$$V_B = 16 \text{ V} \times 15 \text{ k}\Omega/(50 \text{ k}\Omega + 15 \text{ k}\Omega) = 3.7 \text{ V}$$
$$V_E = V_B - V_{BE} = 3.7 \text{ V} - 0.7 \text{ V} = 3 \text{ V}$$
$$I_E = V_E/R_{E(total)} = 3 \text{ V}/1.5 \text{ k}\Omega = 2 \text{ mA} \quad (R_{E(total)} = R_E \text{ in this case.})$$
$$I_C \approx I_E = 2 \text{ mA}$$
$$r'_e = 25 \text{ mV}/I_E = 25 \text{ mV}/2 \text{ mA} = 12.5 \text{ }\Omega$$

EXAMPLE 5.9 continued

> **Step 2.** Calculate the input impedance.
> $$r_b = R_{b1} \parallel R_{b2} = 50 \text{ k}\Omega \parallel 15 \text{ k}\Omega = 11.5 \text{ k}\Omega$$
> $$r_e = r'_e = 12.5 \ \Omega \ (R_E \text{ is bypassed.})$$
> $$z_{\text{in}} = r_b \parallel (\beta \times r_e) = 11.5 \text{ k}\Omega \parallel (100 \times 12.5 \ \Omega) = 1.13 \text{ k}\Omega$$
> **Step 3.** Calculate the output impedance.
> $$z_{\text{out}} = R_c = 4 \text{ k}\Omega$$
> **Step 4.** Calculate the voltage gain.
> $$r_c = R_c \parallel R_L = 4 \text{ k}\Omega \parallel 10 \text{ k}\Omega = 2.9 \text{ k}\Omega$$
> $$A_v = r_c / r_e = 2.9 \text{ k}\Omega / 12.5 \ \Omega = 232$$

The circuits in Examples 5.8 and 5.9 are identical except for the addition of the bypass capacitor. The addition of the bypass capacitor makes a dramatic change to the voltage gain and input impedance. The voltage gain goes up from 1.9 to 232, and the input impedance goes down from 10.7 kΩ to 1.13 kΩ. The increase in voltage gain is generally considered a plus that would probably outweigh the disadvantages of lowering the input impedance. However, the voltage gain is now dependent on r'_e, which makes it less predictable. The input impedance is not only lower but now is dependent on r'_e and β, which makes it less predictable than it was in the unbypassed circuit. Signal distortion is another disadvantage of the fully bypassed circuit. Figure 5–13 shows an example of signal distortion that is easily seen when the output signal has a large swing. The distortion is caused by fluctuations in r'_e and directly affects the voltage gain of the amplifier. The fluctuations in r'_e are caused by the wide variations in emitter current. On the positive swing of the input, the emitter current increases, causing r'_e to decrease and the voltage gain to increase. On the negative swing of the input, the emitter current decreases, causing r'_e to increase and the voltage gain to decrease.

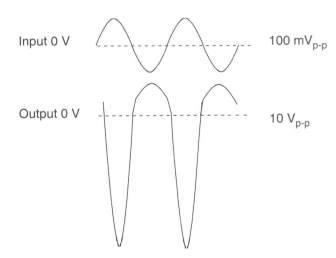

FIGURE 5–13 Signal distortion

The fully bypassed circuit is used in low cost circuitry where the high voltage gain reduces the number of stages needed, therefore, cutting cost. The price paid for this lower cost is distortion, unpredictable voltage gains, and input impedances that fluctuate with changes in environmental conditions and device parameters.

THE SPLIT-EMITTER AMPLIFIER

The split-emitter amplifier is a compromise between the fully bypassed and the unbypassed voltage divider amplifier. Figure 5–14 shows a split-emitter amplifier. The resistor in the emitter leg has been split into two parts: R_e and R_E. Resistor R_E is bypassed with a capacitor and is considered to be zero ohms when calculating signal parameters. Resistor R_e is unbypassed and must be considered when calculating signal parameters. The total value of the DC resistance $(R_e + R_E)$ in the emitter leg is the same as it was in the circuits in Figures 5–11 and 5–12, and the DC currents and voltages are also the same. Example 5.10 will show the differences in signal parameters.

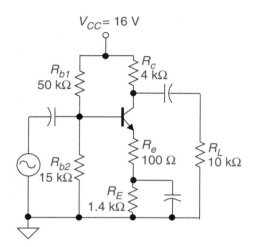

FIGURE 5–14   Split-emitter voltage divider amplifier

EXAMPLE 5.10

Calculate the input impedance, output impedance, and voltage gain for the circuit in Figure 5–14.

Step 1. Calculate r'_e.

$$V_B = V_{cc} \times R_{b2}/(R_{b1} + R_{b2})$$
$$V_B = 16\ \text{V} \times 15\ \text{k}\Omega/(50\ \text{k}\Omega + 15\ \text{k}\Omega) = 3.7\ \text{V}$$
$$V_E = V_b - V_{be} = 3.7\ \text{V} - 0.7\ \text{V} = 3\ \text{V}$$
$$I_E = V_E/R_{E(\text{total})} = 3\ \text{V}/1.5\ \text{k}\Omega = 2\ \text{mA} \quad (R_{E(\text{total})} = R_e + R_E.)$$
$$I_C \approx I_E = 2\ \text{mA}$$
$$r'_e = 25\ \text{mV}/I_E = 25\ \text{mV}/2\ \text{mA} = 12.5\ \Omega$$

Step 2. Calculate the input impedance.

$$r_b = R_{b1} \| R_{b2} = 50\ \text{k}\Omega \| 15\ \text{k}\Omega = 11.5\ \text{k}\Omega$$
$$r_e = r'_e + R_e = 12.5\ \Omega + 100 = 112.5\ \Omega$$
$$z_{\text{in}} = r_b \| (\beta \times r_e) = 11.5\ \text{k}\Omega \| (100 \times 112.5\ \Omega) = 5.7\ \text{k}\Omega$$

Step 3. Calculate the output impedance.

$$z_{\text{out}} = r_c = 4\ \text{k}\Omega$$

Step 4. Calculate the voltage gain.

$$r_c = R_c \| R_L = 4\ \text{k}\Omega \| 10\ \text{k}\Omega = 2.9\ \text{k}\Omega$$
$$A_v = r_c/r_e = 2.9\ \text{k}\Omega/112.5\ \Omega = 26$$

The split-emitter amplifier has values of voltage gain and input impedance between those of the fully bypassed and the unbypassed circuits. The predictability of the signal parameters is also between the other two circuits. The amount of the emitter resistor left unbypassed determines the voltage gain and predictability of the circuit. Because the designer has the ability to control voltage gain and signal parameter predictability, the split-emitter voltage divider circuit has become a popular circuit in electronic systems.

SUMMARY OF VOLTAGE DIVIDER BIASING

In order for the transistor amplifier to be a practical circuit, it is necessary to use a biasing circuit that will stabilize the Q-point. In order to make the Q-point stable, the DC collector current must be independent of beta. This can be accomplished by using voltage divider biasing. There are three variations of voltage divider biasing circuits: the unbypassed circuit with voltage gains in the range of 2 to 10 but with excellent signal parameter predictability, the fully bypassed circuit with voltage gains in the range of 50 to 500 but with unpredictable signal parameters, and the split-emitter circuit with voltage gains in the range of 5 to 50 with good signal parameter predictability.

5.5 EMITTER BIASED AMPLIFIER

Figure 5–15 shows circuits using emitter bias for stabilization. The base is connected to ground through a relatively low resistance (R_b). The base/emitter is forward biased by a negative power supply (V_{EE}) connected to the emitter resistor (R_e). Since the base current is small, the voltage drop across the base resistor (R_b) is negligible, and the base is considered to be at zero volts. The forward-biased base/emitter junction drops to 0.7 V, so the emitter is at –0.7 V. This means the current through the emitter is determined by the value of the negative supply and the value of the emitter resistor. Since the emitter current is independent from changes in beta, the circuit obtains stability. One disadvantage of the emitter biased amplifier is the need for an emitter supply voltage (V_{EE}). Example 5.11 shows how the DC voltage of the emitter biased amplifier can be calculated.

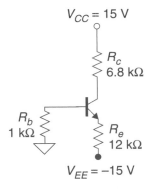

FIGURE 5–15   Emitter biased amplifier

EXAMPLE 5.11

Calculate the DC voltage V_B, V_E, and V_C of the circuit shown in Figure 5–15.

Step 1. Determine the base voltage. The base is at zero voltage because the current through R_b is considered negligible.

$$V_B = 0 \text{ V}$$

EXAMPLE 5.11 continued

> **Step 2.** Determine the emitter voltage. The base/emitter of the NPN transistor is forward biased, so the emitter must be 0.7 V negative with respect to the base.
>
> $$V_E = -0.7 \text{ V}$$
>
> **Step 3.** Calculate the collector current.
>
> $$I_E = (V_{EE} - V_{BE})/R_e = (-15 \text{ V} - (-0.7 \text{ V}))/12 \text{ k}\Omega = -1.2 \text{ mA}$$
>
> (The negative sign can be dropped.)
>
> $$I_C \approx I_E = 1.2 \text{ mA}$$
>
> **Step 4.** Calculate the collector voltage.
>
> $$V_{Rc} = I_C \times R_c = 1.2 \text{ mA} \times 6.8 \text{ k}\Omega = 8.2 \text{ V}$$
> $$V_c = V_{cc} - V_{Rc} = 15 \text{ V} - 8.2 \text{ V} = 6.8 \text{ V}$$

Figure 5–16 shows the same emitter biased amplifier with an input signal and a load connected. The signal parameters are calculated using the same procedure as the voltage divider amplifier. The emitter can be fully bypassed for high voltage gain or partly bypassed, which reduces the gain but improves predictability and reduces distortion. Example 5.12 shows how the signal parameters can be calculated.

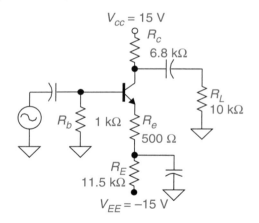

FIGURE 5–16   Emitter biased amplifier with split emitter

EXAMPLE 5.12

> Calculate the input impedance, output impedance, and voltage gain for the circuit in Figure 5–16.
>
> **Step 1.** Calculate r'_e.
>
> $$I_E = 1.2 \text{ mA (Calculated in Example 5.11.)}$$
> $$r'_e = 25 \text{ mV}/I_E = 25 \text{ mV}/1.2 \text{ mA} = 20.8 \text{ }\Omega$$
>
> **Step 2.** Calculate the input impedance.
>
> $$r_b = R_b = 1 \text{ k}\Omega$$
> $$r_e = r'_e + R_e = 20.8 \text{ }\Omega + 500 = 520.8 \text{ }\Omega \quad (r'_e \text{ is small and may be disregarded.})$$
> $$z_{\text{in}} = r_b \parallel \beta r_e = 1 \text{ k}\Omega \parallel (100 \times 521 \text{ }\Omega) = 981 \text{ }\Omega$$

EXAMPLE 5.12 continued

> **Step 3.** Calculate the output impedance.
> $$z_{out} = R_c = 6.8 \text{ k}\Omega$$
> **Step 4.** Calculate the voltage gain.
> $$r_c = R_c \parallel R_L = 6.8 \text{ k}\Omega \parallel 10 \text{ k}\Omega = 4 \text{ k}\Omega$$
> $$A_v = r_c/r_e = 4 \text{ k}\Omega/521 \ \Omega = 7.7$$

5.6 VOLTAGE-MODE FEEDBACK BIASED AMPLIFIER

Figure 5–17 shows circuits using voltage feedback for stabilization. Note that there is not a resistor in the emitter leg. Voltage divider biasing using current-mode feedback is normally preferred over voltage-mode feedback. However, for low voltage power supplies or when maximum output voltage swings are necessary, voltage-mode feedback may be preferred.

Figure 5–17(a) shows a circuit operating from a 4 V supply. A resistor (R_b) is connected between the collector and base. The base current is controlled by the voltage dropped across R_b, which is equal to the collector-to-base voltage. If the collector current increases, the voltage on the collector will decrease, reducing the voltage across R_b. A reduced voltage across R_b causes less base current, which then causes the collector current to decrease. The opposite is also true. A decrease in collector current causes the collector voltage to increase, which increases the base current and causes the collector current to rise. The voltage across the transistor (V_{CE}) stabilizes at approximately 2 V, and the remaining supply voltage is dropped across R_c. The voltage gain of the amplifier is the same as the fully bypassed voltage divider circuit and can be calculated using the formula $A_v = r_c/r_e$. Since there is no emitter resistor in the circuit, $r_e = r'_e$ causing the voltage gain to be high. The input impedance is equal to R_b divided by the voltage gain plus one in parallel with r'_e times beta $[z_{in} = (R_b/A_v + 1) \parallel (r'_e\beta)]$. The reason R_b appears to be much smaller than its actual size is because as the voltage goes up on the base end of R_b, the voltage is forced down on the collector end of R_b. Miller's theorem states that the effective impedance of a component in the negative feedback loop of an amplifier is equal to the actual impedance of the component divided by the voltage gain plus one. The main advantage of the circuit in Figure 5–17(a) is its ability to operate with a low supply voltage.

The circuit in Figure 5–17(b) uses voltage-mode feedback. The second resistor in the base circuit (R_{b2}) permits a portion of the current flowing through R_{b1} to bypass the base. This permits the circuit

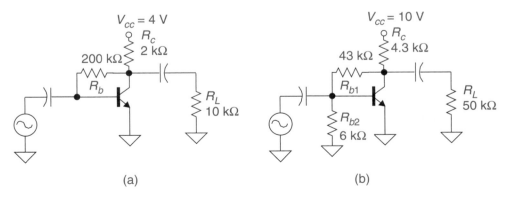

(a)   (b)

FIGURE 5–17   Voltage-mode feedback amplifiers

to operate with a higher value of collector-to-emitter voltage (V_{CE}). The circuit can be designed for ideal midpoint biasing, permitting maximum output signal swings. This type of voltage-mode feedback is often used to drive power amplifier stages when maximum swing is of utmost importance.

5.7 MULTISTAGE RC COUPLED AMPLIFIERS

In order to obtain the gain an electronic system needs, it is often necessary to connect amplifiers in series. When amplifiers are connected in series, they are said to be cascaded. Figure 5–18 shows a two-stage cascaded amplifier circuit. The two stages are both voltage divider biased amplifiers.

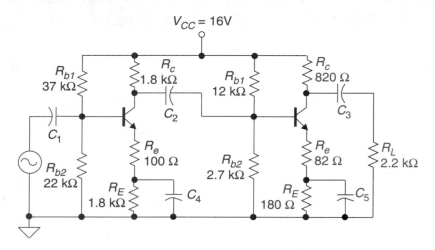

FIGURE 5–18   RC coupled amplifier

One method for coupling transistor stages together is by using coupling capacitors between stages. This method of coupling is called RC, or resistance-capacitor, coupling. Coupling capacitors act like an open for biased currents but are a short for signal currents. Because of this, the biased circuitry in each stage is independent. The biased currents and voltages can be calculated in the same manner used for a single-stage amplifier. You simply consider each stage individually.

The signal parameters of a multistage amplifier are easy to calculate if you remember the following facts. The input impedance of the total circuit is equal to the input impedance of the first stage. The output impedance of the total circuit is equal to the output impedance of the final stage of the system. The voltage gain of the total circuit is equal to the product of the individual stages. The output signal from the first stage will see the input impedance of the second stage as its load resistance.

EXAMPLE 5.13

Analyze the circuit in Figure 5–18.

Step 1. Analyze each stage independently for the biased voltages.

First stage:

$$V_B = V_{cc} \times R_{b2}/(R_{b1}+R_{b2}) = 16 \text{ V} \times 22 \text{ k}\Omega/(37 \text{ k}\Omega + 22 \text{ k}\Omega) = 6 \text{ V}$$

$$V_E = V_B - V_{BE} = 6 \text{ V} - 0.7 \text{ V} = 5.3 \text{ V}$$

$$I_E = V_E/R_{E(total)} = 5.3 \text{ V}/1.9 \text{ k}\Omega = 2.8 \text{ mA}$$

EXAMPLE 5.13 continued

$$V_{RC} = I_C \times R_c = 1.8 \text{ k}\Omega \times 2.8 \text{ mA} = 5 \text{ V}$$
$$V_C = V_{CC} - V_{RC} = 16 \text{ V} - 5 \text{ V} = 11 \text{ V}$$
$$V_{CE} = V_C - V_E = 11 \text{ V} - 5.3 \text{ V} = 5.7 \text{ V}$$

Second stage:
$$V_B = V_{cc} \times R_{b2}/(R_{b1} + R_{b2}) = 16 \text{ V} \times 2.7 \text{ k}\Omega/(12 \text{ k}\Omega + 2.7 \text{ k}\Omega) = 2.9 \text{ V}$$
$$V_E = V_B - V_{BE} = 2.9 \text{ V} - 0.7 \text{ V} = 2.2 \text{ V}$$
$$I_E = V_E/R_{E(\text{total})} = 2.2 \text{ V}/262 \text{ }\Omega = 8.4 \text{ mA}$$
$$V_{RC} = I_C \times R_c = 820 \text{ }\Omega \times 8.4 \text{ mA} = 6.9 \text{ V}$$
$$V_C = V_{CC} - V_{RC} = 16 \text{ V} - 6.9 \text{ V} = 9.1 \text{ V}$$
$$V_{CE} = V_C - V_E = 9.1 \text{ V} - 2.2 \text{ V} = 6.9 \text{ V}$$

Step 2. Calculate the input impedance and output impedance of each stage. Assume beta to equal 100 for both transistors.

First stage:
$$r_b = R_{b1} \| R_{b2} = 37 \text{ k}\Omega \| 22 \text{ k}\Omega = 13.8 \text{ k}\Omega$$
$$r'_e = 25 \text{ mV}/I_E = 25 \text{ mV}/2.8 \text{ mA} = 9 \text{ }\Omega$$
$$r_e = r'_e + R_e = 9 \text{ }\Omega + 100 \text{ }\Omega = 109 \text{ }\Omega$$
$$z_{\text{in}} = r_b \| (\beta \times r_e) = 13.8 \text{ k}\Omega \| (100 \times 109 \text{ }\Omega) = 6 \text{ k}\Omega$$
$$z_{\text{out}} = R_c = 1.8 \text{ k}\Omega$$

Second stage:
$$r_b = R_{b1} \| R_{b2} = 12 \text{ k}\Omega \| 2.7 \text{ k}\Omega = 2.2 \text{ k}\Omega$$
$$r'_e = 25 \text{ mV}/I_E = 25 \text{ mV}/8.4 \text{ mA} = 3 \text{ }\Omega$$
$$r_e = r'_e + R_e = 3 \text{ }\Omega + 82 \text{ }\Omega = 85 \text{ }\Omega$$
$$z_{\text{in}} = r_b \| (\beta \times r_e) = 2.2 \text{ k}\Omega \| (100 \times 85 \text{ }\Omega) = 1.7 \text{ k}\Omega$$
$$z_{\text{out}} = R_c = 820 \text{ }\Omega$$

Step 3. Calculate the voltage gain of each stage.

First stage:
$$r_c = R_c \| R_L = 1.8 \text{ k}\Omega \| 1.7 \text{ k}\Omega = 874 \text{ }\Omega \text{ } (R_L = z_{\text{in}} \text{ of next stage.})$$
$$A_v = r_c/r_e = 874 \text{ }\Omega/109 \text{ }\Omega = 8$$

Second stage:
$$r_c = R_c \| R_L = 820 \text{ }\Omega \| 2.2 \text{ k}\Omega = 597 \text{ }\Omega$$
$$A_v = r_c/r_e = 597 \text{ }\Omega/85 \text{ }\Omega = 7$$

Step 4. Calculate the input impedance, output impedance, and voltage gain of the total amplifier circuit.

$$z_{\text{in}} = z_{\text{in}} \text{ of first stage} = 6 \text{ k}\Omega$$
$$z_{\text{out}} = z_{\text{out}} \text{ of last stage} = 820 \text{ }\Omega$$
$$A_{v(\text{total})} = A_{V1} \times A_{V2} = 8 \times 7 = 56$$

5.8 COUPLING AND BYPASS CAPACITORS

Up to this point we have assumed the capacitors in the circuits to have zero impedance (X_c) for signal currents and infinite impedance for biased currents. The impedance of a capacitor is dependent on its capacitance and the frequency of the signal and is expressed as $X_c = 1/(2\pi FC)$. Any size capacitor will exhibit some ohms of impedance at any frequency. Engineers design circuits so X_c is relatively small at the lowest operating frequency. Normally, technicians will not have to calculate the size of the capacitors, but it is helpful to have a rule of thumb to determine the approximate size. As a rule, X_c, at the lowest frequency, should be equal to one-tenth or less of the series impedance being driven by the signal passing through the capacitor.

The signal passing through the input coupling capacitor is driving z_{in} of the first stage of the circuit, and the impedance of the coupling capacitor should be one-tenth of z_{in}. The coupling capacitor between stages should have an impedance of one-tenth of z_{in} of the stage being driven. The output coupling capacitor should have an impedance of one-tenth of R_L.

The purpose of the emitter capacitor is to bypass signal currents to ground. The parallel combination of R_E and the bypass capacitor should be one-tenth of r'_e for the fully bypassed circuit or one-tenth of r_e ($r'_e + R_e$) for the split-emitter circuit. Example 5.14 will demonstrate how this rule of thumb can be used to calculate the size of coupling and bypass capacitors. It should be emphasized that the procedure to be used in Example 5.14 will result in estimated values that permit the technician to select capacitors for experimental circuits. Engineers use other calculation methods that result in smaller capacitor values that save money on commercially produced circuits.

EXAMPLE 5.14

Calculate the size of the coupling and bypass capacitors needed in the circuit in Figure 5–18 if the lowest signal frequency to be amplified is 300 Hz.

Step 1. Determine the series resistance being driven by each capacitor. In Example 5.13, the input impedance and the value of signal resistance in the emitter leg (r_e) was calculated for each stage. These values are listed as the driven resistance in Table 5–1.

TABLE 5–1

Capacitor	Driven resistance	$X_{C(max)}$	Capacitor value
C_1	6 kΩ (z_{in} Q_1)	600 Ω	0.9 µF
C_2	1.7 kΩ (z_{in} Q_2)	170 Ω	3.1 µF
C_3	2.2 kΩ (R_L)	220 Ω	2.4 µF
C_4	109 Ω (r_e Q_1)	10.9 Ω	49 µF
C_5	85 Ω (r_e Q_2)	8.5 Ω	63 µF

Step 2. Calculate the maximum impedance of the capacitor by assuming the capacitive reactance (X_c) to equal one-tenth of the driven resistance value shown in Table 5–1.

Step 3. Using 300 Hz, calculate the value of each capacitor using the following formula. The calculation for the first capacitor is shown. Table 5–1 shows the calculated values for all capacitors for the circuit shown in Figure 5–18.

$$C_1 = 1/(2\pi X_c F) = 1/(2\pi \times 600\ \Omega \times 300\ \text{Hz}) = 0.9\ \mu F$$

The results in Table 5–1 show the minimum size capacitor using our rule of thumb for calculation. The next larger standard size capacitor can be used to construct the circuit. Since coupling and bypass capacitors with larger capacitance values will function fine, often circuits will use the same value for all coupling capacitors.

5.9 DIRECT COUPLED AMPLIFIERS

Capacitor coupled (RC coupled) amplifiers are popular because each stage has its own independently biased circuit. Fluctuations in the DC operating point in one stage is not amplified by the next stage. Occasionally, however, it is necessary to have an amplifier that is capable of amplifying DC currents and voltages. In order to couple a DC signal from one stage to the next, it is necessary to remove the coupling capacitor and connect the output of one stage directly to the input of the next. Emitter-bypass capacitors are removed; thus, the DC signal gain will be the same as the dynamic signal gain.

Figure 5–19 shows a two-stage direct coupled amplifier. The first stage uses emitter biasing. With no input signal, the 4.3 kΩ emitter resistor connected to the negative 5 V supply sets the emitter and collector current of Q_1 to 1 mA. The 1 mA of collector current flowing through Q_1 causes a 15 V drop across the 15 kΩ collector resistor setting the collector voltage at 5 V. The second stage uses a PNP transistor, and the bias voltage on the base is held constant at 5 V by the output of the first stage. The emitter of Q_2 is connected to a 10 V supply through a 4.3 kΩ emitter resistor. Since 4.3 V will be dropped across the emitter, the emitter current and collector current of Q_2 will equal 1 mA. The 1 mA of collector current flows through the 20 kΩ collector resistor of Q_2, dropping 20 V. This causes the output to be zero volts at quiescence.

One advantage of DC coupling is the high input impedance of the second stage. The high input impedance is a result of the absence of base resistors at the input of the second stage. This reduces loading on the first stage and permits a higher voltage gain. Example 5.15 shows how the direct coupled amplifier can be analyzed.

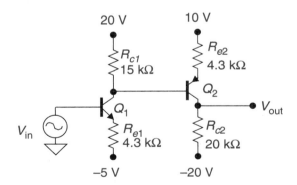

FIGURE 5–19 Direct coupled amplifier

EXAMPLE 5.15

For the circuit in Figure 5–19, calculate V_B, V_E, V_C, V_{CE}, z_{in}, z_{out}, and A_v for each transistor. Then calculate z_{in}, z_{out}, and A_v for the total circuit.

Step 1. Calculate the DC voltage for the first stage.

$$V_B = 0 \text{ V (Assume the input signal has zero DC offset.)}$$
$$V_E = V_B - V_{BE} = 0 \text{ V} - 0.7 \text{ V} = -0.7 \text{ V}$$

EXAMPLE 5.15 continued

$$I_E = (-5 \text{ V} - (-0.7 \text{ V}))/4.3 \text{ k}\Omega = -1 \text{ mA}$$
$$I_E \approx I_C$$
$$V_{Rc} = I_c \times R_{c1} = 1 \text{ mA} \times 15 \text{ k}\Omega = 15 \text{ V}$$
$$V_C = V_{CC} - V_{Rc} = 20 \text{ V} - 15 \text{ V} = 5 \text{ V}$$
$$V_{CE} = V_C - V_E = 5 \text{ V} - (-0.7) \text{ V} = 5.7 \text{ V}$$

Step 2. Calculate the DC voltage for the second stage.
$$V_B = V_C \text{ of first stage} = 5 \text{ V}$$
$$V_E = V_B + V_{BE} = 5 \text{ V} + 0.7 \text{ V} = 5.7 \text{ V} \quad \text{(PNP transistor } V_{BE} \text{ is added.)}$$
$$I_E = (10 \text{ V} - 5.7 \text{ V})/4.3 \text{ k}\Omega = 1 \text{ mA}$$
$$I_E \approx I_C$$
$$V_{Rc} = I_c \times R_c = 1 \text{ mA} \times 20 \text{ k}\Omega = 20 \text{ V}$$
$$V_C = V_{CC} - V_{Rc} = -20 \text{ V} - (-20 \text{ V}) = 0 \text{ V}$$
$$V_{CE} = V_C - V_E = 0 \text{ V} - 5.7 \text{ V} = -5.7 \text{ V}$$

Step 3. Calculate the input impedance and output impedance for each stage. Assume beta to equal 100 for both transistors.

First stage:
$$r'_e = 25 \text{ mV}/I_E = 25 \text{ mV}/1 \text{ mA} = 25 \text{ }\Omega$$
$$r_e = r'_e + R_e = 25 \text{ }\Omega + 4.3 \text{ k}\Omega = 4.3 \text{ k}\Omega \text{ } (r'_e \text{ is too small to consider.)}$$
$$z_{in} = \beta \times r_e = 100 \times 4.3 \text{ k}\Omega = 430 \text{ k}\Omega$$
$$z_{out} = R_c = 15 \text{ k}\Omega$$

Second stage:
$$r'_e = 25 \text{ mV}/I_E = 25 \text{ mV}/1 \text{ mA} = 25 \text{ }\Omega$$
$$r_e = r'_e + R_e = 25 \text{ }\Omega + 4.3 \text{ k}\Omega \approx 4.3 \text{ k}\Omega \text{ } (r'_e \text{ is too small to consider.)}$$
$$z_{in} = \beta \times r_e = 100 \times 4.3 \text{ k}\Omega = 430 \text{ k}\Omega$$
$$z_{out} = R_c = 20 \text{ k}\Omega$$

Step 4. Calculate the voltage gain of each stage.

First stage:
$$r_c = R_c \parallel R_L = 15 \text{ k}\Omega \parallel 430 \text{ k}\Omega = 14.5 \text{ k}\Omega \text{ } (R_L = z_{in} \text{ of second stage (430 k}\Omega\text{).)}$$
$$A_v = r_c/r_e = 14.5 \text{ k}\Omega/4.3 \text{ k}\Omega = 3.37$$

Second stage:
$$r_c = R_c = 20 \text{ k}\Omega \text{ } \text{(No load resistance connected.)}$$
$$A_v = r_c/r_e = 20 \text{ k}\Omega/4.3 \text{ k}\Omega = 4.65$$

Step 5. Calculate z_{in}, z_{out}, and $A_{V(total)}$ for the total amplifier circuit.
$$z_{in} = z_{in} \text{ first stage} = 430 \text{ k}\Omega$$
$$z_{out} = z_{out} \text{ last stage} = 20 \text{ k}\Omega$$
$$A_{V(total)} = A_{V1} \times A_{V2} = 3.37 \times 4.65 = 15.7$$

Direct coupled amplifiers have the advantage of being able to amplify DC voltages. However, in order to amplify DC, coupling capacitors and bypass capacitors must be removed. The removal of coupling capacitors means that stages cannot be individually biased, and any Q-point drift in one stage will be amplified by the next stage. The removal of the bypass capacitor causes the gain of each stage to be low. The system will require more stages to obtain the needed gain. Even though direct coupled amplifiers have these disadvantages, they still are used in high quality circuits where low frequency and DC amplification are required.

Another way to obtain DC and low frequency amplification is to use a capacitor coupled amplifier and chop (turning the signal off and on at a high frequency) the DC and low frequency signals so that they can be passed by the coupling capacitors. Chopper circuits will be discussed in a future chapter.

5.10 TROUBLESHOOTING TRANSISTOR CIRCUITS

Troubleshooting transistor circuits can be divided into two areas: troubleshooting the circuit for correct bias voltages, and troubleshooting the circuit for correct signal responses.

BIAS TROUBLESHOOTING

If the stages are capacitor coupled, the procedure used in troubleshooting single-stage circuits can be applied. Each stage in a capacitor coupled amplifier is biased independently, and therefore, troubleshooting the biased circuitry is done one stage at a time. In direct coupled amplifiers, the biased voltage on the base is the collector voltage of the previous stage. Troubleshooting the biased circuit becomes a multistage problem.

SIGNAL TROUBLESHOOTING

Signal troubleshooting is performed by inserting a signal from a signal generator through a capacitor into the circuit, and then tracing the signal through the system. In multistage amplifiers, remember, the load on one stage is the input impedance of the next. So when the gain of a stage is not as estimated, it may be because the next stage is loading it down. This can be checked by disconnecting the next stage and replacing it with a resistor equal to its input impedance. Open capacitors can cause signal problems while not affecting the biased voltages. The bypass capacitor in the emitter leg, if it is open, will cause the voltage gain of the stage to go down. Open coupling capacitors can be detected by noting that the signal is not being coupled from one stage to the next. One of the best ways to check for open capacitors is to bypass these capacitors with a known good capacitor. If the circuit returns to proper working order, the capacitor needs replacing.

Voltage Divider Biased Amplifiers

1. Calculate and record V_B, V_E, V_C, z_{in}, z_{out}, A_v, and maximum output voltage swings without distortion for each circuit in Figure 5–20.

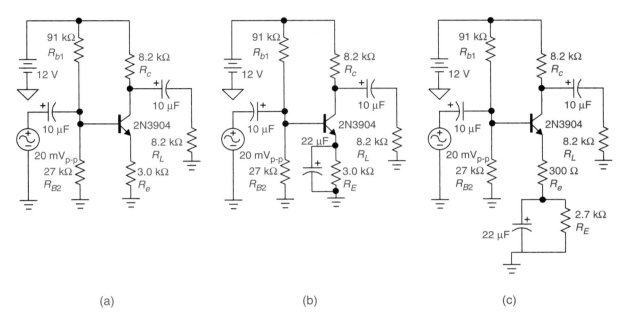

(a)   (b)   (c)

FIGURE 5–20

Circuit a:

$V_B =$ _____ $V_E =$ _____ $V_C =$ _____

$z_{in} =$ _____ $z_{out} =$ _____ $A_v =$ _____

Maximum output voltage swing = _____

Circuit b:

$V_B =$ _____ $V_E =$ _____ $V_C =$ _____

$z_{in} =$ _____ $z_{out} =$ _____ $A_v =$ _____

Maximum output voltage swing = _____

Circuit c:

$V_B =$ _____ $V_E =$ _____ $V_C =$ _____

$z_{in} =$ _____ $z_{out} =$ _____ $A_v =$ _____

Maximum output voltage swing = _____

2. For a minimum frequency of 1 kHz, calculate the minimum capacitor size for each capacitor in the circuit in Figure 5–20.

$C_1 =$ _____ $C_2 =$ _____ $C_3 =$ _____ $C_4 =$ _____

$C_5 =$ _____ $C_6 =$ _____ $C_7 =$ _____ $C_8 =$ _____

3.   Simulate and record V_B, V_E, V_C, z_{in}, z_{out}, A_v, and maximum output voltage swings without distortion for each circuit in Figure 5–20.

Circuit a:

$V_B =$ _____ $V_E =$ _____ $V_C =$ _____

$z_{in} =$ _____ $z_{out} =$ _____ $A_v =$ _____

Maximum output voltage swing = _____

Circuit b:

$V_B =$ _____ $V_E =$ _____ $V_C =$ _____

$z_{in} =$ _____ $z_{out} =$ _____ $A_v =$ _____

Maximum output voltage swing = _____

Circuit c:

$V_B =$ _____ $V_E =$ _____ $V_C =$ _____

$z_{in} =$ _____ $z_{out} =$ _____ $A_v =$ _____

Maximum output voltage swing = _____

Voltage Divider Biased Amplifiers

I. Objective

To analyze and construct three variations of voltage divider biased amplifiers.

II. Test Equipment

(1) sine wave generator
(1) dual trace oscilloscope
(1) 12 V power supply

III. Components

Resistors: (1) 200 Ω, (1) 620 Ω, (1) 820 Ω, (1) 3.9 kΩ, (1) 5.6 kΩ, (1) 8.2 kΩ, (1) 39 kΩ
Capacitors: (2) 10 μF and (1) 22 μF
Transistors: (1) 2N3904

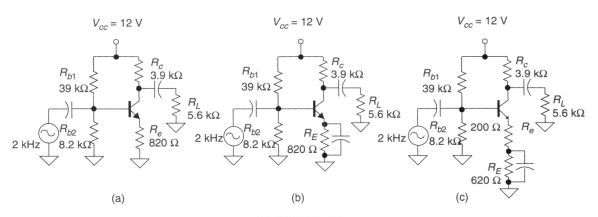

FIGURE 5–21

IV. Procedure

1. Calculate and record $V_B, V_E, V_C, z_{in}, z_{out}, A_v$, and maximum output voltage swing without distortion for each circuit in Figure 5–21.

   Circuit a:

$V_B =$ _____	$V_E =$ _____	$V_C =$ _____
$z_{in} =$ _____	$z_{out} =$ _____	$A_v =$ _____

   Maximum output voltage swing = _____

   Circuit b:

$V_B =$ _____	$V_E =$ _____	$V_C =$ _____
$z_{in} =$ _____	$z_{out} =$ _____	$A_v =$ _____

   Maximum output voltage swing = _____

   Circuit c:

$V_B =$ _____	$V_E =$ _____	$V_C =$ _____
$z_{in} =$ _____	$z_{out} =$ _____	$A_v =$ _____

   Maximum output voltage swing = _____

2. Construct each circuit in Figure 5–21 and measure the values of V_B, V_E, V_C, z_{in}, z_{out}, A_v, and maximum output voltage swing without distortion. Set the input signal frequency to 2 kHz and the signal magnitude as needed.

Circuit a:

$V_B =$ _____ $V_E =$ _____ $V_C =$ _____

$z_{in} =$ _____ $z_{out} =$ _____ $A_v =$ _____

Maximum output voltage swing = _____

Circuit b:

$V_B =$ _____ $V_E =$ _____ $V_C =$ _____

$z_{in} =$ _____ $z_{out} =$ _____ $A_v =$ _____

Maximum output voltage swing = _____

Circuit c:

$V_B =$ _____ $V_E =$ _____ $V_C =$ _____

$z_{in} =$ _____ $z_{out} =$ _____ $A_v =$ _____

Maximum output voltage swing = _____

V. Points to Discuss

1. Analyze and discuss the percent of difference in calculated and measured values.

2. Explain the difference in voltage gain (A_v) among the three circuits.

3. Explain the difference in input impedance (z_{in}) among the three circuits.

4. Why were z_{out} and the DC voltage readings approximately the same for all three circuits?

Multistage Amplifier

1. Calculate and record all values listed below for the amplifier circuit in Figure 5–22.

   Stage 1:

   $V_B =$ _____ $V_E =$ _____ $V_C =$ _____

   $A_v =$ _____ $z_{in} =$ _____ $z_{out} =$ _____

   Stage 2:

   $V_B =$ _____ $V_E =$ _____ $V_C =$ _____

   $A_v =$ _____ $z_{in} =$ _____ $z_{out} =$ _____

   Total Circuit:

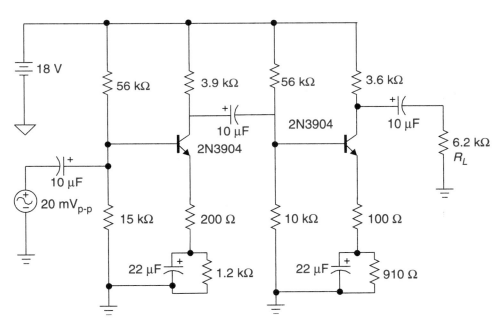

FIGURE 5–22

   $A_v =$ _____ $z_{in} =$ _____ $z_{out} =$ _____

2. For a minimum frequency of 1 kHz, calculate the minimum capacitor size for each capacitor in the circuit in Figure 5–22.

   $C_1 =$ _____ $C_2 =$ _____ $C_3 =$ _____ $C_4 =$ _____

   $C_5 =$ _____

3. Simulate and record $V_B, V_E, V_C, A_v, z_{in},$ and $z_{out},$ for the amplifier circuit in Figure 5–22.

   Stage 1:

   $V_B =$ _____ $V_E =$ _____ $V_C =$ _____

   $A_v =$ _____ $z_{in} =$ _____ $z_{out} =$ _____

Stage 2:

$V_B =$ _____ $V_E =$ _____ $V_C =$ _____

$A_v =$ _____ $z_{in} =$ _____ $z_{out} =$ _____

Total Circuit:

$A_v =$ _____ $z_{in} =$ _____ $z_{out} =$ _____

Multistage Amplifier

I. Objective

To analyze a multistage amplifier to determine
the bias and signal values at test points.

II. Test Equipment

(1) sine wave generator
(1) dual-trace oscilloscope
(1) 12 V power supply

III. Components

Resistors: (1) 100 Ω, (1) 180 Ω, (1) 680 Ω, (1) 820 Ω, (2) 3.9 kΩ, (3) 8.2 kΩ, (2) 39 kΩ
Capacitors: (3) 2.2 µF and (2) 10 µF
Transistors: (2) 2N3904

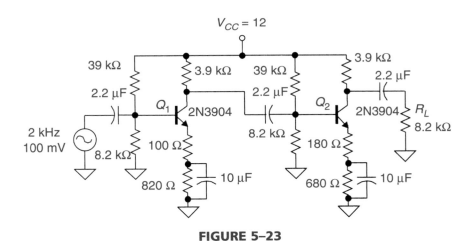

FIGURE 5–23

IV. Procedure

1. Calculate and record all values listed below for the amplifier circuit in Figure 5–23.

Stage 1:

$V_B =$ _____	$V_E =$ _____	$V_C =$ _____
$A_v =$ _____	$z_{in} =$ _____	$z_{out} =$ _____

Stage 2:

$V_B =$ _____	$V_E =$ _____	$V_C =$ _____
$A_v =$ _____	$z_{in} =$ _____	$z_{out} =$ _____

Total Circuit:

$A_v =$ _____	$z_{in} =$ _____	$z_{out} =$ _____

2. Construct the circuit in Figure 5–23 and measure the values listed below for the amplifier circuit.

    Stage 1:

    $V_B =$ _____ $V_E =$ _____ $V_C =$ _____

    $A_v =$ _____ $z_{in} =$ _____ $z_{out} =$ _____

    Stage 2:

    $V_B =$ _____ $V_E =$ _____ $V_C =$ _____

    $A_v =$ _____ $z_{in} =$ _____ $z_{out} =$ _____

    Total Circuit:

    $A_v =$ _____ $z_{in} =$ _____ $z_{out} =$ _____

V. Points to Discuss

1. Explain any differences between calculated and measured values.

2. Explain why z_{in} of the first stage equals z_{in} of the total circuit.

3. Explain why z_{out} of the last stage equals z_{out} of the total circuit.

4. Explain why input impedance of the second stage affected the voltage gain of the first stage.

5. The total voltage gain of this circuit could be obtained by using a single-stage amplifier. Explain the advantages of using two stages.

QUESTIONS

5.1 Introduction

____ 1.   Why is it necessary to stabilize the bipolar transistor amplifier against changes in beta?
   a.   Beta changes with temperature.
   b.   Beta changes with changes in coupling capacitors.
   c.   Beta is different among transistors of the same type.
   d.   Both a and c.

____ 2.   The typical beta of a transistor should be considered to be _____.
   a.   +50% and −50%
   b.   +50% and −100%
   c.   +100% and −50%
   d.   +100% and −100%

____ 3.   If beta changes, how will the lack of bias stability show up in an amplifier circuit?
   a.   The collector voltage will change.
   b.   The collector current will change.
   c.   The emitter current will change.
   d.   All of the above.

5.2 Voltage Divider Biasing

____ 4.   In voltage divider biasing, why is the voltage at the junction of R_{b1} and R_{b2} considered to be independent of the transistor base current?
   a.   The base current does not flow through R_{b1} or R_{b2}.
   b.   The base current is small in comparison to the bleeder current through R_{b1} and R_{b2}.
   c.   Only the emitter current has an effect on the current flow through R_{b1} and R_{b2}.
   d.   The coupling capacitor blocks base current through the voltage divider.

____ 5.   In voltage divider biased amplifiers, the difference in voltage between the emitter and base is always _____.
   a.   0 V
   b.   0.2 V
   c.   0.7 V
   d.   2 V

____ 6.   In voltage divider biased amplifiers, once the DC emitter voltage is calculated, the quiescence collector current can be approximated by dividing the emitter voltage by _____ .
   a.   the resistance in the base leg
   b.   the resistance in the emitter leg
   c.   the resistance in the collector leg
   d.   the resistance of the load

_____ 7.   In voltage divider biased amplifiers, the collector voltage is calculated by _____.

     a.   multiplying the collector current times the collector resistor

     b.   multiplying the collector current times the load resistor

     c.   adding the base voltage and the emitter voltage

     d.   subtracting the voltage dropped across the collector resistor from the supply voltage

5.3   Signal Parameters in Voltage Divider Circuits

_____ 8.   Voltage divider biased amplifiers are beta independent, but what is the price paid for this independence?

     a.   loss of stability

     b.   low output impedance

     c.   loss of voltage gain

     d.   both a and c

_____ 9.   When calculating input impedance, the two base resistors (R_{b1} and R_{b2}) appear in _____ with each other.

     a.   series

     b.   series/parallel

     c.   parallel

     d.   opposing series

_____ 10.   The dynamic resistance of the base/emitter junction is in _____.

     a.   series with the signal resistance in the base leg

     b.   parallel with the signal resistance in the base leg

     c.   parallel with the signal resistance in the emitter leg

     d.   series with the signal resistance in the emitter leg

_____ 11.   The output impedance of the common emitter amplifier is equal to _____.

     a.   the collector resistor

     b.   the load resistance

     c.   the collector resistor in parallel with the load resistance

     d.   the collector resistor times beta

5.4   Variations of Voltage Divider Biased Amplifiers

_____ 12.   Which type of voltage divider biased amplifier has the highest input impedance?

     a.   fully bypassed

     b.   split-emitter

     c.   unbypassed

     d.   all the same

_____ 13.   Which type of voltage divider biased amplifier has the highest output impedance?

     a.   fully bypassed

     b.   split-emitter

     c.   unbypassed

     d.   all the same

____ 14. Which type of voltage divider biased amplifier has the highest voltage gain?
   a. fully bypassed
   b. split-emitter
   c. unbypassed
   d. all the same

____ 15. Which type of voltage divider biased amplifier has the least distortion?
   a. fully bypassed
   b. split-emitter
   c. unbypassed
   d. all the same

5.5 Emitter Biased Amplifier

____ 16. The quiescent base voltage of the emitter biased amplifier is normally _____.
   a. 0 V
   b. 0.7 V
   c. 2 V
   d. V_{cc}

____ 17. A disadvantage of the emitter biased amplifier when compared to the voltage divider biased amplifier is the emitter biased amplifier requires _____.
   a. transistors with higher beta
   b. two power supply voltages
   c. higher value of V_{cc}
   d. none of the above

____ 18. The voltage gain of the emitter biased amplifier is _____.
   a. dependent on beta
   b. calculated using the same general formula as that used for the voltage divider biased amplifier
   c. equal to $\beta \times r_c$
   d. always higher than that of the voltage divider biased amplifier

5.6 Voltage-Mode Feedback Biased Amplifier

____ 19. Voltage-mode feedback biased amplifiers are particularly suitable for operations with _____.
   a. high frequency signals
   b. low voltage power supplies
   c. circuits requiring an exceedingly high input impedance
   d. circuits requiring an exceedingly low output impedance

____ 20. The input impedance of a voltage-mode feedback biased amplifier is affected by the _____.
   a. wattage value of the collector resistor
   b. amplifier voltage gain
   c. resistance of the feedback resistor
   d. both b and c

5.7 Multistage RC Coupled Amplifiers

____ 21. Why is it important to know the input impedances of each stage in a multistage amplifier?
    a. The total input impedance is the product of each stage's input impedance.
    b. The voltage gain of a stage is affected by the input impedance of the next stage.
    c. The input impedance of a stage is the load resistance on the previous stage.
    d. Both b and c.

____ 22. What is one of the main advantages of using coupling capacitors between stages?
    a. They permit the multistage amplifier to pass DC signals.
    b. They permit the bias circuitry in each stage to be independent.
    c. They bypass the emitter resistor and increase the gain.
    d. Both b and c.

____ 23. The total voltage gain of a multistage amplifier is equal to the _____.
    a. sum of each stage's voltage gain
    b. product of each stage's voltage gain
    c. first stage voltage gain
    d. last stage voltage gain

____ 24. The input impedance of the total multistage amplifier is equal to the _____.
    a. sum of each stage's input impedance
    b. product of each stage's input impedance
    c. first stage input impedance
    d. last stage input impedance

____ 25. The output impedance of the total multistage amplifier is equal to the _____.
    a. sum of each stage's output impedance
    b. product of each stage's output impedance
    c. first stage output impedance
    d. last stage output impedance

5.8 Coupling and Bypass Capacitors

____ 26. The size of coupling and bypass capacitors is one of the main factors determining _____.
    a. low frequency cutoff
    b. voltage gain
    c. current gain
    d. high frequency cutoff

____ 27. If the input impedance of the second stage is 1 kΩ, the coupling capacitor connecting the first stage to the second stage should have an X_c of approximately _____ for the lowest frequency to be amplified.
    a. 1 Ω
    b. 10 Ω
    c. 100 Ω
    d. 1 kΩ
    e. 10 kΩ

5.9 Direct Coupled Amplifiers

____ 28. Direct coupled amplifiers have an advantage over RC coupled amplifiers in that they can amplify _____.

    a. larger signals

    b. high frequency signals

    c. smaller signals

    d. low frequency signals

____ 29. Direct coupled amplifiers are particularly susceptible to _____ problems.

    a. gain

    b. saturation

    c. DC drift

    d. impedance

5.10 Troubleshooting Transistor Circuits

____ 30. The collector of Q_1 in Figure 5–24 measures approximately 20 V DC.

    a. The circuit is functioning correctly.

    b. Capacitor C_2 is shorted.

    c. Capacitor C_2 is open.

    d. Resistor R_1 is open.

____ 31. The collector of Q_2 in Figure 5–24 measures 13.8 V DC.

    a. The circuit is functioning correctly.

    b. Transistor Q_2 is open between the collector and the emitter.

    c. Capacitor C_5 is shorted.

    d. Resistor R_8 is shorted.

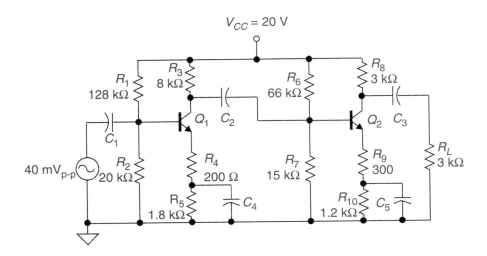

FIGURE 5–24

____ 32. The DC voltage at the junction of resistors R_4 and R_5 in Figure 5–24 is zero volts.

    a. The circuit is functioning correctly.

    b. Transistor Q_1 has a collector to emitter short.

      c.   Capacitor C_4 is open.

      d.   Resistor R_2 is shorted.

___ 33. The signal voltage gain of Q_2 in Figure 5–24 is approximately two times the calculated gain.

      a.   The circuit is functioning correctly.

      b.   Capacitor C_3 is open.

      c.   Capacitor C_3 is shorted.

      d.   Capacitor C_5 is open.

___ 34. The signal voltage gain of Q_1 is approximately three.

      a.   The circuit is functioning correctly.

      b.   Capacitor C_4 is open.

      c.   Capacitor C_4 is shorted.

      d.   Capacitor C_2 is open.

PROBLEMS

1. For the circuit in Figure 5–25, find:

    $V_c =$ _____ $V_B =$ _____ $V_E =$ _____ $A_v =$ _____

    $V_{CE} =$ _____ $z_{in} =$ _____ $z_{out} =$ _____ $V_{out} =$ _____

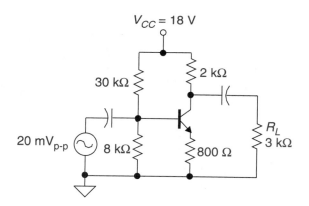

FIGURE 5–25

2. For the circuit in Figure 5–26, find:

    $V_c =$ _____ $V_B =$ _____ $V_E =$ _____ $A_v =$ _____

    $V_{CE} =$ _____ $z_{in} =$ _____ $z_{out} =$ _____ $V_{out} =$ _____

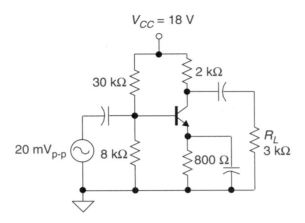

FIGURE 5–26

3. For the circuit in Figure 5–27, find:

   $V_c =$ _____ $V_B =$ _____ $V_E =$ _____ $A_v =$ _____

   $V_{CE} =$ _____ $z_{in} =$ _____ $z_{out} =$ _____ $V_{out} =$ _____

4. What is the largest signal that can be accommodated at the input of the circuit in Figure 5–27 before the output begins to clip?

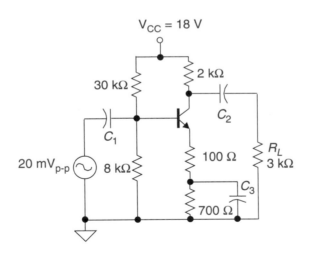

FIGURE 5–27

5. If the lowest frequency of operation is to be 100 Hz, select a value for C_1, C_2, and C_3 in the circuit in Figure 5–27.

   $C_1 =$ _____ $C_2 =$ _____ $C_3 =$ _____

6. For the circuit in Figure 5–28, find:

   $V_c =$ _____ $V_B =$ _____ $V_E =$ _____ $A_v =$ _____

   $V_{CE} =$ _____ $z_{in} =$ _____ $z_{out} =$ _____ $V_{out} =$ _____

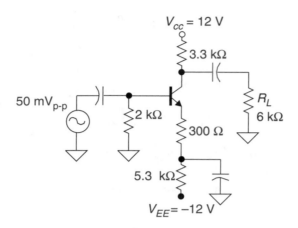

FIGURE 5–28

7. Find the values requested below for the circuit in Figure 5–29.

Stage 1:

$V_B =$ _____ $V_E =$ _____ $V_C =$ _____

$A_v =$ _____ $z_{in} =$ _____ $z_{out} =$ _____

Stage 2:

$V_B =$ _____ $V_E =$ _____ $V_C =$ _____

$A_v =$ _____ $z_{in} =$ _____ $z_{out} =$ _____

Total Circuit:

$A_v =$ _____ $z_{in} =$ _____ $z_{out} =$ _____

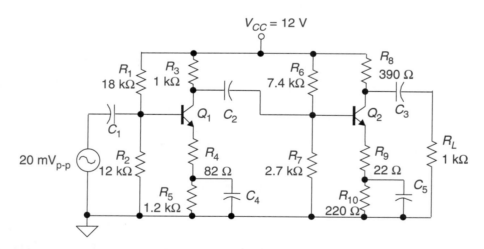

FIGURE 5–29

8. If the lowest frequency of operation is to be 1 kHz, select a value for C_1, C_2, C_3, C_4, and C_5 in the circuit in Figure 5–29.

$C_1 =$ _____ $C_2 =$ _____ $C_3 =$ _____

$C_4 =$ _____ $C_5 =$ _____

9. Find the values requested below for the circuit in Figure 5–30.
    Stage 1:

    $V_B =$ _____ $V_E =$ _____ $V_C =$ _____

    $A_v =$ _____ $z_{in} =$ _____ $z_{out} =$ _____

    Stage 2:

    $V_B =$ _____ $V_E =$ _____ $V_C =$ _____

    $A_v =$ _____ $z_{in} =$ _____ $z_{out} =$ _____

    Total Circuit:

    $z_{in} =$ _____ $z_{out} =$ _____ $A_v =$ _____

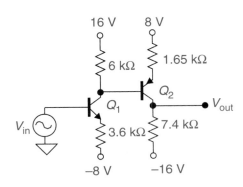

FIGURE 5–30

CHAPTER 6

Other Transistor Circuits

OBJECTIVES

After studying the material in this chapter, you will be able to describe and/or analyze:

○ common-collector amplifiers,
○ power and current gain,
○ Darlington pairs,
○ the common-collector stage in the multistage amplifier,
○ common-base amplifiers,
○ the advantages of different amplifier configurations,
○ current sources,
○ the differential amplifier, and
○ troubleshooting procedures for transistor circuits.

6.1 COMMON-COLLECTOR AMPLIFIERS

INTRODUCTION

Figure 6–1 shows a common-collector circuit that at first glance may look like a common-emitter circuit, but further investigation will reveal a difference. Notice that the collector terminal is connected directly to the power bus (V_{cc}). The power bus has a low signal impedance (assumed to be zero ohms) to ground, so the collector is at signal ground. The input signal is applied between the base of the transistor and ground. The collector is at signal ground, so the input signal is actually between the base and the collector. The output signal of the circuit is taken between the emitter and ground. Since the terminal common to both the input and output is the collector, the circuit is a common-collector amplifier.

Common-collector amplifiers often are used as buffers to prevent loading of a signal source. All signal sources have an output impedance. This output impedance acts like an internal resistance (R_{int}) in series with the signal source. For example, the function generator in your lab probably has an output impedance of 50 Ω. Figure 6–2 shows how this 50 Ω of internal resistance can affect the terminal voltage of the function generator. When a 5 kΩ load is connected across the

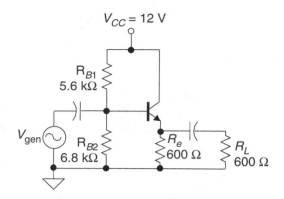

FIGURE 6–1   Common-collector amplifier

output of the generator, as shown in Figure 6–2(a), the 50 Ω of internal resistance has little effect on the output voltage of the generator. However, in Figure 6–2(b), a 50 Ω load has been connected across the output of the generator, causing the output voltage to be cut in half to 0.5 V. Many signal sources have internal resistances considerably higher than 50 Ω. Loading the output of these sources, therefore, becomes a problem that engineers must consider. In previous chapters we considered the ideal signal source driving our circuits to be zero ohms of internal resistance. In this chapter, we will consider the internal resistance of the signal source and examine how the common-collector amplifier can help prevent loading.

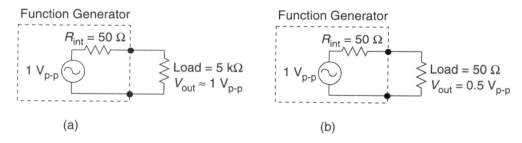

FIGURE 6–2   Internal resistance

BIASING THE COMMON-COLLECTOR AMPLIFIER

The most popular method of biasing the common-collector amplifier is voltage divider biasing. Voltage divider biasing provides independence from variations in beta. The circuit can be analyzed for the DC operating point (Q-point) by the same method used for the common-emitter amplifier. In Example 6.1, the Q-point values will be calculated.

EXAMPLE 6.1

For the circuit in Figure 6–1, calculate the Q-point values of V_B, V_E, V_C, V_{CE}, and I_E.

$$V_B = V_{cc} \times R_{b2}/(R_{b1} + R_{b2}) = 12 \text{ V} \times 6.8 \text{ k}\Omega/5.6 \text{ k}\Omega + 6.8 \text{ k}\Omega = 6.6 \text{ V}$$

$$V_E = V_B - V_{BE} = 6.6 \text{ V} - 0.7 \text{ V} = 5.9 \text{ V}$$

$$I_E = V_{RE}/R_E = 5.9 \text{ V}/600 \text{ }\Omega = 9.8 \text{ mA}$$

$$V_c = V_{cc} = 12 \text{ V (The collector is connected directly to } V_{cc}.)$$

$$V_{CE} = V_c - V_E = 12 \text{ V} - 5.9 \text{ V} = 6.1 \text{ V}$$

SIGNAL PARAMETERS OF THE COMMON-COLLECTOR AMPLIFIER_____

The signal parameters of the common-collector amplifier that are important to the technician are input impedance, output impedance, and voltage gain.

INPUT IMPEDANCE _____

The input impedance of the common collector is calculated in approximately the same manner as the input impedance of the common emitter. The value of r'_e is normally negligible because the emitter resistor is large in comparison and is never bypassed. The load resistor (R_L) is in parallel with the emitter resistor (R_e), and the resistance of the load affects the input impedance.

OUTPUT IMPEDANCE _____

The common-collector amplifier has extremely low output impedance, and this characteristic makes the circuit very useful. To understand the low output impedance of the common-collector, it is helpful to use Thevenin's theorem. First, remove the load (see Figure 6–3), then look back into the circuit from where the load was removed for possible signal current paths. The emitter resistor (R_e) is directly across the load and provides a signal current path. There is, however, another parallel path for current flow. This parallel path is through the forward-biased base/emitter junction and back through the signal resistance on the base leg. The base/collector junction is reverse biased and acts like an open.

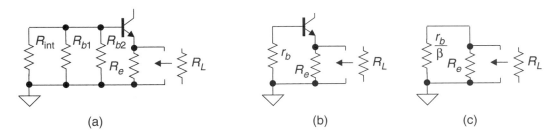

(a)   (b)   (c)

FIGURE 6–3 Common-collector output impedance

Figure 6–3(a) shows the base leg resistors seen when viewed through the forward-biased emitter/base. The three resistors (R_{b1}, R_{b2}, and R_{int}) appear in parallel. Resistor R_{int} is the internal resistance of the source. For simplification, in Figure 6–3(b) all three resistors are combined into one equivalent resistance called r_b.

Recall that when the signal resistance in the emitter leg is viewed from the base leg, it appears beta times larger. This magnification of resistance is caused by a magnification of current. This magnification of resistance can be compared to viewing objects through a telescope with a magnification factor equal to beta. If the signal resistance on the base is viewed from the perspective of the emitter, the opposite effect takes place. The signal resistance in the base leg is divisible by beta. This can be compared to looking through a telescope backward, causing everything you see to appear smaller by a factor equal to beta. Figure 6–3(c) shows the output impedance equal to R_e in parallel with the base equivalent resistance (r_b) divided by beta. The formula for the output impedance of the common-collector amplifier is $z_{out} = R_e \parallel (r_b/\beta)$. Normally, R_e is large in comparison to r_b/β, so the output impedance can be approximated by solving for r_b/β.

VOLTAGE GAIN

Figure 6–4(a) shows a common-collector amplifier, and Figure 6–4(b) shows a detailed drawing of the signal path through the transistor. The signal applied to the base of the transistor sees two resistances in series: r'_e and r_e. Resistance r'_e is the signal resistance of the forward-biased base/emitter junction, and resistance r_e is the signal resistance in the emitter leg. The signal resistance in the emitter leg consists of R_e in parallel with R_L ($r_e = R_e \parallel R_L$).

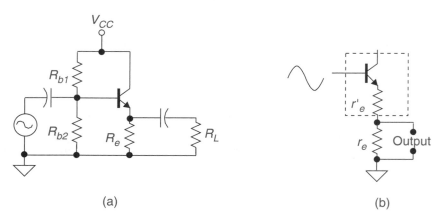

(a)   (b)

FIGURE 6–4   Common-collector voltage gain

The signal voltage out (V_o) is equal to the voltage developed across r_e. The signal output voltage can be calculated by using the voltage divider formula [$V_o = V_i \times r_e/(r'_e + r_e)$], since the input voltage at the base is divided between r'_e and r_e. In a common-collector amplifier, the value of r_e is always many times r'_e; therefore, the value of $r_e/(r'_e + r_e)$ is always very close to unity. In most cases, the circuit is considered to have a voltage gain of unity ($A_v = 1$).

The output signal is in phase with the input signal. If the voltage on the base changes in a positive direction, the voltage of the emitter will follow. On the other hand, if the voltage on the base decreases, the emitter voltage will also decrease. The magnitude and phase of the output signal on the emitter follows the input. This is why common-collector amplifiers are often referred to as emitter followers.

EXAMPLE 6.2

Calculate the value of z_in, z_out, and A_v of the circuit in Figure 6–5.

Step 1. Calculate the input impedance. (Assume beta to be 100.)

$$z_\text{in} = R_{b1} \parallel R_{b2} \parallel \beta \, (R_e \parallel R_L)$$
$$z_\text{in} = 5.6 \text{ k}\Omega \parallel 6.8 \text{ k}\Omega \parallel (100(600 \parallel 100)) = 2.26 \text{ k}\Omega$$

Step 2. Calculate the output impedance.

$$z_\text{out} = R_e \parallel ((R_{b1} \parallel R_{b2} \parallel R_\text{int})/\beta)$$
$$z_\text{out} = 600 \text{ }\Omega \parallel ((5.6 \text{ k}\Omega \parallel 6.8 \text{ k}\Omega \parallel 6 \text{ k}\Omega)/100) = 19.7 \text{ }\Omega$$

Step 3. The voltage gain of the common-collector amplifier is unity.

$$A_v = 1$$

EXAMPLE 6.2 continued

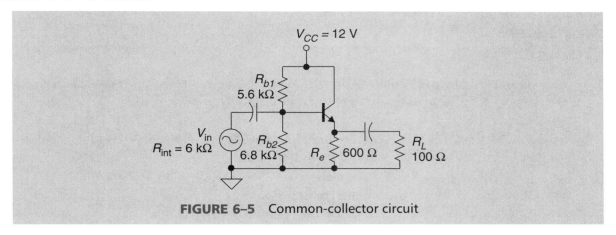

FIGURE 6–5 Common-collector circuit

6.2 POWER AND CURRENT GAIN

You may be asking yourself, "What good is an amplifier that has a voltage gain of one?" The answer is that it is useless as a voltage amplifier, but let us examine current and power gain. Current gain is equal to the signal current out divided by the signal current in (i_{out}/i_{in}). Power gain is equal to the signal power out divided by the signal power in (P_{out}/P_{in}). In Example 6.3, the current and power gains are calculated for the circuit in Figure 6–5.

EXAMPLE 6.3

For the circuit in Figure 6–5, calculate the current gain (A_i) and power gain (A_p).

Step 1. Assume a convenient RMS value for signal input voltage. Using this value, calculate the input and output signal currents.

$$i_{in} = V_{in}/z_{in} \text{ (Assume } V_{in} = 2.26 \text{ V}_{rms}.)$$

$$i_{in} = 2.26 \text{ V}/2.26 \text{ k}\Omega = 1 \text{ mA}_{rms}$$

$$V_{out} = A_V \times V_{in} = 1 \times 2.26 \text{ V} = 2.26 \text{ V}$$

$$i_{out} = V_{out}/R_L$$

$$i_{out} = 2.26 \text{ V}/100 \text{ }\Omega = 22.6 \text{ mA}_{rms}$$

Step 2. Calculate the current gain.

$$A_i = i_{out}/i_{in}$$

$$A_i = 22.6 \text{ mA}/1 \text{ mA} = 22.6$$

Step 3. Calculate the signal power in and the signal power out using the RMS current and voltage values calculated above.

$$P_{in} = i_{in} \times V_{in} = 1 \text{ mA} \times 2.26 \text{ V} = 2.26 \text{ mW}$$

$$P_{out} = i_{out} \times V_{out} = 22.6 \text{ mA} \times 2.26 \text{ V} = 51.1 \text{ mW}$$

Step 4. Calculate the circuit power gain.

$$A_p = P_{out}/P_{in} = 51.1 \text{ mW}/2.26 \text{ mW} = 22.6$$

Even though the emitter follower does not amplify voltage, it does have current gain and power gain. In addition, the high input impedance and low output impedance prevent loading of the output signal. Example 6.4 will demonstrate how a common-collector amplifier can reduce loading and, therefore, increase power to the load.

EXAMPLE 6.4

Calculate the signal power delivered to the load in both circuits in Figure 6–6. Each circuit is driven by a 3 V_{rms} signal source with an internal impedance of 1 kΩ.

Circuit a:

Step 1. Calculate the V_{out} in the circuit in Figure 6–6(a). The output voltage is calculated easily by using the voltage divider formula.

$$V_{out} = v_{gen} \times R_L/(R_{int} + R_L) = 3 \text{ V} \times 300 \text{ }\Omega/(1 \text{ k}\Omega + 300) = 0.69 \text{ V}$$

Step 2. Calculate the power delivered to the load.

$$P = V^2/R = 0.69 \text{ V}^2/300 \text{ }\Omega = 1.6 \text{ mW}$$

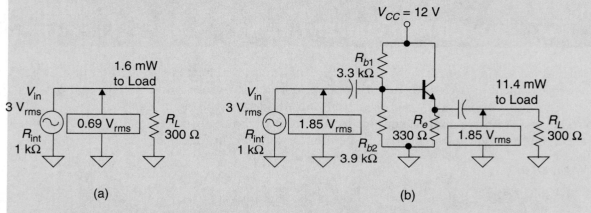

FIGURE 6–6   The common-collector used to reduce loading

Circuit b:

Step 1. Calculate the input impedance of the amplifier in the circuit in Figure 6–6(b).

$$z_{in} = R_{b1} \| R_{b2} \| \beta(R_e \| R_L)$$
$$z_{in} = 3.3 \text{ k}\Omega \| 3.9 \text{ k}\Omega \| (100(330 \| 300)) = 1.6 \text{ k}\Omega$$

Step 2. Calculate the signal voltage present at the input of the amplifier. The input impedance forms a voltage divider with the internal impedance of the signal source, so the input voltage of the circuit can be calculated using the voltage divider formula.

$$V_{in} = v_{gen} \times z_{in}/(R_{int} + z_{in})$$
$$V_{in} = 3 \text{ V} \times 1.6 \text{ k}\Omega/(1 \text{ k}\Omega + 1.6 \text{ k}\Omega) = 1.85 \text{ V}$$

Step 3. Calculate the signal voltage out. The voltage gain of the common-collector amplifier is unity ($A_v = 1$); therefore, the signal voltage out will equal the voltage present at the input of the amplifier.

$$V_{out} = V_{in} = 1.85 \text{ V}$$

Step 4. Calculate the power delivered to the load.

$$P = V^2/R = 1.85 \text{ V}^2/300 \text{ }\Omega = 11.4 \text{ mW}$$

When a 3 V_{rms} signal source with an internal resistance of 1 kΩ is connected directly to the 300 Ω load, only 1.6 mW of power is delivered to the load. If the common-collector amplifier is placed between the signal source and the load, power delivered to the load increases to 11.4 mW. The common-collector amplifier is effectively making the 300 Ω load appear as a 1.6 kΩ load to the signal source.

6.3 DARLINGTON PAIRS

In Example 6.4, the power to the load was enhanced by using a common-collector amplifier as a buffer between the signal source and the load. This stage helps raise the low impedance of the load. However, the impedance was only raised to 1.6 kΩ, which still caused considerable loading of the signal source. If the input impedance of the buffer stage could be raised, the signal source would load less, and greater power could be delivered to the load.

Two transistors connected in a Darlington pair configuration is the answer. Figure 6–7 shows two transistors connected in the Darlington pair configuration. The two collectors are connected together, and the emitter of Q_1 is connected to the base of Q_2. The base of Q_1 functions as the base for the pair, and the emitter of Q_2 functions as the emitter for the pair. The base current of Q_1 is multiplied by the beta of Q_1 (actually $\beta + 1$) and becomes the base current of Q_2. The base current of Q_2 is multiplied by the beta of Q_2 to become the collector current of Q_2. The two transistors form a super transistor whose overall beta is equal to the product of the beta of Q_1 times the beta of Q_2.

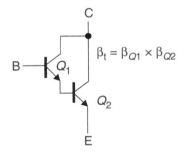

FIGURE 6–7   Darlington pair

EXAMPLE 6.5

Assume a beta of 100 for Q_1 and a beta of 50 for Q_2. What would be the beta of the Darlington pair shown in Figure 6–7?

Total beta:

$$\beta_t = \beta_{Q1} \times \beta_{Q2} = 100 \times 50 = 5000$$

Figure 6–8 shows a common-collector amplifier using a Darlington pair as a buffer between a 3 V_{rms} signal source with 1 kΩ of internal resistance and a 300 Ω load. The Darlington pair can be analyzed as a single device with a high beta; however, remember there are two 0.7 V junction drops between the emitter and the base.

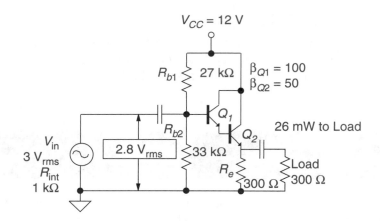

FIGURE 6–8   Common-collector circuit using Darlington pairs

EXAMPLE 6.6

Calculate the bias voltages (V_{BQ1} and V_{EQ2}) and the signal power output for the circuit in Figure 6–8.

Step 1. Calculate the bias voltages (V_{BQ1} and V_{EQ2}). The DC voltage at the base of Q_1 can be calculated using the voltage divider formula. Once the DC voltage is known on the base of Q_1, the DC voltage on the emitter of Q_2 can be found by subtracting 1.4 V from the base voltage of Q_1.

$$V_{BQ1} = V_{cc} \times R_{b2}/(R_{b1} + R_{b2})$$
$$V_{BQ1} = 12 \text{ V} \times 33 \text{ k}\Omega/(27 \text{ k}\Omega + 33 \text{ k}\Omega) = 6.6 \text{ V}$$
$$V_{EQ2} = V_{BQ1} - 1.4 \text{ V} = 6.6 \text{ V} - 1.4 \text{ V} = 5.2 \text{ V}$$

Step 2. Calculate the input impedance of the amplifier.

$$z_{in} = R_{b1} \parallel R_{b2} \parallel (\beta_{Q1} \times \beta_{Q2}) (R_e \parallel R_L)$$
$$z_{in} = 27 \text{ k}\Omega \parallel 33 \text{ k}\Omega \parallel (100 \times 50) \times (300 \parallel 300) = 14.8 \text{ k}\Omega$$

Step 3. Calculate the signal voltage present at the input of the amplifier.

$$V_{in} = v_{gen} \times z_{in}/(R_{int} + z_{in})$$
$$V_{in} = 3 \text{ V} \times 14.8 \text{ k}\Omega/(1 \text{ k}\Omega + 14.8 \text{ k}\Omega) = 2.8 \text{ V}$$

Step 4. Calculate the signal voltage out. The voltage gain of the common-collector amplifier is unity ($A_v = 1$); therefore, the signal voltage out will equal the voltage present at the input of the amplifier.

$$V_{out} = V_{in} = 2.8 \text{ V}$$

Step 5. Calculate the power delivered to the load.

$$P = V^2/R = 2.8 \text{ V}^2/300 \ \Omega = 26 \text{ mW}$$

The Darlington pair amplifier was able to raise the 300 Ω load to 14.8 kΩ. This high impedance had little loading effect on the signal source. Since the common-collector amplifier, using a Darlington pair, still has a voltage gain of one, the signal voltage present on the base will be present across the load. This increase of output voltage will enhance the output power.

The Darlington pair configuration is so popular that manufacturers provide single packages containing two transistors already connected as Darlington pairs. These commercially available packages can be analyzed as a single transistor with super high beta and a 1.4 voltage drop between the base and emitter junction.

6.4 COMMON-COLLECTOR STAGE IN THE MULTISTAGE AMPLIFIER

The common-collector amplifier finds use in many output circuits because of its relatively high input impedance and low output impedance. Using a common-collector stage between the final voltage amplifier and a low resistance load will greatly reduce loading. Figure 6–9(a) shows a common-emitter amplifier directly driving a 300 Ω load. The voltage gain of the common-emitter stage is only 1.36 because of the severe loading. The same common-emitter stage is used in Figure 6–9(b); however, a common-collector stage is used as a buffer between the 300 Ω load and the output of the common-emitter stage. The common-emitter stage now has a gain of 12.5. Example 6.7 shows how the gain of each circuit is calculated.

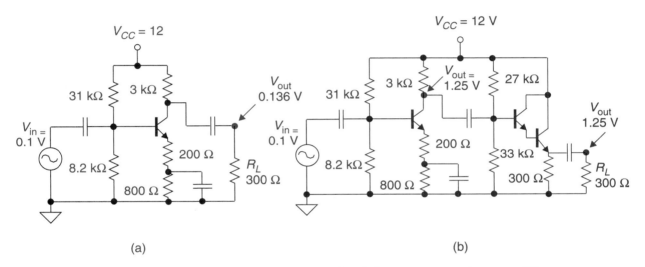

(a)   (b)

FIGURE 6–9   Buffer used to increase gain of the common-emitter amplifier

EXAMPLE 6.7

Calculate the voltage gain of both circuits shown in Figure 6–9.

Step 1. Calculate the voltage gain of the circuit in Figure 6–9(a).

$$r_c = R_c \| R_L = 3 \text{ k}\Omega \| 300 \text{ }\Omega = 273 \text{ }\Omega$$
$$r_e = r'_e + R_e \approx R_e = 200 \text{ }\Omega \text{ } (r'_e \text{ is negligible.})$$
$$A_v = r_c/r_e = 273 \text{ }\Omega/200 \text{ }\Omega = 1.36$$

Step 2. Calculate the voltage of the first stage in Figure 6–9(b).

$R_L = z_{in}$ of next stage $= 14.8 \text{ k}\Omega$ (The input impedance was calculated in Example 6.6.)

$$r_c = R_c \| R_L = 3 \text{ k}\Omega \| 14.8 = 2.5 \text{ k}\Omega$$
$$r_e = r'_e + R_e \approx R_e = 200 \text{ }\Omega \text{ } (r'_e \text{ is negligible.})$$
$$A_v = r_c/r_e = 2.5 \text{ k}\Omega/200 \text{ }\Omega = 12.5$$

Step 3. Calculate the total voltage gain for the circuit in Figure 6–9(b). The common-collector stage has a voltage gain of approximately one.

$$A_{v(total)} = A_{v(stage \text{ } 1)} \times A_{v(stage \text{ } 2)} = 12.5 \times 1 = 12.5$$

By using a common-collector stage between the final voltage amplifier and a low resistance load (Figure 6–10), the power output of the system can be greatly increased because the output voltage is not loaded down by the low impedance of the load. The power gain is equal to the product of the voltage and current gains. In a multistage amplifier system, the power gain can be in the millions, as shown in Example 6.8.

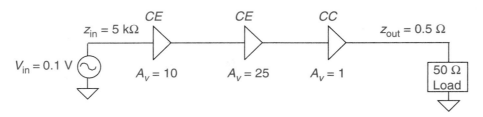

FIGURE 6–10   Multi-staged amplifier

EXAMPLE 6.8

Calculate the voltage gain, current gain, and power gain for the system shown in Figure 6–10.

Step 1. Calculate the voltage gain of the system. The total voltage gain is equal to the products of the gains of each stage.

$$A_{V(total)} = A_{V1} \times A_{V2} \times A_{V3}$$
$$A_{V(total)} = 10 \times 25 \times 1 = 250$$

Step 2. Calculate the output signal voltage.

$$V_{out} = V_{in} \times A_{V(total)}$$
$$V_{out} = 0.1 \text{ V} \times 250 = 25 \text{ V}$$

Step 3. Calculate the input signal current.

$$i_{in} = V_{in}/z_{in}$$
$$i_{in} = 0.1 \text{ V}/5 \text{ k}\Omega = 20 \text{ μA}$$

Step 4. Calculate the output signal current.

$$i_{out} = V_{out}/R_L$$
$$i_{out} = 25 \text{ V}/50 \text{ }\Omega = 0.5 \text{ A}$$

Step 5. Calculate the current gain of the system.

$$A_i = i_{out}/i_{in}$$
$$A_i = 0.5 \text{ A}/20 \text{ μA} = 25,000$$

Step 6. Calculate the power gain of the system.

$$A_p = A_v \times A_i$$
$$A_p = 250 \times 25,000 = 6,250,000$$

6.5 COMMON-BASE AMPLIFIERS

Figure 6–11 shows a block diagram of a common-base amplifier circuit. The input signal is connected between the emitter and base, and the output signal is taken between the collector and base. All of the circuit parameters of the common-base amplifier are similar to those of the common-emitter with the exception of phase shift and input impedance. The output signal is in phase with the input. The input impedance is equal to r'_e ($25 \text{ mV}/I_E$). This means that input impedance is extremely low (less than 50 Ω). In most cases, a low input impedance is not desirable, so the common-emitter is often preferred over the common-base.

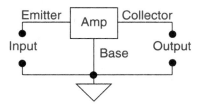

FIGURE 6–11 Block diagram of common-base amplifier

The common-base configuration does find use in high frequency amplifier applications (RF or radio frequencies). In the common-emitter configuration, the input signal sees the reverse-biased base/collector junction as a capacitance that connects the input to the output. At high frequencies, the reactance (X_c) of the base/collector junction drops to a value low enough to permit the signal to bypass the transistor. In the common-base configuration, the signal is applied to the forward-biased emitter/base junction, and there is no capacitive path bypassing the transistor. This permits the common-base amplifier to function at frequencies higher than a common-emitter amplifier using the same type of transistor.

The common-base amplifier can be stabilized by using voltage divider biasing or emitter biasing. Figure 6–12(a) shows a common-based amplifier using voltage divider biasing. The base is connected to signal ground through a capacitor, and the input signal is capacitively coupled to the emitter. Figure 6–12(b) is the same circuit with the transistor rotated, which is often done when common-base circuits are drawn in schematics. Example 6.9 shows how DC and signal values for the common-base circuit can be calculated.

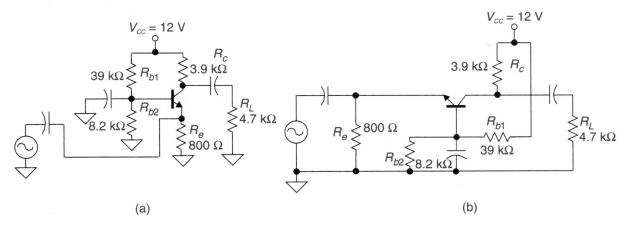

FIGURE 6–12 Common-base amplifier

EXAMPLE 6.9

Analyze the circuit in Figure 6–12 and calculate (a) the DC bias voltages for V_B, V_E, V_c, and V_{CE}, and (b) the signal voltage gain (A_v), input impedance (z_{in}), output impedance (z_{out}), and the approximate maximum signal output voltage swing.

EXAMPLE 6.9 continued

Step 1. Calculate the DC biased voltages and currents.
$$V_B = V_{cc} \times R_{b2}/(R_{b1} + R_{b2})$$
$$V_B = 12 \text{ V} \times 8.2 \text{ k}\Omega/(39 \text{ k}\Omega + 8.2 \text{ k}\Omega) = 2.1 \text{ V}$$
$$V_E = V_B - V_{BE} = 2.1 \text{ V} - 0.7 \text{ V} = 1.4 \text{ V}$$
$$I_E = V_E/R_E = 1.4 \text{ V}/800 \text{ }\Omega = 1.75 \text{ mA}$$
$$I_C \approx I_E = 1.75 \text{ mA}$$
$$V_{RC} = I_C \times R_c = 1.75 \text{ mA} \times 3.9 \text{ k}\Omega = 6.8 \text{ V}$$
$$V_c = V_{cc} - V_{RC} = 12 \text{ V} - 6.8 \text{ V} = 5.2 \text{ V}$$
$$V_{CE} = V_c - V_E = 5.2 \text{ V} - 1.4 \text{ V} = 3.8 \text{ V}$$

Step 2. Calculate the input impedance.
$$z_{in} = r'_e = 25 \text{ mV}/I_E = 25 \text{ mV}/1.75 \text{ mA} = 14.3 \text{ }\Omega$$

Step 3. Calculate the output impedance.
$$z_{out} = R_c = 3.9 \text{ k}\Omega$$

Step 4. Calculate the voltage gain.
$$r_c = R_c \| R_L = 3.9 \text{ k}\Omega \| 4.7 \text{ k}\Omega = 2.1 \text{ k}\Omega$$
$$A_v = r_c/r_e = 2.1 \text{ k}\Omega/14.3 \text{ }\Omega = 147 \ (r_e = r'_e)$$

Step 5. Calculate the maximum output signal swing. An approximation of the output peak-to-peak signal swing equals $2 \times V_{CE}$.
$$V_{CE} = 3.8 \text{ V}$$

Therefore, the maximum output signal swing equals
$$2 \times V_{CE}$$
$$2 \times 3.8 \text{ V} = 7.6 \text{ V}_{\text{p-p}}$$

6.6   COMPARISON OF AMPLIFIER CONFIGURATIONS

Bipolar transistors are three-terminal devices. One of the terminals must be common to both the input and the output signal when the device is used as an amplifier. The three configurations (common-emitter, common-collector, and common-base) that can function as amplifier circuits are shown in Figure 6–13. The name of each circuit configuration identifies which terminal is common to both the input and the output.

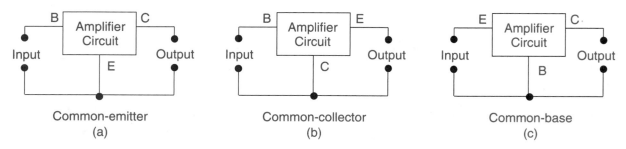

FIGURE 6–13 Transistor circuit configurations

An amplifier circuit has six important parameters: input impedance (z_{in}), output impedance (z_{out}), phase relationship, voltage gain (A_v), current gain (A_i), and power gain (A_p). The ideal amplifier would have the following circuit parameters: very high input impedance, very low output impedance, and high voltage, current, and power gain. The phase relationship between the input and the output signals should be known, but it is usually not necessary that they be in phase. Unfortunately, none of the circuit configurations as seen in Table 6–1 is the best with respect to all of the parameters.

TABLE 6–1   Amplifier comparisons

	z_{in}	z_{out}	Phase input to output	A_v	A_i	A_p
CE	MED	MED	180	YES	YES	YES
CC	HIGH	LOW	0	UNITY	YES	YES
CB	LOW	MED	0	YES	UNITY	YES

Note that the common-emitter is the best configuration if all the parameters are considered. For this reason, most circuitry uses the common-emitter configuration. The common-collector configuration does not have voltage gain, but it does approach the ideal amplifier in the area of input and output impedance, and it finds extensive use in the area of impedance matching circuits. The common-base configuration has a low input impedance, which is usually undesirable, and is used in only a few special applications where high frequency response is needed.

6.7 CURRENT SOURCES

By this time in your study of electronics, you are comfortable with voltage sources. For example, the ideal constant 12 V source will provide 12 V output regardless of the current flowing out of its terminals. There are, of course, limits that real voltage sources must operate within, and in a future chapter on power supplies, we will study these limitations. You are probably not so comfortable with current sources. The ideal current source is a circuit that will supply a constant current output regardless of the voltage across its terminals. There are, of course, limits to any real current source. A voltage source can be converted to a current source with a transistor and a few other components. Figure 6–14(a) shows the schematic symbol for a current source. The arrow in the symbol points in the direction of conventional current flow.

Figure 6–14(b) shows how a constant current source can be constructed using a bipolar transistor with voltage divider biasing. The voltage on the base of the transistor is held constant at –4 V by the voltage divider formed by R_1 and R_2. The transistor is forward biased so the emitter is held constant at –4.7 V. Since one end of R_3 is connected to the emitter and the other end is connected to –12 V, the voltage across R_3 is constant at 7.3 V. Using Ohm's law, the current through R_3 is calculated to equal 2 mA. The collector current approximately equals the emitter current, so the output of the circuit is a constant 2 mA. The circuit will maintain a constant 2 mA output unless the transistor is driven into saturation. Since the emitter of the transistor is at –4.7 V, the circuit should function properly as long as the collector does not drop below –4.5 V. The top of the load is connected to 12 V or a maximum difference voltage of 16.5 V. The circuit will maintain current regulation at 2mA as long as the voltage across the load does not excess 16.5 V.

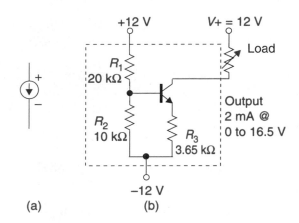

(a)   (b)

FIGURE 6–14   Current source symbol and current source circuitry

Figure 6–15(b) shows a constant current source using a current mirror. The operation of the current mirror circuit depends on having two PN junctions with identical current-versus-voltage curves. This means that if the two junctions have the same voltage drop across them, they will have the same current flowing through them. Let us assume the diode PN junction and the base/emitter PN junction are identically matched in the circuit shown in Figure 6–15(b). Since the junctions are in parallel, the voltage across both junctions is the same; therefore, the current flowing through each junction is equal. The current through the diode is controlled by the applied voltage and the value of the resistor. In this case, the current will equal 2 mA (15.3 V/7.65 kΩ). The emitter current will also equal 2 mA. Since the emitter is approximately equal to the collector current, the output current will also be 2 mA.

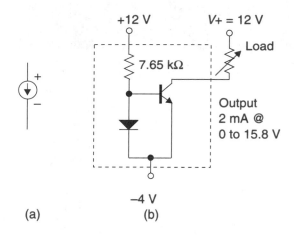

(a)   (b)

FIGURE 6–15   Current mirror

The circuit will maintain a constant 2 mA output unless the transistor is driven into saturation. Since the emitter of the transistor is at –4 V, the circuit should function properly as long as the collector does not drop below –3.8 V. The top of the load is connected to 12 V or a maximum difference voltage of 15.8 V. The circuit will maintain current regulation as long as the voltage across the load does not excess 15.8 V.

The advantages of a current mirror include fewer components needed for construction and lower voltage requirements. The main disadvantage of the current mirror is the need for matched PN junctions. Matched junctions are difficult to obtain in discrete circuitry but are easy to obtain in integrated circuitry. For this reason, current mirrors are very popular with engineers designing integrated circuits.

6.8 DIFFERENTIAL AMPLIFIERS

Figure 6–16 shows the circuit of a differential amplifier. The name comes from the fact that there are two inputs and the signal amplified is the difference of the signals between the two inputs. An output with respect to ground can be taken from the collector of either transistor. A signal with respect to ground is called single-ended. An output signal taken from the collector of one transistor to the collector of the other transistor is also available. This signal is not with respect to ground and is called a *differential output* or *balanced output*.

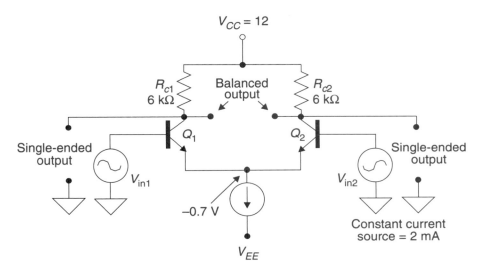

FIGURE 6–16 Differential amplifier

    The key to understanding the operation of the differential amplifier is to realize that the emitter of each transistor is fed by a single constant current source. The current out of the current source is constant. The way the current divides between the two transistors is dependent on the signal present at the base of each transistor. Let us assume that the current source is outputting 2 mA and the input signals to the base of each transistor is 0 V. The base/emitter junction of each transistor is forward biased, so the voltage at the top of the current source is −0.7 V. The transistors are matched so the current divides evenly, and 1 mA flows through each transistor. The collector current flowing through each transistor produces a 6 V drop across the collector resistors, causing the voltage on the collector of each transistor to be at 6 V with respect to ground. The difference voltage between the two collectors is zero volts, so the balanced output is zero. If the voltage on the base of each transistor is increased, both transistors will attempt to conduct harder. However, the current is limited to 2 mA, and since the transistors are matched, the current through each transistor stays at 1 mA, and the balanced output stays at zero volts.

    Figure 6–17 shows how a constant current source can be constructed using a bipolar transistor with voltage divider biasing. The voltage divider resistors (R_1 and R_2) and the emitter resistor (R_3) set the current source output equal to 2 mA. The 2 mA divides between transistors Q_1 and

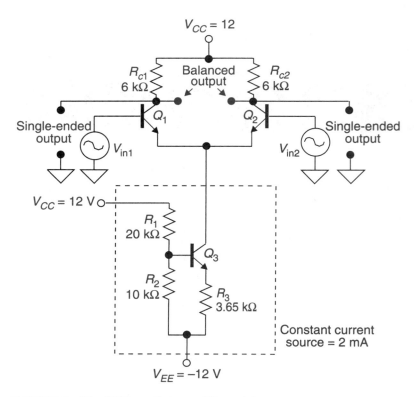

FIGURE 6–17   Differential amplifier with constant current source

Q_2. Example 5.10 shows the calculations of all biased voltages for the circuit in Figure 6–17 if the base voltage of both transistors is zero volts.

EXAMPLE 6.10

Calculate the voltage on the collector, base, and emitter of all three transistors for the circuit shown in Figure 6–17.

Step 1. Calculate the voltage on the base of Q_3.
$$V_{R2} = (V_{cc} - V_{EE}) \times R_2/(R_1 + R_2)$$
$$V_{R2} = (12\ V - (-12\ V)) \times 10\ k\Omega/(20\ k\Omega + 10\ k\Omega) = 8\ V$$
$$V_{B3} = V_{EE} + 8\ V = -12\ V + 8\ V = -4\ V$$

Step 2. Calculate the voltage on the emitter of Q_3.
$$V_{E3} = V_{B3} - 0.7\ V = -4\ V - 0.7\ V = -4.7\ V$$

Step 3. Calculate the emitter and collector current of Q_3.
$$V_{R3} = V_{EE} - V_{E3} = -12\ V - (-4.7\ V) = -7.3\ V$$
$$I_{EQ3} = V_{R3}/R_3 = -7.3\ V/3.65\ k\Omega = -2\ mA \text{ (Negative indicates direction of current flow.)}$$
$$I_{CQ3} \approx I_{EQ3} = 2\ mA$$

Step 4. Calculate the emitter voltage of Q_1 and Q_2. With no signal input, the base voltage of Q_1 and Q_2 are both at zero volts with respect to ground. The emitters of both transistors are connected together, and the base/emitter junctions of both transistors are forward biased, so the voltage on the emitters will be one junction drop below ground, or -0.7 V.

EXAMPLE 6.10 continued

$$V_{EQ1} = V_{BQ1} - 0.7 \text{ V} = 0 \text{ V} - 0.7 \text{ V} = -0.7 \text{ V}$$
$$V_{EQ2} = V_{EQ1} = -0.7 \text{ V}$$

Step 5. Calculate the collector voltage for Q_1 and Q_2. Since the transistors are balanced and have equal base signals applied, the current from the current source (Q_3) will divide equally and 1 mA will flow through both transistors, Q_1 and Q_2.

$$V_{RC1} = I_{C1} \times R_{C1} = 1 \text{ mA} \times 6 \text{ k}\Omega = 6 \text{ V}$$
$$V_{CQ1} = V_{CC} - V_{RC1} = 12 \text{ V} - 6 \text{ V} = 6 \text{ V}$$
$$V_{CQ2} = V_{CQ1} = 6 \text{ V}$$

If one input of a differential amplifier is referenced to ground and a signal is applied to the other input, the circuit will amplify the difference between the inputs. For example, if the base of Q_2, in Figure 6–17, has an input of 0 V and the base of Q_1 has an input of 10 mV, then Q_1 will be turned on harder than Q_2. This will create more current flow through Q_1, causing its collector voltage to drop. The reduced current flowing through Q_2 will cause its collector voltage to rise. The output signal on the collector of Q_1 is 180° out-of-phase with respect to the input signal on the base of Q_1. The output signal on the collector of Q_2 is in phase with respect to the input signal on the base of Q_1. Since the voltage on one collector increases and the voltage on the other collector decreases, there is a difference voltage between the collectors (balanced output). Since a small base current controls a large collector current, the circuit will have gain.

The signal parameters of the differential amplifier can be best studied by considering the differential amplifier as a two-stage circuit, as shown in Figure 6–18. We will apply a signal to the base of Q_1 and reference the base of Q_2 to ground. The first stage (Q_1) behaves as a common-collector stage. The constant current source in the emitter acts as a large emitter resistor for Q_1. The output of the common-collector stage drives the second stage (Q_2), which acts as a common-base amplifier. Since the input stage functions as a common-collector stage, the differential amplifier has high input impedance and the second stage functions as a common-base stage, giving the differential amplifier a high voltage gain. If the base of Q_1 is grounded and the signal is applied to the base of Q_2, the transistors reverse their roles. Example 6.11 shows the signal parameters.

EXAMPLE 6.11

Estimate the input impedance, calculate the voltage gain for the single-ended output, and calculate the voltage gain for the balanced output for the circuit in Figure 6–18.

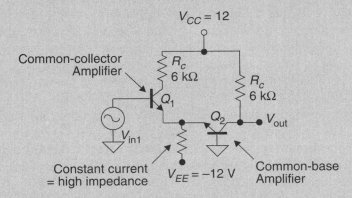

FIGURE 6–18 Signal equivalent circuit of the differential amplifier

EXAMPLE 6.11 continued

> **Step 1.** Estimate the input impedance of the differential amplifier. The input impedance of the differential amplifier is equal to beta of the transistor times the signal resistance of the current source. The ideal current source appears as an open. While our current source will not be near the ideal, it will have a large impedance. This impedance times beta will cause the differential amplifier to have a very high input impedance.
>
> **Step 2.** Calculate the voltage gain for the single-ended output. The gain of the transistor functioning as a common-emitter will be unity, and the gain of the transistor functioning as the common-base is equal to r_c/r_e.
>
> $r_c = R_c$ (If the differential amplifier has a load connected $r_c = R_c \parallel R_L$.)
>
> $$r_c = 6 \text{ k}\Omega$$
>
> $r_e = r'_e = 25 \text{ mV}/I_E = 25 \text{ mV}/1 \text{ mA} = 25 \Omega$ (I_E was calculated in Example 5.10.)
>
> $$A_v = r_c/r_e = 6 \text{ k}\Omega/25 \Omega = 240$$
>
> **Step 3.** Calculate the voltage gain for the balanced output. The signal input will cause the collector voltages to change in equal amounts but in opposite directions. This means the differential output voltage will be twice the single-ended output voltage. Hence, the differential voltage gain is twice the single-ended voltage gain.
>
> $$A_{v(\text{diff})} = 2 \times A_v = 2 \times 240 = 480$$

The differential amplifier has excellent qualities: high input impedance, high voltage gain, differential inputs, and good common mode rejection ratio. The common mode rejection ratio (CMRR) is the ability of an amplifier to reject noise present at the differential inputs. CMRR will be discussed in greater detail in Chapter 11. One problem in using the differential amplifier is the requirement for matched transistors. The economical solution is to incorporate the circuitry into integrated circuits (IC). All transistors inside an integrated circuit are identically matched because they are all made from the same block of silicon. The excellent qualities of differential amplifiers and the availability of matched transistors in integrated circuits have made the differential amplifier a popular circuit used in ICs.

6.9 TROUBLESHOOTING OTHER TRANSISTOR CIRCUITS

To troubleshoot common-collector, common-base, or differential amplifiers, the technician uses the same techniques used in troubleshooting the common-emitter amplifier. The first step is to calculate or estimate the DC bias voltages. These voltages are then measured, and if correct, the bias circuitry is assumed to be functioning properly. Then the technician proceeds to evaluate the signal parameters of the circuit by using signal insertion and checking the circuit for correct gains.

With common-emitter, common-base, and differential amplifier circuits, the technician can make a quick check of circuit operation by checking the DC voltage on the collector terminal of the transistor. The voltage at this point should measure approximately one-half of the DC supply voltage if the amplifier is designed to operate class-A. The common-collector is a different case. The collector connects directly to the DC supply voltage; however, the emitter terminal should measure approximately one-half of the DC supply voltage.

Remember that the common-collector amplifier has unity voltage gain; therefore, the input signal voltage on the base will measure the same as the output signal voltage on the emitter. The common-collector is designed to have a high input impedance. If the previous stage driving the common-collector stage is loading down, there is a good possibility the common-collector stage is defective.

The key to proper operation of a differential amplifier circuit is balanced components and a constant current source. If the signals at both inputs can be set to equal magnitudes, the differential amplifier should be balanced, and the voltage between the collectors of the two output transistors should be zero. If there is an imbalance, the paths through the two transistors are not dividing the current from the current source equally, and the components need to be checked. If there is no current (both collectors measure V_{cc}), the current source is suspect.

Emitter Follower Buffer

1. With the load (75 Ω) removed from the circuit in Figure 6–19(a), calculate and record the following. (Assume a minimum beta of 100.)

    $V_B =$ _____ $V_E =$ _____ $V_C =$ _____ $V_{CE} =$ _____

    $z_{in} =$ _____ $z_{out} =$ _____ $A_v =$ _____ $V_{out} =$ _____

2. With the load connected to the circuit in Figure 6–19(a), calculate the following.

    $V_B =$ _____ $V_E =$ _____ $V_C =$ _____ $V_{CE} =$ _____

    $z_{in} =$ _____ $A_v =$ _____ $V_{out} =$ _____

3. Calculate and record the current gain and power gain for the circuit in Figure 6–19(a).

    $A_i =$ _____ $A_p =$ _____

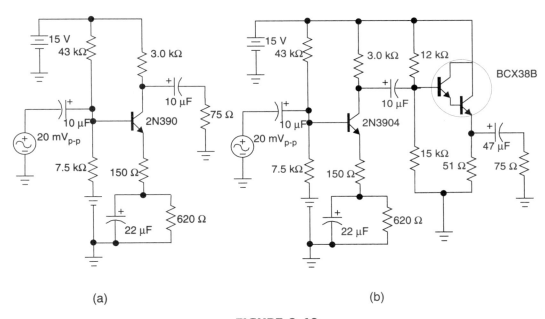

(a)   (b)

FIGURE 6–19

4. Calculate and record the signal output power for the circuit in Figure 6–19(a).

    $P_{out} =$ _____

5. Simulate and record the signal output power for the circuit in Figure 6–19(a).

    $P_{out} =$ _____

6. With the load (75 Ω) removed from the circuit in Figure 6–19(b), calculate the following. (Assume the minimum beta of the transistor is 100 and the minimum beta of the Darlington pair is 5000.)

    $V_{B(Q1)} =$ _____ $V_{E(Q1)} =$ _____ $V_{C(Q1)} =$ _____ $V_{CE(Q1)} =$ _____

    $z_{in(Q1)} =$ _____ $z_{out(Q1)} =$ _____ $A_{v(Q1)} =$ _____ $V_{out(Q1)} =$ _____

    $V_{B(Q2)} =$ _____ $V_{E(Q2)} =$ _____ $V_{C(Q2)} =$ _____ $V_{CE(Q2)} =$ _____

    $z_{in(Q2)} =$ _____ $z_{out(Q2)} =$ _____ $A_{v(Q2)} =$ _____ $V_{out(Q2)} =$ _____

    $z_{in(total)} =$ _____ $z_{out(total)} =$ _____ $A_{v(total)} =$ _____ $V_{out(total)} =$ _____

7. With the load connected to the circuit in Figure 6–19(b), calculate the following.

$V_{B(Q1)} = $ _____ $V_{E(Q1)} = $ _____ $V_{C(Q1)} = $ _____ $V_{CE(Q1)} = $ _____

$z_{in(Q1)} = $ _____ $z_{out(Q1)} = $ _____ $A_{v(Q1)} = $ _____ $V_{out(Q1)} = $ _____

$V_{B(Q2)} = $ _____ $V_{E(Q2)} = $ _____ $V_{C(Q2)} = $ _____ $V_{CE(Q2)} = $ _____

$z_{in(Q2)} = $ _____ $A_{v(Q2)} = $ _____ $V_{out(Q2)} = $ _____

$z_{in(total)} = $ _____ $A_{v(total)} = $ _____ $V_{out(total)} = $ _____

8. Calculate the current gain and power gain for the circuit in Figure 6–19(b).

$A_i = $ _____ $A_p = $ _____

9. Calculate and record the signal output power for the circuit in Figure 6–19(b).

$P_{out} = $ _____

10. Simulate and record the signal output power for the circuit in Figure 6–19(b).

$P_{out} = $ _____

Emitter Follower Buffer

I. Objective

To experimentally show how the common-collector (emitter follower) circuit can be used to buffer the output of a common-emitter amplifier to prevent loading.

II. Test Equipment

(1) sine wave generator   (1) dual-trace oscilloscope   (1) 12 V power supply

III. Components

Transistors: (1) 2N3904, (1) MPSA13

Resistors: (2) 100 Ω, (1) 220 Ω, (1) 820 Ω, (1) 2.7 kΩ, (1) 10 kΩ, (1) 27 kΩ, (2) 33 kΩ

Capacitors: C_1, C_2, C_3, and C_4 = 10 µf

IV. Procedure

1. With the load (100 Ω) removed from the circuit in Figure 6–20(a), calculate the following. (Assume a minimum beta of 100.)

   $V_B = $ _____ $V_E = $ _____ $V_C = $ _____ $V_{CE} = $ _____

   $z_{in} = $ _____ $z_{out} = $ _____ $A_v = $ _____ $V_{out} = $ _____

2. With the load connected to the circuit in Figure 6–20(a), calculate the following.

   $V_B = $ _____ $V_E = $ _____ $V_C = $ _____ $V_{CE} = $ _____

   $z_{in} = $ _____ $A_v = $ _____ $V_{out} = $ _____

3. Calculate the current gain and power gain for the circuit in Figure 6–20(a).

   $A_i = $ _____ $A_p = $ _____

4. Construct the circuit in Figure 6–20(a), and measure and record the following with the load removed. (Set the input signal to 100 mV$_{p-p}$ at a frequency of 5 kHz.)

   $V_B = $ _____ $V_E = $ _____ $V_C = $ _____ $V_{CE} = $ _____

   $z_{in} = $ _____ $z_{out} = $ _____ $V_{out} = $ _____

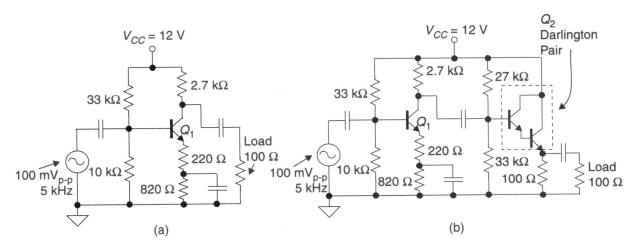

FIGURE 6–20

5. With the load connected, measure and record the following values for the circuit in Figure 6–20(a).

$V_B =$ _____ $V_E =$ _____ $V_C =$ _____ $V_{CE} =$ _____

$z_{in} =$ _____ $V_{out} =$ _____

6. Using measured values, calculate the voltage, current, and power gain for the circuit in Figure 6–20(a).

$A_v =$ _____ $A_i =$ _____ $A_p =$ _____

7. With the load (100 Ω) removed from the circuit in Figure 6–20(b), calculate the following. (Assume the minimum beta of the transistor is 100 and the minimum beta of the Darlington pair is 5000.)

$V_{B(Q1)} =$ _____ $V_{E(Q1)} =$ _____ $V_{C(Q1)} =$ _____ $V_{CE(Q1)} =$ _____

$z_{in(Q1)} =$ _____ $z_{out(Q1)} =$ _____ $A_{v(Q1)} =$ _____ $V_{out(Q1)} =$ _____

$V_{B(Q2)} =$ _____ $V_{E(Q2)} =$ _____ $V_{C(Q2)} =$ _____ $V_{CE(Q2)} =$ _____

$z_{in(Q2)} =$ _____ $z_{out(Q2)} =$ _____ $A_{v(Q2)} =$ _____ $V_{out(Q2)} =$ _____

$z_{in(total)} =$ _____ $z_{out(total)} =$ _____ $A_{v(total)} =$ _____ $V_{out(total)} =$ _____

8. With the load connected to the circuit in Figure 6–20(b), calculate the following.

$V_{B(Q1)} =$ _____ $V_{E(Q1)} =$ _____ $V_{C(Q1)} =$ _____ $V_{CE(Q1)} =$ _____

$z_{in(Q1)} =$ _____ $z_{out(Q1)} =$ _____ $A_{v(Q1)} =$ _____ $V_{out(Q1)} =$ _____

$V_{B(Q2)} =$ _____ $V_{E(Q2)} =$ _____ $V_{C(Q2)} =$ _____ $V_{CE(Q2)} =$ _____

$z_{in(Q2)} =$ _____ $A_{v(Q2)} =$ _____ $V_{out(Q2)} =$ _____

$z_{in(total)} =$ _____ $A_{v(total)} =$ _____ $V_{out(total)} =$ _____

9. Calculate the current gain and power gain for the circuit in Figure 6–20(b).

$A_i =$ _____ $A_p =$ _____

10. Construct the circuit in Figure 6–20(b), and measure and record the following with the load removed. The calculated output impedance of this circuit is less than 1 Ω. This circuit will not drive a 1 Ω load. However, the output impedance can be calculated from measured values by carefully measuring V_{out} without the 100 Ω load connected and then carefully measuring V_{out} with the 100 Ω load connected. The small change in the output voltage can be used to calculate the output impedance. The procedure to calculate z_{out} was covered in Chapter 4.

$V_{B(Q1)} =$ _____ $V_{E(Q1)} =$ _____ $V_{C(Q1)} =$ _____ $V_{CE(Q1)} =$ _____

$A_{v(Q1)} =$ _____ $V_{out(Q1)} =$ _____

$V_{B(Q2)} =$ _____ $V_{E(Q2)} =$ _____ $V_{C(Q2)} =$ _____ $V_{CE(Q2)} =$ _____

$A_{v(Q2)} =$ _____ $V_{out(Q2)} =$ _____

$z_{in(total)} =$ _____ $z_{out(total)} =$ _____ $A_{v(total)} =$ _____ $V_{out(total)} =$ _____

11. With the load connected, measure and record the following values for the circuit in Figure 6–20(b).

$V_{B(Q1)} =$ _____ $V_{E(Q1)} =$ _____ $V_{C(Q1)} =$ _____ $V_{CE(Q1)} =$ _____

$A_{v(Q1)} =$ _____ $V_{out(Q1)} =$ _____

$V_{B(Q2)} =$ _____ $V_{E(Q2)} =$ _____ $V_{C(Q2)} =$ _____ $V_{CE(Q2)} =$ _____

$V_{out(Q2)} =$ _____

$z_{in(total)} =$ _____ $V_{out(total)} =$ _____

12. Using measured values, calculate the voltage, current, and power gain for the circuit in Figure 6–20(b).

$A_v =$ _____ $A_i =$ _____ $A_p =$ _____

V. Points to Discuss

1. Explain why the voltage gain was affected when the load was connected to the circuit in Figure 6–20(a).

2. Explain why the voltage gain was affected only slightly when the load was connected to the circuit in Figure 6–20(b).

3. Why was the power gain of the second circuit so much higher than that of the first circuit?

QUESTIONS

6.1 Common-Collector Amplifiers

____ 1. In the common-collector amplifier, the collector terminal _____.
   a. connects to DC ground
   b. connects to signal ground
   c. connects to V_{cc}
   d. both b and c

____ 2. The common-collector amplifier can use _____ biasing.
   a. emitter
   b. collector voltage feedback
   c. voltage divider
   d. both a and c

____ 3. The common-collector configuration has a _____.
   a. high input impedance and low output impedance
   b. low input impedance and output impedance equal to R_c
   c. high input impedance and output impedance equal to R_c
   d. none of the above

____ 4. Voltage gain of the common-collector circuit is _____.
   a. more than 1
   b. exactly 1
   c. slightly less than 1
   d. equal to ß

6.2 Power and Current Gain

____ 5. The power gain of a circuit can be calculated by multiplying _____.
   a. beta times the voltage gain
   b. the load current squared times the load voltage
   c. the voltage gain times the current gain
   d. input power times output power

____ 6. Which transistor configuration has power gain?
   a. common-emitter
   b. common-base
   c. common-collector
   d. all of the above

6.3 Darlington Pairs

_____ 7. The Darlington pair that replaces a normal transistor in a common-collector circuit will _____.

    a. raise the input impedance of the circuit

    b. permit the circuit to drive a lower resistance load

    c. change the emitter bias voltage by 0.7 V

    d. all of the above

_____ 8. If a Darlington pair is constructed using two transistors, each with a beta of 40, the Darlington pair will have a beta equal to _____.

    a. 40

    b. 80

    c. 60

    d. 1600

_____ 9. A commercially packaged Darlington pair is operating in a class-A amplifier. What is the voltage between the base leg and the emitter leg?

    a. 0.2 V

    b. 0.7 V

    c. 1.4 V

    d. 2 V

6.4 Common-Collector Stage in the Multistage Amplifier

_____ 10. Often a common-collector will be the last stage before the load. The main function of this stage is to _____.

    a. provide voltage gain

    b. buffer the voltage amplifiers from the low resistance of the load

    c. provide phase inversion

    d. provide a high frequency path that improves high-end frequency response

_____ 11. An amplifier has three stages with voltage gains of 32, 16, and 1. What is the voltage gain of the total amplifier?

    a. 49 V

    b. 49

    c. 512 V

    d. 512

6.5 Common-Base Amplifiers

_____ 12. The common-base configuration has a _____.

    a. high input impedance and low output impedance

    b. low input impedance and output impedance equal to R_c

    c. high input impedance and output impedance equal to R_c

    d. none of the above

____ 13. The current gain of the common-base circuit is _____.
    a. more than 1
    b. exactly 1
    c. slightly less than 1
    d. equal to β

6.6 Comparison of Amplifier Configurations

____ 14. The common-emitter configuration has a _____.
    a. high input impedance and low output impedance
    b. low input impedance and output impedance equal to R_c
    c. medium input impedance and output impedance equal to R_c
    d. none of the above

____ 15. When a bipolar transistor is connected in the common-base configuration, it is capable of providing _____.
    a. voltage and power gain
    b. current and power gain
    c. current, voltage, and power gain
    d. voltage gain only

____ 16. When a bipolar transistor is connected in the common-collector configuration, it is capable of providing _____.
    a. voltage and power gain
    b. current and power gain
    c. current, voltage, and power gain
    d. voltage gain only

____ 17. When a bipolar transistor is connected in the common-emitter configuration, it is capable of providing _____.
    a. voltage and power gain
    b. current and power gain
    c. current, voltage, and power gain
    d. voltage gain only

6.7 Current Sources

____ 18. Which of the following is the symbol for a constant current source?

    a. ⊘   b. ⟍   c. (I)   d. ⊙

____ 19. A 10 mA constant current source is driving a 1 kΩ load. What is the voltage across the load?
    a. 1 V
    b. 5 V
    c. 10 V
    d. 20 V

_____ 20. A 10 mA constant current source is driving a 1 kΩ load. If the load resistance changes
to 2 kΩ, what will the current through the load be?

    a.   1 mA

    b.   5 mA

    c.   10 mA

    d.   40 mA

6.8 Differential Amplifiers

_____ 21. A differential amplifier is designed with a 10 mA constant current source, a 12 V power
source, and a 1 kΩ resistor (R_c) in the collector leg of both output transistors. Both
inputs are set at 100 mV. What will be the voltage difference measured between the
collector of the output transistors?

    a.   0 V

    b.   100 mV

    c.   5 V

    d.   7 V

_____ 22. A differential amplifier is designed with a 10 mA constant current source, a 12 V power
source, and a 1 kΩ resistor (R_c) in the collector leg of both output transistors. Both
inputs are set at 100 mV. What will be the voltage measured on the collector of one of
the output transistors with respect to ground?

    a.   0 V

    b.   100 mV

    c.   5 V

    d.   7 V

6.9 Troubleshooting Other Transistor Circuits

_____ 23. The signal voltage across the load in Figure 6–21 measures 1 $V_{p\text{-}p}$. What is the most
likely problem with the circuit?

    a.   The beta of the transistor is low.

    b.   The output coupling capacitor is shorted.

    c.   The transistor has a shorted base/emitter junction.

    d.   The circuit is operating correctly.

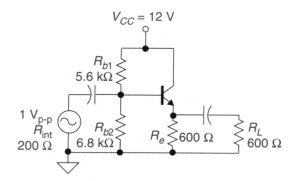

FIGURE 6–21

_____ 24. The voltage on the base in Figure 6–21 measures 1.2 V DC. What is the most likely problem with the circuit?

    a.   The beta of the transistor is low.

    b.   The output coupling capacitor is shorted.

    c.   The transistor has a shorted base/emitter junction.

    d.   The circuit is operating correctly.

_____ 25. The output signal in Figure 6–21 is riding on a 6 V DC level. What is the most likely problem with the circuit?

    a.   The beta of the transistor is low.

    b.   The output coupling capacitor is shorted.

    c.   The transistor has a shorted base/emitter junction.

    d.   The circuit is operating correctly.

_____ 26. The input signal is set to zero in Figure 6–22. The voltage on the collector of Q_1 and Q_2 measures 12 V. What is the most likely problem with the circuit?

    a.   The transistor Q_1 is open.

    b.   The transistor Q_2 is open.

    c.   The current source is defective.

    d.   The circuit is operating correctly.

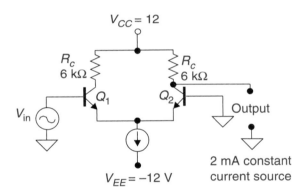

FIGURE 6–22

_____ 27. The input signal is set to zero in Figure 6–22. The voltage on the collector of Q_1 measures 12 V and the voltage at Q_2 measures 0.2 V. What is the most likely problem with the circuit?

    a.   The transistor Q_1 is open.

    b.   The transistor Q_2 is open.

    c.   The current source is defective.

    d.   The circuit is operating correctly.

_____ 28. The input signal is set to zero in Figure 6–22. The voltage on the collector of Q_1 and Q_2 measures 6 V. What is the most likely problem with the circuit?

    a.   Transistor Q_1 is open.

    b.   Transistor Q_2 is open.

    c.   The current source is defective.

    d.   The circuit is operating correctly.

PROBLEMS

1. For the circuit in Figure 6–23, calculate the Q-point values of V_B, V_E, V_C, V_{CE}, and I_E.

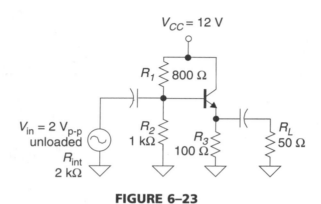

FIGURE 6–23

2. Calculate the values of z_{in}, z_{out}, and A_v for the circuit in Figure 6–23.

3. For the circuit in Figure 6–23, calculate the current gain (A_i) and power gain (A_p).

4. Assume a beta of 150 for Q_1 and a beta of 30 for Q_2. What would be the beta of the Darlington pair shown in Figure 6–24?

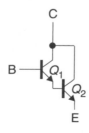

FIGURE 6–24

5. For the circuit in Figure 6–25, calculate the Q-point values of V_{BQ1}, V_{EQ1}, V_{CQ1}, V_{BQ2}, V_{EQ2}, V_{CQ2}, and I_{EQ2}. (Assume $\beta_{Q1} = 90$ and $\beta_{Q2} = 40$.)

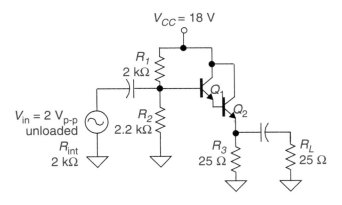

FIGURE 6–25

6. Calculate the values of z_{in}, z_{out}, and A_v for the circuit in Figure 6–25. (Assume $\beta_{Q1} = 90$ and $\beta_{Q2} = 40$.)

7. For the circuit in Figure 6–25, calculate the current gain (A_i) and power gain (A_p). (Assume $\beta_{Q1} = 90$ and $\beta_{Q2} = 40$.)

8. Calculate the voltage gain, current gain, and power gain for the system shown in Figure 6–26.

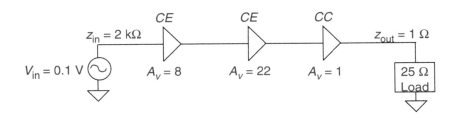

FIGURE 6–26

9. Calculate the signal power delivered to the load in both circuits in Figure 6–27. Each circuit is driven by a 2 $V_{p\text{-}p}$ (unloaded) signal source with an internal impedance of 2 kΩ. (Assume $\beta_{Q1} = 90$ and $\beta_{Q2} = 40$.)

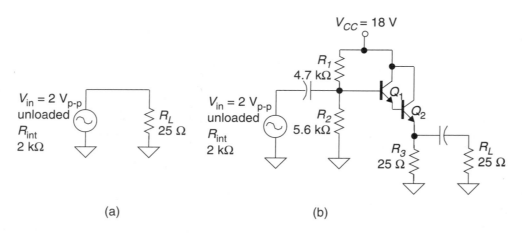

(a)   (b)

FIGURE 6–27

10. Calculate the DC bias voltages V_B, V_E, V_c, and V_{CE} for the circuit in Figure 6–28.

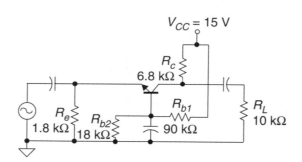

FIGURE 6–28

11. Calculate the signal voltage gain (A_v), input impedance (z_{in}), output impedance (z_{out}), and the approximate maximum output voltage swing for the circuit in Figure 6–28.

12. Calculate the voltage on the collector, base, and emitter of all three transistors for the circuit shown in Figure 6–29.

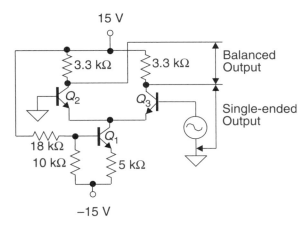

FIGURE 6–29

13. Calculate the single-ended voltage gain and the differential voltage gain for the circuit shown in Figure 6–29.

CHAPTER 7

Junction Field Effect Transistors

OBJECTIVES

After studying the material in this chapter, you will be able to describe and/or analyze:

○ the theory of JFETs,

○ the difference between a JFET and a bipolar transistor,

○ JFET characteristics,

○ JFET biasing circuits,

○ JFET signal parameters,

○ JFET common-source circuits,

○ common-drain circuits,

○ P-channel JFETs,

○ JFET switching circuits, and

○ troubleshooting procedures for JFET circuits.

7.1 INTRODUCTION TO JFETs

The junction field effect transistor (JFET) is a three-terminal device. Like the bipolar transistor, the JFET can be used for switching and signal amplification. The internal structure of the JFET, however, differs greatly from the bipolar transistor. This gives JFET circuits some advantages and, of course, some disadvantages. The three terminals of the JFET are the source, drain, and gate. The schematic symbols for an N-channel JFET are shown in Figure 7–1. The symbol in Figure 7–1(a) does not distinguish between the drain and source terminal, since the arrow points to the middle of the bar. Some JFETs are symmetrical, permitting the source and drain to be interchanged. However, some are not symmetrical and the source and drain cannot be interchanged. It would be nice if companies would use the symbol in Figure 7–1(b) for nonsymmetrical JFETs, but most companies use the symbol in Figure 7–1(a) for all JFETs. This leaves it up to the technician to examine the circuitry to determine the source and drain.

The JFET utilizes a reverse-biased PN junction for its operation. Recall these facts about a reverse-biased PN junction: the depletion region is located between the N and P regions, the

FIGURE 7–1   N-channel JFET symbol

depletion region cannot support current flow, and the width of the depletion region can be controlled by the magnitude of the reverse-biased voltage.

A JFET is constructed by taking a channel of N-type semiconductor material and surrounding it by a collar of P-type material. The source terminal is connected to one end of the channel, and the drain terminal is connected to the other end of the channel. The gate terminal connects to the collar. You could compare this to a water hose with a clamp connected around the hose. The water flows into one end of the hose and flows out the other end. The rate of water flow can be controlled by tightening or loosening the clamp. In the case of the JFET, electrons flow into one end of the channel (source) and out the other end (drain). The rate of current flowing through the JFET is controlled by the magnitude of the signal applied to the gate.

Figure 7–2 shows a cross-sectional drawing of the JFET. The source and drain are connected at opposite ends of the channel made from N-type material. The gate is made from P-type material and surrounds the channel. The magnitude of current flow through the channel is dependent on the size of the channel. When a negative voltage is applied to the gate, it reverse biases the junction, and a depletion region is formed between the gate and the channel material. This depletion region cuts the size of the channel and reduces the current from the source to the drain. The size of the channel can be controlled by controlling the reverse bias of the gate. The greater the reverse-biased voltage, the larger the depletion region and the smaller the channel.

In Figure 7–2(a), there is a small reverse-biased voltage (V_{GS}) and the depletion region is relatively small, causing the channel to be large and the drain current to be large. In Figure 7–2(b), V_{GS} is increased, which causes the depletion region to increase, thus causing the channel to decrease and the drain current to decrease. In summary, the voltage on the gate controls the current through the channel.

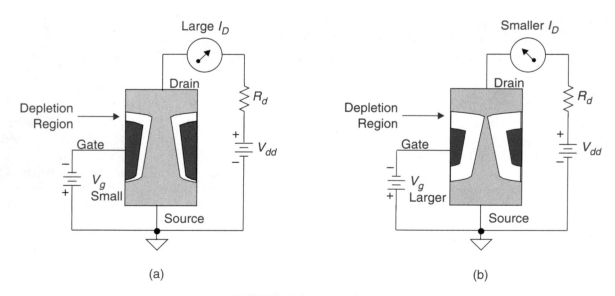

FIGURE 7–2   JFET theory

7.2 THE JFET VERSUS THE BIPOLAR TRANSISTOR

The JFET and the bipolar transistor can be used for signal amplification. The difference between them can be seen by examining the diagram of each device shown in Figure 7–3. The control terminal of the bipolar transistor is the base. A change in base current (I_B) causes a change in the output collector current (I_c). The gain of the bipolar transistor device is called beta (β) and is equal to the ratio of change in collector current to change in base current ($\beta = \Delta I_C/\Delta I_B$). Since the device gain is a ratio of two currents, the units cancel, and beta has no units. The output current is collector current ($I_C = \beta \times I_B$).

FIGURE 7–3 Function diagram of the bipolar transistor and the JFET

The control terminal of the JFET is the gate. A change in gate voltage causes a change in the output drain current (I_D). The gate voltage is measured with reference to the source terminal (V_{GS}). The gain of the JFET device is equal to the ratio of change in drain current (ΔI_D) to change in gate voltage (ΔV_{GS}). Since the device gain is a ratio of output current to input voltage, the gain is a measurement of conductance in siemens. The gain of a JFET device is called *transconductance* (*gm*), which means a transfer of conductance. The formula for transconductance is *gm* = $\Delta I_D/\Delta V_{GS}$.

The main difference between the bipolar transistor and the JFET is that the input control signal in a bipolar transistor is a current (I_B), while the input control signal in a JFET is a voltage (V_{GS}).

7.3 JFET CHARACTERISTICS

TRANSCONDUCTANCE CURVES

The transconductance (abbreviated *gm* or Y_{fs}) of a JFET is one characteristic that is extremely useful to the technician. Transconductance (*gm*) depends on two factors: the JFET and the bias voltage applied to the gate. The bias voltage is the DC difference voltage between the gate and the source terminal. In normal operation, this voltage will reverse bias the PN junction inside the JFET. The magnitude of the reverse-biased gate to source voltage (V_{GS}) determines the magnitude of drain current (I_D). Figure 7–4 shows a graph of V_{GS} versus I_D. This graph is called a transconductance curve. The drain current (I_D) is at maximum (6 mA) when the gate voltage (V_{GS}) equals zero. As V_{GS} is adjusted toward the negative cutoff voltage (–6 V), I_D continues to decrease. When V_{GS} reaches the cutoff voltage, I_D drops to zero. The graph shown in Figure 7–4 is not linear. Notice that the drain current is more sensitive to changes in gate voltage when V_{GS} is near zero.

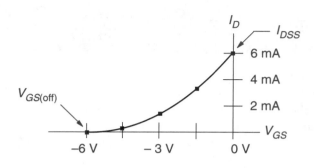

V_{GS} (V)	I_D (mA)
−6.0	0.0
−4.6	0.375
−3.0	1.5
−1.5	3.375
0.0	6.0

FIGURE 7–4   Transconductance curve

I_{DSS} and $V_{GS(off)}$ are shown on the transconductance curve in Figure 7–4. I_{DSS} is equal to the drain current with zero volts of gate bias ($V_{GS} = 0$ V). $V_{GS(off)}$ is equal to the value of the gate-bias voltage needed to cause the drain current to drop to zero amps.

The general formula for calculating the slope of a curve at any point is *slope* $= \Delta_{rise}/\Delta_{run}$, where Δ_{rise} and Δ_{run} are small changes around the point of interest. The slope of the transconductance curve equals the transconductance (gm) at a point on the curve (gm $= \Delta I_d/\Delta V_{GS}$). It is apparent that the slope of the transconductance curve changes as V_{GS} goes from $V_{GS(off)}$ to zero volts. Example 7.1 shows that transconductance varies greatly as we change the operating point on the curve.

EXAMPLE 7.1

Calculate the transconductance for the JFET (from the curve in Figure 7–5) at a bias of:
(a) $V_{GS} = -1$ V and (b) $V_{GS} = -3$ V.

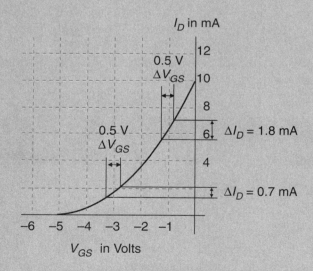

FIGURE 7–5   Transconductance at two Q-points

Solution a:

Step 1. For a positive change of 0.25 V above the bias point of −1 V (−0.75 V), read the drain current (I_D) from the graph.

$$\text{For } V_{GS} = -0.75 \text{ V} \qquad I_D = 7.2 \text{ mA}$$

EXAMPLE 7.1 continued

> **Step 2.** For a negative change of –0.25 V below the bias point of –1 V (–1.25 V), read the drain current (I_D) from the graph.
>
> $$\text{For } V_{GS} = -1.25 \text{ V} \qquad I_D = 5.4 \text{ mA}$$
>
> **Step 3.** Calculate the change in drain current (ΔI_D).
>
> $$\Delta I_D = 7.2 \text{ mA} - 5.4 \text{ mA} = 1.8 \text{ mA}$$
>
> **Step 4.** Calculate the transconductance at the bias point of –1 V.
>
> $$\Delta V_{GS} = 0.5 \text{ V}$$
> $$\Delta I_D = 1.8 \text{ mA}$$
> $$gm = \Delta I_D / \Delta V_{GS}$$
> $$gm = 1.8 \text{ mA} / 0.5 \text{ V}$$
> $$gm = 3.6 \text{ mS}$$
>
> **Solution b:**
>
> **Step 1.** For a positive change of 0.25 V above the bias point of –3 V (–2.75 V), read the drain current (I_D) from the graph.
>
> $$\text{For } V_{GS} = -2.75 \text{ V} \qquad I_D = 2.1 \text{ mA}$$
>
> **Step 2.** For a negative change of –0.25 V below the bias point of –3 V (–3.25 V), read the drain current (I_D) from the graph.
>
> $$\text{For } V_{GS} = -3.25 \text{ V} \qquad I_D = 1.4 \text{ mA}$$
>
> **Step 3.** Calculate the change in drain current (ΔI_D).
>
> $$\Delta I_D = 2.1 \text{ mA} - 1.4 \text{ mA} = 0.7 \text{ mA}$$
>
> **Step 4.** Calculate the transconductance at the bias point of –3 V.
>
> $$\Delta V_{GS} = 0.5 \text{ V}$$
> $$\Delta I_D = 0.7 \text{ mA}$$
> $$gm = \Delta I_D / \Delta V_{GS}$$
> $$gm = 0.7 \text{ mA} / 0.5 \text{ V}$$
> $$gm = 1.4 \text{ mS}$$

Example 7.1 shows that transconductance (gm) is a function of the bias voltage (V_{GS}). The less negative the bias voltage, the greater the transconductance. Data sheets will not list a value for transconductance, since it depends on circuit design. Data sheets do give a value for gm_0, which is the transconductance when the bias voltage is zero volts. If $V_{GS(\text{off})}$ and gm_0 are known, the transconductance (gm) for a given value of V_{GS} can be calculated using the following formula.

$$gm = gm_0 \times [1 - (V_{GS}/V_{GS(\text{off})})]$$

A transconductance curve is part of a parabola curve and is often referred to as a square-law curve. The mathematical relationship that describes the curve is:

$$I_D = I_{DSS}(1 - V_{GS}/V_{GS(\text{off})})^2$$

From this equation, you can see that all points on the curve are a function of I_{DSS}, V_{GS}, and $V_{GS(\text{off})}$. If I_{DSS} and $V_{GS(\text{off})}$ are constant for a particular type of JFET, then for a given bias voltage (V_{GS}), the drain current (I_D) can be calculated. However, as with other semiconductor devices, variations in manufacturing cause a varying set of values for a given type of JFET. Figure 7–6 shows three transconductance curves for the same type of JFET. The important thing to note is that the basic shape of the curve is the same because I_{DSS} and $V_{GS(\text{off})}$ vary together, and the shape of the curve is a function of their relationship.

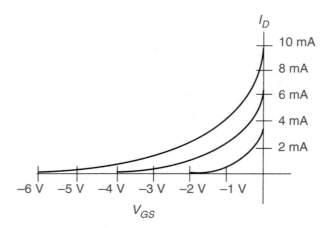

FIGURE 7–6   Transconductance curves

OUTPUT CHARACTERISTIC CURVES

A graph showing drain current (I_D) versus drain-to-source voltage (V_{DS}) with gate voltage (V_{GS}) held constant is called an output characteristic curve. Figure 7–7 shows an experimental circuit used to obtain the needed data to draw an output characteristic curve. The data needed to draw the output characteristic curve shown in Figure 7–7 is obtained by holding the gate voltage constant at 0 V and monitoring the drain current as the drain-to-source voltage is varied from 0 V to 25 V.

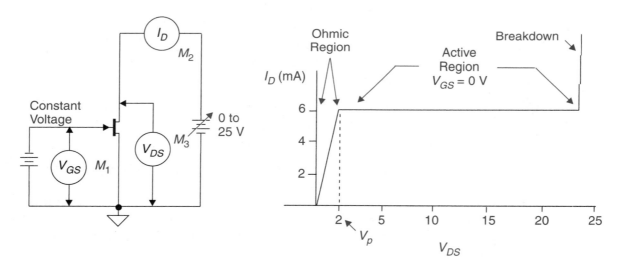

FIGURE 7–7   Test circuit and characteristic curve

Examine the characteristic curve in Figure 7–7. As the drain-to-source voltage (V_{DS}) increases from 0 V, the drain current increases until 2 V is reached. The drain current then levels off, and further increases in the drain-to-source voltage causes very little increase in drain current. The magnitude of drain-to-source voltage that causes the drain current to level off is called the *pinch-off voltage* (V_p).

The characteristic curve shows two distinct regions. To the left of the pinch-off voltage (V_p), the JFET is in the ohmic region and acts as a voltage-controlled resistor. To the right of the pinch-off voltage, the JFET is in the active region and acts as a voltage-controlled current source. If the drain voltage (V_{DS}) is increased above normal operating voltages, a point will be reached where the PN junction inside the JFET breaks down. This voltage is called the *breakdown voltage* and should be avoided.

If the gate voltage is adjusted to –0.5 V and the procedure to obtain the first curve is repeated, data can be obtained for a second curve. This process can be repeated for gate voltages of –1 V, –1.5 V and –2 V, and a family of output characteristic curves can be obtained as shown in Figure 7–8.

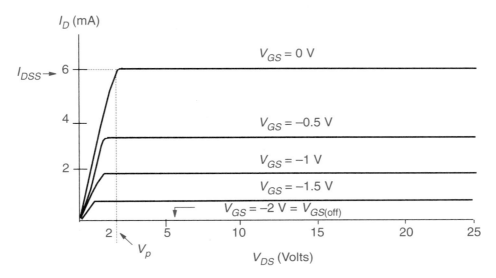

FIGURE 7–8   Family of output characteristic curves

As the gate voltage becomes more negative, the drain current decreases. The output characteristic curves in Figure 7–8 show that I_{DSS} = 6 mA (drain current with zero gate voltage) and the cutoff voltage ($V_{GS(off)}$) equals –2 V (gate voltage that causes zero drain current). Observe that the absolute magnitude of pinch-off voltage is equal to the absolute magnitude of cutoff voltage. Figure 7–8 shows that pinch-off voltage is equal to +2 V and cutoff voltage is equal to –2 V. Knowing $|V_p| = |V_{GS(off)}|$ is important because normally only one of these values is given in data sheets.

Engineers make use of characteristic curves for designing circuits. The technician does not normally utilize characteristic curves in circuit troubleshooting. Understanding them, however, is useful for technicians when using a curve tracer. By using a curve tracer, the technician can display a family of output characteristic curves to check the dynamic performance of JFETs.

Transconductance curves and output characteristic curves are useful in understanding the JFET parameters and how they interact. Let us review a list of parameters and their definitions. Lowercase letters denote signal values, and uppercase letters denote bias values.

> I_D = drain-bias current.
>
> i_d = drain-signal current.
>
> I_{DSS} = drain current with the gate voltage equal to zero (the maximum drain current for a JFET).
>
> V_{GS} = gate-to-source bias voltage.
>
> v_{gs} = gate-to-source signal voltage.
>
> $V_{GS(off)}$ = gate-to-source voltage that stops drain current (I_D) from flowing.
>
> V_p = pinch-off voltage (the drain-to-source voltage with zero gate voltage applied where the drain current becomes approximately constant).

gm = transconductance. A measurement of the sensitivity of the JFET, it is equal to the change in drain current divided by the corresponding change in gate voltage ($gm = \Delta I_D / \Delta V_{GSD}$).

gm₀ = transconductance when bias voltage (V_{GS}) is zero volts.

7.4 BIASING THE COMMON-SOURCE JFET AMPLIFIER

The gate terminal of a JFET is connected to a reverse-biased PN junction inside the JFET; consequently, there is zero current flowing in or out of the gate terminal. There is a conductive channel inside the JFET, connecting the source to the drain. All the current flowing into the source terminal flows out the drain terminal; therefore, the source current always equals the drain current ($I_S = I_D$). In order to use a JFET as an amplifier, it is necessary to design a bias circuit that will set the quiescent point (Q-point).

The transconductance curve in Figure 7–9 shows that the amount of drain current flowing will be a percentage of I_{DSS} and is dependent on the bias voltage (V_{GS}). If V_{GS} is equal to one-fourth of $V_{GS(off)}$, the drain current (I_D) will equal approximately one-half of I_{DSS} (actually nine-sixteenths).

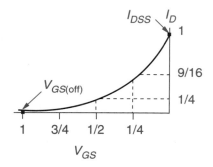

FIGURE 7–9 I_{DSS} versus V_{GS}

I_{DSS} and $V_{GS(off)}$ are not constant from one JFET to the next. This is illustrated in Figure 7–10 where three curves for the same type of JFET are graphed. The top curve uses the maximum values for I_{DSS} and $V_{GS(off)}$. The bottom curve uses the minimum values for I_{DSS} and $V_{GS(off)}$. The middle curve uses midrange values for I_{DSS} and $V_{GS(off)}$.

The graph in Figure 7–10 shows how the Q-point changes as the transconductance curve of the JFET changes. A bias circuit designed to maintain a constant bias voltage will cause large changes in drain current. A bias circuit designed to maintain a constant drain current causes the *gm* value to have large variations with changes in JFETs.

In the following sections, we will investigate three circuits that are used for biasing the JFET: self-biasing, voltage divider biasing, and source biasing. Self-biasing gives the best results. It maintains the drain current relatively constant, but more important, it holds the *gm* near a constant value.

SELF-BIASING

Figure 7–10 shows a line labeled *self-biasing*. This line intersects all three transconductance curves. When a JFET circuit is biased using the self-biasing method, the JFET is biased at the

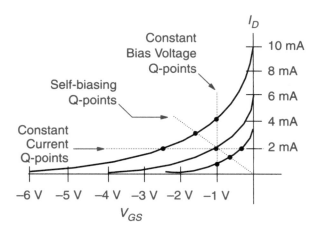

FIGURE 7–10 Bias considerations

intersection of this line and the transconductance curve. Notice that I_D is not constant for the three self-biasing Q-points; nevertheless, variations in I_D are considerably less than if the biased voltage (V_{GS}) is held constant.

Observe in Figure 7–10 that the slopes of the curves at each Q-point on the self-biasing line are approximately the same. Transconductance (gm) is equal to the slope of the curve at the Q-point; therefore, self-biasing maintains a nearly constant gm with changing transconductance curves. A constant gm will result in a constant voltage gain.

Figure 7–11(a) shows a JFET amplifier circuit using self-biasing. To analyze the biasing circuitry, a bias equivalent circuit is drawn as illustrated in Figure 7–11(b). This bias equivalent circuit is drawn by assuming all capacitors are open for bias currents.

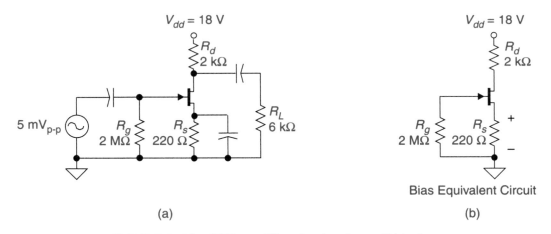

FIGURE 7–11 JFET amplifier circuit using self-biasing

Study the bias equivalent circuit in Figure 7–11(b). The source current (I_S) flowing through R_s causes a voltage drop across R_s, raising the voltage at the source above ground. There is virtually no current flowing through R_g and no voltage drop across R_g, so the gate is referenced to ground. The source is positive with respect to the gate, or stated differently, the gate is negative with respect to the source. The amount of bias voltage (V_{GS}) is a function of the size of R_s and the amount of source current. Using a transconductance curve for the JFET used in the circuit, bias voltage (V_{GS}) and drain current (I_D) can be calculated.

Figure 7–12 shows the transconductance curve for the JFET used in the circuit in Figure 7–11. The Q-point of the circuit can be determined by plotting R_s on the graph and locating the intersection of the R_s line and the transconductance curve. The origin is one point needed in order to plot R_s. A second point can be calculated by assuming a current through R_s and calculating the value of voltage across R_s. For our example, assume the current through R_s is equal to 10 mA. The voltage across R_s is then equal to 2.2 V (10 mA × 220 Ω = 2.2 V). The second point needed to plot R_s is at the intersection of 2.2 V and 10 mA on the graph. A line connecting the origin and this point is drawn. The Q-point is located at the intersection of the line drawn and the transconductance curve. Using the Q-point as a reference, the bias voltage (V_{GS}) and drain current (I_D) can be read from the graph. In this example, $V_{GS} = -1.25$ V and $I_D = 5.6$ mA.

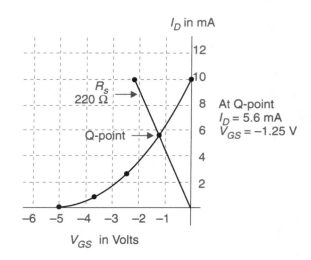

FIGURE 7–12   Transconductance curve with plot of R_s

Engineers must consider several factors when selecting the value of R_s. If midpoint biasing is the most important factor, the engineer will pick a value for R_s that will cause the quiescent drain voltage (V_D) to be halfway between V_{DD} and ground. Many circuits using JFETs have small output signal swings, and the engineer may trade off midpoint biasing in order to obtain a higher transconductance. Example 7.2 shows how easy the DC voltages can be calculated for a self-biasing JFET amplifier if the value of V_{GS} is known.

EXAMPLE 7.2

Calculate V_G, V_S, V_D, and V_{DS} for the circuit in Figure 7–13. (Assume $V_{GS} = -2$ V.)

Step 1. Determine the value of V_G. Since there is zero current in the gate leg, there is no voltage drop across the 250 kΩ resistor; therefore, the gate is at zero volts.

$$V_G = 0 \text{ V}$$

Step 2. Calculate V_S.

$$V_S = V_G - V_{GS} = 0 \text{ V} - (-2 \text{ V}) = 2 \text{ V}$$

Step 3. Calculate V_D.

$$I_D = I_S = V_S/R_s = 2 \text{ V}/1 \text{ k}\Omega = 2 \text{ mA}$$
$$V_{Rd} = I_D \times R_d = 2 \text{ mA} \times 2.7 \text{ k}\Omega = 5.4 \text{ V}$$
$$V_D = V_{DD} - V_{Rd} = 12 \text{ V} - 5.4 \text{ V} = 6.6 \text{ V}$$

Step 4. Calculate V_{DS}.

$$V_{DS} = V_D - V_S = 6.6 \text{ V} - 2 \text{ V} = 4.6 \text{ V}$$

EXAMPLE 7.2 continued

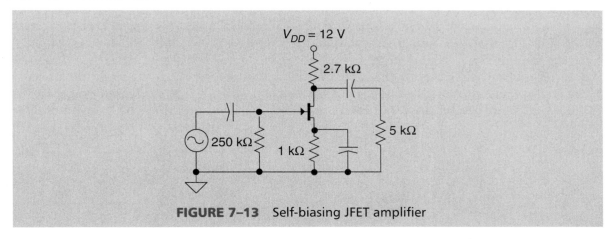

FIGURE 7–13 Self-biasing JFET amplifier

Self-biasing does not hold the drain constant, but it does a good job of maintaining the drain current (I_D) to a constant percentage of maximum drain current (I_{DSS}). If the actual value of V_{GS} is higher than the assumed value (–2 V), the drain current will increase. However, JFETs that have greater values of V_{GS} will also have higher values of I_{DSS}, so drain current maintains approximately the same percentage of maximum drain current (I_{DSS}).

VOLTAGE DIVIDER BIASING

Self-biasing is effective at maintaining a constant transconductance (gm), but voltage divider biasing is superior in maintaining a constant drain current (I_D). Voltage divider biasing is shown in Figure 7–14(a).

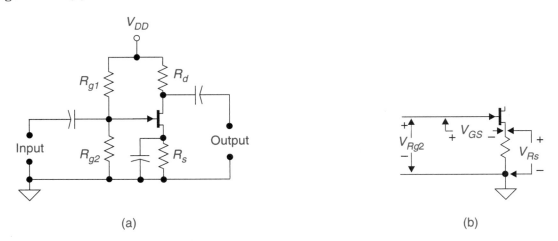

(a)   (b)

FIGURE 7–14 Voltage divider biasing

Voltage divider biasing used in JFET circuitry is similar to voltage divider biasing used for bipolar transistors. The idea is to hold the voltage on the gate constant and to make the voltage drop across resistor R_s large enough to swamp out changes in V_{GS}. Mathematically expressed, $V_{Rg2} = V_{GS} + V_{Rs}$. This formula can be explained by examining Figure 7–14(b). You can see that this is a Kirchhoff's loop equation for the loop made up of V_{Rg2}, V_{GS}, and V_{Rs}. If V_{Rg2} is constant and V_{Rs} is large in comparison to V_{GS}, variations in V_{GS} will not have much efect on the current through R_s. The current flowing through R_s is the quiescent current (I_D). This method works well in bipolar transistors because V_{BE} equals approximately 0.7 V, and changes in V_{BE} from device to device are normally less than 0.1 V. The situation is different in JFETs, however, because V_{GS} is normally in the range of 2 V, with variations of a volt or more from device to device. In order for voltage divider biasing to be effective in JFET circuitry, it is necessary to make the voltage at the junction of the voltage divider larger than in bipolar circuits. This is accomplished by selecting R_{g1} and R_{g2} so that a larger part of the power supply voltage is across R_{g2}. The ratio of R_{g1} and R_{g2} sets the voltage on the gate, but the actual values of R_{g1} and R_{g2} are not critical. Since no gate current flows, the values of these resistors can be in the megohm range. Example 7.3 illustrates how quiescent values can be calculated for a voltage divider biased amplifier.

EXAMPLE 7.3

Figure 7–15 shows a voltage divider biased JFET amplifier. Calculate the following quiescent values: I_D, V_G, V_S, V_D, and V_{GS}. (V_{GS} = –2 V is assumed.)

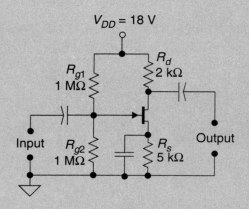

FIGURE 7–15   Voltage divider biased amplifier

$$V_G = V_{DD} \times R_{g2}/(R_{g1} + R_{g2}) = 18\ \text{V} \times 1\ \text{M}\Omega/(1\ \text{M}\Omega + 1\ \text{M}\Omega) = 9\ \text{V}$$
$$V_s = V_G - V_{GS} = 9\ \text{V} - (-2\ \text{V}) = 11\ \text{V}\quad (\text{VGS} = -2\ \text{V is assumed.})$$
$$I_D = I_S = V_S/R_s = 11\ \text{V}/5\ \text{k}\Omega = 2.2\ \text{mA}$$
$$V_{Rd} = I_D \times R_d = 2.2\ \text{mA} \times 2\ \text{k}\Omega = 4.4\ \text{V}$$
$$V_D = V_{DD} - V_{Rd} = 18\ \text{V} - 4.4\ \text{V} = 13.6\ \text{V}$$
$$V_{DS} = V_D - V_S = 13.6 - 11\ \text{V} = 2.6\ \text{V}$$

If the –2 V assumed for V_{GS} were a volt higher or lower, it would not have greatly affected the value of the drain current.

SOURCE BIASING

Source biasing uses a constant current source to hold the drain current (I_D) constant. The current source will maintain a constant drain current even with changes in devices. Figure 7–16 shows three means of obtaining constant current sources.

Figure 7–16(a) uses a high ohm value source resistor (R_s) connected to a large negative power supply voltage to obtain a constant source current. The source current in the circuit is determined by V_{ss}/R_s, if V_{ss} is much larger than variations in V_{GS}. The main disadvantage of this type of biasing is the need for a large negative power supply.

The circuits in Figures 7–16(b) and 7–16(c) use a negative supply voltage, but the magnitude of the negative supply can be much smaller than that required for the circuit in Figure 7–16(a).

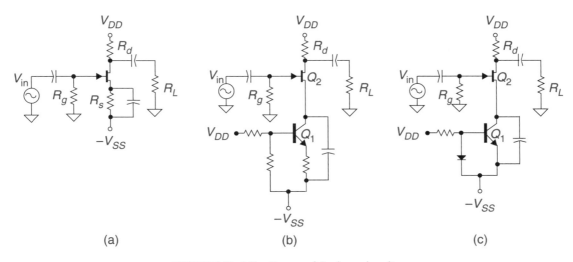

FIGURE 7–16 Source biasing circuits

The constant current source in circuit "b" consists of a voltage divider biased bipolar transistor circuit. The constant current source in circuit "c" is a current mirror consisting of a resistor, diode, and a bipolar transistor. In all three circuits, the current source holds the source current constant. Source current, however, always equals drain current ($I_S = I_D$); therefore, the drain current is constant. Example 7.4 shows how to calculate quiescent current and voltage for a JFET amplifier circuit using source biasing.

EXAMPLE 7.4

Calculate the following quiescent values for the circuit in Figure 7–17: I_D, V_D, V_S, V_G, and V_{DS}. (Assume $V_{GS} = -2$ V.)

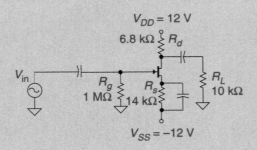

FIGURE 7–17 Source biasing JFET amplifier

$$V_G = 0 \text{ V (No current flow in gate leg.)}$$
$$V_S = V_G - V_{GS} = 0 \text{ V} - (-2 \text{ V}) = 2 \text{ V}$$
$$V_{Rs} = V_{SS} - V_S = -12 \text{ V} - 2 \text{ V} = -14 \text{ V}$$
$$I_S = V_{Rs}/R_s = 14 \text{ V}/14 \text{ k}\Omega = 1 \text{ mA}$$
$$I_S = I_D$$
$$V_{Rd} = I_D \times R_d = 1 \text{ mA} \times 6.8 \text{ k}\Omega = 6.8 \text{ V}$$
$$V_D = V_{DD} - V_{Rd} = 12 \text{ V} - 6.8 \text{ V} = 5.2 \text{ V}$$
$$V_{DS} = V_D - V_S = 5.2 \text{ V} - 2 \text{ V} = 3.2 \text{ V}$$

If the −2 V assumed for V_{GS} were a volt higher or lower, it would not have greatly affected the value of the drain current.

SUMMARY OF BIASING CIRCUITS

Source biasing is the best method for maintaining a constant drain current, and voltage divider biasing is the second best method for maintaining a constant drain current. However, in JFET amplifiers, it is usually more important to maintain a constant gm so the voltage gain of the amplifier will be constant. The circuit that does the best job of maintaining a constant gm is the self-biasing circuit. The self-biasing circuit also has the advantage of containing few components. For these reasons, most JFET amplifiers use self-biasing.

7.5   JFET SIGNAL PARAMETERS

VOLTAGE GAIN

Figure 7–18(a) shows a common-source JFET amplifier. Analyzing this circuit will give you an understanding of how the JFET can function as an amplifier. Figure 7–18(b) shows the signal equivalent circuit of the amplifier. The coupling and bypass capacitors and DC power supply (V_{DD}) appear as shorts to the signal current. This causes R_s to disappear and R_d and R_L to appear in parallel in the signal equivalent circuit. The input signal voltage (V_{in}) is applied directly to the gate.

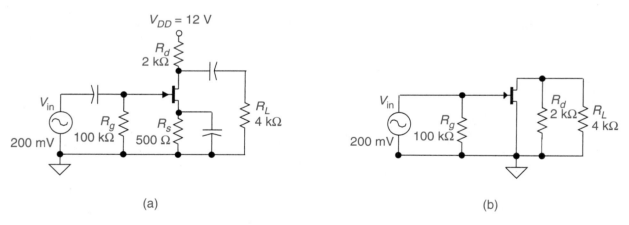

(a)   (b)

FIGURE 7–18   JFET amplifier circuit

In Figure 7–19, the JFET symbol is replaced with the function diagram of the JFET, and resistors R_D and R_L are combined into a parallel equivalent resistor, r_d. The input signal voltage at the gate controls the output current (i_d) flowing through the signal resistance (r_d) in the drain leg. The signal current through the signal drain resistance develops the output signal voltage (V_{out}).

Voltage gain by definition is V_{out}/V_{in}. The input signal (V_{in}) is connected between the gate and signal ground. Since the source is at signal ground, $V_{in} = v_{gs}$. The output signal (V_{out}) is equal to $\Delta i_d \times r_d$. Using this information and the formula for transconductance ($gm = \Delta i_d/\Delta v_{gs}$), the voltage gain formula for the JFET amplifier can be derived as follows:

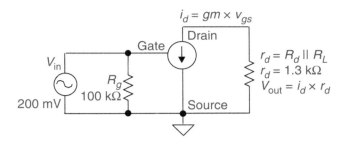

FIGURE 7–19 Signal equivalent circuit

$A_v = V_{out}/V_{in}$

$V_{in} = \Delta v_{gs}$

$V_{out} = \Delta i_d \times r_d$

Therefore:

$A_v = \Delta i_d \times r_d/\Delta v_{gs}$

Or:

$A_v = (\Delta i_d /\Delta v_{gs}) \times r_d$

$gm = \Delta i_d/\Delta v_{gs}$

Therefore:

$A_v = gm \times r_d$

In order for the voltage gain to equal $gm \times r_d$, the signal resistance in the source leg must equal zero. If there is signal resistance in the source leg (r_s), the formula for voltage gain becomes $A_v = r_d/(1/gm + r_s)$. There normally is a bypassed capacitor in the source leg that returns the source to signal ground. If the source bypass capacitor opens, the voltage gain of the circuit will be greatly reduced. Example 7.5 illustrates that the voltage gain of a JFET amplifier is easy to calculate if the transconductance (gm) is known.

EXAMPLE 7.5

Calculate the A_v and V_{out} for the circuit in Figure 7–18(a). Assume $gm = 3.6$ mS.

$$r_d = R_d \parallel R_L = 2\text{ k}\Omega \parallel 4\text{ k}\Omega = 1.3\text{ k}\Omega$$
$$A_v = gm \times r_d = 3.6\text{ mS} \times 1.3\text{ k}\Omega = 4.7$$
$$V_{out} = V_{in} \times A_v = 200\text{ mV} \times 4.7 = 940\text{ mV}$$

INPUT IMPEDANCE

The input impedance of a JFET amplifier equals the resistance the signal current sees looking into the amplifier. The junction between the gate and the channel is always reverse biased, so the input impedance is a function of the gate circuitry. The combination of resistors the signal current sees looking into the gate circuitry depends on the type of bias circuit. Figure 7–20 shows the circuits and formulas for z_{in} in self-biasing, voltage divider biasing, and constant current biasing.

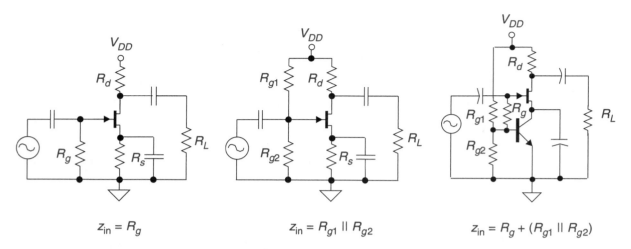

$$z_{in} = R_g \qquad\qquad z_{in} = R_{g1} \parallel R_{g2} \qquad\qquad z_{in} = R_g + (R_{g1} \parallel R_{g2})$$

FIGURE 7–20   Input impedance

OUTPUT IMPEDANCE

In order to obtain a value for signal output impedance, it is necessary to look back into the circuit from the output. By looking back from the output, there are two possible signal paths to ground [see Figure 7–21(a)]. One path is through R_d and the power supply to ground, and the other path is through the JFET to ground. These two paths are shown in Figure 7–21(b). The power supply is a short for the signal, so one path is equal to the impedance of R_d. The impedance through the JFET is equal to $\Delta v_d / \Delta i_{ds}$. In the active region, the change in drain current is very small in comparison to the change in drain voltage. Because of this, the impedance back through the JFET is in megohms, which is many times the normal value of R_d; hence, the output impedance is equal to the value of R_d for all common-source JFET amplifiers.

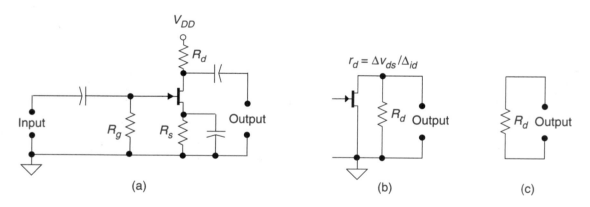

FIGURE 7–21   Output impedance

CURRENT AND POWER GAIN

Since the input impedance can be extremely high and the output impedance is equal to R_d, which is normally in the 1 kΩ to 10 kΩ range, the common-source JFET amplifier is capable of having high current gain. With a high current gain (A_i) and modest voltage gain (A_v), the common-source JFET amplifier has a high power gain (A_p).

7.6 ANALYZING JFET AMPLIFIER CIRCUITS

As a technician, it will be your job to look at the schematic of a circuit and make logical approximations of voltage readings, impedance, and gains. This section will cover procedures you can use in analyzing JFET amplifier circuits. The good news is JFET circuits are easy to analyze if you know two values: the gate-to-source voltage (V_{GS}) and the transconductance (gm) at the Q-point. The bad news is, the technician will almost never know the exact value of these two variables. So, let us talk about an approximation of these values.

The bias voltage between the gate and source (V_{GS}) normally is in the range of 0.25 V to 4 V. A good guess for troubleshooting purposes is 2 V. The polarity of V_{GS} depends on whether the JFET is an N-channel or a P-channel. (P-channel JFETs will be discussed later.) The first measurement a technician normally takes is V_{GS}. If the measured value of V_{GS} is within reason, the technician can use the measured value to calculate other circuit values.

Normally, the gm_0 of a JFET ranges from about 2 mS to 10 mS, so use 5 mS as an estimation for gm_0. The gm value is reduced from gm_0, depending on the Q-point of the amplifier. A good starting estimation of gm for troubleshooting purposes is 3 mS.

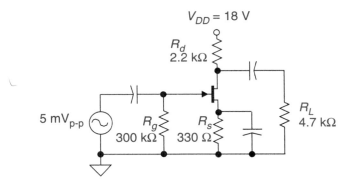

FIGURE 7–22   Self-biasing amplifier

We now have values for V_{GS} and gm. We will assume V_{GS} equals 2 V and gm equals 3 mS. These are guesses, but they should put us in the ballpark. When troubleshooting, just remember that actual V_{GS} and gm have ranges, so do not take our approximations as absolute values.

Figure 7–22 shows a self-biasing amplifier. The self-biasing amplifier is a popular circuit because of its simplicity and its ability to maintain a constant voltage gain. Using circuit components, you can roughly estimate DC voltage readings and signal values. Example 7.6 will demonstrate the procedure for estimating readings you would expect to observe with your test equipment.

EXAMPLE 7.6

Calculate V_G, V_S, V_D, V_{DS}, A_v, V_{out}, z_{in}, and z_{out} for the circuit in Figure 7–22 using the values of the components in the circuit.

Step 1. Determine V_G.

There is no gate current, so the voltage drop across R_g is zero, causing the gate to be referenced to ground.

$$V_G = 0 \text{ V}$$

EXAMPLE 7.6 continued

Step 2. Calculate V_S.

The voltage on the source will equal the gate voltage minus V_{GS} (Assume $V_{GS} = -2$ V.)

$$V_S = V_G - V_{GS} = 0 \text{ V} - (-2 \text{ V}) = 2 \text{ V}$$

Step 3. Calculate V_D.

With the value of V_S known, the source current can be calculated. The source current equals the drain current. With the drain current known, the voltage drop across R_d can be calculated. If the voltage drop across R_d is subtracted from the supply voltage, the value of V_D is obtained.

$$I_S = V_S/R_s$$

$$I_S = 2 \text{ V}/330 \ \Omega = 6 \text{ mA}$$

$$I_S = I_D = 6 \text{ mA}$$

$$V_{Rd} = I_D \times R_d$$

$$V_{Rd} = 6 \text{ mA} \times 2.2 \text{ k}\Omega = 13.2 \text{ V}$$

$$V_D = V_{DD} - V_{Rd}$$

$$V_D = 18 \text{ V} - 13.2 \text{ V} = 4.8 \text{ V}$$

Step 4. Calculate V_{DS}.

The voltage from drain to source is equal to the difference between V_D and V_S.

$$V_{DS} = V_D - V_S$$

$$V_{DS} = 4.8 \text{ V} - 2 \text{ V} = 2.8 \text{ V}$$

Step 5. Calculate A_v. The voltage gain is equal to gm times the signal resistance in the drain leg. There are two paths to signal ground from the drain terminal. One is through R_d, and the other is through R_L.

$$r_d = R_d \parallel R_L$$

$$r_d = 2.2 \text{ k}\Omega \parallel 4.7 \text{ k}\Omega = 1.5 \text{ k}\Omega$$

$$A_v = gm \times r_d$$

$$A_v = 3 \text{ mS} \times 1.5 \text{ k}\Omega = 4.5 \text{ (Assume } gm = 3 \text{ mS.)}$$

Step 6. Calculate V_{out}.

$$V_{\text{out}} = V_{\text{in}} \times A_v$$

$$V_{\text{out}} = 5 \text{ mV} \times 4.5 = 22.5 \text{ mV}$$

Step 7. Calculate z_{in} and z_{out}.

$$z_{\text{in}} = R_g = 300 \text{ k}\Omega$$

$$z_{\text{out}} = R_d = 2.2 \text{ k}\Omega$$

The two assumptions, $V_{GS} = -2$ V and $gm = 3$ mS, used in Example 7.5 made the obtained results rough approximations. A technician can use these approximations to evaluate the operation of JFET amplifiers. Since these are only approximations, the actual values read may vary considerably. As you make circuit readings, keep asking yourself, "does the value read make sense?" If the value read seems to be logical, move to the next reading. If the value read seems wrong, determine what could cause the incorrect reading.

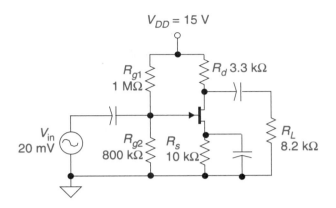

FIGURE 7–23   Voltage divider biased amplifier

Figure 7–23 illustrates a voltage divider biased amplifier. One advantage the voltage divider biased amplifier has over the self-biasing amplifier is that calculated DC circuit values will be close to the values read. The voltage gain will still be a rough approximation.

EXAMPLE 7.7

Calculate V_G, V_S, V_D, V_{DS}, A_v, V_{out}, z_{in}, and z_{out} for the circuit in Figure 7–23 using the values of the components in the circuit.

Step 1. Calculate the values of DC voltages V_G, V, V_S, V_D, and V_{DS}.

$$V_G = V_{DD} \times R_{g2}/(R_{g1} + R_{g2}) = 15 \text{ V} \times 800 \text{ k}\Omega/(1 \text{ M}\Omega + 800 \text{ k}\Omega) = 6.7 \text{ V}$$
$$V_S = V_G - V_{GS} = 6.7 \text{ V} - (-2 \text{ V}) = 8.7 \text{ V} \ (V_{GS} = -2 \text{ V is assumed.})$$
$$I_D = I_S = V_S/R_s = 8.7 \text{ V}/10 \text{ k}\Omega = 0.87 \text{ mA}$$
$$V_{Rd} = I_D \times R_d = 0.87 \text{ mA} \times 3.3 \text{ k}\Omega = 2.9 \text{ V}$$
$$V_D = V_{DD} - V_{Rd} = 15 \text{ V} - 2.9 \text{ V} = 12.1 \text{ V}$$
$$V_{DS} = V_D - V_S = 12.1 \text{ V} - 8.7 \text{ V} = 3.4 \text{ V}$$

Step 2. Calculate A_v.

$$r_d = R_d \parallel R_L$$
$$r_d = 3.3 \text{ k}\Omega \parallel 8.2 \text{ k}\Omega = 2.4 \text{ k}\Omega$$
$$A_v = gm \times r_d$$
$$A_v = 3 \text{ mS} \times 2.4 \text{ k}\Omega = 7.2 \text{ (Assume } gm = 3 \text{ mS.)}$$

Step 3. Calculate V_{out}.

$$V_{out} = V_{in} \times A_v$$
$$V_{out} = 20 \text{ mV} \times 7.2 = 144 \text{ mV}$$

Step 4. Calculate z_{in} and z_{out}.

$$z_{in} = R_{g1} \parallel R_{g2} = 444 \text{ k}\Omega$$
$$z_{out} = R_d = 3.3 \text{ k}\Omega$$

Figure 7–24 shows a source biased circuit using a current mirror as a constant source current. The calculated DC circuit values will always be close to values read. The voltage gain will still be a rough approximation.

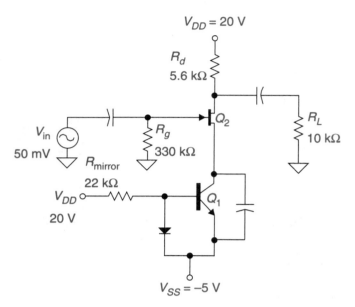

FIGURE 7–24   Source biased circuit using current mirror

EXAMPLE 7.8

Calculate V_G, V_S, V_D, V_{DS}, A_v, V_{out}, z_{in}, and z_{out} for the circuit in Figure 7–24 using the values of the components in the circuit.

Step 1. Determine V_G. There is no gate current, so the voltage drop across R_g is zero, causing the gate to be referenced to ground.

$$V_G = 0 \text{ V}$$

Step 2. Calculate V_S. The voltage on the source will equal the gate voltage minus V_{GS}. (Assume $V_{GS} = -2$ V.)

$$V_S = V_G - V_{GS} = 0 \text{ V} - (-2 \text{ V}) = 2 \text{ V}$$

Step 3. Calculate I_S. The source current is set by the current mirror output. The current output of the current mirror equals the current flowing through the diode.

$$I_{diode} = (V_{DD} - V_D - V_{SS})/R_{mirror} = (20 - 0.7 - (-5 \text{ V}))/22 \text{ k}\Omega = 1.1 \text{ mA}$$

$$I_s = I_{diode} = I_D = 1.1 \text{ mA}$$

Step 4. Calculate V_D.

$$V_{Rd} = I_D \times R_d = 1.1 \text{ mA} \times 5.6 \text{ k}\Omega = 6.2 \text{ V}$$
$$V_D = V_{DD} - V_{Rd} = 20 \text{ V} - 6.2 \text{ V} = 13.8 \text{ V}$$

Step 5. Calculate V_{DS}.

$$V_{DS} = V_D - V_s = 13.8 \text{ V} - 2 \text{ V} = 11.8 \text{ V}$$

Step 6. Calculate A_v.

$$r_d = R_d \parallel R_L$$
$$r_d = 5.6 \text{ k}\Omega \parallel 10 \text{ k}\Omega = 3.6 \text{ k}\Omega$$
$$A_v = gm \times r_d$$
$$A_v = 3 \text{ mS} \times 3.6 \text{ k}\Omega = 10.8 \text{ (Assume } gm = 3 \text{ mS.)}$$

EXAMPLE 7.8 continued

Step 7. Calculate V_{out}.

$$V_{\text{out}} = V_{\text{in}} \times A_v$$
$$V_{\text{out}} = 50 \text{ mV} \times 10.8 = 540 \text{ mV}$$

Step 8. Calculate z_{in} and z_{out}.

$$z_{\text{in}} = R_g = 330 \text{ k}\Omega$$
$$z_{\text{out}} = R_d = 5.6 \text{ k}\Omega$$

7.7 COMMON–DRAIN AMPLIFIER

Like bipolar transistors, JFETs can be used in different configurations. The JFET circuits discussed in this chapter so far have used common-source configurations. The common-source is the most popular configuration. It is possible, however, to use JFETs in the common-drain and the common-gate configurations. The common-gate configuration is sometimes used in high frequency circuits but is limited in use at lower frequencies. We will not discuss the common-gate configuration. The common-drain configuration, which is also known as the source follower, is used as a buffer amplifier and will be covered in this section.

The purpose of a buffer amplifier is to prevent a low impedance load from loading a signal source. In order to do this, the buffer must have a high input impedance and a relatively low output impedance. The common-drain circuit (shown in Figure 7–25) has a high input impedance, a relatively low output impedance, and a voltage gain of slightly less than one.

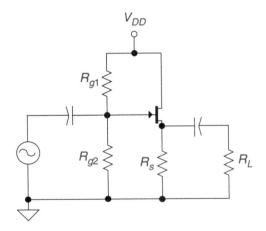

FIGURE 7–25 Common-drain circuit

VOLTAGE GAIN OF THE COMMON-DRAIN AMPLIFIER

The voltage gain of the common-drain amplifier is approximately equal to unity and can be derived as follows:

$$V_{\text{out}} = \Delta i_s \times r_s \, (r_s = R_s \parallel R_L)$$
$$gm = \Delta i_d / \Delta v_{gs}$$

Or:

$$\Delta i_d = \Delta v_{gs} \times gm$$
$$\Delta i_d = \Delta i_s$$

Therefore:

$$\Delta i_s = \Delta v_{gs} \times gm$$

Substituting $\Delta v_{gs} \times gm$ for Δi_s:

$$V_{out} = \Delta v_{gs} \times gm \times r_s$$
$$V_{in} = \Delta v_{gs} + V_{out} \text{ (Kirchhoff's law)}$$

Substituting $\Delta i_s \times r_s$ for V_{out}:

$$V_{in} = \Delta v_{gs} + (\Delta i_s \times r_s)$$

Substituting $\Delta v_{gs} \times gm$ for Δi_s:

$$V_{in} = \Delta v_{gs} + ((\Delta v_{gs} \times gm) \times r_s)$$

Factoring:

$$V_{in} = \Delta v_{gs} \times (1 + gm \times r_s)$$
$$A_v = V_{out}/V_{in}$$
$$A_v = \Delta v_{gs} \times gm \times r_s/\Delta v_{gs} \times (1 + gm \times r_s)$$
$$A_v = gm \times r_s/(1 + gm \times r_s)$$

Multiplying by $1/gm$ gives:

$$A_v = r_s/(1/gm + r_s)$$

If:

$$r_s \gg 1/gm$$

Then:

$$A_v = r_s/r_s$$

Or:

$$A_v = 1$$

In most cases, the engineer designs the circuit so the value of r_s is made much larger than $1/gm$. Therefore, the technician can consider the voltage gain of the common drain to be near unity. The output voltage is in phase and the same magnitude as input voltage. For this reason, the circuit is often called a source follower.

INPUT AND OUTPUT IMPEDANCE OF THE COMMON-DRAIN AMPLIFIER

The input impedance of the common-drain circuit in Figure 7–25 is easy to calculate. The circuit is using voltage divider biasing. The gate is reverse biased, so input impedance is equal to the parallel equivalent of R_{g1} and R_{g2}.

The output impedance can be analyzed by looking back into the circuit from the load. Since the circuit has unity voltage gain, the signal present on the source is equal in magnitude to the signal present on the gate ($v_s = v_g$). The signal current flow in the source is equal to the signal drain current ($i_s = i_d$). Therefore, looking back from the load, output impedance appears as v_g/i_d, which equals the reciprocal of gm ($1/gm$). Since gm is normally in the range of 1 mS to 10 mS, the output is in the range of 1 kΩ to 100 Ω. This is relatively low impedance when you consider the input impedance of the common drain can be over 1 MΩ.

SUMMARY OF THE COMMON-DRAIN AMPLIFIER

Looking at the equivalent circuit of the common-drain circuit in Figure 7–26 will help you summarize the important facts. It fulfills the requirements of a buffer circuit with a high input impedance and low output impedance. The input equivalent circuit in Figure 7–26(a) shows that the input impedance is equal to $R_{g1} \| R_{g2}$. Since the gate current is approximately zero, the values of R_{g1} and R_{g2} can be made extremely high. The output equivalent circuit in Figure 7–26(b) shows that the voltage gain is equal to one, and the output impedance is equal to $1/gm$, which is normally under 1 kΩ. Example 7.9 illustrates the value of using a common-drain as a buffer amplifier. The calculations are easy if the approximations developed in this section are used.

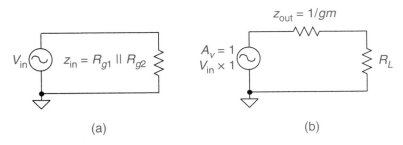

(a)   (b)

FIGURE 7–26 Common-drain equivalent circuits

EXAMPLE 7.9

Calculate the signal voltage (V_{out}) across the load and the power delivered to the load for the circuit shown in Figure 7–27(a). Calculate $V_G, V_S, V_D, A_v, z_{in}, z_{out}$ and the signal voltage (V_{out}) across the load, and the power delivered to the load for the circuit shown in Figure 7–27(b). (Assume $gm = 3$ mS and $V_{GS} = -2$ V.)

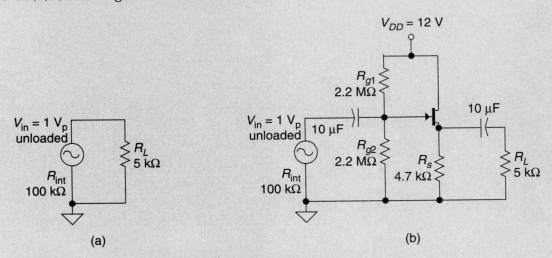

(a)   (b)

FIGURE 7–27 Common-drain circuit

Step 1. Calculate V_{out} for the circuit in Figure 7–27(a).
$$V_{out} = V_{in} \times R_L/(R_{int} + R_L) = 1 \text{ V}_p \times 5 \text{ k}/(100 \text{ k} + 5 \text{ k}) = 48 \text{ mV}_p$$
Step 2. Calculate the signal P_{out} for the circuit in Figure 7–27(a).
$$v_{rms} = v_p \times 0.707$$
$$v_{rms} = 48 \text{ mV}_p \times 0.707 = 34 \text{ mV}_{rms}$$
$$P_{out} = (V_{rms})^2/R_L$$
$$P_{out} = (34 \text{ mV})^2/5 \text{ k}\Omega = 0.2 \text{ }\mu\text{W}$$

EXAMPLE 7.9 continued

Step 3. Calculate the DC voltages V_G, V_S, V_D, and V_{DS} for the circuit in Figure 7–27(b).

$$V_G = V_{DD} \times (R_{g2}/(R_{g1} + R_{g2})) = 12\text{ V }(2.2\text{ M}\Omega/(2.2\text{ M}\Omega + 2.2\text{ M}\Omega)) = 6\text{ V}$$

$$V_S = V_G - V_{GS} = 6\text{ V} - (-2)\text{ V} = 8\text{ V (Assume }-2\text{ V for }V_{GS}.)$$

$$V_D = V_{DD} = 12\text{ V}$$

$$V_{DS} = V_D - V_S = 12\text{ V} - 8\text{ V} = 4\text{ V}$$

Step 4. Calculate the input impedance for the circuit in Figure 7–27(b).

$$z_{\text{in}} = R_{g1} \parallel R_{g2} = 2.2\text{ M}\Omega \parallel 2.2\text{ M}\Omega = 1.1\text{ M}\Omega$$

Step 5. Calculate the output impedance for the circuit in Figure 7–27(b). (Assume $gm = 3$ mS.)

$$z_{\text{out}} = 1/gm = 1/3\text{ mS} = 333\ \Omega$$

Step 6. Calculate the output signal voltage for the circuit in Figure 7–27(b).

$$V_{\text{out}} = V_{\text{in}} \times A_v$$

$$V_{\text{out}} = 1\text{ V}_{\text{p}} \times 1 = 1\text{ V}_{\text{p}}$$

Step 7. Draw the input and output equivalent circuits for the circuit in Figure 7–27(b). (The equivalent circuits with values calculated are shown in Figure 7–28.)

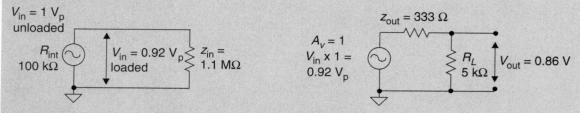

FIGURE 7–28   Equivalent circuits

Step 8. Calculate the input signal voltage. The signal source will be loaded slightly by the input impedance of the circuit.

$$V_{\text{in}} = V_{\text{in(unloaded)}} \times (z_{\text{in}}/(R_{\text{int}} + z_{\text{in}})) = 1\text{ V}_{\text{p}} \times (1.1\text{ M}/(0.1\text{ M} + 1.1\text{ M})) = 0.92\text{ V}_{\text{p}}$$

Step 9. Calculate the output signal voltage. The signal will be loaded slightly by the resistance of the load.

$$V_{\text{out}} = V_{\text{in}} \times R_L/(z_{\text{out}} + R_L) = 0.92\text{ V}_{\text{p}} \times 5\text{ k}\Omega/(333\ \Omega + 5\text{ k}\Omega) = 0.86\text{ V}_{\text{p}}$$

Step 10. Calculate the signal P_{out} for the circuit in Figure 7–27(b).

$$v_{\text{rms}} = v_p \times 0.707$$

$$v_{\text{rms}} = 0.86\text{ V}_{\text{p}} \times 0.707 = 0.608\text{ V}_{\text{rms}}$$

$$P_{\text{out}} = (V_{\text{rms}})^2/R_L$$

$$P_{\text{out}} = (0.608\text{ V})^2/5\text{ k}\Omega = 74\ \mu\text{W}$$

In circuit "a" in Example 7.9, only 0.2 μW was delivered to the load from the high impedance source. By using the JFET buffer amplifier between the high impedance source and the load [Figure 7–27(b)], 74 μW is delivered to the load. By using the JFET buffer amplifier, the power delivered to the load is increased by a factor of 370.

Higher input impedance can be obtained by using JFET common-drain amplifiers rather than bipolar common-collector amplifiers. Another advantage of JFET circuits is less internal noise. Bipolar transistors depend on the recombining of electrons and holes at the PN junction for operation. This action causes random noise in current flow through the bipolar device. When signal levels are low, this noise can mask out signal currents. JFET devices do not have this recombining of electrons and holes and, therefore, have much less internal noise. This makes JFETs desirable as input devices where signal levels are often small. JFET common-drain amplifiers often are used as the input stage of communication and test equipment to prevent circuit loading and provide a front end with low internal noise. They also are used to isolate transducers and oscillators so their operation will not be affected by loading.

7.8 N-CHANNEL AND P-CHANNEL JFETs

The JFETs discussed in this chapter have been N-channel JFETs. The channel connecting the source to the drain is made of N-type material. If the channel is made of P-type material, the device is a P-channel JFET. The gate region of a P-channel JFET is made of N-type material. The symbols for both types of JFETs are shown in Figure 7–29. By looking at the symbols, you can tell if the device is an N-channel or P-channel JFET. The direction of the arrow in the symbol always points toward the N-type material. So, if the arrow points toward the channel, the device is an N-channel device, and if the arrow points away from the channel, the device is a P-channel device. The bias voltage will be of opposite polarity when the circuit is designed to operate using P-channel devices. N-channel devices are most popular, but P-channel JFETs are used, and some circuits will use both N-channel and P-channel devices. The calculations for all the circuits are the same; the only difference is the polarity of the voltages. Example 7.10 calculates the circuit values for a P-channel self-biasing common-source amplifier.

N-channel   P-channel

FIGURE 7–29   JFET symbols

EXAMPLE 7.10

Calculate V_G, V, V_S, V_D, V_{DS}, A_v, V_{out}, z_{in}, and z_{out} for the circuit in Figure 7–30 using the values of the components in the circuit.

Step 1. Determine V_G.

There is no gate current, so the voltage drop across R_g is zero.

$$V_G = 0 \text{ V}$$

Step 2. Calculate V_S.

The voltage on the source will equal the gate voltage minus V_{GS}. (Assume $V_{GS} = 2$ V P-channel device.)

$$V_S = V_G - V_{GS} = 0 \text{ V} - 2 \text{ V} = -2 \text{ V}$$

Step 3. Calculate V_D.

$$I_S = V_s / R_s$$

EXAMPLE 7.10 continued

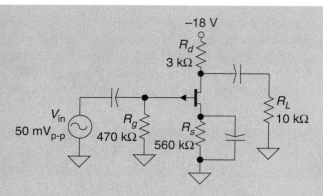

FIGURE 7–30   P-channel self-biasing amplifier

$$I_S = -2 \text{ V}/560 \ \Omega = -3.57 \text{ mA}$$
$$I_S = I_D = -3.57 \text{ mA}$$
$$V_{Rd} = I_D \times R_d$$
$$V_{Rd} = -3.57 \text{ mA} \times 3 \text{ k}\Omega = -10.7 \text{ V}$$
$$V_D = V_{DD} - V_{Rd}$$
$$V_D = -18 \text{ V} - (-10.7 \text{ V}) = -7.3 \text{ V}$$

Step 4. Calculate V_{DS}.

The voltage from drain to source is equal to the difference between V_D and V_S.

$$V_{DS} = V_D - V_S$$
$$V_{DS} = -7.3 \text{ V} - (-2 \text{ V}) = -5.3 \text{ V}$$

Step 5. Calculate A_v.

The voltage gain is equal to gm times the signal resistance in the drain leg. There are two paths to signal ground from the drain terminal. One is through R_d, and the other is through R_L.

$$r_d = R_d \parallel R_L$$
$$r_d = 3 \text{ k}\Omega \parallel 10 \text{ k}\Omega = 2.3 \text{ k}\Omega$$
$$A_v = gm \times r_d$$
$$A_v = 3 \text{ mS} \times 2.3 \text{ k}\Omega = 6.9$$

Step 6. Calculate V_{out}.

$$V_{out} = V_{in} \times A_v$$
$$V_{out} = 50 \text{ mV} \times 6.9 = 345 \text{ mV}$$

Step 7. Calculate z_{in} and z_{out}.

$$z_{in} = R_g = 470 \text{ k}\Omega$$
$$z_{out} = R_d = 3 \text{ k}\Omega$$

7.9 JFET SWITCHING CIRCUITS

JFETs can be used as analog switching devices. Figure 7–31 shows a JFET used as a series switch. The load is connected between the source and ground. The input signal is connected to the drain, and the control signal is connected to the gate. When the JFET switch is turned on, the JFET operates in the ohmic region of the characteristic curve and acts like a small resistor in series with the load. Switching JFETs will have specifications called $R_{DS(on)}$. This is the maximum resistance between the drain and the source when the device is turned on. The $R_{DS(on)}$ for the JFET in Figure 7–31 is 50 Ω. This means that when the JFET is on, its resistance will not exceed 50 Ω. When the JFET is cut off, its resistance will be in megohms. If a signal is applied to the gate that switches between 0 V and $V_{GS(off)}$, the signal to the load will be cut on and off as shown in Figure 7–31. For small signals (less than a few hundred mV), no DC bias voltage is needed. For larger signals, however, the JFET must be biased so the gate-to-channel junction is never forward biased.

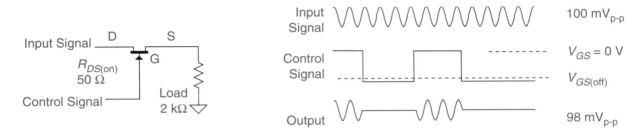

FIGURE 7–31   Series switching circuit

Figure 7–32 shows a JFET shunt switching circuit. The JFET is in parallel with the load. When the JFET is on, the input signal is dropped across resistor R_s, and the signal to the load is near zero. If the JFET is cut off by applying a negative control signal to the gate, a voltage divider is formed between R_s and the load. Since R_s is much smaller than the load, most of the input signal will be delivered to the load (Figure 7–32).

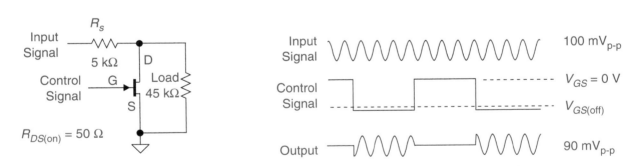

FIGURE 7–32   Shunt switching circuit

7.10   TROUBLESHOOTING JFET CIRCUITS

When troubleshooting a JFET amplifier, the first question is, what should the DC voltages be? The DC voltage on the gate usually can be quickly calculated. A shorted JFET could cause the gate voltage to be incorrect, but an open JFET would not show up by measuring the gate voltage, since the gate is effectively an open anyway.

The source voltage should be approximately 2 V (V_{GS} = 0.25 to 4 V) different from the voltage measured on the gate. With the source voltage known, the source current can be quickly determined. The source current equals the drain current. The drain current can be used to quickly determine the voltage at the drain. A high voltage reading on the drain indicates no current flow, and possibly an open JFET.

If the DC measurements are wrong, the circuit has problems and repairs can be made accordingly. If the DC readings are correct, however, the circuit may still have signal problems, and signal insertion and tracing is the next step.

Input and output impedances are easy to calculate and can be used to estimate loading between stages. The voltage gain per stage is another matter. In order to calculate the voltage gain, the value of gm must be known, but it is usually an unknown to the technician in the field. The value of gm_0 usually ranges from 1 mS to 10 mS; therefore, a good approximation of gm for troubleshooting purposes is 3 mS. Voltage gain can be approximated by using the formula 3 mS × r_d. The exact voltage gain will not be known. By making a reading of the signal on the input and comparing it to the output signal, a reasonable judgment can be made of the performance of the device.

In field service, it is typically necessary to troubleshoot equipment with less than complete information. Even a schematic is sometimes considered a luxury! The important point here is that general knowledge of how a JFET works will enable the technician to troubleshoot circuits containing a JFET. In every instance, the art of troubleshooting consists of first deciding (calculating or estimating) what values you expect to find, and then verifying those values. When your measurements do not agree with your predictions, you have either located the area of the problem or made an erroneous prediction. Experience will minimize erroneous predictions.

JFET Amplifiers

1. Calculate and record the DC voltages and signal values for the circuit in Figure 7–33.

   $V_G =$ _____ $V_S =$ _____ $V_D =$ _____ $V_{DS} =$ _____

   $z_{in} =$ _____ $z_{out} =$ _____ $A_v =$ _____ $V_{out} =$ _____

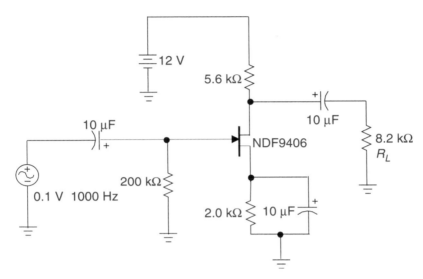

FIGURE 7–33

2. Simulate and record the DC voltages and signal values for the circuit in Figure 7–33.

   $V_G =$ _____ $V_S =$ _____ $V_D =$ _____

   $V_{DS} =$ _____ $A_v =$ _____ $V_{out} =$ _____

3. Calculate and record the DC voltages and signal values for the circuit in Figure 7–34.

   $V_G =$ _____ $V_S =$ _____ $V_D =$ _____ $V_{DS} =$ _____

   $z_{in} =$ _____ $z_{out} =$ _____ $A_v =$ _____ $V_{out} =$ _____

4. Simulate and record the DC voltages and signal values for the circuit in Figure 7–34.

   $V_G =$ _____ $V_S =$ _____ $V_D =$ _____

   $V_{DS} =$ _____ $A_v =$ _____ $V_{out} =$ _____

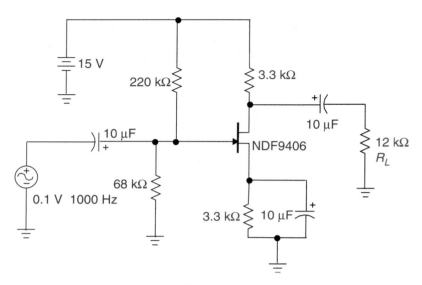

FIGURE 7–34

JFET Amplifiers

I. Objective

To design and construct a JFET amplifier.

II. Test Equipment

    (1) sine wave generator   (1) dual-trace oscilloscope

    (1) DC power supply

III. Components

    Transistors: (1) MPF102 JFET

    Resistors: (1) 1.5 kΩ, (2) 4.7 kΩ, (1) 10 kΩ, (1) 100 kΩ, (1) 330 kΩ

    Capacitors: C_1, C_2, and $C_3 = 10$ µF

IV. Procedure

1. Calculate the DC voltages and signal values for the circuit in Figure 7–35. Calculated values:

    $V_G =$ _____ $V_S =$ _____ $V_D =$ _____ $V_{DS} =$ _____

    $z_{in} =$ _____ $z_{out} =$ _____ $A_v =$ _____ $V_{out} =$ _____

2. Construct the circuit in Figure 7–35 and measure the DC voltages and signal values. Measured values:

    $V_G =$ _____ $V_S =$ _____ $V_D =$ _____ $V_{DS} =$ _____

    $z_{in} =$ _____ $z_{out} =$ _____ $A_v =$ _____ $V_{out} =$ _____

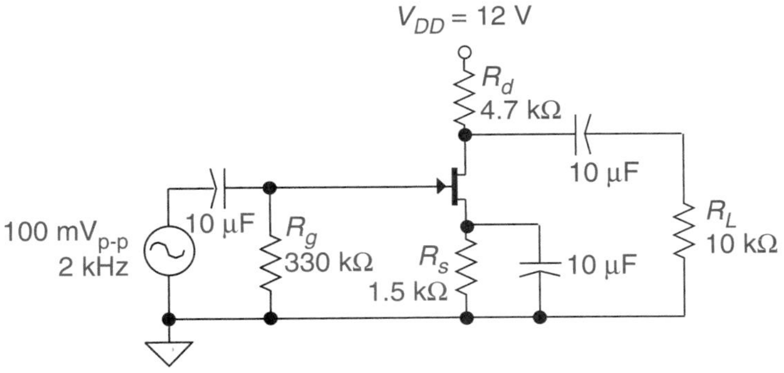

FIGURE 7–35

3. Calculate the DC voltages and signal values for the circuit in Figure 7–36. Calculated values:

    $V_G =$ _____ $V_S =$ _____ $V_D =$ _____ $V_{DS} =$ _____

    $z_{in} =$ _____ $z_{out} =$ _____ $A_v =$ _____ $V_{out} =$ _____

4. Construct the circuit in Figure 7–36 and measure the DC voltages and signal values. Measured values:

    $V_G =$ _____ $V_S =$ _____ $V_D =$ _____ $V_{DS} =$ _____

    $z_{in} =$ _____ $z_{out} =$ _____ $A_v =$ _____ $V_{out} =$ _____

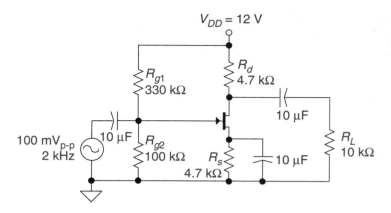

FIGURE 7–36

V. Points to Discuss

1. How could the voltage gain of the circuit in Figure 7–35 be increased?

2. How could the input impedance of the circuit in Figure 7–35 be increased?

3. How could the voltage gain of the circuit in Figure 7–36 be increased?

4. How could the input impedance of the circuit in Figure 7–36 be increased?

QUESTIONS

7.1 Introduction to JFETs

____ 1. The three terminals of the JFET are referred to as the ____.
   a. source, gate, and drain
   b. emitter, base, and collector
   c. emitter, drain, and base
   d. source, base, and drain

____ 2. Which of the following is the schematic symbol for an N-channel JFET?

____ 3. The current through the channel of a JFET is controlled by the amount of ____.
   a. forward bias between the gate and source
   b. forward bias between the gate and drain
   c. reverse bias between the gate and channel
   d. gate current flowing through the gate/source junction

____ 4. The drain current of an N-channel JFET will decrease when ____.
   a. the gate voltage changes in a negative direction
   b. the gate voltage changes in a positive direction
   c. the gate current increases
   d. the gate current decreases

____ 5. Some JFETs are constructed in a symmetrical manner, making it possible to interchange their ____.
   a. gate and drain leads
   b. gate and source leads
   c. source and drain leads
   d. two leads

7.2 The JFET Versus the Bipolar Transistor

____ 6. JFETs are different from bipolar transistors because they ____.
   a. can only amplify voltage
   b. can only amplify current
   c. are voltage-controlled devices
   d. have a low input impedance

____ 7. The gain of a bipolar transistor device is ____ and the gain of a JFET device is ____.
   a. beta, transconductance
   b. transconductance, beta
   c. beta, beta
   d. transconductance, transconductance

7.3 JFET Characteristics

____ 8. The drain current with the gate voltage equal to zero is _____.
  a. i_d
  b. I_D
  c. I_{DSS}
  d. none of the above

____ 9. The gate-to-source voltage that stops drain current from flowing is _____.
  a. V_{GS}
  b. $V_{GS(off)}$
  c. $V_{GS(stop)}$
  d. V_p

____ 10. The formula $\Delta i_d / \Delta v_{gs}$ is the mathematical expression for _____.
  a. I_{DSS}
  b. $V_{GS(off)}$
  c. $V_{GS(stop)}$
  d. gm

____ 11. The _____ voltage is the drain-to-source voltage where the drain current becomes approximately constant.
  a. pinch-off
  b. cutoff
  c. saturation
  d. constant rate

____ 12. Transconductance when gate-to-source voltage equals zero is___.
  a. gm_0
  b. βgm
  c. $gm_{(gate)}$
  d. gm

7.4 Biasing the Common-Source JFET Amplifier

____ 13. What type of biasing JFET circuit is not discussed in this chapter?
  a. voltage divider biasing
  b. voltage-feedback biasing
  c. source biasing
  d. self-biasing

____ 14. Self-biasing circuits require a minimum number of components and also have the advantage of _____.
  a. holding drain current constant with changes in JFET characteristics
  b. holding gate current constant with changes in JFET characteristics
  c. holding V_{GS} constant with changes in JFET characteristics
  d. holding gm constant with changes in JFET characteristics

____ 15. What type of JFET biasing circuit is best at holding the drain constant with changes in JFET characteristics?

  a.   voltage divider biasing

  b.   voltage-feedback biasing

  c.   source biasing

  d.   self-biasing

7.5 JFET Signal Parameters

____ 16. The signal voltage gain of a JFET amplifier is equal to ____.

  a. $\beta \times r_d$

  b. r_d / r_s

  c. r_s / r_d

  d. $gm \times r_d$

____ 17. A decrease in load resistance (R_L) will cause _____ in the voltage gain of the JFET amplifier.

  a.   a decrease

  b.   no change

  c.   a small increase

  d.   a big increase

____ 18. The input impedance of the common-source JFET amplifier is _____.

  a.   normally higher than the input impedance of a bipolar transistor amplifier

  b.   equal to the signal impedance to ground for the gate terminal

  c.   easily adjusted by changing the resistors connected to the gate terminal

  d.   all of the above

____ 19. The output impedance of the common-source JFET amplifier is equal to _____.

  a. R_L

  b. R_d

  c. $R_L \| R_d$

  d. R_d / R_s

7.6 Analyzing JFET Amplifier Circuits

____ 20. If _____ and _____ are known, analyzing the JFET amplifier is an easy task.

  a. $I_{DSS}, V_{GS(off)}$

  b. gm, I_{DSS}

  c. gm, V_{GS}

  d. $V_{GS}, V_{GS(off)}$

____ 21. A reasonable approximation of V_{GS} for analyzing JFET circuits is _____.

  a.   0 V

  b.   0.3 V

  c.   0.7 V

  d.   2 V

_____ 22. A reasonable approximation of *gm* for analyzing JFET circuits is _____.
    a. 100 mS
    b. 10 mS
    c. 3 mS
    d. 0.2 mS

7.7 Common-Drain Amplifier

_____ 23. The voltage gain of the common-drain amplifier is approximately _____.
    a. 100
    b. 10
    c. 3
    d. 1

_____ 24. The common-drain amplifier has a _____.
    a. high input impedance and low output impedance
    b. low input impedance and high output impedance
    c. low input impedance and low output impedance
    d. high input impedance and high output impedance

7.8 N-Channel and P-Channel JFETs

_____ 25. Which of the following is the schematic symbol for a P-channel JFET?

7.9 JFET Switching Circuits

_____ 26. When the JFET switch is on, the resistance between the drain and source is less than _____.
    a. $1 \, \Omega$
    b. $R_{DS(on)}$
    c. R_d
    d. $10 \, \Omega$

_____ 27. Series JFET switches are not perfect, but as long as the load resistance is much larger than the _____ resistance, the switch will function properly.
    a. R_g
    b. $R_{DS(on)}$
    c. R_d
    d. $R_{DS(off)}$

_____ 28. In a shunt JFET switching circuit, the JFET is _____.
    a. in series with the load and in parallel with R_s
    b. in series with R_s and in parallel with the load
    c. in series with the load and in series with R_s
    d. in parallel with the load and in parallel with R_s

7.10 Troubleshooting JFET Circuits

____ 29. In Figure 7–37, the DC voltage on the gate terminal measures 0 V. What is wrong with the circuit?

    a. Resistor R_g is shorted.

    b. Resistor R_g is open.

    c. The JFET is shorted between gate and source.

    d. The circuit is functioning properly.

____ 30. In Figure 7–37, the DC voltages are correct, but the signal voltage gain is low. What is wrong with the circuit?

    a. Capacitor C_3 is shorted.

    b. Capacitor C_3 is open.

    c. Capacitor C_2 is shorted.

    d. Capacitor C_2 is open.

____ 31. In Figure 7–37, the DC voltage on the drain terminal measures 12 V. What is wrong with the circuit?

    a. Capacitor C_3 is shorted.

    b. Capacitor C_3 is open.

    c. The JFET is shorted.

    d. The JFET is open.

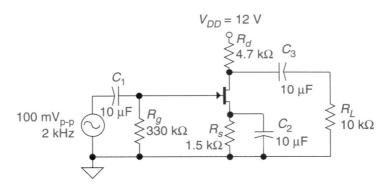

FIGURE 7–37

____ 32. In Figure 7–38, if the voltage on the gate is 2.8 V and the voltage on the source and drain is 6 V, what is wrong with the circuit?

    a. The JFET is shorted drain to source.

    b. The JFET is open drain to source.

    c. Resistor R_{g2} is open.

    d. The circuit is functioning properly.

____ 33. In Figure 7–38, if the voltage on the drain is 6 V and the voltage on the source is 5.9 V, what is wrong with the circuit?

    a. The JFET is open drain to source.

    b. Resistor R_{g1} is open.

    c. Resistor R_{g2} is open.

    d. The circuit is functioning properly.

____ 34. If the generator has an internal impedance of 5 kΩ and connecting it to the circuit in
          Figure 7–38 causes the generator to load considerably, what is wrong with the circuit?
   a. The JFET is open drain to source.
   b. The JFET is shorted gate to source.
   c. Resistor R_{g2} is open.
   d. The circuit is functioning properly.

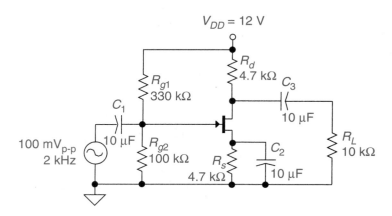

FIGURE 7–38

____ 35. In Figure 7–39, if the control voltage equals 0 V and the signal voltage across the load
          equals 80 mV$_{p-p}$, what is wrong with the circuit?
   a. The JFET is open drain to source.
   b. The JFET is shorted drain to source.
   c. Resistor R_g is open.
   d. The circuit is functioning properly.

____ 36. In Figure 7–39, if the control voltage equals –8 V and the signal voltage across the load
          equals 100 mV$_{p-p}$, what is wrong with the circuit?
   a. The JFET is open drain to source.
   b. The JFET is shorted drain to source.
   c. Resistor R_g is open.
   d. The circuit is functioning properly.

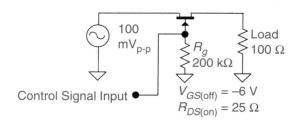

FIGURE 7–39

PROBLEMS

1. Calculate the transconductance for the JFET using the curve in Figure 7–40 at a bias of $V_{GS} = -2$ V (use a 0.5 V change in V_{GS} for your calculations).

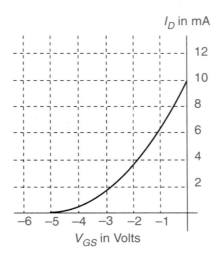

FIGURE 7–40

2. Calculate the DC voltages for the circuit in Figure 7–41.
    $V_G =$ _____ $V_s =$ _____ $V_D =$ _____ $V_{DS} =$ _____
3. Calculate z_{in}, z_{out}, A_v, and V_{out} for the circuit in Figure 7–41.
    $z_{in} =$ _____ $z_{out} =$ _____ $A_v =$ _____ $V_{out} =$ _____

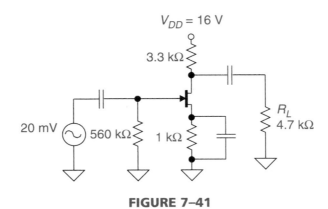

FIGURE 7–41

4. If the source voltage in Figure 7–41 measures 1 V, calculate the gate and drain voltages.
    $V_G =$ _____ $V_D =$ _____
5. Figure 7–42 shows a self-biasing JFET amplifier. Calculate the following values:
    $V_G =$ _____ $V_s =$ _____ $V_D =$ _____
    $z_{in} =$ _____ $z_{out} =$ _____ $A_v =$ _____ $V_{out} =$ _____

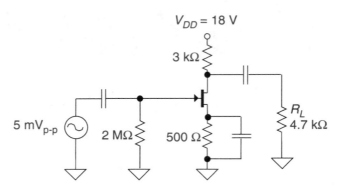

FIGURE 7–42

6. Figure 7–43 shows a voltage divider biased JFET amplifier. Calculate the following quiescent values: V_G, V_S, V_D, and V_{DS}.

$V_G =$ _____ $V_S =$ _____ $V_D =$ _____ $V_{DS} =$ _____

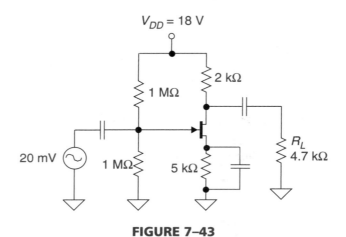

FIGURE 7–43

7. Calculate z_{in}, z_{out}, A_v, and V_{out} for the circuit in Figure 7–43.

   $z_{in} =$ _____ $z_{out} =$ _____ $A_v =$ _____ $V_{out} =$ _____

8. Calculate I_D, V_G, V_S, V_D, and V_{DS} for the circuit in Figure 7–44.

   $I_D =$ _____ $V_G =$ _____ $V_S =$ _____ $V_D =$ _____ $V_{DS} =$ _____

9. Calculate z_{in}, z_{out}, A_v, and V_{out} for the circuit in Figure 7–44.

   $z_{in} =$ _____ $z_{out} =$ _____ $A_v =$ _____ $V_{out} =$ _____

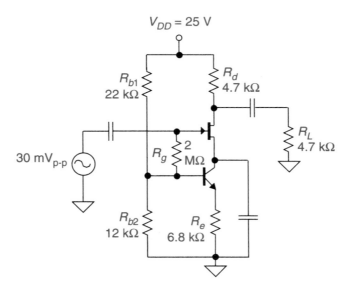

FIGURE 7–44

10. Calculate the following quiescent values for the circuit in Figure 7–45.

$V_D =$ _____ $V_S =$ _____ $V_G =$ _____ $V_{DS} =$ _____

11. Calculate $z_{in}, z_{out}, A_v,$ and V_{out} for the circuit in Figure 7–45.

$z_{in} =$ _____ $z_{out} =$ _____ $A_v =$ _____ $V_{out} =$ _____

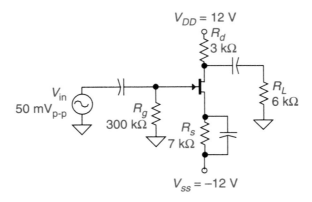

FIGURE 7–45

12. Calculate the following values for the circuit in Figure 7–46.

$V_G =$ _____ $V_s =$ _____ $V_D =$ _____ $V_{DS} =$ _____

$z_{in} =$ _____ $z_{out} =$ _____ $A_v =$ _____ $V_{out} =$ _____

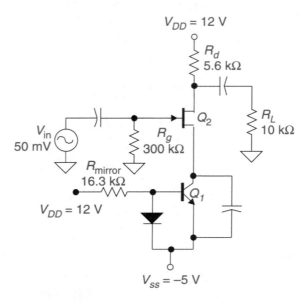

FIGURE 7–46

13. Figure 7–47 shows a self-biasing JFET amplifier. Calculate the following values:

$V_G =$ _____ $V_s =$ _____ $V_D =$ _____ $V_{DS} =$ _____

$z_{in} =$ _____ $z_{out} =$ _____ $A_v =$ _____ $V_{out} =$ _____

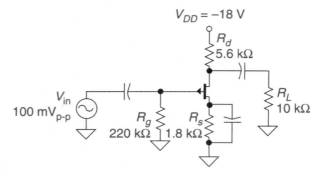

FIGURE 7–47

14. Figure 7–48 shows a common-drain JFET amplifier, calculate the following values:

$V_G =$ _____ $V_s =$ _____ $V_D =$ _____

$z_{in} =$ _____ $z_{out} =$ _____ $A_v =$ _____

$V_{out} =$ _____ $A_i =$ _____ $A_p =$ _____

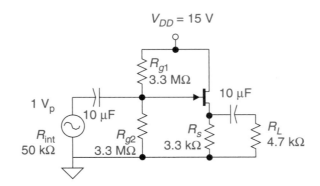

FIGURE 7–48

15. Draw the output waveform for the circuit in Figure 7–49.

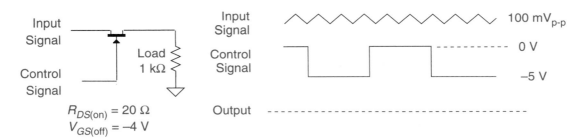

FIGURE 7–49

16. Draw the output waveform for the circuit in Figure 7–50.

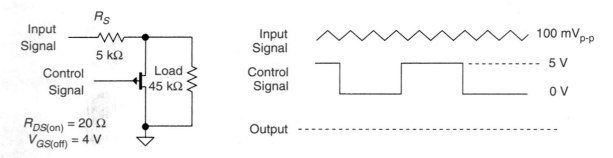

FIGURE 7–50

CHAPTER 8

MOSFETs

OBJECTIVES

After studying the material in this chapter, you will be able to describe and/or analyze:

○ the theory of the D-MOSFET,

○ D-MOSFET amplifier circuits,

○ the theory of the E-MOSFET,

○ E-MOSFET switching circuits,

○ E-MOSFET amplifier circuits,

○ handling MOSFET procedures, and

○ troubleshooting procedures for MOSFET circuits.

8.1 INTRODUCTION TO MOSFETs

MOSFETs, like the JFET, are three-terminal devices having a source, a gate, and a drain. The term *MOSFET* is an acronym for metal-oxide semiconductor field-effect transistor. Like JFETs, MOSFETs have a channel between the source and the drain. The main difference between a JFET and a MOSFET is the way the gate terminal is isolated from the channel. The JFET depends on a reverse-biased PN junction for isolation, whereas the MOSFET uses a thin layer of silicon dioxide as an insulator between the gate and the channel. The two types of MOSFETs are the depletion and the enhancement type. The depletion MOSFET is called the *D-MOSFET* and the enhancement MOSFET is called the *E-MOSFET*.

8.2 D-MOSFET

Figure 8–1 shows the symbol and construction of an N-channel D-MOSFET. A channel made of N-type material runs through a substrate made of P-type material. The source terminal is connected to one end of the channel, and the drain terminal is connected at the other end of the channel. The gate is separated from the channel by a thin insulator made of silicon dioxide.

Silicon dioxide has extremely high resistance, measured in megohms. The schematic symbol shows a gap between the gate and the channel that represents the high impedance insulator.

The substrate is normally internally connected to the source, making the MOSFET a three-terminal device, as the schematic symbol shows in Figure 8–1(b). There are a few MOSFETs on the market that connect the substrate to a separate terminal, making the MOSFET a four-terminal device, as shown in Figure 8–1(c).

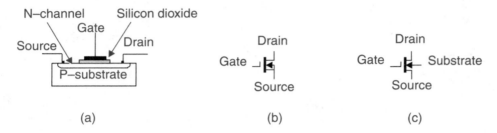

(a)   (b)   (c)

FIGURE 8–1   Depletion MOSFETs

The voltage applied to the gate can control the size of the channel even though the gate is insulated from the channel. If a negative voltage is applied to the gate, the electrons under the gate will be repelled, causing the channel to become smaller. If the negative voltage is increased, a voltage will be reached where the channel under the gate will be totally depleted of all current carriers and the device will be turned off. When the voltage on the gate is negative and reduces current carriers, the MOSFET is said to be working in the depletion mode. If a positive voltage is applied to the gate, electrons are attracted to the channel under the gate, and the size of the channel is increased. This mode of operation is called the *enhancement mode.*

The depletion MOSFET can operate in both the enhancement and depletion modes. Figure 8–2 shows the transconductance curve of a D-MOSFET. If a negative voltage is applied to gate, the drain current decreases from I_{DSS}. If a positive voltage is applied to the gate, the drain current increases from I_{DSS}. Since drain current can be controlled above and below I_{DSS}, the D-MOSFET amplifier can operate class-A with a bias of zero volts (V_{GS}).

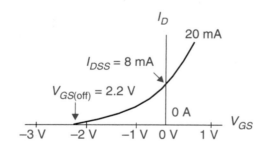

FIGURE 8–2   Transconductance curve

D-MOSFET AMPLIFIER

Since the D-MOSFET can function in the depletion mode, all forms of standard JFET biases can be used. But since D-MOSFETs can also work in the enhancement mode, it is possible to operate a D-MOSFET with zero bias.

Figure 8–3 shows a simple MOSFET amplifier circuit. The gate is simply tied to ground through a large resistor, and the source is connected directly to ground. With zero bias ($V_{GS} = 0$ V), the drain current will equal I_{DSS} with no input signal. Since the circuit is operating with zero bias, gm at the Q-point is equal to gm_0. Example 8.1 shows how to calculate the circuit values.

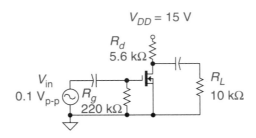

FIGURE 8–3   D-MOSFET amplifier

EXAMPLE 8.1

Calculate V_G, V_S, V_D, z_{in}, z_{out}, A_v, V_{out}, and draw the output waveform for the D-MOSFET amplifier circuit in Figure 8–3, where $gm_0 = 3.3$ mS and $I_{DSS} = 1.2$ mA.

Step 1. Calculate the DC voltages for V_G, V_S, and V_D. The gate is connected to ground through resistor R_g. Since zero bias current flows through R_g, the DC voltage drop across R_g is zero; therefore, the DC voltage on the gate is zero. The source terminal is connected directly to ground; hence, the voltage on the source is zero. The voltage on the drain is calculated by first calculating the voltage drop across R_d, and then subtracting this voltage from the supply voltage.

$$V_G = 0 \text{ V}$$
$$V_s = 0 \text{ V}$$
$$V_{Rd} = I_{DSS} \times R_d = 1.2 \text{ mA} \times 5.6 \text{ k}\Omega = 6.7 \text{ V}$$
$$V_D = V_{DD} - V_{Rd} = 15 \text{ V} - 6.7 \text{ V} = 8.3 \text{ V}$$

Step 2. Calculate the input impedance. Since the gate is insulated from the channel, the D-MOSFET device has a resistance of many megohms between the gate and the channel; therefore, the input impedance of the circuit is simply the value of the gate resistor.

$$z_{in} = R_g = 220 \text{ k}\Omega$$

Step 3. Calculate the output impedance. MOSFETs are like JFETs in that the output impedance is found by looking back from the output. The impedance through the MOSFET is in the megohms; therefore, the only signal path to ground is through R_d, and the output impedance is equal to the value of R_d.

$$z_{out} = R_D = 5.6 \text{ k}\Omega$$

Step 4. Calculate the voltage gain. The voltage gain of a MOSFET is calculated using the same formula used for JFETs ($A_v = gm \times r_d$). If the D-MOSFET is operating at zero bias, then $gm = gm_0$.

$$r_d = R_d \| R_L = 5.6 \text{ k}\Omega \| 10 \text{ k}\Omega = 3.6 \text{ k}\Omega$$

$gm = gm_0 = 3.3$ mS   (If the value of gm_0 is unknown, use 3 mS as an approximation.)
$$A_v = gm \times r_d = 3.3 \text{ mS} \times 3.6 \text{ k}\Omega = 11.9$$

Step 5. Calculate the signal out.
$$V_{out} = V_{in} \times A_v = 0.1 \text{ V}_{p\text{-}p} \times 11.9 = 1.19 \text{ V}_{p\text{-}p}$$

Step 6. Draw the output waveform. Figure 8–4 shows the output waveform with respect to the input signal. The output signal is 180º out-of-phase with respect to the input signal.

EXAMPLE 8.1 continued

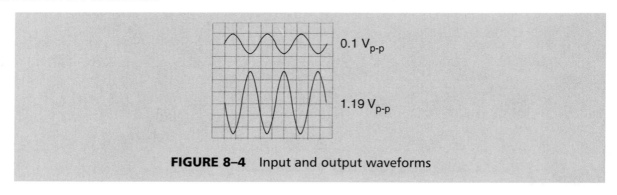

FIGURE 8–4   Input and output waveforms

P-CHANNEL D-MOSFET

A P-channel D-MOSFET is shown in Figure 8–5. An N-type substrate has a P-channel connecting the source to the drain. Notice that the symbol for the P-channel MOSFET shows the arrow pointing away from the channel. The P-channel MOSFET works the same as the N-channel MOSFET except a positive voltage on the gate will reduce current flow, while a negative voltage on the gate will increase current flow. This is because the current carriers in the channel are holes that are positively charged.

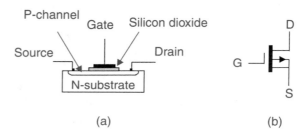

FIGURE 8–5   P-channel D-MOSFET

8.3   E-MOSFET

Figure 8–6 shows the symbols and construction of an N-channel E-MOSFET. It is similar to the D-MOSFET, but note that without bias, there is no channel between the source and drain. This means the E-MOSFET will act like an open between the source and drain until a positive signal is applied to the gate, which will create a channel. Drain current will start to flow when the gate

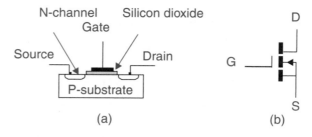

FIGURE 8–6   E-MOSFET

is made positive enough to enhance electrons to move into the area below the gate and complete the channel.

Figure 8–7 shows the transconductance curve for an E-MOSFET. The shape of the curve is the same as for other FETs, but a positive bias voltage must be applied to the gate before drain current starts to flow. The amount of gate-to-source voltage needed to cause drain current to start to flow is called the *threshold voltage* ($V_{GS(\text{TH})}$). Further increases in V_{GS} above the threshold voltage will cause increases in the drain current.

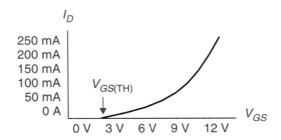

FIGURE 8–7   Transconductance curve for E-MOSFET

E-MOSFET SWITCH

E-MOSFETs are often used in switching circuits. With zero gate voltage applied, the E-MOSFET is an open circuit. A positive voltage applied to the gate of an N-channel E-MOSFET can cause the device to act like a closed switch. The E-MOSFET will have a specification called $R_{DS(\text{on})}$, which is the maximum resistance between the drain and source in the on state. $R_{DS(\text{on})}$ is given in ohms at a stated value of drain current. For example, an E-MOSFET could have a rating of $R_{DS(\text{on})(\text{max})} = 8\ \Omega$ at $I_D = 200$ mA.

Figure 8–8(a) shows an E-MOSFET in series with a 250 Ω resistor connected to a 50 V source. When the gate voltage is zero, the E-MOSFET is off, and no current flows in the circuit. The 50 V is dropped across the E-MOSFET. When the gate voltage switches to a positive value, the E-MOSFET conducts, and the current in the circuit is limited by the series resistance. The current in Figure 8–8(a) will equal 200 mA (50 V/250 Ω). If the E-MOSFET is rated for $R_{DS(\text{on})(\text{max})} = 8\ \Omega$, the equivalent circuit for the on state is shown in Figure 8–8(b).

Two additional items must be considered to ensure the proper operation of the circuit in Figure 8–8(a). The E-MOSFET must be rated to withstand the voltage across its drain-to-source terminals when in the off state. E-MOSFETs have a specification called $V_{(\text{BR})DSS}$, which is the voltage the device can withstand across the drain-to-source terminals with the gate shorted to the source. The E-MOSFET used in Figure 8–8(a) would need to have a minimum $V_{(\text{BR})DSS}$ rating of 50 V.

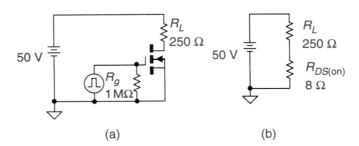

(a)   (b)

FIGURE 8–8   E-MOSFET switching circuit

The second consideration is the magnitude of the gate control voltage. The control voltage applied to the gate (V_{GS}) must be sufficient to drive the E-MOSFET into saturation. The load in Figure 8–8 will limit the current to 200 mA; however, if the gate voltage applied causes only 100 mA of drain current, then current through the load will be 100 mA. Figure 8–9 is the transconductance curve for the E-MOSFET used in the circuit in Figure 8–8. By examining the transconductance curve, you can see that a gate voltage of plus 10 V is needed to cause 200 mA. Therefore, for the switching circuit in Figure 8–8 to operate properly, the gate control voltage needed to turn the E-MOSFET on is plus 10 V or greater. Example 8.2 shows how an E-MOSFET switching circuit can be designed for a given load.

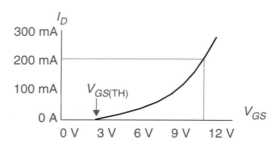

FIGURE 8–9   Transconductance curve

EXAMPLE 8.2

A 200 Ω load is designed to operate with 40 V applied. Design a switching circuit that will switch the load on and off. The control source should see the switching circuit as a 200 kΩ impedance. The transconductance curve of the E-MOSFET to be used in the circuit is shown in Figure 8–9.

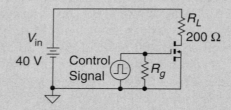

FIGURE 8–10   E-MOSFET switching circuit

Step 1. Draw a schematic of the switching circuit. The schematic is shown in Figure 8–10.

Step 2. Calculate the load current when the circuit is operating in the on state.

$$I_{(on)} = V_{in}/R_L = 40 \text{ V}/200 \text{ Ω} = 200 \text{ mA}$$

Step 3. Calculate the minimum value of gate voltage needed to turn on the switch. Examine the transconductance curve in Figure 8–9 and determine the minimum gate voltage necessary to cause 200 mA of drain current.

$$V_{GS} = 11 \text{ V}   \text{(Value read from graph.)}$$

Step 4. Calculate the value of R_g. The value of R_g sets the input impedance to the gate circuitry.

$$R_g = 200 \text{ kΩ}$$

E-MOSFET AMPLIFIER

While E-MOSFETs are popular in switching circuits, they can be used in amplifier circuits. Figure 8–11 shows two methods of biasing an E-MOSFET amplifier. Figure 8–11(a) shows a voltage divider biasing method. The designer selected the value of R_{g1} and R_{g2} to provide sufficient gate voltage to bias the E-MOSFET in the conduction region. No bias gate current flows; hence, the voltage at the gate is easy to calculate using the voltage divider formula. The problem with holding the gate voltage constant is that variations in FET characteristics will cause large variations in the Q-point.

E-MOSFETs do not have a value for gm_0 because at zero gate-to-source voltage, the device is off and no drain current flows. Once biased in the conductance region, they do have a value of gm, which depends on the Q-point. Getting a value for gm of an E-MOSFET is next to impossible for the technician. There are formulas and curves that the engineer can use to calculate the gm, but the technician in the field is left making a good guess. Since gm values are normally in the range of 1 mS to 10 mS, a reasonable guess can be made. We will use 3 mS to be consistent with other FETs. Once a value is assumed for gm, the voltage gain can be calculated in the same manner as used for the JFET and D-MOSFET. There are, however, some E-MOSFETs on the market with transconductance in the range of 100 mS to 500 mS. If one of these is used in an amplifier circuit, the voltage gain would be much higher than our approximation.

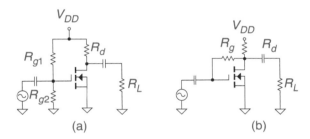

FIGURE 8–11 E-MOSFET amplifier circuits

Figure 8–11(b) shows an E-MOSFET amplifier using drain-feedback biasing. This type of biasing is often used with E-MOSFETs and is especially well suited when operating with a low voltage power supply (V_{DD}). The voltage on the gate is equal to the voltage on the drain ($V_{DS} = V_{GS}$), since no bias current flows through R_g. As the drain current increases, the gate voltage drops. The decreased gate voltage causes the drain current to decrease, which causes the gate voltage to increase. The feedback loop reaches an equilibrium that is the bias point for the circuit. Some data sheets for E-MOSFETs give a value for $I_{D(on)}$, where $V_{GS} = V_D$. If $I_{D(on)}$ is known, the circuit component can be easily calculated, as shown in Example 8.3. The input impedance of a circuit using drain feedback biasing is equal to the value of R_g divided by the voltage gain plus one.

EXAMPLE 8.3

> Calculate V_G, V_S, V_D, A_v, V_{out}, z_{in}, and z_{out} for the circuit in Figure 8–12. Assume $I_{D(on)} = 2$ mA and gm at the Q-point equals 3 mS.
>
> **Step 1.** Calculate V_G, V_S, and V_D.
> $$V_D = V_{DD} - (R_d \times I_{D(on)}) = 9 \text{ V} - (2.5 \text{ k}\Omega \times 2 \text{ mA}) = 4 \text{ V}$$
> $$V_G = V_D = 4 \text{ V}$$
> $$V_S = 0 \text{ V}$$

EXAMPLE 8.3 continued

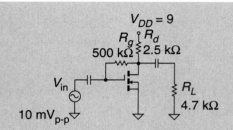

FIGURE 8–12   Amplifier using drain-feedback biasing

Step 2. Calculate A_v.

$$r_d = R_d \| R_L = 2.5 \text{ k}\Omega \| 4.7 \text{ k}\Omega = 1.6 \text{ k}\Omega$$
$$A_v = gm \times r_d = 3 \text{ mS} \times 1.6 \text{ k}\Omega = 4.8$$

Step 3. Calculate z_in.

$$z_\text{in} = R_g/(A_v + 1) = 500 \text{ k}\Omega/4.8 + 1 = 86 \text{ k}\Omega$$

Step 4. Calculate z_out.

$$z_\text{out} = R_d = 2.5 \text{ k}\Omega$$

P-CHANNEL E-MOSFET

A P-channel E-MOSFET is shown in Figure 8–13. An N-type substrate has two P-type pockets: one connected to the source, and the other connected to the drain. Notice that the symbol for the P-channel E-MOSFET shows the arrow pointing away from the channel. The P-channel E-MOSFET works the same way as the N-channel except a negative voltage on the gate is needed to complete the channel and turn the device on.

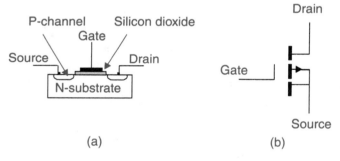

FIGURE 8–13   P-channel E-MOSFET

8.4 HANDLING MOSFETs

The layer of silicon dioxide between the gate and the channel is very thin and is subject to puncture by static electricity. When working with MOSFETs, special handling precautions should be taken as follows:

1. MOS devices should have their leads shorted together for shipment and storage. Conductive foam is often used for this purpose.
2. Avoid unnecessary handling, and pick up the device by the case, not the leads.
3. All test equipment and tools should be connected to earth ground.
4. Workers handling MOSFET devices should have grounding straps attached to their wrists.
5. Never remove or insert MOSFET devices with the power on.

Many MOSFET devices are now available with built-in zener diode protection. Zener diodes are placed between the gate and the channel, as shown in Figure 8–14. The voltage rating of the zener diode is below the voltage that would cause punctures to the silicon dioxide insulator. If the voltage between the gate and the channel exceeds the voltage rating of the zener, the diode conducts, thus protecting the MOSFET from damage. At normal operating voltages, zener diodes have no effect on the operation of the MOSFET.

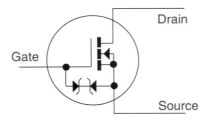

FIGURE 8–14   MOSFET with zener diode protection

8.5 MOSFET PARAMETERS

Most of the parameters you learned when studying JFETs apply to MOSFETs; however, there are a few differences. Lowercase letters denote signal values, and uppercase letters denote bias values:

I_D = the drain-bias current.

i_d = the drain-signal current.

I_{DSS} = the drain current with the gate equal to zero. I_{DSS} is the maximum drain current for a JFET, but not for a D-MOSFET. I_{DSS} is often used as the bias point for D-MOSFETs, since the drain current can be controlled above and below this value. I_{DSS} is always zero for E-MOSFETs.

$V_{(BR)DSS}$ = the drain-to-source voltage. If exceeded, it will cause the device to break down and conduct. If breakdown occurs, the FET will be destroyed. This parameter is the same for all three FETs.

gm = transconductance, which is a measurement of the sensitivity of the MOSFET. It is equal to the change in drain current divided by the corresponding change in gate voltage ($gm = \Delta i_d / \Delta v_{gs}$).

gm_0 = transconductance when bias voltage (V_{GS}) is zero volts. When D-MOSFET amplifiers operate at zero bias, $gm_0 = gm$ of the amplifier.

$R_{DS(on)}$ = the resistance of a FET from the drain to source when in the on state. The value of $R_{DS(on)}$ is normally given for a stated drain current.

V_{GS} = gate-to-source bias voltage.

v_{gs} = gate-to-source signal voltage.

$V_{GS(off)}$ = the gate-to-source voltage that stops drain current (I_D) from flowing. This term has the same meaning for JFETs and D-MOSFETs but is replaced by $V_{GS(TH)}$ for E-MOSFETs.

$V_{GS(TH)}$ = the gate-to-source voltage that permits drain current (I_D) to start flowing in E-MOSFETs.

V_p = pinch-off voltage is the drain-to-source voltage with zero gate voltage applied where the drain current becomes approximately constant.

8.6 MOSFET APPLICATIONS

There are few discrete systems that are made completely of FETs, but many applications take advantage of the high input impedance and low internal noise of the FET. This is especially true for low power circuits such as the front end of receivers or the input circuitry of instruments. FETs are used as analog switches as well as in numerous other applications.

FRONT END OF RECEIVERS

When signals are at low power levels, it is possible that electrical noise can cover up the signal and make it undetectable. There are two types of noise: external noise and internal noise. External noise is generated outside of the circuit, and internal noise is generated by the circuit. Internal noise can be caused by random movement of current carriers. Bipolar transistors depend on the combining of electrons and holes in the base region. This causes random movement of current carriers and is a source of internal noise. This small electrical noise is no problem when signal levels are large, but at low signal levels, it causes serious problems. The FET does not depend on the combining of holes and electrons and generates much less internal noise. This fact makes FETs ideal for use in the front end of receivers where input signal levels are low and noise levels need to be kept to a minimum.

Another feature of the FET that makes it desirable for use in the front end of receivers is the ability to control the gain of the FET by controlling its bias point. By changing the bias point, the value of gm is changed, and therefore, the gain is changed ($A_v = gm \times r_d$). This characteristic is put to work in the front end of receivers. Most receivers use automatic gain control (AGC) to maintain a constant output even though the input signal levels change. This is accomplished by controlling the gain of the amplifier in relation to the strength of the input signal. By using a FET and a feedback loop, the gain of the FET can be controlled.

Many receivers use the super-heterodyne technique. This technique involves mixing the incoming signal with a signal from a local oscillator in a nonlinear device to develop an IF (intermediate frequency) signal. MOSFETs are available with two gates, and these are excellent devices to use for the mixer circuit. The two gates permit signals to be mixed in the output circuit without interaction between the two input circuits. Figure 8–15 shows the front end of a receiver using a JFET as an RF (radio frequency) amplifier and a dual-gate D-MOSFET as a mixer stage.

The signal input from the antenna is applied to the JFET gate through a coupling capacitor. The JFET circuit is configured as a common-source amplifier, using voltage divider biasing, and will amplify the small signal from the antenna adding a minimum amount of internal noise. The voltage gain of the JFET stage equals $gm \times r_d$. The gm of the JFET amplifier, and thus the gain, will depend on the bias voltage applied to the gate. The AGC signal is generated by circuitry monitoring the

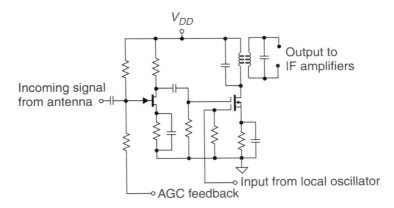

FIGURE 8–15 Receiver front end

receiver output. When the output signal is large, the AGC generates a large negative voltage, which decreases the gain of the JFET amplifier. When the output signal is small, the AGC generates a small negative voltage, which increases the gain of the JFET amplifier. AGC is particularly useful in portable receivers when the signal selected at the antenna varies continuously. AGC permits the output of the receiver to stay at a constant volume even with changes in input signal.

The output of the JFET amplifier is connected to one of the gates of the D-MOSFET. The signal from the JFET amplifier will control the drain current of the D-MOSFET; however, the other gate of the D-MOSFET is connected to a sine wave signal from the local oscillator. The signal from the local oscillator will also control the drain current of the D-MOSFET. The drain current flowing at any instant in time is a function of the independent voltages on both gates. The result is that the drain current represents a mixing of both signals. The transconductance curve is not a straight line, or stated another way, the FET is a nonlinear device. If two signals are mixed in a nonlinear device, the resulting signal will contain four frequencies: the two original frequencies, the sum of the two frequencies, and the difference of the two frequencies. The difference frequency is called the *intermediate frequency* (IF).

The gain of the D-MOSFET stage is equal to $gm \times r_d$. In this case, r_d is equal to the impedance of the parallel LC circuit. The LC circuit is designed to have a resonance frequency equal to the intermediate frequency (IF). Since parallel resonance circuits have a high impedance, the gain of the D-MOSFET amplifier for the IF signal will be high; however, the gain of the amplifier for the other three frequencies will be low. This means that only the IF signal will be present at the output of the transformer. To select a station, the local oscillator is adjusted to a frequency that is offset from the desired station signal frequency by an amount equal to IF. In this way, any station can be selected, but IF will always remain constant.

E-MOSFET SWITCHING CIRCUITS

Figure 8–16(a) shows the E-MOSFET used as a series switch. If the E-MOSFET is in series with the load and is off, all the signal voltage is dropped across the high impedance of the E-MOSFET, and the output is zero. If the series E-MOSFET is turned on, the signal voltage will divide between R_L and $R_{DS(on)}$. Since $R_{DS(on)}$ is normally less than 10 Ω, most of the voltage will be dropped across R_L (assuming $R_L \gg R_{DS(on)}$).

Figure 8–16(b) shows an E-MOSFET parallel switching circuit. If R_s is much smaller than R_L but much larger than $R_{DS(on)}$, the circuit can control the signal voltage on the load. When the E-MOSFET is off, most of the voltage will be delivered to the load because R_L is much larger than R_s. However, when the E-MOSFET is on, it shunts R_L, causing the input voltage to be dropped across R_s.

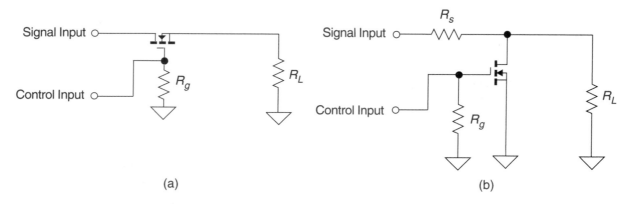

FIGURE 8–16   E-MOSFET switches: (a) series and (b) parallel

Since an E-MOSFET is not a perfect switch, neither the series nor the parallel switching circuit is perfect. When the switch is off, a small percentage of the input signal may reach the load, and when the switch is on, a small percentage of the input signal does not reach the load. An improved switching circuit can be designed by using one E-MOSFET in series and a second in parallel. Figure 8–17 shows a switching circuit using a pair of E-MOSFETs: one N-channel and one P-channel. The E-MOSFET in series with the load is an N-channel and turns on when a positive voltage is applied to the gate. The E-MOSFET in parallel with the load is a P-channel and turns off when a positive voltage is applied to the gate. The waveforms shown in Figure 8–17 demonstrate how the output signal can be switched off and on by the control signal.

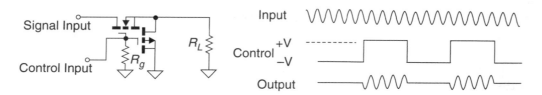

FIGURE 8–17   Switching circuit using two E-MOSFETs

Often in electronic systems, it is desirable to use a single line to carry several signals to distant loads. If the distant loads do not require all input signals to be present at all times, a method called time multiplexing can permit one line to transport several signals. Figure 8–18 shows how E-MOSFET switches can be used to time multiplex two signals. When the control voltage is positive, N-channel FETs (Q_1 and Q_3) are on, connecting signal #1 to load #1 through the transmission line. When the control is negative, P-channel FETs (Q_2 and Q_4) are on, connecting signal #2 to load #2 through the transmission line.

Q_1 and Q_2 form a time multiplexing circuit, combining the two signals on one transmission line (see waveform TP_4). Transistors Q_3 and Q_4 form a de-multiplexing circuit that separates the signals and delivers the signals to the correct load (see waveforms TP_5 and TP_6). The control signal must synchronize the de-multiplexing with the multiplexing. In Figure 8–18, the same control signal is applied to both circuits. The advantage of time multiplexing becomes apparent when you consider that a single line could support a dozen signals or more and that the control synchronization can also be sent down the same line.

E-MOSFETs are often used in chopper circuits. A chopper circuit converts a DC signal or a low frequency signal into a high frequency signal. To accomplish this, a switching circuit alternately connects and disconnects the input source to the load at some periodic frequency. This process is called *chopping*. The chopping signal may be amplified by a conventional RC

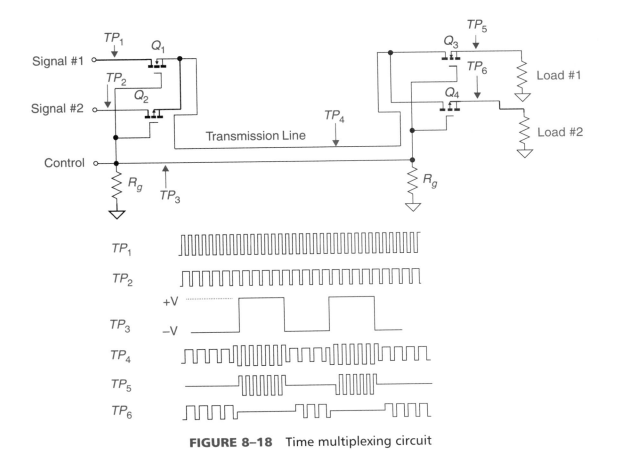

FIGURE 8–18 Time multiplexing circuit

coupled amplifier, thus avoiding the use of direct coupled amplifiers, which have limited gain and are prone to bias drift.

Figure 8–19 shows a chopping circuit using two E-MOSFETs. When the chopping signal is positive, the series E-MOSFET is on, the parallel E-MOSFET is off, and an input is connected to the load. When the chopping signal is negative, the series E-MOSFET is off, and the parallel E-MOSFET is on; thus, no signal is developed across the load. Figure 8–19 shows the waveforms. The input can be a DC signal or a low frequency signal that would see coupling capacitors as opens or high impedance. The output signal has a high frequency component at the frequency of

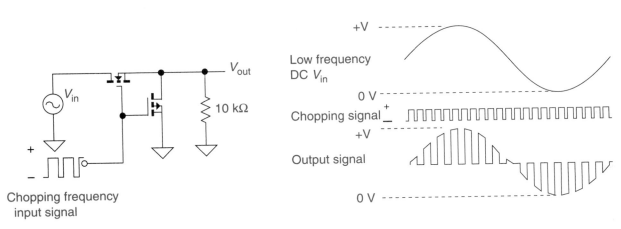

FIGURE 8–19 Chopping circuit

the chopping signal. The amplifier circuitry should be designed to amplify the signal at the chopping frequency.

Figure 8–20 shows an E-MOSFET being used as an active load resistor. Q_1 is selected to have a value of $R_{DS(on)}$ ten times or greater than the value of Q_2 $R_{DS(on)}$. When Q_2 is off, the output voltage will be equal to or near V_{DD}, and when Q_2 is on, the output voltage will drop to near zero. Transistor Q_1 performs the function of a drain resistor. In integrated circuitry, less room is required for an active component than a passive resistor, so most ICs use active loads. Example 8.4 shows how the output voltage can be calculated if $R_{DS(on)}$ is known for both E-MOSFETs.

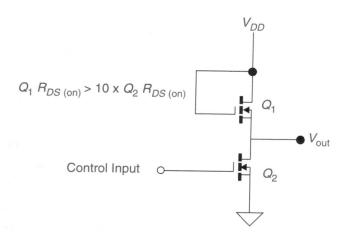

FIGURE 8–20   FET digital switch

EXAMPLE 8.4

Calculate the output voltage for the circuit in Figure 8–20, if $V_{DD} = 12$ V, Q_1 on resistance ($R_{DS(on)}$) is 120 Ω, and Q_2 on resistance ($R_{DS(on)}$) is 10 Ω: (a) with Q_2 turned on, and (b) with Q_2 turned off.

Step 1. If Q_2 is turned on, the voltage divider formula can be used to calculate the output signal.

$$V_{out} = V_{DD} \times Q_2 R_{DS(on)} / (Q_1 R_{DS(on)} + Q_2 R_{DS(on)})$$
$$V_{out} = 12 \text{ V} \times 10 \text{ }\Omega / 120 \text{ }\Omega + 10 \text{ }\Omega) = 0.9 \text{ V}$$

Step 2. If Q_2 is turned off, the E-MOSFET acts like an open circuit, and the output voltage is equal to the supply voltage (V_{DD}).

$$V_{out} = V_{DD}$$
$$V_{out} = 12 \text{ V}$$

CMOS CIRCUIT

Circuit designs using both types of MOSFETs are called CMOS circuits. The "C" in the acronym stands for complementary. CMOS circuits are popular in digital circuitry. The circuit in Figure 8–21 is a CMOS inverter. A P-channel and an N-channel are connected in series between the positive supply and ground. The gates of both devices are connected together. If the input equals V_{DD}, the N-channel E-MOSFET is on and connects the output to ground. If the input equals zero volts, the P-channel E-MOSFET is on and connects the output to V_{DD}.

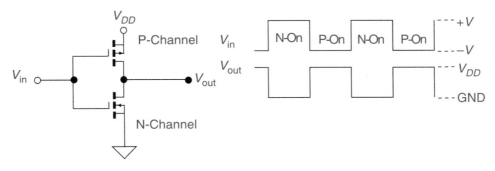

FIGURE 8–21 CMOS inverter

8.7 TROUBLESHOOTING FET CIRCUITS

AMPLIFIER CIRCUITS

Like all amplifier circuits, the first step in troubleshooting is estimating the bias voltage in the circuit. To estimate the bias voltage, start by determining the type of FET and the biasing method being used. The schematic symbol for the device quickly identifies the FET type. By examining the schematic, you should be able to determine the biasing method. For example, Figure 8–22 shows three FET circuits. Circuit "a" is a JFET device using self-biasing, circuit "b" is a D-MOSFET using zero biasing, and circuit "c" is an E-MOSFET device using drain-feedback biasing.

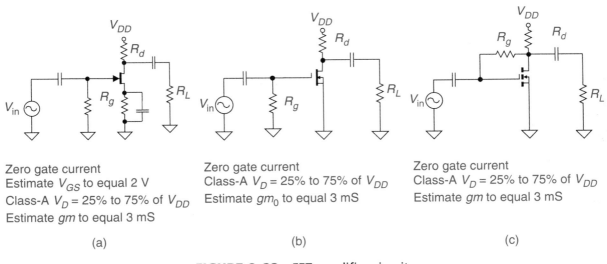

Zero gate current
Estimate V_{GS} to equal 2 V
Class-A V_D = 25% to 75% of V_{DD}
Estimate gm to equal 3 mS

(a)

Zero gate current
Class-A V_D = 25% to 75% of V_{DD}
Estimate gm_0 to equal 3 mS

(b)

Zero gate current
Class-A V_D = 25% to 75% of V_{DD}
Estimate gm to equal 3 mS

(c)

FIGURE 8–22 FET amplifier circuits

Remember, the one thing all FET amplifier circuits have in common is zero current flow between the gate and the channel. Knowing this fact makes it easy to estimate gate voltages. Both the JFET and the D-MOSFET, in Figure 8–22, are operating with zero volts on the gate. The E-MOSFET is operating with a positive voltage equal to V_D.

Next, based on the type of FET being used, estimate the bias between the gate and source. The JFET must be reverse biased. Depending on the JFET, this voltage ranges from 0.25 V to

4 V. For troubleshooting purposes, start by estimating 2 V between the gate and the source. Since the JFET in Figure 8–22 is an N-channel, the gate would be negative with respect to the source. The D-MOSFET can operate in the depletion and enhancement modes. For this reason, it is often biased with zero volts between the gate and the source. For the E-MOSFET to function as an amplifier, the bias voltage must be above $V_{GS(TH)}$. Depending on the type of E-MOSFET, $V_{GS(TH)}$ ranges from 0.5 V to 5 V. For troubleshooting purposes, start by estimating 2 V between the gate and the source. The E-MOSFET in Figure 8–22 is an N-channel, so the gate will be positive with respect to the source.

Since the circuits in Figure 8–22 are functioning as class-A amplifiers, the drain voltage should be between 25% of V_{DD} to 75% of V_{DD}, so estimate the drain voltage to equal 50% of V_{DD} for a starting point. If the drain voltage measured is equal to V_{DD}, no current is flowing, and there may be an open FET. A low drain voltage reading means that too much current is flowing and there may be a shorted FET. If the bias voltage (DC) measurements are wrong, the circuit has problems, and repairs can be made accordingly. If the bias readings are correct, however, the circuit may still have dynamic problems, and signal insertion and tracing is the next step.

In order to calculate voltage gain, you must know the value of gm, which is usually an unknown to the technician in the field. If necessary, the technician can research the technical data, find values for gm_0 and $V_{GS(off)}$, and then, using the measured value of V_{GS}, calculate the value of gm at the operating point [$gm = gm_0(1 - (V_{GS}/V_{GS(off)}))$]. This is usually more trouble than necessary. The value for gm_0 usually ranges from 1 mS to 10 mS. The value for gm depends on the Q-point. To be consistent, use 3 mS as an approximation of gm at the Q-point for all FETs. Voltage gains can then be approximated by using the formula 3 mS $\times r_d$. The voltage gain can be checked by reading the input signal and comparing it to the output signal.

The input impedance of a FET amplifier equals the resistance the signal current sees looking into the amplifier. The combination of resistors the signal current sees looking into the gate circuitry depends on the type of bias circuit used. The output impedance for all common source FET amplifier circuits is equal to the value of R_d. Examples 8.5 through 8.7 will review how to calculate the values for three amplifier circuits.

EXAMPLE 8.5

Estimate V_G, V_S, V_D, A_v, V_{out}, z_{in}, and z_{out} for the circuit shown in Figure 8–23.

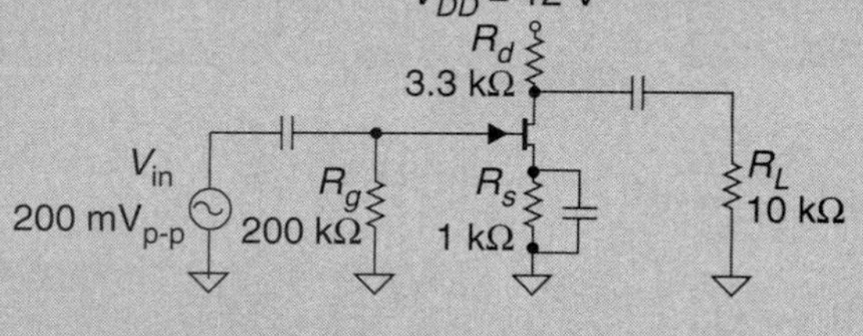

FIGURE 8–23   JFET amplifier

Step 1. Calculate V_G. Since zero gate current flows, the voltage drop across R_g is zero.
$$V_G = 0 \text{ V}$$
Step 2. Estimate V_S. V_S is in the range of 0.5 V to 5 V.
$$V_S = 2 \text{ V}$$

EXAMPLE 8.5 continued

Step 3. Estimate V_D. V_D should have a range from 25% to 75% of V_{DD}. Once the source voltage is measured, the drain voltage can be calculated. However, for a starting point, estimate the drain voltage to be 50% of V_{DD}.

$$V_D = 6 \text{ V}$$

Step 4. Estimate the value of gm. The value of gm_0 ranges from 1 mS to 10 mS. A starting point for gm is 3 mS.

$$gm = 3 \text{ mS}$$

Step 5. Calculate the voltage gain.

$$r_d = R_d \parallel R_L = 3.3 \text{ k}\Omega \parallel 10 \text{ k}\Omega = 2.5 \text{ k}\Omega$$
$$A_v = gm \times r_d = 3 \text{ mS} \times 2.5 \text{ k}\Omega = 7.5$$

Step 6. Calculate the signal out.

$$V_{\text{out}} = V_{\text{in}} \times A_v = 200 \text{ mV}_{\text{p-p}} \times 7.5 = 1.5 \text{ V}_{\text{p-p}}$$

Step 7. Calculate the input impedance.

$$z_{\text{in}} = R_g = 200 \text{ k}\Omega$$

Step 8. Calculate the output impedance.

$$z_{\text{out}} = R_d = 3.3 \text{ k}\Omega$$

EXAMPLE 8.6

Estimate V_G, V_S, V_D, A_v, V_{out}, z_{in}, and z_{out} for the circuit shown in Figure 8–24.

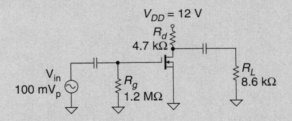

FIGURE 8–24   D-MOSFET amplifier

Step 1. Calculate V_G. Since zero gate current flows, the voltage drop across R_g is zero.

$$V_G = 0 \text{ V}$$

Step 2. Estimate V_S.

$$V_S = 0 \text{ V}$$

Step 3. Estimate V_D. V_D should have a range from 25% to 75% of V_{DD}.

$$V_D = 6 \text{ V (50\% of } V_{DD})$$

Step 4. Estimate the value of gm. The value of gm_0 ranges from 1 mS to 10 mS. Since V_{GS} is zero, gm will equal gm_0. The value of gm_0 is unknown, so use 3 mS as an approximation.

$$gm = gm_0 = 3 \text{ mS}$$

EXAMPLE 8.6 continued

Step 5. Calculate the voltage gain.
$$r_d = R_d \,\|\, R_L = 4.7 \text{ k}\Omega \,\|\, 8.6 \text{ k}\Omega = 3 \text{ k}\Omega$$
$$A_v = gm \times r_d = 3 \text{ mS} \times 3 \text{ k}\Omega = 9$$

Step 6. Calculate the signal out.
$$V_{\text{out}} = V_{in} \times A_v = 100 \text{ mV}_\text{p} \times 9 = 0.9 \text{ V}_\text{p}$$

Step 7. Calculate the input impedance.
$$z_{\text{in}} = R_g = 1.2 \text{ M}\Omega$$

Step 8. Calculate the output impedance.
$$z_{\text{out}} = R_d = 4.7 \text{ k}\Omega$$

EXAMPLE 8.7

Estimate V_G, V_S, V_D, A_v, V_{out}, z_{in}, and z_{out} for the circuit shown in Figure 8–25.

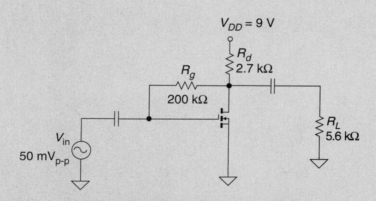

FIGURE 8–25 E-MOSFET amplifier

Step 1. Calculate V_G. The value of V_G equals V_D. The problem is knowing a value for V_D. As with other class-A amplifiers, assume a starting point of 50% of V_{DD}.
$$V_G = V_D = 4.5 \text{ V}$$

Step 2. Determine V_S. (The source is connected to ground.)
$$V_S = 0 \text{ V}$$

Step 3. Estimate V_D.
$$V_D = 4.5 \text{ V (50\% of } V_{DD})$$

Step 4. Estimate the value of gm. The gm at the Q-point is normally in a range from 1 mS to 10 mS. Let us be conservative and guess 3 mS.
$$gm = 3 \text{ mS}$$

Step 5. Calculate the voltage gain.
$$r_d = R_d \,\|\, R_L = 2.7 \text{ k}\Omega \,\|\, 5.6 \text{ k}\Omega = 1.8 \text{ k}\Omega$$
$$A_v = gm \times r_d = 3 \text{ mS} \times 1.8 \text{ k}\Omega = 5.4$$

Step 6. Calculate the signal out.
$$V_{\text{out}} = V_{in} \times A_v = 50 \text{ mV}_\text{p-p} \times 5.4 = 270 \text{ mV}_\text{p-p}$$

EXAMPLE 8.7 continued

> **Step 7.** Calculate the input impedance.
> $$z_{in} = R_g/(A_v + 1) = 200 \text{ k}\Omega/(5.4 + 1) = 31.25 \text{ k}\Omega$$
> **Step 8.** Calculate the output impedance.
> $$z_{out} = R_d = 2.7 \text{ k}\Omega$$

SWITCHING CIRCUITS

Switching circuits are inherently easier to troubleshoot than amplifier circuits. When the switching device is off, it can be considered an open, and when the switching device is on, it can be considered a short. If it functions as an open when on or short when off, the FET is bad or the control voltage applied to the gate is not correct.

Figure 8–26 shows a JFET used in a series switching circuit and an E-MOSFET used in a shunt switching circuit. For most troubleshooting, the FET can be considered an ideal switch. However, if the value of $R_{DS(on)}$ is known, the actual output signal can be calculated, as shown in Examples 8.8 and 8.9.

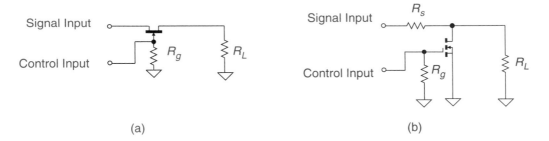

FIGURE 8–26   FET switching circuits

EXAMPLE 8.8

Calculate the signal output of the JFET switching circuit in Figure 8–27 with the JFET off and with the JFET on.

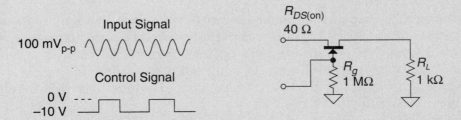

FIGURE 8–27   JFET switching circuit

> **Step 1.** Calculate the signal output with the JFET off. The load (R_L) is in series with the off JFET, which has a resistance in the megohms. Therefore, the signal output would be zero.
> $$V_{out} = 0 \text{ V}$$

EXAMPLE 8.8 continued

> **Step 2.** Calculate the signal output with the JFET on. The load is in series with the on JFET, which has a resistance of 40 Ω ($R_{DS(on)}$). Therefore, the signal output can be calculated using the voltage divider formula.
>
> $V_{out} = V_{in} \times R_L/(R_L + R_{DS(on)}) = 100\ \text{mV}_{p\text{-}p} \times (1\ \text{k}\Omega/1\ \text{k}\Omega + 40\ \Omega) = 96\ \text{mV}_{p\text{-}p}$

EXAMPLE 8.9

> Calculate the signal output of the E-MOSFET switching circuit in Figure 8–28 with the E-MOSFET off and with the E-MOSFET on.
>
>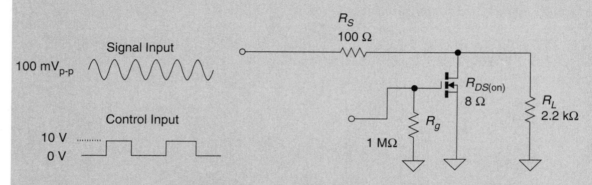
>
> **FIGURE 8–28**   E-MOSFET switching circuit
>
> **Step 1.** Calculate the signal output with the E-MOSFET off. The load forms a voltage divider with R_s when the E-MOSFET is off. The shunt switching circuit will have an output signal when the FET is off.
>
> $V_{out} = V_{in} \times R_L/(R_s + R_L) = 100\ \text{mV}_{p\text{-}p} \times (2.2\ \text{k}\Omega/100\ \Omega + 2.2\ \text{k}\Omega) = 96\ \text{mV}_{p\text{-}p}$
>
> **Step 2.** Calculate the signal output with the E-MOSFET on. Resistor R_s is in series with the on E-MOSFET, which has a resistance of 8 Ω ($R_{DS(on)}$).
>
> $V_{out} = V_{in} \times R_{DS(on)}/(R_s + R_{DS(on)}) = 100\ \text{mV}_{p\text{-}p} \times (8\ \Omega/(100\ \Omega + 8\ \Omega) = 7\ \text{mV}_{p\text{-}p}$

The troubleshooting procedure for predicting circuit values does not require the technician to look up device specifications. The technician is working with estimates, and the measured values may vary considerably. However, with a working knowledge of FETs and some common sense, the technician can troubleshoot FET circuits using estimated values. In every instance, the art of troubleshooting consists of first deciding (calculating or estimating) what values you expect to find, and then measuring the values.

MOSFET Amplifier and Switching Circuits

1. Calculate and record DC voltages, z_{in}, z_{out}, A_v, and V_{out} for the circuit in Figure 8–29. (Assume I_{DSS} = 1.5 mA.)

   $V_G =$ _____ $V_S =$ _____ $V_D =$ _____ $V_{DS} =$ _____
   $z_{in} =$ _____ $z_{out} =$ _____ $A_v =$ _____ $V_{out} =$ _____

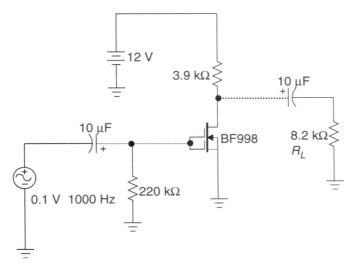

FIGURE 8–29

2. Simulate and record the DC voltages and signal values for the circuit in Figure 8–29.

   $V_G =$ _____ $V_S =$ _____ $V_D =$ _____
   $V_{DS} =$ _____ $A_v =$ _____ $V_{out} =$ _____

3. Calculate, draw, and label the output waveform for the circuit in Figure 8–30. (Assume $R_{DS(on)}$ = 5 Ω.)

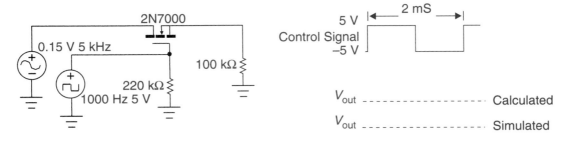

FIGURE 8–30

4. Simulate, draw, and label the output waveform for the circuit in Figure 8–30.

MOSFET Amplifier and Switching Circuits

I. Objectives

1. To construct a D-MOSFET amplifier circuit and measure circuit values.
2. To construct an E-MOSFET switching circuit and measure circuit values.

II. Test Equipment

      (1) sine wave generator   (1) dual-trace oscilloscope
      (1) DC power supply

III. Components

      Transistors: (1) 40673 D-MOSFET, (1) 2N7000 E-MOSFET
      Resistors: (1) 4.7 Ω, (2) 47 Ω, (1) 1 kΩ, (1) 5.6 kΩ, (1) 100 kΩ
      Capacitors: (2) 10 μF

IV. Procedure

1. Calculate DC voltages, z_{in}, z_{out}, A_v, and V_{out} for the circuit in Figure 8–31. (Assume I_{DSS} = 10 mA.)

   $V_G =$ _____ $V_S =$ _____ $V_D =$ _____ $V_{DS} =$ _____
   $z_{in} =$ _____ $z_{out} =$ _____ $A_v =$ _____ $V_{out} =$ _____

2. Construct the MOSFET amplifier shown in Figure 8–31. (Note the 40673 D-MOSFET is a dual-gate device; however, this experiment requires only a single-gate device. The two gates can be tied together and the device will function as a single-gate D-MOSFET.)

3. Measure DC voltages, z_{in}, z_{out}, V_{out}, and the A_v of the circuit in Figure 8–31.

   $V_G =$ _____ $V_S =$ _____ $V_D =$ _____ $V_{DS} =$ _____
   $z_{in} =$ _____ $z_{out} =$ _____ $A_v =$ _____ $V_{out} =$ _____

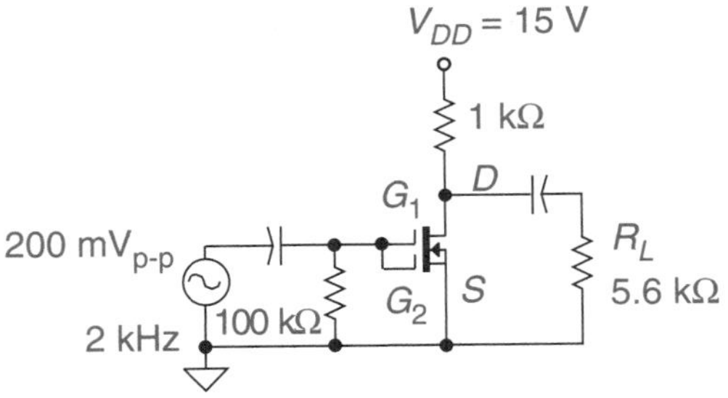

FIGURE 8–31

4. Calculate the output voltage for the circuit in Figure 8–32 with 5 V applied to the control
    signal and with 0 V applied to the control signal. (Assume $R_{DS(on)} = 6\ \Omega$.)

    $V_{out(control = 5\ V)} =$ _____ $V_{out(control = 0\ V)} =$ _____

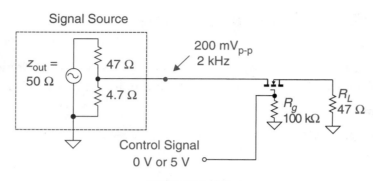

FIGURE 8–32

5. Construct a MOSFET switch as shown in Figure 8–32.

6. Measure the input and output signal voltages of the circuit in Figure 8–32 with 5 V applied
    to the control signal and with 0 V applied to the control signal. (Assume $R_{DS(on)} = 5\ \Omega$.)

    $V_{in(control = 5\ V)} =$ _____ $V_{in\ (control = 0\ V)} =$ _____
    $V_{out(control = 5\ V)} =$ _____ $V_{out(control = 0\ V)} =$ _____

7. Using measured values, calculate $R_{DS(on)}$.

    $R_{DS(on)} =$ _____

V. Points to Discuss

1. What is the actual value of I_{DSS} of the D-MOSFET used in the circuit in Figure 8–31?

2. What is the actual value of gm_0 of the D-MOSFET used in the circuit in Figure 8–31?

3. Explain how circuitry can be added to the amplifier circuit in Figure 8–31 to adjust the
    DC drain voltage in a positive direction.

4. Explain how circuitry can be added to the amplifier circuit in Figure 8–31 to adjust the DC drain voltage in a negative direction.

5. Why is it necessary to add circuitry to the function generate to lower its output impedance in order to drive the circuit in Figure 8–32?

6. Could the value of R_g in Figure 8–32 be adjusted up or down without affecting the operation of the circuit? Explain.

QUESTIONS

8.1 Introduction to MOSFETs

____ 1. The MOS in MOSFET stands for _____.

    a. material of semiconductor

    b. metal-oxide semiconductor

    c. most standard semiconductor

    d. none of the above

____ 2. The JFET depends on a reverse-biased PN junction for isolation, but the MOSFET uses thin layers of _____ as an insulator between the gate and the channel.

    a. silicon dioxide

    b. glass

    c. rubber

    d. enhanced glass

8.2 D-MOSFET

____ 3. The schematic symbol for an N-channel D-MOSFET is _____.

____ 4. When the voltage on the gate of an N-channel D-MOSFET is negative and reduces current carriers, the MOSFET is said to be working in the _____ mode.

    a. depletion

    b. enhancement

    c. conductive

    d. local

____ 5. Drain current can flow through a D-MOSFET operating in _____.

    a. the depletion mode only

    b. the enhancement mode only

    c. either the depletion or the enhancement mode

    d. the local mode only

____ 6. If an amplifier using a D-MOSFET is designed to have zero bias between the gate and source, the drain current at quiescent will equal _____.

    a. 0 mA

    b. 10 mA

    c. I_{DSS}

    d. R_d/gm

____ 7. The schematic symbol for a P-channel D-MOSFET is _____.

8.3 E-MOSFET

____ 8. The schematic symbol for an N-channel E-MOSFET is _____.

   a. ⊣⊢   b. ⊔⊢   c. ⊔⊢   d. ⊔⊢

____ 9. Drain current can flow through an E-MOSFET operating in _____.
   a. the depletion mode only
   b. the enhancement mode only
   c. either the depletion or the enhancement mode
   d. the local mode only

____ 10. The amount of gate-to-source voltage needed to cause drain current to start to flow is called the _____ voltage.
   a. cutoff
   b. cut-on
   c. saturation
   d. threshold

____ 11. With zero gate voltage applied, the E-MOSFET switching circuit acts as a/an _____ circuit.
   a. conductive path
   b. short
   c. open
   d. low resistance

____ 12. When the E-MOSFET switching circuit is turned on, the resistance between the drain and source is less than _____.
   a. $1\,\Omega$
   b. $R_{DS(on)}$
   c. R_g/gm
   d. R_d/gm

____ 13. For class-A operation, N-channel E-MOSFETs must have _____.
   a. the source voltage positive with respect to the gate voltage
   b. the source voltage positive with respect to the drain voltage
   c. the gate voltage positive with respect to the source voltage
   d. the gate voltage negative with respect to the source voltage

____ 14. A type of biasing circuit often used with E-MOSFET amplifier circuits is the_____.
   a. drain-feedback bias
   b. source bias
   c. self-bias
   d. constant current bias

____ 15. The schematic symbol for a P-channel E-MOSFET is _____.

   a. ⊣⊢   b. ⊔⊢   c. ⊔⊢   d. ⊔⊢

8.4 Handling MOSFETs

_____ 16. When working with MOSFETs, special handling precautions should be taken. Which of the following is not one of these precautions?

   a.   All test equipment and tools should be connected to earth ground.

   b.   MOSFET devices should have their leads shorted together for shipment and storage.

   c.   Never remove or insert MOSFET devices with the power off.

   d.   Workers handling MOSFET devices should have grounding straps attached to their wrists.

8.5 MOSFET Parameters

_____ 17. If exceeded, it will cause the FET to break down and conduct.

   a. $V_{(BR)DSS}$

   b. I_{DSS}

   c. $V_{GS(TH)}$

   d. $V_{GS(on)}$

_____ 18. The gate-to-source bias that permits drain current (I_D) to start flowing in E-MOSFETs is _____.

   a. $V_{(BR)DSS}$

   b. I_{DSS}

   c. $V_{GS(TH)}$

   d. $V_{GS(on)}$

8.6 MOSFET Applications

_____ 19. One reason FETs are ideal for use in the front end of receivers is _____.

   a.   the fact they generate less internal noise than bipolar transistors

   b.   the inability to control the gain of the FET by controlling its bias point

   c.   their ability to operate at temperatures above 500°C

   d.   both a and b

_____ 20. E-MOSFETs are excellent devices to use in switching circuits because _____.

   a. $R_{DS(on)}$ is normally less than 10 Ω

   b.   they are normally open until a control signal turns the device on

   c.   only a small amount of power is needed to drive the control circuitry

   d.   all of the above

_____ 21. More than one signal can be transmitted over a single line by utilizing _____.

   a.   time multiplexing

   b.   single-line production (SLP)

   c.   series MOSFET control

   d.   noise separation

____ 22. A _____ circuit converts a DC signal or a low frequency signal into a high frequency signal.

    a. multiplexing

    b. clipper

    c. chopper

    d. demultiplexing

____ 23. Circuit designs using both N-channel and P-channel MOSFETs are called _____ circuits.

    a. NP

    b. PN

    c. two-channel

    d. CMOS

8.7 Troubleshooting FET Circuits

____ 24. If the generator has an internal impedance of 100 kΩ and connecting it to the circuit in Figure 8–33(a) causes the generator to load, what is probably wrong with the circuit?

    a. Resistor R_g is open.

    b. Resistor R_d is shorted.

    c. Resistor R_d is open.

    d. The MOSFET is defective.

____ 25. For the circuit in Figure 8–33(a), if the voltage on the drain is 9 V and the voltage on the source and gate is 0 V, what is wrong with the circuit?

    a. Resistor R_g is open.

    b. Resistor R_g is shorted.

    c. The MOSFET is defective.

    d. The circuit is operating properly.

____ 26. For the circuit in Figure 8–33(a), if the voltage on the drain is 16 V and the voltage on the gate is 0 V, what is wrong with the circuit?

    a. Resistor R_d is open.

    b. Resistor R_d is shorted.

    c. Resistor R_g is shorted.

    d. The circuit is operating properly.

____ 27. For the circuit in Figure 8–33(b), if the voltage on the drain is 6 V and the voltage on the gate is 6 V, what is wrong with the circuit?

    a. Resistor R_g is open.

    b. Resistor R_g is shorted.

    c. The MOSFET is defective.

    d. The circuit is operating properly.

____ 28. All DC voltages measure correctly in the circuit in Figure 8–33(b), but the signal voltage on the drain is near zero. What is wrong with the circuit?

    a. Resistor R_d is open.

    b. Resistor R_d is shorted.

    c. Resistor R_L is shorted.

    d. Resistor R_L is open.

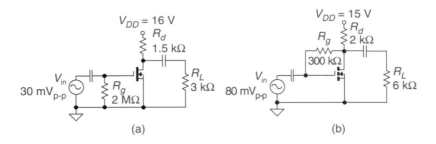

FIGURE 8–33

____ 29. For the circuit in Figure 8–34(a), if the voltage on the control input is 5 V and the signal voltage across the load is zero, what is wrong with the circuit?

a. Resistor R_g is open.

b. Resistor R_g is shorted.

c. The MOSFET is defective.

d. The circuit is operating properly.

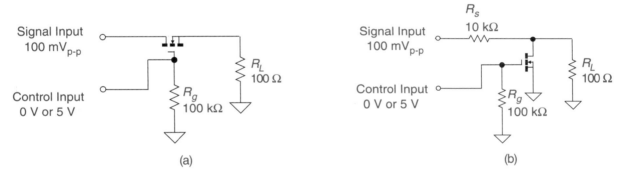

FIGURE 8–34

____ 30. For the circuit in Figure 8–34(b), if the voltage on the control input is 5 V and the signal voltage across the load is near zero, what is wrong with the circuit?

a. Resistor R_g is open.

b. Resistor R_g is shorted.

c. The MOSFET is defective.

d. The circuit is operating properly.

PROBLEMS

1. Figure 8–35 shows the transconductance curve of a D-MOSFET. Using the curve, determine the values of $V_{GS(\text{off})}$ and I_{DSS}.

   $V_{GS(\text{off})} =$ _____ $I_{DSS} =$ _____

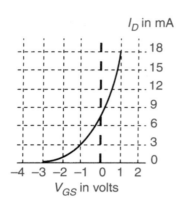

FIGURE 8–35

2. Calculate V_G, V_S, V_D, and V_{DS} for the circuit in Figure 8–36, if $I_{Dss} = 10$ mA.

   $V_G =$ _____ $V_S =$ _____ $V_D =$ _____ $V_{DS} =$ _____

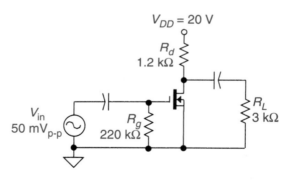

FIGURE 8–36

3. Calculate z_{in}, z_{out}, A_v, and V_{out} for the circuit in Figure 8–36.

   $z_{\text{in}} =$ _____ $z_{\text{out}} =$ _____ $A_v =$ _____ $V_{\text{out}} =$ _____

4. Calculate z_{in}, z_{out}, A_v, and V_{out} for the circuit in Figure 8–36, if gm_0 is known to equal 6 mS.

   $z_{\text{in}} =$ _____ $z_{\text{out}} =$ _____ $A_v =$ _____ $V_{\text{out}} =$ _____

5. Calculate V_G, V_S, V_D, and V_{DS} for the circuit in Figure 8–37, if $I_{Dss} = 2.5$ mA.

   $V_G =$ _____ $V_S =$ _____ $V_D =$ _____ $V_{DS} =$ _____

6. Calculate z_{in}, z_{out}, A_v, and V_{out} for the circuit in Figure 8–37.

   $z_{\text{in}} =$ _____ $z_{\text{out}} =$ _____ $A_v =$ _____ $V_{\text{out}} =$ _____

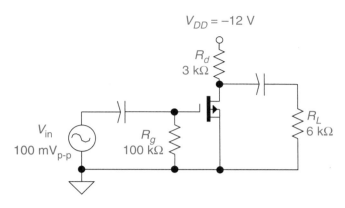

FIGURE 8–37

7. Figure 8–38 shows the transconductance curve of an E-MOSFET. Using the curve, determine the value of $V_{GS(Th)}$.

$V_{GS(Th)} = $ _____

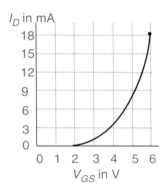

FIGURE 8–38

8. Calculate the V_G, V_S, V_D, and V_{DS} for the circuit in Figure 8–39. Assume $I_{D(on)} = 2$ mA.

$V_G = $ _____ $V_S = $ _____ $V_D = $ _____ $V_{DS} = $ _____

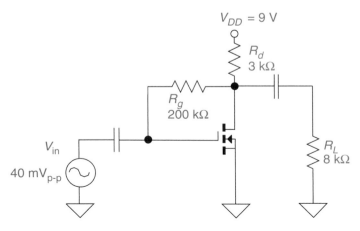

FIGURE 8–39

9. Calculate z_{in}, z_{out}, A_v, and V_{out} for the circuit in Figure 8–39. Assume gm at the Q-point equals 3 mS.

    z_{in} = _____ z_{out} = _____ A_v = _____ V_{out} = _____

10. Calculate the signal output of the switching circuit in Figure 8–40 with the MOSFET off and with the MOSFET on.

    Switch on = _____   Switch off = _____

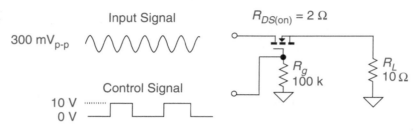

FIGURE 8–40

11. Calculate the signal output of the switching circuit in Figure 8–41 with the MOSFET off and with the MOSFET on.

    MOSFET off = _____   MOSFET on = _____

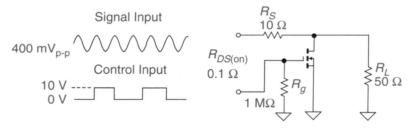

FIGURE 8–41

12. Design and draw a switching circuit that will switch a 15 V supply on and off to a 100 Ω load. The control source should see the switching circuit as a 100 kΩ impedance. The transconductance curve of the E-MOSFET to be used in the circuit is shown in Figure 8–42.

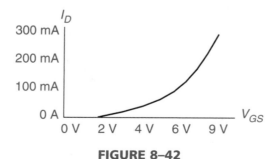

FIGURE 8–42

CHAPTER 9

Basics of Operational Amplifiers

OBJECTIVES

After studying the material in this chapter, you will be able to describe and/or analyze:
○ the two basic rules used for analyzing op-amp circuits,
○ a voltage follower circuit,
○ a noninverting amplifier circuit,
○ negative feedback in amplifier circuits,
○ an inverting amplifier circuit,
○ a comparator circuit, and
○ troubleshooting procedures for op-amp circuits.

9.1 INTRODUCTION TO OP-AMPS

Operational amplifiers, op-amps, are general purpose amplifiers with several specific characteristics. The name *operational amplifier* comes from the use of these amplifiers in performing mathematical operations in analog computers. Today most analog computers have been replaced with digital computers, but op-amps have become the basic building blocks for electronics circuitry. The characteristics of op-amps are as follows:

1. High gain—an internal voltage gain of 100,000 or more is not uncommon.
2. High input impedance—an input impedance of 1 MΩ or more is typical.
3. Low output impedance—an output impedance of less than 100 Ω is typical. Once connected in a circuit, the output impedance of the op-amp approaches zero.
4. Differential input—differential input means that the output is determined by the difference voltage between the two inputs.

OP-AMP BASICS

The op-amp has five basic terminals: two power supply inputs, two signal inputs, and one signal output. Figure 9–1 shows the symbol for the op-amp with its five basic terminals. The power supply terminals are usually connected to a positive and a negative voltage in the range of 5 to 18 V. However, op-amps can operate off of a single voltage supply and ground.

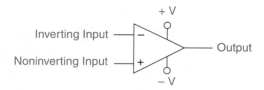

FIGURE 9–1   Op-amp symbol with five basic terminals

There are two differential signal inputs: one is called the *inverting input* (denoted by –), and the other is called the *noninverting input* (denoted by +). A positive voltage at the inverting input (with respect to the noninverting input) will drive the output negative, whereas a positive voltage at the noninverting input (with respect to the inverting input) will drive the output positive. The output voltage is the difference between the inputs times the voltage gain of the op-amp. The output voltage has an approximate range equal to the voltage between the negative and positive supplies.

Two assumptions can be made that simplify the analysis of op-amp circuits. The first assumption is that there is no current flowing in or out of the input terminals of the op-amp. This assumption is possible because of the high input impedance of op-amps.

The second assumption is that if the output is not in saturation, the voltage between the inverting and noninverting input terminals is essentially zero. This assumption is possible because of the high voltage gain of the op-amp. The output voltage swing is limited by the size of the power supplies; therefore, the output signal divided by the high voltage gain results in a negligible input voltage between the input terminals of the op-amp. Example 9.1 shows the small input voltage needed to drive the output of the op-amp into saturation.

EXAMPLE 9.1

> If the op-amp is using 15 V supplies, the output signal would have a range of a little less than –15 V to +15 V. What voltage difference between the inputs would drive the output to +15 V?
>
> $$A_v = V_{out}/V_{in}$$
> $$V_{in} = V_{out}/A_v = 15\ V/100,000\ (\text{Op-amp gain} = 100\ \text{k.})$$
> $$V_{in} = 150\ \mu V$$

The actual voltage between the input terminals is so small it normally cannot be measured and can be assumed to be zero.

THE TWO BASIC RULES OF OP-AMPS

The previously stated assumptions are so useful in analyzing op-amp circuits that we will call them Rule 1 and Rule 2. They are restated below:

> **Rule 1:** The current in or out of either input is negligible.
>
> **Rule 2:** If the output is *not* in saturation, the voltage between the two input terminals is essentially zero.

9.2 VOLTAGE FOLLOWER

A voltage follower circuit is constructed by connecting the output of the op-amp back to the inverting input as shown in Figure 9–2. A signal applied to the noninverting input will drive the output to precisely the same voltage of the input signal. In other words, the output signal follows the input signal—hence, the name *voltage follower*.

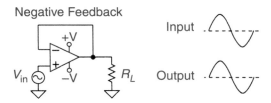

FIGURE 9–2 Op-amp voltage follower

The operation of the voltage follower circuit can be explained using Rule 2 of op-amps. Assuming that the output of the voltage follower is not in saturation, the voltage between the two input terminals is zero. The input signal voltage present on the noninverting input is also present on the inverting input. The inverting input is connected to the output; therefore, the voltage at the output is the same as that at the input.

Of what benefit is a circuit that has an output signal the same as its input signal? The input and output voltage may be the same, but the output impedance of the voltage follower circuit is very low. An op-amp has a relatively low output impedance. For example, the 741 op-amp has a typical output impedance of 75 Ω. Once the op-amp is connected in a circuit with negative feedback, the output impedance drops to near zero ohms. Example 9.2 shows the advantage of using a voltage follower circuit as a buffer between a high impedance signal source and a low impedance load.

EXAMPLE 9.2

Calculate the signal output in both circuits in Figure 9–3.

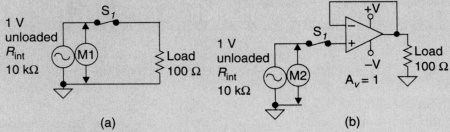

FIGURE 9–3 Advantage of using a voltage follower circuit

Step 1. Calculate the voltage the meter will read in both circuits if switch S_1 is open. In both circuits, if switch S_1 is open, the meter will read the unloaded generator output of 1 V.

$$V_{M1} = V_{M2} = 1 \text{ V (switch open)}$$

EXAMPLE 9.2 continued

Step 2. Calculate the voltage the meter will read in the circuit in Figure 9–3(a) if switch S_1 is closed. In the circuit in Figure 9–3(a), the 10 kΩ internal resistance of the generator (R_{int}) is connected in series with the 100 Ω load. The meter will read the voltage across the load.

$$V_{M1} = V_{in} \times R_L/(R_{int} + R_L)$$
$$V_{M1} = 1 \text{ V} \times 100 \ \Omega/(10 \text{ k}\Omega + 100)$$
$$V_{M1} = 10 \text{ mV (switch closed)}$$

Step 3. Calculate the voltage the meter will read in the circuit in Figure 9–3(b) if switch S_1 is closed. In the circuit in Figure 9–3(b), the 10 kΩ internal resistance of the generator (R_{int}) is connected in series with the input impedance of the op-amp. The input impedance of the op-amp is so high it acts like an open circuit; thus, the output of the generator is not loaded down, and the meter still reads 1 V with S_1 closed.

$$V_{M2} = 1 \text{ V (switch closed)}$$

Step 4. Calculate the signal voltages out of both circuits in Figure 9–3. The circuit in Figure 9–3(a) delivers only 10 mV to the load because the generator is severely loaded down. The circuit in Figure 9–3(b) uses a voltage follower as a buffer. The high input impedance of the voltage follower does not load down the generator, so there is 1 V present at the input of the voltage follower. Since the voltage gain of the voltage follower is one, the signal across the load will be 1 V.

$$V_{out} = 10 \text{ mV (circuit a)} \qquad V_{out} = 1 \text{ V (circuit b)}$$

The benefits of the voltage follower circuit should now be clear. Its high input impedance and low output impedance makes it the ideal circuit for matching low impedance loads with high impedance signal sources. It should be noted that in our calculations, we have assumed the input impedance of the op-amp to be pure resistance (phase angle of zero degrees). For most applications, assuming input and output impedance of devices to be pure resistance gives acceptable results.

9.3 NONINVERTING AMPLIFIER

Op-amp applications are not limited to simple voltage followers. Instead of feeding back all of the output to the inverting input, we can feed back only a part of the output.

In Figure 9–4, resistors R_f and R_i form a voltage divider providing negative feedback. One-third of the output voltage will be present at the inverting input. The voltages at the inverting and noninverting inputs are essentially the same. Therefore, the signal input voltage is equal to one-third of the output voltage. Since the output is three times the input, the circuit has a voltage gain of three. By changing the feedback ratio, any voltage gain is possible.

Let us examine the current flow in the circuit in Figure 9–4. If there is +1 V present at the noninverting input, there is also +1 V at the inverting input. Positive one volt at the inverting input causes 1 mA through R_i (1 kΩ). The current flowing in or out of the input of the op-amp is essentially zero, so the current flowing through resistor R_i is forced to flow through R_f (2 kΩ). The 1 mA flowing through R_f causes the output to be two volts more positive (1 mA $\times$ 2 kΩ) than the inverting input, so the output is equal to three volts. The voltage gain of the circuit is equal to three.

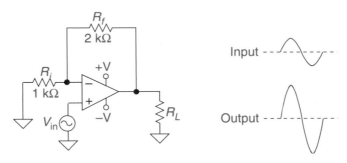

FIGURE 9–4 Noninverting amplifier

$A_v = V_{out}/V_{in}$

$A_v = 3\text{ V}/1\text{ V} = 3$

If the input is changed to –1 V, the current flow will be reversed, and the output will be driven to –3 V. Note that the output is in phase with the input as shown in Figure 9–4. The voltage gain of the circuit is dependent on the ratio of the two resistors and can be derived as follows:

$A_v = V_{out}/V_{in}$

$V_{out} = V_{Ri} + V_{Rf}$ (R_i and R_f form an unloaded voltage divider.)

$V_{in} = V_{Ri}$ (Voltage on the inverting input equals voltage on the noninverting input.)

$A_v = (V_{Ri} + V_{Rf})/V_{Ri}$

$A_v = 1 + (V_{Rf}/V_{Ri})$

$A_v = 1 + (IR_f/IR_i)$ (The same current flows through R_f and R_i.)

Therefore:

$A_v = 1 + (R_f/R_i)$

EXAMPLE 9.3

Calculate the input impedance and the voltage gain of the circuit in Figure 9–5.

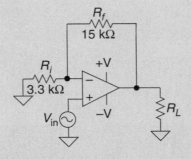

FIGURE 9–5 Noninverting amplifier

Step 1. Input impedance of the op-amp is very high, and when negative feedback is used, the impedance is increased even further. The input impedance of a noninverting amplifier, therefore, can be thought of as infinite. In this text, if you are asked to calculate the input impedance of a noninverting amplifier or voltage follower,

EXAMPLE 9.3 continued

> do not attempt to give your answer as a numeral. Instead respond by stating
> that the input impedance of the noninverting amplifier is very high.
>
> $$z_{in} = \text{very high}$$
>
> **Step 2.** Voltage gain:
>
> $$A_v = 1 + (R_f/R_i) = 1 + (15 \text{ k}\Omega/3.3 \text{ k}\Omega) = 5.55$$

9.4 INSIDE THE OP-AMP AND NEGATIVE FEEDBACK

In this section, we will look inside the op-amp and develop the voltage gain formula for the
noninverting amplifier using negative feedback theory. It is not necessary to know what is inside
the op-amp to work with op-amp circuits, since we already know the voltage gain of the
noninverting amplifier is equal to $A_v = 1 + (R_f/R_i)$. Some people, however, will want to know
what is in the little package and how it can be controlled by negative feedback. Others may prefer
to leave it as a mystery.

There is no magic inside the op-amp—just circuitry. Figure 9–6 shows a simplified schematic
(the actual circuit is considerably more complex) of the circuitry inside op-amp circuits. The first
and second stages are differential amplifiers, and the third stage is a common-collector amplifier.
The circuit has excellent qualities: high impedance differential inputs, high voltage gain, and
low output impedance.

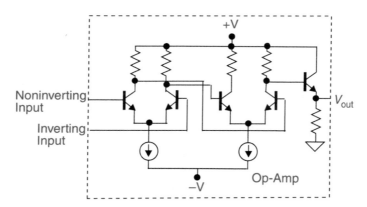

FIGURE 9–6   Simplified schematic of an op-amp

There are two problems. First, while the voltage gain is high, it is not predictable. Second, in
order for the circuit to function properly, the transistors used in differential amplifiers must be
matched exactly. The first problem can be corrected by using negative feedback to control the
gain. The solution to the second problem is to incorporate the circuitry into an integrated circuit
(IC) called an op-amp. All transistors inside an integrated circuit are identically matched because
they are all made from the same block of silicon. The excellent qualities of differential amplifiers
and the availability of matched transistors in integrated circuits have made the differential
amplifier the most popular circuit used in op-amps.

The circuitry in Figure 9–6 is shown as a block in Figure 9–7. A negative feedback loop has
been added to control the voltage gain of the circuit. The high voltage gain of the amplifier without
feedback is called the open-loop voltage gain (A_{ol}). The feedback circuit takes a small portion of

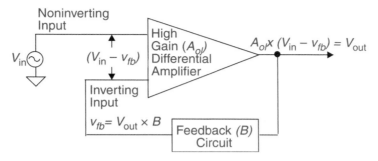

FIGURE 9–7 Feedback control of voltage gain

the output signal and feeds it back to the inverted differential input. The ratio of the feedback voltage to the output voltage is called the feedback factor and is represented by the letter "B" ($B = V_{fb}/V_{out}$). The differential amplifier amplifies the difference voltage between the input signal and the feedback signal ($V_{in} - V_{fb}$). The difference voltage is extremely small—up to this point, we have considered it essentially zero. In this section, however, we will acknowledge a small difference across the inputs of the op-amp.

Let us examine the block diagram and develop a formula for the closed-loop voltage gain (A_v). The signal out is equal to the difference between the inputs times the open-loop gain of the amplifier.

$$V_{out} = A_{ol} \times (V_{in} - V_{fb})$$

The feedback signal is equal to the output signal times the feedback factor.

$$V_{fb} = V_{out} \times B$$

Therefore:

$$V_{out} = A_{ol} \times (V_{in} - (V_{out} \times B))$$

Expanding:

$$V_{out} = (A_{ol} \times V_{in}) - (A_{ol} \times V_{out} \times B)$$

Rearranging:

$$V_{out} + (A_{ol} \times V_{out} \times B) = A_{ol} \times V_{in}$$

Factoring:

$$V_{out} \times (1 + (A_{ol} \times B)) = A_{ol} \times V_{in}$$

Rearranging:

$$V_{out}/V_{in} = A_{ol}/(1 + (A_{ol} \times B))$$

The closed-loop voltage gain (A_v) is equal to V_{out}/V_{in}

Therefore:

$$A_v = A_{ol}/(1 + (A_{ol} \times B))$$

If $A_{ol} \times B$ is much larger than 1, the equation can be simplified to:

$$A_v = A_{ol}/(A_{ol} \times B)$$

Which simplifies to:

$$A_v = 1/B$$

The voltage gain of the complete amplifier system is simply equal to the reciprocal of the feedback factor if $A_{ol} \times B \gg 1$, where $\gg$ stands for much larger than. Rearranging the "if" statement gives $A_{ol} \gg 1/B$. Since $A_v = 1/B$, it can be stated that if an open-loop gain (A_{ol}) of an amplifier is much larger (ten times) than the desired closed-loop gain (A_v), the gain of the complete circuit is equal to the reciprocal of the feedback factor ($1/B$).

Figure 9–8 shows that a feedback circuit can consist of two resistors: R_f and R_i. These two resistors form a voltage divider circuit that is divided by the output signal. Because of the high impedance of the amplifier, the voltage divider is virtually an unloaded voltage divider. The feedback signal can be calculated by multiplying the output signal times the ratio of the resistors. Recall that the feedback signal is also equal to the output signal times the feedback factor; therefore, the feedback factor must equal the ratio of the resistors. Since the voltage gain of the circuit is equal to the reciprocal of the feedback factor, the closed-loop voltage gain is equal to the ratio of the two resistors.

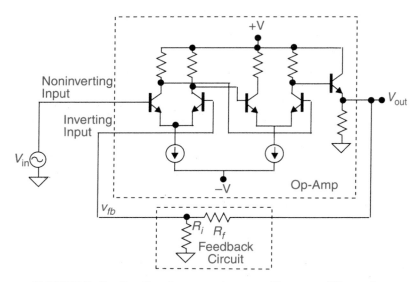

FIGURE 9–8   Feedback network controlling amplifier gain

The formula for the closed-loop voltage gain of the circuit is developed as follows:

$$V_{fb} = V_{out} \times B$$
$$V_{fb} = V_{out} \times R_i/(R_i + R_f)$$

Therefore:

$$B = R_i/(R_i + R_f)$$

and

$$1/B = (R_i + R_f)/R_i$$

Simplified to:

$$1/B = 1 + R_f/R_i$$

since

$$A_v = 1/B$$

Then:

$$A_v = 1 + R_f/R_i$$

The voltage gain of a system (integrated or discrete) using a high gain amplifier circuit can be controlled by taking part of the output and feeding it back to the input in such a manner that this feedback signal cancels part of the input signal. This is referred to as negative feedback. The feedback circuit can be a simple two-resistor voltage divider if the following two conditions are met:

1. The feedback circuit is not loaded.
2. The open-loop voltage gain (A_{ol}) is much higher than the closed-loop gain (A_v).

The use of negative feedback makes it easy to adjust the voltage gain. Also, the voltage gain obtained is predictable. The disadvantages are the loss of voltage gain and the need for a high impedance differential amplifier circuit. Today, with the use of op-amp integrated circuits, these disadvantages are no longer valid, and most amplifier systems are designed using op-amps with a negative feedback loop to control the gain.

9.5 THE INVERTING AMPLIFIER

Another application of the op-amp is as an inverting amplifier as shown in Figure 9–9. When a signal is applied to the inverting input, the output is driven opposite in polarity from the input.

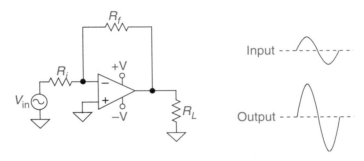

FIGURE 9–9 Inverting amplifier

Since the noninverting input is connected to ground and there is essentially zero volts between the two inputs, the inverting input is also at zero volts. This means the current flowing through R_i is equal to V_{in}/R_i. The current flowing into the op-amp is negligible; consequently, the current that flows through R_i also flows through R_f. This makes the output voltage a function of the input current times the value of R_f. The current flow is such that it causes the output voltage to be inverted. The voltage gain of the circuit can be derived as follows:

$$A_v = V_{out}/V_{in}$$

$$V_{in} = V_{Ri} = I_i \times R_i \text{ (Same current flows through } R_i \text{ and } R_f.)$$

Therefore:

$$-V_{out} = -V_{Rf} = -I_i \times R_f \text{ (The direction of current flow causes a negative output voltage.)}$$

$$A_v = -I_i \times R_f / I_i \times R_i$$

Therefore:

$$A_v = -R_f/R_i$$

The circuit gain is given by the relationship $A_v = -R_f/R_i$ and can be controlled simply by selecting the ratio of the two resistors. The minus sign denotes phase inversion and is sometimes dropped from the equation.

How does the inverting amplifier differ from the noninverting amplifier? Obviously one inverts, and the gain formulas are slightly different, but there is a more significant difference: the circuit input impedances are different. With the noninverting amplifier, the input is connected directly to the high input impedance of the op-amp. The input impedance of the noninverting amplifier, for most applications, is so high that it can be considered infinite. By comparison, the input impedance of the inverting amplifier is simply equal to the value of R_i. Resistor R_i is connected to the high input impedance terminal, but the action of the feedback resistor keeps the inverting terminal at zero volts or virtual ground. Consequently, input impedance of the inverting amplifier is equal to R_i.

EXAMPLE 9.4

Calculate the input impedance and voltage gain of the circuit in Figure 9–10.

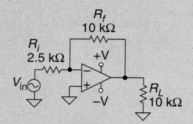

FIGURE 9–10   Inverting amplifier

Step 1. Calculate input impedance.
$$z_{in} = R_i = 2.5 \text{ k}\Omega$$
Step 2. Calculate the voltage gain.
$$A_v = R_f/R_i = 10 \text{ k}\Omega/2.5 \text{ k}\Omega = 4$$

VIRTUAL GROUND

An inverting amplifier has the noninverting input connected to ground. Op-amp Rule 1 tells us that if one input is at ground level, the other input would be at ground level if the output is not in saturation. The inverting input pin will be at zero volts, but this point is not a true ground. It is referred to as *virtual ground*. As we study op-amp circuits, we will see several applications that make use of virtual ground.

9.6 COMPARATORS

The circuits we have discussed so far have used op-amps in closed-loop configurations. In closed-loop configurations, a portion of the output signal is fed back to the input of the op-amp. Amplifier circuits use negative feedback to stabilize the circuit and make gains predictable.

The comparator circuit we will examine in this section uses an op-amp in an open-loop configuration. In the open-loop configuration, the output signal is not fed back to the input. Without negative feedback, the extremely high voltage gain of the op-amp will force the output into saturation. The output of the comparator will always be in one of two states: positive saturation or negative saturation. The saturation state will depend on the signals present at the inputs.

The saturation voltage levels are mainly a function of the voltage of the power supplies. Positive and negative saturation normally occur within one volt of the power supplies voltages. For example, if a circuit is operating with +15 V and –15 V supplies, saturation levels would be approximately +14 V and –14 V. The saturation levels are sometimes referred to as *power supply rails*. If the output is at the positive rail, it is in positive saturation; the negative rail would mean negative saturation. If the output of the op-amp is between the rails, its output is in the linear operating region. The output of an amplifier circuit is normally between the power supply rails and operates in the linear region. The output of a comparator circuit is normally at the positive or negative power supply rail. In this text, we will assume the saturation voltages to be one volt below the voltages of the power supplies unless otherwise stated.

Figure 9–11(a) shows a zero crossover comparator. There is no feedback from the output to the input. The inverting input of the op-amp is connected to ground. When the noninverting input goes above ground (zero volts), the output will be driven to positive saturation by the high voltage gain of the op-amp. When the noninverting input drops below ground, the output switches to negative saturation. These waveforms are shown in Figure 9–11(b).

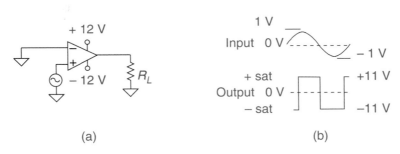

(a)   (b)

FIGURE 9–11   Zero crossover comparator

Example 9.5 shows how only a small difference voltage is needed at the inputs of a comparator to drive the output into saturation.

EXAMPLE 9.5

If the op-amp in Figure 9–11 has an open-loop gain of 100,000, calculate the output voltage when the input voltage is 1 mV.

Step 1. Calculate the output voltage using the voltage gain formula.

$$V_{out} = V_{in} \times A_v$$
$$V_{out} = 1 \text{ mV} \times 100,000$$
$$V_{out} = 100 \text{ V}$$

Step 2. Examine the voltage level calculated in step 1. If the calculated level exceeds the saturation voltage of the circuit, the circuit is in saturation. In this case, 11 V is the saturation level, so the circuit is in saturation, and the output is 11 V.

$$V_{out} = \text{saturation}$$
$$V_{out} = 11 \text{ V}$$

If an oscilloscope is used to measure the signal between the inputs in Figure 9–11, a 2 $V_{p\text{-}p}$ signal will be measured. One may think this is in conflict with basic op-amp Rule 2; however, Rule 2 states that if the output is *not* in saturation, the voltage between the two input terminals is essentially zero. Since the comparator circuit normally operates in saturation, there is usually a measurable signal present between the input terminals.

9.7 TROUBLESHOOTING OP-AMP CIRCUITS

Op-amp circuits can normally be analyzed for troubleshooting by using two rules: (1) the current in or out of either input is negligible, and (2) if the output is *not* in saturation, the voltage between the two input terminals is essentially zero. Armed with these two rules and a pin identification (pinout) of the IC, the next step is to analyze the circuit to determine the configuration of the op-amp: buffer, inverting amplifier, noninverting amplifier, comparator, and so on.

Once the function of the circuit is known, the next step is to estimate and measure the DC and signal values in the circuit. When working with inverting circuits, remember that the inverting input pin is at virtual ground, if the noninverting pin is connected to ground. Therefore, the signal input needs to be measured before R_i on inverting amplifiers. In order to not overlook important test points, ask yourself what DC and signal values should be on each pin of the IC. Always check power supply voltages at the IC pins. If one of the inputs is found to be bad, this input should be traced back to the previous stage. Remember that this stage could be loading the previous stage, causing the input to be incorrect.

If all the inputs to an IC are correct but the output is incorrect, the IC may be defective. Before replacing the IC, check for loading from the next stage or check for bad components. It is sometimes practical to disconnect the next stage to check for loading. Most of the components associated with op-amps are resistors and capacitors. A good physical inspection should be performed on these associated components. Resistors often show signs of heat damage if they are defective. Capacitors can be bypassed by a known good capacitor to check for opens and checked with an ohmmeter for shorts. Parallel paths for ohmmeter readings must be considered when making in-circuit checks.

The amount of time spent on analyzing associated circuitry before replacing the IC depends on the difficulty of replacement. If the IC is in a socket, it can be easily replaced. If the IC is soldered directly onto the PC board, replacement is a big job, in which case more time is justified in analyzing the circuit before assuming the IC to be defective.

Basic Op-Amp Circuits

1. Calculate the input impedance and voltage gain, and draw the output waveform, for the circuit in Figure 9–12.

    Z_{in} = _____ A_v = _____

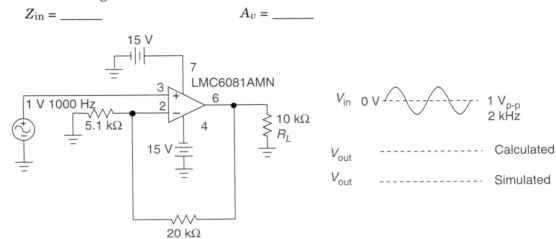

FIGURE 9–12

2. Simulate and draw the output waveform for the circuit in Figure 9–12.

3. What value would resistor R_f in Figure 9–12 need to be changed to in order to make the gain of the circuit equal to 12?

    New value of R_f = _____

4. Calculate the input impedance and voltage gain, and draw the output waveform, for the circuit in Figure 9–13.

    Z_{in}= _____ A_v = _____

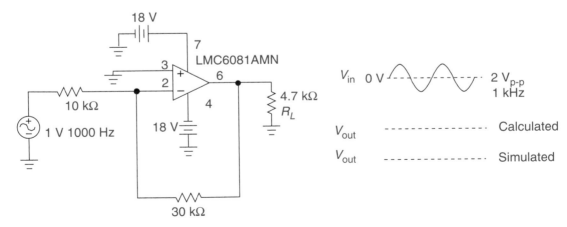

FIGURE 9–13

5. Simulate and draw the output waveform for the circuit in Figure 9–13.

6. What value would resistor R_f in Figure 9–13 need to be changed to in order to make the gain of the circuit equal to 7?

    New value of $R_f =$ _____

7. Calculate and draw the output waveform for the circuit in Figure 9–14.

    $Z_{in} =$ _____ $A_v =$ _____

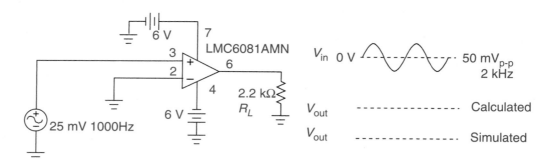

FIGURE 9–14

8. Simulate and draw the output waveform for the circuit in Figure 9–14.

Basic Op-Amp Circuits

I. Objectives

1. To design a noninverting amplifier with a desired gain and to experimentally verify the circuit design.
2. To design an inverting amplifier with a desired gain and to experimentally verify the circuit design.
3. To experimentally determine how the output of an op-amp voltage comparator will respond to changes in input voltages.

II. Test Equipment

+12 V power supply   −12 V power supply

Voltmeter   Oscilloscope

Function generator

III. Component List

Op-amp: (1) 741

Resistors: (1) 1 kΩ, (1) 2.2 kΩ, (1) 4.7 kΩ, (1) 10 kΩ

IV. Procedure

1. Calculate the voltage gain for the circuit in Figure 9–15.

   $A_v =$ _____

2. Draw the calculated output waveform for the circuit in Figure 9–15.

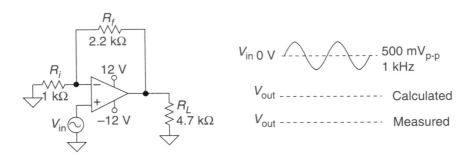

FIGURE 9–15

3. Determine the input impedance for the circuit in Figure 9–15.

   $z_{in} =$ _____

4. Construct the circuit in Figure 9–15 and measure and draw the output waveform.

5. Calculate the voltage gain using measured values.

    $A_v = $ _____

6. Attempt to measure the input impedance. (Use ×10 oscilloscope probe.)

    $z_{in} = $ _____

7. Make the necessary modifications to the circuit in Figure 9–15 to change the voltage gain of the circuit to 11. Verify that the voltage gain of the modified circuit is 11.

8. Calculate the voltage gain and the input impedance for the circuit in Figure 9–16.

    $A_v = $ _____ $z_{in} = $ _____

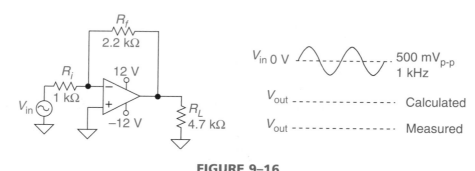

FIGURE 9–16

9. Draw the calculated output waveform for the circuit in Figure 9–16.

10. Construct the circuit in Figure 9–16 and measure and draw the output waveform.

11. Calculate the voltage gain using measured values.
$A_v =$ _____

12. Measure the input impedance.
$z_{in} =$ _____

13. Make the necessary modifications to the circuit in Figure 9–16 to change the voltage gain of the circuit to 10. Verify that the voltage gain of the modified circuit is 10.

14. Draw the calculated output waveform for the circuit in Figure 9–17.

FIGURE 9–17

15. Construct the circuit in Figure 9–17 and measure and record the output waveform.

V. Points to Discuss

1. How does the input impedance of the inverting amplifier compare to that of the noninverting amplifier?

2. What would be the voltage gain of the noninverting amplifier (Figure 9–15) if the value of R_f was changed to 500 Ω? $A_v =$ _____

3. What would be the voltage gain of the inverting amplifier (Figure 9–16) if the value of R_f was changed to 500 Ω? $A_v =$ _____

4. Explain why the noninverting amplifier cannot have a voltage gain of less than one.

5. Explain the phase relationships between the inputs and outputs of all three circuits.

6. Explain why there are only two output levels for the circuit in Figure 9–17.

QUESTIONS

9.1 Introduction to Op-Amps

____ 1. Which of the following is not a general characteristic of an op-amp?
   a. High input impedance.
   b. Low output impedance.
   c. Differential output.
   d. High voltage gain.

____ 2. What are the five basic terminals of the op-amp?
   a. Two signal outputs, two power supplies, and one signal input.
   b. Two signal inputs, two signal outputs, and one power supply.
   c. Two signal inputs, two power supplies, and one signal output.
   d. One signal input, three power supplies, and one signal output.

____ 3. The current flowing in or out of either input of an op-amp is _____.
   a. in the milliamps
   b. in the 1 to 5 amp range
   c. hole current flow only
   d. negligible

____ 4. If the output is *not* in saturation, the voltage between the two input terminals is _____.
   a. near zero
   b. approximately equal to +V
   c. equal to V_{in}
   d. equal to $I_{out} \times R_f$

9.2 Voltage Follower

____ 5. The voltage gain of the voltage follower circuit is equal to _____.
   a. near zero
   b. one
   c. R_f/R_i
   d. $1 + R_f/R_i$

____ 6. The voltage follower circuit has a _____.
   a. high input impedance
   b. low output impedance
   c. unity voltage gain
   d. all of the above

____ 7. A voltage follower circuit is an excellent _____ between a high impedance signal source and a low impedance load.
   a. voltage amplifier
   b. frequency multiplier
   c. comparator
   d. buffer

9.3   Noninverting Amplifier

____ 8.   The voltage gain of the noninverting amplifier is equal to _____.
     a.   near zero
     b.   one
     c. $-R_f/R_i$
     d. $1 + R_f/R_i$

____ 9.   The gain of the noninverting amplifier can be controlled by changing the ratio of _____ resistors.
     a.   two
     b.   three
     c.   four
     d.   none of the above

____ 10.   The input impedance of the noninverting amplifier is _____ .
     a.   equal to R_i
     b.   equal to R_i in parallel with R_f
     c.   equal to $R_L/(A_v + 1)$
     d.   very high

9.4   Inside the Op-Amp and Negative Feedback

____ 11.   Differential amplifiers are popular in integrated circuits because of the availability of _____ transistors.
     a.   power
     b.   current controlled
     c.   matched
     d.   FET and bipolar

____ 12.   The ratio of the feedback voltage to the output voltage is called the _____.
     a.   feedback factor
     b.   back ratio
     c.   transconductance ratio
     d.   feed-around factor

____ 13.   The long formula for the voltage gain of a noninverting amplifier using negative feedback is ____.
     a. $A_v = (1/B) \times R_f$
     b. $A_v = A_{ol}/(1 + (A_{ol} \times B))$
     c. $A_v = (1 + (A_{ol} \times B))/A_{ol}$
     d. $A_v = A_{ol}/(A_{ol} \times B)$

____ 14.   If _____ is much larger than the desired closed-loop gain, the gain of the complete circuit is equal to the reciprocal of the feedback factor.
     a.   the ratio of R_f/R_i
     b.   the ratio of $(R_f \| R_L)/R_i$
     c.   virtual gain
     d.   the open-loop gain

9.5 The Inverting Amplifier

_____ 15. The voltage gain of the inverting amplifier is equal to _____.
   a. near zero
   b. one
   c. $-R_f/R_i$
   d. $1 + R_f/R_i$

_____ 16. The input impedance of the inverting amplifier is _____ .
   a. equal to R_i
   b. equal to R_i in parallel with R_f
   c. equal to $R_L/(A_v + 1)$
   d. very high

_____ 17. One of the greatest differences between the inverting amplifier and the noninverting amplifier is the difference in _____.
   a. voltage gain
   b. input impedance
   c. output impedance
   d. output signal swing

_____ 18. If an inverting amplifier has the noninverting pin of the op-amp connected to ground, the inverting pin of the op-amp will be _____.
   a. at ground
   b. at virtual ground
   c. equal to V_{in}
   d. equal to V_{out}

9.6 Comparators

_____ 19. A comparator circuit makes use of the op-amp without using _____.
   a. power supply pins
   b. negative feedback
   c. the inverting input
   d. the noninverting input

_____ 20. The output of the comparator will always be _____.
   a. at positive saturation
   b. at negative saturation
   c. at positive or negative saturation
   d. in the linear operating region

9.7 Troubleshooting Op-Amp Circuits

_____ 21. The signal voltage on the inverting input pin of the op-amp in Figure 9–18(a) measures 1 $V_{p\text{-}p}$. What is wrong with the circuit?
   a. R_i is open.
   b. R_f is open.
   c. R_L is open.
   d. The circuit is functioning properly.

_____ 22. The signal voltage on the inverting input pin of the op-amp in Figure 9–18(a) measures zero. What is wrong with the circuit?

a. R_i is open.

b. R_f is open.

c. R_L is open.

d. The circuit is functioning properly.

_____ 23. The output signal of the circuit in Figure 9–18(a) switches between positive and negative saturation. What is wrong with the circuit?

a. R_i is open.

b. R_f is open.

c. R_L is open.

d. The circuit is functioning properly.

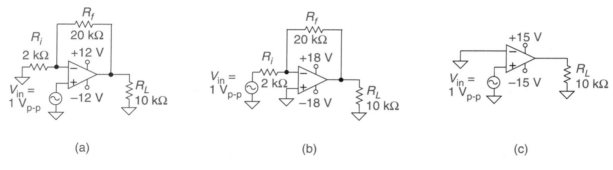

(a)　　　　　　　　　　　　　　(b)　　　　　　　　　　　　　　(c)

FIGURE 9–18

_____ 24. The signal voltage on the inverting input pin of the op-amp in Figure 9–18(b) measures 1 $V_{p\text{-}p}$. What is wrong with the circuit?

a. R_i is open.

b. R_f is open.

c. R_L is open.

d. The circuit is functioning properly.

_____ 25. The signal voltage on the inverting input pin of the op-amp in Figure 9–18(b) measures zero. What is wrong with the circuit?

a. R_i is open.

b. R_f is open.

c. R_L is open.

d. The circuit is functioning properly.

_____ 26. The output signal of the circuit in Figure 9–18(c) switches between positive and negative saturation. What is wrong with the circuit?

a. The op-amp is defective.

b. R_L is shorted.

c. R_L is open.

d. The circuit is functioning properly.

PROBLEMS

1. Calculate z_{in}, A_v, and V_{out} for the circuit in Figure 9–19. Draw the output waveform.

   z_{in} _____ $A_v =$ _____ $V_{out} =$ _____

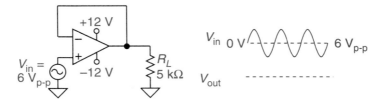

FIGURE 9–19

2. Calculate the signal output in both circuits in Figure 9–20.

   V_{out} (a) = _____ V_{out} (b) = _____

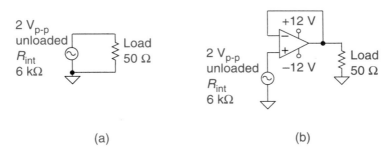

(a)   (b)

FIGURE 9–20

3. Calculate z_{in}, A_v, and V_{out} for the circuit in Figure 9–21. Draw the output waveform.

   z_{in} _____ $A_v =$ _____ $V_{out} =$ _____

4. In a noninverting amplifier, if R_f is 10 kΩ and R_i is 2 kΩ, find A_v and the input impedance of the circuit.

   $A_v =$ _____ $z_{in} =$_____

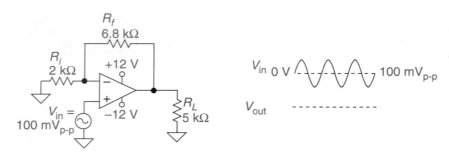

FIGURE 9–21

5. A noninverting amplifier has a gain of 100 and R_i equals 1 kΩ. What is the value of R_f?
   $R_f = $ _____

6. Sketch an amplifier that has a voltage gain of 25 and a high input impedance.

7. Calculate z_{in}, A_v, and V_{out} for the circuit in Figure 9–22. Draw the output waveform.

   $z_{in} = $ _____ $A_v = $ _____ $V_{out} = $ _____

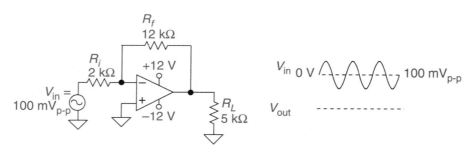

FIGURE 9–22

8. In an inverting amplifier, if R_f is 10 kΩ and R_i is 2 kΩ, find A_v and the input impedance of the circuit.

   $A_v = $ _____ $z_{in} = $ _____

9. An inverting amplifier has a gain of 100 and R_i equals 1 kΩ. What is the value of R_f?
   $R_f = $ _____

10. Sketch an amplifier that has a voltage gain of 10, a phase shift of 180 degrees, and an input impedance of 600 Ω.

11. Calculate the following values for the circuits in Figure 9–23:

    Circuit a: $z_{in} =$ _____ $A_v =$ _____ $V_{out} =$ _____

    Circuit b: $z_{in} =$ _____ $A_v =$ _____ $V_{out} =$ _____

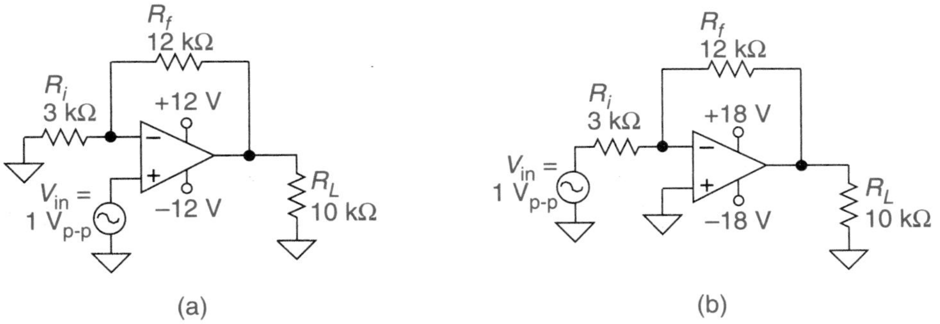

(a)   (b)

FIGURE 9–23

12. Draw and label the output waveform for the circuit in Figure 9–24.

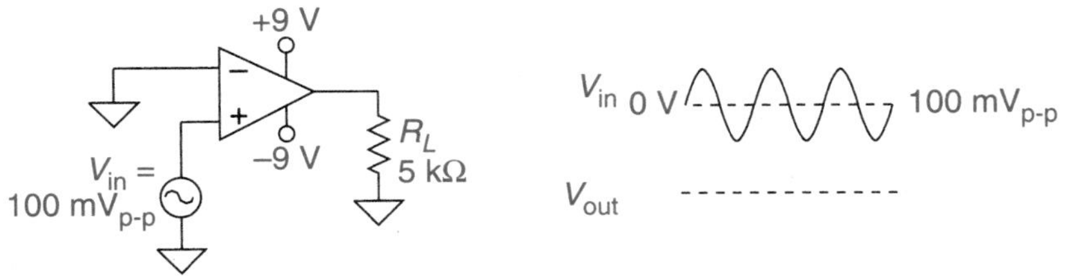

FIGURE 9–24

CHAPTER

Op-Amp Limitations

10

OBJECTIVES

After studying the material in this chapter, you will be able to describe and/or analyze:

○ input bias current (I_{IB}),

○ input offset current (I_{IO}),

○ input offset voltage (V_{IO}),

○ output voltage swing (V_o) limitations,

○ the effects of output short-circuitry current (I_{sc}) limiting,

○ the frequency response limits of the op-amp,

○ circuit gain problems using decibels (dBs),

○ slew rate (SR) limitations, and

○ troubleshooting procedures for IC op-amp circuits.

INTRODUCTION TO OP-AMP LIMITATIONS

Voltage followers, noninverting amplifiers, inverting amplifiers, and comparators are four basic op-amp circuits. Before exploring numerous op-amp applications, a closer look at the limitations of op-amps is in order. The op-amp is close to being the perfect amplifier, but it has some limitations. It is important that you know what the limitations are and how these limitations affect the operation of op-amp circuits.

10.1 INPUT BIAS CURRENT (I_{IB})

A previously stated rule indicated that the current in or out of either input is negligible. Indeed, the small current flow through the input pins of an op-amp is in the nanoamps and is normally negligible, but not zero. This small current is called *input bias current* (I_{IB}) and is one of the parameters given on op-amp data sheets. The lower the input bias current, the better the op-amp.

Figure 10–1 shows a test circuit that can be used to measure input bias current. The circuit is a voltage follower with 1 MΩ in the feedback loop. Ideally, there should be zero current flowing through the 1 MΩ resistor. With no current through the resistor, the voltage across the resistor would equal zero. However, if there is input bias current (I_{IB}) flowing, the current flow will cause a voltage drop across the 1 MΩ resistor. Example 10.1 shows how the input bias current can be calculated.

EXAMPLE 10.1

Calculate the input bias current (I_{IB}) for the op-amp shown in Figure 10–1 if the meter reads 0.12 V.

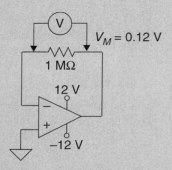

FIGURE 10–1   Test circuit for measuring input bias current

Step 1.   Calculate input bias current. The current that flows through the inverting input pin of the op-amp must also flow through the 1 MΩ resistor. Therefore, Ohm's law can be used to calculate the input bias current for the op-amp.

$$(I_{IB}) = 0.12 \text{ V}/1 \text{ M}\Omega = 120 \text{ nA}$$

Since the input bias currents are small, they normally are not a problem. However, for circuits with large resistance paths to ground, these small currents can cause the output to be offset from zero volts. The engineer can design circuitry to compensate for input bias current. To understand this circuitry, we need to understand the next parameter called input offset current.

10.2 INPUT OFFSET CURRENT (I_{IO})

Both op-amp inputs have a bias current (I_{IB}). The manufacturer strives to keep the magnitude of the input bias currents as small as possible and equal on both inputs. If the inverting and noninverting inputs have equal bias currents, a circuit can be designed to cancel out the effects of the input bias currents. The difference in the absolute values of bias input currents is called *input offset current* (I_{IO}) and is calculated by the formula $I_{IO} = |I_{IB} + |-|I_{IB}-|$.

The rule for compensation of input bias current is to have the resistance from the noninverting input pin to ground the same as the resistance from the inverting input pin to ground. For this calculation, the output terminal is considered at ground potential. Example 10.2 shows how an inverting amplifier circuit can be modified to eliminate the effects of input bias current.

EXAMPLE 10.2

Modify the circuit in Figure 10–2(a) so it compensates for input bias current.

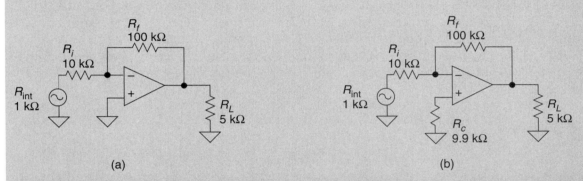

(a)   (b)

FIGURE 10–2   Inverting amplifier with compensation resistor

Step 1. Calculate the value of R_c. Make the resistance to ground in each input pin the same. Looking from the inverting input pin to ground, the resistance is equal to $R_f \| (R_i + R_{int})$. This assumes the output is at ground potential.

$$R_c = R_f \| (R_i + R_{int}) = 100 \text{ k}\Omega \| (10 \text{ k}\Omega + 1 \text{ k}\Omega) = 9.9 \text{ k}\Omega$$

Step 2. Insert the compensation resistor (R_c) calculated in step 1 between the noninverting input and ground. Figure 10–2(b) shows the inverting amplifier circuit with the compensation resistor added.

If the op-amp in Figure 10–2(b) has an offset bias current of zero, the output will be fully compensated for input bias current. The compensation network depends on the input bias currents being equal. With equal bias currents, both inputs will develop the same voltage, and the difference voltage across the inputs will be zero. Since the output is a function of the difference voltage between the inputs, the output will also remain zero.

Ideally, the offset bias current would equal zero, but in practice, it is usually kept below 25% of the average input bias currents. This permits the compensation to be at least 75% effective.

10.3   INPUT OFFSET VOLTAGE (V_{IO})

If op-amps are connected at random, as in Figure 10–3(a), some will be driven into positive saturation, and some will be driven into negative saturation. The op-amp functions as though an input voltage is present as shown in Figure 10–3(b). Actually, what is driving the op-amp into

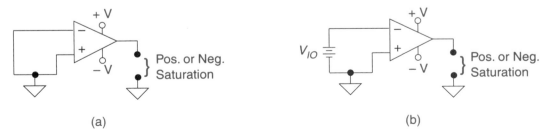

(a)   (b)

FIGURE 10–3   Offset voltage

saturation is the imbalance of the internal circuitry. When considering the effects of this internal imbalance, it is best to use the model shown in Figure 10–3(b). In the model, all internal voltage imbalances are lumped into one external voltage source called *input offset voltage* (V_{IO}).

Figure 10–4 shows a test circuit that can be used to determine the input offset voltage (V_{IO}). The circuit has been compensated for input offset current, and the inputs are both connected to ground. The output voltage will still be offset from ground. The output voltage offset is caused by input offset voltage (V_{IO}). The input offset voltage can be calculated by measuring the output offset voltage and dividing by the voltage gain of the circuit as shown in Example 10.3.

EXAMPLE 10.3

The output voltage of the circuit in Figure 10–4 is 1.5 V. What is the input offset voltage (V_{IO}) of the op-amp?

Step 1. Calculate the voltage gain (A_v) of the circuit.
$$A_v = R_f/R_i = 100 \text{ k}\Omega/100 \text{ }\Omega = 1000$$

Step 2. Calculate the input offset voltage.
$$V_{\text{in}} = V_{\text{out}}/A_v = 1.5 \text{ V}/1000 = 1.5 \text{ mV}$$

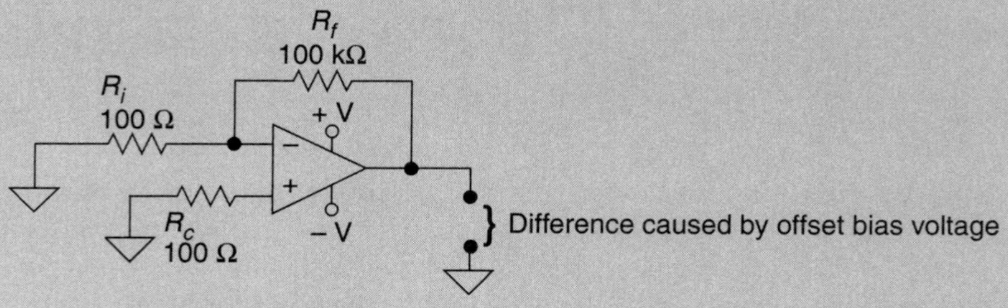

FIGURE 10–4   Test circuit for input offset voltage

Correction for offset voltage should be according to the manufacturer's specifications in the technical data. This usually consists of connecting a variable resistor between the two pins which is called *offset null*, or connecting the variable resistor between the power supply and a trim pin. In general, the manufacturer's suggested method is the most economical and effective way to compensate for input offset voltage. Figure 10–5 shows an inverting amplifier using a 741 op-amp. This circuit has been compensated for both input bias currents and input offset voltage.

The null potentiometer needs to be adjusted for the specific op-amp in the circuit. This means if an op-amp is replaced in the field, it will be the responsibility of the technician to null the circuit. The procedure is simple: all signal sources are adjusted to zero. If a source cannot be adjusted to zero, it may be necessary to apply a jumper to disable the input signal. Caution should be taken to make sure the jumper does not damage the circuit. The potentiometer is then adjusted for zero volts at the output.

Engineers often design op-amp circuits that do not include correction circuitry for input offset current (I_{IB}) or input offset voltage (V_{IO}). Manufacturers have improved op-amps over the years. Op-amps are available with maximum input offset currents of 100 pA (0.1 nA). This compares to 500 nA for the 741 op-amp. There are, in fact, many op-amps on the market that do not have pins available for adding correction circuits for input offset voltage.

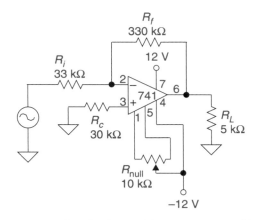

FIGURE 10–5   Compensation for input bias currents and input offset voltage

10.4 OUTPUT VOLTAGE SWING (V_o)

The output voltage swing (V_o) of an op-amp is limited by the DC supply voltage applied to the op-amp. If the output signal is driven too close to the DC supply, clipping will occur. The voltage levels at which clipping occurs are referred to as the power supply rails. The output signal of most op-amps can swing within approximately a volt of the supply voltages without being clipped. To estimate the power supply rails, subtract 1 V from the DC supply voltages. This is only an approximation, but it is close enough for troubleshooting circuits.

EXAMPLE 10.4

What is the maximum peak-to-peak signal that can be developed without distortion at the output of the circuit in Figure 10–6? Draw the output waveform for the input signal shown.

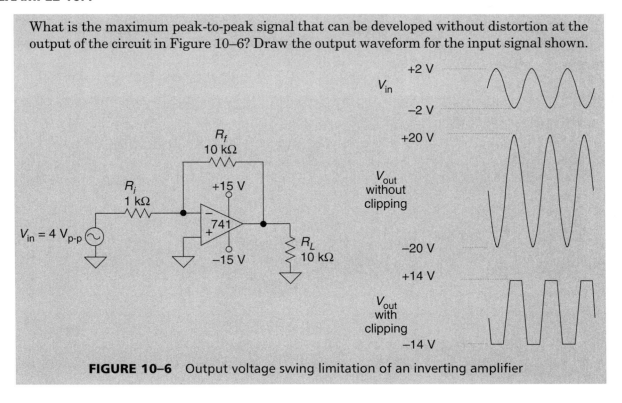

FIGURE 10–6   Output voltage swing limitation of an inverting amplifier

EXAMPLE 10.4 continued

> **Step 1.** The positive rail equals the plus supply minus 1 V.
> $$+V_{rail} = +V - 1\ V$$
> $$+V_{rail} = 15\ V - 1\ V = 14\ V$$
> **Step 2.** The negative rail equals the negative supply minus 1 V.
> $$-V_{rail} = -V - (-1\ V)$$
> $$-V_{rail} = -15\ V - (-1\ V) = -14\ V$$
> **Step 3.** The maximum peak-to-peak signal is equal to the difference between the rails.
> $$V_o = +V_{rail} - (-V_{rail})$$
> $$V_o = 14\ V - (-14\ V) = 28\ V_{p\text{-}p}$$
> **Step 4.** Draw the output waveform. Calculate and draw the output waveform without considering clipping levels. Then apply the levels calculated and redraw the output with clipping as shown in Figure 10–6.

10.5   OUTPUT SHORT-CIRCUIT CURRENT (I_{SC})

Op-amps are designed with built in current limiting circuitry. This protection circuit limits the output current to a value that will not cause damage to the internal circuitry of the op-amp. The maximum output current is called output short-circuit current (I_{sc}).

The output pin of an op-amp must deliver current to the load and the feedback loop. Normally, the current flowing into the feedback loop is so small that it can be considered negligible. The load current depends on the resistance of the load and the peak voltage present on the output. If the load current demand exceeds the output short-circuit current (I_{sc}) rating of the op-amp, the built-in current limiting of the op-amp will take over and cause clipping of the output signal. Example 9.5 shows that this clipping can occur at voltages considerably below the rails of the power supply.

EXAMPLE 10.5

> If the op-amp in Figure 10–7 has an output short-circuit current (I_{sc}) of 20 mA, what is the maximum peak-to-peak output signal without distortion? Draw the output waveform for the input signal shown.
> **Step 1.** Calculate and draw the output waveform without considering clipping levels.
> $$A_v = R_f/R_i = 10\ k/1\ k = 10$$
> $$V_{out(peak)} = V_{in} \times A_v = 2\ V_{p\text{-}p} \times 10 = 20\ V_{p\text{-}p}\ (\text{See Figure 10–7.})$$
> **Step 2.** Calculate the maximum peak signal output that current limiting will permit.
> $$V_{out(peak)} = I_{sc} \times R_L$$
> $$V_{out(peak)} = 20\ mA \times 100$$
> $$V_{out(peak)} = 2\ V_p$$
> **Step 3.** If the signal calculated in step 1 exceeds the maximum peak signal calculated in step 2, apply the levels and redraw the output with clipping. (See Figure 10–7.)

EXAMPLE 10.5 continued

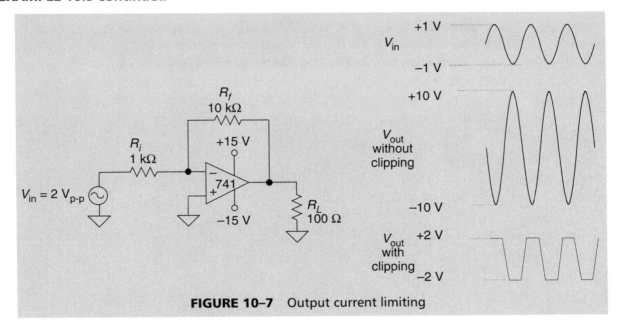

FIGURE 10-7 Output current limiting

10.6 FREQUENCY RESPONSE

The noninverting amplifier circuit in Figure 10–8(a) is designed for a voltage gain (A_v) of 100. The frequency response curve for the amplifier is shown in Figure 10–8(b). The graph shows that DC and low frequency signals are amplified with a gain of 100, but as the signal frequency increases, the gain of the circuit drops off. Cutoff frequency occurs when the gain drops to 70.7% of the DC gain. All frequencies below the cutoff frequency are in the bandpass, and those above the cutoff frequency are in the bandstop. In this section, we will examine what causes the voltage gain of an op-amp to drop off as signal frequency increases.

Figure 10–8(a) shows resistors R_i and R_f forming a voltage divider that feeds a portion of the output voltage back to the inverting input. The feedback voltage (V_f) is calculated using the formula $V_f = V_{out} \times (R_i/R_i + R_f)$. The ratio of $R_i/R_i + R_f$ is equal to the feedback ratio (B). The feedback ratio (B) is the fractional part of the output voltage fed back to the input.

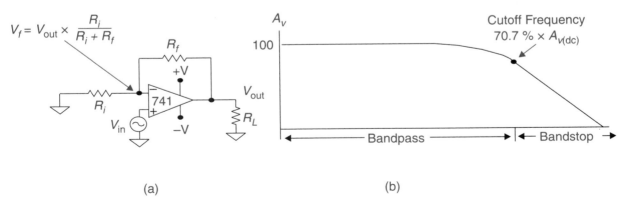

(a)

(b)

FIGURE 10-8 Noninverting amplifier frequency response

The simple formula $A_v = 1 + R_f/R_i$, used for calculating the voltage gain, is a ratio of two resistors and does not consider op-amp characteristics. This simple formula is derived from the voltage gain formula $A_v = A_{ol}/(1 + A_{ol} \times B)$, where A_{ol} is the open-loop gain of the op-amp and B is the feedback factor. Reviewing how the simple gain formula is derived will help you understand why the gain of the amplifier circuit drops off at higher frequencies.

$$A_v = A_{ol}/(1 + A_{ol} \times B)$$

If:

$$A_{ol} \times B \gg 1$$

Then:

$$A_v = A_{ol}/(A_{ol} \times B)$$
$$A_v = 1/B$$
$$B = R_i/R_i + R_f$$

Therefore:

$$A_v = 1/(R_i/(R_i + R_f))$$
$$A_v = (R_i + R_f)/R_i$$
$$A_v = 1 + R_f/R_i$$

In order for the simplified formula $A_v = 1 + R_f/R_i$ to be valid, the assumption $A_{ol} \times B \gg 1$ must be true. Rearranging the equation $A_{ol} \times B \gg 1$ gives $A_{ol} \gg 1/B$. The desired closed-loop voltage gain is equal to the reciprocal of the feedback factor ($A_v = 1/B$). In order for the simplified gain formula to be valid, the open-loop voltage gain must be much larger than the desired closed-loop voltage gain ($A_{ol} \gg A_v$).

Up to this point, we have considered the op-amp open-loop voltage gain (A_{ol}) to be 100,000 or more. This is true at low frequencies. As frequency increases, the open-loop gain of the op-amp rolls off. This roll-off in gain is normally built into the op-amp circuitry to prevent oscillations at high frequencies. Figure 10–9 shows a graph of the open-loop frequency response for the 741 op-amp.

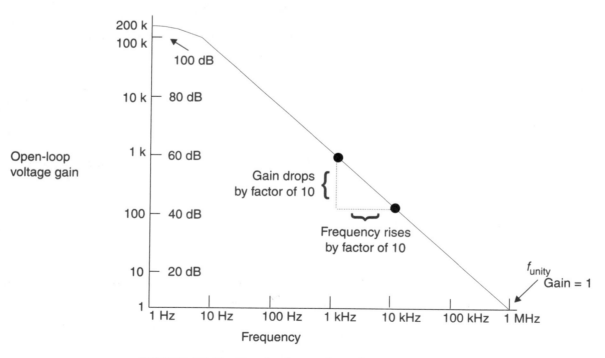

FIGURE 10–9   Graph of open-loop frequency response

The graph shows that the op-amp has an open-loop gain (A_{ol}) of 200,000 when amplifying DC signals. As the frequency of the signal goes up, the open-loop gain drops off. If the frequency goes up by a factor of 10, the gain goes down by a factor of 10. At 1 MHz, the voltage gain of the 741 op-amp drops to unity (one). This frequency is called the unity gain frequency (f_{unity}). Example 10.6 shows how the roll-off of the open-loop gain (A_{ol}) affects the closed-loop gain (A_v) of the amplifier circuit.

EXAMPLE 10.6

Calculate the voltage gain of the circuit in Figure 10–10 for a signal with a frequency of (a) 100 Hz, (b) 4.17 k, and (c) 10 kHz.

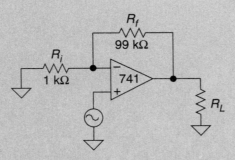

FIGURE 10–10 Noninverting amplifier

Solution "a" A_v **at 100 Hz:**

Step 1. Calculate the circuit feedback factor for the circuit in Figure 10–10.
$$B = R_i/(R_i + R_f)$$
$$B = 1\ \text{k}/(1\ \text{k} + 99\ \text{k})$$
$$B = 0.01$$

Step 2. Find the open-loop voltage gain (A_{ol}) of the op-amp at 100 Hz using the graph in Figure 10–11. Draw a vertical line up from 100 Hz to intersect the curve. Then extend a line horizontally from the intersection to the vertical axis and read the open-loop gain.
$$A_{ol} = 10,000\ @\ 100\ \text{Hz}$$

Step 3. Calculate the voltage gain using the long formula.
$$A_v = A_{ol}/(1 + (A_{ol} \times B))$$
$$A_v = 10\ \text{k}/(1 + (10\ \text{k} \times 0.01))$$
$$A_v = 10\ \text{k}/(1 + 100)$$
$$A_v = 10\ \text{k}/101 = 99.1$$

Solution "b" A_v **at 4.17 kHz:**

Step 1. Find the open-loop voltage gain (A_{ol}) of the op-amp at 4.17 kHz using the graph in Figure 10–10.
$$A_{ol} = 240\ @\ 4.17\ \text{kHz}$$

Step 2. Calculate the voltage gain using the long formula.
$$A_v = A_{ol}/(1 + (A_{ol} \times B))$$
$$A_v = 240/(1 + (240 \times 0.01))$$
$$A_v = 240/(1 + 2.4)$$
$$A_v = 240/3.4 = 70.6$$

EXAMPLE 10.6 continued

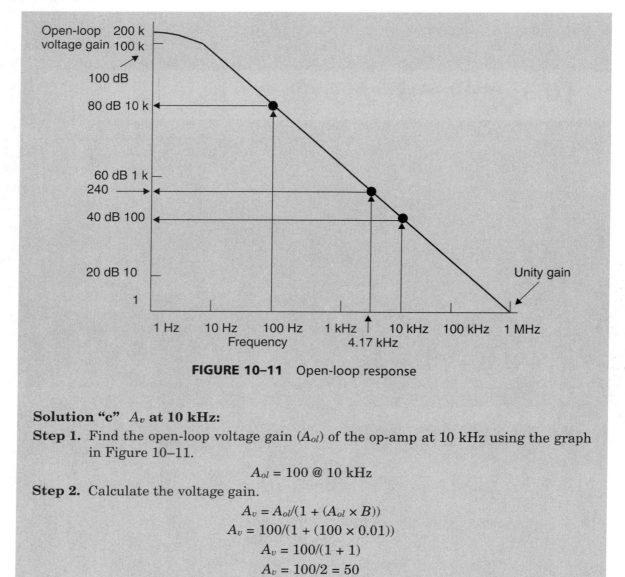

FIGURE 10–11   Open-loop response

Solution "c" A_v **at 10 kHz:**

Step 1. Find the open-loop voltage gain (A_{ol}) of the op-amp at 10 kHz using the graph in Figure 10–11.

$$A_{ol} = 100 @ 10 \text{ kHz}$$

Step 2. Calculate the voltage gain.

$$A_v = A_{ol}/(1 + (A_{ol} \times B))$$
$$A_v = 100/(1 + (100 \times 0.01))$$
$$A_v = 100/(1 + 1)$$
$$A_v = 100/2 = 50$$

Using the simplified gain formula ($A_v = 1 + R_f/R_i$), the voltage gain of the circuit in Figure 10–10 is 100. The value calculated using the simplified formula is an approximation. The actual voltage gain is calculated by using the long formula, which considers the open-loop gain of the op-amp. Table 10–1 compares the results of calculating voltage gain for the three different frequencies using both formulas. When the open-loop gain is large in comparison to the desired closed-loop gain, the simplified voltage gain formula gives acceptable results. As the open-loop gain decreases, the error resulting from using the simple formula increases.

Engineers design amplifier circuits so the desired voltage gain of the circuit is not dependent upon the open-loop voltage gain of the op-amp. If the circuit is designed properly, the voltage gain of the circuit can be calculated using the simplified formula ($A_v = 1 + R_f/R_i$). Technicians must realize, however, that gain does roll off with increases in frequency. They should be able to estimate the cutoff frequency of an amplifier.

TABLE 10–1

Signal Frequency	Long Formula	Simplified Formula	Op-Amp Open-Loop	% Error
100 Hz	99.1	100	10,000	0.9%
4.17 kHz	70.6	100	240	41.6%
10 kHz	50	100	100	100%

The frequency response curve for a closed-loop amplifier is plotted on the same graph (Figure 10–12) as the open-loop gain response curve of the 741 op-amp. The closed-loop gain is held constant at 100 and intersects the open-loop gain curve, and the two curves roll off together. The point where the open-loop gain curve intersects the closed-loop gain line is the approximate cutoff frequency of the circuit. In our example, the approximate cutoff frequency is equal to 10 kHz.

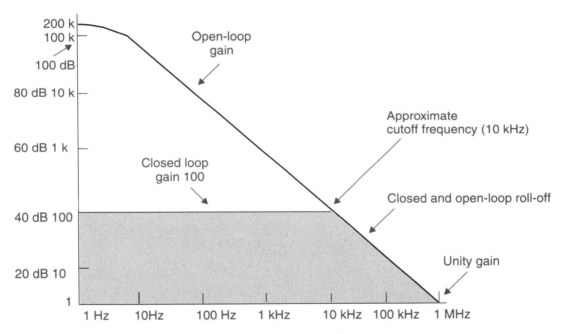

FIGURE 10–12 Open-loop and closed-loop response curve

Figure 10–13 shows the actual frequency response curve and a Bode plot response curve for the amplifier circuit in Figure 10–10. The actual response curve starts to roll off before the closed-loop curve meets the open-loop curve. If you review Example 10.6, the calculations verify that the gain roll-off starts before 10 kHz. The voltage gain of 100 drops to 70.6 at a frequency of 4.17 kHz, and at 10 kHz, the gain has dropped to 50.

The Bode plot is a simplified response curve. The Bode plot assumes the gain of the circuit is constant until it intersects the open-loop gain curve. The intersection point is assumed to be the cutoff frequency. Beyond the cutoff frequency, the circuit gain rolls off at the same rate as the open-loop gain. In Figure 10–13, the Bode plot shows the cutoff frequency occurring at 10 kHz, but the actual cutoff frequency occurs at 4.17 kHz. The difference between 4.17 kHz and 10 kHz may seem large at first, but since the frequency is dimensioned using logarithmic scaling, the two cutoff frequencies are reasonably close. The Bode plot curve is normally used when examining the band response of an amplifier because calculations are easier and the results are acceptable.

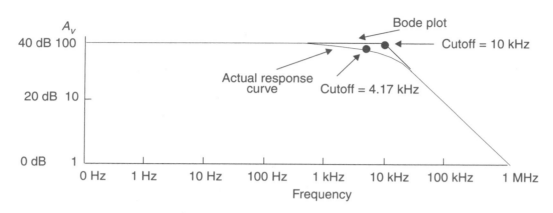

FIGURE 10–13   Actual frequency response curve versus Bode plot response curve

Op-amp data sheets often give a value for the gain bandwidth product (GBW) or unity gain bandwidth (UGB). Unity gain bandwidth (UGB) is equal to the range of frequencies amplified with a gain greater than one. Since the op-amp amplifies all frequencies from DC up to the unity gain frequency (f_{unity}), the unity gain bandwidth equals the unity gain frequency. Sometimes the unity gain bandwidth is called the *small signal bandwidth*, or simply *bandwidth*. The gain bandwidth product (GBW) equals the closed-loop gain times the cutoff frequency ($GBW = A_v \times F_{cutoff}$). The gain bandwidth product (GBW) is equal to unity gain frequency for op-amps having a single pole roll-off curve like the one shown for the 741 op-amp in Figure 10–12. Example 10.7 demonstrates these relationships.

EXAMPLE 10.7

A 741 op-amp circuit has a gain of 100 and a cutoff frequency of 10 kHz. What is the op-amp gain bandwidth product (GBW), unity gain bandwidth, and unity gain frequency (f_{unity})?

Step 1. Calculate the gain bandwidth product (GBW).
$$GBW = A_V \times F_{cutoff} = 100 \times 10 \text{ kHz} = 1 \text{ MHz}$$
Step 2. Calculate the unity gain frequency (f_{unity}).
$$GBW = UGB = f_{unity} = 1 \text{ MHz}$$

When working with circuitry, the technician can calculate the voltage gain of the circuit by using circuit values and then looking up the gain bandwidth product (GBW). Using these values, the cutoff frequency of the circuit can be calculated. Example 10.8 shows this procedure.

EXAMPLE 10.8

Calculate the voltage gain (A_v) and cutoff frequency for the circuit in Figure 10–14. The GBW of the 741 op-amp is 1 MHz.

Step 1. Calculate the voltage gain using circuit values.
$$A_v = 1 + R_f/R_i = 1 + 120 \text{ k}\Omega/2.2 \text{ k}\Omega = 55.5$$
Step 2. Calculate the cutoff frequency.
$$f_{cutoff} = GBW/A_v = 1 \text{ Mz}/55.5 = 18 \text{ kHz}$$

EXAMPLE 10.8 continued

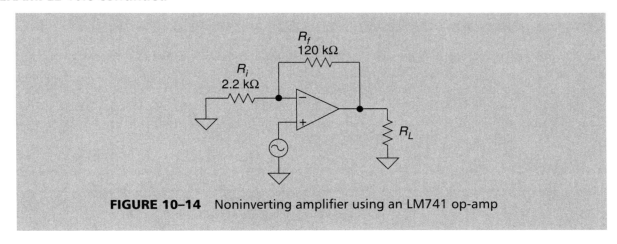

FIGURE 10–14   Noninverting amplifier using an LM741 op-amp

Throughout this section we have been using noninverting amplifiers. The procedure for calculating the cutoff frequency is the same for the inverting amplifier as shown in Example 10.9.

EXAMPLE 10.9

Calculate the voltage gain (A_v) and cutoff frequency for the circuit in Figure 10–15. The GBW of the LF441C op-amp is 2 Mz.

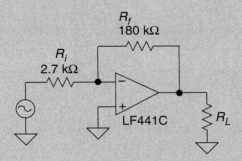

FIGURE 10–15   Inverting amplifier using an LF441C op-amp

Step 1. Calculate the voltage gain using circuit values.

$A_v = -R_f/R_i = -180\ \text{k}\Omega/2.7\ \text{k}\Omega = -66.7$ (Minus sign indicates phase inversion.)

Step 2. Calculate the cutoff frequency.

$f_{\text{cutoff}} = GBW/A_v = 2\ \text{Mz}/66.7 = 30\ \text{kHz}$

In this section, you examined the frequency response of the op-amp circuit. You learned that the op-amp has a bandpass area that starts at DC and goes up in frequency until the cutoff frequency is reached. In the bandpass area, the gain of the amplifier can be considered flat. At the cutoff frequency, the gain of the amplifier rolls off. The gain of the amplifier rolls off as the signal frequency increases because of the decrease in the open-loop gain of the op-amp. If we know the closed-loop gain (A_v) and the gain bandwidth product (GBW), the cutoff frequency can be easily calculated.

10.7   WORKING WITH LOGARITHMIC SCALES

Graphs of frequency response curves use logarithmic scaling on the vertical gain axis and on the horizontal frequency axis. The wide range of gains and frequencies makes log scaling necessary. The voltage gain on the graph we have been working with has gain given as a voltage ratio and also as a dB value. The dB value is equal to twenty times the log of the voltage ratio (V_{out}/V_{in}). Examples 10.10 and 10.11 show how a voltage ratio gain can be converted to a dB gain.

EXAMPLE 10.10

An amplifier has a voltage gain of 134. Express this gain in dB.
$$A_v \text{ in dB} = 20 \times \log A_v \text{ ratio}$$
$$A_v \text{ in dB} = 20 \times \log 134$$
$$A_v \text{ in dB} = 20 \times 2.127$$
$$A_v \text{ in dB} = 42.5 \text{ dB}$$

EXAMPLE 10.11

At the cutoff frequency, the voltage gain of an amplifier drops to 0.707. Express the gain at cutoff in dB.

$$A_v \text{ in dB} = 20 \times \log A_v \text{ ratio}$$
$$A_v \text{ in dB} = 20 \times \log 0.707$$
$$A_v \text{ in dB} = 20 \times -0.15$$
$$A_v \text{ in dB} = -3 \text{ dB}$$

The gain roll-off of amplifiers above the cutoff frequency is rated in dB/decade or dB/octave. Understanding how to work with logarithmic scales permits you to calculate the size of the output signal at frequencies above cutoff. Example 10.12 will show the procedure for calculating the output voltage at frequencies above the cutoff frequency.

EXAMPLE 10.12

For the circuit in Figure 10–16, calculate the following values: A_v expressed as a ratio, A_v expressed in dB, f_{cutoff}, V_{out} of signals below cutoff, the output signal ($V_{out(1dec)}$) at one decade above cutoff, and the output signal ($V_{out(2dec)}$) at two decades above cutoff. Draw and label a Bode plot of the frequency response curve for the circuit shown in Figure 10–17. The 741 op-amp has a GBW of 1 MHz, and its open-loop gain rolls off at a rate of 20 dB/decade.

Step 1.   Calculate the voltage gain using circuit values.
$$A_v = -R_f/R_i = 220 \text{ k}\Omega/1.2 \text{ k}\Omega = 183$$

Step 2.   Express the voltage gain in dB.
$$A_v \text{ in dB} = 20 \log A_v \text{ ratio}$$
$$A_v \text{ in dB} = 20 \log 183$$

EXAMPLE 10.12 continued

$$A_v \text{ in dB} = 20 \times 2.262$$
$$A_v = 45.25\text{dB}$$

Step 3. Calculate the cutoff frequency.

$$f_{cutoff} = GBW/A_v$$
$$f_{cutoff} = 1 \text{ MHz}/183$$
$$f_{cutoff} = 5.46 \text{ kHz}$$

Step 4. Calculate V_{out} in the bandpass region (below cutoff).

$$V_{out} = V_{in} \times A_v$$
$$V_{out} = 25 \text{ mV} \times 183$$
$$V_{out} = 4.6 \text{ V}$$

Step 5. Calculate V_{out} at one decade above cutoff. Since the gain rolls off at 20 dB/decade, the signal output $V_{out(1dec)}$ will be –20 dB below the value V_{out} (4.6 V) in the bandpass.

$$-20 \text{ dB} = 20 \log V_{out(1dec)}/V_{out(ref)}$$
$$-1 \text{ dB} = \log V_{out(1dec)}/V_{out(ref)}$$
$$\text{Antilog} - 1 \text{ dB} = V_{out(1dec)}/V_{out(ref)}$$
$$0.1 = V_{out(1dec)}/V_{out(ref)}$$
$$V_{out(1dec)} = V_{out(ref)} \times 0.1$$
$$V_{out(1dec)} = 4.6 \text{ V} \times 0.1$$
$$V_{out(1dec)} = 0.46 \text{ V}$$

Step 6. Calculate V_{out} at two decades above cutoff. Since the gain rolls off at 20 dB/decade, the signal output $V_{out(2dec)}$ will be –40dB below the value V_{out} (4.6 V) in the bandpass.

$$-40 \text{ dB} = 20 \log V_{out(2dec)}/V_{out(ref)}$$
$$-2 \text{ dB} = \log V_{out(2dec)}/V_{out(ref)}$$
$$\text{Antilog} - 2 \text{ dB} = V_{out(2dec)}/V_{out(ref)}$$
$$0.01 = V_{out(2dec)}/V_{out(ref)}$$
$$V_{out(2dec)} = V_{out(ref)} \times 0.01$$
$$V_{out(2dec)} = 4.6 \text{ V} \times 0.01$$
$$V_{out(2dec)} = 0.046 \text{ V}$$

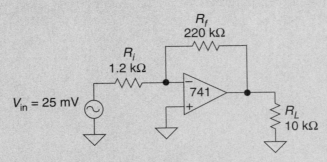

FIGURE 10–16 Circuit for Example 9.12

Step 7. Draw and label a Bode plot graph of the frequency response curve. Figure 10–17 shows the graph.

EXAMPLE 10.12 continued

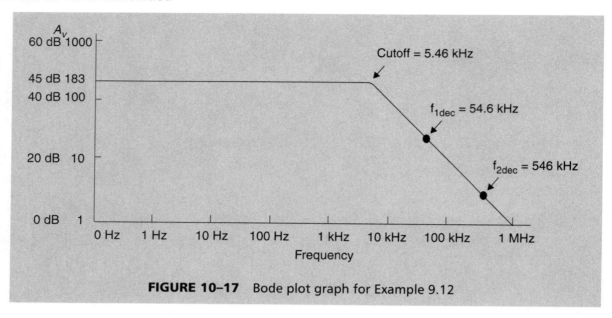

FIGURE 10–17   Bode plot graph for Example 9.12

The use of decibels is popular when working with power gain. In fact, the formula for dB was developed for power gain. In power calculation, dB is equal to ten times the log of the ratio of output power to input power. The formula for working with decibels using voltage ratio was developed from the basic decibel power formula as follows:

$$dB = 10 \log P_{out}/P_{in}$$
$$dB = 10 \log (V_{out}^2/R)/(V_{in}^2/R)$$

If the resistance values are equal, then:

$$dB = 10 \log V_{out}^2/V_{in}^2$$
$$dB = 10 \log (V_{out}/V_{in})^2$$
$$dB = 2 \times (10 \log V_{out}/V_{in})$$
$$dB = 20 \log V_{out}/V_{in}$$

One advantage of expressing power gain and attenuation in dB is that the total system power gain can be quickly calculated. If power gains are expressed in dB, the total gain is equal to the sum of the gain and attenuation. If power gain and attenuation are expressed as ratios, the total gain is equal to the product of gain and attenuation. Example 10.13 shows this advantage.

EXAMPLE 10.13

Calculate the total power gain of the system in Figure 10–18 using (a) dB and (b) ratio. Validate that the total gains are equal.

Step 1. Solve for total system power gain using dB and ratio.

Solution "a":

$$A_{p(total)} = +19.5 \text{ dB} - 1 \text{ dB} + 20.7 \text{ dB} - 2.2 \text{ dB} + 15.3 \text{ dB} = 52.3 \text{ dB}$$

Solution "b":

$$A_{p(total)} = 90 \times 0.8 \times 117 \times 0.6 \times 34 = 171,850$$

Step 2. Verify that the two gains are equal by converting the total dB gain to a ratio.

$$\text{Ratio gain} = \text{antilog dB}/10$$

EXAMPLE 10.13 continued

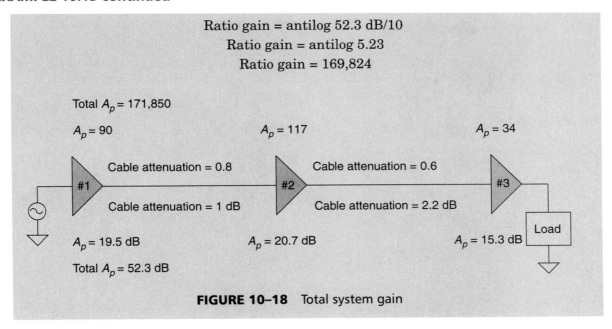

Ratio gain = antilog 52.3 dB/10
Ratio gain = antilog 5.23
Ratio gain = 169,824

Total A_p = 171,850

A_p = 90 A_p = 117 A_p = 34

Cable attenuation = 0.8   Cable attenuation = 0.6

#1   #2   #3

Cable attenuation = 1 dB   Cable attenuation = 2.2 dB

Load

A_p = 19.5 dB A_p = 20.7 dB A_p = 15.3 dB

Total A_p = 52.3 dB

FIGURE 10–18   Total system gain

The small differences in calculated gains are due to round-off errors. Cables are normally rated in dB attenuation per 1000 ft. Using dB makes it easy to combine cable losses with amplifier gains.

10.8 SLEW RATE (SR)

Slew rate is a measurement of the maximum rate at which the output of an op-amp can change. The slew rate is measured in volts/microsecond (V/µS). Figure 10–19 shows an op-amp voltage follower circuit. The input signal is a 1 V peak square wave with a frequency of 250 kHz. Since this circuit is designed for unity gain, the output signal should look identical to the input signal.

Figure 10–19 shows an output signal not identical to the input signal. The difference in input and output signals is caused by the slew rate limitation of the op-amp. When the input signal steps from 0 V to 1 V, the output signal should respond, but op-amps have a maximum rate (slew

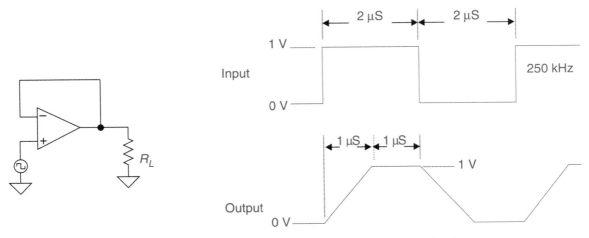

Input

2 µS   2 µS

1 V

250 kHz

0 V

R_L

Output

1 µS   1 µS

1 V

0 V

FIGURE 10–19   Voltage follower with slew rate limitation

rate) at which output voltages can change. The op-amp shown in Figure 10–19 has a slew rate of 1 V/µS. The slew rate for most op-amps is considered to be the same on both the rising and falling edges; thus, only one slew rate is given in the data sheets.

If the slew rate of an op-amp is known, the output waveform can be calculated when the input signal to the op-amp is a square wave of a known frequency. Example 10.14 shows how the frequency of the input signal determines the amount of slew rate distortion in the output signal.

EXAMPLE 10.14

An op-amp with a slew rate of 2 V/µS is used in a voltage follower circuit. Draw the output waveform with respect to the input signal shown in Figure 10–20: (a) 10 kHz square wave and (b) 50 kHz square wave.

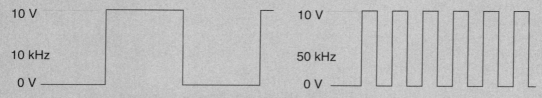

FIGURE 10–20   Input signals for Example 9.14

Solution "a":

Step 1. Calculate the period of the 10 kHz input signal.
$$P = 1/f = 1/10 \text{ kHz}$$
$$P = 100 \text{ µS}$$

Step 2. Draw one cycle of the input waveform showing timing (Figure 10–21).

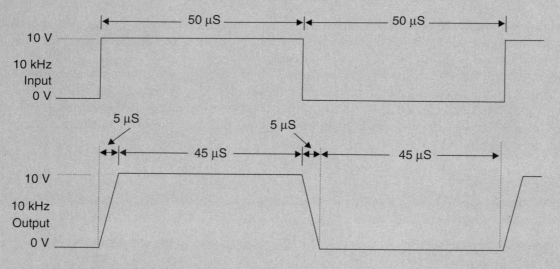

FIGURE 10–21   10 V$_{p\text{-}p}$ output signal with 10 kHz input signal

Step 3. Calculate the time required for the output to rise from 0 V to 10 V.
$$T_R = \Delta V/SR$$
$$T_R = 10 \text{ V}/2 \text{ V/µS}$$
$$T_R = 5 \text{ µS}$$

EXAMPLE 10.14 continued

Step 4. Calculate the time required for the output to fall from 10 V to 0 V.
$$T_F = \Delta V/SR$$
$$T_F = 10 \text{ V}/2 \text{ V/}\mu\text{S}$$
$$T_F = 5 \ \mu\text{S}$$

Step 5. Draw the output signal under the input signal as shown in Figure 10–21.

Solution "b":

Step 1. Calculate the period for the 50 kHz input signal.
$$P = 1/f = 1/50 \text{ kHz}$$
$$P = 20 \ \mu\text{S}$$

Step 2. Draw one cycle of the input waveform showing timing (Figure 10–22).

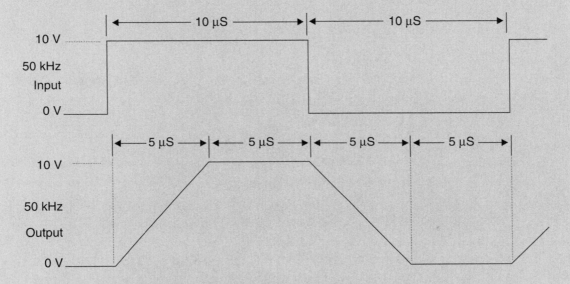

FIGURE 10–22   10 V$_{p-p}$ output signal with 50 kHz input signal

Step 3. Calculate the time required for the output to rise from 0 V to 10 V.
$$T_R = \Delta V/SR$$
$$T_R = 10 \text{ V}/2 \text{ V/}\mu\text{S}$$
$$T_R = 5 \ \mu\text{S}$$

Step 4. Calculate the time required for the output to fall from 10 V to 0 V.
$$T_F = \Delta V/SR$$
$$T_F = 10\text{V}/2 \text{ V/}\mu\text{S}$$
$$T_F = 5 \ \mu\text{S}$$

Step 5. Draw the output signal under the input signal as shown in Figure 10–22.

In Example 10.14, notice that with a 10 kHz square-wave input, the output waveform shows only a slight slew rate distortion. When the input was raised to 50 kHz, the slew rate distortion became more apparent in the output signal. This example demonstrates that the higher the frequency, the greater the distortion caused by the slew rate. In Example 10.15, we will demonstrate that the larger the output signal swing, the greater the distortion caused by the slew rate.

EXAMPLE 10.15

An op-amp with a slew rate of 2 V/µS is used in a voltage follower circuit. Draw the output waveform with respect to the input signal shown in Figure 10–23: (a) 10 V peak at 50 kHz square wave, and (b) 1 V peak at 50 kHz square wave.

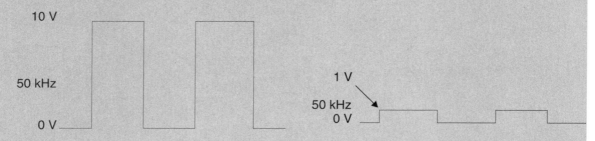

FIGURE 10–23   Input signals for Example 9.15

Solution "a"   10 V peak signal:

This is the same circuit and input used in Example 10.14. The input and output waveforms are shown in Figure 10–24.

Solution "b"   1 V peak signal:

Step 1. Calculate the period of the 50 kHz input signal.
$$P = 1/f = 1/50 \text{ kHz}$$
$$P = 20 \text{ µS}$$

Step 2. Draw one cycle of the input waveform showing timing (Figure 10–24).

Step 3. Calculate the time required for the output to rise from 0 V to 1 V.
$$T_R = \Delta V/SR$$
$$T_R = 1 \text{ V}/2 \text{ V/µS}$$
$$T_R = 0.5 \text{ µS}$$

Step 4. Calculate the time required for the output to fall from 1 V to 0 V.
$$T_F = \Delta V/SR$$
$$T_F = 1 \text{ V}/2 \text{ V/µS}$$
$$T_F = 0.5 \text{ µS}$$

Step 5. Draw the output signal under the input signal as shown in Figure 10–24.

Figure 10–24 shows the two output signals for Example 10.15. Both are the same frequency (50 kHz). The only difference is in amplitude. When the output signal has a 10 V swing, the slew rate distortion is considerable. It requires 50% of the alternation time for the output signal to reach the desired output level. When the output signal has a 1 V swing, the slew rate distortion is hardly noticeable. The output reaches the desired level in 5% of the alternation time. This example demonstrates why slew rate distortion is usually not a problem for circuits with small output voltage swings but often is a problem for circuits with large output voltage swings.

In this section, you have been examining the slew rate using a voltage follower circuit with unity gain. Slew rate limitations affect the output signals of all types of op-amp circuits. To determine the output for an amplifier circuit, first calculate the ideal output, then consider how

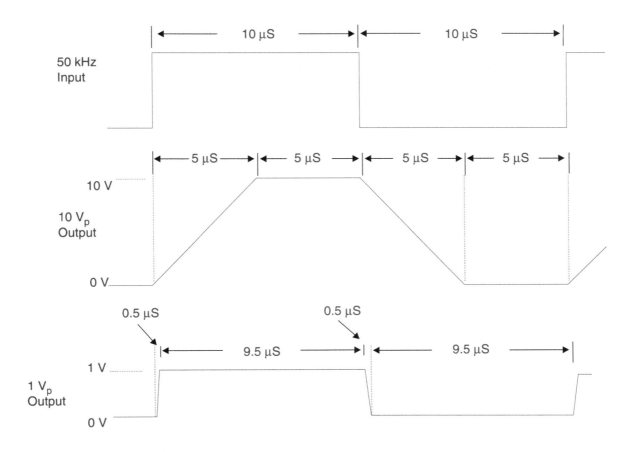

FIGURE 10–24 10 V and 1 V output swings with 50 kHz input signal

this signal would be affected by the slew rate limitations of the op-amp. Example 10.16 shows how to examine an inverting amplifier for slew rate problems.

EXAMPLE 10.16

Draw the output waveform for the circuit shown in Figure 10–25. The input is a 2 V_{p-p} square wave with a frequency of 12.5 kHz. The 741 op-amp has a slew rate of 0.5 V/μS.

Step 1. Calculate the period of the signal.
$$P = 1/f$$
$$P = 1/12.5 \text{ kHz} = 80 \text{ μS}$$

Step 2. Calculate the gain of the circuit.
$$A_v = R_f/R_i = 200 \text{ k}\Omega/20 \text{ k}\Omega = 10$$

Step 3. Using the information from steps 1 and 2, draw the ideal output waveform (Figure 10–26).

Step 4. Calculate the time required for the output to rise from –10 V to +10 V.
$$T_R = \Delta V/SR$$
$$T_R = 20 \text{ V}/0.5 \text{ V/μS}$$
$$T_R = 40 \text{ μS}$$

EXAMPLE 10.16 continued

Step 5. Calculate the time required for the output to fall from +10 V to –10 V. It also will take 40 μS, since slew rate is the same for rise time and fall time.

$$T_F = 40 \ \mu S$$

Step 6. Draw the output signal under the input signal as shown in Figure 10–26.

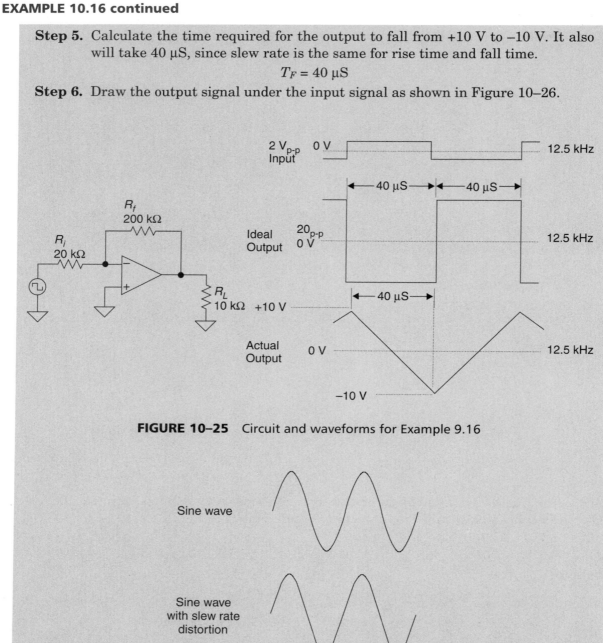

FIGURE 10–25 Circuit and waveforms for Example 9.16

FIGURE 10–26 Slew rate distortion

SLEW RATE DISTORTION IN SINE WAVES

You have been examining slew rates for circuits with rectangular waveform outputs. Slew rate limitations also affect circuits with sine wave outputs, but the output distortion is harder to detect. The rate of change of a sine wave is not constant. A sine wave has its greatest rate of change at zero crossover, and then the rate of change tapers off as the waveform nears its peak.

This makes it difficult to detect low levels of slew rate distortion in sine waves. Figure 10–26 shows how slew rate distortion starts near zero crossover, causing the slope of the waveform to take on a linear rise and fall.

Engineers have derived a formula that can be used to determine the maximum sine wave output frequency without slew rate distortion. The formula shows that the maximum undistorted signal is directly related to the slew rate of the op-amp and indirectly related to the output swing. The formula is:

$$f_{\max} = SR \times 10^6/(2\pi \times V_p)$$

where SR is the slew rate in V/μS and V_p is the peak voltage of the output sine wave.

Example 10.17 shows how this formula can be used to determine at what frequency a sine wave will be affected by slew rate distortion.

EXAMPLE 10.17

A circuit using a 741 op-amp has a 16 $V_{p\text{-}p}$ sine wave present at its output. What is the highest frequency signal that can pass through the amplifier without slew rate distortion? The 741 op-amp has a slew rate of 0.5 V/μS.

$$f_{\max} = SR \times 10^6/(2\pi \times V_p)$$
$$f_{\max} = (0.5 \text{ V/μS} \times 10^6)/(2\pi \times 8 \text{ V}_p)$$
$$f_{\max} = 9.95 \text{ kHz}$$

10.9 TROUBLESHOOTING IC OP-AMP CIRCUITS

It is helpful to mentally remove compensation components that confuse the basic function of the circuit. It is often useful to sketch the circuit, leaving out components that are not needed to determine the basic function of the circuit. Another valuable tool is the manufacturer's manual, which often contains a section on the theory of operation. Time can be saved by reading this technical information.

While the op-amp is close to perfect, there are a few limitations that should be remembered. Input bias currents and offset voltages are bias problems, and their effects can be offset with the use of external components. DC offset output voltage can be checked by grounding the inverting and noninverting inputs and measuring the output for zero DC offset.

The frequency handling ability of an op-amp is limited by open-loop frequency responses and slew rates. Normally, frequency response is the limiting factor of small signals, and slew rate limits larger signals. The output swing of the signal is limited by the magnitude of the power supplies and the current drive ability of the op-amp. Clipping will occur if the signal is driven too close to the power supply rails or if the load requires so much current that it drives the op-amp into the current limiting mode.

Op-Amp Limitations

1. Calculate the value of the bias current compensating resistor (R_c) needed for the circuit in Figure 10–27(b).

   $R_c =$ _____

2. Calculate the DC output voltages for the circuits in Figure 10–27.

   $V_{out(a)} =$ _____ (Assume $I_{IB} = 100$ nA and $V_{IO} = 0$ V.)
   $V_{out(b)} =$ _____ (Assume $V_{IO} = 1$ mV.)
   $V_{out(c)} =$ _____ (Assume R_{null} adjusted correctly.)

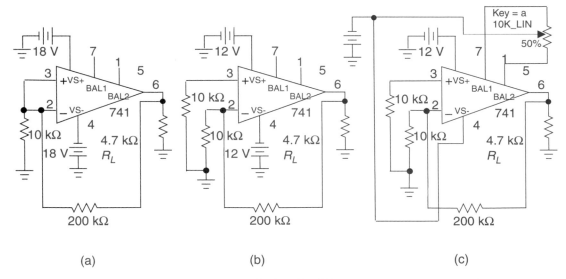

(a)   (b)   (c)

FIGURE 10–27

3. Simulate and record the DC output voltages for the circuits in Figure 10–27.

   $V_{out(a)} =$ _____
   $V_{out(b)} =$ _____
   $V_{out(c)} =$ _____ (With R_{null} adjusted for best reading.)

4. Calculate the following values for the two inverting amplifiers in Figure 10–28. The LM607 has a bandwidth of 1.8 MHz, slew rate of 0.7 V/µS, and current limiting of 20 mA. The LM627 has a bandwidth of 14 MHz, slew rate of 4.5 V/µS, and current limiting of 25 mA.

LM607	Calculated		Simulated
a. Voltage gain =	_____	_____	_____
b. Maximum signal output with 5 kΩ load =	_____	_____	_____
c. Maximum signal output with 50 Ω load =	_____	_____	_____
d. Circuit bandwidth ($V_{out} = 1$ V$_{p-p}$) =	_____	_____	_____

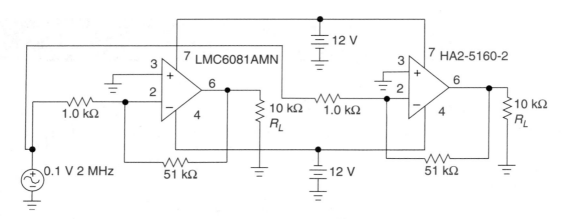

FIGURE 10–28

e.   If V_{in} is the following waveform, draw V_{out}.

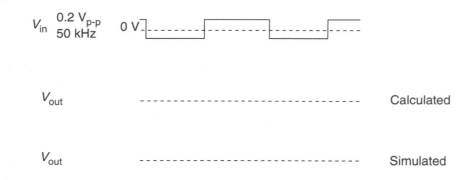

LM627

		Calculated	Simulated
a.	Voltage gain =		
b.	Maximum signal output with 10 kΩ load =		
c.	Maximum signal output with 100 Ω load =		
d.	Circuit bandwidth (V_{out} = 1 $V_{p\text{-}p}$) =		
e.	If V_{in} is the following waveform, draw V_{out}.		

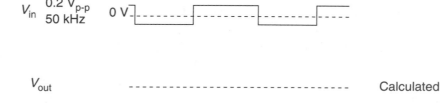

f.   If V_{in} is the following waveform, draw V_{out}.

V_{in} 0.2 $V_{p\text{-}p}$ 0 V
        1 MHz

V_{out} -   Calculated

V_{out} -   Simulated

5.   Simulate and record the values in step 4 above.

Op-Amp Limitations

I. Objectives

1. To experimentally examine the effects of compensating circuitry for input bias current and offset voltage.
2. To experimentally compare the signal responses of the LM741 op-amp to the LM318 op-amp.

II. Test Equipment

+12 V power supply   −12 V power supply

Voltmeter   Oscilloscope

Function generator   Bread boarding system

III. Components

ICs: (1) LM741, (1) LM318

Resistors: (2) 100 Ω, (2) 1 k, (2) 22 k, (1) 91 kΩ, (2) 10 kΩ, (1) 100 kΩ, (1) 1 MΩ

Potentiometer: (1) 10 kΩ

Capacitors: 0.1 μF

IV. Procedure

1. Calculate the value of the bias current compensating resistor (R_c) needed for the circuit in Figure 10–29(b). If the driving signal source has a high internal impedance (R_{int}), the value of R_c should include R_{int} in series with R_i. However, if the circuit is to be driven by a function generator with a low impedance of 50 Ω or 600 Ω, there is no need to include this small value in the calculation. (Assume the signal source will be a low impedance function generator.)

   $R_c = \underline{\hspace{2cm}}$

2. Calculate the output voltage for each of the circuits in Figure 10–29.

   $V_{out(a)} = \underline{\hspace{1cm}}$ $V_{out(b)} = \underline{\hspace{1cm}}$ $V_{out(c)} = \underline{\hspace{1cm}}$

3. Construct the circuits in Figure 10–29(a) and measure the output voltage using a sensitive voltmeter.

   $V_{out(a)} = \underline{\hspace{1cm}}$

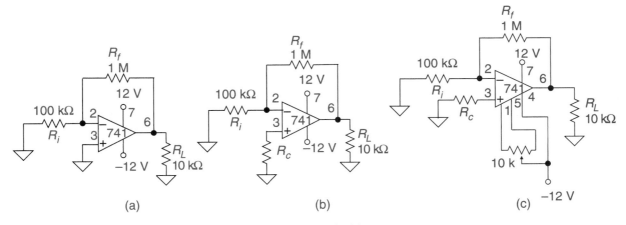

(a)   (b)   (c)

FIGURE 10–29

4. Construct the circuits in Figure 10–29(b) using the value of R_c calculated in step 1 and measure the output voltage using a sensitive voltmeter.

    $V_{out(b)} = $ _____

5. Construct the circuits in Figure 10–29(c) and monitor the output voltage. Adjust the null potentiometer full range, recording the voltage at each extreme.

    $V_{out(min)} = $ _____ $V_{out(max)} = $ _____

6. Null out the circuits in Figure 10–29(c) by adjusting the null potentiometer for zero output voltage. The circuits are now ready to receive a signal input.

7. Calculate the following values for the two inverting amplifiers in Figure 10–30. The LM741 has a bandwidth of 1 MHz, slew rate of 0.5 V/μS, and current limiting of 25 mA. The LM318 has a bandwidth of 15 MHz, slew rate of 50 V/μS, and current limiting of 22 mA.

LM741

a. Voltage gain = _____
b. Maximum signal output with 10 kΩ load = _____
c. Maximum signal output with 100 Ω load = _____
d. Circuit bandwidth ($V_{out} = 1$ V$_{p-p}$) = _____
e. If V_{in} is the following waveform, draw V_{out}.

LM318

a. Voltage gain = _____
b. Maximum signal output with 10 kΩ load = _____
c. Maximum signal output with 100 Ω load = _____
d. Circuit bandwidth ($V_{out} = 1$ V$_{p-p}$) = _____
e. If V_{in} is the following waveform, draw V_{out}.

f. If V_{in} is the following waveform, draw V_{out}.

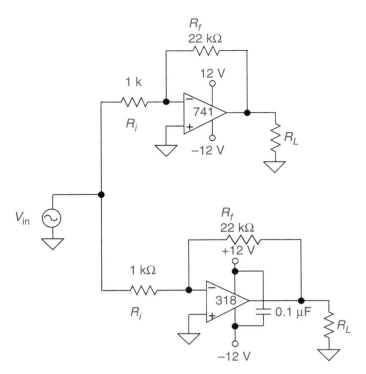

FIGURE 10-30

8. Construct the circuit shown in Figure 10–30. The LM318 is a high performance op-amp and is sensitive to noise on the DC supply bus. The 0.1 µF capacitor connected between +V and −V will prevent the LM318 from breaking into oscillations.

9. Set the signal generator for 50 mV$_{p-p}$ at a frequency of 1 kHz. Connect a 10 kΩ load resistor to each circuit. Observe the two outputs while increasing the magnitude of the input signal. Record the maximum signal each op-amp can deliver to the load without distortion.

   LM741 $V_{out(max)}$ = _____   LM318 $V_{out(max)}$ = _____

10. Set the signal generator for 50 mV$_{p-p}$ at a frequency of 1 kHz. Connect a 100 Ω load resistor to each circuit. Observe the two outputs while increasing the magnitude of the input signal. Record the maximum signal each op-amp can deliver to the load without distortion.

   LM741 $V_{out(max)}$ = _____   LM318 $V_{out(max)}$ = _____

11. Remove the 100 Ω load resistors. Connect a 10 kΩ load resistor to each circuit.

12. Set the signal generator for 50 mV$_{p-p}$ at a frequency of 1 kHz. Measure the output signals and adjust the generator as required to obtain a 1 V$_{p-p}$. Since each amplifier has the same gain, both outputs should equal 1 V$_{p-p}$. Observe the outputs of both amplifiers while increasing the frequency. Record the frequency ($f_{(cutoff)}$) at which the output signal drops to 0.7 V$_{p-p}$. (Note that the LM318 uses a frequency compensation circuit that actually causes the output to increase just before decreasing.)

   LM741 $f_{(cutoff)}$ = _____   LM318 $f_{(cutoff)}$ = _____

13. Set the signal generator for a 0.9 V$_{p-p}$ square-wave signal at a frequency of 10 kHz and measure and draw the output waveform for each op-amp. Measure the slew rate of the LM741 op-amp.

   Slew rate LM741 = _____

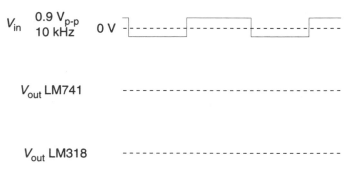

14. Set the signal generator for a 0.9 V$_{p\text{-}p}$ square-wave signal at a frequency of 500 kHz and measure and draw the output waveform for each op-amp. Measure the slew rate of the LM318 op-amp.

Slew rate LM318 = _____

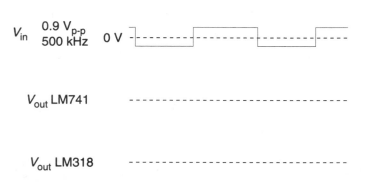

V. Points to Discuss

1. Explain why adding the compensation resistor (R_c) in Figure 10–29(b) did not cause the DC output voltage to equal exactly zero?

2. In Figure 10–29(c), a potentiometer was added that permitted the output voltage to be adjusted to exactly zero. If the op-amp in the circuit were replaced, would the potentiometer need to be adjusted for zero voltage output? Explain.

3. Explain what caused the clipping of the output signals in step 9.

4. Explain what caused the clipping of the output signals in step 10.

5. With a starting output voltage of 1 $V_{p\text{-}p}$, which op-amp dropped to an output of 0.707 $V_{p\text{-}p}$ first as the frequency was increased? Explain why.

6. Explain how you were able to measure the slew rate of each op-amp.

QUESTIONS

10.1 Input Bias Current (I_{IB})

____ 1. The small current that flows through the input pins of an op-amp is called the _____ current.
  a. input leakage
  b. input reverse junction
  c. input bias
  d. input

____ 2. Normally, the magnitude of input bias currents are _____.
  a. approximately the same in both the inverting and noninverting inputs
  b. ten times greater in the inverting input than in the noninverting input
  c. ten times greater in the noninverting input than in the inverting input
  d. in the 0.1 mA to 1 mA range

10.2 Input Offset Current (I_{IO})

____ 3. The difference in the absolute values of bias input currents is called _____ .
  a. input difference current
  b. input offset current
  c. absolute current
  d. none of the above

____ 4. The input bias current compensation resistor can be calculated for an inverting amplifier using which of the following formulas?
  a. $R_c = R_L \parallel (R_i + R_{int})$
  b. $R_c = R_f \parallel R_{int}$
  c. $R_c = R_f \parallel (R_i + R_{int})$
  d. $R_c = R_{int} \parallel R_i$

____ 5. Ideally, the offset bias current should be _____.
  a. zero
  b. 25% of I_{out}
  c. 25% of the input bias current
  d. greater than the input bias current

10.3 Input Offset Voltage (V_{IO})

____ 6. If an op-amp is connected to a plus and a negative supply and the two input pins are grounded, the output of the op-amp will be at _____ .
  a. zero voltage
  b. positive saturation
  c. negative saturation
  d. b or c

_____ 7. The imbalances of the internal circuitry are lumped into one value called the _____.
   a. imbalance factor
   b. input offset voltage
   c. output offset voltage
   d. balance offset voltage

_____ 8. If an op-amp is replaced in the field, the null potentiometer should be adjusted for _____ when input signal sources are set to zero.
   a. zero output voltage
   b. maximum output voltage
   c. maximum input impedance
   d. balance input currents

10.4 Output Voltage Swing (V_o)

_____ 9. The output voltage swing of an op-amp is limited by _____ .
   a. the resistance of R_f
   b. the resistance of R_i
   c. the positive power supply
   d. both the positive and negative power supply voltages

_____ 10. The voltage levels at which clipping occurs is referred to as the _____.
   a. clipping voltage level
   b. power supply rails
   c. peak limiting voltage
   d. top-bottom voltage

10.5 Output Short-Circuit Current (I_{SC})

_____ 11. If the load current demand exceeds the op-amp rating, the built-in current limiting of the op-amp will take over and cause _____ of the output signal.
   a. clipping
   b. clamping
   c. saturation
   d. zenering

_____ 12. The maximum current that can flow through the output pin of an op-amp is called _____.
   a. max output current
   b. output control current
   c. output short-circuit current
   d. none of the above

10.6 Frequency Response

_____ 13. Cutoff frequency occurs when the voltage gain drops to _____ of the DC voltage gain.
   a. 10%
   b. 50%

c. 70%

d. 90%

_____ 14. A formula for the voltage gain of the noninverting amplifier is _____.

a. $A_v = R_i/R_f$

b. $A_v = (R_f/R_i)$

c. $A_v = A_{ol}/(1 + A_{ol} \times B)$

d. none of the above

_____ 15. The desired closed-loop voltage gain of an amplifier using negative feedback is equal to the _____.

a. reciprocal of the feedback ratio

b. open gain (A_{ol}) times the load resistance (R_L)

c. reciprocal of R_i

d. ratio of R_L to R_i

_____ 16. In order for the formula $A_v = 1 + R_f/R_i$ to be valid, the open-loop voltage gain must be _____ the desired closed-loop voltage gain.

a. considerably smaller than

b. equal to

c. twice

d. much larger than

_____ 17. The frequency at which the open-loop gain drops to one is called the _____.

a. unity gain frequency

b. ending frequency

c. bandwidth frequency

d. gain frequency

_____ 18. The _____ response curve is a simplified band response curve that is often used when examining the band response of an amplifier because calculations are easier and the results are acceptable.

a. Miller plot

b. bandwidth

c. Bode plot

d. log

10.7 Working with Logarithmic Scales

_____ 19. Using _____ is especially useful when working with data that spans a wide range of values.

a. compression scaling

b. logarithmic scaling

c. linear scaling

d. micro scaling

_____ 20. The dB value is equal to _____ times the log of the voltage ratio.

a. 1

b. 5

c. 10

d. 20

____ 21.   The dB value is equal to ____ times the log of the power ratio.
   a.   1
   b.   5
   c.   10
   d.   20

____ 22.   One advantage of expressing gain and attenuation in dB is that the total system gain can be calculated by _____ the gains of each stage.
   a.   adding
   b.   subtracting
   c.   multiplying
   d.   averaging

10.8 Slew Rate (SR)

____ 23.   Slew rate is a measurement of the _____.
   a.   maximum rate at which the input of an op-amp can change
   b.   maximum rate at which the output of an op-amp can change
   c.   maximum rate at which the load resistance can change
   d.   ability of the op-amp to handle changes in power supply voltages

____ 24.   The slew rate is measured in _____.
   a.   volts/microseconds
   b.   amps/microseconds
   c.   volts/milliseconds
   d.   amps/milliseconds

____ 25.   Slew rate distortion is more prevalent in circuits with _____.
   a.   small input voltage swings
   b.   large input voltage swings
   c.   small output voltage swings
   d.   large output voltage swings

____ 26.   The formula _____ can be used to determine the maximum sine wave output frequency without slew rate distortion.
   a. $F_{max} = (2\pi \times SR \times 10^6)/V_p$
   b. $F_{max} = SR \times 10^6/V_p$
   c. $F_{max} = SR \times Freq.$
   d. $F_{max} = SR \times 10^6/(2\pi \times V_p)$

10.9 Troubleshooting IC Op-Amp Circuits

____ 27.   The signal generator in Figure 10–31 is adjusted to 0 V. The output voltage changes from −100 mV to 100 mV while the potentiometer is adjusted from minimum to maximum. What is wrong with the circuit?
   a.   The op-amp is defective.
   b.   The potentiometer is defective.
   c.   Resistor R_c is open.
   d.   The circuit is functioning properly.

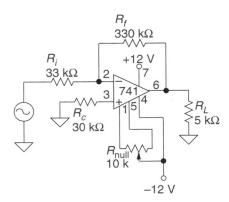

FIGURE 10–31

___ 28. The signal generator in Figure 10–31 is adjusted to 1 V_{p-p} riding on zero volts DC offset at a frequency of 1 kHz. The output voltage measures 10 V_{p-p} riding on a 50 mV DC offset. What is wrong with the circuit?

a. The op-amp is defective.

b. The potentiometer needs to be adjusted.

c. Resistor R_c is open.

d. The circuit is functioning properly.

___ 29. The signal generator in Figure 10–31 is adjusted to 3 V_{p-p} at a frequency of 1 kHz. The output signal is clipped. What is wrong with the circuit?

a. The op-amp is defective.

b. Resistor R_f is shorted.

c. Resistor R_c is shorted.

d. The circuit is functioning properly.

___ 30. The signal generator in Figure 10–31 is adjusted to 1 V_{p-p} at a frequency of 1 kHz. The output signal is clipped. What is wrong with the circuit?

a. The load resistance has decreased.

b. Resistor R_f is shorted.

c. Resistor R_c is shorted.

d. The circuit is functioning properly.

___ 31. The signal generator in Figure 10–31 is adjusted to 0.1 V_{p-p} at a frequency of 100 kHz. The output signal measures approximately 0.7 V_{p-p}. What is wrong with the circuit?

a. The load resistance has decreased.

b. Resistor R_f is shorted.

c. Resistor R_c is shorted.

d. The circuit is functioning properly.

___ 32. The signal generator in Figure 10–31 is adjusted to a 2 V_{p-p} sine wave at a frequency of 15 kHz. The output signal is a triangular waveform. What is wrong with the circuit?

a. The op-amp is defective.

b. Resistor R_f is shorted.

c. Resistor R_c is shorted.

d. The circuit is functioning properly.

PROBLEMS

1. Calculate the input bias current (I_{IB}) for the op-amp shown in Figure 10–32 if the meter reads 0.09 V.

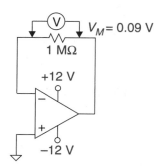

FIGURE 10–32

2. Calculate the value of the compensation resistor (R_c) for the circuit in Figure 10–33.

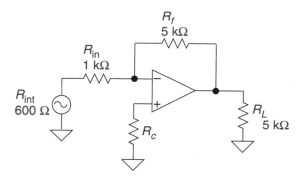

FIGURE 10–33

3. The output voltage of the circuit in Figure 10–34 is 2.5 V. What is the input offset voltage?

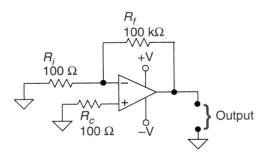

FIGURE 10–34

4. Draw the output waveform for the circuit in Figure 10–35 where the sinusoidal input is 6 V$_{\text{p-p}}$. (I_{sc} = 25 mA, GBW = 1 MHz, SR = 0.5 V/µS, and R_L = 10 kΩ)

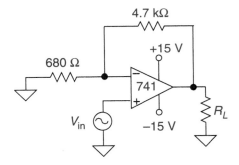

FIGURE 10–35

5. What is the largest input signal at 1 kHz that the circuit in Figure 10–35 can handle without causing distortion at the output? (I_{sc} = 25 mA, GBW = 1 MHz, SR = 0.5 V/µS, and R_L = 10 kΩ)

6. What is the maximum peak-to-peak signal at 500 Hz that can be developed without distortion at the output of the circuit in Figure 10–36? (I_{sc} = 25 mA, GBW = 1 MHz, SR = 0.5 V/μS, and R_L = 10 kΩ)

7. What is the largest input signal that the circuit in Figure 10–36 can handle without causing distortion at the output?

8. What is the maximum peak-to-peak signal at 500 Hz that can be developed without distortion at the output of the circuit in Figure 10–36? (I_{sc} = 25 mA, GBW = 1 MHz, SR = 0.5 V/μS, and R_L = 200 Ω)

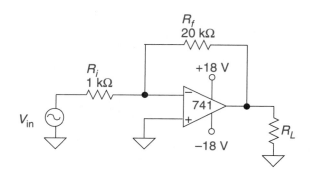

FIGURE 10–36

9. Calculate the voltage gain of the circuit in Figure 10–36 for a input signal of 100 m$V_{p\text{-}p}$ with a frequency of (a) 5 kHz, (b) 50 kHz, and (c) 500 kHz. (I_{sc} = 25 mA, GBW = 1 MHz, SR = 0.5 V/μS, and R_L = 10 kΩ)

10. Calculate the cutoff frequency for the circuit in Figure 10–36. (I_{sc} = 25 mA, GBW = 1 MHz, SR = 0.5 V/μS, and R_L = 10 kΩ)

11. An op-amp circuit has a gain of 200 and a cutoff frequency of 20 kHz. What is the op-amp gain bandwidth product (GBW) and unity gain frequency (f_{unity})?

12. If the op-amp has an open-loop gain of 25,000 and you build a noninverting amplifier circuit with R_f equal to 100 kΩ and R_i equal to 100 Ω, how much will the actual gain differ from 1001?

13. An amplifier has a voltage gain of 254. Express this gain in dB.

14. A 741 op-amp circuit has a voltage gain of 40 and an input signal of 50 mV$_{p-p}$. Calculate the cutoff frequency and the signal present at the output if the frequency of the input signal is adjusted to ten times the cutoff frequency. (I_{sc} = 25 mA, GBW = 1 MHz, SR = 0.5 V/μS, and R_L = 10 kΩ)

15. For the circuit in Figure 10–37, calculate the following values: A_v expressed as a ratio, A_v expressed in dB, F_{cutoff}, V_{out} of signals below cutoff, the output signal ($V_{out(1dec)}$) at one decade above cutoff, and the output signal ($V_{out(2dec)}$) at two decades above cutoff. Draw and label a Bode plot of the frequency response curve for the circuit. (I_{sc} = 22 mA, bandwidth = 15 MHz, and SR = 50 V/μS)

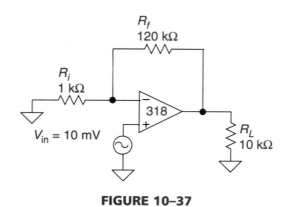

FIGURE 10–37

16. Calculate the total power gain of the system in Figure 10–38 using (a) dB and (b) ratio.

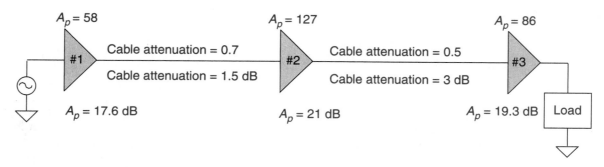

FIGURE 10–38

17. An op-amp with a slew rate of 1.25 V/µS is used in a voltage follower circuit. Draw the output waveform with respect to the input signals shown below:

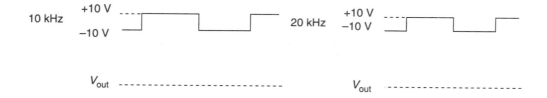

18. An op-amp has a slew rate of 20 V/μS. Sketch the output of a voltage follower whose input is a 10 V$_{p-p}$ square wave with a frequency of (a) 0.2 MHz, (b) 0.5 MHz, and (c) 1 MHz.

19. If the slew rate is 0.5 V/μS, what frequency square wave will give a triangular output of 10 V$_{p-p}$?

20. A circuit using an op-amp has a 10 V$_{p-p}$ sine wave present at its output. What is the highest frequency signal that can pass through the amplifier without slew rate distortion? The op-amp has a slew rate of 2 V/μS.

21. Draw the signal at TP_1 in Figure 10–39 for the input signals given. Label amplitude and show timing with respect to the input (power supplies = +/–15 V DC, SR = 5 V/μS, I_{os} = 30 mA, and bandwidth = 2 MHz).

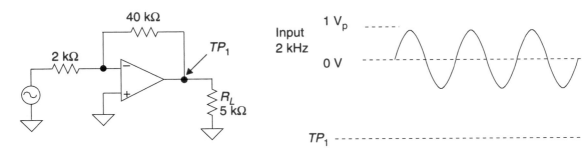

FIGURE 10–39

22. Draw the signal at TP_1 in Figure 10–40 for the input signals given. Label amplitude and show timing with respect to the input (power supplies = +/–15 V DC, SR = 5 V/µS, I_{os} = 30 mA, and bandwidth = 2MHz).

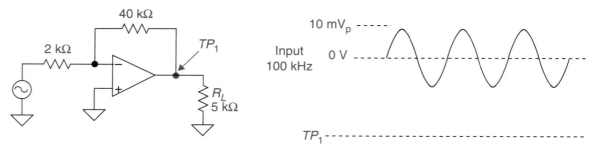

FIGURE 10–40

23. Draw the signal at TP_1 in Figure 10–41 for the input signals given. Label amplitude and show timing with respect to the input (power supplies = +/–15 V DC, SR = 5 V/µS, I_{os} = 30 mA, and bandwidth = 2 MHz).

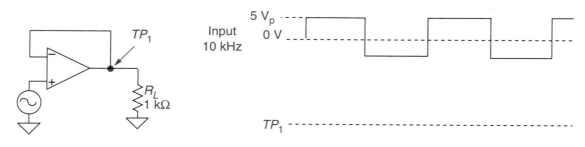

FIGURE 10–41

24. Draw the signal at TP_1 in Figure 10–42 for the input signals given. Label amplitude and show timing with respect to the input (power supplies = +/–15 V DC, SR = 5 V/µS, I_{os} = 30 mA, and bandwidth = 2 MHz).

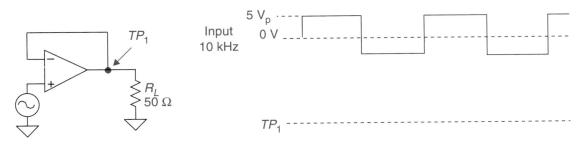

FIGURE 10–42

25. Draw the signal at TP_1 in Figure 10–43 for the input signals given. Label amplitude and show timing with respect to the input (power supplies = +/–15 V DC, SR = 5 V/μS, I_{os} = 30 mA, and bandwidth = 2 MHz).

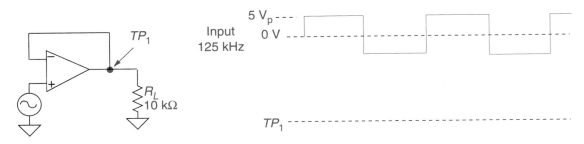

FIGURE 10–43

26. Draw the signal at TP_1 in Figure 10–44 for the input signals given. Label amplitude and show timing with respect to the input (power supplies = +/–15 V DC, SR = 5 V/μS, I_{os} = 30 mA, and bandwidth = 2 MHz).

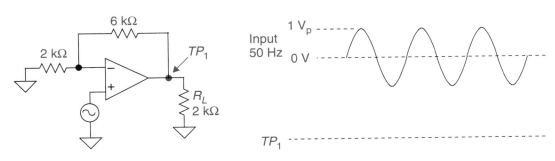

FIGURE 10–44

27. Draw the signal at TP_1 in Figure 10–45 for the input signals given. Label amplitude and show timing with respect to the input (power supplies = +/–15 V DC, SR = 5 V/µS, I_{os} = 30 mA, and bandwidth = 2 MHz).

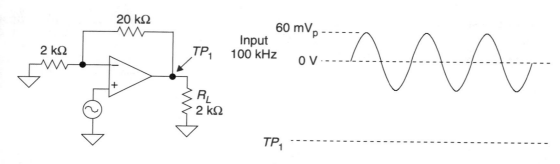

FIGURE 10–45

In Section I, the building blocks that comprise electronic systems were discussed. Engineers use these components to design electronic subsystems. In Section II of the text, electronic circuits (subsystems) will be addressed. All electronic systems are comprised of subsystems.

After completing Section II, you will have gained the needed knowledge to use technical manuals to troubleshoot most electronic systems. Technical manuals are comprised of technical theory, block diagrams, and schematics. A technician with the circuit knowledge covered in Section II can use the information in technical manuals to understand the overall operation of an electronics system. Once the technician understands the theory of operation, he can use the block diagram to isolate the system's problems to a subsystem. Once the problem is isolated to the subsystem, it can be replaced or repaired.

Examining the block diagram of the computer VGA monitor in Figure II–1 will give you an idea of how circuitry can be combined to make a functional system. The circuitry used in the VGA monitor and most other electronics systems are covered in Section II of the text. The vertical and horizontal oscillators are nonsinusoidal oscillators, which are covered in Chapter 14. The output amplifiers (video, vertical, and horizontal) are power amplifiers, which are covered in Chapter 16. Digital-to-analog converters (DACs) are covered in Chapter 19, power supplies are covered in Chapter 18, and the operation of the CRT is covered in Chapter 20. At this point in your study of electronics, you may not have the needed knowledge to follow a detailed theory of operation of a complete electronics system. However, reading the following overview discussion of the block diagram of a computer VGA monitor (see Figure II–1) should emphasize the importance of studying electronics circuitry in Section II.

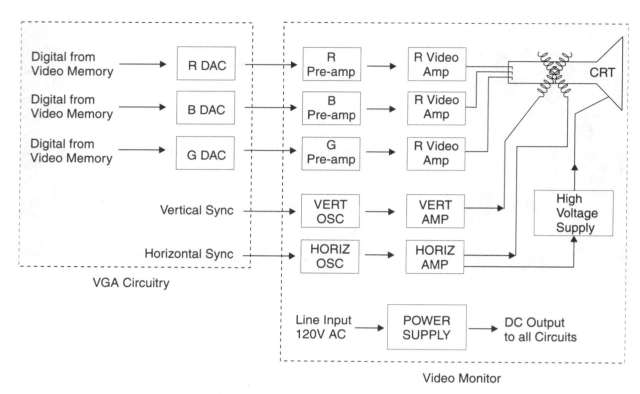

FIGURE II-01   Computer VGA monitor

Overview Discussion of a VGA Monitor

The video monitor is driven by the VGA circuitry located on the computer motherboard or on a video card inside the computer. The VGA circuitry contains three DACs (digital-to-analog converters) that convert digital words stored in video memory to three small analog signals that are sent over the video cable to the monitor. The signals are called the red, blue, and green video signals, and each signal drives a separate electron gun in the CRT (cathode ray tube).

The small signals enter the monitor, so the first step is to use three pre-amps to increase the voltage level of each signal. The three video amps increase the power of the video signals so they have the needed power to drive the electron guns in the CRT. The CRT has three electron guns. Each gun generates an electronic beam that is projected to the front screen of the CRT. Each beam strikes a colored phosphorous coating on the inside of the front screen. The beam from the red gun strikes the red phosphorous, the beam from the blue gun strikes the blue phosphorous, and the beam from the green gun strikes the green phosphorous. The electron beams striking the phosphorous coatings cause red, blue, and green light to be emitted. The magnitude of each electronic beam is controlled by the video signal driving the guns at the back of the CRT. The computer can control the magnitude of each beam; therefore, it can create any color on the screen by mixing different magnitudes of red, blue, and green light.

The horizontal sync pulses from the VGA circuitry are used to synchronize the horizontal oscillator (HORIZ OSC). The output of the horizontal oscillator is a linear ramp that is amplified by the horizontal amplifier (HORIZ AMP). The ramp signal is used to drive the horizontal deflection coil located in the yoke that connects around the neck of the CRT. The ramp current flowing through the horizontal coil of the yoke generates a magnetic field causing the electron beam to be deflected from left to right across the screen. Once the ramp reaches its peak value, it quickly returns to its baseline causing the electron beam to return quickly to the left side of the screen.

Vertical sync pulses from the VGA circuitry are used to synchronize the vertical oscillator (VERT OSC). The vertical oscillator generates a linear ramp that is amplified by the vertical amplifier (VERT AMP) and is used to drive the vertical deflection coil located in the yoke. The ramp current flowing through the vertical coil in the yoke generates a magnetic field causing the electron beam to be deflected from the top of the screen to the bottom of the screen. Once the ramp reaches its peak value, it quickly returns to its baseline causing the electron beam to quickly return to the top of the screen.

The picture on the face of the CRT is generated by a combination of the magnitude of the electron beams and the horizontal and vertical deflections. All three beams are tightly aligned so the human eye sees only one dot of light. The dot of light generated by the three beams is used as a pencil point. The picture is drawn by moving the dot of light across the entire screen. As the dot is moved from left to right, the intensity of the beams controls the brightness and color of the screen. Each horizontal sweep moves the dot of light from left to right. Each time the dot moves left to right, it also moves slightly down because the vertical sweep occurs at the same time. For standard VGA, the result is 480 horizontal lines across the screen. After the dot of light reaches the bottom of the screen, it quickly moves to the top of the screen and the dot starts sweeping again from left to right and from top to bottom. As the dot scans the screen, the computer controls the brightness and color of the light. Thus, the computer system can produce any image on the screen.

In order for the electron beams to be attracted to the screen of the CRT, a high voltage is needed at the front of the CRT. For a color picture, the voltage required is 25,000 V. The high voltage supply circuit makes use of the fast falling edge of the horizontal sweep signal. The falling edge of the horizontal sweep is applied to the primary of a step-up transformer. The rapid change in current in the primary of the transformer causes a large voltage in the secondary of the transformer. The secondary voltage is rectified and filtered and is the source of the high voltage needed at the front of the CRT.

Electronic systems operate on DC voltages. However, most systems receive their energy from a 120 VAC distribution system. The monitor (and most systems) uses a power supply to convert the AC power to DC power. This provides all the needed DC voltage levels to all the subsystems in the monitor.

As you can see from the previous discussion, the VGA monitor system is comprised of multiple subcircuits. There are thousands of electronics systems; however, they all are constructed from common subcircuits. If you understand the operation of the subcircuits, you can read technical material and learn the operation of any electronics system. With this in mind, the remainder of this text does not concentrate on one type of system, but instead focuses on the subcircuits that comprise all electronics systems. An excellent way to learn an electronics system is to start with a block diagram that breaks the system into functional circuits. But in order to troubleshoot the system, the technician must understand the operation of each subcircuit, and that is the goal of Section II.

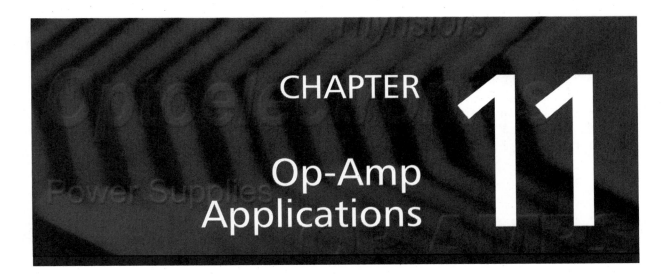

CHAPTER 11

Op-Amp Applications

OBJECTIVES

After studying the material in this chapter, you will be able to describe and/or analyze:

○ high impedance input circuits,
○ basic arithmetic circuits,
○ mixer circuits,
○ periodic signals,
○ integrator circuits,
○ differentiator circuits,
○ the operation of single supply op-amp circuits,
○ precision rectifier circuits,
○ peak detector circuits,
○ comparator circuits, and
○ troubleshooting procedures for op-amp circuits.

INTRODUCTION

The applications of op-amps in electronics are virtually unlimited. With the advent of quad op-amp packages, the cost has dropped so low that it has been stated that if you cannot afford a transistor, use an op-amp! It would be impossible to cover all of the applications of op-amps, but we will discuss some of the most popular ones.

11.1 HIGH INPUT IMPEDANCE CIRCUITS

The first stage of the electronic system must be compatible with the signal source driving the system. A preamplifier is a circuit designed to amplify a signal from a signal source without loading it down. The output of the preamplifier is designed to drive the next stage of the system.

The high input impedance and low output impedance of op-amps make them ideal preamplifiers. Figure 11–1 shows an op-amp functioning as a preamplifier for a high impedance crystal microphone. The crystal microphone has a high internal resistance, and if connected to an amplifier with a low input impedance, the signal voltage level out of the crystal microphone will drop. The input impedance of the voltage follower is in the megohms and does not load the microphone. The output of the preamplifier can easily drive the next voltage amplifier stage. If voltage gain is needed, the preamplifier can use a noninverting amplifier, which provides voltage gain and the needed high input impedance.

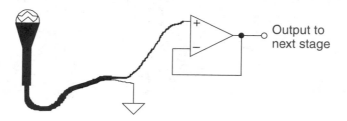

FIGURE 11–1   Crystal microphone preamplifier

Figure 11–2 shows another example of how the noninverting amplifier can be used as a buffer to prevent loading. The high impedance voltmeter in Figure 11–2 uses a noninverting amplifier and a 1 mA meter movement. The voltage present at the noninverting input is also felt at the inverting input and causes current to flow through the resistor connected between the inverting input and ground. The magnitude of the current is equal to V_{in}/R_i. This current flows through the 1 mA meter movement, causing meter deflection. By using various sizes of resistors, a multiscale voltmeter can be made as shown in Figure 11–2. The size of the resistor needed for a particular range can be calculated by dividing the desired full-range voltage reading by the current needed for full-scale deflection of the meter movement. Example 11.1 shows how to calculate the current through the meter for a given voltage input. Example 11.2 shows how to calculate the needed current limiting resistor to add a 50 V range.

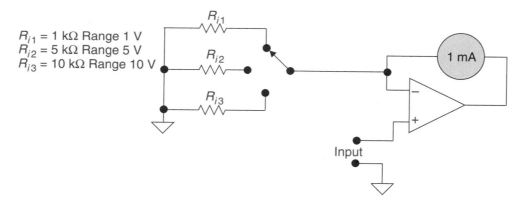

FIGURE 11–2   High input impedance voltmeter

EXAMPLE 11.1

If the meter in Figure 11–2 is in the 5 V range and 2 V is applied to the input, what would be the current flow through the 1 mA meter movement, and what percent of full scale would the meter read?

EXAMPLE 11.1

Step 1. Since 2 V is applied to the noninverting input, 2 V is also present at the inverting input. The current limit in the 5 V range is 5 kΩ, so the current flow can be calculated.

$$I = V_{in}/R_i = 2 \text{ V}/5 \text{ k}\Omega = 0.4 \text{ mA}$$

Step 2. Once the current flow is known, the percent of full-scale deflection can be calculated.

$$\% \text{ of full scale} = I/I_{(full\ scale)} \times 100\%$$
$$\% \text{ of full scale} = (0.4 \text{ mA}/1 \text{ mA}) \times 100\%$$
$$\% \text{ of full scale} = 40\%$$

EXAMPLE 11.2

What would be the value of the current limiting resistor needed to add a 50 V range to the meter in Figure 11–2?

Step 1. Assume a 50 V input and calculate the value of the resistor that will limit the current flow through the meter to the full-scale current.

$$R_{(limiting\ resistor)} = V_{in}/I_{(full\ scale)} = 50 \text{ V}/1 \text{ mA} = 50 \text{ k}\Omega$$

11.2 BASIC ARITHMETIC CIRCUITS

The name *op-amp* traces its origin to amplifiers used to perform mathematical operations in analog computers. An amplifier with a controlled gain provides multiplication, a gain of less than one provides division, and a summing amplifier can be used to add and subtract. Op-amp circuits using capacitors can perform integration and differentiation, which we will cover in a later section.

MULTIPLICATION CIRCUITS

Analog multiplication can be performed in an amplifier. The output is equal to the input times the voltage gain of the circuit. The circuit shown in Figure 11–3 multiplies the input by ten.

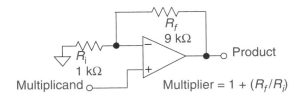

FIGURE 11–3 Multiplier circuit

DIVISION CIRCUITS

Analog division can be performed by designing an amplifier with a gain of less than unity. The output is then equal to the input divided by the reciprocal of the gain. The circuit in Figure 11–4

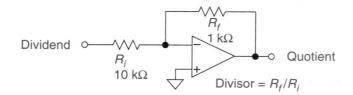

FIGURE 11–4   Divider circuit

has a voltage gain of 0.1. It divides the input by 10 (divider = $1/A_v$). An inverting amplifier must be used, since the noninverting amplifier cannot have a voltage gain of less than one.

ADDING CIRCUITS

Figure 11–5 shows an adder circuit, also called a *summing circuit*. If all the resistors are equal, the output will be the inverted sum of the individual inputs ($V_{out} = -(V_1 + V_2 + V_3)$). The noninverting input is grounded, and the voltage level at the inverting input is at virtual ground. Consequently, the current flowing in each input resistor is totally dependent on the input voltage divided by the input resistor in each leg. The current flowing through each input resistor is independent of the other inputs. Since the input impedance of the op-amp is high, virtually all of the current from all inputs flows through the feedback resistor, producing an output voltage equal to the sum of the inputs.

Be aware that it is the virtual ground at the inverting input that makes the summing circuit possible. Since noninverting circuits do not have a virtual ground, all summing circuits will use the inverting configuration, and the output will equal the inverted sum of the inputs. Example 11.3 shows how the output voltage can be calculated if the input voltages are known.

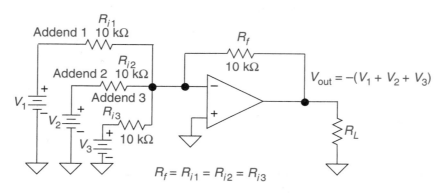

FIGURE 11–5   Adder circuit

EXAMPLE 11.3

What is the output voltage of the circuit in Figure 11–5 if $V_1 = 1$ V, $V_2 = 2$ V, $V_3 = 3$ V and all resistors equal 10 kΩ?

Step 1. Find the current through each input resistor.
$$I_{in1} = V_{in1}/R_{i1} = 1 \text{ V}/10 \text{ k}\Omega = 0.1 \text{ mA}$$
$$I_{in2} = V_{in2}/R_{i2} = 2 \text{ V}/10 \text{ k}\Omega = 0.2 \text{ mA}$$
$$I_{in3} = V_{in3}/R_{i3} = 3 \text{ V}/10 \text{ k}\Omega = 0.3 \text{ mA}$$

Step 2. Find the current through the feedback resistor.
$$I_{Rf} = I_1 + I_2 + I_3 = 0.1 \text{ mA} + 0.2 \text{ mA} + 0.3 \text{ mA} = 0.6 \text{ mA}$$

EXAMPLE 11.3 continued

> **Step 3.** Find the output voltage.
> $V_{out} = R_f \times I_{Rf} = 10 \text{ k}\Omega \times 0.6 \text{ mA} = -6 \text{ V}$ (Negative because of phase inversion.)

A summing circuit with voltage gain can be designed by making resistor R_f larger than the input resistors. Figure 11–6(a) shows a summing circuit with a voltage gain of two. For example, if the input voltages to the circuit in Figure 11–6(a) are 1 V, 2 V, and 3 V, the output voltage would equal –12 V.

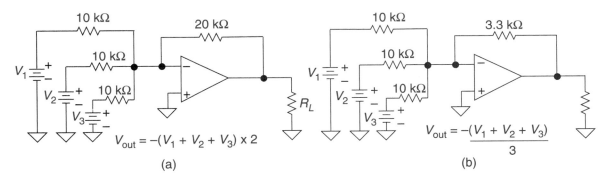

(a)

(b)

FIGURE 11–6 Modified summing circuits

Figure 11–6(b) shows a voltage averaging circuit. The value of resistor R_f is equal to the value of one input resistor divided by the number of inputs. All input resistors are equal in value. If the input voltages to the voltage averaging circuit in Figure 11–6(b) are 1 V, 2 V, and 3 V, the output voltage would equal –2 V.

It is sometimes desirable to have a weighted adder, which has greater gain for some inputs than for others. This is done by adjusting the values of the input resistors for the gain desired. A weighted adder is shown in Figure 11–7. The output voltage can still be found by finding each input current, then adding the currents together to find the current through the feedback resistor. The direction of current must be considered when adding current. Notice that inputs V_1 and V_3 are positive voltage inputs; however, input V_2 is a negative voltage input. The current produced by V_2 will flow in an opposite direction and must be subtracted from the current produced by V_1 and V_3. Example 11.4 demonstrates this procedure.

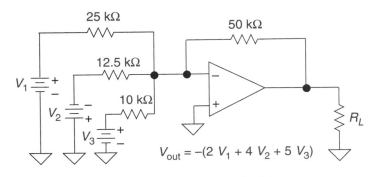

FIGURE 11–7 Weighted adder

EXAMPLE 11.4

What is the output voltage of the circuit in Figure 11–7 if $V_1 = 4$ V, $V_2 = -3$ V, $V_3 = 2$ V?

Step 1. Find the current through each input resistor.

$$I_{in1} = V_{in1}/R_{i1} = 4\text{ V}/25\text{ k}\Omega = 0.16\text{ mA}$$
$$I_{in2} = V_{in2}/R_{i2} = -3\text{ V}/12.5\text{ k}\Omega = -0.24\text{ mA}$$
$$I_{in3} = V_{in3}/R_{i3} = 2\text{ V}/10\text{ k}\Omega = 0.2\text{ mA}$$

Step 2. Find the current through the feedback resistor.

$$I_{Rf} = I_1 + I_2 + I_3 = 0.16\text{ mA} + (-0.24\text{ mA}) + 0.2\text{ mA} = 0.12\text{ mA}$$

Step 3. Find the output voltage.

$$V_{out} = R_f \times I_{Rf} = 50\text{ k}\Omega \times 0.12\text{ mA} = -6\text{ V}$$

SUBTRACTING CIRCUITS

Figure 11–8 shows a subtracting circuit comprised of two op-amps. Op-amp U_1 is an inverter with unity gain, and U_2 is an adder. Signal V_1 is connected directly to the adder, and signal V_2 is first inverted and then connected to the adder. The circuit performs subtraction, and the output signal is equal to V_1 minus V_2. A second way subtracting can be accomplished is by using a differential amplifier. We will study differential amplifiers in Chapter 15.

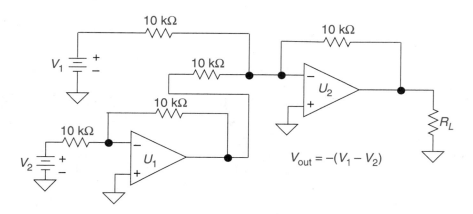

FIGURE 11–8   Subtracting circuit

11.3   MIXERS AND PERIODIC SIGNALS

Sometimes adders are referred to as mixers. Figure 11–9 shows a weighted adder being used as an audio mixer. The variable resistor on the inputs lets the mixing be controlled, and signals can be faded in and out by adjusting the resistors. The sum of instantaneous currents flowing through resistor R_f generates the output voltage.

Audio mixers are used by radio stations to mix audio signals from many sources. For example, one input can be the DJ's microphone, the second input can be a CD player for playing music, and the third can be the cartridge tape machine for playing commercials. Mixer boards also are

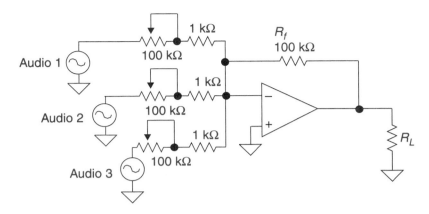

FIGURE 11–9 Audio mixer circuit

used by bands to mix the signals from different instruments and voices into one sound signal that is sent to audio amplifiers.

Fourier analysis is a method engineers use to analyze complex periodic waveforms. It permits any nonsinusoidal period function to be resolved into sine waves. While the mathematical study of Fourier analysis is beyond the scope of this text, we can use a mixer circuit to demonstrate the principle of Fourier analysis. A periodic sine wave contains only one frequency, but all other periodic signals contain multiple frequencies. All periodic signals are comprised of sine waves of different frequencies, magnitudes, and phases.

Figure 11–10 shows three input signals: a 1 kHz sine wave, a 3 kHz sine wave, and a 5 kHz sine wave. Their amplitudes are 15 V$_{p-p}$, 5 V$_{p-p}$, and 3 V$_{p-p}$ respectively. The three signals are in phase, meaning each time the low frequency passes through zero in a positive direction, the other waveforms pass through zero in a positive direction. When these signals are connected to the mixer circuit shown in Figure 11–10, the instantaneous magnitudes of each signal are added, resulting in the square-wave output signal shown.

A square-wave signal is comprised of the odd harmonics. For example, a 1 kHz square wave is comprised of the first harmonic (1 kHz), the third harmonic (3 kHz), the fifth harmonic (5 kHz), the seventh harmonic (7 kHz), and so on. The formula for a square wave is $v = \pi/4\ V_m\ (sin\ \omega t + 1/3\ sin\ 3\ \omega t + 1/5\ sin\ 5\ \omega t + 1/7\ sin\ 7\ \omega t + \ldots + 1/n\ sin\ n\omega t)$. The square wave in Figure 11–10 is not

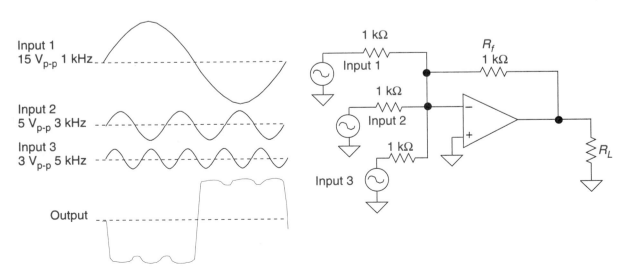

FIGURE 11–10 Mixer circuit

precise because only three harmonics were added. Engineers can write formulas for any periodic wave. It is not important that the technician be able to write or even understand these formulas. What is important is that the technician realize that nonsinusoidal waveforms are comprised of sinusoidal waves of many frequencies. The frequencies present in nonsinusoidal waves are far above the frequencies of fundamental waveforms. For example, a 1 kHz square wave contains a 9 kHz component. Therefore, circuitry and test equipment must be designed for not only the fundamental frequency but also the harmonics present in nonsinusoidal signals.

11.4 INTEGRATION

Integration is a mathematical operation (using calculus) for finding the area under the curve of a graph. Figure 11–11(a) shows a rectangular waveform, and Figure 11–11(b) shows the result of mathematically integrating that waveform (the integral). The amplitude of the waveform in Figure 11–11(b) is proportional to the area under the curve (shaped area) in Figure 11–11(a). Since we are dealing with a rectangular waveform, the area increases linearly as time progresses. Notice that when the waveform in Figure 11–11(a) returns to zero, the waveform in Figure 11–11(b) stays constant until the waveform in Figure 11–11(a) goes high once again. This happens because there is no increase in area when the waveform is at zero. In summary, it can be said that the integral of a waveform is proportional to the total area under the curve.

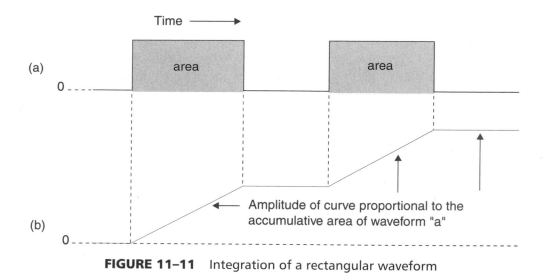

FIGURE 11–11   Integration of a rectangular waveform

Figure 11–12(a) shows an op-amp integrator circuit. If the rectangular waveform shown in Figure 11–12(b) is applied to the input, the output will be a ramp equal to the integral of the input. The circuit operates as follows. The input rectangular wave goes active, causing a constant current to flow through R_i. This same constant current is forced through C_f, causing the voltage across C_f to rise at a constant rate. The output voltage is inverted and is equal to the voltage across the feedback component (C_f). Figure 11–12(b) shows the output waveform. It is a linear ramp until the saturation of the op-amp is reached. If the input drives the output into saturation, it cannot go higher and will remain saturated.

The manner in which the capacitor voltage increases with a linear ramp may seem odd, since you have been taught that the voltage across a capacitor increases according to the universal

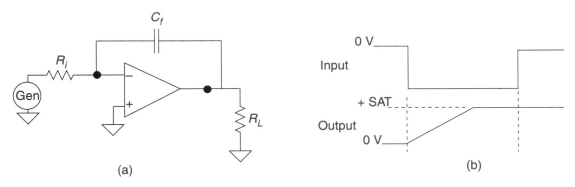

FIGURE 11–12   Integrator circuit

time constant curve. The trick here is the constant charging current. A capacitor charged through a resistor is charged by a current that is continuously decreasing. The formula for the voltage across a capacitor is $V = Q/C$, where Q is equal to current times time ($Q = it$). If all factors in the formula $V = (it)/C$ are held constant, except time, the voltage has to increase linearly with time.

Figure 11–13 shows how an integrator circuit would respond to a series of pulses. You can see that the output is proportional to the total area of all the pulses. Note that when the input changes polarity, the slope of the output of the integrator changes. Integrator circuits are not limited to rectangular-type inputs, but they are the easiest to analyze without the use of calculus. Example 11.5 shows how the output waveform of an integrator circuit can be calculated using the formulas $V = Q/C$ and $Q = it$.

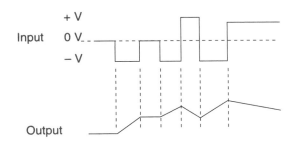

FIGURE 11–13   Integrator input and output signals

EXAMPLE 11.5

Calculate the output waveform for the circuit shown in Figure 11–14.

Step 1. Calculate the input current when the input is –2 V.

$i_{in} = V_{in}/R_i = -2\ V/2\ k\Omega = -1\ mA$ (The sign on current represents current direction.)

Step 2. Calculate the coulombs of charge the capacitor will contain at the end of the first pulse.

$Q = i \times t = 1\ mA \times 1\ mS = 1\ \mu C$ (The sign of current can be dropped.)

Step 3. Calculate the change in voltage at the end of the first pulse. Since the input signal is a rectangular pulse, the capacitor will charge linearly over the period of time the pulse is applied. If the voltage is calculated at the end of the 1 mS pulse, a line can be drawn from zero to the calculated voltage.

$\Delta V = Q/C = 1\ \mu C/1\ \mu F = 1\ V$ (The direction of the current causes a positive output voltage.)

EXAMPLE 11.5

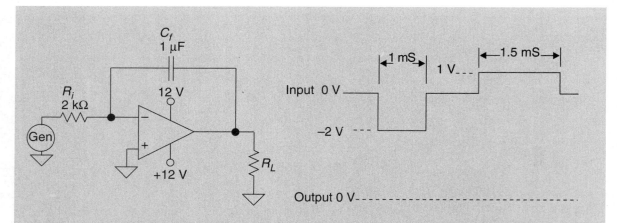

FIGURE 11–14   Integrator circuit

Step 4. Calculate the voltage output between the two pulses. Since the input voltage drops to zero, the current also drops to zero. Since no current flows through the capacitor, it will remain one volt.

Step 5. Calculate the input current when the input is 1 V.
$$i_{in} = V_{in}/R_i = 1\ V/2\ k\Omega = 0.5\ mA$$

Step 6. Calculate the change in charge caused by the second pulse.
$$Q = i \times t = 0.5\ mA \times 1.5\ mS = 0.75\ \mu C$$

Step 7. Calculate the change in voltage at the end of the second pulse. The voltage is at 1 V at the start of the second pulse. The current direction causes the voltage to change in a negative direction.
$$\Delta V_c = Q/C = 0.75\ \mu C/1\ \mu F = 0.75\ V$$

Step 8. Calculate the voltage output after the second pulse. The voltage changes from 1 V to 0.25 V during the second pulse. Since the input voltage drops to zero after the second pulse, there is no current flowing through the capacitor, and it will remain at 0.25 V.

Step 9. Draw the output waveform. The output waveform with respect to the input is shown in Figure 11–15.

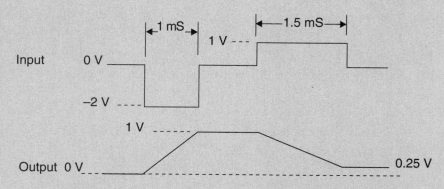

FIGURE 11–15   Output waveform with respect to input

There are many applications for integrator circuits. The missing pulse detector in Figure 11–16(b) is one example. The circuit is reset when switch S_1 is closed by a digital system to discharge C_f. When the switch is open, the integrator circuit charges C_f each time a pulse is received at TP_1. At the end of a time period [see P_1, Figure 11–16(a)], the digital system can check TP_3 for a high level. If TP_3 is at a high level, there were no missing pulses. If TP_3 is at a low level at the end of the time period [see P_2, Figure 11–16(a)], there were missing pulses.

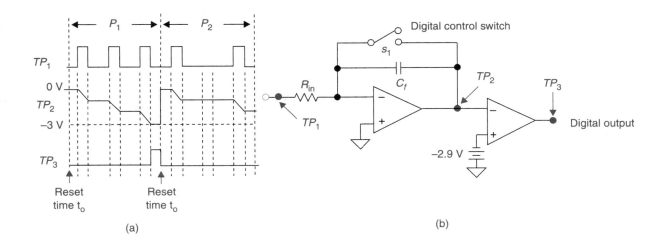

FIGURE 11–16 Missing pulse detector

11.5 DIFFERENTIATION

Differentiation is a mathematical operation (using calculus) that determines the rate of change of a curve at a given point. Figure 11–17 shows a graph of a waveform and a second graph of a waveform that would result by differentiating the first waveform. The second waveform is the derivative of the first waveform. Notice that the magnitude of the derivative waveform is dependent on the rate of change (slope) of the original waveform.

Figure 11–18(a) shows an op-amp differentiator circuit. Figure 11–18(b) shows the input and output waveforms. The circuit functions as follows. First, an input voltage is applied to the inverting input through the capacitor. The linear rise in voltage across the capacitor causes a

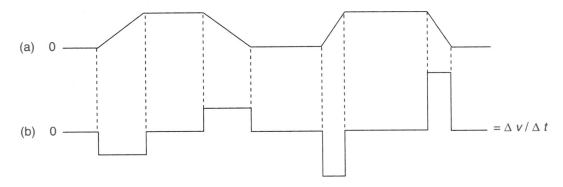

FIGURE 11–17 Differentiation of a waveform

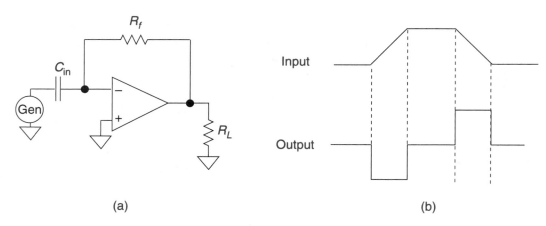

(a)   (b)

FIGURE 11–18   Differentiator circuit

constant current through the capacitor and also through the feedback resistor. This causes a constant voltage output. When the input voltage stays constant, there is no current through the input capacitor and, therefore, no current through the feedback resistor, which results in zero output voltage.

The constant current through the capacitor with a linear change in voltage can be analyzed by reviewing the formula for charges stored in a capacitor. The delta (Δ) sign stands for *change in*.

$$\Delta Q = \Delta v \times C$$
$$\Delta Q = \Delta i \times \Delta t$$
$$\Delta i = \Delta Q / \Delta t$$

Substituting:

$$\Delta i = (\Delta v \times C)/\Delta t$$

Rearranging:

$$\Delta i = (\Delta v / \Delta t) \times C$$

The formula $\Delta i = (\Delta v / \Delta t) \times C$ shows that if the voltage changes at a constant rate with respect to time, the current will be constant. However, if there is no change in voltage, the current will drop to zero.

There are many applications for differentiator circuits. The overacceleration detector circuit in Figure 11–19(b) is an example of one application. The input signal at TP_1 is a DC output from

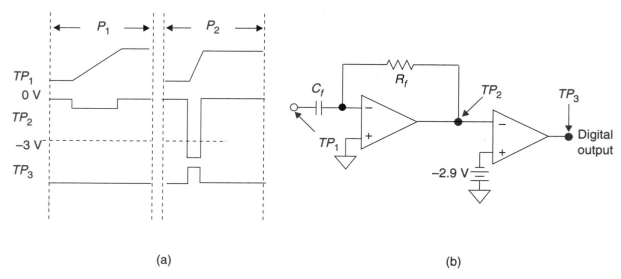

(a)   (b)

FIGURE 11–19   Overacceleration detector

a tachometer. As the speed of rotation changes, the DC voltage at TP_1 varies. The signal at TP_2 is proportional to the rate of speed change or acceleration. In time period P_1 [see Figure 11–19(a)], the acceleration does not exceed the limit set by the circuit, and there is no digital output pulse. However, in time period P_2 [see Figure 11–19(a)], the acceleration does exceed the limit set by the circuit, and the circuit outputs a digital pulse.

11.6 SINGLE SUPPLY OP-AMP CIRCUITS

All of the op-amp circuits we have studied up to this point have used dual power supplies. It is possible, however, to design op-amp circuits that operate from a single supply voltage. Figure 11–20(a) shows an inverting amplifier designed to operate from a single 20 V supply. In order to permit class-A operation, the output pin must be biased halfway between V_{cc} and V_{EE}. In single supply operation, V_{EE} is connected to ground; consequently, the output pin should equal one-half of V_{cc}. The bias is set by resistors R_1 and R_2, which supply one-half of V_{cc} (10 V) to the noninverting input pin. Since the voltage difference between the input pins is negligible, the DC voltage on the inverting input will also equal one-half V_{cc}. Capacitor C_1 blocks DC current flow; therefore, there will be zero bias current flowing through R_f, and the voltage on the output pin will equal the voltage on the input pins or one-half V_{cc}. Resistor R_b provides a bias current path for the output pin.

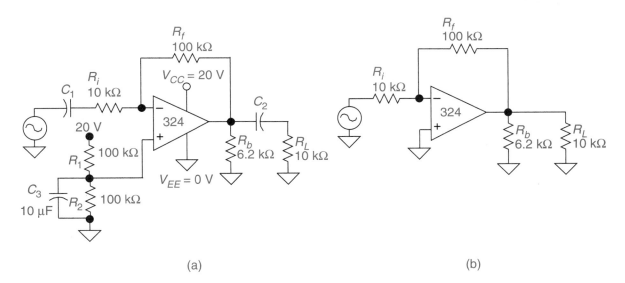

(a)   (b)

FIGURE 11–20 Single-supply inverting amplifier

Coupling capacitors (C_1 and C_2) are required on the input and output to block DC current flow, and capacitor C_3 is required to reference the noninverting input to signal ground. The circuit in Figure 11–20(b) is the signal equivalent obtained if all capacitors are replaced with shorts. Notice that this circuit is a simple inverting amplifier with one exception: resistor R_b is in parallel with R_L. Resistor R_b has no influence on the voltage gain of the circuit, which is equal to $-R_f/R_i$.

Figure 11–21(a) shows a noninverting amplifier circuit operating from a single power supply. The bias voltage is provided by a voltage divider comprised of resistors R_1 and R_2, which provide one-half of V_{cc} to the noninverting input pin through resistor R_3. Since the current flow through R_3 is negligible, the voltage on the noninverting input pin equals the voltage at the junction of

R_1 and R_2. Capacitor C_3 keeps the signal noise off of the bias voltage reference. Resistor R_b provides a biased current path for the output pin. Coupling capacitors (C_2 and C_4) are required on the input and output to block DC current flow. Capacitor C_1 provides a signal path to ground for the negative feedback circuitry. The circuit in Figure 11–21(b) is the signal equivalent obtained if all capacitors are replaced with shorts. This circuit is a simple noninverting amplifier with two exceptions: resistor R_b is in parallel with R_L, and R_3 connects the noninverting input to ground. Resistor R_b has no effect on the signal parameters, and the voltage gain of the circuit is equal to $1 + R_f/R_i$. Resistor R_3 does lower the input impedance from near infinite to the value of R_3.

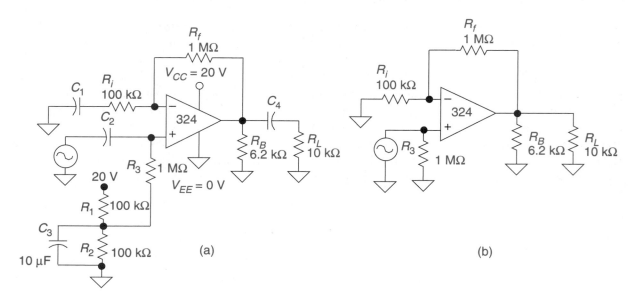

FIGURE 11–21   Single-supply noninverting amplifier

All op-amps can be used in circuit design to operate from a single supply. However, some op-amps have been optimized by the manufacturer for single supply operation. For example, the LM324 is designed to operate with single supply voltages in the range of 3 V to 32 V or dual supplies of –16 V to 16 V. When operating from a single supply, the output signal of an LM324 can swing near ground compared to the LM741, which saturates at approximately 1.5 V. The LM324 is a quad op-amp containing four op-amps in a single IC package and costs approximately forty cents. This makes the cost of an op-amp less than the cost of a single transistor.

11.7 PRECISION RECTIFIER CIRCUITS

The silicon diode has a 0.7 V drop in forward bias. This voltage drop is no problem when working with signals with large voltage swings. However, if a signal has a voltage swing of less than a volt, the normal rectifier circuit will not function. The precision rectifier circuit can rectify signals with voltage swings in the millivolts.

Figure 11–22 shows a precision half-wave rectifier circuit. On the positive alternation of the input, the output pin of the op-amp swings positive, forward biasing the diode. The forward-biased diode completes the feedback loop. The voltage between the two inputs of an op-amp must be the same if the output is not in saturation. Thus, the voltage on the inverting input is equal to the input signal on the positive alternation. The voltage on the output pin of the op-amp is approximately 0.7 V higher than the voltage across the load on the positive swing.

During the negative alternation, the diode is reverse biased, and the feedback loop is open. The output pin of the op-amp is driven into negative saturation, reverse biasing the diode. No current flows through the load so the output voltage is zero. The output waveform is shown for an input signal of 0.5 V peak. If the diode is reversed in Figure 11–22, the circuit will rectify the negative alternation.

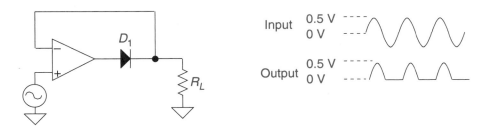

FIGURE 11–22 Precision half-wave rectifier

Figure 11–23 shows a precision half-wave rectifier using an inverting amplifier circuit. During the positive swing of the input, the output pin of the op-amp is driven negative, which forward biases D_2 and completes the feedback loop through R_f. The voltage gain is one ($R_f = R_i$), so the output signal equals the inverted input signal.

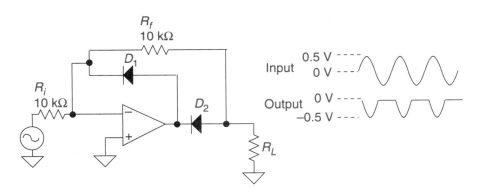

FIGURE 11–23 Precision high frequency half-wave rectifier

During the negative swing of the input, the output pin of the op-amp is driven positive, which reverse biases D_2, causing the output voltage to be zero. Diode D_1 is forward biased during this alternation, which clamps the voltage on the output pin of the op-amp to 0.7 V. Without D_1, the output pin of the op-amp would swing to positive saturation. The rectifier circuit in Figure 11–23 can operate at higher frequencies than the rectifier circuit in Figure 11–22, since the op-amp does not swing to saturation. If the diodes are reversed in Figure 11–23, the circuit will rectify the positive alternation.

Figure 11–24 is a precision full-wave rectifier. The circuit is comprised of a precision negative half-wave rectifier followed by a voltage adder circuit. The inputs to the adder stage are the sine wave source signal and the signal from the rectifier stage. The adder is weighted; the signal from the rectifier is amplified by two, and the sine wave source signal is amplified by one. The output signal at any instant equals $V_{out} = -(V_{TP1} + 2 V_{TP2})$. Example 11.6 calculates the output voltage at the positive and negative peaks of the sine wave.

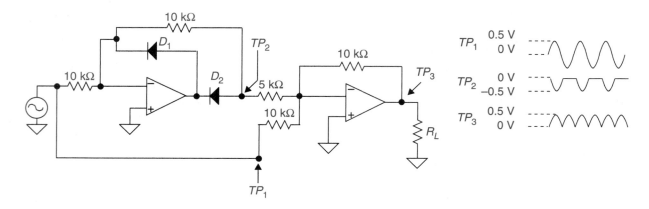

FIGURE 11–24   Precision full-wave rectifier

EXAMPLE 11.6

Calculate the output voltage at the positive and negative peaks of the sine wave signal (TP_1) shown in Figure 11–24.

Step 1. The first stage is a negative rectifier. The output is a $-0.5\ V_p$ signal as shown in Figure 11–24 (TP_2).

Step 2. Calculate the adder output on the positive swing of the input signal.

$$V_{out} = -(V_{TP1} + 2\ V_{TP2})$$
$$V_{out} = -(0.5\ V + (2 \times -0.5\ V)) = 0.5\ V$$

Step 3. Calculate the adder output on the negative swing of the input signal.

$$V_{out} = -(V_{TP1} + 2\ V_{TP2})$$
$$V_{out} = -(-0.5\ V + 2 \times 0\ V) = 0.5\ V$$

11.8 PEAK DETECTOR

Figure 11–25 shows a peak detector circuit. The output of a peak detector is always equal to the magnitude of the maximum peak level received since the last reset pulse. Op-amp U_2 is a voltage follower circuit, so the voltage across the capacitor (C_1) is equal to the output voltage. When a reset pulse (TP_1) is received, the E-MOSFET switch turns on and discharges the capacitor, setting the output voltage to zero volts. The output stays at zero until a positive input voltage is received (TP_2).

The positive voltage at input U_1 will drive the output pin of U_1 positive, forward biasing the diode and quickly charging the capacitor to the level of the input. The voltage across the capacitor is connected to the inverting input of U_1 and to the noninverting input of U_2. When the input voltage level at TP_2 drops below the capacitor voltage, the output pin of U_1 is driven into negative saturation, reverse biasing the diode. With the diode reverse biased and the E-MOSFET switch cut off, there is no discharge path for capacitor C_1, so the capacitor maintains its charge. The voltage level on the capacitor is transferred to the output of U_2, which drives the load.

The circuit is now in a static condition until the input receives a voltage level more positive than the capacitor voltage being fed back to the inverting input. If a more positive voltage appears at the input, the capacitor is quickly charged through the diode to the new voltage level. The

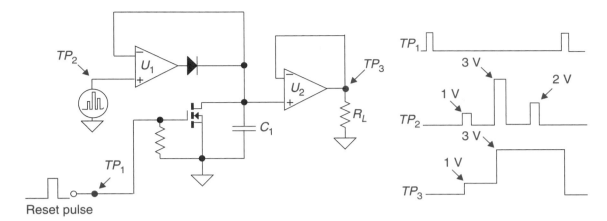

FIGURE 11–25 Peak detector

circuit then returns to the static state until a more positive voltage appears at the input or until a reset pulse is received. Since components are not perfect, the technician should realize that voltage levels cannot be maintained indefinitely because leakage currents will cause the capacitor voltage to drop after a period of time.

Figure 11–25 depicts the output waveforms for three consecutive input pulses of 1 V, 3 V, and 2 V. On the first pulse, the output jumps to 1 V, and on the next pulse the output jumps to 3 V, but note that on the third pulse the output stays at 3 V because the third pulse was less than the current output voltage.

11.9 COMPARATOR CIRCUITS

INTRODUCTION

General purpose op-amps can function as comparators, but they have been designed to operate in the linear mode, and not the switching mode. Manufacturers have designed special op-amp IC circuits to operate in a switching environment. These ICs are called *voltage comparators* and are designed to switch rapidly.

Figure 11–26 shows the response time for a comparator. Response time is the time it takes the output to switch from one state to another after a step voltage is received on the input. Typical response times for comparators are in the range from 200 nS to 1500 nS. High speed comparators, however, are available with response times in the range from 5 nS to 100 nS.

The circuit in Figure 11–27(a) shows a typical output circuit of the normal op-amp. The output swing is limited by supply voltages V_{cc} and V_{EE}. If the circuit is used as a comparator, the output

FIGURE 11–26 Response time

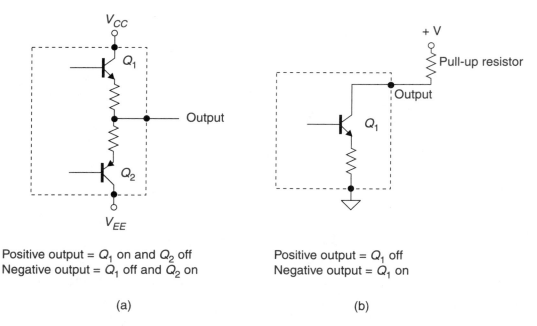

Positive output = Q_1 on and Q_2 off   Positive output = Q_1 off
Negative output = Q_1 off and Q_2 on   Negative output = Q_1 on

(a)   (b)

FIGURE 11–27   Output circuitry of comparators

will be in positive saturation or negative saturation. The internal resistances are relatively small; hence, the output voltage is near V_{cc} or V_{EE}. Some comparators use the type of circuitry shown in Figure 11–27(a), but most comparators use the type of circuitry shown in Figure 11–27(b). The output circuit in Figure 11–27(b) is called an *open collector output*. An external pull-up resistance is required for the circuit to function. If the output transistor is "on," the voltage on the output pin drops to near zero. The internal resistance is small, and the voltage drop is normally negligible. If the transistor is "off," the output pin rises to the value of the external positive voltage (+V) connected to the pull-up resistor.

Figure 11–28 shows a pin-out of the LM311, which is a flexible voltage comparator in an 8-pin package. The LM311 has differential inputs, an open collector output, V_{CC} input, V_{EE} input, and a ground pin. The LM311 can operate from split power supplies or from a single supply. There is also a strobe input pin that can be used to enable or disable the output.

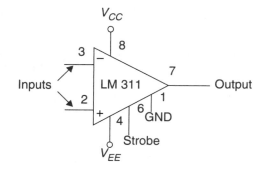

FIGURE 11–28   LM311 voltage comparator

The circuit in Figure 11–29 shows a zero crossover comparator circuit designed to control an AC load. Examining this circuit will demonstrate the flexibility of the LM311. The output pin of the comparator is connected to 24 V through a relay coil. When the output circuitry of the

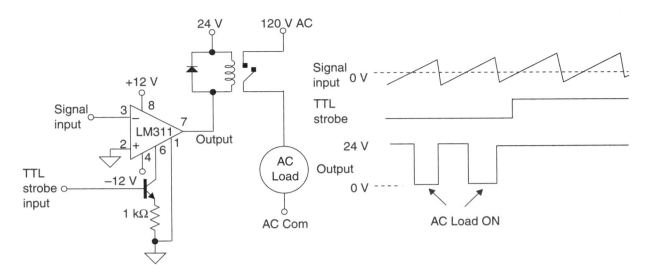

FIGURE 11–29 Relay driving circuit

comparator is not conducting, the voltage on the output pin rises to 24 V, and the relay is de-energized. When the output circuitry of the comparator is conducting, the voltage on the output pin drops to 0 V, causing the 24 V to energize the relay (see waveforms in Figure 11–29). The energized relay connects the AC load to the 120 V AC. The diode in parallel with the relay coil prevents a large induced voltage from being developed across the coil, which could damage the circuitry.

The LM311 is operating on plus and minus 12 V. The strobe is returned to ground through an external transistor and resistor. The off/on condition of the strobe is controlled by a TTL input (0 V or 5 V) connected to the base of the transistor. When the transistor is off, the circuit is enabled, and the output signal will be a function of the input signal (see Figure 11–29). If the strobe is active (high state), the output will be disabled and will be pulled up to the high output level. If the strobe pin is left open, it has no effect on the output.

THE SCHMITT TRIGGER

Because of the high open-loop gain, comparators often have a problem if the two signals at their inputs are close to the same magnitude. The problem is that noise on one of the inputs can cause the output to continuously switch. This problem is illustrated in Figure 11–30. In Figure

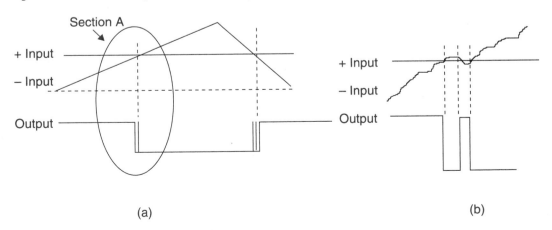

(a)

(b)

FIGURE 11–30 Noise causing output switching

11–30(a), the inverting input exceeds the noninverting input, and the output switches. There are several pulses present on the output. Figure 11–30(b) is an enlargement of section "A," and it can be seen that these pulses are caused by noise on the inverting input.

Figure 11–31 shows a Schmitt trigger circuit that overcomes the problem of switching caused by noise. The circuit has a positive feedback loop that causes electrical hysteresis. Electrical hysteresis causes the circuit to switch at one voltage level on the positive slope and switch at another voltage level on the negative slope. The switching voltage level on the positive slope is called the *upward trip point*. The switching voltage level on the negative slope is called the *downward trip point*. Example 11.7 shows how trip points can be calculated for the Schmitt trigger zero crossover comparator circuit in Figure 11–31.

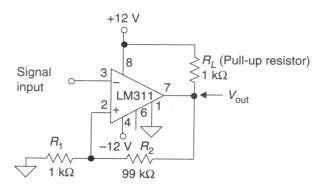

FIGURE 11–31   Schmitt trigger zero crossover comparator

EXAMPLE 11.7

Calculate the upward trip point voltage and the downward trip point voltage for the circuit in Figure 11–31.

Step 1. Start by assuming the input signal to the inverting input is less than zero volts; therefore, the output will be in its positive state. The voltage present on the noninverting input will equal the upward trip point voltage. The upward trip point voltage equals the voltage across R_1 and can be calculated by using the voltage divider formula. Notice that since the circuit is using an open collector output, the high level output voltage can be considered equal to the external pull-up voltage. However, this assumes that the resistance of the positive feedback loop ($R_1 + R_2 = 100$ k) is high in comparison to the load resistance ($R_L = 1$ k). If this is not the case, the value of V_{RL} would have to be included in the calculations.

$$V_{(UTP)} = V_{R1} = V_{cc} \times R_1/(R_1 + R_2)$$
$$V_{(UTP)} = 12 \text{ V} \times 1 \text{ k}/(1 \text{ k} + 99 \text{ k})$$
$$V_{(UTP)} = 120 \text{ mV}$$

Step 2. Now assume the input signal to the inverting input is greater than 120 mV. Therefore, the output will be in its negative state. The negative state output voltage for the LM311 is zero volts. The downward trip point voltage equals the voltage across R_1.

$$V_{(DTP)} = V_{R1} = V_{out} \times R_1/(R_1 + R_2)$$
$$V_{(DTP)} = 0 \text{ V} \times 1 \text{ k}/(1 \text{ k} + 98 \text{ k})$$
$$V_{(DTP)} = 0 \text{ V}$$

Figure 11–32 pictures the input and output waveforms for the circuit in Figure 11–31. An enlargement of section "A" shows that on the positive slope, the signal crosses the upward trip point (120 mV), then drops below the upward trip point. Notice that the output switches only once because the output will switch on a negative slope only when the input drops below the downward trip point (0 V). An enlargement of section "B" shows that on the negative slope, the signal crosses the downward trip point (0 V), then rises above the downward trip point. Also, note that the output switches only once.

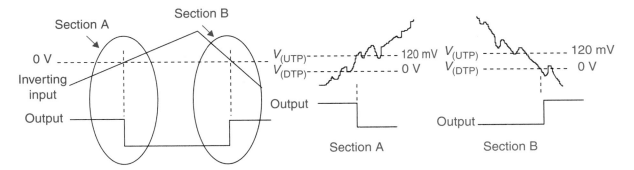

FIGURE 11–32   Switching with hysteresis

A comparator circuit designed to switch at any reference voltage can be designed with hysteresis. The circuit will switch just above the reference voltage on the positive slope and just below the reference voltage on the negative slope. Example 11.8 illustrates how switching voltages can be calculated.

EXAMPLE 11.8

Calculate the upward trip point voltage and the downward trip point voltage for the circuit in Figure 11–33.

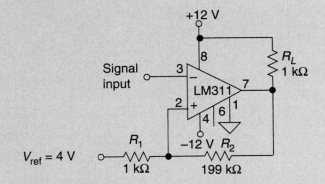

FIGURE 11–33   Schmitt trigger comparator circuit

Step 1. Start by assuming that the input signal to the inverting input is less than the reference voltage. Therefore, the output will be in its positive state. The voltage present on the noninverting input will equal the upward trip point voltage. The upward trip point voltage equals the voltage across R_1 plus the reference voltage (V_{ref}).

EXAMPLE 11.8 continued

$$V_{R1} = (V_{cc} - V_{ref}) \times R_1/(R_1 + R_2)$$
$$V_{R1} = (12 \text{ V} - 4 \text{ V}) \times 1 \text{ k}/(1 \text{ k} + 199 \text{ k})$$
$$V_{R1} = 40 \text{ mV}$$
$$V_{(UTP)} = V_{R1} + V_{ref}$$
$$V_{(UTP)} = 0.04 \text{ V} + 4 \text{ V} = 4.04 \text{ V}$$

Step 2. Now assume the input signal to the inverting input is greater than 4.04 V. Therefore, the output will be in its negative state. The negative state output voltage for the LM311 is zero volts. The downward trip point voltage equals the voltage across R_1 plus the reference voltage (V_{ref}).

$$V_{R1} = (V_{out} - V_{ref}) \times R_1/(R_1 + R_2)$$
$$V_{R1} = (0 \text{ V} - 4 \text{ V}) \times 1 \text{ k}/(1 \text{ k} + 198 \text{ k})$$
$$V_{R1} = -20 \text{ mV}$$
$$V_{(DTP)} = V_{R1} + V_{ref}$$
$$V_{(DTP)} = -0.020 \text{ V} + 4 \text{ V} = 3.98 \text{ V}$$

The circuit in Figure 11–33 is a Schmitt trigger circuit designed to switch at approximately 4 V. On the positive slope, the circuit actually switches at 4.04 V, and on the negative slope, the circuit switches at 3.98 V. The electrical hysteresis built into the circuit will keep the circuit from switching on noise as long as the noise level is below 60 mV. The trip points can be adjusted by changing the ratio of the feedback resistors.

WINDOW DETECTOR

The open collector output circuitry of the LM311 makes it easy to connect the outputs of two or more comparators together through use of a common pull-up resistor. The outputs are combined in an AND function. If any one of the outputs goes low, the combined output is forced low. Figure 11–34 shows a window detector circuit designed to light an LED when the 5 V TTL power supply

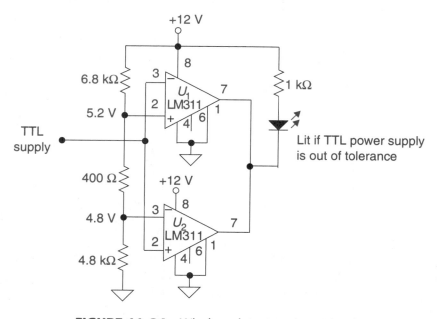

FIGURE 11–34 Window detector alarm circuit

is out of limits. The circuit uses two LM311s operating from a single 12 V supply. When the LM311 is operated from a single supply, V_{EE} and ground pins are connected together and tied to ground.

The high input impedance of the comparators does not load the input voltage divider consisting of three resistors. The input voltage divider sets the limits to 5.2 V on the high side and 4.8 V on the low side. If the input is between 4.8 V to 5.2 V, both comparators have a high output, and the LED is off. When the 5 V supply goes outside the window (4.8 V to 5.2 V), the output will go low, causing the LED to energize. The circuit is called a *window detector* because it detects when the voltage is outside of the limits, or outside of the window. If the LM311 did not have open collector outputs, a logic circuit would have to be used at the output of the two comparators.

11.10 TROUBLESHOOTING OP-AMP APPLICATIONS

We have covered a variety of op-amp applications in this chapter, but there are thousands of op-amp applications. It is impossible to cover them all. How does the technician troubleshoot the almost unlimited number of circuits? By understanding the basics and reading all the technical information available on the circuitry.

Let me share with you a story that illustrates the importance of reading the technical information. My first job out of school was working with a military contractor in the field engineering division. After three months of on-the-job training at the home plant, I was sent into the field to be the sole representative and expert on an airborne radar system. A few weeks after arriving at the base, I was called at 2:00 A.M. to give technical assistance to technicians who were replacing a defective antenna. I had never seen the antenna and had only a general idea where the antenna was located on the aircraft. Fortunately, two things worked in my favor: first, I had the technical manual, and second, the first step in the manual was to remove 44 screws from the antenna enclosure. While the technicians were removing the screws (thank goodness many required drilling), I read the remaining 83 steps on removing and replacing the antenna system. When the enclosure was open, I was ready. I glanced at the manual and stated step two in my own words. After a couple of hours and 82 steps later, the job was completed. The chief of maintenance thanked me and said they could not have done the job without me. My immediate thought was, if you could read, you would not have needed me. Fortunately, I did not verbalize my thoughts. Actually, the only thing I did that evening was read the technical manual, and lucky for me it was correct. The moral of this story is: Always read the technical material. It may be excellent, or it may be horrible, but it is the best place to start.

Many circuits you will encounter are variations of the basic circuits covered in this text, so an understanding of basic circuits will help you separate an electronic system into small blocks you can analyze. Remember that circuits containing op-amps can be analyzed by using two fundamental rules. Let us add a third rule for comparators:

Rule 1: The current in or out of either input is negligible.

Rule 2: If the output is *not* in saturation, the voltage between the two input terminals is essentially zero.

Rule 3: The output of a comparator is one of two states.

Pulse-Activated Switching System

1. The circuit in Figure 11–35 is a sound-activated switching system. Once activated, the LED will turn on for a time period, and then automatically shut off. After the LED is turned off, the system is reset and will respond to the next sound signal of sufficient amplitude. The system is comprised of six subcircuits. At the output of each subcircuit is a test point (*TP*). Analyze the system and list the name of each subcircuit after its corresponding test point listed below (all circuits were covered in this chapter).

    $TP_1 =$ _____ $TP_2 =$ _____ $TP_3 =$ _____
    $TP_4 =$ _____ $TP_5 =$ _____ $TP_6 =$ _____

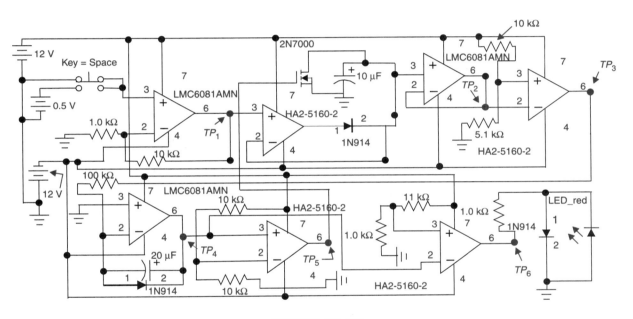

FIGURE 11–35

2. The momentary switch has not been activated in Figure 11–35. Calculate and record the voltages at the following test points.

    $TP_1 =$ _____ $TP_2 =$ _____ $TP_3 =$ _____
    $TP_4 =$ _____ $TP_5 =$ _____ $TP_6 =$ _____

3. Once the momentary switch is activated in Figure–35, calculate and record the voltages at the following test points for the times given below. Time T_0 is at the activation of the switch.

    0.2 S after time T_0

    $TP_1 =$ _____ $TP_2 =$ _____ $TP_3 =$ _____
    $TP_4 =$ _____ $TP_5 =$ _____ $TP_6 =$ _____

    1 S after time T_0

    $TP_1 =$ _____ $TP_2 =$ _____ $TP_3 =$ _____
    $TP_4 =$ _____ $TP_5 =$ _____ $TP_6 =$ _____

2 S after time T_0

$TP_1 = $ _____ $TP_2 = $ _____ $TP_3 = $ _____
$TP_4 = $ _____ $TP_5 = $ _____ $TP_6 = $ _____

4 S after time T_0

$TP_1 = $ _____ $TP_2 = $ _____ $TP_3 = $ _____
$TP_4 = $ _____ $TP_5 = $ _____ $TP_6 = $ _____

4. The momentary switch has not been activated in Figure 11–35. Simulate and record the voltages at the following test points.

$TP_1 = $ _____ $TP_2 = $ _____ $TP_3 = $ _____
$TP_4 = $ _____ $TP_5 = $ _____ $TP_6 = $ _____

5. Once the momentary switch is activated in Figure 11–35, simulate and record the voltages at the following test points for the times given below. Time T_0 is at the activation of the switch. Use the simulated time, not real time.

0.2 S after time T_0

$TP_1 = $ _____ $TP_2 = $ _____ $TP_3 = $ _____
$TP_4 = $ _____ $TP_5 = $ _____ $TP_6 = $ _____

1 S after time T_0

$TP_1 = $ _____ $TP_2 = $ _____ $TP_3 = $ _____
$TP_4 = $ _____ $TP_5 = $ _____ $TP_6 = $ _____

2 S after time T_0

$TP_1 = $ _____ $TP_2 = $ _____ $TP_3 = $ _____
$TP_4 = $ _____ $TP_5 = $ _____ $TP_6 = $ _____

4 S after time T_0

$TP_1 = $ _____ $TP_2 = $ _____ $TP_3 = $ _____
$TP_4 = $ _____ $TP_5 = $ _____ $TP_6 = $ _____

Sound-Activated Switching System

I. Objective

To construct and analyze a sound-activated switching system.

II. Test Equipment

+12 V power supply

Voltmeter

Bread boarding system

−12 V power supply

Oscilloscope

III. Components

ICs: (2) LM311, (1) LM324, (1) LM741

Diodes: (2) 1N4148, (1) LED

E-MOSFET: (1) 2N7000

Resistors: (5) 1 kΩ, (1) 2.2 kΩ, (3) 10 kΩ, (3) 100 kΩ

Capacitors: (1) 1 µF, (1) 100 µF

IV. Procedure

1. The circuit in Figure 11–36 is designed to respond to sounds. If the circuit is functioning properly, it will respond as follows: once a sound input of sufficient level is received, the system delays approximately 1 second, then the LED lights for approximately 20 seconds. Assuming the system is reset, calculate and record the voltages at the following test points.

   $TP_1 = $ _____ $TP_2 = $ _____ $TP_3 = $ _____
   $TP_4 = $ _____ $TP_5 = $ _____ $TP_6 = $ _____

2. A one-second sound burst is generated, causing the microphone to output a 30 mV$_{p-p}$ signal for one second. Using the start of the sound burst as time zero, calculate and record the voltages at the test points for the following times.

   0.5 S after time zero

   $TP_1 = $ _____ $TP_2 = $ _____ $TP_3 = $ _____
   $TP_4 = $ _____ $TP_5 = $ _____ $TP_6 = $ _____

   2 S after time zero

   $TP_1 = $ _____ $TP_2 = $ _____ $TP_3 = $ _____
   $TP_4 = $ _____ $TP_5 = $ _____ $TP_6 = $ _____

   10 S after time zero

   $TP_1 = $ _____ $TP_2 = $ _____ $TP_3 = $ _____
   $TP_4 = $ _____ $TP_5 = $ _____ $TP_6 = $ _____

   20 S after time zero

   $TP_1 = $ _____ $TP_2 = $ _____ $TP_3 = $ _____
   $TP_4 = $ _____ $TP_5 = $ _____ $TP_6 = $ _____

3. Construct the circuit in Figure 11–36. The voltage gain of the buffer amplifier stage (LM741) may need to be adjusted for the output of the microphone being used.

4. Activate the system with a sound input. Using the start of the sound burst as time zero, measure and record the voltages at the test points for the following times.

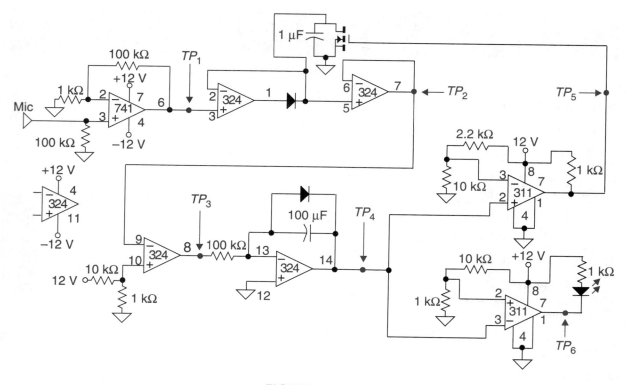

FIGURE 11–36

0.5 S after time zero

$TP_1 = $ _____	$TP_2 = $ _____	$TP_3 = $ _____
$TP_4 = $ _____	$TP_5 = $ _____	$TP_6 = $ _____

2 S after time zero

$TP_1 = $ _____	$TP_2 = $ _____	$TP_3 = $ _____
$TP_4 = $ _____	$TP_5 = $ _____	$TP_6 = $ _____

10 S after time zero

$TP_1 = $ _____	$TP_2 = $ _____	$TP_3 = $ _____
$TP_4 = $ _____	$TP_5 = $ _____	$TP_6 = $ _____

20 S after time zero

$TP_1 = $ _____	$TP_2 = $ _____	$TP_3 = $ _____
$TP_4 = $ _____	$TP_5 = $ _____	$TP_6 = $ _____

V. Points to Discuss

1. If the microphone being used in the circuit has an output of 100 mV_{p-p}, what should be the gain of the input amplifier for the desired sound level to cause it to trigger?

2. Explain why the sound input only needs to be present for an instant to trigger the system.

3. Once the sound input is present, the system delays approximately one second before the LED lights. Explain how this delay could be adjusted.

4. Explain how the on time of the LED could be adjusted.

QUESTIONS

11.1 High Input Impedance Circuits

____ 1. The high input impedance and low output impedance of noninverting amplifiers make them _____.
   a. ideal preamplifiers
   b. only useful in high frequency circuits
   c. likely to have low voltage gain
   d. likely to have low current gain

____ 2. It is often a good idea to buffer the output of a high impedance signal source to _____.
   a. provide 1:1 impedance matching
   b. prevent negative feedback
   c. reduce the input signal size
   d. prevent loading

____ 3. A high impedance voltmeter can be designed using a noninverting amplifier and _____.
   a. two capacitors
   b. an inductor
   c. a meter movement
   d. an LED

11.2 Basic Arithmetic Circuits

____ 4. An amplifier with a gain greater than one can be used to perform analog _____.
   a. addition
   b. subtraction
   c. multiplication
   d. division

____ 5. An amplifier with a gain less than one can be used to perform analog _____.
   a. addition
   b. subtraction
   c. multiplication
   d. division

____ 6. The _____ input makes the summing circuit possible.
   a. virtual ground at the noninverting
   b. virtual ground at the inverting
   c. low voltage
   d. high voltage

____ 7. If all the resistors in a summing circuit are equal, the output will be equal to the _____.
   a. average of the individual inputs
   b. inverted average of the individual inputs
   c. sum of the individual inputs
   d. inverted sum of the individual inputs

____ 8. If the value of resistor R_f in a summing circuit is equal to the value of one input resistor divided by the number of inputs, the output will be equal to the ____.

   a. average of the individual inputs

   b. inverted average of the individual inputs

   c. sum of the individual inputs

   d. inverted sum of the individual inputs

____ 9. A subtracting circuit can be designed using two op-amps: one as _____, and the other as an adder.

   a. an inverter

   b. a buffer

   c. an amplifier

   d. a multiplier

11.3 Mixers and Periodic Signals

____ 10. When audio signals are connected to the inputs of an adder, the circuit is sometimes referred to as _____.

   a. an audio circuit

   b. a blender circuit

   c. a mixer circuit

   d. a combiner circuit

____ 11. A periodic signal _____ .

   a. repeats its wave shape continuously

   b. is always sinusoidal in shape

   c. always contains only one harmonic

   d. always contains more than one harmonic

____ 12. A periodic sine wave _____.

   a. repeats its wave shape continuously

   b. is always sinusoidal in shape

   c. contains only one harmonic

   d. all of the above

____ 13. A square-wave signal is comprised of _____ harmonics.

   a. no

   b. all

   c. even

   d. odd

11.4 Integration and 11.5 Differentiation

____ 14. A mathematical operation that determines the rate of change of a curve is called ____.

   a. differentiation

   b. integration

   c. curve averaging

   d. none of the above

____ 15. A mathematical operation for finding the area under the curve of a graph is called _____.
   a. differentiation
   b. integration
   c. curve averaging
   d. none of the above

____ 16. The formula for the voltage across a capacitor is $V = Q/C$, where Q is equal to current times time ($Q = it$). If all factors, in the formula $V = (it)/C$, are held constant, except time, the voltage will _____ with time.
   a. increase linearly
   b. decrease linearly
   c. increase logarithmically
   d. decrease logarithmically

____ 17. The formula $\Delta i = (\Delta v/\Delta t) \times C$ shows that for a given capacitor, if the voltage changes at a constant rate with respect to time, then the current will _____.
   a. increase
   b. decrease
   c. be constant
   d. be dynamic

11.6 Single Supply Op-Amp Circuits

____ 18. In order to permit class-A operation, the output pin must be biased halfway between V_{cc} and V_{EE}. This means in single supply operations, the output pin should equal _____.
   a. zero volts
   b. one-half of V_{cc}
   c. one-half of V_{EE}
   d. V_{EE}

____ 19. Op-amp circuits using single-supply operation cannot amplify DC signals because _____.
   a. a coupling capacitor must be used on the input
   b. a coupling capacitor must be used on the output
   c. they use RC coupling
   d. all of the above

____ 20. The signal voltage gain of a noninverting amplifier using single supply operation is equal to _____.
   a. $1 + R_f/R_i$
   b. R_f/R_i
   c. $(1 + R_f/R_i) \times DC_{(offset)}$
   d. $(1 + R_f/R_i)/DC_{(offset)}$

11.7 Precision Rectifier Circuits

____ 21. The precision rectifier circuit is designed to _____.
   a. rectify precision waveforms
   b. amplify and rectify waveforms

    c.   rectify waveforms with small voltage swings

    d.   rectify waveforms with large voltage swings

____ 22.   A precision half-wave rectifier circuit with a diode in the feedback loop can operate at higher frequencies than the rectifier circuit without a diode in the feedback loop. The diode keeps the op-amp from _____.

    a.   oscillating

    b.   swinging into saturation

    c.   filtering high frequencies

    d.   all of the above

____ 23.   A precision full-wave rectifier circuit can be designed by using a precision half-wave rectifier followed by a _____.

    a.   second half-wave rectifier

    b.   voltage adder circuit

    c.   voltage subtracter circuit

    d.   buffer

11.8   Peak Detector

____ 24.   The output of a peak detector is always equal to the magnitude of the _____.

    a.   peak level of the current input signal

    b.   maximum peak level received

    c.   average of all peak levels received

    d.   last peak received

____ 25.   The output stage of a peak detector is _____ circuit.

    a.   an inverting amplifier

    b.   a voltage follower

    c.   a switching

    d.   a rectifier

11.9   Comparator Circuits

____ 26.   The time it takes the output of a comparator to switch from one state to another after a step voltage is received on the input is called _____.

    a.   switching time

    b.   operating time

    c.   falling time

    d.   response time

____ 27.   Most comparator ICs use _____ circuitry.

    a.   open collector output

    b.   open base input

    c.   open emitter output

    d.   differential output

____ 28. Some comparator ICs have a strobe input. The strobe is used to _____.
   a.   amplify the output
   b.   disable one input
   c.   disable the output
   d.   split the output

____ 29. The _____ circuit overcomes the problem of switching caused by noise on the inputs.
   a.   input buffer
   b.   Schmitt trigger
   c.   input noise rejecter
   d.   none of the above

____ 30. Comparator circuits designed to switch one voltage level on the positive swing and a second voltage level on the negative swing _____.
   a.   are designed with hysteresis
   b.   use positive feedback
   c.   tend not to switch on input noise
   d.   all of the above

____ 31. A _____ circuit has one output level for a defined range of input voltages and a second output level for input voltages above or below the defined range.
   a.   range comparator
   b.   range detector
   c.   window detector
   d.   dual comparator

11.10   Troubleshooting Op-Amp Applications

____ 32. The circuit in Figure 11–37 is a sound control switch; however, the microphone has been replaced by a function generator and a switch so the system can be activated without having to make a loud sound. With S_1 closed, the signal at TP_1 measures 40 mV$_{p-p}$. What is wrong with the circuit?
   a.   The 1 kΩ resistor connected to pin 2 of the LM741 is open.
   b.   The 1 kΩ resistor connected to pin 2 of the LM741 is shorted.
   c.   The 100 kΩ resistor connected between pin 2 and pin 6 of the LM741 is open.
   d.   a or c
   e.   The circuit is operating properly.

____ 33. Switch S_1 is closed but then opens, the signal at TP_1 measures 0 V, the signal at TP_2 measures 4 V, and the signal at TP_6 measures 0 V. What is wrong with the circuit?
   a.   The LM741 is defective.
   b.   The diode connected to the pin of the LM324 is shorted.
   c.   The 2 µF capacitor is open.
   d.   The E-MOSFET is open.
   e.   The circuit is operating properly.

____ 34. Switch S_1 is closed but then opens, the signal at TP_1 measures 0 V, and at first the signal at TP_2 measures 4 V and the signal at TP_6 measures 0 V. After a period of time, the signal at TP_6 changes to 12 V, but the signal at TP_2 remains at 4 V. What is wrong with the circuit?

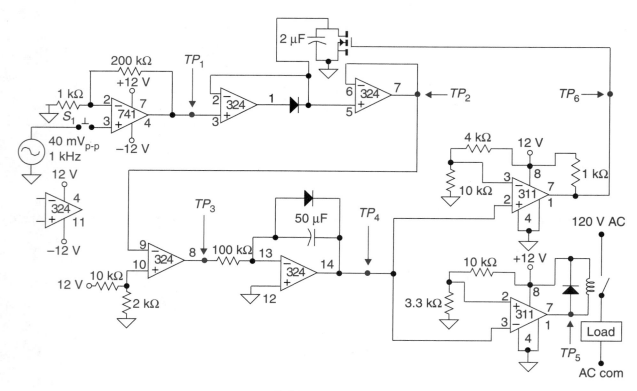

FIGURE 11–37

a. The LM741 is defective.

b. The diode connected between pin 1 and pin 5 of the LM324 is shorted.

c. The 2 μF capacitor is open.

d. The E-MOSFET is open.

e. The circuit is operating properly.

_____ 35. The function generator is adjusted for a 15 mV$_{p-p}$ output, switch S_1 is closed then opens, the signal at pin 9 of the LM324 measures 1.5 V, and the signal at TP_3 measures 11 V. What is wrong with the circuit?

a. The 2 kΩ resistor connected to pin 10 of LM324 is open.

b. The 10 kΩ resistor connected to pin 10 of LM324 is open.

c. The 100 kΩ resistor connected to pin 8 of LM324 is shorted.

d. The circuit is operating properly.

_____ 36. The function generator is adjusted for a 30 mV$_{p-p}$ output, switch S_1 is closed then opens, the signal at pin 9 of the LM324 measures 3 V, and the signal at TP_3 measures 11 V. What is wrong with the circuit?

a. The 2 kΩ resistor connected to pin 10 of LM324 is open.

b. The 10 kΩ resistor connected to pin 10 of LM324 is open.

c. The 100 kΩ resistor connected to pin 8 of LM324 is shorted.

d. The circuit is operating properly.

_____ 37. The system is sent in for repair. The LM311 comparator connected to the relay is found defective and replaced. The unit is returned to service but soon returns for repair. The second LM311 comparator connected to the relay is found defective and replaced. What else is wrong with the circuit?

a. The diode in parallel with the relay coil is shorted.

b. The diode in parallel with the relay coil is open.

c. The relay coil is shorted.

d. The relay coil is open.

PROBLEMS

1. Calculate the values of the resistors needed to make the circuit in Figure 11–38 function as a high input impedance voltmeter with 10 V, 30 V, and 100 V scales.

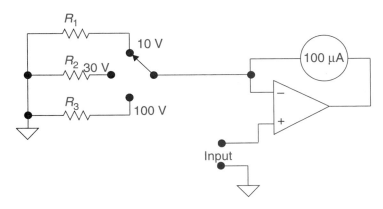

FIGURE 11–38

2. If the meter in Figure 11–38 is in the 10 V range and 7 V is applied to the input, what would be the current flow through the 100 μA meter movement ?

3. A 2 V signal source has an internal resistance of 20 kΩ. (a) Calculate the output signal if the source is connected directly across a 500 Ω load. (b) Calculate the output signal if an op-amp buffer is connected between the source and the 500 Ω load.

4. What is the output voltage of the circuit in Figure 11–39 if $V_1 = 2$ V, $V_2 = 4$ V, $V_3 = 3$ V, and all resistors equal 10 kΩ?

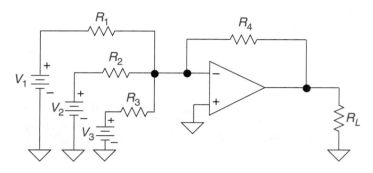

FIGURE 11–39

5. In Figure 11–39, resistors R_1, R_2, and R_3 all equal 6 kΩ, and resistor R_4 equals 2 kΩ. What is the output voltage of the circuit if $V_1 = 2$ V, $V_2 = 4$ V, and $V_3 = 3$ V?

6. In Figure 11–39, $R_1 = 1$ kΩ, $R_2 = 2$ kΩ, $R_3 = 4$ kΩ, and $R_4 = 2$ kΩ. What is the output voltage of the circuit if $V_1 = 2$ V, $V_2 = 4$ V, and $V_3 = 3$ V?

7. Configure and draw an op-amp circuit with three inputs (a, b, and c) that will provide an output (x) for the equation $x = -(a + 2b + 4c)$.

8. Draw the output waveform for the circuit in Figure 11–40. (Show timing with respect to input and label amplitude.)

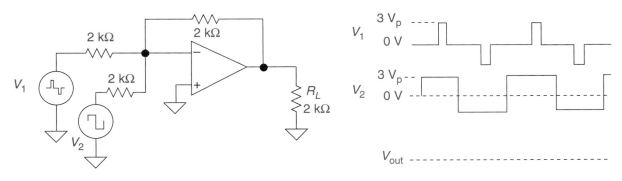

FIGURE 11–40

9. What is the output voltage of the circuit in Figure 11–41 if $V_1 = 3$ V and $V_2 = 3$ V?

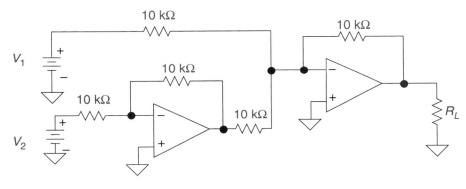

FIGURE 11–41

10. What is the output voltage of the circuit in Figure 11–41 if $V_1 = 3$ V and $V_2 = 5$ V?

11. Draw the output waveform for the circuit in Figure 11–42. (Show timing with respect to input and label amplitude.)

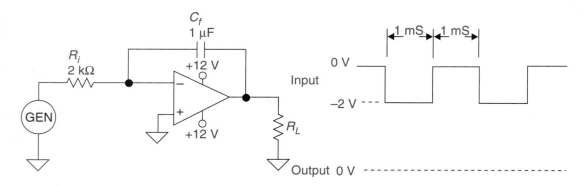

FIGURE 11–42

12. Calculate the output waveform for the circuit shown in Figure 11–43. (Show timing with respect to input and label amplitude.)

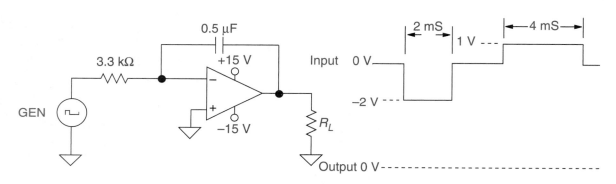

FIGURE 11–43

13. The capacitor in Figure 11–43 is discharged, and the generator is set for 3.3 V DC. How long after the generator is switched on before the op-amp reaches saturation?

14. Calculate the size of the capacitor required in an integrator circuit so that the output will rise 10 V in 0.1 seconds with an input of 1 V and an input resistance of 1 kΩ.

15. Draw the output waveform for the circuit in Figure 11–44. (Show timing with respect to input and label amplitude.)

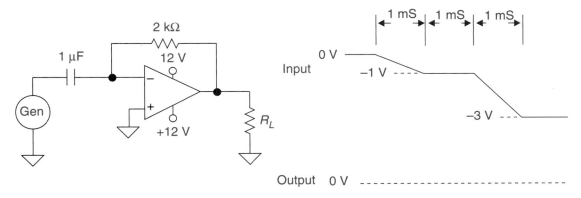

FIGURE 11–44

16. A differentiation circuit has an input capacitor of 10 μF. Calculate the feedback resistor required so that the output will be 5 V when the rate of change at the input is 100 V/S.

17. Draw the output waveforms for the circuit in Figure 11–45. (Show timing with respect to input and label amplitude.)

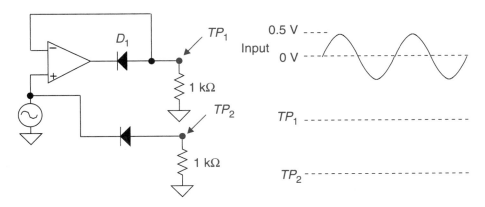

FIGURE 11–45

18. Calculate and draw the waveforms at TP_1 and TP_2 for the circuit shown in Figure 11–46. (Show timing with respect to inputs and label amplitude.)

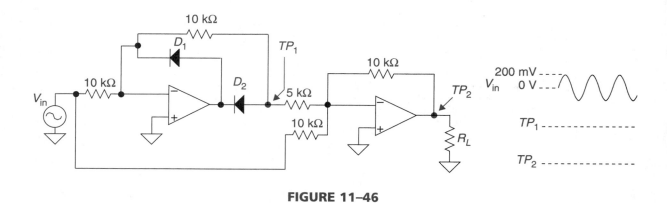

FIGURE 11–46

19. Draw the output waveform (TP_3) for the circuit in Figure 11–47. (Show timing with respect to inputs and label amplitude.)

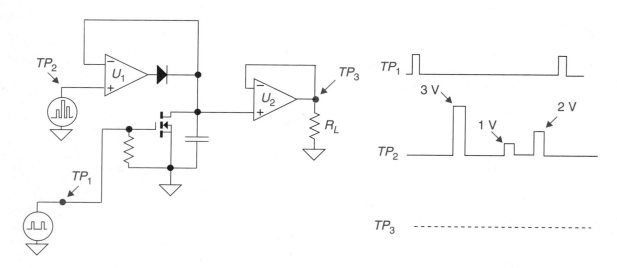

FIGURE 11–47

20. Draw the output waveform for the circuit in Figure 11–48. (Show timing with respect to inputs and label amplitude.)

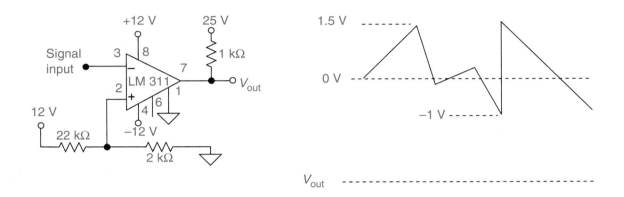

FIGURE 11–48

21. Draw the output waveform for the circuit in Figure 11–49. (Show timing with respect to inputs and label amplitude.)

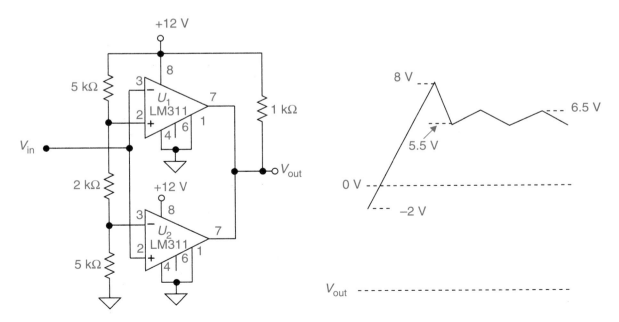

FIGURE 11–49

22. Configure a window comparator that will detect any input between 4.9 V and 5.1 V.

23. Draw the output waveform for the circuit in Figure 11–50. (Show timing with respect to inputs and label amplitude.)

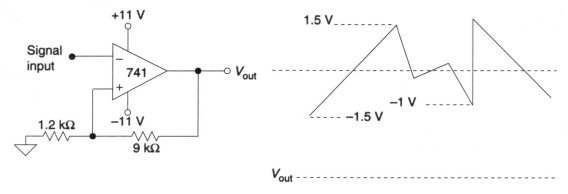

FIGURE 11–50

24. Calculate the upward trip point voltage, $V_{(UTP)}$, and the downward trip point voltage, $V_{(DTP)}$, for the circuit in Figure 11–51.

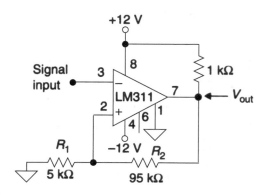

FIGURE 11–51

CHAPTER
Filter Circuits
12

OBJECTIVES

After studying the material in this chapter, you will be able to describe and/or analyze:
- ⭘ the types of filters,
- ⭘ RC passive filters,
- ⭘ the Bode plot,
- ⭘ first-order active filters,
- ⭘ higher-order active filters,
- ⭘ bandpass active filters,
- ⭘ bandstop active filters,
- ⭘ state-variable filters,
- ⭘ switched capacitor filters,
- ⭘ the operation of LC tuned amplifiers,
- ⭘ crystal filters, and
- ⭘ troubleshooting procedures for filter circuits.

12.1 INTRODUCTION TO FILTER CIRCUITS

A filter is a circuit that is designed to select electronic signals based on their frequencies. Figure 12–1 shows the four classes of filters based on their frequency responses. It would be ideal if the frequency response curves of filters were as depicted by the dotted lines. In practice, however, response curves do not cut off sharply but roll off as shown by the solid lines in Figure 12–1. By definition, a frequency is considered in the bandpass if it passes through the filter with voltage amplitudes of 70.7% or greater as compared to signals at the midband.

If the signal is attenuated more than 70.7%, it is considered to be in the bandstop. The frequency at which the voltage amplitude drops to the 70.7% point is called the *cutoff frequency* and is also referred to as the *half-power point*. If the voltage drops to 70.7%, the current will also drop to 70.7%. Since the power is the product of the voltage times current, the power will drop

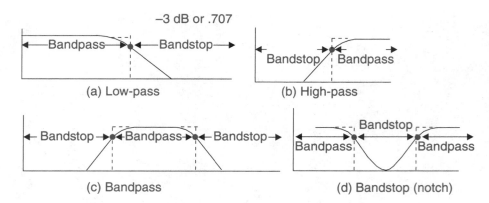

FIGURE 12–1   Frequency response of filters

to 50% ($0.5 = 0.707 \times 0.707$). The cutoff frequency is often described in decibels (dB). Negative 3 dB is equal to 70.7% using the voltage formula for dB [($dB = 20 \log (V_x/V_{ref})$] and is equal to 50% using the power formula for dB [($dB = 10 \log (P_x/P_{ref})$].

The low-pass filter, shown in Figure 12–1(a), passes all frequencies from zero up to the cutoff frequency. Figure 12–1(b) shows that the high-pass filter has a low cutoff frequency and passes all frequencies above that cutoff frequency. The bandpass filter has an upper and lower cutoff frequency, and the frequencies that pass are between the two cutoff points as shown in Figure 12–1(c). The bandstop or notch filter passes a lower band of frequencies up to a cutoff frequency and an upper band of frequencies above a second cutoff frequency. It rejects the frequencies between the two cutoff points. This is shown in Figure 12–1(d).

12.2 PASSIVE RC FILTER CIRCUITS

Figure 12–2 shows passive low-pass and high-pass filters using RC networks. Recall from your study of basic electronics, reactance of the capacitor is dependent on frequency ($X_c = 1/2\pi fc$). As the frequency goes up, the reactance (X_c) of the capacitor goes down. The resistance of a resistor is constant with changes in frequency. By examining the low-pass filter in Figure 12–2, you can see that it is a voltage divider made up of R and X_c. The output is taken across X_c. At low frequencies, X_c is much larger than R, and most of the input voltage is across the output. At higher frequencies, X_c becomes smaller than R, and most of the input voltage is dropped across R, causing the output to be small. For the high-pass filter in Figure 12–2, notice that the components have been reversed. In this case, the high frequencies will be passed, and the low frequencies will be attenuated.

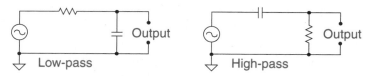

FIGURE 12–2   Passive filter circuits

One important point that must be remembered when analyzing RC filters is that all calculations must be performed using phasor math because of the phase shift. The cutoff frequency is equal to the frequency where the output voltage equals 0.707 of the input voltage; this occurs when $X_c = R$. Example 12.1 demonstrates that in a series circuit containing a resistor and a capacitor of equal impedance, the voltage across each component equals 70.7% of the input voltage.

EXAMPLE 12.1

Calculate V_C and V_R for the circuit in Figure 12–3.

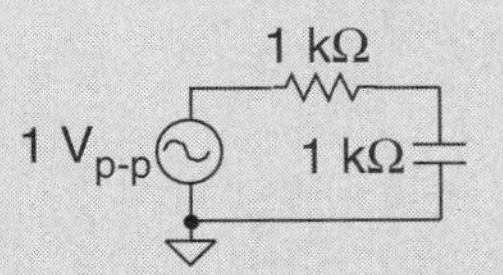

FIGURE 12–3 Simple RC filter circuit

$$Z = \sqrt{R^2 + X_C^2} = \sqrt{1\text{k}^2 + 1\text{k}^2} = 1.414\text{k}$$
$$I = V/Z = 1\ \text{V}_{\text{p-p}}/1.414\ \text{k}\Omega = 0.707\ \text{mA}$$
$$V_R = I \times R = 0.707\ \text{mA} \times 1\ \text{k}\Omega = 0.707\ \text{V}_{\text{p-p}}$$
$$V_c = I \times X_c = 0.707\ \text{mA} \times 1\ \text{k}\Omega = 0.707\ \text{V}_{\text{p-p}}$$

In a series circuit, where $R = X_c$, the voltage across each component is equal to 70.7% of the input signal. An RC filter is a series circuit containing a resistor and a capacitor. The output signal is taken across one of the components. The cutoff frequency of a filter is defined as the frequency where the output signal drops to 70.7% of the input signal. Therefore, at the cutoff frequency, $R = X_c$. Using the fact that R equals X_c at cutoff, a formula can be derived for the cutoff frequency of an RC filter.

$R = X_c$ (at cutoff)

$R = 1/(2\pi fc)$

$f_{(\text{cutoff})} = 1/(2\pi RC)$

Example 12.2 demonstrates how easy it is to calculate the cutoff frequency of an RC filter.

EXAMPLE 12.2

Calculate the cutoff frequency for both circuits in Figure 12–4, and draw a sketch of the frequency-response curve for each circuit.

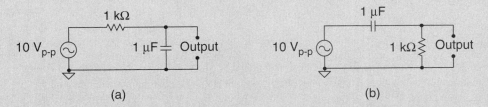

(a)   (b)

FIGURE 12–4 Low-pass and high-pass RC filters

Step 1. Since both circuits are using a 1 kΩ resistor and a 1 μF capacitor, the cutoff frequency will be the same for both circuits.

EXAMPLE 12.2 continued

$$f_{(\text{cutoff})} = 1/(2\pi RC)$$
$$f_{(\text{cutoff})} = 1/(2\pi 1 \text{ k}\Omega \times 1 \text{ }\mu\text{F})$$
$$f_{(\text{cutoff})} = 159 \text{ Hz}$$

Step 2. Figure 12–5 shows the frequency-response curve for both circuits. Circuit "a" is a low-pass filter, and circuit "b" is a high-pass filter.

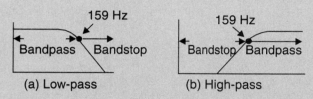

FIGURE 12–5   Frequency-response curves

The graph in Figure 12–5 shows that frequency cutoff occurs at 159 Hz in both circuits. Also, there is a sizable signal well beyond cutoff in both circuits. The ideal filter would totally reject signals out of the bandpass, but RC filters are far from ideal. Example 12.3 shows how a simple RC filter responds to frequencies above the cutoff frequency.

EXAMPLE 12.3

The low-pass filter in Figure 12–6 has a cutoff frequency of 1 kHz. Find the amplitude of the output voltage at cutoff (1 kHz), at one decade above cutoff (10 kHZ), and at two decades above cutoff (100 kHz).

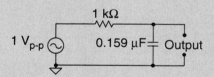

FIGURE 12–6   Low-pass RC filter

Step 1. Find the value of X_c for a frequency of 1 kHz.
$$X_c = 1/(2\pi fc) = 1/(2\pi \times 1 \text{ kHz} \times 0.159 \text{ }\mu\text{F}) = 1 \text{ k}\Omega$$

Step 2. Find the output voltage at a frequency of 1 kHz using the voltage divider formula.
$$V_{\text{out}} = V_{\text{in}} \times X_c/(R + X_c) = 1 < 0 \times 1 \text{ k}\Omega < -90º/(1 \text{ k}\Omega < 0º + 1 \text{ k}\Omega < -90º) = 0.707 \text{ V} < -45º$$

Step 3. Find the value of X_c for a frequency of 10 kHz.
$$X_c = 1/(2\pi fc) = 1/(2\pi \times 10 \text{ kHz} \times 0.159 \text{ }\mu\text{F}) = 100 \text{ }\Omega$$

Step 4. Find the output voltage at a frequency of 10 kHz using the voltage divider formula.
$$V_{\text{out}} = V_{\text{in}} \times X_c/(R + X_c) = 1 < 0 \times 100 < -90º/(1 \text{ k}\Omega < 0º + 100 < -90º) = 0.1 \text{ V} < -84º$$

Step 5. Find the value of X_c for a frequency of 100 kHz.
$$X_c = 1/(2\pi fc) = 1/(2\pi \times 100 \text{ kHz} \times 0.159 \text{ }\mu\text{F}) = 10 \text{ }\Omega$$

EXAMPLE 12.3 continued

> **Step 6.** Find the output voltage at a frequency of 100 kHz using the voltage divider formula.
>
> $V_{out} = V_{in} \times X_c/(R + X_c) = 1 < 0 \times 10 < -90°/(1\ \text{k}\Omega < 0° + 10 < -90°) = 0.01\ \text{V} < -89.4°$

12.3 ROLL-OFF

Figure 12–7 shows a logarithmic graph of the frequency-response curve for the low-pass filter circuit in Example 12.3. In this example, the voltage range of interest is from 1 mV to 1 V, and the frequency-response range of interest is from 1 Hz to 10 MHz. In order to graph values with such wide ranges, it is necessary to dimension the graph logarithmically on both the horizontal and vertical axis. The frequency-response curve shows that in the bandpass area, the output is relatively constant, and the voltage gain of the circuit is one. The output voltage gradually drops from 1 V to 0.707 V at the cutoff frequency of 1 kHz. Above the cutoff frequency, the output voltage drops off at a constant rate.

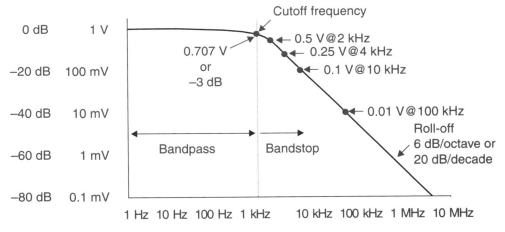

FIGURE 12–7 Bandpass response curve

How effective a filter is in rejecting signals beyond cutoff frequency is a function of the filter roll-off rating. Roll-off is measured in dB/octave or dB/decade. An octave is an increase or decrease in frequency by a factor of two. A decade is an increase or decrease in frequency by a factor of ten. Roll-off is calculated by measuring the output signal at a known frequency, and this value becomes the reference output signal. The frequency of the signal is adjusted by an octave or decade, and the amplitude of the output signal is measured at this new frequency. The dB value is calculated by taking the ratio of the two signal values measured. Example 12.4 shows how to calculate the roll-off of a filter for an octave change in frequency and a decade change in frequency.

EXAMPLE 12.4

The graph in Figure 12–7 shows voltage amplitude readings at one octave above cutoff, two octaves above cutoff, one decade above cutoff, and two decades above cutoff. Express the roll-off in dB/octave and dB/decade.

Step 1. Calculate the roll-off in dB/octave. Take the ratio of the output signal at 4 kHz to the output signal at 2 kHz.

$$dB/octave = 20 \log V_{out(4\ kHz)}/V_{out(2\ kHz)}$$
$$dB/octave = 20 \log 0.25\ V/0.5\ V = -6\ dB/octave$$

Step 2. Calculate the roll-off in dB/decade. Take the ratio of the output signal at 100 kHz to the output signal at 10 kHz.

$$dB/decade = 20 \log V_{out(100\ kHz)}/V_{out(10\ kHz)}$$
$$dB/decade = 20 \log 0.01\ V/0.1\ V = -20\ dB/decade$$

The frequency-response curve in Figure 12–7 shows the roll-off rate expressed as –6 dB/octave or –20 dB/decade. Normally, the negative sign is dropped when referring to roll-off of a filter. The word *roll-off* infers a negative direction.

12.4 BODE PLOT

Figure 12–8 shows a Bode plot for the filter circuit in Example 12.3. The Bode plot is a simplification of the frequency-response curve. The Bode plot shows that the output of the filter is constant in the bandpass area. Above the cutoff frequency, the roll-off is shown at a constant rate. The Bode plot (Figure 12–8) shows that the filter has a cutoff frequency of 1 kHz and then rolls off at 20 dB/decade (6 dB/octave). Using a Bode plot in place of the actual frequency-response curve makes calculating circuit values easy; therefore, we will use the Bode plot whenever possible.

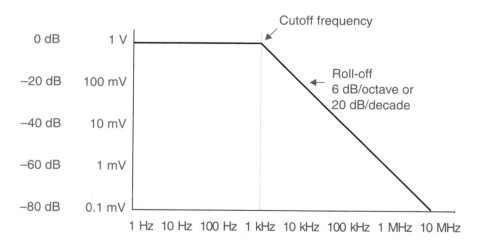

FIGURE 12–8   Bode plot

The ideal filter circuit would totally reject all signals above the cutoff frequency. The roll-off of a single-stage RC filter is limited to 20 dB/decade, which is far from the ideal filter. By connecting RC stages in series, the roll-off can be improved. However, when using passive filters, this is complicated because the next stage will load the previous stage. Loading can be overcome by adding an active device (transistor or op-amp) to buffer the output. Filter circuits that use active devices are called *active filters*.

12.5 FIRST-ORDER ACTIVE FILTERS

Figure 12–9 shows a low-pass and a high-pass active filter. The circuits consist of a passive RC filter connected to an op-amp follower stage. The voltage at the output of the follower is the same as the voltage at the noninverting input. These circuits have the same output as single-stage passive filters. Since the RC filter is connected to the high impedance input of the noninverting amplifier, the RC filter will not be loaded by the next stage. It is possible to replace the op-amp follower circuit with an op-amp noninverting amplifier and obtain voltage gain.

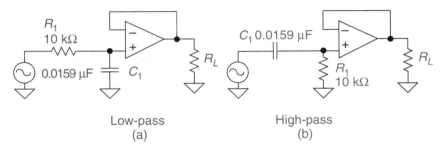

FIGURE 12–9 Active filters

Filters can be classified by the steepness of their roll-off. The roll-off is a measurement of how effective the filter circuit is at rejecting signals beyond the cutoff frequency. The roll-off is a ratio of a reference output signal in the bandpass to the output signal at a given frequency beyond cutoff.

Figure 12–10 shows the Bode plot for three different orders of filters. The passive filters we have studied so far are limited to a roll-off of 20 dB/decade or 6 dB/octave and are called *first-order filters*. A roll-off of 20 dB/decade means that as the frequency is increased by ten times, the amplitude of the output signal is reduced to one-tenth. First-order filters use only one RC network and are also called *single-pole filters*.

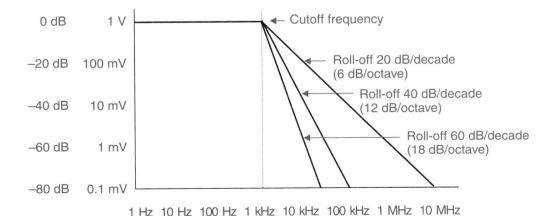

FIGURE 12–10 Bode plot of different order filters

12.6 HIGHER-ORDER ACTIVE FILTERS

Filters can be made closer to the ideal by increasing the number of RC networks or poles. If a filter has a roll-off of –40 dB/decade, it has two RC networks and is referred to as a second-order, or double-pole, filter. If a filter has a roll-off of –60 db/decade, it has three RC networks and is referred to as a third-order, or three-pole, filter. The steepness of the roll-off can be increased by adding additional RC networks.

The roll-off of a filter can be expressed in dB per octave. For example, a first-order filter has a roll-off of 20 dB/decade, which is equal to 6 dB/octave. Figure 12–10 shows a roll-off of 20 dB/decade equal to 6 dB/octave, a roll-off of 40 dB/decade equal to 12 dB/octave, and a roll-off of 60 dB/decade equal to 18 dB/octave.

By putting two first-order active filters in series, a second-order filter can be formed. However, this requires two op-amps. Figure 12–11 shows a method of obtaining a second-order filter using only one op-amp. Figure 12–11(a) is a low-pass filter, and Figure 12–11(b) is a high-pass filter. The gain of the circuit is unity, and the filtering components must have the relationship shown in Figure 12–11 for proper operation. The cutoff frequency of both circuits can be found using the following formula:

$$f_{(\text{cut off})} = \frac{1}{2\pi\sqrt{R_1 R_2 C_1 C_2}}$$

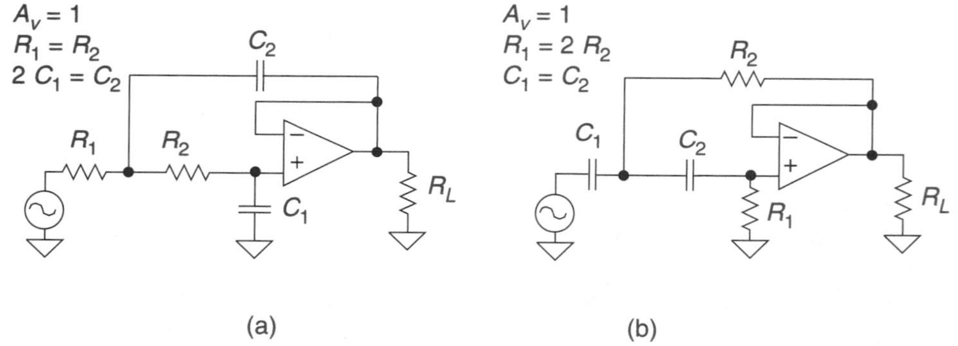

FIGURE 12–11   Second-order filter circuits with unity gain

Figure 12–12 shows a second-order filter that uses equal value filtering capacitors and resistors. In order for this circuit to work properly, the gain must be set to 1.586. The gain is calculated using the normal voltage gain formula for a noninverting amplifier ($A_v = 1 + R_f/R_{in}$). When equal value filtering components are used, the cutoff frequency for the total circuit is calculated by calculating the cutoff frequency for one pole using the formula $f_{(\text{cutoff})} = 1/2\pi RC$.

Higher-order filters can be made by cascading stages of first- and second-order filters. For example, Figure 12–13(a) shows a third-order low-pass filter constructed by using a second-order low-pass filter followed by a first-order low-pass filter. The roll-off of the first stage is –40 db/decade, and the second stage has a roll-off of –20 db/decade. The total roll-off is equal to –60 db/decade.

Figure 12–13(b) shows a high-pass active filter with a total roll-off of 60 db/decade. If steeper roll-offs are needed, they can be obtained by adding more stages. If all the filtering capacitors and resistors are equal, the formula $f_{(\text{cutoff})} = 1/2\pi RC$ can be used to calculate the cutoff frequency of the filter circuit. In order to use equal value filtering components, the voltage gain of each stage must be set to a specific value using resistors R_f and R_i in each stage. The calculations for these gains are beyond the scope of this text. Table 12–1 shows the voltage gains needed to design up to six-order filter circuits using equal value components.

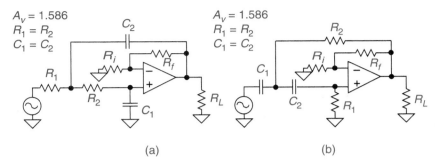

FIGURE 12–12 Second-order filter circuits with 1.59 gain

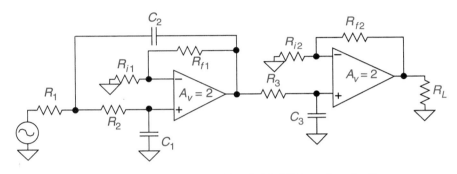

(a) Third-order low-pass filter (–60 dB/decade roll-off)

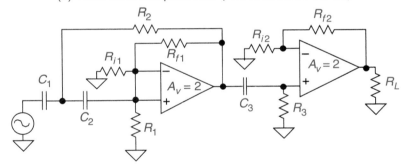

(b) Third-order high-pass filter (60 dB/decade roll-off)

FIGURE 12–13 Third-order filters

TABLE 12–1

Order	Roll-off	1st stage		2nd stage		3rd stage		Total
	dB/decade	Poles	Av	Poles	Av	Poles	Av	Av
1	20 dB	1	Optional					Optional
2	40 dB	2	1.586					1.586
3	60 dB	2	2.000	1	2.000			4.000
4	80 dB	2	1.152	2	2.235			2.575
5	100 dB	2	2.000	2	1.382	1	2.382	6.583
6	120 dB	2	1.068	2	1.586	2	2.482	4.204

The technician should not expect high voltage gains when working with filter circuits. The voltage of the total circuit is equal to the product of each stage. The voltage gain of a three-stage six-order filter, as shown in Table 12–1, is only 4.2.

12.7 BANDPASS FILTERS

Bandpass filters are designed to pass a band of frequencies between a low cutoff frequency (f_1) and a high cutoff frequency (f_2). The frequency in the center of the band is called the *center frequency* (f_o). Both response curves in Figure 12–14 have an output signal of 1 V at the center frequency (f_o) of 100 kHz. Since they are bandpass filters, each response curve has a low cutoff frequency (f_1) and a high cutoff frequency (f_2). The magnitude of the output signal at the cutoff frequency is equal to 70.7% of the magnitude at center frequency, or in this case 0.707 V. The bandwidth (BW) is defined as the difference between the high and low cutoff frequencies ($BW = f_2 - f_1$). The response curve in Figure 12–14(a) shows a bandwidth of 20 kHz, whereas the response curve in Figure 12–14(b) shows a bandwidth of 2 kHz.

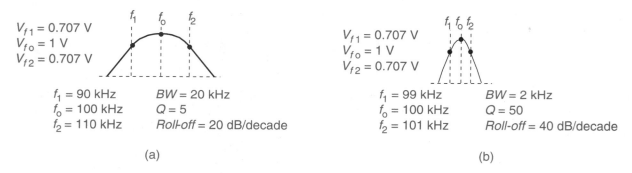

(a)   (b)

FIGURE 12–14   Bandwidth comparison

Q is equal to the center frequency divided by the bandwidth ($Q = f_o/BW$). The Q value is useful when comparing filter circuits. For example, both response curves in Figure 12–14 have a center frequency of 100 kHz, but the Q in Figure 12–14(a) is 5, and the Q in Figure 12–14(b) is 50. The higher the Q, the more selective the filter. Filter circuits with a Q below 10 are called *low-Q filters*, and filter circuits with a Q above 10 are called *high-Q filters*. Low-Q filters are wide bandpass circuits, and high-Q filters are narrow bandpass circuits.

Figure 12–15(a) shows the frequency-response curve for a bandpass filter circuit, and Figure 12–15(b) shows the Bode plot for the same circuit. The Bode plot shows the output signal as a constant 2 V across the bandwidth, and then a constant rate of roll-off beyond cutoff. The filter has a low-end roll-off and high-end roll-off. Normally both ends roll off at the same rate (as shown in Figure 12–15), but it is possible to have a bandpass filter circuit with different roll-off rates on the low and high ends. To calculate the high-end roll-off in dB/decade, the center frequency output signal is divided into the output signal at a frequency of one decade above cutoff. To calculate the low-end roll-off, the center frequency output signal is divided into the output signal at a frequency of one decade below cutoff. Example 12.5 shows how to evaluate a bandpass filter by examining its frequency-response curve.

EXAMPLE 12.5

Calculate the center frequency, bandwidth, Q, and roll-off for the filter response curve shown in Figure 12.15(b).

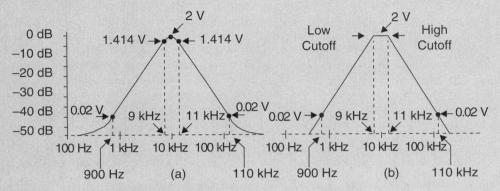

FIGURE 12–15 Band response and Bode plot curves

Step 1. From the Bode plot, determine the low cutoff frequency.
$$f_1 = 9 \text{ kHz}$$

Step 2. From the Bode plot, determine the high cutoff frequency.
$$f_2 = 11 \text{ kHz}$$

Step 3. Calculate the center frequency. For a circuit with a Q of two or higher, the center frequency can be calculated by dividing the sum between f_2 and f_1 by two.
$$f_o = (f_1 + f_2)/2 = (9 \text{ kHz} + 11 \text{ kHz})/2 = 10 \text{ kHz}$$

Step 4. Calculate the bandwidth of the filter.
$$BW = f_2 - f_1 = 11 \text{ kHz} - 9 \text{ kHz} = 2 \text{ kHz}$$

Step 5. Calculate the Q of the filter.
$$Q = f_o/BW = 10 \text{ kHz}/2 \text{ kHz} = 5$$

Step 6. Calculate the low-end roll-off in dB/decade. The low cutoff frequency is 9 kHz. Therefore, read the magnitude of the signal at one-tenth of the cutoff (900 Hz) from Figure 12–15(b). The output at 900 Hz is 0.02 V.
Low-end roll-off $= 20 \log (V_{(f1/10)}/V_{(fo)}) = 20 \log (0.02 \text{ V}/2 \text{ V}) = -40 \text{ dB/decade}$

Step 7. Calculate the high-end roll-off in dB/decade. The high cutoff frequency is 11 kHz; therefore, read the magnitude to the signal at ten times the cutoff (110 kHz) from Figure 12–15(b). The output at 110 kHz is 0.02 V.
High-end roll-off $= 20 \log(V_{(10f2)}/V_{(fo)}) = 20 \log (0.02 \text{ V}/2 \text{ V}) = -40 \text{ dB/decade}$

Figure 12–16 shows a bandpass filter using a second-order low-pass filter followed by a second-order high-pass filter. The upper cutoff frequency (f_2) is determined by calculating the cutoff frequency of the low-pass filter, and the lower cutoff frequency (f_1) is determined by calculating the cutoff frequency of the high-pass filter. This type of filter is used for wide bandpass filters. The Q of this filter will be below two; consequently, the center frequency should be calculated using the formula

$$f_o = \sqrt{f_1 \times f_2} \,.$$

This formula is valid for calculating the center frequency of all bandpass filters; however, for bandpass filters with Qs higher than two, the formula $f_o = (f_1 + f_2)/2$ gives approximately the same results. In Example 12.6, the cutoff frequencies and the bandwidth are calculated.

EXAMPLE 12.6

Calculate the low and high cutoff frequencies, the bandwidth, the center frequency, and the Q of the filter circuit shown in Figure 12–16. Draw a Bode plot of the frequency response curve.

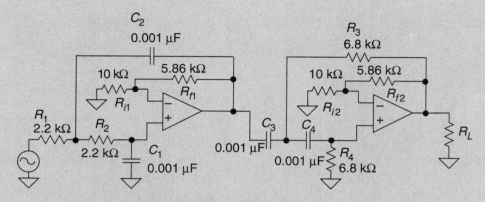

FIGURE 12–16   Bandpass filter

Step 1. Calculate the low cutoff frequency. The low cutoff frequency is a function of the high-pass filter stage.

$$f_2 = 1/(2\pi R_3 C_3) = 1/(2\pi \times 6.8\ \text{k} \times 0.001\ \mu\text{F}) = 23\ \text{kHz}$$

Step 2. Calculate the high cutoff frequency. The high cutoff frequency is a function of the low-pass filter stage.

$$f_2 = 1/(2\pi R_1 C_2) = 1/(2\pi \times 2.2\ \text{k} \times 0.001\ \mu\text{F}) = 72\ \text{kHz}$$

Step 3. Calculate the center frequency.

$$f_o = \sqrt{f_1 \times f_2} = \sqrt{23\ \text{kHz} \times 72\ \text{kHz}} = 40.7\ \text{kHz}$$

Step 4. Calculate the bandwidth.

$$BW = f_2 - f_1 = 73\ \text{kHz} - 23\ \text{kHz} = 49\ \text{kHz}$$

Step 5. Calculate the Q.

$$Q = f_o/BW = 40.7\ \text{kHz}/49\ \text{kHz} = 0.83$$

Step 6. Draw a Bode plot of the frequency-response curve. Figure 12–17 shows the Bode plot for the high-pass, low-pass, and bandpass. Since both the low-pass and high-pass are second-order filters, the bandpass will have a roll-off of 40 dB on both the low and high ends.

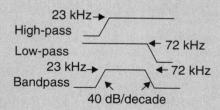

FIGURE 12–17   Frequency response curves

Figure 12–18 shows a multiple-feedback bandpass filter. The circuit is essentially an inverting amplifier with two feedback paths. At frequencies in the bandpass, capacitors C_i and C_f are at moderate impedance values, permitting the circuit to function as an inverting amplifier. In the bandpass area, the main factors controlling voltage gain are resistors R_f and R_i ($A_v \approx R_f/R_i$). At frequencies below the bandpass, the main factors controlling the voltage gain are the resistance of R_f and the impedance of capacitor C_i ($A_v \approx R_f/C_i$). Since C_i has a larger impedance at low frequencies, the gain is low, resulting in a small output signal. At frequencies above the bandpass, the main factors controlling the voltage gain are the impedance of capacitor C_f and the resistance of R_i ($A_v \approx C_f/R_i$). Since C_f has a low impedance at high frequencies, the gain is low, resulting in a small output signal.

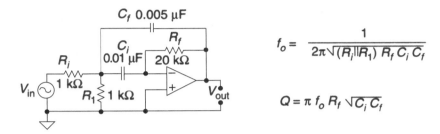

FIGURE 12–18   Multiple-feedback bandpass filter

The formula for calculating the center frequency of the circuit is shown in Figure 12–18. The Q of the filter can be calculated using the other formula in Figure 12–18. Once the Q and center frequencies are known, the bandwidth can be calculated by using the formula $BW = f_c/Q$. Since approximately half of the band is above the center frequency and half below, the upper and lower cutoff frequencies can be approximated. Example 12.7 shows how to analyze the multiple-feedback filter.

EXAMPLE 12.7

Calculate the center frequency, the Q, the bandpass, the low cutoff frequency, and the high cutoff frequency for the circuit in Figure 12–18.

Step 1. Calculate the center frequency.

$$f_o = \frac{1}{2\pi\sqrt{(R_i\|R_1)R_f\,C_i\,C_f}}$$

$$f_0 = \frac{1}{2\pi\sqrt{(1k\|1k)20k \times 0.01\mu F \times 0.005\mu F}}$$

$$f_o = 7.1 \text{ kHz}$$

Step 2. Calculate the Q.

$$Q = \pi f_o\, R_f\, \sqrt{C_i\, C_f}$$

$$Q = \pi 7.1\,\text{kHz}\; 20k\sqrt{0.01\mu F \times 0.005\mu F}$$

$$Q = 3$$

Step 3. Calculate the bandwidth.

$$BW = f_o/Q = 7.1 \text{ kHz}/3 = 2.36 \text{ kHz}$$

Step 4. Calculate low cutoff frequency (f_1).

$$f_1 = f_o - (BW/2)$$

$$f_1 = 7.1 \text{ kHz} - (2.36 \text{ kHz}/2)$$

$$f_1 = 5.92 \text{ kHz}$$

EXAMPLE 12.7 continued

Step 5. Calculate high cutoff frequency (f_2).
$$f_2 = f_o + (BW/2)$$
$$f_2 = 7.1 \text{ kHz} + (2.36 \text{ kHz}/2)$$
$$f_2 = 8.28 \text{ kHz}$$

12.8   BANDSTOP FILTERS

A bandstop filter filters out a particular band of frequencies while passing those frequencies below and above the bandstop. Figure 12–19 shows a bandstop filter (notch) designed by using a summer circuit to add the outputs of a low-pass and a high-pass filter. The low-pass filter permits frequencies below its cutoff to reach the input of the summer. The high-pass filter permits frequencies above its cutoff to reach the input of the summer. Frequencies above the low-pass cutoff but below the high-pass cutoff will fall into the band reject area and will not be present at the output of the summer. Example 12.8 shows how to analyze the circuit.

EXAMPLE 12.8

Calculate the low and high cutoff frequencies and the bandwidth of the reject band of the filter circuit shown in Figure 12–19. Draw a Bode plot of the frequency-response curve.

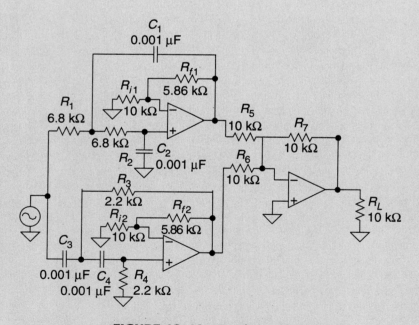

FIGURE 12–19   Bandstop filter

Step 1. Calculate the cutoff frequency of the low-pass filter.
$$f_1 = 1/(2\pi R_1 C_1) = 1/(2\pi \times 6.8 \text{ k} \times 0.001 \text{ μF}) = 23 \text{ kHz}$$

EXAMPLE 12.8 continued

> **Step 2.** Calculate the cutoff frequency of the high-pass filter.
> $$f_2 = 1/(2\pi R_3 C_3) = 1/(2\pi \times 2.2 \text{ k} \times 0.001 \text{ μF}) = 72 \text{ kHz}$$
>
> **Step 3.** Calculate the bandwidth of the reject band. The signals in the range of frequencies from 23 kHz to 72 kHz will be rejected by this bandstop filter.
> $$BW = f_2 - f_1 = 72 \text{ kHz} - 23 \text{ kHz} = 49 \text{ kHz}$$
>
> **Step 4.** Draw a Bode plot of the frequency-response curve. Figure 12–20 shows the Bode plot for the high-pass, low-pass, and bandstop.
>
>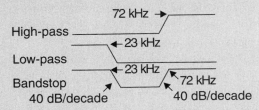
>
> **FIGURE 12–20**   Frequency-response curve

An example of a multiple-feedback bandstop filter is shown in Figure 12–21. The feedback from the output to the inverting input is similar to the multiple-feedback bandpass filter, but the input signal is also connected to the noninverting input. Resistors R'_f and R'_i set the voltage gain on the noninverting input side to a constant value ($A_v = R'_f/R'_i + 1$). The voltage gain on the inverting input side is determined by the multiple-feedback network. At low and high frequencies, the gain is low, and the output is a function of the signal passing through the noninverting side of the circuit. At the center frequency, the inverting side of the circuit has a voltage gain equal to the voltage gain of the noninverting side. Since the op-amp has differential inputs, the gains cancel, causing the output signal to equal zero at center frequency. The center frequency of the bandstop and the bandwidth can be calculated using the formulas in Figure 12–21.

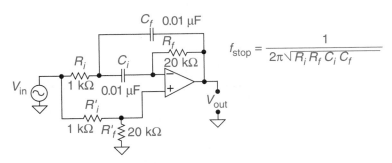

FIGURE 12–21   Bandstop filter

12.9   STATE-VARIABLE FILTERS

Figure 12–22 shows how three op-amps can be connected into a filter network called a *state-variable filter*, or sometimes called a *universal filter*. The circuit has three outputs: a second-order high-pass, a second-order low-pass, and a one-pole bandpass. Op-amp 1 is an inverting summer

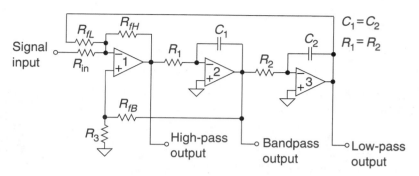

FIGURE 12–22 State-variable filter

that combines the input signal with feedback signals. Op-amp 2 and op-amp 3 are integrator circuits. Each integrator provides voltage gain and a 90° phase shift. There are two feedback paths: one from the output of op-amp 2 to the noninverting input of op-amp 1, and a second from the output of op-amp 3 to the inverting input of op-amp 1.

At frequencies below cutoff, the integrators have a high voltage gain ($A_v = X_c/R$), causing op-amp 3 to have a large output signal. The feedback signals at these frequencies tend to cancel the input signal so the signal present at the output of op-amp 1 is small. At frequencies above cutoff, the integrators have a low voltage gain ($A_v = X_c/R$), causing op-amp 3 to have a small output signal. The negative feedback signals are small at these frequencies, permitting the input signal to pass through op-amp 1 with little attenuation. Consequently, at high frequencies, the signal is large at the output of op-amp 1. At center frequency, the gain of the integrators are unity, and the feedback signal aids the input signal. This permits a fairly large signal at the outputs of all three op-amps. The signal at the output of op-amp 2, however, will roll off if the frequency is increased or decreased.

The state-variable filter can function as a low-pass, high-pass, and bandpass filter. Center frequency, low cutoff, and high cutoff are all calculated by using the formula $f_c = 1/2\pi R_1 C_1$. The Q of the circuit is controlled by the ratio of R_{fB} and R_3. For the low-pass and high-pass outputs to have a flat response curve, the ratio must equal one or less ($1 \geq R_{fB}/R_3$). If the circuit is to be used as a bandpass filter, the Q will need to be increased by changing the ratio of R_{fB} and R_3 (normally $10 < R_{fB}/R_3$). For this reason, the circuit is not useful as a bandpass filter when optimized for low-pass and high-pass filtering.

12.10 SWITCHED CAPACITOR FILTERS

Switched capacitor filters are based on the state-variable filter. One of the basic building blocks of the state-variable filter is the integrator circuit as shown in Figure 12–23(a). The RC time constant of the integrator is the determining factor of the frequency response of the filter circuit. Figure 12–23(b) shows the same integrator circuit, but the resistor has been replaced with a switch and capacitor (C_{in}) combination.

In the integrator circuit in Figure 12–23(a), the rate of charge of the feedback capacitor is a function of the input signal voltage (V_{in}) and the resistance of the input resistor. The capacitor is charged with a continuous current set by V_{in}/R. In the integrator circuit in Figure 12–23(b), the rate of charge of the feedback capacitor is a function of the signal input voltage (V_{in}), the capacitance of the input capacitor (C_{in}), and the switching speed of the switch. When the switch is connected to the input voltage, C_{in} charges. When the switch is switched to the input of the op-amp, C_{in} dumps this charge into C_f. The coulombs of charge stored by C_{in} is a function of the size of

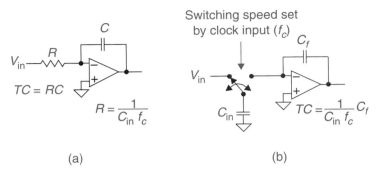

(a)   (b)

FIGURE 12–23   Integrator circuits

C_{in} and the signal input voltage (V_{in}). Input capacitor C_{in} is much smaller than the feedback capacitor (C_f); therefore, it takes many dumps of C_{in} to charge C_f. The rate of charging C_f is a function of the size of the charge stored in the input capacitor (C_{in}) and the frequency at which these coulombs of charge are dumped into the capacitor. The formula for the time constant of the circuit is $TC = (1/C_{in}f_c)C_f$.

Figure 12–24 shows a switched capacitor filter circuit. It is essentially a state-variable filter, but the input resistors on the two integrators have been replaced with an input capacitor (C_{in}) and a switch. The speed of switching is controlled by an input clock signal. The time constants of the integrators determine the frequency responses of the filter circuit. Since the time constant of the integrator is a function of the input frequency of the clock [$TC = (1/C_{in}f_c)C_f$], the bandpass of the filter can be controlled by varying the input frequency of the clock.

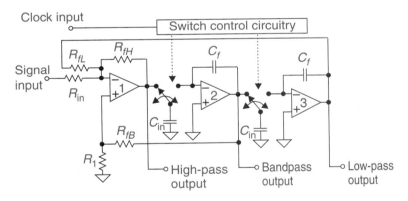

FIGURE 12–24   Switched capacitor filter

Manufacturers have integrated complete switched capacitor filter circuits into single IC packages. National Semiconductor's MF5 is an example of an integrated switched capacitor filter circuit as shown in Figure 12–24. The cutoff frequency of the high-pass and low-pass filter is equal to 1/100 of the input frequency of the clock.

12.11   LC TUNED AMPLIFIER

Figure 12–25 shows two tuned LC amplifier circuits. Recall that the voltage gain of the common-emitter amplifier is equal to the signal resistance in the collector leg divided by the signal resistance

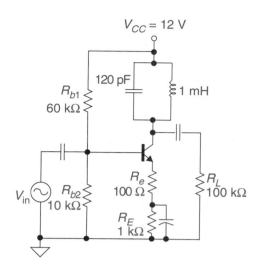

FIGURE 12–25   LC tuned amplifier

in the emitter leg ($A_v = r_c/r_e$). The parallel tuned circuit (tank circuit) is the signal resistance in the collector leg, and the signal resistance in the emitter is equal to R_e (assume r'_e is negligible). In electronic fundamentals, you learned that the impedance of the ideal parallel tuned circuit at resonance is infinite ohms. At frequencies below or above resonance, the impedance of the tuned circuit quickly decreases. Actually, tuned circuits have a high impedance at resonance, but not infinite. Therefore, the voltage gain of the circuit in Figure 12–25 is high at resonance. If the input frequency increases or decreases, the gain of the circuit decreases. The tuned LC amplifier circuit is a bandpass filter, so the center frequency can be calculated using the resonant frequency formula.

$$f_o = \frac{1}{2\pi\sqrt{LC}}$$

The Q of the filter circuit is equal to the Q of the LC tuned circuit. The Q of the LC tuned circuit is equal to the value of the inductive reactance (X_L) at resonance divided by the resistance in the tuned circuit. The bulk of the resistance in the tuned circuit is the resistance of the coil (R_L); therefore, $Q = X_L/R_L$. Once the Q is known, the bandwidth, low cutoff, and high cutoff can be calculated as shown in Example 12.9.

EXAMPLE 12.9

The coil in Figure 12–25 has an effective resistance of 20 Ω. Calculate V_B, V_E, V_C, V_{CE}, center frequency, Q, bandwidth, low cutoff, and high cutoff.

Step 1. The bias voltages are calculated using the procedures discussed in previous chapters. Since the coil has a resistance of only 20 Ω, the collector voltage will approximately equal V_{cc}.

$$V_B = V_{cc} \times R_{b2}/(R_{b1} + R_{b2}) = 12\text{ V} \times 10\text{ k}\Omega/(60\text{ k}\Omega + 10\text{ k}\Omega) = 1.7\text{ V}$$
$$V_E = V_B - 0.7\text{ V} = 1.7\text{ V} - 0.7\text{ V} = 1\text{ V}$$
$$I_c = I_E = V_E/R_e = 1\text{ V}/1.1\text{ k} = 0.91\text{ mA}$$
$$V_{RC} = I_c \times R_c = 0.91\text{ mA} \times 20\text{ }\Omega = 18.2\text{ mV}$$
$$V_C = V_{cc} - V_{RC} = 12\text{ V} - 18.2\text{ mV} = 11.9818\text{ or } 12\text{ V}$$
$$V_{CE} = V_c - V_E = 12\text{ V} - 1\text{ V} = 11\text{ V}$$

Step 2. Calculate the center frequency.

EXAMPLE 12.9 continued

$$f_o = \frac{1}{2\pi\sqrt{LC}} = \frac{1}{2\pi\sqrt{1m\text{ H}120pF}} = 459\,\text{kHz}$$

Step 3. Calculate the Q of the filter.
$$X_L = 2\pi \times f_o \times L = 2\pi \times 459\text{ kHz} \times 1\text{ mH} = 2.88\text{ k}\Omega$$
$$Q = X_L/R_L = 2.88\text{ k}\Omega/20\ \Omega = 144$$

Step 4. Calculate the filter bandwidth.
$$BW = f_o/Q = 459\text{ kHz}/144 = 3.18\text{ kHz}$$

Step 5. Calculate the low cutoff frequency.
$$f_{(\text{low cutoff})} = f_o - (BW/2) = 459\text{ kHz} - (3.18\text{ kHz}/2) = 457.4\text{ kHz}$$

Step 6. Calculate the high cutoff frequency.
$$f_{(\text{high cutoff})} = f_o + (BW/2) = 459\text{ kHz} + (3.18\text{ kHz}/2) = 460.6\text{ kHz}$$

LC tuned bandpass circuits are often used at RF (radio frequencies). At these frequencies, the size and cost of inductors are reasonable. It is common to use RF transformers in place of a single coil. Transformers coupling between stages provide DC isolation. In addition, the turns ratio of transformers are designed to provide impedance matching.

The circuit in Figure 12–26 is a bandstop filter circuit using a series LC circuit. The circuit is an inverting amplifier with a voltage gain of ten. The feedback resistor is shunted by a series LC circuit. At resonance, the series LC circuit will effectively become a short, causing the gain of the circuit to drop to zero. At frequencies below or above resonance, the LC circuit has a relatively high impedance, and the gain of the circuit is a function of the resistor ratio ($A_v = R_f/R_i$). The center frequency of the rejected band can be calculated using the resonant frequency formula.

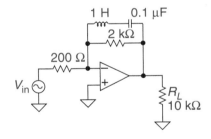

FIGURE 12–26 LC bandstop filter

EXAMPLE 12.10

Calculate the bandstop frequency for the circuit in Figure 12–26.
$$f_{(\text{stop})} = \frac{1}{2\pi\sqrt{LC}} = \frac{1}{2\pi\sqrt{1\text{H }0.1\,\mu\text{F}}} = 503\text{ Hz}$$

12.12 CRYSTAL AND OTHER PIEZOELECTRIC FILTERS

A crystal is a piece of quartz. The dimensions and cut determine the frequency at which it will mechanically resonate. The cut is the angle, relative to the crystalline structure of the bulk quartz, at which the crystal is sliced. The thinner the slice, the higher the resonant frequency. Frequencies from 100 kHz to over 100 MHz are possible. Crystals are piezoelectric, meaning a mechanical force across a crystal causes a small voltage to be generated. When metal contacts are attached to the crystal, the application of a small electrical signal causes the transfer of electrical energy to mechanical energy and back. Figure 12–27(a) shows the electrical equivalent of a crystal. The crystal itself is equivalent to the series LC circuit, but when leads are installed, they form a capacitance that is in parallel with the crystal. Figure 12–27(b) shows the symbol for a crystal.

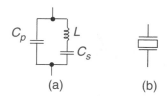

(a)   (b)

FIGURE 12–27   Crystal equivalent circuit and crystal symbol

A crystal has two resonate frequencies: one for series resonance, and a second for parallel resonance. At series resonance, the impedance is near zero, and the crystal functions as a short. At parallel resonance, the impedance of the crystal is extremely high, and the crystal functions as an open. Figure 12–28 shows an impedance versus frequency graph for the crystal. The formulas for calculating the resonant frequencies for both series and parallel resonance are shown in Figure 12–28.

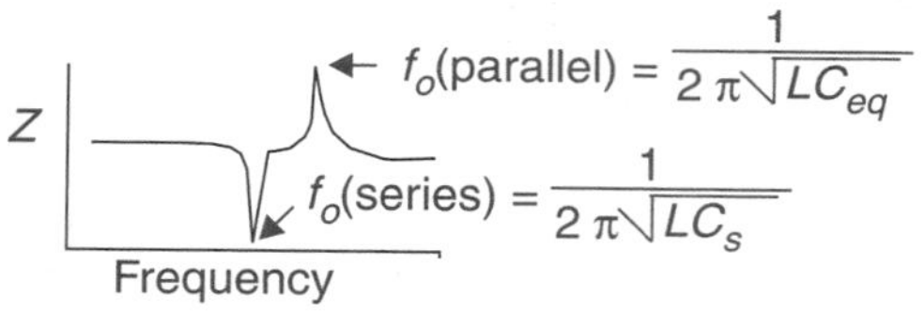

$$f_o(\text{parallel}) = \frac{1}{2\pi\sqrt{LC_{eq}}}$$

$$f_o(\text{series}) = \frac{1}{2\pi\sqrt{LC_s}}$$

Z

Frequency

FIGURE 12–28   Impedance versus frequency

At parallel resonance, a loop current flows, causing capacitors (C_s and C_p) to appear in series. Capacitor C_p is typically one hundred times the capacitance of C_s; hence, the effective equivalent capacitance (C_{eq}) is approximately equal to the capacitance of C_s. The series and parallel resonance frequencies of the crystal, however, are approximately the same, with the parallel resonance frequency being only slightly higher because C_{eq} is slightly lower than C_s.

The high Q of the crystal is an advantage of using a crystal filter over an LC tuned circuit. The Q of an LC tuned circuit is equal to X_L/R_L. With discrete components, it is difficult to obtain Qs of over a few hundred. Crystals have a very high value of inductance and a low resistance, and Qs can be obtained as high as 100,000. The higher the Q of the tuned circuit, the better the ability of the filter circuit to be frequency selective.

The ceramic filter is another piezoelectric device that has become popular in communication circuits. Ceramic filters are made from lead zinconate-titanate and formed into small disks. The small disks have electrical properties similar to crystals, but the Q for ceramic filters is considerably lower (normally not exceeding 2,000). Manufacturers offer ceramic filters for the popular frequencies used in communications circuitry. For example, ceramic filters are readily available for 455 kHz and 10.7 MHz. These are the IF (intermittent frequencies) for AM and FM radio receivers.

12.13 TROUBLESHOOTING FILTER CIRCUITS

A filter is a circuit that responds to changes in frequency, so the first step is to figure out how the filter will respond. Answering the following questions will give you the needed information to start troubleshooting: (1) what is the filter type—low-pass, high-pass, bandpass, or bandstop; (2) what is the cutoff frequency or frequencies; and (3) what is the roll-off rate? Use these answers to sketch the frequency-response curve. Now apply power to the circuit and check the DC voltages. Most filter circuits are modified amplifier circuits, so the method developed to calculate DC voltages for amplifier circuits often can be applied to determine the DC readings for filter circuits. If the DC voltages seem correct, it is time to use signal insertion to check the frequency response of the filter.

Figure 12–29 shows a test configuration that can be used to check the frequency response of a filter circuit. A function generator is connected to the input of the filter. An oscilloscope is used to measure the input and output of the filter. If available, a frequency counter should be used to monitor the frequency output of the generator. Set the function generator for a sine wave signal. Adjust the frequency control for a frequency in the bandpass area. If using a bandpass filter, the frequency should be set for the calculated center frequency. If a low, high, or bandstop filter is used, the frequency setting should be in the bandpass area and a minimum of one decade from the cutoff frequency. Adjust the function generator amplitude for a convenient output level. Record the output signal level. This level will be your reference level. The level should be below saturation but as large as possible without distortion, which will permit measurements beyond cutoff and allow the roll-off rate to be checked. With the output signal level set, measure and record the input signal. The input level must stay at this amplitude throughout the rest of the test. Monitor the input signal throughout the test to ensure it stays at the recorded level.

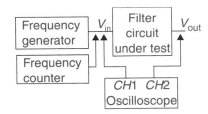

FIGURE 12–29 Test circuit for measuring frequency response

To measure the cutoff frequency, adjust the frequency control of the function generator until the output level reaches 70.7% of the original reference level. (The input should stay constant.) The cutoff frequency can be read on the frequency counter. If a frequency counter is not available, the frequency can be read from the dial on the function generator. On a bandpass or bandstop circuit, there will be two cutoff frequencies.

To check the roll-off rate, increase (decrease) the frequency of the function generator one decade or one octave above (below) the measured cutoff and record the amplitude of the output signal. Using the amplitude measurement and the original reference voltage amplitude, the roll-off can be calculated in dB/decade or dB/octave as follows:

$$\text{dB/decade} = 20 \log V_{(1 \text{ decade from cutoff})}/V_{(\text{reference})}$$

$$\text{dB/octave} = 20 \log V_{(1 \text{ octave from cutoff})}/V_{(\text{reference})}$$

The procedure just covered is a manual frequency sweep of the filter circuit. Many test generators have a sweep mode. In this mode, the generator outputs a constant amplitude signal but automatically adjusts the frequency. The generator has a control that permits starting and ending frequencies to be set for the desired frequency range. If this signal is connected to the input of a filter, the output of the filter will be a sine wave that varies in amplitude as the sweep generator sweeps across its frequency range. The generator will have a sweep output, which is a linear sawtooth signal. The oscilloscope can be operated in the xy mode. The x axis is connected to the sweep output of the generator, and the y axis is connected to the output of the filter. The result is a graph of amplitude versus frequency.

Active RC Filters

1. Calculate the cutoff frequency of the filter circuit in Figure 12–30.

   $f_{(cutoff)} =$ _____

2. What is the roll-off rate of the filter circuit in Figure 12–30? Give the answer in dB/octave and dB/decade.

   Roll-off = _____dB/octave   Roll-off = _____dB/decade

3. Draw and label a graph of the frequency-response curve for the filter circuit in Figure 12–30.

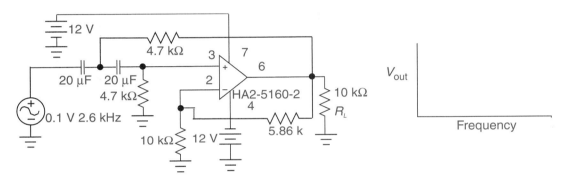

FIGURE 12–30

4. If the input signal is 1 V_{p-p}, what will be the output signal at 20 kHz and at the cutoff frequency for the circuit in Figure 12–30?

   $V_{out(20\ kHz)} =$ _____ $V_{out(cutoff)} =$ _____

5. Simulate and record the cutoff frequency and the roll-off rate for the filter circuit in Figure 12–30.

   $f_{(cutoff)} =$ _____   Roll-off = _____

6. Simulate the circuit in Figure 12–30 with an input signal of 1 V_{p-p}. Record the output voltage at 20 kHz.

   $V_{out(20\ kHz)} =$ _____

7. Calculate the cutoff frequency of the filter circuit in Figure 12–31.

   $f_{(cutoff)} =$ _____

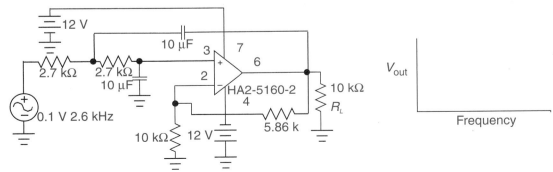

FIGURE 12–31

8. What is the roll-off rate of the filter circuit in Figure 12–31? Give the answer in dB/octave and dB/decade.

    Roll-off = _____dB/octave   Roll-off = _____dB/decade

9. Draw and label a graph of the frequency-response curve for the filter circuit in Figure 12–31.

10. If the input signal is 1 V_{p-p}, what will be the output signal at 500 Hz and at the cutoff frequency for the circuit in Figure 12–31?

    $V_{out(500\ Hz)}$ = _____ $V_{out(cutoff)}$ = _____

11. Simulate and record the cutoff frequency and the roll-off rate for the filter circuit in Figure 12–31.

    $f_{(cutoff)}$ = _____   Roll-off = _____

12. Simulate the circuit in Figure 12–31 with an input signal of 1 V_{p-p}. Record the output voltage at 500 Hz.

    $V_{out(500\ Hz)}$ = _____

13. Calculate the cutoff frequencies, the bandwidth, and the center frequency for the filter circuit in Figure 12–32.

    $f_{(low\ cutoff)}$ = _____ $f_{(high\ cutoff)}$ = _____
    BW = _____ f_o = _____

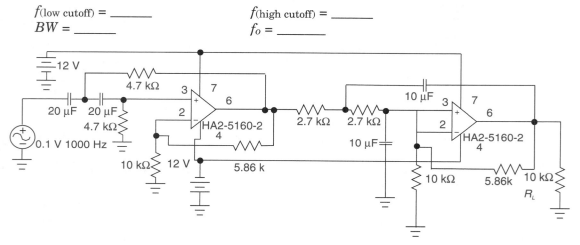

FIGURE 12–32

14. What are the low end and high end roll-off rates of the filter circuit in Figure 12–32? Give the answers in dB/octave and dB/decade.

    Low end: Roll-off = _____dB/octave   Roll-off = _____dB/decade
    High end: Roll-off = _____dB/octave   Roll-off = _____dB/decade

15. If the input signal is 1 V_{p-p}, what will be the output signal at center frequency and at the cutoff frequencies for the circuit in Figure 12–32?

    $V_{out(fo)}$ = _____ $V_{out(low\ cutoff)}$ = _____ $V_{out(high\ cutoff)}$ = _____

16. Draw and label a graph of the frequency-response curve for the filter circuit in Figure 12–32.

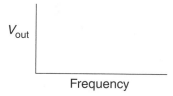

17. Simulate and record the cutoff frequencies, the bandwidth, and the center frequency for the filter circuit in Figure 12–32.

$f_{\text{(low cutoff)}}$ = _____ $f_{\text{(high cutoff)}}$ = _____

BW = _____ f_0 = _____

18. Simulate the circuit in Figure 12–32 with an input signal of 1 $V_{\text{p-p}}$. Record the output voltage at the center frequency.

$V_{\text{out}(f_0)}$ = _____

Active RC Filters

I. Objective

To experimentally examine an RC active filter.

II. Test Equipment

±12 V power supplies	Oscilloscope
Function generator	Frequency counter

III. Components

ICs: (2) LM318 op-amp

Capacitors: (2) 0.01 µF, (2) 0.022 µF

Resistors: (2) 2.2 kΩ, (4) 5.6 kΩ, (4) 10 kΩ

IV. Procedure

1. Calculate the cutoff frequency of the filter circuit in Figure 12–33.

    $f_{(cutoff)}$ = _____

2. What is the roll-off rate of the filter circuit in Figure 12–33? Give the answer in dB/decade.

    Roll-off = _____ dB/decade

3. If the input signal is 1 V_{p-p}, what will be the output signal at 15 kHz and at the cutoff frequency for the circuit in Figure 12–33?

    $V_{out(15\ kHz)}$ = _____ $V_{out(cutoff)}$ = _____

4. Construct the circuit in Figure 12–33.

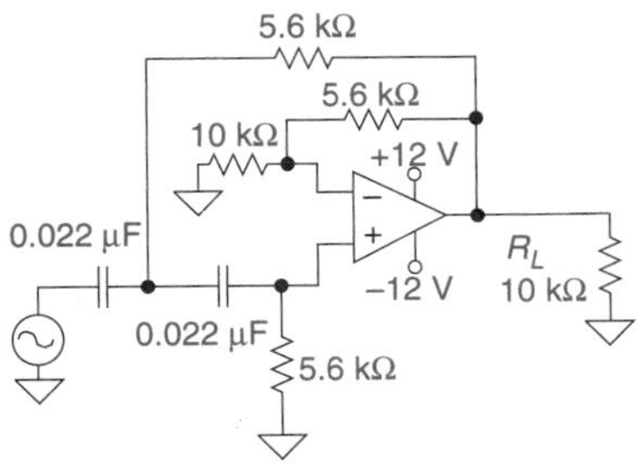

Freq.	V_{out}	Δ dB
100 Hz		
$f_{(cutoff)/10}$		
250 Hz		
500 Hz		
750 Hz		
1 kHz		
$f_{(cutoff)}$		
2.5 kHz		
5 kHz		
7.5 kHz		
10 kHz		
25 kHz		
50 kHz		
75 kHz		

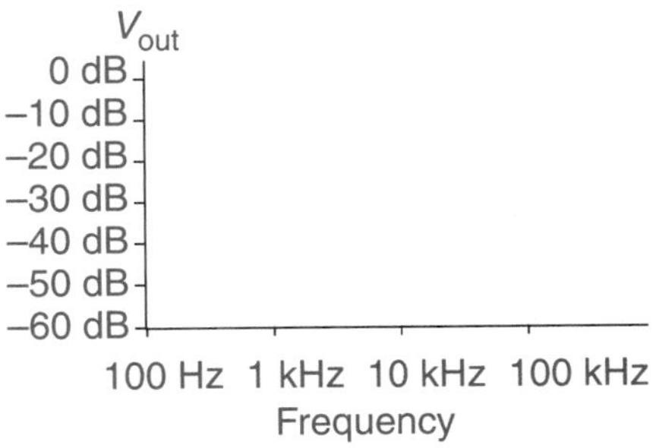

FIGURE 12–33

5. Adjust the input amplitude for an output signal of 10 $V_{p\text{-}p}$ at 100 kHz. Complete the table in Figure 12–33 by adjusting the frequency of the generator to the values shown. Be sure to monitor the input to make sure it maintains a constant amplitude. Calculate and record the changes in dB from the reference voltage level (10 $V_{p\text{-}p}$).

6. Calculate the roll-off rate of the filter circuit using measured values.
   Roll-off = _____

7. Draw and label a graph of the frequency-response curve using measured values.

8. Calculate the cutoff frequency of the filter circuit in Figure 12–34.
   $f_{(cutoff)}$ = _____

9. What is the roll-off rate of the filter circuit in Figure 12–34? Give the answer in dB/decade.
   Roll-off = _____dB/decade

10. If the input signal is 1 $V_{p\text{-}p}$, what will be the output signal at 200 Hz and at the cutoff frequency for the circuit in Figure 12–34?
    $V_{out(200\ Hz)}$ = _____ $V_{out(cutoff)}$ = _____

11. Construct the circuit in Figure 12–34.

12. Adjust the input amplitude for an output signal of 10 $V_{p\text{-}p}$ at 100 Hz. Complete the table in Figure 12–34 by adjusting the frequency of the generator to the values shown. Be sure to monitor the input to make sure it maintains a constant amplitude. Calculate and record the changes in dB from the reference voltage level (10 $V_{p\text{-}p}$).

13. Calculate the roll-off rate of the filter circuit using measured values.
    Roll-off = _____

14. Draw and label a graph of the frequency-response curve using measured values.

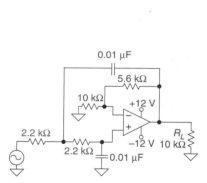

Freq.	V_{out}	Δ dB
100 Hz	10 $V_{p\text{-}p}$	0 dB
250 Hz		
500 Hz		
750 Hz		
1 kHz		
2.5 kHz		
5 kHz		
$f_{(cutoff)}$		
7.5 kHz		
10 kHz		
25 kHz		
50 kHz		
10 $f_{(cutoff)}$		
75 kHz		

FIGURE 12–34

15. Calculate the cutoff frequencies and center frequency of the filter circuit in Figure 12–35.

$f_{(\text{cut-off})}$ = _____

16. What are the low-end and high-end roll-off rates of the filter circuit in Figure 12–35? Give the answers in dB/decade.

Low-end roll-off = _____   High-end roll-off = _____

17. If the input signal is 1 V_{p-p}, what will be the output signal at center frequency and at the cut-off frequencies for the circuit in Figure 12–35?

$V_{\text{out(center freq.)}}$ = _____ $V_{\text{out(low cutoff)}}$ = _____ $V_{\text{out(high cutoff)}}$ = _____

18. Construct the circuit in Figure 12–35.

19. Adjust the input amplitude for an output signal of 10 V_{p-p} at center frequency. Complete the table in Figure 12–35 by adjusting the frequency of the generator to the values shown. Be sure to monitor the input to make sure it maintains a constant amplitude. Calculate and record the changes in dB from the reference voltage level (10 V_{p-p}).

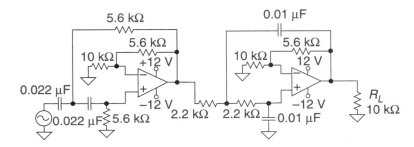

Freq.	V_{out}	Δ dB
100 Hz		
$f_{1/10}$		
250 Hz		
500 Hz		
750 Hz		
1 kHz		
f_1		
2.5 kHz		
f_0	10 V_{p-p}	0 dB
5 kHz		
f_2		
7.5 kHz		
10 kHz		
25 kHz		
50 kHz		
$f_{2\times10}$		
75 kHz		
100 kHz		

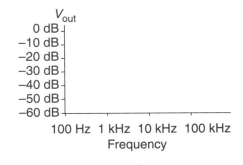

FIGURE 12–35

20. Calculate the low-end and high-end roll-off rates of the filter circuit using measured values.

Low-end roll-off = _____   High-end roll-off = _____

21. Draw and label a graph of the frequency-response curve using measured values.

V. Points to Discuss

1. Discuss the roll-off. Did each filter roll off as expected?

2. What was the gain of each filter? Could the gain be changed? Explain.

3. Was the bandpass filter in this experiment a high-Q or low-Q filter circuit? Explain.

4. The center frequency of the bandpass filter should have been calculated using the formula $f_o = \sqrt{f_1 \times f_2}$, and not the formula $f_o = (f_1 + f_2)/2$. Explain why.

QUESTIONS

12.1 Introduction to Filter Circuits

____ 1. Filter circuits separate electronic signals based on their _____.
    a. amplitude
    b. frequency
    c. logic level
    d. wave shape

____ 2. A signal is considered in the bandpass if it passes through the filter with voltage amplitudes of _____ or greater as compared to signals at the midband.
    a. 10%
    b. 50%
    c. 70.7%
    d. 90.9%

____ 3. The cutoff frequency is also referred to as the _____.
    a. half-power point
    b. low-end frequency
    c. high-end frequency
    d. both b and c

____ 4. A negative 3 dB drop causes the output to drop to _____.
    a. 97% of its original voltage level
    b. 70.7% of its original voltage level
    c. 50% of its original power level
    d. both b and c

12.2 Passive RC Filter Circuits

____ 5. As the frequency goes up, the reactance (X_c) of the capacitor _____.
    a. does not change
    b. doubles for every 1 kHz increase in frequency
    c. increases
    d. decreases

____ 6. As the frequency goes up, the resistance of the resistor _____.
    a. does not change
    b. doubles for every 1 kHz increase in frequency
    c. increases
    d. decreases

____ 7. The cutoff frequency of a passive RC filter occurs when _____.
    a. the power drops to 50%
    b. the voltage drops to 70.7%
    c. $X_c = R$
    d. all the above

12.3 Roll-Off

_____ 8.   Above the cutoff frequency of a low-pass filter, the output voltage _____ .
   a.   does not change
   b.   doubles for every 1 kHz increase in frequency
   c.   increases
   d.   decreases

_____ 9.   How effective a filter is in rejecting signals beyond cutoff frequency is a function of the _____.
   a.   signal amplitude
   b.   signal wave shape
   c.   filter roll-off rating
   d.   filter center frequency

_____ 10.   _____ is an increase or decrease in frequency by a factor of two.
   a.   A decade
   b.   An octave
   c.   A doubler
   d.   A duo

12.4 Bode Plot

_____ 11.   The Bode plot is a simplification of the _____.
   a.   filter schematic diagram
   b.   vertical curve
   c.   horizontal curve
   d.   frequency-response curve

_____ 12.   The Bode plot shows the output of the filter to _____ in the bandpass area.
   a.   be constant
   b.   be decreased at a constant rate
   c.   be increased at a constant rate
   d.   equal near zero volts

_____ 13.   The Bode plot shows the output of the filter to _____ in the roll-off area.
   a.   be constant
   b.   be decreased at a constant rate
   c.   be increased at a constant rate
   d.   equal near zero volts

12.5 First-order Active Filters

_____ 14.   A first-order active filter circuit consists of _____ connected to an op-amp follower stage.
   a.   a filtered IC
   b.   a crystal
   c.   an inductor
   d.   a passive RC filter

_____ 15. A first-order active filter has a roll-off of _____.
   a. 20 dB/decade
   b. 20 dB/octave
   c. 3 dB/decade
   d. 3 dB/octave

_____ 16. First-order filters are also called _____ filters.
   a. 20 dB
   b. one op-amp
   c. single-pole
   d. none of the above

12.6 Higher-Order Active Filters

_____ 17. Filters can be made closer to the ideal by increasing the number of _____.
   a. op-amps
   b. RC networks
   c. diodes
   d. transistors

_____ 18. A second-order active filter has a roll-off of _____.
   a. 20 dB/decade
   b. 20 dB/octave
   c. 3 dB/decade
   d. none of the above

_____ 19. A fourth-order filter is constructed by using a second-order filter followed by a _____.
   a. first-order filter
   b. second-order filter
   c. third-order filter
   d. fourth-order filter

12.7 Bandpass Filters

_____ 20. Bandpass filters are designed to pass a band of frequencies between _____ .
   a. a low cutoff and a high cutoff
   b. a bandstart and bandstop
   c. 1 kHz and 10 kHz
   d. 1 kHz and 1 MHz

_____ 21. The frequency in the middle of the band is called the _____.
   a. middle frequency
   b. Q-point frequency
   c. center frequency
   d. none of the above

_____ 22. The _____ is defined as the difference between the high and low cut-off frequencies.
   a. Q frequency range
   b. bandwidth

     c. filter width

     d. none of the above

____ 23. The higher the Q, the _____ the filter.

     a. lower the center frequency of

     b. greater the bandwidth of

     c. less selective

     d. more selective

12.8 Bandstop Filters

____ 24. A bandstop filter filters out _____.

     a. odd harmonics

     b. even harmonics

     c. a band of frequencies

     d. signals with large voltage swings

____ 25. The bandstop filter is also called _____.

     a. a notch filter

     b. a harmonics filter

     c. an open-loop filter

     d. a frequency stop filter

____ 26. A bandstop filter can be designed by using ____, a low-pass filter, and a high-pass filter.

     a. a bandpass filter

     b. a summer circuit

     c. an amplifier circuit

     d. none of the above

12.9 State-Variable Filters

____ 27. The state-variable filter is also called _____.

     a. a notch filter

     b. a harmonics filter

     c. a universal filter

     d. a feedback filter

____ 28. The state-variable filter circuit uses three op-amp stages. The first stage is a summer, and the other two stages are _____.

     a. amplifiers

     b. high-pass filters

     c. bandpass filters

     d. integrators

____ 29. The state-variable filter functions as a _____.

     a. low-pass filter

     b. high-pass filter

     c. bandpass filter

     d. all of the above

12.10 Switched Capacitor Filters

____ 30. In the normal integrator circuit, the rate of charge of the feedback capacitor is a function of the input voltage and the _____.
   a. resistance of the input resistor
   b. switching speed of the switch
   c. resistance of the output resistor
   d. input impedance of the op-amp

____ 31. In switched capacitor filters, the integrator circuit is modified so the rate of charge of the feedback capacitor is a function of the input voltage, the capacitance of the input capacitor, and the _____.
   a. switching speed of the switch
   b. resistance of the input resistor
   c. resistance of the output resistor
   d. input impedance of the op-amp

____ 32. In switched capacitor filters, the cutoff frequency is controlled by the _____.
   a. switching speed of the switch
   b. resistance of the input resistor
   c. RC network
   d. LC network

12.11 LC Tuned Amplifier

____ 33. The center frequency of an LC tuned amplifier is controlled by the _____.
   a. resonant frequency of the tank circuit
   b. RC network
   c. gain of the amplifier
   d. signal resistance in the emitter

____ 34. The impedance of an ideal parallel tuned circuit is _____.
   a. zero
   b. low
   c. high
   d. infinite

____ 35. The impedance of an ideal series tuned circuit is _____.
   a. zero
   b. low
   c. high
   d. infinite

12.12 Crystal and Other Piezoelectric Filters

____ 36. The electrical equivalent of a crystal is _____.
   a. a series LC circuit
   b. a parallel LC circuit

c.   both a series LC circuit and a parallel LC circuit

d.   an amplifier connected to an RC network

_____ 37.   Which of the following is the schematic symbol for a crystal?

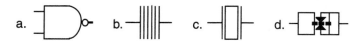

_____ 38.   The advantage of using a crystal filter over an LC tuned circuit filter is the _____ of
the crystal.

a.   high gain

b.   low Q

c.   high Q

d.   wide bandwidth

12.13   Troubleshooting Filter Circuits

_____ 39.   The circuit in Figure 12–36 has a 1 V_{p-p} input signal at a frequency of 4 kHz, and the
output is 2.4 V_{p-p}. What is wrong with the circuit?

a. C_1 is open.

b. C_1 is shorted.

c.   Op-amp #1 is defective.

d.   The circuit is functioning properly.

_____ 40.   The circuit in Figure 12–36 has a 1 V_{p-p} input signal at a frequency of 4 kHz, the output
of op-amp #1 is 1.6 V_{p-p}, and the output of op-amp #2 is swinging between plus and
negative saturation at a 4 kHz rate. What is wrong with the circuit?

a.   Op-amp #1 is defective.

b. R_5 is open.

c. R_8 is shorted.

d. R_8 is open.

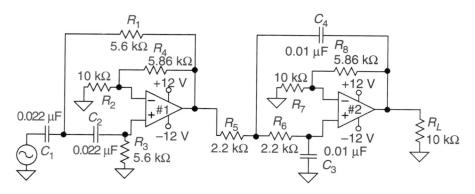

FIGURE 12–36

____ 41. The circuit in Figure 12–36 has a 1 V_{p-p} input signal at a frequency of 50 Hz, and the output measures only a few millivolts. What is wrong with the circuit?

a. C_1 is open.

b. C_1 is shorted.

c. Op-amp #1 is defective.

d. The circuit is functioning properly.

____ 42. The low end roll-off measures 20 dB/decade. What is wrong with the circuit?

a. C_1 is open.

b. C_1 is shorted.

c. Op-amp #1 is defective.

d. The circuit is functioning properly.

PROBLEMS

1. Calculate V_C and V_R for the circuit in Figure 12–37.

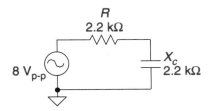

FIGURE 12–37

2. Calculate the cutoff frequency for both circuits in Figure 12–38, and draw a sketch of the frequency-response curve for each circuit.

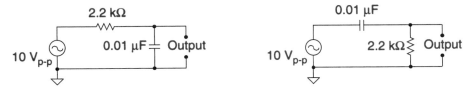

FIGURE 12–38

3. The low-pass filter in Figure 12–39 has a cutoff frequency of 10 kHz. Find the amplitude of the output voltage at cutoff (10 kHz), at one decade above cutoff (100 kHz), and at two decades above cutoff (1 MHz).

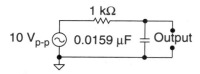

FIGURE 12–39

4. If a low-pass filter has a voltage output of 0.1 V at 10 kHz and 0.001 V at 100 kHz, express the roll-off in dB/decade.

5. Calculate the cutoff frequency and roll-off rate, and draw a sketch of the frequency-response curve for the circuit in Figure 12–40(a).

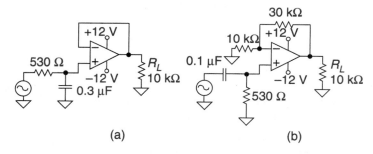

FIGURE 12–40

6. If the input signal to the circuit in Figure 12–40(a) is 2 V_{p-p}, what is the output at the following frequencies: (a) 100 Hz, (b) 1 kHz, (c) 2 kHz, and (d) 10 kHz?

7. Calculate the cutoff frequency and roll-off rate, and draw a sketch of the frequency-response curve for the circuit in Figure 12–40(b).

8. The input signal to the circuit in Figure 12–40(b) is 1 V$_{p\text{-}p}$. What is the output at the following frequencies: (a) 300 Hz, (b) 3 kHz, (c) 6 kHz, and (d) 30 kHz?

9. Calculate the cutoff frequency and roll-off rate, and draw a sketch of the frequency-response curve for the circuit in Figure 12–41.

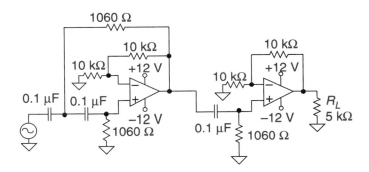

FIGURE 12–41

10. The input signal to the circuit in Figure 12–41 is 1.5 V$_{p\text{-}p}$. What is the output at the following frequencies: (a) 150 Hz, (b) 750 Hz, (c) 1.5 kHz, and (d) 15 kHz?

11. Figure 12–42 shows a Bode plot response curve for a filter circuit. Use the graph to determine the center frequency, bandwidth, Q, and roll-off for the filter.

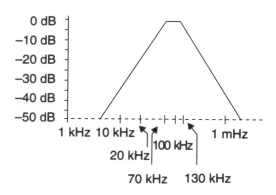

FIGURE 12–42

12. Calculate the low and high cutoff frequencies, the bandwidth, the center frequency, and the Q of the filter circuit shown in Figure 12–43. Draw a Bode plot of the frequency-response curve.

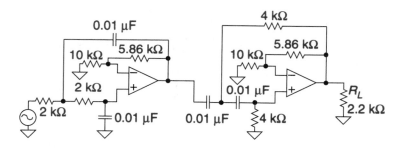

FIGURE 12–43

13. The input voltage of the generator is set so the output voltage equals 4 V when the frequency of the generator is adjusted to f_o. When the generator frequency is adjusted to $f_{(low\ cutoff)}$, what will be the output voltage of the circuit in Figure 12–43? _____

14. The frequency of the generator is adjusted to f_o, and the amplitude of the generator is adjusted so the output equals 6 V. When the generator frequency is increased to $10 \times f_{(high\ cutoff)}$, what will be the output voltage of the circuit in Figure 12–43? _____

Questions 15 through 20 refer to Figure 12–44 and are matching with the waveforms shown in Figure 12–44.

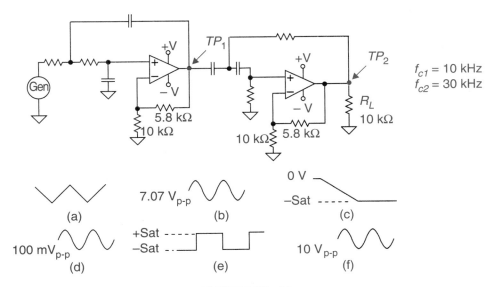

FIGURE 12–44

15. What is the waveform present at TP_2 with the input shown below? _____

$$4\ V_{p-p}$$
$$20\ kHz$$

16. What is the waveform present at TP_2 with the input shown below? _____

$$4\ V_{p-p}$$
$$10\ kHz$$

17. What is the waveform present at TP_2 with the input shown below? _____

$$4\ V_{p-p}$$
$$30\ kHz$$

18.   What is the waveform present at TP_2 with the input shown below? _____

4 V$_{p-p}$
1 kHz

19.   What is the waveform present at TP_2 with the input shown below? _____

4 V$_{p-p}$
300 kHz

20.   What is the waveform present at TP_1 with the input shown below? _____

6.33 V$_{p-p}$
1 kHz

21.   Calculate the center frequency, the Q, the bandpass, the low cutoff frequency, and the high cutoff frequency for the circuit in Figure 12–45.

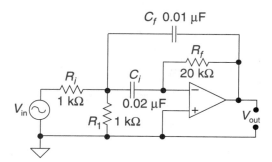

FIGURE 12–45

22.   Calculate the low and high cutoff frequencies of the filter circuit shown in Figure 12–46. Draw a Bode plot of the frequency-response curve.

23.   The coil in Figure 12–47 has an effective resistance of 30 Ω. Calculate V_B, V_E, V_C, V_{CE}, center frequency, Q, bandwidth, low cutoff, and high cutoff.

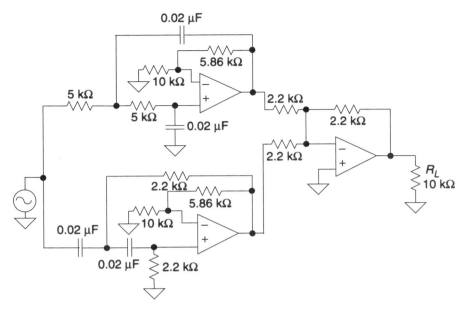

FIGURE 12–46

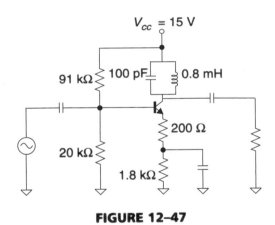

FIGURE 12–47

24. Calculate the bandstop frequency for the circuit in Figure 12–48.

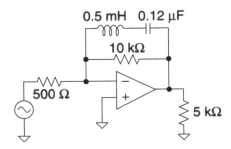

FIGURE 12–48

CHAPTER 13

Sine Wave Oscillator Circuits

OBJECTIVES

After studying the material in this chapter, you will be able to describe and/or analyze:

- ○ basic oscillator theory,
- ○ a Wien-bridge oscillator circuit,
- ○ a phase-shift oscillator circuit,
- ○ a Colpitts oscillator circuit,
- ○ a Hartley oscillator circuit,
- ○ a crystal oscillator circuit, and
- ○ troubleshooting procedures for sine wave oscillator circuits.

13.1 INTRODUCTION TO BASIC OSCILLATOR THEORY

Oscillators are circuits that produce a periodic waveform, with the only input requirement being a supply of DC power. There are three general categories of sine wave oscillators: RC, LC, and crystal controlled. The RC group finds applications at relatively low frequencies where often it is desirable to have the ability to adjust the output frequency and the stability of the output frequency is not a critical requirement. LC oscillators are also adjustable and are usually used at higher frequencies (radio frequency) where reasonably priced inductors are available. The crystal oscillator finds applications where accurate and stable outputs are required.

There are many types of sine wave oscillators, and most operate on the same basic theory. When an electronics circuit is activated, there are electrical noises present that span the electrical frequency spectrum. If the noise is amplified and the output of the amplifier is connected to a positive feedback network, an oscillator can be created. The positive feedback network is designed to select a signal at one frequency. This signal is coupled back to the input in proper phase so it reinforces the signal present at the input. As this process of amplifying and positive feedback continues, the selected frequency becomes the dominant signal in the circuit. Figure 13–1 shows the three essentials needed for an oscillator to function: an amplifier, a frequency selection network, and positive feedback.

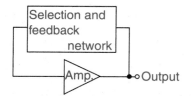

FIGURE 13–1   Essentials needed for oscillation

Amplification is required to change DC energy into signal energy and overcome circuit losses so the selected signal can be sustained. It should be noted that an oscillator does not generate energy but only transforms the DC supply energy into signal energy. The frequency of oscillation is determined by the selection network. The stability of oscillation is determined by how well the network can select a given frequency. Not only does the desired signal have to be selected from the output, but the feedback must also provide the proper phase shift to obtain positive feedback. Positive feedback must be present to support oscillation. Negative feedback is used in amplifier circuits to stabilize and prevent oscillations.

13.2 RC SINE WAVE OSCILLATOR CIRCUITS

WIEN-BRIDGE OSCILLATOR

Figure 13–2 shows an RC network. The input is divided between two impedances (Z_1 and Z_2), and the output is taken across Z_2. The output voltage can be expressed by the following formula: $V_o = V_{in} \times Z_2/(Z_1 + Z_2)$. The output will always be less than the input, and the magnitude of the output will depend on the frequency of the input signal.

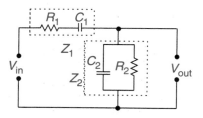

FIGURE 13–2   RC network

At high frequencies, the capacitive reactance of capacitors C_1 and C_2 approaches zero, causing the impedance of Z_1 to approach the value of R_1 and the impedance of Z_2 to approach zero. Since the output is taken across Z_2, the output voltage will be near zero. At low frequencies, the capacitive reactance of capacitors C_1 and C_2 approach infinity, causing the impedance of Z_1 to approach infinity and the impedance of Z_2 to approach the value of R_2. The output voltage will again be near zero because most of the input signal is dropped across the high impedance of Z_1. If $R_1 = R_2$ and $C_1 = C_2$, the RC network in Figure 13–2 will have maximum output when $R = X_C$. This will be shown in Example 13.1.

EXAMPLE 13.1

> For the circuit in Figure 13–2, find the output voltage when the input voltage is 10 V with a frequency of (a) 159 Hz, (b) 1.59 kHz, and (c) 15.9 kHz. ($R_1 = R_2 = 1\,k\Omega$ and $C_1 = C_2 = 0.1\,\mu F$.)

EXAMPLE 13.1 continued

Solution:

(a) Frequency = 159 Hz

$$V_o = V_{in} \times Z_2/(Z_1 + Z_2)$$
$$X_C = 1/(2\pi f C) = 1/(2\pi \times 159 \text{ Hz} \times 0.1 \text{ } \mu\text{F}) = 10 \text{ k}\Omega$$
$$Z_1 = R_1 - jX_C = 1 \text{ k}\Omega - j10 \text{ k}\Omega = 10.05 \text{ k}\Omega < -84.3^\circ$$
$$Z_2 = 1/Y_2 = 1/(G_2 + jB) = 1/(1 \text{ mS} + j0.1 \text{ mS}) = 0.99 \text{ k}\Omega < -5.7^\circ$$
$$V_o = V_{in} \times Z_2/(Z_1 + Z_2)$$
$$V_o = 10 \text{ V}/0^\circ \times (0.99 \text{ k}\Omega < -5.7^\circ)/(10.05 \text{ k}\Omega < -84.3^\circ + 0.99 \text{ k}\Omega < -5.7^\circ) = 0.96 \text{ V} < 73^\circ$$

(b) Frequency = 1.59 kHz

$$V_o = V_{in} \times Z_2/(Z_1 + Z_2)$$
$$X_C = 1/(2\pi f C) = 1/(2\pi \times 1.59 \text{ kHz} \times 0.1 \text{ } \mu\text{F}) = 1 \text{ k}\Omega$$
$$Z_1 = R_1 - jX_C = 1 \text{ k}\Omega - j1 \text{ k}\Omega = 1.41 \text{ k}\Omega < -45^\circ$$
$$Z_2 = 1/Y_2 = 1/(G_2 + jB) = 1/(1 \text{ mS} + j1 \text{ mS}) = .707 \text{ k}\Omega < -45^\circ$$
$$V_o = V_{in} \times Z_2/(Z_1 + Z_2)$$
$$V_o = 10 \text{ V}/0^\circ \times (.707 \text{ k}\Omega < -45^\circ)/(1.41 \text{ k}\Omega < -45^\circ + .707 \text{ k}\Omega < -45^\circ) = 3.3 \text{ V} < 0^\circ$$

(c) Frequency = 15.9 kHz

$$V_o = V_{in} \times Z_2/(Z_1 + Z_2)$$
$$X_C = 1/(2\pi f C) = 1/(2\pi \times 15.9 \text{ kHz} \times 0.1 \text{ } \mu\text{F}) = 100 \text{ } \Omega$$
$$Z_1 = R_1 - jX_C = 1 \text{ k}\Omega - j100 \text{ } \Omega = 1 \text{ k}\Omega < -5.7^\circ$$
$$Z_2 = 1/Y_2 = 1/(G_2 + jB) = 1/(1 \text{ mS} + j10 \text{ mS}) = 0.1 \text{ } \Omega < -84.3^\circ$$
$$V_o = V_{in} \times Z_2/(Z_1 + Z_2)$$
$$V_o = 10 \text{ V} < 0^\circ \times (0.1 \text{ } \Omega < -84.3^\circ)/(1 \text{ k}\Omega < -5.7^\circ + 0.1 < -84^\circ) = 0.98 \text{ V} < -73^\circ$$

Two things should be noted. First, maximum output occurs when the input frequency causes X_C to equal R, and the output voltage drops off when the frequency goes above or below this value. Second, the output is in phase with the input only at the frequency where R is equal to X_C. These relationships are shown graphically in Figure 13–3.

The frequency that has maximum output and zero phase shift is equal to that frequency where $R = X_C$. The formula for this frequency is derived as follows:

$$R = X_C$$
$$X_C = 1/2\pi f C$$

Therefore:

$$R = 1/2\pi f C$$

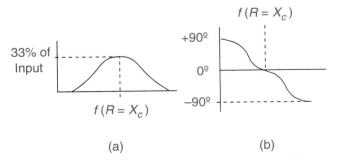

FIGURE 13–3 Amplitude and phase of output

Solving for frequency:

$$f = 1/2\pi RC$$

Figure 13–4 shows a Wien-bridge oscillator that uses the RC circuit of Figure 13–2 to select the frequency and provide positive feedback. The RC network will produce one-third of the output voltage across Z_2 at a frequency where $R = X_C$. The voltage across Z_2 will be in phase with the output. Since the feedback is connected to the noninverting input of the op-amp, it will be in phase with the input and reinforce the input signal at the selected frequency. The op-amp serves as a noninverting amplifier and has a gain of three $(1 + R_f/R_i)$.

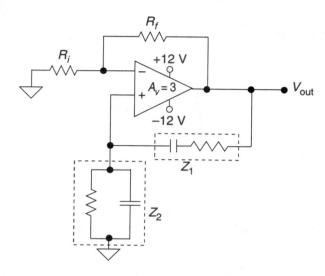

FIGURE 13–4   Wien-bridge oscillator

In order for the circuit to oscillate, the product of the voltage gain (A_v) times the feedback factor (B) must be one or greater. In this case, the product is equal to one ($A_v = 3$, $B = 1/3$); consequently, the circuit will oscillate. Even though this circuit will oscillate, it may not be self-starting. This means that when DC power is applied to the circuit, it will not go into oscillation without an external signal. The circuit can be made self-starting by making the product of $A_v \times B$ greater than one. However, this causes the output of the oscillator to be clipped, which makes the output nonsinusoidal. In order to make the oscillator self-starting and still have a sinusoidal output, it is necessary to have the amplifier start with a high gain. Once the selected signal builds up, the gain needs to be reduced so the product of $A_v \times B$ equals one. The circuits in Figure 13–5 show two ways this can be accomplished.

In Figure 13–5(a), a resistor in series with the two back-to-back diodes shunts the negative feedback resistor (R_f). When the circuit is first turned on, the gain of the amplifier is equal to ten (9 kΩ/1 kΩ + 1) and causes the oscillator to be self-starting ($A_v \times B > 1$). As the signal builds up, greater than 0.7 V peak, the diodes are turned on, shunting the 9 kΩ resistor with 2.57 kΩ. This reduces the gain to three, thus preventing clipping of the output.

In Figure 13–5(b), a tungsten lamp is used to control the gain of the amplifier. When the circuit is first turned on, the lamp is at room temperature and has a low resistance. This causes the amplifier to have a high gain. As current flows through the lamp, the filament is heated and the resistance of the lamp goes up. This reduces the gain of the amplifier. The product of $A_v \times B$ equals 1, which keeps the output from clipping.

The Wien-bridge oscillator is a popular RC oscillator. It can be used to generate frequencies up to 1 MHz, and by varying the values of the resistances or capacitances, the output frequency can be varied. Example 13.2 reviews the Wien-bridge oscillator.

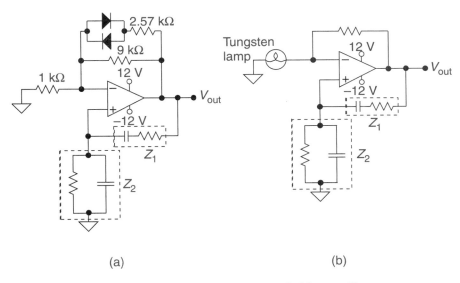

(a)   (b)

FIGURE 13–5   Self-starting Wien-bridge oscillators

EXAMPLE 13.2

Calculate the output frequency and the operating resistance of the tungsten lamp for the circuit in Figure 13–6.

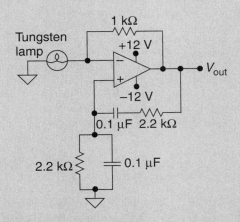

FIGURE 13–6   Wien-bridge oscillator with tungsten lamp

Step 1. Calculate the output frequency.
$$f = 1/(2\pi RC) = 1/(2\pi \times 2.2 \text{ k} \times 0.1 \text{ μF}) = 724 \text{ Hz}$$

Step 2. Calculate the resistance of the tungsten lamp. The gain of the circuit must be three to permit oscillations but prohibit clipping. Therefore, the resistance of the lamp (R_i) can be calculated by solving the gain formula for the resistance of R_i.

$$A_v = R_f/R_i + 1$$
$$R_i = R_f/(A_v - 1)$$
$$R_i = 1 \text{ k}/(3 - 1) = 500 \text{ Ω}$$

PHASE-SHIFT OSCILLATOR

Figure 13–7(a) shows an RC circuit. The voltage at the output is 0º to 90º out of phase with the input voltage. The actual phase is dependent upon the component values and the frequency of the input signal. Figure 13–7(b) shows an RC network circuit containing three stages. At one frequency, this circuit provides a 180º phase shift and an output signal magnitude equal to 1/29 of the input signal. If the components are of equal value, each stage will provide a 60º phase shift.

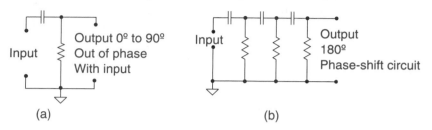

FIGURE 13–7   Phase-shift circuits

Figure 13–8 shows a phase-shift oscillator made up of an inverting amplifier and an RC phase-shift network that provides a 180º phase shift. If the components are of equal value, the frequency of oscillations can be calculated using the formula shown in Figure 13–8. The virtual ground on the inverting input pin of the op-amp permits resistor R_i in Figure 13–8 to have two functions. It serves as the last resistor in the feedback network and also functions as the input resistor for the amplifier. The amplifier must have a gain equal to 29 in order for the product of $A_v \times B$ to equal 1. A gain greater than 29 will permit oscillations but will cause clipping.

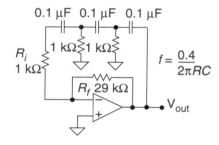

FIGURE 13–8   Phase-shift oscillator circuit

13.3 OSCILLATIONS IN AMPLIFIER CIRCUITS

Wien-bridge and phase-shifts are examples of how an oscillator can be designed using an amplifier and a positive feedback network. The higher the gain of the amplifier, the smaller the feedback factor required for oscillation. It is important to realize that any amplifier will oscillate any time there is positive feedback, and the product of the gain times the feedback factor ($A_v \times B$) is equal to or greater than 1. This is why circuits that have high gain are prone to break into oscillations.

There are several things designers do to prevent unwanted oscillations. Amplifier circuits are normally designed using negative feedback networks to prevent oscillations. Since all the stages in a system often share the power bus, the power bus can cause unwanted positive feedback. To prevent feedback, the power bus should be at signal ground. Throughout the system, small capacitors are connected from the power bus to ground to ensure the power bus is maintaining at signal ground.

The physical layout of the system becomes important because paths of high power output signals are physically separated from the low power input circuitry. Shielding is used to prevent electrostatic or electromagnetic coupling. High frequency signals find capacitive feedback paths; hence, high gain amplifiers, if not designed properly, will break into high frequency oscillations. The technician should understand the limitations of breadboard systems. Many circuits designed for high frequencies and high gain operations cannot use breadboards because the physical layout of the PC board is a critical part of the design.

13.4 LC OSCILLATOR CIRCUITS

When a capacitor and an inductor are connected in parallel, they form a parallel tuned circuit (tank circuit). If energy is introduced into this tuned circuit, it will oscillate at a resonant frequency where $X_C = X_L$. The resonant frequency can be calculated by using the formula

$$f_r = \frac{1}{2\pi\sqrt{LC}}.$$

Figure 13–9 shows a tuned circuit connected to a DC source through a push-button switch. If the push button is momentarily closed, energy is applied to the tuned circuit. The energy applied to the circuit will cause the tuned circuit to oscillate. This can be explained by recalling that inductors and capacitors are theoretically wattless devices. They store energy but do not consume it. When the switch closes, energy is stored in the capacitor in the form of an electrostatic field. When the switch is opened, the capacitor discharges its energy, causing a current to flow through the inductor. The current flowing through the inductor causes energy to be stored in the inductor in the form of an electromagnetic field. Theoretically, all of the energy in the electrostatic field of the capacitor will be transferred to the electromagnetic field of the inductor. At this point, the inductor takes over and causes current to continue to flow in the same direction. This current flow causes the capacitor to charge in the opposite direction. All the energy stored in the electromagnetic field of the inductor will be transferred back to the capacitor. This transferring back and forth of energy in the tuned circuit is known as the flywheel effect and would, theoretically, continue forever. In practical circuits, however, there is always some resistance present, so the energy in the circuit dissipates as the current flows back and forth, and the oscillations are dampened out as shown in Figure 13–9. If energy can be fed back into the circuit to overcome the losses, the oscillations can be sustained.

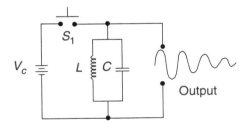

FIGURE 13–9 Flywheel effect

The Colpitts oscillator circuit is constructed with an amplifier and a tuned LC circuit. Figure 13–10(a) shows a Colpitts oscillator circuit that uses an op-amp for the amplifier and an LC tuned circuit consisting of an inductor and two capacitors. A ground reference is connected to the junction of the two capacitors in the tuned circuit. The signal at the top of the tuned circuit is 180° out of phase with respect to the signal at the bottom of the tuned circuit when measured from the ground reference. The signal at the bottom of the tuned circuit is fed back to the inverting input of the op-amp. The op-amp provides an additional 180º phase shift, which supplies the top of the tuned circuit with the proper signal to sustain oscillation. Figure 13–10(b) shows a Colpitts oscillator using a bipolar transistor as the amplifier stage.

In order for a Colpitts oscillator to maintain oscillations, the amplifier must provide enough gain so the product of the feedback factor times the gain is greater than 1. The feedback factor is determined by the voltage divider formed by the two capacitors (C_1 and C_2). The feedback voltage will equal the voltage developed across capacitor C_2 in Figure 13–10. Capacitor C_1 equals 0.002 μF, and capacitor C_2 equals 0.02 μF, so the feedback factor equals 1/11; therefore, the voltage gain of the amplifier must be 11 or greater to sustain oscillation. The Q (X_L/R_{coil}) of the tuned circuit is also an important factor in designing an LC oscillator. A tuned circuit with a high Q will ensure stable oscillations, whereas a circuit with a low Q may not oscillate or may oscillate off frequency.

The Colpitts oscillator frequency is determined by the resonant frequency of the LC tuned circuit. The two capacitors are in series for the resonant current flow. The bias voltage can be calculated in the same manner used for amplifiers. Example 13.3 shows how to calculate for circuit values using the components in the circuit.

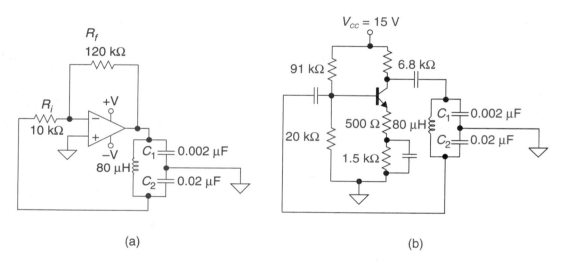

FIGURE 13–10   Colpitts oscillators

EXAMPLE 13.3

Calculate V_B, V_E, V_C, V_{CE}, A_v, and the oscillating frequency for the circuit in Figure 13–10(b).

Step 1. Calculate V_B, V_E, V_C, and V_{CE}. The amplifier is a common-emitter amplifier.

$$V_B = V_{cc} \times R_{B2}/(R_{B1} + R_{B2}) = 15 \text{ V} \times 20 \text{ k}\Omega/(91 \text{ k}\Omega + 20 \text{ k}\Omega) = 2.7 \text{ V}$$
$$V_E = V_B - 0.7 \text{ V} = 2.7 \text{ V} - 0.7 \text{ V} = 2 \text{ V}$$
$$I_C = I_E = V_E/R_e = 2 \text{ V}/2 \text{ k} = 1 \text{ mA}$$
$$V_{Rc} = I_c \times R_c = 1 \text{ mA} \times 6.8 \text{ k} = 6.8 \text{ V}$$
$$V_C = V_{cc} - V_{Rc} = 15 \text{ V} - 6.8 \text{ V} = 8.2 \text{ V}$$
$$V_{CE} = V_c - V = 8.2 \text{ V} - 2 \text{ V} = 6.2 \text{ V}$$

Step 2. Calculate A_v. The impedance of the tuned circuit at resonance will be high and cause little loading; therefore, the signal resistance in the collector leg will equal the value of R_c. The signal in the emitter leg equals R_e plus r'_e. Since R_e is 500 Ω, r'_e is negligible.

EXAMPLE 13.3 continued

$A_v = r_c/r_e = 6.8\ \text{k}\Omega/500\ \Omega = 13.6$ (The circuit will oscillate, since the gain is over 11.)

Step 3. Calculate the equivalent of C_1 and C_2.

$$C_{eq} = \frac{1}{\dfrac{1}{C_1} + \dfrac{1}{C_2}} = \frac{1}{\dfrac{1}{0.002\ \mu\text{F}} + \dfrac{1}{0.02\ \mu\text{F}}} = 0.0018\ \mu\text{F}$$

(Since C_2 is ten times C_1, the equivalent value equals approximately C_1.)

Step 4. Calculate the oscillating frequency.

$$f_r = \frac{1}{2\pi\sqrt{LC}}$$

$$f_r = \frac{1}{2\pi\sqrt{80\ \mu\text{H} \times 0.0018\ \mu\text{F}}}$$

$$f_r = 420\ \text{kHz}$$

The Hartley oscillator is similar to the Colpitts oscillator; however, in the Hartley oscillator, the split capacitors are replaced with a split inductor. Figure 13–11(a) shows a Hartley oscillator using an op-amp, and Figure 13–11(b) shows a Hartley oscillator using a bipolar amplifier stage. The size of the feedback signal is determined by how the inductor is split. Normally only a small portion of the signal developed across the tank is fed back to the input of the amplifier. The smaller the feedback signal, the higher the needed gain of the amplifier. However, designing LC oscillators where the feedback signal is a small percentage of the voltage across the tank reduces the loading on the tuned circuit and helps frequency stability.

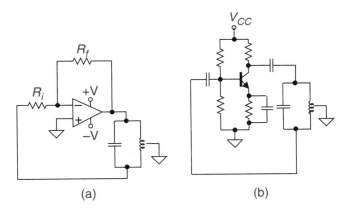

(a)   (b)

FIGURE 13–11 Hartley oscillators

13.5 CRYSTAL OSCILLATOR CIRCUITS

A crystal is a small wafer of quartz that has a natural resonant frequency. Crystals have the piezoelectric effect where a mechanical force across a crystal causes a small voltage to be generated. The reverse is also true. A voltage across the crystal will cause the material to mechanically bend.

The dimension and cut of a crystal determine the resonant frequency of the crystal. Crystals are available with resonant frequencies from a few hundred kilohertz to upwards of 100 MHz. Figure 13–12(a) shows the electrical equivalent of a crystal. The crystal by itself is equivalent to

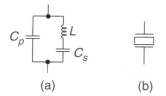

FIGURE 13–12   Crystal equivalent circuit and symbol

a series LC circuit, but when leads are installed, a parallel LC circuit is created. Figure 13–12(b) shows the symbol for a crystal.

Be aware that the crystal has two resonate frequencies: one for series resonance, and a second for parallel resonance. The series resonate frequency is equal to

$$f_r = \frac{1}{2\pi\sqrt{LC_s}}$$

The parallel resonance frequency is equal to

$$f_r = \frac{1}{2\pi\sqrt{LC_{eq}}}$$

where C_{eq} is the series equivalent of C_s and C_p. At parallel resonance, there is a loop current that flows through both capacitors so they appear to be in series. C_p is typically one hundred times the capacitance of C_s. When two capacitors are connected in series, their effective capacitance is less than the smaller one. If one capacitor is many times larger than the other capacitor, the effective capacitance is approximately equal to the capacitance of the smaller one. So, C_{eq} is approximately equal to C_s, and the series and parallel resonance frequencies of the crystal are approximately the same. Actually, the parallel resonance frequency is slightly higher because C_{eq} is slightly lower than C_s.

The advantage of using a tuned crystal circuit over a tuned circuit made of discrete components is the high Q of the crystal. The higher the Q of the tuned circuit, the better the ability of the circuit to be frequency selective. When used in an oscillator, the ability of a tuned circuit to select one frequency determines the stability of the oscillator. Crystal controlled oscillators are extremely stable oscillators. Typical crystal performance permits an oscillator circuit to maintain a stable output frequency within 0.0001%. This is demonstrated in inexpensive electronic watches that use crystals that provide an accuracy of fifteen seconds per month, or better than 0.001% accuracy.

If a crystal is placed in an amplifier circuit so that it provides positive feedback, oscillation occurs. Figure 13–13 shows two possible crystal oscillator circuits. The circuit in Figure 13–13(a) is a modified Colpitts oscillator. The crystal replaces the tuned circuit. The two capacitors, C_1 and C_2, form a voltage divider shunting the crystal but have very little effect on the frequency of operation. The circuit in Figure 13–13(b) is also a modification of the Colpitts, called a Pierce

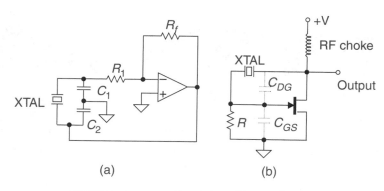

FIGURE 13–13   Crystal oscillator circuits

oscillator. It is dependent on the internal capacitance of the field effect transistor to provide a capacitive voltage divider across the crystal.

13.6 TROUBLESHOOTING SINE WAVE OSCILLATOR CIRCUITS

The techniques used in troubleshooting amplifier circuits can be applied to troubleshooting oscillator circuits. The procedure is to calculate or estimate what signals you expect to observe and then use test equipment to measure the signals. If the signals are as expected, move on to the next stage. If they are not as expected, then the oscillator stage must be repaired.

The first thing the technician needs to determine when troubleshooting an oscillator is its output frequency. There are many types of oscillators, and often an oscillator will have a unique formula for determining its frequency. Be sure to read the technical manual on the equipment. Good manuals normally contain information on the oscillating frequencies of oscillators used in the system. Most RC sine waves use some form of the formula

$$f = \frac{1}{2\pi RC}$$

where R and C are values of components in the positive feedback loop. Once the feedback loop is determined, using the formula

$$f = \frac{1}{2\pi RC}$$

should give you an estimate of the oscillating frequency. The operating frequency of LC oscillators is set by the resonant frequency of the tuned LC circuit. Once the components in the tuned circuit are determined, the formula

$$f_r = \frac{1}{2\pi\sqrt{LC}}$$

can be used to calculate the oscillating frequency. The crystal used in a crystal oscillator determines the frequency of oscillation. Normally, the label on crystals will state the operating frequency.

Often the oscillator stage is followed by a buffer stage. The buffer prevents the oscillator from being loaded. The technician should realize oscillators are sensitive to loading and must take precautions when connecting test equipment to prevent loading. For example, the oscilloscope should be connected to the oscillator under test with a times-ten probe. This is especially true when troubleshooting high frequency oscillators. The standard oscilloscope using a times-one probe has input resistance of 1 MΩ and input capacitance of 30 pF. The input resistance is normally not a problem, but the input capacitance can easily load high frequency oscillators. At 10 MHz, the capacitive reactance of 30 pF is only 530 Ω. A times-ten probe has input resistance of 10 MΩ and input capacitance of 3 pF. The reduction in input capacitance is the main factor in reducing loading. A good rule of thumb is to use a times-ten probe when measuring signals above 1 MHz.

Sine Wave Oscillators

1. Calculate the output frequency of the circuit in Figure 13–14. _____

2. What resistance will the potentiometer need to be adjusted to in order to permit oscillations? _____

3. Calculate the output of the circuit in Figure 13–14 if the 2 kΩ resistors are changed to 5.1 kΩ resistors. _____

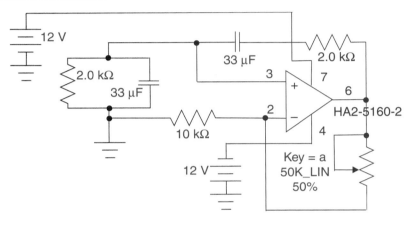

FIGURE 13–14

4. Simulate the circuit in Figure 13–14 and record the output frequency with 2 kΩ resistors and with 5.1 kΩ resistors.

   $f_{(2\,k)}$ = _____ $f_{(5.1\,k)}$ = _____

5. Calculate the output frequency of the circuit in Figure 13–15. _____

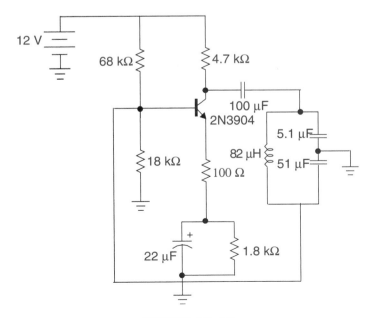

FIGURE 13–15

6. Calculate the output frequency of the circuit in Figure 13–15 if the 82 μH inductor is changed to a 51 μH inductor.

7. Simulate the circuit in Figure 13–15 and record the output frequency with an 82 μH inductor and with a 51 μH inductor.

$f_{(82\,\mu H)}$ = _____ $f_{(51\,\mu H)}$ = _____

Sine Wave Oscillators

I. Objectives

1. To calculate and experimentally determine the output frequency of a Wien-bridge oscillator.
2. To modify the design of the Wien-bridge oscillator so it is self-starting but outputs an undistorted sine wave.
3. To calculate and experimentally determine the output frequency of a Colpitts oscillator.
4. To calculate and experimentally determine the output frequency of a crystal oscillator.

II. Test Equipment

±12 V power supply   Frequency counter
Oscilloscope

III. Components

Resistors: (1) 100 Ω, (3) 1 kΩ, (1) 1.8 kΩ, (2) 2.2 kΩ, (1) 4.7 kΩ, (3) 10 kΩ, (1) 18 kΩ, (1) 67 kΩ, (1) 220 kΩ, and (1) 5 kΩ potentiometer
Capacitors: (1) 0.001 μF, (1) 0.0047 μF, (1) 0.01 μF, (2) 0.047 μF, (2) 0.1 μF
Inductor: (1) 100 μH, (1) Lamp (5 V, 60 mA)
Transistors: (1) 2N3904 JFET, (1) MPF102 Crystal, (1) 4 MHz
IC: (1) LM741 op-amp

IV. Procedure

1. Calculate the output frequency of the circuit in Figure 13–16.

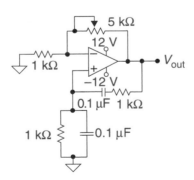

FIGURE 13–16

2. Construct the circuit and adjust the potentiometer until the circuit oscillates with minimum distortion.
3. Use an oscilloscope to examine the wave shape and a frequency counter to measure the frequency of the output signal.
4. Adjust the potentiometer until the output is sinusoidal. Turn the circuit off and then back on. Note if the circuit was self-starting.
5. Measure the resistance of the potentiometer and calculate the gain of the op-amp.

6. Modify the circuit so it is like the circuit in Figure 13–17. Adjust the potentiometer until the circuit oscillates without distortion.

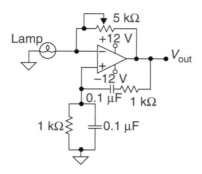

FIGURE 13–17

7. Turn the circuit off and then back on. Note if the circuit was self-starting.
8. Measure the resistance of the potentiometer and calculate the operating resistance of the lamp.
9. Calculate the output of the circuit in Figure 13–17 if the 0.1 µF capacitors are changed to 0.047 µF capacitors.
10. Change the capacitors to 0.047 µF and measure the output frequency and waveform.
11. Calculate the output frequency of the circuit in Figure 13–18.
12. Construct the circuit in Figure 13–18.
13. Using an oscilloscope, examine the waveform and measure the voltage gain. Use a frequency counter to measure the output frequency.
14. Calculate the output frequency of the circuit in Figure 13–18 if the 0.001 µF capacitor is changed to a 0.0047 µF capacitor.

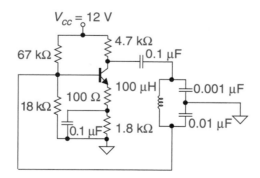

FIGURE 13–18

15. Change the capacitors to 0.0047 µF and measure the output frequency of the circuit in Figure 13–18.
16. Calculate the output frequency of the circuit in Figure 13–19.
17. Construct the circuit in Figure 13–19.

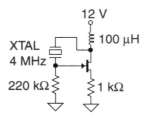

FIGURE 13–19

18. Using an oscilloscope, examine the waveform, and measure the output frequency using a frequency counter. (Use a times-ten probe.)

V. Points to Discuss

1. Why did the output of the circuit in Figure 13–16 output a rectangular waveform until the potentiometer was adjusted?

2. What was the gain of the circuit in Figure 13–16 when its output was sinusoidal? Was this what you expected? Explain.

3. Explain why the circuit in Figure 13–16 was or was not self-starting.

4. Explain why the circuit in Figure 13–17 was self-starting.

5. What was the gain of the circuit in Figure 13–18? Is voltage gain needed for the Colpitts oscillator to operate? Explain.

6. Explain how the Colpitts oscillator obtains positive feedback.

7. What determines the oscillating frequency of the circuit in Figure 13–19? Can the output frequency be varied in the circuit? Explain.

8. Explain why it is a good idea to use a times-ten probe when measuring the output of a high frequency oscillator.

QUESTIONS

13.1 Introduction to Basic Oscillator Theory

____ 1. Oscillators are circuits that produce a periodic waveform output with the only input being _____.
  a. an AC signal
  b. a DC supply voltage
  c. a sine wave
  d. a square wave

____ 2. Three categories of oscillators are RC, LC, and crystal oscillators. Which of these types is the most stable?
  a. RC oscillators
  b. LC oscillators
  c. Crystal oscillators
  d. They are all equal in stability.

____ 3. Which of the following is not an essential requirement of a sine wave oscillator?
  a. positive feedback network
  b. negative feedback network
  c. frequency selecting circuit
  d. amplifier circuit

13.2 RC Sine Wave Oscillator Circuits

____ 4. In order for a Wien-bridge to oscillate, the product of the feedback factor times the voltage gain must be _____.
  a. zero
  b. less than one
  c. one or greater than one
  d. 0.9

____ 5. The RC feedback network used in the Wien-bridge oscillator has a maximum output when ___.
  a. $X_C = X_L$
  b. $X_L = R$
  c. $X_C = R$
  d. $R = 0\ \Omega$

____ 6. The RC feedback network used in the Wien-bridge oscillator causes a _____ phase shift.
  a. 0º
  b. 45º
  c. 90º
  d. 180º

____ 7. The amplifier of a Wien-bridge oscillator has a voltage gain of ____.
  a. 0
  b. 1

c. 3

d. 10

____ 8. The tungsten lamp in a Wien-bridge oscillator is used to control the _____.

a. light output of the circuit

b. phase of the feedback signal

c. gain of the amplifier

d. none of the above

____ 9. Which of the following formulas is used to calculate the oscillating frequency of a Wien-bridge oscillator?

a. $f = \dfrac{0.4}{2\pi RC}$

b. $f = \dfrac{1}{2\pi RX_C}$

c. $f = \dfrac{1}{2\pi RC}$

d. $f = \dfrac{1}{2\pi\sqrt{LC}}$

____ 10. The RC feedback network used in the phase shift oscillator causes a _____ phase shift.

a. 0º

b. 45º

c. 90º

d. 180º

____ 11. The amplifier of a phase shift oscillator has a voltage gain of ____.

a. 3

b. 10

c. 29

d. 100

____ 12. Which of the following formulas is used to calculate the oscillating frequency of a phase shift oscillator?

a. $f = \dfrac{0.4}{2\pi RC}$

b. $f = \dfrac{1}{2\pi RX_C}$

c. $f = \dfrac{1}{2\pi RC}$

d. $f = \dfrac{1}{2\pi\sqrt{LC}}$

13.3 Oscillations in Amplifier Circuits

____ 13. _____ amplifiers are more likely to break into oscillation than _____ amplifiers.

a. Bipolar transistor, FET

b. Low gain, high gain

c. High gain, low gain

d. Inverting, noninverting

____ 14. Which of the following will help limit unwanted oscillations in amplifier circuits?

a. positive feedback

b. negative feedback

c. increased voltage gain

d. increased power gain

13.4 LC Oscillator Circuits

____ 15. If energy is transferred to a parallel LC circuit, it will oscillate at resonant frequency. However, these oscillations will dampen out because of the _____ in the circuit.

a. resistance

b. capacitive reactance

c. inductive reactance

d. impedance

____ 16. What is the main frequency determining factor in Colpitts and Hartley oscillator circuits?

a. the capacitance in the tuned circuit

b. the inductance in the tuned circuit

c. the resonant frequency of the tuned circuit

d. the RC feedback network

____ 17. Which of the following formulas is used to calculate the oscillating frequency of an LC oscillator?

a. $f = \dfrac{0.4}{2\pi\,RC}$

b. $f = \dfrac{1}{2\pi\,RX_C}$

c. $f = \dfrac{1}{2\pi\,RC}$

d. $f = \dfrac{1}{2\pi\sqrt{LC}}$

____ 18. The transferring back and forth of energy in the tuned circuit is known as the _____ effect.

a. Colpitts

b. Hartley

c. piezoelectric

d. flywheel

____ 19. The Colpitts oscillator has _____ in the tuned LC circuit.

a. a split inductor

b. two capacitors

c. two capacitors and a split inductor

d. a capacitor in series with an inductor

____ 20. The Hartley oscillator has _____ in the tuned LC circuit.
    a. a split inductor
    b. two capacitors
    c. two capacitors and a split inductor
    d. a capacitor in series with an inductor

13.5 Crystal Oscillator Circuits

____ 21. Crystals demonstrate the _____ effect where a mechanical force across a crystal causes a small voltage to be generated.
    a. Colpitts
    b. Hartley
    c. piezoelectric
    d. flywheel

____ 22. The crystal itself is equivalent to _____ circuit, but when leads are installed, _____ circuit is created.
    a. an open, a shorted
    b. a high impedance, a low impedance
    c. a parallel LC, a series LC
    d. a series LC, a parallel LC

____ 23. The advantage of using a crystal over a tuned circuit made of discrete components is _____.
    a. the high Q of the crystal
    b. the low Q of the crystal
    c. the wide bandpass
    d. lower frequency operation

____ 24. Which of the following formulas is used to calculate the oscillating frequency of a crystal oscillator?
    a. $f = \dfrac{0.4}{2\pi RC}$
    b. $f = \dfrac{1}{2\pi RC}$
    c. $f = \dfrac{1}{2\pi \sqrt{LC}}$
    d. none of the above

13.6 Troubleshooting Sine Wave Oscillator Circuits

____ 25. The loading effect of an oscilloscope can be reduced by using _____.
    a. a times-ten probe
    b. the 20 mV/cm scale
    c. the 10 μS/cm scale
    d. AC coupling

____ 26. The output signal of the circuit in Figure 13–20(a) measures 12 $V_{p\text{-}p}$, and the signal at
the inverting input measures 4 $V_{p\text{-}p}$. What is wrong with the circuit?

   a. The lamp is open.

   b. R_1 is open.

   c. C_2 is open.

   d. The circuit is operating properly.

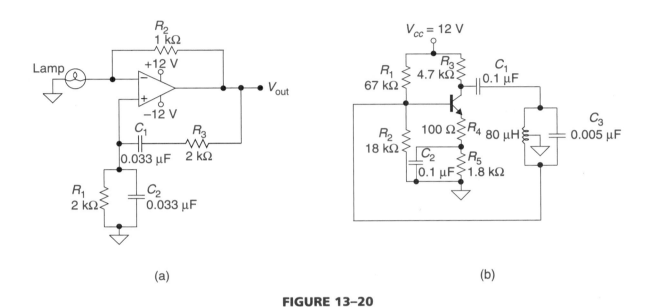

(a)   (b)

FIGURE 13–20

____ 27. The output signal of the circuit in Figure 13–20(a) measures 12 $V_{p\text{-}p}$, and the signal at
the noninverting input measures 4 $V_{p\text{-}p}$. What is wrong with the circuit?

   a. The lamp is open.

   b. R_1 is open.

   c. C_2 is open.

   d. The circuit is operating properly.

____ 28. The output signal of the circuit in Figure 13–20(a) measures 12 V DC. What is wrong
with the circuit?

   a. The lamp is open.

   b. R_2 is shorted.

   c. C_1 is shorted.

   d. The circuit is operating properly.

____ 29. The signal on the collector of the transistor in Figure 13–20(b) measures 12 V DC.
What is wrong with the circuit?

   a. C_3 is shorted.

   b. R_1 is open.

   c. R_1 is shorted.

   d. The circuit is operating properly.

_____ 30. The output signal (top of tank circuit) of the circuit in Figure 13–20(b) measures 10 V_{p-p}, and the signal on the base of the transistor measures 0.22 V_{p-p}. What is wrong with the circuit?

    a. C_3 is shorted.

    b. R_1 is open.

    c. The inductor is open.

    d. The circuit is operating properly.

PROBLEMS

1. Show the waveforms for the circuit in Figure 13–21 (DC and signal components) at test points 1 through 3. Assume the signal at TP_1 is 4 V_{p-p}.

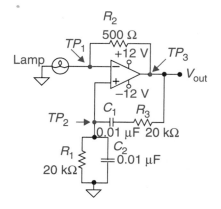

FIGURE 13–21

2. What is the resistance of R_x in Figure 13–22 when the circuit is operating properly?

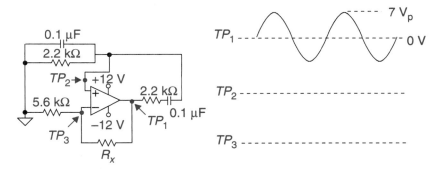

FIGURE 13–22

3. Draw the waveforms for the signals at TP_2 and TP_3 for the circuit in Figure 13–22. (Show amplitude and timing.)

4. What is the frequency of the output signal in the circuit in Figure 13–22?

5. What is the resistance of R_x in Figure 13–23 when the circuit is operating properly?

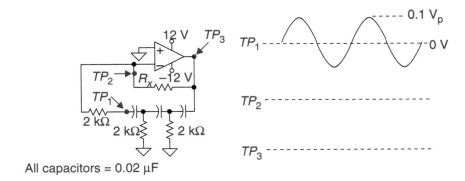

FIGURE 13–23

6. Draw the waveforms for the signals at TP_2 and TP_3 for the circuit in Figure 13–23. (Show amplitude and timing.)

7. What is the frequency of the output signal in the circuit in Figure 13–23?

8. What is the frequency of the output signal in the circuit in Figure 13–24?

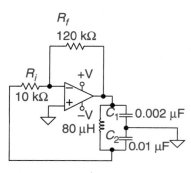

FIGURE 13–24

9. What is the minimum voltage gain needed for the amplifier in Figure 13–24 to permit oscillations?

10. What is the frequency of the output signal in the circuit in Figure 13–25?

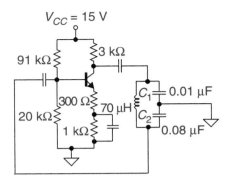

FIGURE 13-25

11. Calculate V_B, V_E, V_C, and V_{CE} for the circuit in Figure 13–25.

12. What is the minimum voltage gain needed for the amplifier in Figure 13–25 to permit oscillations?

13. What is the frequency of the output signal in the circuit in Figure 13–26?

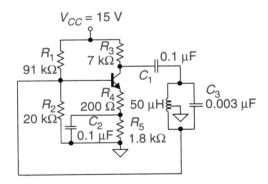

FIGURE 13-26

14. Calculate V_B, V_E, V_C, and V_{CE} for the circuit in Figure 13–26.

15. If the coil has 200 turns, 190 above the tap and 10 below the tap, what is the minimum voltage gain needed for the amplifier to permit oscillations?

16. What is the frequency of the output signal in the circuit in Figure 13–27?

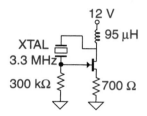

FIGURE 13–27

CHAPTER 14

Nonsinusoidal Oscillators

OBJECTIVES

After studying the material in this chapter, you will be able to describe and/or analyze:

○ the operation of the 555 astable circuit,

○ the operation of the 555 monostable circuit,

○ inverter oscillators,

○ the Schmitt trigger RC oscillator,

○ crystal controlled oscillators,

○ triangular wave oscillators,

○ the procedure for wave shaping a triangular wave into a sine wave,

○ sawtooth oscillators, and

○ troubleshooting procedures for oscillator circuits.

14.1 INTRODUCTION TO RECTANGULAR WAVE OSCILLATORS

Multivibrators are circuits whose outputs are bistate; their outputs are always at one of two voltage levels. Multivibrators can be divided into three groups: astable, monostable, and bistable. Figure 14–1 shows block diagrams with the waveforms for each group. Each circuit has a nickname that is derived from its output signal. The astable multivibrator is called a *free-running circuit*, the monostable multivibrator is called a *one-shot circuit*, and the bistable multivibrator is called a *flip-flop circuit*.

Astable stands for not stable. As shown in Figure 14–1(a), the output constantly switches between two states. The output of the astable circuit is a rectangular waveform, and no trigger input is required. The monostable circuit is stable in one of two voltage levels. When a trigger input is received, it switches to the unstable state for a period of time and then returns to the stable state. The bistable circuit has two stable states and will stay in either one indefinitely. When a trigger input is received, it will switch to the other state and remain in that state until the next trigger input is received.

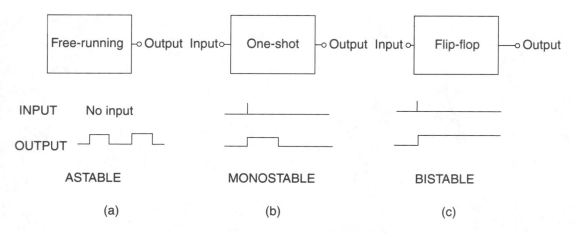

FIGURE 14-1   Types of multivibrator circuits

Astable circuits are used as rectangular oscillators, monostable circuits are used as timers, and bistable circuits are used as digital storage registers. In the next section, we will examine a popular IC that can be used as an astable or monostable circuit. Since bistable multivibrator circuits (flip-flops) are explained in digital electronics, they will not be covered in this text.

14.2 THE 555 ASTABLE CIRCUIT

An integrated circuit that can function as an astable oscillator or a monostable timer and is widely used is the inexpensive 555 timer. A typical application of the 555 is shown in Figure 14–2. In this configuration, capacitor C charges through R_A and R_B until an upper threshold voltage is reached. At that point, pin 7 pulls to ground and C discharges through R_B to a lower threshold voltages. Then pin 7 is released from ground, and C charges until the upper threshold is reached; and this cycle continues.

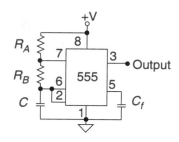

FIGURE 14-2   Astable circuit

A better understanding of the 555 in the astable oscillator configuration can be gained by examining the simplified circuit in Figure 14–3. All components inside the dotted lines are internal to the 555. The internal circuitry consists of two comparators, an RS flip-flop, a transistor switch, an output buffer amplifier, and a voltage divider made up of three resistors. There are four external components: R_A, R_B, C, and C_f. Capacitor C_f is a noise filter designed to keep noise off the internal voltage divider.

The three resistors comprising the internal voltage divider are of equal values. This means two-thirds of +V will be present at the inverting input of comparator #1 and one-third of +V will be

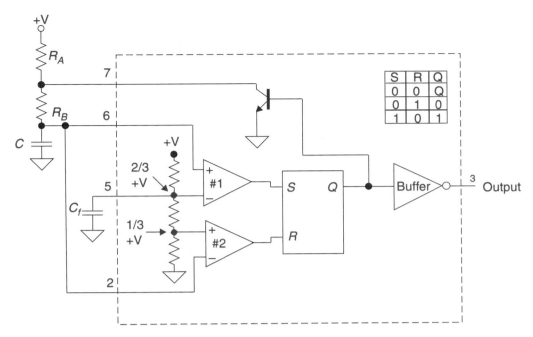

FIGURE 14–3 Internal circuitry of a 555 in the astable mode

present at the noninverting input of comparator #2. When power is first applied, external capacitor (C) has zero volts across it. The voltage applied to both the noninverting input of comparator #1 (pin 6) and the inverting input of comparator #2 (pin 2) is zero. Initially, comparator #1 has a more positive input at its inverting input, and its output will be driven low. Comparator #2 has a more positive input at its noninverting input, driving its output high.

The outputs of the comparators are applied to the RS flip-flop, which responds as shown in the truth table in Figure 14–3. With a low on the S input and a high on the R input, the Q output will be reset (low state). This low is fed back to the transistor turning it off and permitting the external capacitor to charge through resistors R_A and R_B. As the external capacitor charges, its voltage will reach one-third of +V, and at this point, comparator #2 will switch, causing the R input of the flip-flop to go low. Now both inputs to the RS flip-flop are lows. As the truth table shows, two lows will cause the output to stay constant. This means the output of the flip-flop will remain low and keeps the transistor off, thus permitting the external capacitor (C) to continue to charge toward +V.

As the voltage across the external capacitor increases to two-thirds of +V, comparator #1 switches. This causes the S input of the RS flip-flop to go high, and with $S = 1$ and $R = 0$, the Q output of the flip-flop sets (high state), causing the transistor to conduct. This provides a discharge path through R_B and the transistor to ground. As the voltage across the external capacitor drops below two-thirds of +V, comparator #1 switches back to a low output. This causes the flip-flop to have inputs of $S = 0$ and $R = 0$, and the output of the flip-flop stays high, keeping the transistor on. The discharge continues until the voltage across the external capacitor drops below one-third of +V. At this point, comparator #2 switches, causing the flip-flop to have inputs of $S = 0$ and $R = 1$. This drives the Q output of the flip-flop low, and the transistor turns off. Then the capacitor starts its charge cycle again. This pattern of charging to two-thirds of +V and then discharging to one-third of +V is continuous.

The output of the flip-flop is connected to an inverting buffer amplifier, which drives the output pin. Figure 14–4 shows the waveforms at pins 2 and 6, at the output of the internal flip-flop, and at the output of the IC. The output swings between +V and ground. The 555 is designed to operate with a positive supply voltage from 4 V to 18 V.

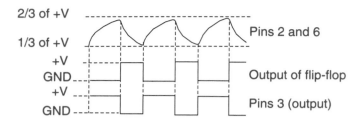

FIGURE 14–4   Waveforms of the astable 555 circuit

The frequency of oscillations can be calculated by the formula $f = 1.44/(R_A + 2R_B)C$. The duty cycle is defined as the time the output is active divided by the total period of the output signal. The question is: which state is active? For a 555, the low state is considered active. Since the capacitor charges through both R_A and R_B and discharges only through R_B, the discharge time will always be shorter than the charge time. The amount of time the output signal is in the high state (inactive) will always be longer; consequently, the duty cycle is always less than 50%. The relationship between the values of R_A and R_B determines the relationship between the high and low states of the output signal. The duty cycle can be calculated using the formula $DC = R_B/(R_A + 2R_B) \times 100\%$. Figure 14–5 shows the output waveform and formulas for calculating the duty cycle with measured values, the duty cycle using component values, and the output frequency using component values. Example 14.1 shows how output waveforms can be calculated.

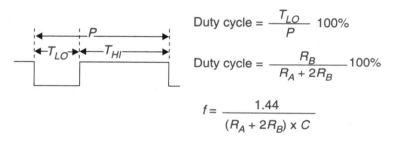

$$\text{Duty cycle} = \frac{T_{LO}}{P} \; 100\%$$

$$\text{Duty cycle} = \frac{R_B}{R_A + 2R_B} \; 100\%$$

$$f = \frac{1.44}{(R_A + 2R_B) \times C}$$

FIGURE 14–5   Duty cycle and frequency formulas

EXAMPLE 14.1

Calculate and draw the waveforms at TP_1 and TP_2 for the circuit in Figure 14–6. Capacitor C_f is a noise filter and has no effect on the waveforms.

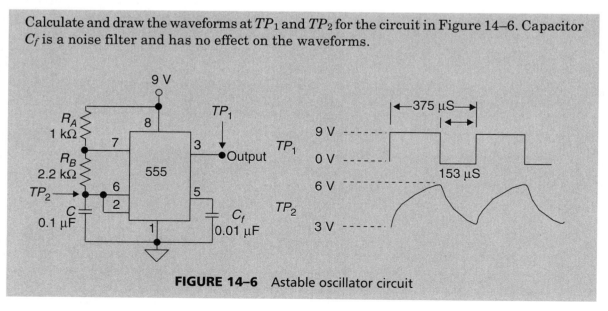

FIGURE 14–6   Astable oscillator circuit

EXAMPLE 14.1 continued

> **Step 1.** Calculate the frequency and period of the output waveform.
>
> $$f = 1.44/(R_A + 2R_B) \times C = 1.44/(1 \text{ k}\Omega + 4.4 \text{ k}\Omega) \times 0.1 \text{ }\mu F = 2.67 \text{ kHz}$$
> $$P = 1/f = 1/2.67 \text{ kHz} = 375 \text{ }\mu S$$
>
> **Step 2.** Calculate the duty cycle of the output waveform.
>
> $$DC = R_B/(R_A + 2R_B)100\% = 2.2 \text{ k}\Omega/(1 \text{ k}\Omega + 4.4 \text{ k}\Omega)100\% = 40.7\%$$
> $$\text{Output low} = 153 \text{ }\mu S$$
> $$\text{Output high} = 222 \text{ }\mu S$$
>
> **Step 3.** Using the information calculated in steps 1 and 2, draw the waveform at TP_1. (See waveform in Figure 14–6.)
>
> **Step 4.** Draw the waveform at TP_2. When the output is high, the internal transistor is off, thus permitting the external capacitor to charge to two-thirds of the supply voltage (6 V). When the output goes low, the internal transistor turns on, thus discharging the external capacitor until it reaches one-third of the supply voltage (3 V). (See waveform in Figure 14–6.)

Figure 14–7 shows two ways to design a 555 astable oscillator with a 50% duty cycle; this is called a square-wave oscillator. In the circuit shown in Figure 14–7(a), resistor R_A equals resistor R_B, and resistor R_B is bypassed with a diode. The current charging the capacitor will flow through R_A and the diode. When the voltage across the capacitor reaches two-thirds of the supply, the internal circuitry grounds pin 7. This causes the capacitor to discharge through R_B until the voltage across the capacitor drops to one-third of the supply. The direction of the discharge current reverses biases the diode. Since the capacitor is charged and discharged through the same value of resistance, the charge and discharge times will be equal, and the circuit will have a 50% duty cycle.

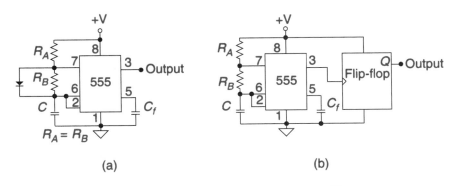

(a)   (b)

FIGURE 14–7   Astable square-wave oscillators

Figure 14–7(b) uses a flip-flop to obtain a 50% duty cycle. The 555 oscillator is designed to operate at twice the desired output frequency. The duty cycle of the 555 is not important. The flip-flop is edge-triggered and will toggle on each positive edge. This action will divide the frequency by two, but the output of the flip-flop will be a waveform with a 50% duty cycle.

14.3 THE 555 AS A MONOSTABLE CIRCUIT

Figure 14–8(a) shows a 555 connected as a monostable multivibrator. The 555 monostable multivibrator circuit is not an oscillator, since it requires a trigger input to obtain an output signal. Figure 14–8(b) shows the input trigger and the output waveforms. The output stays at a low level until the trigger input goes low. The edge of the trigger input going low causes the output switch to go high for a period of time determined by the value of the RC network.

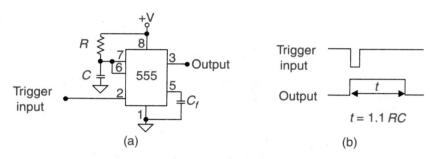

(a)   (b)

FIGURE 14–8   Monostable multivibrator

Figure 14–9 shows an external RC network and the internal workings of the 555 IC. At power up, assume the flip-flop is set. The positive voltage (two-thirds of +V) connected to the inverting input of comparator #1 drives the output of comparator #1 low. The positive voltage on the trigger input (pin 2) drives the output of comparator #2 low. The flip-flop will remain set because both inputs are low ($S = 0$, $R = 0$) and the transistor remains on, thereby keeping the external capacitor discharged. This is the stable state of the circuit, and it will maintain this state until a trigger input is received.

If at power up the flip-flop is reset, the transistor is off, and the external capacitor will charge to two-thirds of the supply voltage, causing comparator #1 to switch to a high setting of the flip-flop. The set flip-flop will turn on the transistor and discharge the capacitor. The circuit is now in its stable state, and it will maintain this state until a trigger input is received.

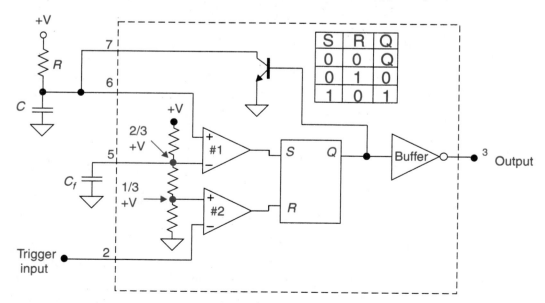

FIGURE 14–9   Internal circuitry of a 555 in the monostable mode

When a trigger input causes pin 2 to drop below one-third of +V, the output of comparator #2 is driven high. This resets the flip-flop causing the transistor to turn off. The circuit is now in its unstable state. With the transistor off, the external capacitor will charge through resistor (R). When the voltage across the capacitor reaches two-thirds of +V, the output of comparator #1 goes high and sets the flip-flop. The transistor turns on and quickly discharges the capacitor, causing the circuit to return to its stable state.

Since the capacitor must charge to 67% (two-thirds) of + V to return the circuit back to its stable state, the circuit is in its unstable state for approximately one time constant as determined by the external RC network. Actually, a capacitor in an RC circuit only charges to 63% in one time constant, so a formula that gives a closer approximation is $t = 1.1RC$. By selecting the proper external components, delays of up to 100 seconds are possible.

If the trigger input is connected directly to pin 2 of the 555, the trigger must be returned to high voltage levels (above one-third of +V) before the circuit times out. In other words, the trigger input pulse must be of shorter duration than the output pulse. A trigger conditioning circuit has been added to the circuit in Figure 14–10 consisting of a pull-up resistor (1 k), a coupling capacitor (0.01 µF), and a diode. This circuitry permits the trigger input pulse to be longer in duration than the output pulse. Figure 14–10 shows the waveforms. With the trigger input high, both sides of the coupling capacitor (TP_1 and TP_2) are at 12 V; consequently, the coupling capacitor is discharged. When the trigger input switches to 0 V (TP_1), the voltage at TP_2 also drops to 0 V. However, the coupling capacitor is quickly charged through the pull-up resistor so the voltage at TP_2 returns to 12 V after five time constants. Hence, the voltage on pin 2 of the 555 is a short duration pulse regardless of the pulse width of the trigger input. During the time the trigger input is low, the capacitor remains charged at 12 V. When the trigger input switches from 0 V to 12 V, the voltage at TP_2 would (for an instant) increase to 24 V. This high voltage could damage the 555. The diode prevents damage by clamping the positive swing to 0.7 V above the supply (12.7 V). Example 14.2 shows how the waveforms for the circuit in Figure 14–10 can be determined.

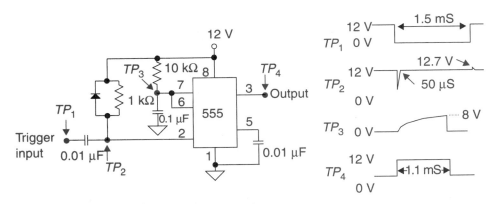

FIGURE 14–10 Monostable with trigger conditioning circuitry

EXAMPLE 14.2

A trigger input is shown in Figure 14–10 (TP_1). Calculate and draw the waveforms at TP_2, TP_3, and TP_4.

Step 1. Calculate the magnitude and width of the pulse at TP_2 on the falling edge of the input. The pulse will be 12 V in magnitude, dropping from a 12 V level to ground. If viewed on an oscilloscope, it may appear not to reach ground level; however, theoretically, it does but the pulse width at ground level is too narrow to detect. The pulse width at the 12 V level will equal five time constants of the capacitor and resistor in the conditioning circuit.

EXAMPLE 14.2 continued

$$TC = RC = 1\ k\Omega \times 0.01\ \mu F = 10\ \mu S$$
$$\text{Pulse width at 12 V level} = 5TC = 5 \times 10\ \mu S = 50\ \mu S$$

Step 2. Calculate the magnitude of the pulse at TP_2 on the rising edge of the input. The diode will forward bias and clamp the voltage at TP_2 to a maximum of 12.7 V. The waveform for TP_2 is shown in Figure 14–10.

Step 3. Calculate the pulse width of the output.

$$t = 1.1RC = 1.1 \times 10\ k\Omega \times 0.1\ \mu F = 1.1\ mS$$

Step 4. Draw the waveform at TP_3. The external capacitor starts charging on the falling edge of the trigger and continues charging for 1.1 mS. At that time, the voltage across the capacitor will equal two-thirds of +V. The waveform is shown in Figure 14–10.

Step 5. Draw the waveform at TP_4. The output switches to the value of +V on the falling edge of the trigger. It stays at +V for 1.1 mS. The waveform is shown in Figure 14–10.

The circuit in Figure 14–11 is a pulse generator that will output continuous pulses. The frequency and pulse width are adjustable. The potentiometer in the circuitry of the first 555 provides a frequency adjustment but has no effect on the pulse width of the output of the second 555. The pulse width of the output is controlled by the potentiometer in the circuitry of the second 555.

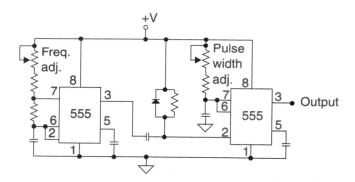

FIGURE 14–11   Pulse generator

14.4   INVERTER OSCILLATORS

Any odd number of digital inverters can be connected to form an oscillator. Figure 14–12 shows three inverters that will output a rectangular wave. Digital inverters are high gain inverting amplifiers whose outputs are always driven to positive (high level) or negative (low level) saturation by the signals at their inputs.

Assume that when power is initially applied, the output of inverter "a" is high. This high level is fed to the input of inverter "b," causing its output to be driven low. The low level output of inverter "b" is connected to the input of inverter "c," causing its output to go high. The high level output of inverter "c" is fed back to the input of inverter "a," causing its output to switch to a low level. The outputs of the inverters continue to switch as shown in Figure 14–12.

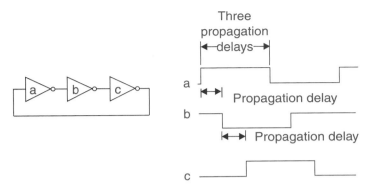

FIGURE 14–12 Inverter oscillator

The frequency of oscillation depends on the speed of the inverters used. Each inverter has a propagation time delay (t_p) that is the time it takes the output to respond to a change in input. Since the circuit in Figure 14–12 has three inverters in series, it will take three propagation time delays for the signal to progress around the loop one time. There are an odd number of inverters in the loop, so when the signal returns to inverter "a," it will force the output to switch to a low state. The output of inverter "a" will stay low for three propagation time delays, then switch back to high. The period of the waveform can be calculated by multiplying the propagation time delay times twice the number of inverters in the loop. This circuit is completely symmetrical, and it makes no difference where the output is taken. Simply increasing the number of inverters to any larger odd number will decrease the output frequency.

EXAMPLE 14.3

Calculate the period and frequency of the inverter oscillator in Figure 14–12 if the propagation time delay (t_p) for each inverter is 20 nS.

Step 1. Calculate the period of the output.

$$P = t_p \times 2 \times \text{no. of inverters} = 20 \text{ nS} \times 2 \times 3 = 120 \text{ nS}$$

Step 2. Calculate the frequency of the output.

$$f = 1/P = 1/120 \text{ nS} = 8.3 \text{ MHz}$$

14.5 SCHMITT TRIGGER RC OSCILLATORS

The circuit in Figure 14–13(a) is a Schmitt trigger RC oscillator. The trip points are set on the noninverting input by the positive feedback network consisting of R_2 and R_3. The output is connected to capacitor C_1 through resistor R_1. If the output is in positive saturation, the capacitor connected to the inverting input will charge in a positive direction until the upper trip point (UTP) is reached as shown in Figure 14–13(b). At this point, the output switches to negative saturation, causing the capacitor to start charging in a negative direction. When the downward trip point (DTP) is reached, the output switches to positive saturation, and the cycle starts over. The period of the output square wave depends on the saturation voltages, the trip point voltage levels, and the RC time constant (R_1C_1) of the negative feedback loop. If all these factors are known, the period of the output square wave can be calculated using the formula for the universal time constant curve. For troubleshooting procedures, the technician can approximate the period of the output square wave by calculating the RC time constant of the negative feedback loop.

The circuit in Figure 14–13(c) is a Schmitt trigger RC oscillator using a digital Schmitt trigger inverter gate. The digital Schmitt trigger gate has a built-in hysteresis of approximately 1 V and does not require an external positive feedback loop to set the trip points. The period of the output square wave can still be approximated by calculating the RC time constant of the negative feedback loop.

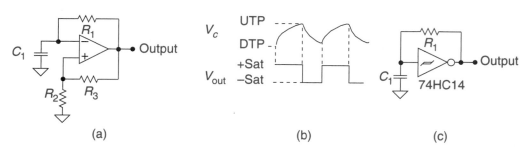

(a)   (b)   (c)

FIGURE 14–13   Schmitt trigger RC oscillator circuits

EXAMPLE 14.4

Draw and label the waveforms at TP_1 and TP_2 for the circuit shown in Figure 14–14.

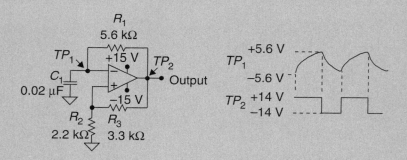

FIGURE 14–14   Schmitt trigger RC oscillator and waveforms

Step 1. Approximate the period of the output waveform.
$$P = R_1 C_1 = 5.6 \text{ k}\Omega \times 0.02 \text{ μF} = 112 \text{ μS}$$

Step 2. Calculate the frequency using the approximate period.
$$f = 1/P = 1/112 \text{ μS} = 8.9 \text{ kHz}$$

Step 3. Calculate the amplitude at TP_1 and TP_2. The trip points set the waveform amplitude at TP_1 and the saturation voltages set the waveform amplitude at TP_2. The saturation voltages are equal to ±14 V.

$$UTP = +\text{Sat } R_2/(R_2 + R_3) = 14 \text{ V} \times 2.2 \text{ k}\Omega/(2.2 \text{ k}\Omega + 3.3 \text{ k}\Omega) = 5.6 \text{ V}$$
$$UTP = -\text{Sat } R_2/(R_2 + R_3) = -14 \text{ V} \times 2.2 \text{ k}\Omega/(2.2 \text{ k}\Omega + 3.3 \text{ k}\Omega) = -5.6 \text{ V}$$

Step 4. Draw the waveform using the information calculated. (See Figure 14–14.)

14.6 CRYSTAL CONTROLLED OSCILLATORS

The circuit in Figure 14–15 is a crystal controlled Schmitt trigger RC oscillator. The first inverter and the RC network form a standard Schmitt trigger RC oscillator. The second inverter provides a 180° phase shift, and the crystal provides a positive feedback path for the signal at the resonant frequency of the crystal. If the RC circuit is designed to set the frequency of oscillations slightly above the resonant frequency of the crystal, the positive feedback through the crystal will force the circuit to oscillate at the resonant frequency of the crystal. The inverter oscillator, therefore, becomes a crystal controlled oscillator, and the output frequency is extremely stable.

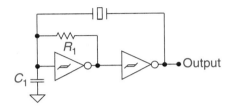

FIGURE 14–15   Crystal controlled inverter oscillator circuit

14.7 TRIANGULAR WAVE OSCILLATORS

Figure 14–16 shows a triangular wave oscillator. The circuit consists of a Schmitt trigger square-wave oscillator followed by an integrator circuit. The frequency of the square-wave oscillator is determined by the values of R_1 and C_1. The amplitude of the square-wave signal (TP_2) switches between plus saturation and negative saturation. The integrator circuit produces a positive ramp when the signal at TP_2 is in negative saturation and a negative ramp when the signal at TP_2 is in positive saturation. The circuit is designed so the square-wave input to the integrator switches at a high enough frequency to prohibit the integrator from reaching saturation. Example 14.5 shows how the waveforms for the test points shown in Figure 14–16 can be determined.

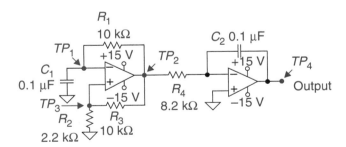

FIGURE 14–16   Triangular wave oscillator circuit

EXAMPLE 14.5

Determine the waveforms for the test points shown in Figure 14–16.

Step 1. Determine the approximate frequency and the amplitude of the square wave at TP_2.

EXAMPLE 14.5 continued

$$P \approx R_1 \times C_1 = 10 \text{ k}\Omega \times 0.1 \text{ }\mu\text{F} = 1 \text{ mS}$$
$$f = 1/P = 1/1 \text{ mS} = 1 \text{ kHz}$$
$$+V_{p(TP2)} = +V - 1 \text{ V} = 15 \text{ V} - 1 \text{ V} = 14 \text{ V}$$
$$-V_{p(TP2)} = -V - (-1 \text{ V}) = -15 \text{ V} - (-1 \text{ V}) = -14 \text{ V}$$
$$V_{p\text{-}p(TP2)} = 28 \text{ V}_{p\text{-}p}$$

Step 2. Determine the wave shape and amplitude of the signal at TP_3. The signal at TP_3 is the same wave shape (square wave) as the signal at TP_2. Its amplitude is reduced by the ratio of the voltage divider consisting of resistors R_2 and R_3.

$$V_{p\text{-}p(TP3)} = V_{p\text{-}p(TP1)} \times R_2/(R_2 + R_3) = 28 \text{ V}_{p\text{-}p} \times 2.2 \text{ k}\Omega/(2.2 \text{ k}\Omega + 10 \text{ k}\Omega) = 5 \text{ V}_{p\text{-}p}$$

Step 3. Determine the wave shape and amplitude of the signal at TP_1. Capacitor C_1 charges through R_1 in a positive direction until the upper trip point is reached (+2.5 V). Capacitor C_1 then charges in the opposite direction until the downward trip point is reached (–2.5 V). The wave shape at TP_1 is that of a charging and discharging capacitor with an amplitude of 5 $V_{p\text{-}p}$.

Step 4. Determine the wave shape and amplitude of the signal at TP_4. The output of the integrator stage will be a triangular wave. Capacitor C_2 will ramp positive and then negative. The amplitude of the signal at TP_4 is a function of the frequency of the square-wave input, the amplitude of the square-wave input, the resistance value of R_4, and the capacitance value of C_2. The square wave was calculated in step 1 to be a 1 kHz signal with an amplitude of 14 V_p (28 $V_{p\text{-}p}$). The signal has a period of 1 mS and a duty cycle of 50%; therefore, the signal at the input of the integrator will be +14 V for 0.5 mS, and then will switch to –14 V for 0.5 mS. The amplitude of the ramp can be calculated using two formulas: $Q = it$ and $V = Q/C$.

$$i = V_{p(\text{in})}/R_4 = 14 \text{ V}/8.2 \text{ k}\Omega = 1.7 \text{ mA}$$
$$Q = it = 1.7 \text{ mA} \times 0.5 \text{ mS} = .85 \text{ }\mu\text{C}$$
$$V = Q/C = 0.85 \text{ }\mu\text{C}/0.1 \text{ }\mu\text{F} = 8.5 \text{ V}$$

Step 5. Using the information calculated, draw the waveforms at the test points shown in Figure 14–16. (See Figure 14–17.)

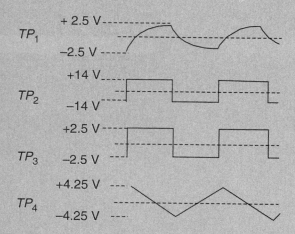

FIGURE 14–17 Waveforms for triangular wave oscillator

Figure 14–18 shows a two-stage feedback triangular wave oscillator circuit. The first stage of the circuit is a zero crossover comparator, and the second stage is an integrator. The input signal to the comparator is taken from a voltage divider consisting of resistors R_1 and R_2. One end of the voltage divider is connected to the output of the comparator, and the other end is connected to the output of the integrator. Assume the output of the comparator is at positive saturation. This positive voltage at the input of the integrator will cause the output of the integrator to ramp negative. At this point in time, the voltage divider has a positive voltage (+Sat.) on one end and a negative going voltage on the other end. The voltage level at TP_1 will decrease as the ramp goes negative until it reaches 0 V. At this point, the output of the comparator switches to negative saturation. The negative voltage at the output of the comparator drives TP_1 to a negative voltage level and causes the integrator to start ramping positive. The positive ramp of the integrator will, in time, cause the voltage at TP_1 to reach 0 V, which will drive the output of the comparator into positive saturation and start the cycle over.

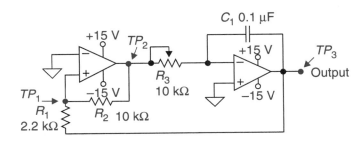

FIGURE 14–18 Two-stage feedback triangular wave oscillator circuit

Figure 14–19 shows the waveforms at the three test points shown in Figure 14–18. The amplitude of the output signal is a function of the voltage divider network (R_1 and R_2). The peak output can be determined by calculating the output needed to cause the voltage at the input of the comparator (TP_1) to equal zero. Once the output voltage swing is known, the period of the triangular wave can be calculated. The timing is a function of the integrator stage and is set by the values of R_3 and C_1. Example 14.6 calculates the amplitude and frequency of the output signal.

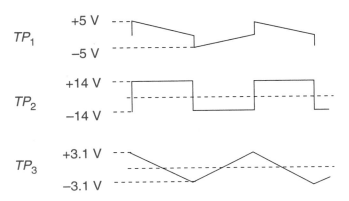

FIGURE 14–19 Waveforms for two-stage feedback triangular wave oscillator

EXAMPLE 14.6

Calculate the amplitude and frequency of the output signal for the circuit in Figure 14–18. Assume the potentiometer (R_3) is set to 5 kΩ.

Step 1. Calculate the output voltage swing. Assume one end of the voltage divider is at positive saturation (14 V), then calculate the voltage needed at the other end of the voltage divider to cause the voltage at TP_1 to equal zero. Figure 14–20 shows a simplified schematic of the voltage divider network.

$$I = V_{R2}/R_2 = 14 \text{ V}/10 \text{ k}\Omega = 1.4 \text{ mA}$$
$$V_{R1} = I_{R1} = 1.4 \text{ mA} \times 2.2 \text{ k}\Omega = 3.1 \text{ V}$$
$$V_{p(TP3)} = -3.1 \text{ V (The voltage at } TP_3 \text{ must be negative to cause 0 V at } TP_1.)$$
$$V_{p\text{-}p(TP3)} = 6.2 \text{ V}$$

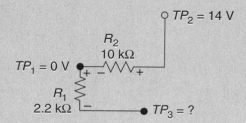

FIGURE 14–20   Voltage divider network

Step 2. Calculate the output frequency of the circuit in Figure 14–18. The formula $Q = IT$ and $Q = VC$ can be used to calculate the time it takes the signal to ramp to 6.2 V. This time equals half the period of the waveform, since the signal must ramp 6.2 V in each direction.

$$Q = VC = 6.2 \text{ V} \times 0.1 \text{ }\mu\text{F} = 0.62 \text{ }\mu\text{C}$$
$$I = V_{(sat)}/R_3 = 14 \text{ V}/5 \text{ k}\Omega = 2.8 \text{ mA}$$

Since the output of the comparator is the input to the integrator, the input to the integrator will always be $\pm V_{(sat)}$.

$$T = Q/I = 0.62 \text{ }\mu\text{F}/2.8 \text{ mA} = 221 \text{ }\mu\text{S}$$
$$P = 2T = 2 \times 221 \text{ }\mu\text{S} = 442 \text{ }\mu\text{S}$$
$$f = 1/P = 1/442 \text{ }\mu\text{S} = 2.26 \text{ kHz}$$

A quick approximation of the period of the output can be made by calculating the RC time constant of the resistor and capacitor used in the integrator. In this case, it would have resulted in a period of 500 μS. This is only an approximation but often is close enough for troubleshooting purposes.

The main advantage of using the two-stage feedback triangular wave oscillator is the ability to adjust the output frequency without affecting the amplitude of the output. A secondary advantage is that the two-stage feedback triangular wave oscillator requires fewer components than the triangular wave oscillator designed with a square-wave oscillator followed by an integrator.

14.8 WAVE SHAPING A TRIANGULAR WAVE INTO A SINE WAVE

Square- and triangular wave oscillators are easy to construct and are always self-starting. If a sine wave is needed, an alternative to using a sine wave oscillator is to convert a triangular wave into a sine wave using a wave shaping circuit similar to the one shown in Figure 14–21. The circuit changes the slope of the triangular wave input as it increases in amplitude. The plus and minus 12 V provide the required bias voltages for the diodes to turn on at the correct voltage levels to modify the slope.

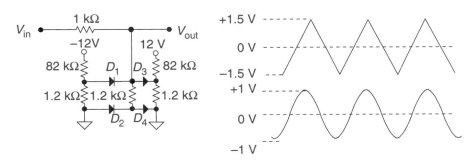

FIGURE 14–21 Wave shaping circuit

Figure 14–22 shows the equivalent circuits the signal sees on the positive alternation. At signal levels below 0.7 V, all diodes are off, and the output signal tracks the input signal. The equivalent circuit is shown in Figure 14–22(a). A sine wave with a 1 V peak will exceed 0.7 V at 45º. Between 45º and 60º, diode D_4 will turn on, and the output signal will be a function of the voltage divider shown in Figure 14–22(b). The input voltage will reach an amplitude of 1 V, but the output will be limited to an amplitude of 0.86 V. This is the magnitude of a 1 V peak sine wave at 60º. Between 60º and 120º, both diodes (D_4 and D_3) are turned on, and the output signal will be a function of the voltage divider as shown in Figure 14–22(c). Effectively, the two 1.2 kΩ resistors are in parallel, causing a resistance of 600 Ω in series with a diode. The input signal will reach 1.5 V, but the voltage divider will limit the output signal to 1 V peak. The diodes will turn off as the input drops back toward ground. The other half of the circuit consists of diodes D_1, D_2, and associated resistors performing the same function on the negative alternation.

0º to 45º	45º to 60º	60º to 120º
135º to 180º	120º to 135º	

V_{in} ●–ww–● V_{out}
0 V to 0.7 V | 0 V to 0.7 V
1 kΩ

(a)

V_{in} ●–ww–● V_{out}
0.7 V to 1 V 0.7 V to 0.86 V
1 kΩ
D_4
1.2 kΩ

(b)

V_{in} ●–ww–● V_{out}
1 V to 1.5 V 0.86 V to 1 V
1 kΩ
D_3 1.2 kΩ
1.2 kΩ D_4

(c)

FIGURE 14–22 Equivalent circuits for positive alternation

The circuit shown in Figure 14–21 uses only two diodes on each alternation, and the resulting sine wave is far from perfect. Additional diodes and resistors can be added, and this type of shaping circuit can produce excellent sine wave signals. The circuit is amplitude sensitive and must be designed for a triangular wave of known amplitude. The nice thing about the circuit is that it is not frequency sensitive. This makes it useful in function generators that operate over a wide range of frequencies.

14.9 SAWTOOTH OSCILLATORS

Figure 14–23 shows how an integrator circuit can be used to generate a sawtooth waveform. Resistor R_1 and the zener diode provide a constant negative voltage to the input of the integrator. If the transistor is off, the negative voltage causes the output of the integrator to ramp positive. A trigger input turns on the transistor and quickly discharges the capacitor, causing the output to drop to 0 V. As soon as the transistor is turned off, the output starts another positive ramp. The trigger input can be used to synchronize the sawtooth signal with the rest of the system. If a stand alone sawtooth oscillator is desired, the output ramp is monitored, and when its amplitude reaches the desired level, a trigger is generated to reset the circuit.

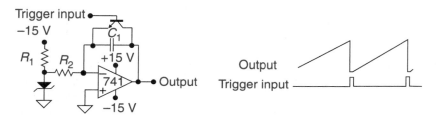

FIGURE 14–23   Sawtooth oscillator with external trigger input

The circuit in Figure 14–24 is a complete stand-alone sawtooth oscillator consisting of a comparator, a 555 one-shot, and integrator. The output of the integrator ramp is in a positive direction. When the ramp voltage exceeds the positive reference voltage connected to the noninverting input of the comparator, the output of the comparator switches from high to low. This falling edge triggers the one-shot, which outputs a narrow pulse that resets the circuit.

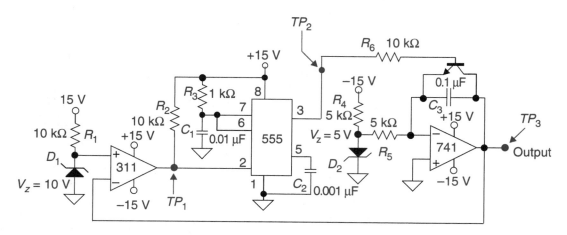

FIGURE 14–24   Sawtooth oscillator circuit

When the output of the 555 (pin 3) returns to a low value, the cycle starts over. Example 14.7 shows how to evaluate the circuit.

EXAMPLE 14.7

Determined the waveforms at TP_1, TP_2, and TP_3 for the circuit in Figure 14–24.

Step 1. Calculate the two reference voltages set by the two zener diodes. Both zener circuits are within regulation, so the voltage across each zener will equal the voltage rating of the zener diode.

$$V_{D1} = 10 \text{ V}$$
$$V_{D2} = 5 \text{ V}$$

Step 2. Calculate the period and the amplitude of the signal at TP_3. The signal will ramp in a positive direction until reaching the reference voltage connected to the comparator; hence, its amplitude will reach 10 V. The time it takes to reach the peak voltage is a function of reference voltage V_{D2}, resistor R_5, and capacitor C_3.

$$V_{p(TP3)} = 10 \text{ V}$$
$$I = V_{D2}/R_5 = 5 \text{ V}/5 \text{ k}\Omega = 1 \text{ mA}$$
$$Q = VC = 10 \text{ V} \times 0.1 \text{ }\mu\text{F} = 1 \text{ }\mu\text{C}$$
$$T = Q/I = 1 \text{ }\mu\text{C}/1 \text{ mA} = 1 \text{ mS}$$

Step 3. Calculate the pulse width of the reset pulse at TP_2. The width of the reset pulse is set by resistor R_3 and capacitor C_1 in the 555 stage.

$$T = 1.1R_3C_1 = 1.1 \times 1 \text{ k}\Omega \times 0.01 \text{ }\mu\text{F} = 11 \text{ }\mu\text{S}$$

Step 4. Using the information calculated, draw the waveforms. (See Figure 14–25.) The pulse at TP_1 will be short. The comparator switches low and will generate the reset pulse at TP_2. The reset pulse will turn on the transistor, causing the ramp voltage at the input of the comparator to decrease. This decrease in voltage causes the comparator to switch to a high level. The width of the pulse at TP_1 is a function of the propagation time around the loop.

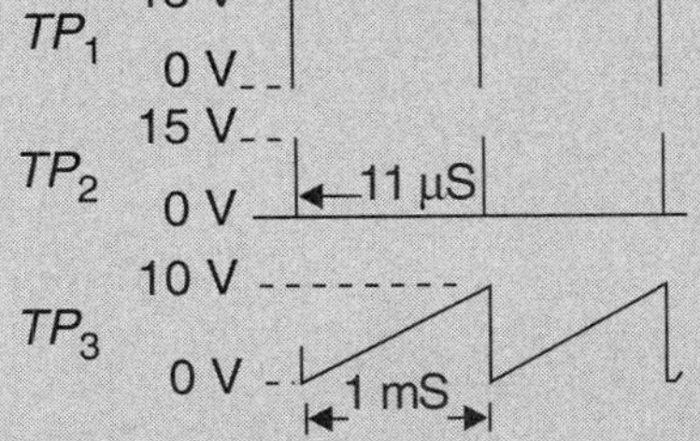

FIGURE 14–25 Waveforms for Example 14.7

The circuit in Figure 14–26 is a sawtooth oscillator using a Shockley diode. The Shockley diode is a thyristor device that will be explained in a later chapter. In essence, the Shockley diode acts like an open until its rated break-over voltage is reached. At that point, the Shockley diode will conduct like a short until the current drops to near zero. The Shockley diode then returns to its open condition, and the cycle starts over.

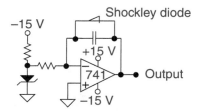

FIGURE 14–26   Sawtooth oscillator using a Shockley diode

14.10   TROUBLESHOOTING OSCILLATOR CIRCUITS

In this chapter, you have examined a number of periodic waveform oscillators. As in all circuits, the first step in troubleshooting is to estimate the voltage and waveforms you expect to measure. It is always best to check the obvious things first. For example, always measure the DC supply voltages. All the circuits studied in this chapter depend on an RC network for the frequency of operation. If the timing components can be determined, a rough approximation of the period of the waveform can be made by assuming the period equals one time constant. The technician must realize that this approach gives only a ballpark approximation. Some of the circuits in this chapter charged a capacitor through a resistance, and the signal developed across the capacitor followed the time constant curve. Other circuits in this chapter charged a capacitor with a constant current. Capacitors charged with a constant current developed a voltage with a linear slope.

Periodic Waveform Oscillators

1. Determine, draw, and label the waveforms for the four test points shown in Figure 14–27 if both potentiometers are set to minimum ohms.

   TP_1 ─────────────────────────────

   TP_2 ─────────────────────────────

   TP_3 ─────────────────────────────

   TP_4 ─────────────────────────────

2. Determine, draw, and label the waveforms for the four test points shown in Figure 14–27 if both potentiometers are set to maximum ohms.

   TP_1 ─────────────────────────────

   TP_2 ─────────────────────────────

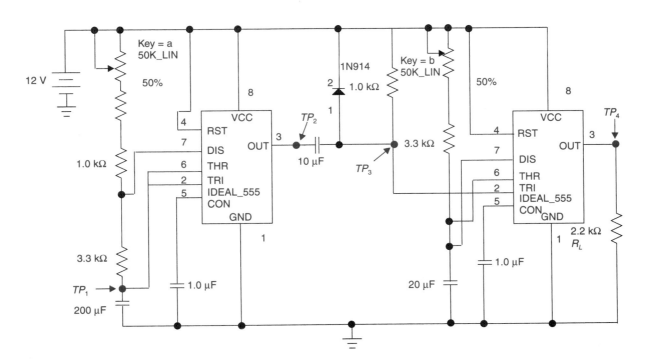

FIGURE 14–27

TP_3 ————————————————————————

TP_4 ————————————————————————

3. Simulate the circuit in Figure 14–27. Draw and label the waveforms at the four test points when both potentiometers are set to minimum ohms.

    TP_1 ————————————————————————

    TP_2 ————————————————————————

    TP_3 ————————————————————————

    TP_4 ————————————————————————

4. Simulate the circuit in Figure 14–27. Draw and label the waveforms at the four test points when both potentiometers are set to maximum ohms.

    TP_1 ————————————————————————

    TP_2 ————————————————————————

    TP_3 ————————————————————————

    TP_4 ————————————————————————

5. Determine, draw, and label the waveforms for the following points shown in Figure 14–28 if the potentiometer is set to minimum ohms.

    TP_1 ————————————————————————

    TP_2 ————————————————————————

    TP_3 ————————————————————————

6. Determine, draw, and label the waveforms for the following points shown in Figure 14–28 if the potentiometer is set to maximum ohms.

    TP_1 ————————————————————————

    TP_2 ————————————————————————

    TP_3 ————————————————————————

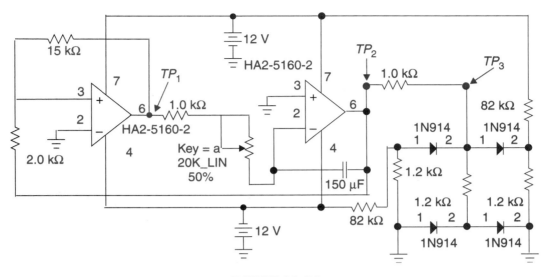

FIGURE 14–28

7. Simulate the circuit in Figure 14–28. Draw and label the waveforms at the three test points when the potentiometer is set to minimum ohms.

TP_1 ———————————————————————

TP_2 ———————————————————————

TP_3 ———————————————————————

8. Simulate the circuit in Figure 14–28. Draw and label the waveforms at the three test points when the potentiometer is set to maximum ohms.

TP_1 ———————————————————————

TP_2 ———————————————————————

TP_3 ———————————————————————

Periodic Waveform Oscillators

I. Objectives

1. To analyze a pulse generator circuit constructed from a 555 astable oscillator and a 555 monostable timer.
2. To analyze generator circuits with triangular and sine wave outputs.

II. Test Equipment

±12 V power supply　　　　　　Oscilloscope

III. Components

Resistors: (4) 1 kΩ, (3) 1.2 kΩ, (1) 2 kΩ, (1) 2.2 kΩ, (2) 10 kΩ, (1) 15 kΩ, (2) 82 kΩ, (2) 10 kΩ potentiometer

Capacitors: (2) 0.001 μF, (1) 0.01 μF, (1) 0.1 μF, (1) 0.22 μF

Diodes: (4) 1N4148

ICs: (2) LM555 timer, (2) LM741

IV. Procedure

1. Determine, draw, and label the waveforms for the four test points shown in Figure 14–29 if both potentiometers are set to minimum ohms.

TP_1 ————————————————————————

TP_2 ————————————————————————

TP_3 ————————————————————————

TP_4 ————————————————————————

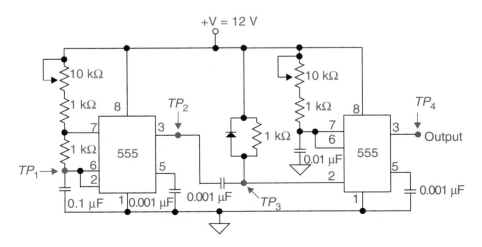

FIGURE 14–29

2. Determine, draw, and label the waveforms for the four test points shown in Figure 14–29 if both potentiometers are set to maximum ohms.

TP_1 ——————————————————————

TP_2 ——————————————————————

TP_3 ——————————————————————

TP_4 ——————————————————————

3. Construct the circuit in Figure 14–29.

4. Set both potentiometers for minimum ohms. Using an oscilloscope, measure and record the waveforms at the four test points.

TP_1 ——————————————————————

TP_2 ——————————————————————

TP_3 ——————————————————————

TP_4 ——————————————————————

5. Set both potentiometers for maximum ohms. Using an oscilloscope, measure and record the waveforms at the four test points.

TP_1 ——————————————————————

TP_2 ——————————————————————

TP_3 ——————————————————————

TP_4 ——————————————————————

6. Determine, draw, and label the waveforms for the following points shown in Figure 14–30 if the potentiometer is set to minimum ohms.

TP_1 ——————————————————————

TP_2 ——————————————————————

TP_3 ——————————————————————

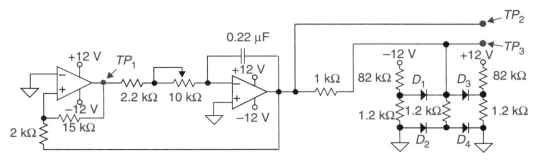

FIGURE 14–30

7. Determine, draw, and label the waveforms for the following points shown in Figure 14–30 if the potentiometer is set to maximum ohms.

TP_1 ————————————————————————————

TP_2 ————————————————————————

TP_3 ————————————————————————

8. Construct the circuit in Figure 14–30.

9. With the adjustment to the potentiometer set for minimum ohms, use an oscilloscope to measure and record the waveforms at the following points.

TP_1 ————————————————————————————

TP_2 ————————————————————————

TP_3 ————————————————————————

10. With the adjustment to the potentiometer set for maximum ohms, use an oscilloscope to measure and record the waveforms at the following points.

TP_1 ————————————————————————————

TP_2 ————————————————————————

TP_3 ————————————————————————

V. Points to Discuss

1. For the circuit in Figure 14–29, explain why the duty cycle of the signal at TP_2 does not affect the pulse width of the signal at TP_4.

2. What controls the amplitude of the output in the circuit in Figure 14–29?

3. Why is trigger conditioning necessary between the stages in the circuit in Figure 14–29?

4. What controls the amplitude of the output in the circuit in Figure 14–30?

5. If the triangular wave circuit in Figure 14–30 is modified so the triangular wave output is 6 $V_{p\text{-}p}$, would the sine wave shaping circuit function properly? Explain.

QUESTIONS

14.1 Introduction to Rectangular Wave Oscillators

____ 1. What do all multivibrator circuits have in common?
   a. They all have a trigger input.
   b. Their output is always a sine wave.
   c. Their output is always a triangular wave.
   d. Their output is always at one of two voltage levels.

____ 2. The _____ is not a type of multivibrator circuit.
   a. free-run
   b. one-shot
   c. flip-flop
   d. multi-shot

14.2 The 555 Astable Circuit

____ 3. The output of the astable circuit _____.
   a. is low until a trigger input is received
   b. is high until a trigger input is received
   c. constantly switches between two states
   d. none of the above

____ 4. An integrated circuit that can function as an astable oscillator or a monostable timer is the widely used and inexpensive _____.
   a. 555 timer
   b. 74LS555 timer
   c. 555 astable circuit
   d. 555 monostable circuit

____ 5. The internal circuitry of the 555 consists of _____, an RS flip-flop, a transistor switch, an output buffer amplifier, and a voltage divider.
   a. a comparator
   b. two comparators
   c. a voltage amplifier
   d. a peak detector

____ 6. The output frequency of a 555 astable oscillator can be calculated by the formula _____.
   a. $f = 1/(R_A + R_B)C$
   b. $f = (R_A + 2R_B)C$
   c. $f = 1/(2\pi RC)$
   d. $f = 1.44/(R_A + 2R_B)C$

____ 7. The _____ is defined as the time the output is active divided by the total period of the output signal.
   a. on time
   b. off time

    c.   duty cycle

    d.   active ratio

14.3 The 555 as a Monostable Circuit

____ 8. The 555 monostable multivibrator circuit is not an oscillator because _____.

    a.   its output switches between two states

    b.   it requires a trigger input to obtain an output signal

    c.   it requires a sine wave input signal

    d.   the circuit does not require a DC power supply

____ 9. On the falling edge of the trigger input, the output switches high for a period of time determined by the _____ .

    a.   value of the RC network

    b.   manufacturing in the design of the 555

    c.   amplitude of the input trigger

    d.   magnitude of the DC supply voltage

____ 10. The output pulse width of the 555 one-shot circuit can be calculated using the formula _____.

    a. $t = RC$

    b. $t = 1/RC$

    c. $t = 1.1RC$

    d. $t = 2\pi RC$

____ 11. The trigger input pulse at pin 2 of the 555 one-shot circuit must be _____.

    a.   a positive going pulse

    b.   a negative going pulse

    c.   of shorter duration than the output pulse

    d.   both b and c

14.4 Inverter Oscillators

____ 12. An inverter oscillator consists of _____ connected in series with the output of the last inverter connected to the input of the first inverter.

    a.   an even number of inverters

    b.   an odd number of inverters

    c.   two inverters

    d.   four inverters

____ 13. The frequency of oscillations of an inverter oscillator depends on the number of inverters and _____.

    a.   the time constant of the RC network

    b.   the resonant frequency

    c.   the propagation time delay of each inverter

    d.   only the number of inverters

14.5 Schmitt Trigger RC Oscillators

____ 14. When troubleshooting a Schmitt trigger RC oscillator, a technician can approximate the period of the output square wave by calculating the _____.

a. RC time constant of the negative feedback loop

b. RC time constant of the positive feedback loop

c. hysteresis

d. feedback factor

____ 15. A Schmitt trigger oscillator using a digital Schmitt trigger inverter gate does not require an external positive feedback loop to set the trip points because _____.

a. electrical hysteresis is built into the IC

b. trip points are not needed

c. the circuit uses negative feedback

d. none of the above

14.6 Crystal Controlled Oscillators

____ 16. The RC network in a crystal Schmitt trigger oscillator is designed to set the frequency of oscillation _____.

a. at electrical hysteresis

b. equal to the resonant frequency of the crystal

c. slightly below the resonant frequency of the crystal

d. slightly above the resonant frequency of the crystal

14.7 Triangular Wave Oscillators

____ 17. A triangular wave oscillator can consist of a Schmitt trigger square-wave oscillator followed by _____.

a. an amplifier

b. an integrator

c. a differentiator

d. a one-shot

____ 18. A two-stage feedback triangular wave oscillator circuit consists of a zero crossover comparator followed by _____.

a. an amplifier

b. an integrator

c. a differentiator

d. a one-shot

____ 19. One advantage of the two-stage feedback triangular wave oscillator circuit is the _____.

a. frequency can be adjusted without affecting the amplitude of the output

b. two-stage feedback circuit requires fewer components

c. output amplitude can be adjusted without affecting the frequency of the output

d. both a and b

14.8 Wave Shaping a Triangular Wave into a Sine Wave

____ 20. An alternative to using a sine wave oscillator is to convert a _____ wave into a sine wave by using a wave shaping circuit.
    a. square
    b. triangular
    c. sawtooth
    d. none of the above

____ 21. The sine wave shaping circuit examined in this chapter is sensitive to _____ changes but is not sensitive to _____ changes.
    a. amplitude, frequency
    b. frequency, amplitude
    c. current, voltage
    d. none of the above

14.9 Sawtooth Oscillators

____ 22. A sawtooth oscillator consists of _____ and a switching circuit.
    a. an amplifier
    b. an integrator
    c. a differentiator
    d. a one-shot

____ 23. The Shockley diode acts like ___ until its rated break-over voltage is reached. At that point, the Shockley diode will conduct like _____.
    a. a low impedance, a high impedance
    b. $10 k\Omega$, 100Ω
    c. a short, an open
    d. an open, a short

14.10 Troubleshooting Oscillator Circuits

____ 24. The voltage at TP_1 in Figure 14–31 measures 0 V. What is wrong with the circuit?
    a. C_1 is open.
    b. R_1 is shorted.
    c. U_1 is defective.
    d. The circuit is operating properly.

____ 25. The signal at TP_1 in Figure 14–31 is observed to be a waveform changing amplitude between 4 V and 8 V. What is wrong with the circuit?
    a. C_1 is open.
    b. R_1 is shorted.
    c. U_1 is defective.
    d. The circuit is operating properly.

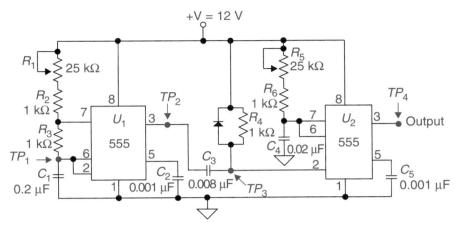

FIGURE 14–31

_____ 26. The signals at TP_2 and TP_3 are identical in Figure 14–31. What is wrong with the circuit?

    a. C_3 is open.

    b. C_3 is shorted.

    c. The diode is open.

    d. The circuit is operating properly.

_____ 27. The frequency of the output signal (TP_4) in Figure 14–31 cannot be adjusted. What is wrong with the circuit?

    a. R_1 is defective.

    b. R_2 is shorted.

    c. R_5 is defective.

    d. R_6 is shorted.

PROBLEMS

1. Calculate and draw the waveforms at TP_1 and TP_2 for the circuit in Figure 14–32.

2. If the resistor values are not changed, calculate the capacitor value needed to modify the circuit in Figure 14–32 so the circuit will have an output frequency of 3.3 kHz.

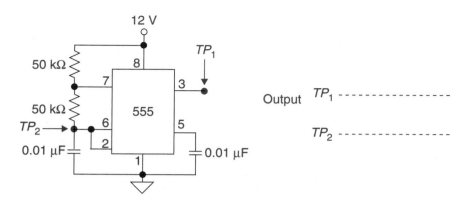

FIGURE 14–32

3. If the capacitor value is not changed, calculate the resistor values needed to modify the circuit in Figure 14–32 so the circuit will have an output signal with a frequency of 2 kHz and a duty cycle of 25%.

4. A trigger input is shown in Figure 14–33 (TP_1). Calculate, label, and draw the waveforms at TP_2 and TP_3.

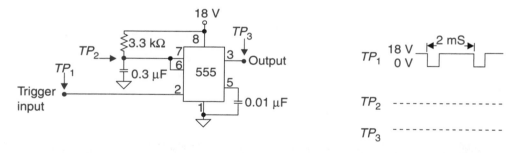

FIGURE 14–33

5. A trigger input is shown in Figure 14–34 (TP_1). Calculate, label, and draw the waveforms at TP_2, TP_3, and TP_4.

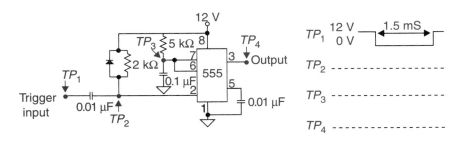

FIGURE 14–34

6. Modify the circuit in Figure 14–34 by changing the value of the 5 kΩ resistor to obtain an output pulse width of 100 μS. What is the new value of the resistor?

7. Show the waveforms at TP_1 through TP_5. Label timing and amplitude for the circuit shown in Figure 14–35.

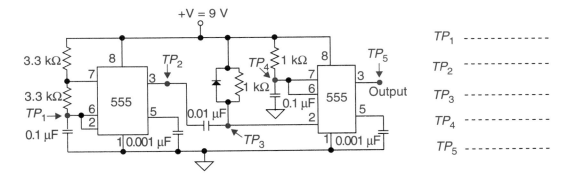

FIGURE 14–35

8. If the propagation time delay (t_p) for each inverter is 35 nS, calculate the period and frequency and draw the waveforms at the test points for the inverter oscillator in Figure 14–36.

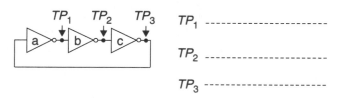

FIGURE 14–36

9. Sketch a diagram using five inverters as an oscillator and draw the waveform. Assume the inverter used has a propagation delay of 15 nS.

10. Draw and label the waveforms at TP_1 and TP_2 for the circuit shown in Figure 14–37.

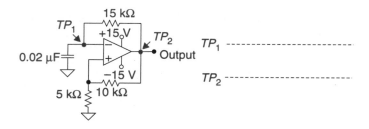

FIGURE 14–37

11. What needs to be changed to modify the circuit in Figure 14–37 so the output signal swing would equal 20 $V_{p\text{-}p}$?

12. Choose a new value for the 15 kΩ resistor in Figure 14–37 that would permit the oscillator to oscillate at approximately 25 kHz.

13. Calculate the frequency of the output signal for the oscillator shown in Figure 14–38.

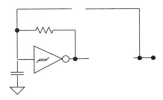

FIGURE 14–38

14. Determine the waveforms for the test points shown in Figure 14–39.

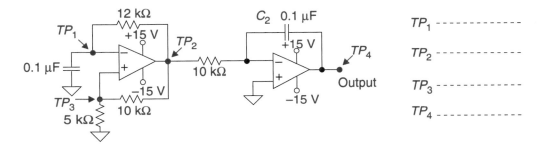

FIGURE 14–39

15. Determine the waveforms for the test points shown in Figure 14–40.

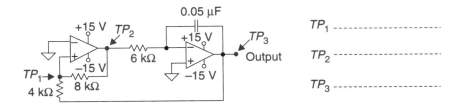

FIGURE 14–40

16. Determine the waveforms at TP_1, TP_2, and TP_3 for the circuit in Figure 14–41.

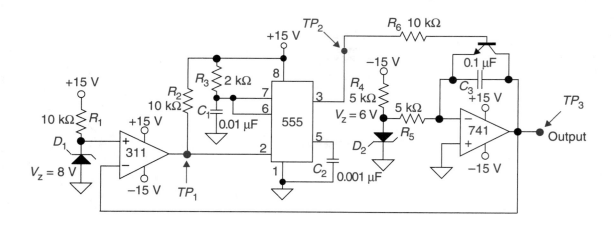

TP_1 - - - - - - - - - - - - -

TP_2 - - - - - - - - - - - - -

TP_3 - - - - - - - - - - - - -

FIGURE 14–41

OBJECTIVES

After studying the material in this chapter, you will be able to describe and/or analyze:
- ○ a differential amplifier circuit,
- ○ an instrumentation amplifier circuit,
- ○ an operational transconductance amplifier (OTA) circuit,
- ○ an optoisolator circuit,
- ○ a voltage controlled oscillator (VCO),
- ○ a phase-lock loop (PLL),
- ○ VCO and PLL applications, and
- ○ troubleshooting techniques used with special ICs.

15.1 DIFFERENTIAL AMPLIFIERS

The circuits you have worked with so far have operated with single-ended input signals. This means that one side of the source signal is referenced to ground. Figure 15–1(a) shows an inverting amplifier circuit. The positive terminal of the 2 V signal source is connected to ground, and the negative terminal of the signal source is connected to the inverting input of the amplifier through resistor R_i. The input signal (V_x) is –2 V with respect to ground. The inverting amplifier in Figure 15–1(a) has a voltage gain of 5, so the output voltage is 10 V.

Figure 15–1(b) shows a differential amplifier with a 2 V signal source connected across its inputs. Neither terminal of the 2 V signal source is connected to ground. The positive terminal is connected to the noninverting input through resistor R_i', and the negative terminal is connected to the inverting input through resistor R_i. The input signal at point x is –2 V with respect to point z. The input signal (V_{xz}) is the voltage across the input and is not referenced to ground. Example 15.1 will show that the voltage gain of the differential amplifier is equal to R_f/R_i.

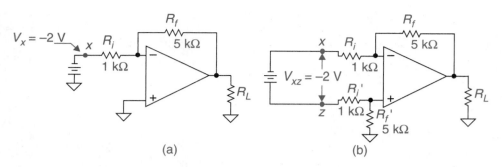

(a) (b)

FIGURE 15–1   Single-ended and differential amplifiers

EXAMPLE 15.1

Calculate V_{out} and A_v for the circuit in Figure 15–1(b).

Step 1. Calculate the voltage across resistors R_i and R_i'. Assume the output of the op-amp is not in saturation; therefore, the voltage between the inverting and noninverting inputs is equal to zero ($V_{-+} = 0$ V). Kirchoff's loop equation for the input loop is:

$$V_{xz} - V_{Ri} - V_{-+} - -V_{Ri'} = 0 \text{ V (Since } V_{-+} = 0 \text{ V.)}$$
$$V_{xz} - V_{Ri} - V_{Ri'} = 0 \text{ V}$$
$$-2 \text{ V} - V_{Ri} - V_{Ri'} = 0 \text{ V}$$
$$2 \text{ V} = -V_{Ri} - V_{Ri'} \text{ (Since } R_i = R_i'.)$$
$$V_{Ri} = -1 \text{ V}$$
$$V_{Ri'} = -1 \text{ V}$$

Step 2. Calculate the current through R_i and R_i'.

$$I_{Ri} = V_{Ri}/R_i = 1 \text{ V}/1 \text{ k}\Omega = 1 \text{ mA}$$
$$I_{Ri'} = V_{Ri'}/R_i = 1 \text{ V}/1 \text{ k}\Omega = 1 \text{ mA}$$

Step 3. Calculate the current through R_f and R_f'. Since no current flows in or out of the op-amp terminals, $I_{Rf} = I_{Ri}$ and $I_{Rf'} = I_{Ri'}$.

$$I_{Rf} = I_{Ri} = 1 \text{ mA}$$
$$I_{Rf'} = I_{Ri'} = 1 \text{ mA}$$

Step 4. Calculate the voltage drop across R_f and R_f' using the calculated currents.

$$V_{Rf} = I_{Rf} \times R_f = 1 \text{ mA} \times 5 \text{ k}\Omega = 5 \text{ V}$$
$$V_{Rf'} = I_{Rf'} \times R_f' = 1 \text{ mA} \times 5 \text{ k}\Omega = 5 \text{ V}$$

Step 5. Calculate V_{out}. The output voltage equals the voltage dropped across R_f plus the voltage dropped across R_f'.

$$V_{out} = V_{Rf} + V_{Rf'} = 5 \text{ V} + 5 \text{ V} = 10 \text{ V}$$

Step 6. Show the polarity of all voltage drops in the circuit in Figure 15–2.

Step 7. Calculate the voltage gain. The 2 V input signal across the inputs of the differential amplifier causes an output signal of 10 V; therefore, the circuit has a voltage gain of 5. The differential voltage gain of the differential amplifier can be calculated using circuit components and the formula $A_{v(\text{diff})} = R_f/R_i$.

$$A_{v(\text{diff})} = R_f/R_i$$
$$A_{v(\text{diff})} = 5 \text{ k}\Omega/1 \text{ k}\Omega = 5$$

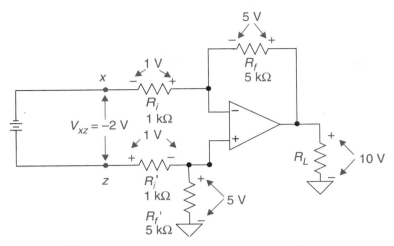

FIGURE 15–2 Voltage reading in a differential amplifier circuit

Figure 15–3(a) shows a differential amplifier with a differential input signal; the signal is connected across the inputs of the amplifier. Figure 15–3(b) shows a differential amplifier with a single-ended input signal. The single-ended input signal is common to both inputs; therefore, zero voltage is present between the inputs. With zero differential signal input, the output of the amplifier is zero. A signal connected to both inputs of a differential amplifier is called a common-mode signal. The ideal differential amplifier has a common-mode voltage gain of zero ($A_{v(cm)} = 0$) and totally rejects common-mode signals. Example 15.2 shows how the output waveforms for the circuits in Figure 15–3 can be calculated.

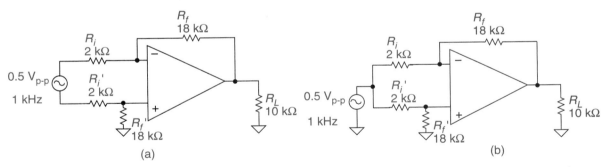

FIGURE 15–3 Differential input signal versus common-mode input signal

EXAMPLE 15.2

Calculate the output waveforms for both circuits in Figure 15–3.

Step 1. Calculate the voltage gain for the circuit in Figure 15–3(a).

$$A_{v(diff)} = R_f/R_i$$

$$A_{v(diff)} = 18 \text{ k}\Omega/2 \text{ k}\Omega = 9$$

Step 2. Calculate the amplitude of the output signal for the circuit in Figure 15–3(a).

$$V_{out} = V_{in} \times A_{v(diff)}$$

$$V_{out} = 0.5 \text{ V}_{p\text{-}p} \times 9 = 4.5 \text{ V}_{p\text{-}p}$$

Step 3. Calculate the voltage gain for the circuit in Figure 15–3(b).

$$A_{v(cm)} = 0$$

EXAMPLE 15.2 continued

Step 4. Calculate the amplitude of the output signal for the circuit in Figure 15–3(b).

$$V_{out} = V_{in} \times A_{v(cm)}$$
$$V_{out} = 0.5 \text{ V}_{p\text{-}p} \times 0 = 0 \text{ V}_{p\text{-}p}$$

Step 5. Draw the input and output waveforms for both circuits in Figure 15–3. Figure 15–4(a) shows the input and output waveforms for the differential amplifier with a differential input signal. Notice that the input signal is not in reference to ground. Figure 15–4(b) shows the input and output waveforms for the differential amplifier with a common-mode input signal. The output is zero because the voltage gain of the differential amplifier is zero for common-mode signals.

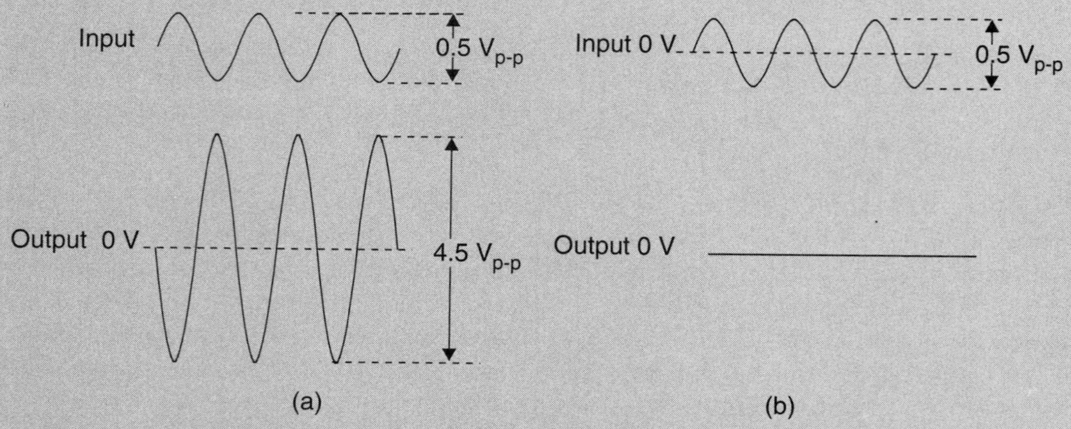

FIGURE 15–4 Waveforms for Example 15.2

COMMON-MODE REJECTION RATIO (CMRR)

The measurement of the ability of a differential amplifier to reject common-mode signals is called *common-mode rejection ratio* (CMRR). The common-mode rejection ratio is the ratio between the differential voltage gain and the common-mode voltage gain. The ideal common-mode voltage gain is zero. This means that regardless of the amplitude of the common-mode signal, the resulting output signal is zero. Actually, a large common-mode signal will result in a small signal, so while the common-mode gain is near zero, it is not exactly zero.

Figure 15–5 gives the needed information to calculate the CMRR for the differential amplifier circuit shown. In Figure 15–5(a), a 100 mV differential signal is applied to the inputs, and the output signal is measured to be 10 V. The differential voltage gain is calculated to be 100. In Figure 15–5(b), a 100 mV common-mode signal is applied to the inputs, and the output signal is measured to be 0.25 mV. The common-mode voltage gain is calculated to be 0.0025. The common-mode rejection ratio (CMRR) is defined as the ratio of differential voltage gain [$A_{v(diff)}$] to common-mode voltage gain [$A_{v(cm)}$] and is normally expressed in dB. The CMRR of the differential amplifier circuit in Figure 15–5 is 92 dB and is calculated as follows:

$$CMRR = 20 \times \log(A_{v(diff)}/A_{v(cm)})$$
$$CMRR = 20 \times \log(100/0.0025)$$
$$CMRR = 20 \times \log 40,000$$
$$CMRR = 92 \text{ dB}$$

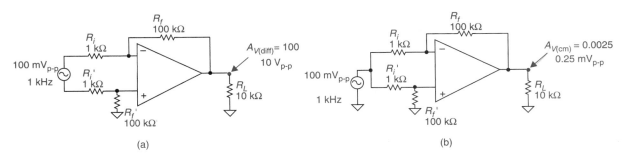

(a)   (b)

FIGURE 15–5 Differential amplifier

The common-mode rejection ratio is normally given on data sheets for op-amps. For example, the 741 op-amp is rated for a typical CMRR equal to 96 dB. A differential amplifier constructed with an op-amp with a CMRR equal to 96 dB would have a somewhat lower CMRR because of the imbalance in resistor values used to construct the circuit.

USING DIFFERENTIAL AMPLIFIERS TO CANCEL NOISE

Noise interference is often encountered in connecting transducers to electronic systems because the voltage generated by the transducer is small and transducers are often located away from the system in noisy environments. The noise problem can be overcome by using a differential amplifier. Figure 15–6 shows a single-ended noninverting amplifier with a transducer as a signal source (v_g). Two more sources are shown in series with the transducer. These represent noise coupled into the cable between the transducer and the noninverting input of the op-amp. The v_{nc} source represents the noise coupled to the cable by capacitive coupling, and the v_{ni} source represents the noise coupled to the cable by inductive coupling. The total input signal at the input of the amplifier is equal to the sum of all three sources ($V_{in} = v_g + v_{ni} + v_{nc}$). The signal and the noise will both be amplified by the gain of the amplifier. If the noise signal is large compared to the signal from the transducer, the signal from the transducer will be lost in the noise.

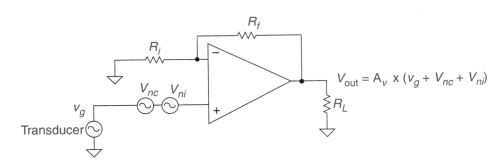

FIGURE 15–6 Single-ended amplifier

Figure 15–7 shows a differential amplifier. The transducer is isolated from the ground, and the signal is applied to the amplifier over a balanced line to the inputs of the op-amp. A balanced line means that both sides of the signal source see the same impedance to ground; in this case, they see $R_i + R_f$. The two wires connecting the transducer are a twisted pair (two wires twisted together), so the noise coupled into each lead is the same.

The signal out of the transducer is a differential input and is amplified by the differential gain ($A_{v(diff)}$) of the circuit. The noise signal is amplified by the common-mode gain ($A_{v(cm)}$) of the circuit. The output will equal the input signal times the differential gain plus the noise input

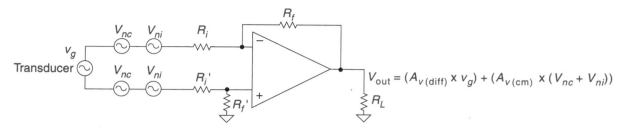

FIGURE 15–7   Differential amplifier

times the common-mode gain. Since the common-mode gain is near zero, the noise present in the output will be extremely small. Example 15.3 demonstrates the advantage of using a differential amplifier to reject noise.

EXAMPLE 15.3

Calculate the signal output and noise output of the single-ended amplifier in Figure 15–8(a) and the differential amplifier in Figure 15–8(b). Assume the circuit has a CMRR of 70 dB.

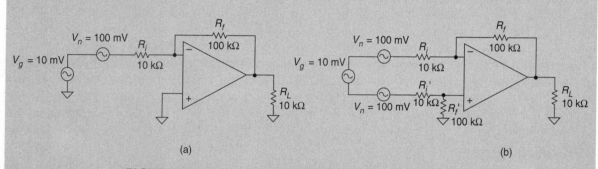

(a)   (b)

FIGURE 15–8   Single-ended and differential amplifiers

Step 1. Calculate the voltage gain of the circuit in Figure 15–8(a).
$$A_v = R_f / R_{in}$$
$$A_v = -100 \text{ k}\Omega / 10 \text{ k}\Omega = 10$$

Step 2. Calculate the output signal and noise of the circuit in Figure 15–8(a).
$$V_{out(signal)} = v_g \times A_v = 10 \text{ mV} \times 10 = 100 \text{ mV}$$
$$V_{out(noise)} = v_n \times A_v = 100 \text{ mV} \times 10 = 1 \text{ V}$$

Step 3. Calculate the differential voltage gain of the circuit in Figure 15–8(b).
$$A_{v(diff)} = R_f / R_{in}$$
$$A_v = -100 \text{ k}\Omega / 10 \text{ k}\Omega = 10$$

Step 4. Calculate the common-mode voltage gain of the circuit in Figure 15–8(b).
$$CMRR = 20 \times \log (A_{v(diff)}/A_{v(cm)})$$
$$70 \text{ dB} = 20 \times \log (10/A_{v(cm)})$$
$$3.5 \text{ dB} = \log (10/A_{v(cm)})$$
$$\text{antilog } 3.5 \text{ dB} = 10/A_{v(cm)}$$
$$3162 = 10/A_{v(cm)}$$

EXAMPLE 15.3 continued

$$A_{v(cm)} = 10/3162$$
$$A_{v(cm)} = 0.0032$$

Step 5. Calculate the output signal and noise of the circuit in Figure 15–8(b).

$$V_{out(signal)} = v_g \times A_{v(diff)} = 10 \text{ mV} \times 10 = 100 \text{ mV}$$
$$V_{out(noise)} = v_n \times A_{v(cm)} = 100 \text{ mV} \times 0.0032 = 0.32 \text{ mV}$$

Step 6. Draw the waveforms for the circuit in Figure 15–8. Figure 15–9(a) shows the waveforms of the single-ended amplifier, and Figure 15–9(b) shows the waveforms of the differential amplifier.

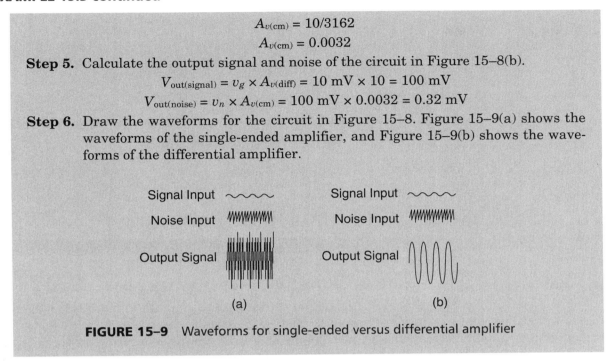

FIGURE 15–9 Waveforms for single-ended versus differential amplifier

An advantage of the differential amplifier can be seen by reviewing the data from Example 15.3. The single-ended amplifier has a signal output of only 100 mV, and the noise output is 1 V. With this much noise present at the output, the signal is lost in the noise, as shown in Figure 15–9(a). On the other hand, the differential amplifier has a signal output of 100 mV, but the noise output is only 0.32 mV, and the signal level at the output is well above the noise level, as shown in Figure 15–9(b).

15.2 INSTRUMENTATION AMPLIFIERS

Even though the differential amplifier has good noise rejection, it still has shortcomings. The gain is hard to adjust because proper adjustment requires exact matching of the two pairs of resistors. A change in one pair for gain purposes would require the other pair to be changed as well. The second problem is the relatively low input impedance of the differential amplifier. The input impedance the signal source sees is the differential input impedance, which is equal to $R_i + R_i'$. There is also an impedance to ground, which is equal to $R_i + R_f$. These problems can be overcome by using three op-amps to form an instrumentation amplifier.

Figure 15–10 shows an instrumentation amplifier. One op-amp is used as a differential amplifier, and the other two are used as input buffers. The input buffers are noninverting amplifiers whose gains are controlled by the variable resistor R_i. The difference voltage between the input terminals of the op-amps is zero; thus, the differential signal applied across the noninverting terminals of U_1 and U_2 is dropped across R_i. The current flowing through R_i flows through R_f and R_f', causing the output voltage. The voltage gain of the buffer stage is equal to $1 + ((R_f + R_f')/R_i)$, or simply $1 + (2R_f/R_i)$. The voltage gain of the total circuit is the product of

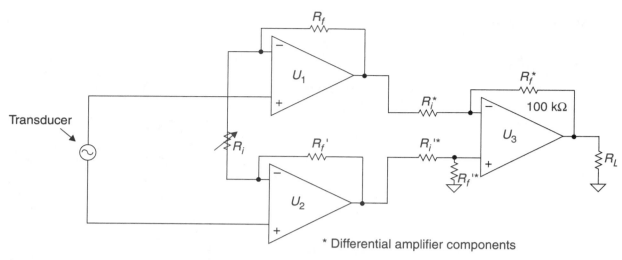

FIGURE 15–10   Instrumentation amplifier

the buffer stage times the differential stage. The gain of the buffer can be easily adjusted by adjusting the value of R_i without upsetting the balance of the differential amplifier.

The input impedance is extremely high because both sides of the signal source are feeding noninverting amplifiers. Instrumentation amplifiers can be constructed from three op-amps; however, because of the popularity of instrumentation amplifiers, they are available in IC packages like National's LH0036.

EXAMPLE 15.4

Calculate the total voltage gain and the circuit input impedance for a circuit like the one in Figure 15–10 with the following components: R_i = 5 kΩ, R_f and R_f' = 35 kΩ, R_i^* and $R_i^{'*}$ = 2 kΩ , and R_f^* and $R_f^{'*}$ = 16 kΩ.

Step 1. Calculate the voltage gain of the first stage.

$$A_v = 1 + (2R_f/R_i)$$
$$A_v = 1 + (2 \times 35 \text{ kΩ}/5 \text{ kΩ})$$
$$A_v = 15$$

Step 2. Calculate the differential voltage gain of the second stage.

$$A_{v(\text{diff})} = R_f^*/R_i^*$$
$$A_{v(\text{diff})} = 16 \text{ kΩ}/2 \text{ kΩ}$$
$$A_{v(\text{diff})} = 8$$

Step 3. Calculate the total voltage gain of the circuit.

$$A_{v(\text{total})} = A_{v(\text{stage 1})} \times A_{v(\text{stage 2})}$$
$$A_{v(\text{total})} = 15 \times 8$$
$$A_{v(\text{total})} = 120$$

Step 4. Calculate the input impedance. The input impedance of the circuit would equal the input impedance of the first stage (which is very high) because both sides of the signal source are feeding noninverting amplifiers.

$$z_{\text{in}} = \text{very high}$$

15.3 OPERATIONAL TRANSCONDUCTANCE AMPLIFIERS (OTAs)

Figure 15–11 shows the schematic symbol for an operational transconductance amplifier (OTA). The OTA has an inverting and noninverting input like the conventional op-amp. The signal input is the difference voltage between the inputs. The output pin of the OTA is a current source. The output current is dependent on the input voltage times the gain of the OTA. Since gain is a ratio of output current to input voltage, the gain is measured in transconductance ($gm = i_{out}/V_{in}$).

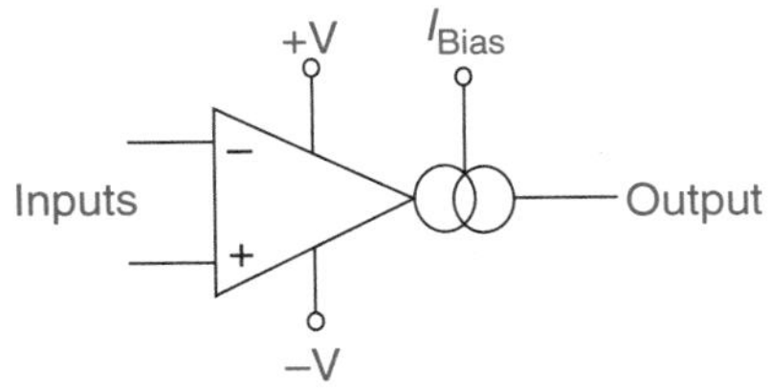

FIGURE 15–11   Operational transconductance amplifier

The transconductance of the OTA is a function of the bias current times a constant ($gm = I_B \times K$). The constant (K) is a characteristic of a given OTA. The bias current (I_B) is set by circuit components controlling the current flowing into the bias current pin.

Figure 15–12 is a voltage amplifier using an OTA. The bias current is determined by calculating the current flowing through R_B. The current bias pin of the OTA connects to the negative supply voltage through an internal diode junction. The voltage across R_B is, therefore, the difference voltage between the voltage applied to R_B and the negative supply. Example 15.5 shows how the voltage gain of the circuit can be calculated if the constant (K) of the OTA is known.

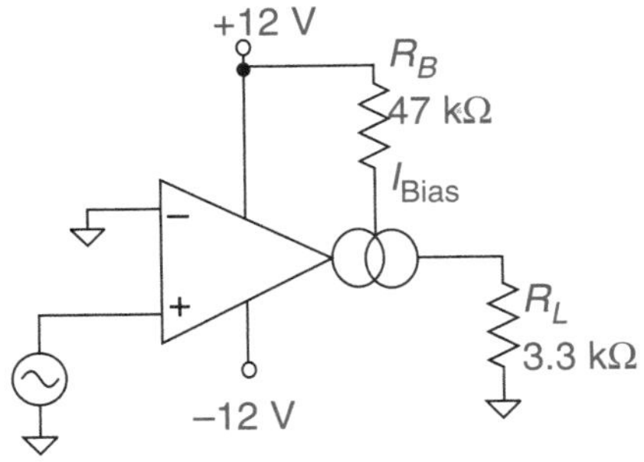

FIGURE 15–12   OTA voltage amplifier

EXAMPLE 15.5

Calculate the voltage gain for the circuit in Figure 15–12 where the OTA constant K is 12.

Step 1. Calculate the bias current. First, determine the voltage across R_B, then calculate the bias current. One end of resistor R_B is connected to +12 V, and the other end of R_B is connected to –12 V through the OTA. There is a PN junction drop of 0.7 V inside the OTA.

$$V_{RB} = 12\ \text{V} - (-12\ \text{V}) - 0.7\ \text{V}$$

EXAMPLE 15.5 continued

$$V_{RB} = 23.3 \text{ V}$$
$$I_B = V_{RB}/R_B$$
$$I_B = 23.3 \text{ V}/47 \text{ k}\Omega$$
$$I_B = 0.5 \text{ mA}$$

Step 2. Calculate *gm*.

$$gm = I_B \times K$$
$$gm = 0.5 \text{ mA} \times 12 = 6 \text{ mS}$$

Step 3. Calculate the voltage gain.

$$A_v = gm \times R_L$$
$$A_v = 6 \text{ mS} \times 3.3 \text{ k}\Omega = 19.8$$

The output of the OTA is a current source and has a high output impedance. For this reason, op-amps are preferred for most applications. However, the voltage gain of an OTA circuit can be easily controlled by a voltage level connected to the biased resistor, making it a useful device to use in circuits that require automatic gain control. Figure 15–13 shows a circuit with a gain control voltage of 1 V to 10 V. If the OTA has a constant *K* value of 12, the gain of the circuit would be 10.4 with 1 V applied to the control input. The circuit would have a gain of 17.9 with 10 V applied to the control input. You can verify these gains by performing the calculations shown in Example 15.5.

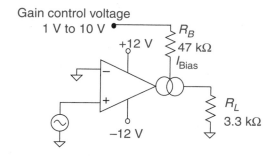

FIGURE 15–13   OTA amplifier with adjustable gain

The circuit in Figure 15–14(a) is a summer circuit with two inputs: a 1 kHz and a 100 kHz. The circuit functions as a linear mixer. The output of the linear mixer is the instantaneous sum of the two inputs and contains two frequency components: 1 kHz and 100 kHz.

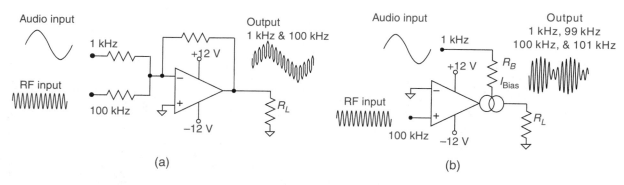

FIGURE 15–14   Linear and nonlinear mixers

Figure 15–14(b) shows an OTA used as a nonlinear mixer circuit. The result is an amplitude modulated (AM) output signal. The output signal contains four frequency components: the two original frequencies (1 kHz and 100 kHz), the sum of the two original frequencies (101 kHz), and the difference of the two original frequencies (99 kHz). A nonlinear mixer is actually an analog multiplier.

Amplitude modulator circuits (nonlinear mixers) are used in transmitters to superimpose the audio signal onto the radio frequency (RF) signal of the transmitter. The OTA permits the audio signal to mix with the RF signal by controlling the gain of the circuit. Another use of the nonlinear mixer is in phase detector circuits. If two signals of identical frequency and phase are the inputs to a nonlinear mixer, the difference signal at the output is zero. As the signals vary in phase, the difference signal can be detected and used to measure the phase difference. Phase detectors are used in phase-lock loops, which are circuits we will cover in a later section.

15.4 OPTOISOLATORS

Optoisolators (also called *optocouplers*) are integrated packages that incorporate a light-emitting diode (LED), a photo diode, and a transistor enclosed in a single package. Figure 15–15(a) shows a 6N135 optoisolator. When current flows through the LED, the light from the LED activates the photodiode, which turns on the transistor. The other optoisolators in Figure 15–15 do not show a photodiode but instead show an LED directly activating the transistor. All optoisolators have a photo-sensitive PN junction that is activated by the light from an LED. In most cases, the photosensitive PN junction is connected between the base and collector of the transistor, creating a phototransistor. Some optoisolators are designed with the base terminal of the transistor connected to a pin of the IC. This design permits the transistor to be biased with external circuitry. Other designs do not have an external base connection and depend totally on the light from the LED to control the conduction of the transistor. Many optoisolators use a Darlington pair, as shown in Figure 15–15(d), to increase the current drive of the output stage.

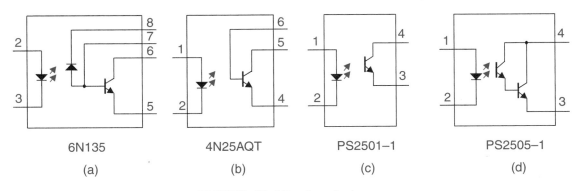

| 6N135 | 4N25AQT | PS2501–1 | PS2505–1 |
| (a) | (b) | (c) | (d) |

FIGURE 15–15 Optoisolators

The function of an optoisolator is to provide electrical isolation between stages of an electronic system but still permit signal coupling. Optoisolators have an isolation voltage rating, which is the voltage difference between the input stage and the output stage, that the device can withstand without breaking down. Optoisolators are available with isolation voltage ratings from 500 V to 5000 V.

Another important specification of optoisolators is the current transfer ratio (CTR), which is the ratio between the collector current of the transistor and the current flowing through the LED. It is expressed in percentages and can be calculated using the formula $CTR = I_C/I_F \times 100\%$, where I_F is the forward current through the diode and I_C is the collector current of the transistor. Optoisolators are available with CTR ratings from 10% to 1000%.

Figure 15–16 shows an optoisolator circuit at the output of an electronic system. If lightning or other factors cause a large voltage surge on the output line, the optoisolator will protect the system by preventing the voltage surge from being coupled back into the system. Example 15.6 shows how the CTR rating of an optoisolator is useful when calculating circuit values.

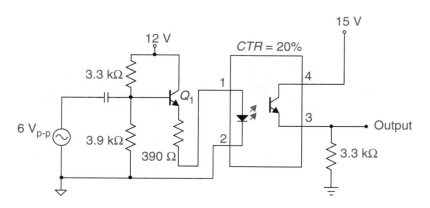

FIGURE 15–16   Optoisolator circuit

EXAMPLE 15.6

If the optoisolator in Figure 15–16 has a current transfer ratio of 20%, calculate the signal at the output.

Step 1. Calculate the current flow through the LED with a signal input of zero.

$$V_B = V_{cc} \times R_{B2}/(R_{B1} + R_{B2}) = 12\ \text{V} \times 3.9\ \text{k}\Omega/(3.3\ \text{k}\Omega + 3.9\ \text{k}\Omega) = 6.5\ \text{V}$$
$$V_E = V_B - 0.7\ \text{V} = 6.5\ \text{V} - 0.7\ \text{V} = 5.8\ \text{V}$$
$$V_{RE} = V_E - V_D = 5.8\ \text{V} - 2\ \text{V} = 3.8\ \text{V}$$
$$I_F = V_{RE}/R_E = 3.8\ \text{V}/390\ \Omega = 9.74\ \text{mA}$$

Step 2. Calculate the current flow in the emitter circuit of the optoisolator.

$$CTR = I_C/I_F \times 100\%$$
$$I_C = CTR \times I_F = 20\% \times 9.74\ \text{mA} = 1.95\ \text{mA}$$
$$I_C \approx I_E = 1.95\ \text{mA}$$

Step 3. Calculate the DC output.

$$V_{\text{out}} = R_E \times I_E = 3.3\ \text{k}\Omega \times 1.95\ \text{mA} = 6.4\ \text{V}$$

Step 4. Calculate the output voltage on the positive swing of the input. The common-emitter amplifier has a voltage gain of one; therefore, if the voltage on the input swing is 3 V positive, the voltage on the emitter will increase by 3 V.

$$V_E = V_{EQ} + 3\ \text{V} = 5.8\ \text{V} + 3\ \text{V} = 8.8\ \text{V}$$
$$V_{RE} = V_E - V_D = 8.8\ \text{V} - 2\ \text{V} = 6.8\ \text{V}$$
$$I_F = V_{RE}/R_E = 6.8\ \text{V}/390\ \Omega = 17.4\ \text{mA}$$
$$CTR = I_C/I_F \times 100\%$$
$$I_C = CTR \times I_F = 20\% \times 17.4\ \text{mA} = 3.48\ \text{mA}$$

EXAMPLE 15.6 continued

$$I_C \approx I_E = 3.48 \text{ mA}$$
$$V_{\text{out}} = R_E \times I_E = 3.3 \text{ k}\Omega \times 3.48 \text{ mA} = 11.5 \text{ V}$$

Step 5. Calculate the output voltage on the negative swing of the input.

$$V_E = V_{EQ} - 3 \text{ V} = 5.8 \text{ V} - 3 \text{ V} = 2.8 \text{ V}$$
$$V_{RE} = V_E - V_D = 2.8 \text{ V} - 2 \text{ V} = 0.8 \text{ V}$$
$$I_F = V_{RE}/R_E = 0.8 \text{ V}/390 \text{ }\Omega = 2 \text{ mA}$$
$$CTR = I_C/I_F \times 100\%$$
$$I_C = CTR \times I_F = 20\% \times 2 \text{ mA} = 0.4 \text{ mA}$$
$$I_C \approx I_E = 0.4 \text{ mA}$$
$$V_{\text{out}} = R_E \times I_E = 3.3 \text{ k}\Omega \times 0.4 \text{ mA} = 1.3 \text{ V}$$

Step 6. Draw the input and output waveforms. The output is a 10.2 $V_{\text{p-p}}$ (11.5 V − 1.3 V) riding on a 6.4 V DC reference. Since both transistors are functioning as common-collector amplifiers, the output will be in phase with the input. The waveforms are shown in Figure 15–17.

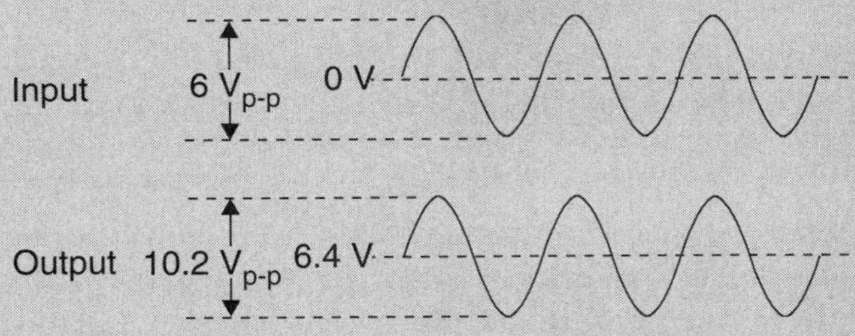

FIGURE 15–17 Optoisolator circuit waveforms

15.5 VOLTAGE CONTROLLED OSCILLATORS (VCOs)

A voltage controlled oscillator (VCO) is a circuit in which the output frequency is a function of input voltage. The 555 timer can function as a simple VCO. Figure 15–18 shows a 555 astable oscillator with pin 5 connected to an input voltage. Under normal operation with a 12 V supply, the internal voltage divider sets the inverting input of comparator one to 8 V and the noninverting input of comparator two to 4 V. If pin 5 has an input of 8 V, then both comparators have the same inputs. The external capacitor will charge and discharge between 4 V and 8 V; therefore, the frequency can be calculated using the normal $f = 1.44/(R_A + 2R_B)C$. As the waveforms in Figure 15–18 show, this results in an output frequency of 1000 Hz.

If the voltage at pin 5 is reduced to 4 V, the inverting input of comparator one has a 4 V input, and the noninverting input of comparator two has a 2 V input. This means the external capacitor will charge and discharge between 2 V and 4 V. Since the voltage across the capacitor is reduced, the time needed for charging and discharging has also been reduced, as shown by the waveform in Figure 15–18. This means the output frequency has increased. The normal formula for calculating the frequency of a 555 astable assumes one-third of +V and two-thirds of +V switching points, so the formula is not valid. The frequency can be calculated by using the time constant curve. With a 4 V input, the output frequency equals 1845 Hz, as shown in Figure 15–18. If the input voltage at pin 5 is varying between 4 V and 8 V, the output frequency will vary between

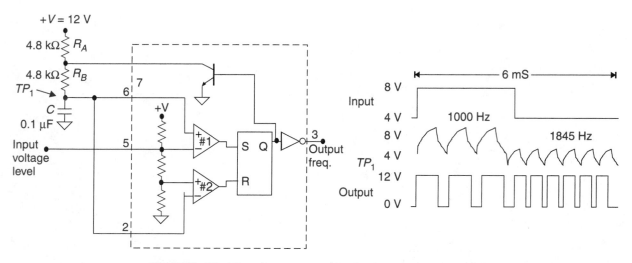

FIGURE 15-18   The 555 astable functioning as a VCO

1845 Hz and 1000 Hz; hence, an input voltage is controlling an output frequency, and the 555 is functioning as a VCO.

High quality VCOs have a linear relationship between the input voltage and the output frequency. While the 555 timer demonstrated the concept of a VCO, it is not very linear. Figure 15–19 shows a VCO circuit using an LM566, which is a popular general purpose VCO integrated circuit. The output frequency at pin 3 is a function of the difference voltage between the supply voltage ($+V$) and the voltage on pin 5. The voltage on pin 5 must be in the range between three-fourths of $+V$ to $+V$. The graph in Figure 15–19 shows that variations in input voltage within this range cause linear changes in output frequency. External components R_1 and C_1 set the frequency range. Capacitor C_2 is a noise filter to prevent unwanted oscillations.

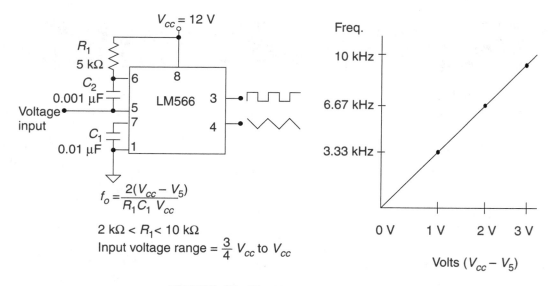

$$f_o = \frac{2(V_{cc} - V_5)}{R_1 C_1 V_{cc}}$$

$2\ k\Omega < R_1 < 10\ k\Omega$

Input voltage range $= \frac{3}{4} V_{cc}$ to V_{cc}

FIGURE 15-19   LM566 VCO circuit

The LM566 VCO has two outputs: a square wave at pin 3 and a triangular wave at pin 4. Both of these signals maintain a 50% duty cycle across the frequency range. Example 15.7 shows how the output frequency can be calculated for the circuit in Figure 15–19.

EXAMPLE 15.7

Calculate the output frequency for the circuit in Figure 15–19 when the input voltage is 10 V.

$$f_o = \frac{2(V_{cc} - V_5)}{R_1 C_1 V_{cc}} = \frac{2(12 \text{ V} - 10 \text{ V})}{5 \text{ k}\Omega \times 0.01 \text{ μF} \times 12 \text{ V}} = 6.7 \text{ kHz}$$

Figure 15–20 shows how an LM566 VCO can be used to obtain frequency modulation (FM). The bias voltage on pin 5 is set by the voltage divider comprised of the 22 kΩ and 150 kΩ resistors. This bias voltage causes the output frequency of the LM566 VCO to be approximately 1 MHz. The audio signal is capacitively coupled to pin 5 and will cause the output to vary above and below 1 MHz. The square-wave output of the LM566 is capacitively coupled to the base of a bipolar transistor, which is biased class-C. The diode prevents the base from being driven negative by clipping the negative swing of the signal. The positive swing of the signal turns on the transistor and causes a current pulse to flow through the tank circuit in the collector leg. The tank circuit is tuned for a center frequency of 1 MHz. The flywheel effect converts the current pulses into a sinusoidal wave that is transformer coupled to the output. An audio sine wave input of 0.2 $V_{p\text{-}p}$ will cause the output frequency to change from 0.950 MHz to 1.050 MHz. This FM radio signal could be amplified and coupled to an antenna. An FM receiver could pick up the signal and convert the frequency deviations back to an audio signal.

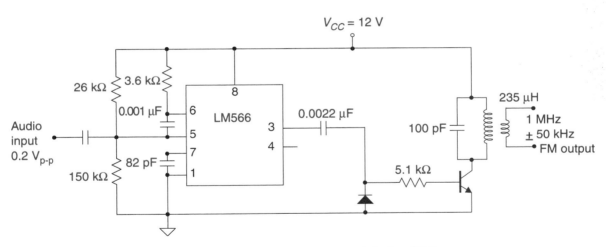

FIGURE 15–20   Frequency modulation (FM)

15.6   PHASE-LOCKED LOOP (PLL)

An important circuit is the phase-locked loop shown in Figure 15–21. It consists of three subcircuits: a phase detector, a low-pass filter, and a voltage controlled oscillator. The phase detector is a nonlinear mixer with two input signals: the input signal and a feedback signal from the VCO. The signal present at the output of the phase detector is comprised of the two original input frequencies plus the sum and difference of these frequencies.

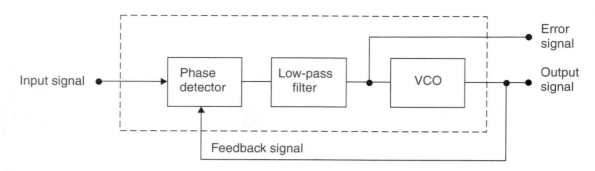

FIGURE 15–21   Block diagram of the phase-locked loop (PLL)

The difference signal is the desired output of the phase detector. If the two input signals to the phase detector are close in frequency, the difference signal will have a low frequency. The phase detector is followed by a low-pass filter. The low-pass filter passes and conditions the difference signal and rejects all others. The signal passed by the low-pass filter is called the error signal. The VCO is biased to an internal reference voltage and will oscillate at a free-running frequency. The error signal from the filter is superimposed on the reference voltage and causes the VCO to change frequency.

CAPTURE AND LOCK RANGE

Figure 15–22 shows a diagram of the capture and lock range. If the frequency of the input signal and the VCO signal are far apart, the difference signal out of the phase detector will be at a high frequency above the low-pass filter bandpass. The low-pass filter will not pass an error signal to the VCO, and the VCO will oscillate at its free-running frequency (f_o). At this point, the PLL is not in lock, and the frequency of the output signal is not affected by the input signal.

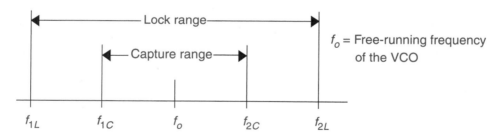

FIGURE 15–22   Capture and lock range

To obtain lock, the input signal frequency must come within the capture range (f_{1C} to f_{2C}). Once the input signal is within the capture range, the frequency of the difference signal decreases until it is passed by the low-pass filter. Once the difference signal is passed by the filter, the error signal will start driving the VCO toward the frequency of the input signal, and the PLL will become locked. The capture range is a function of the low-pass filter. The lower the bandpass of the filter, the narrower the capture range.

Once the incoming signal is captured by the PLL and lock is established, the input signal can vary over a range of frequencies, and the VCO will track. This is called the lock range (f_{1L} to f_{2L}). Since the VCO frequency is constantly changing with input signals, the difference signal is small and does not go beyond the bandpass of the filter. The lock range is a function of the VCO frequency range and is wider than the capture range, as shown in Figure 15–22.

PLL INTEGRATED CIRCUITS

The integrated circuit has made it possible to use PLLs to perform many functions. PLL circuits have been around for years, but many of the functions that they are now used for would not have been cost-effective using discrete components. An example of a low cost general purpose PLL integrated circuit is the National Semiconductor LM565.

Figure 15–23 shows an LM565 and all necessary external components needed to make a functional PLL circuit. The phase detector is totally integrated and requires no external components. The filter utilizes an internal amplifier and resistor (3.6 kΩ) and requires only one external capacitor (C_2). The VCO is totally internal with the exception of R_1 and C_1, which set the free-running frequency of the VCO. The formulas for calculating the free-running frequency (f_o), the lock range (f_{Lock}), and the capture range ($f_{Capture}$) are listed in Figure 15–23. Example 15.8 shows how the operational parameters can be calculated if the component values are known.

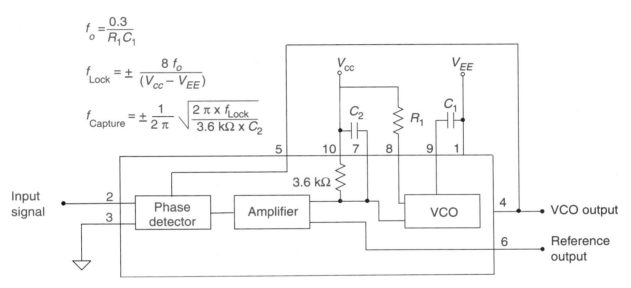

FIGURE 15–23 LM565 PLL circuit

EXAMPLE 15.8

Calculate the free-running frequency, the lock range, and the capture range for the PLL in Figure 15–23 where $R_1 = 33$ kΩ, $C_1 = 100$ pF, $C_2 = 0.01$ µF, and the power supplies are plus and minus 12 V.

Step 1. Calculate the free-running frequency.

$$f_o = \frac{0.3}{R_1 C_1} = \frac{0.3}{33 \text{ kΩ} \times 100 \text{ pF}} = 91 \text{ kHz}$$

Step 2. Calculate the lock range.

$$f_{Lock} = \pm \frac{8 f_o}{(V_{CC} - V_{EE})} = \pm \frac{8 \times 91 \text{ kHz}}{12 \text{ V} - (-12 \text{ V})} = \pm 30 \text{ kHz}$$

Step 3. Calculate the capture range.

$$f_{Capture} = \pm \frac{1}{2\pi} \sqrt{\frac{2\pi \times f_{Lock}}{3.6 \text{ kΩ} \times C_2}} = \pm \frac{1}{2\pi} \sqrt{\frac{2\pi \times 30 \text{ kHz}}{3.6 \text{ kΩ} \times 0.01 \text{ µF}}} = \pm 11 \text{ kHz}$$

Step 4. Draw a diagram showing the free-running frequency, the lock range, and the capture range. (Diagram is shown in Figure 15–24.)

EXAMPLE 15.8 continued

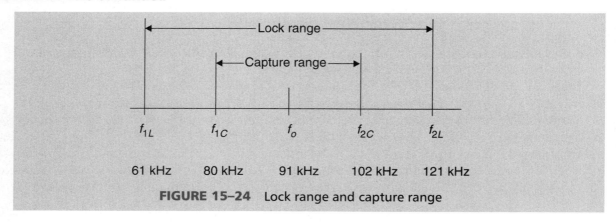

FIGURE 15–24   Lock range and capture range

15.7   PLL APPLICATIONS

The phase-locked loop, like so many other integrated circuits, finds many uses. A few of these include: FM demodulators, automatic frequency controls (AFC), frequency multiplier circuits, frequency synchronizing circuits, and frequency synthesis.

FM DEMODULATOR

Figure 15–25 shows a PLL being used as an FM demodulator. The function of the FM demodulator is to convert frequency deviations in the RF input signal into audio signals. The RF input signal is connected to the phase detector through a clipper circuit consisting of a resistor and two diodes. The clipper circuit removes the amplitude variations by limiting the signal swings to 0.7 V_{peak}. A clipper circuit preceding an FM demodulator is normally called a *limiter*.

The high frequency output signal (TP_2) of the PLL is fed back to the phase detector. The VCO is designed to operate at a free-running frequency near the RF input signal frequency so the PLL will lock on the RF input. Since the PLL is in lock, as the frequency of the RF input signal changes, the error signal will change. The error signal is capacitively coupled to an audio amplifier that

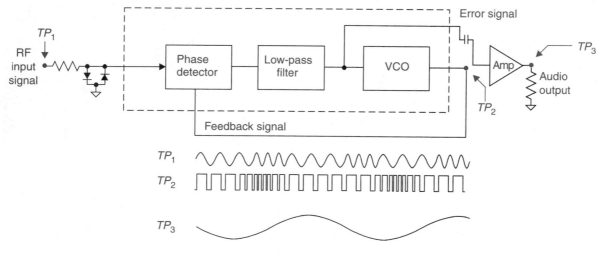

FIGURE 15–25   FM demodulator

drives the load. If the RF signal was originally modulated with a sine wave, the audio output will be a sine wave, as shown (TP_3) in Figure 15–25.

FREQUENCY MULTIPLIER CIRCUITS

There are two ways to use a PLL circuit to obtain frequency multiplication. One is to cause the output of the PLL to operate as a harmonic of the input to the PLL. This is accomplished by inputting a square wave, which is rich in odd harmonics, and setting the free-running frequency of the VCO to the approximate frequency of the desired harmonic. The phase detector will output an error signal within the capture range based on the mixing of the VCO feedback and the desired harmonic. Once locked, the output will be an exact multiple of the input frequency. Figure 15–26(a) shows a block diagram of a frequency multiplication circuit using the harmonic method.

A second method for obtaining frequency multiplication is shown in Figure 15–26(b). In this circuit, the output of the VCO is divided by four using a digital counter. Since both inputs to the phase detector are approximately in the 10 kHz range, capture will occur. Once locked, the output frequency will be exactly four times the input frequency. The digital counter can be set to divide by any desired integer, and any multiple of the input frequency is obtainable.

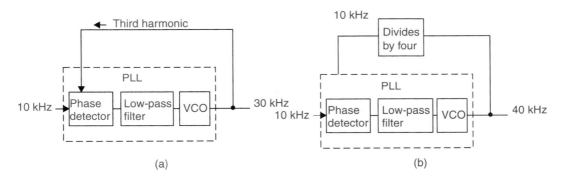

FIGURE 15–26 Frequency multiplier circuits

FREQUENCY SYNTHESIS

Frequency synthesis is a process by which many frequencies can be obtained from one frequency standard (usually a crystal oscillator). Figure 15–27 shows a block diagram of a frequency synthesis circuit. The crystal oscillator is divided by a digital counter, that can divide the crystal

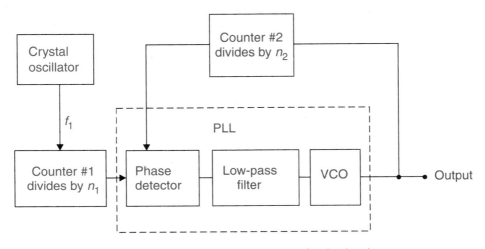

FIGURE 15–27 Frequency synthesis circuit

frequency by any number (n_1). The output of the VCO is also divided by a digital counter, and this counter can divide by any number (n_2).

When the PLL is operating in lock, the two inputs to the phase detector must be of the same frequency. Stated mathematically: $f_1/n_1 = f_{(out)}/n_2$, where f_1 equals the frequency of the crystal oscillator. Solving the formula for the VCO frequency gives: $f_{(out)} = (f_1/n_1) \times n_2$. The digital counters can be set to any number, so the output signal can be adjusted to any frequency by changing the digital counters. The frequency of the output will have the accuracy and stability of the crystal oscillator. This explains how forty-channel CBs with crystal accuracy can fit into such small packages. When you change channels, you are simply changing the values of n_1 or n_2, or both.

EXAMPLE 15.9

Calculate the output frequency for the frequency synthesis circuit shown in Figure 15–27 if the crystal is 4 MHz, counter n_1 is set for 800, and counter n_2 is set for 230.

$$f_{(out)} = (f_1/n_1) \times n_2 = (4 \text{ MHz}/800) \times 230 = 1150 \text{ kHz}$$

15.8   TROUBLESHOOTING ICs

In this chapter we have covered a few special ICs that perform certain applications. If your circuit contains an instrumentation amp, an OTA, an optoisolator, a VCO, or a PLL, you should have an understanding of the function of these circuits. There are numerous ICs on the market that perform these applications, and each IC has unique support circuitry and formulas; therefore, it is necessary to obtain information on the particular IC being used in the circuit.

Obtaining the needed information on ICs is often the most difficult task the technician has to perform when troubleshooting. If you are lucky, the technical manual will give the needed information. If the technical manual does not contain the needed information, the IC can be looked up by the manufacturer's number printed on the IC package. Each manufacturer has its own part number, so a cross-reference book like the *IC Masters* (also available on CD-ROM) must be used to find the manufacturer and the function of the IC. The data sheet normally can be obtained once the manufacturer is known. Some manufacturers have on-line services that permit you to download data sheets.

The advances of integrated circuit technology have made it possible for most applications to be integrated into a single chip. Since there are thousands of applications in electronics, it is not cost-effective to integrate every application into a unique IC. Therefore, general purpose VCOs and PLLs, and many other general purpose ICs, will continue to be used.

Differential Amplifier and PLL Circuit

1. Calculate the signal voltage gain and noise voltage gain for the circuit in Figure 15–28. Assume the op-amp has a CMRR of 90 dB and the resistors are exact values.

$Av_{(signal)} =$ _____ $Av_{(noise)} =$ _____

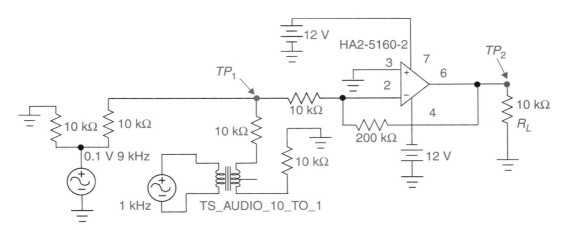

FIGURE 15–28

2. With both the signal and noise generators on, draw the waveforms at TP_1 and TP_2.

   TP_1 — — — — — — — — — —

   TP_2 — — — — — — — — — —

3. Simulate the circuit in Figure 15–28. Draw and label the waveforms at TP_1 and TP_2.

   TP_1 — — — — — — — — — —

   TP_2 — — — — — — — — — —

4. Calculate the signal voltage gain and noise voltage gain for the circuit in Figure 15–29. Assume the op-amp has a CMRR of 90 dB, the transformer has a 1:1 turns ratio, and the resistors are exact values.

   $Av_{(signal)} =$ _____ $Av_{(noise)} =$ _____

5. With both the signal and noise generators on, draw the waveforms at TP_1 and TP_2.

   TP_1 — — — — — — — — — —

   TP_2 — — — — — — — — — —

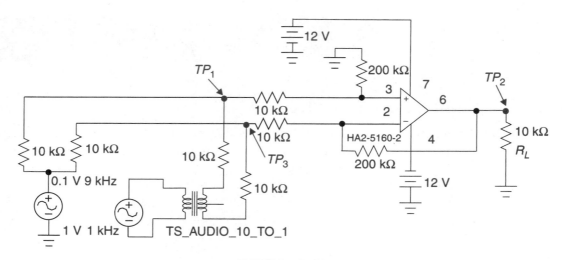

FIGURE 15–29

6. Simulate the circuit in Figure 15–29. Draw and label the waveforms at TP_1 and TP_2.

TP_1 — — — — — — — — — — —

TP_2 — — — — — — — — — — —

7. Calculate the free-running frequency, capture range, and lock range of the PLL circuit in Figure 15–30.

f_o = _____ f_{Lock} = _____ $f_{Capture}$ = _____

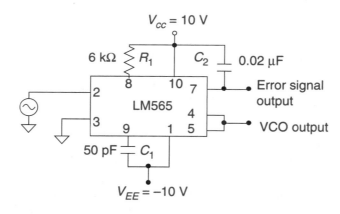

FIGURE 15–30

Differential Amplifier and PLL Circuit

I. Objectives

1. To experimentally examine the ability of a differential amplifier to cancel common-mode noise.
2. To experimentally analyze a PLL and determine the free-running frequency, capture range, and lock range of a PLL circuit.

II. Test Equipment

+/- 5 V power supplies and +/- 12 V power supplies

Dual-trace oscilloscope

Two function generators

Frequency counter

III. Components

(1) LM741

(1) 1:1 isolation transformer

Resistors: (2) 1 kΩ, (2) 2.2 kΩ, (3) 10 kΩ, (2) 100 kΩ

IV. Procedure

1. Calculate the signal voltage gain and noise voltage gain for the circuit in Figure 15–31. Assume the op-amp has a CMRR of 80 dB and the resistors are exact values.

    $A_{v(signal)} =$ _____ $A_{v(noise)} =$ _____

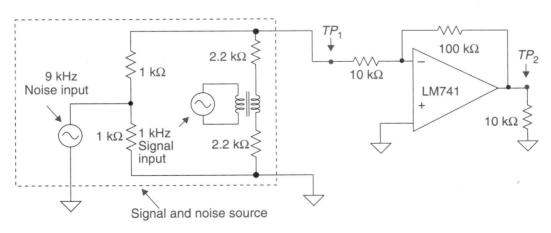

FIGURE 15–31

2. Construct the circuit in Figure 15–31. The circuitry inside the dotted line is the signal and noise source.
3. With the signal generator off, adjust the noise input at TP_1 to 200 mV$_{p-p}$.
4. Measure the output at TP_2 and calculate the noise voltage gain of the circuit.

    $A_{v(noise)} =$ _____

5. With the noise generator off, adjust the signal input at TP_1 to 200 mV$_{p-p}$.

6. Measure the output at TP_2 and calculate the signal gain of the circuit.

    $A_{v(signal)}$ = _____

7. Using the values determined in steps 4 and 6, calculate the *CMRR* of the circuit.

    CMRR = _____

8. With both the signal and noise generators on, measure and draw the waveforms at TP_1 and TP_2.

    TP_1 — — — — — — — — — — —

    TP_2 — — — — — — — — — — —

9. Calculate the signal voltage gain and noise voltage gain for the circuit in Figure 15–32. Assume the op-amp has a CMRR of 80 dB and the resistors are exact values.

    $A_{v(signal)}$ = _____ $A_{v(noise)}$ = _____

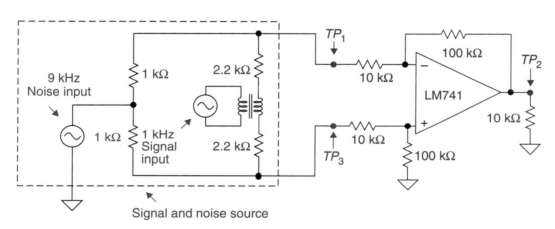

FIGURE 15–32

10. Construct the circuit in Figure 15–32. The circuitry inside the dotted line is the signal and noise source.

11. With the signal generator off, adjust the noise input at TP_1 to 200 mVp-p. The noise input at T_3 will also measure 200 mV.

12. Measure the output at TP_2 and calculate the noise voltage gain of the circuit.

    $A_{v(noise)}$ = _____

13. With the noise generator off, adjust the differential input signal (signal between TP_1 and TP_3) to 200 mV$_{p-p}$. A differential reading must be made by using the oscilloscope. Channel 1 and channel 2 are adjusted to the same scale (50 mV/cm), the vertical mode is set to "both" and "add," and the invert control of channel 2 is activated.

14. Measure the output at TP_2 and calculate the signal gain of the circuit.

    $A_{v(signal)}$ = _____

15. Using the values determined in steps 12 and 14, calculate the *CMRR* of the circuit.

    CMRR = _____

16. With both the signal and noise generators on, measure and draw the waveforms at TP_1 and TP_2.

TP_1 — — — — — — — — — — —

TP_2 — — — — — — — — — — —

17. Calculate the free-running frequency, capture range, and lock range of the PLL circuit in Figure 15–33.

$f_o = $ _____ $f_{Lock} = $ _____ $f_{Capture} = $ _____

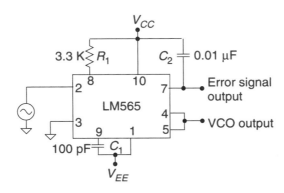

FIGURE 15–33

18. Construct the circuit in Figure 15–33.

19. With no input signal, measure the free-running frequency of the VCO and the voltage amplitude of the error signal.

$f_o = $ _____ $V_{(Error\ signal)} = $ _____

20. Measure the low capture frequency (f_{1C}). Starting with a low input frequency, increase the input frequency until capture occurs. Record the low frequency capture (f_{1C}). The easy way to detect capture is to monitor the error signal. The error signal reads a constant voltage when the VCO is operating at its free-running frequency. When capture occurs, the error signal voltage will switch to a new DC level.

(f_{1C}) = _____

21. Measure the high capture frequency (f_{2C}). Starting with a high input frequency, decrease the input frequency until capture occurs. Record the high frequency capture (f_{2C}).

(f_{2C}) = _____

22. Calculate the capture range using the measured values from steps 4 and 5.

$f_{Capture} = $ _____

23. Measure the low lock dropout frequency. Adjust the input frequency so the PLL is in the lock range, and decrease the input frequency until lock is lost. Record the low lock dropout frequency. Monitor the error signal. When the PLL drops out of lock, the error signal voltage will switch to a new DC level.

(f_{1L}) = _____

24. Measure the high lock dropout frequency. Adjust the input frequency so the PLL is in the lock range and increase the input frequency until lock is lost. Record the high lock dropout frequency.

(f_{2l}) = _____

25. Calculate the capture range using the measured values from steps 7 and 8.

f_{Lock} = _____

V. Points to Discuss

1. Did both circuits have the same signal voltage gain? Explain.

2. Did both circuits have the same noise gain? Explain.

3. What is the ratio expressed in decibels of signal to noise level at the output of the two circuits?

4. Explain the difference in the signal to noise level at the output of the two circuits in this experiment.

5. Explain why twisted pair cabling is often used when connecting the signal to the input of a differential amplifier.

6. When the PLL in Figure 15–33 is operating in lock, the DC level of the error signal varies with changes in input frequency. However, when the circuit is not in lock, the DC level of the error signal stays constant. Why?

7. If a divide by four counter is connected between pin 4 and pin 5 of the PLL in Figure 15–33, what frequency range at the input of the PLL would cause capture? Explain.

QUESTIONS

15.1 Differential Amplifiers

____ 1. The input signal to a differential amplifier is connected between _____.

   a. the inverting input and ground
   b. the noninverting input and ground
   c. the noninverting and the inverting inputs
   d. the inverting input and the reference pin

____ 2. The formula _____ is used to calculate the voltage gain of a differential amplifier.

   a. $A_{v(\text{diff})} = R_f / R_i$
   b. $A_{v(\text{diff})} = 2R_f / R_i$
   c. $A_{v(\text{diff})} = (R_f / R_i) + 1$
   d. $A_{v(\text{diff})} = gm \times R_L$

____ 3. A signal connected to both inputs of a differential amplifier is called a_____.

   a. differential input signal
   b. common-mode signal
   c. biased input signal
   d. double-coupled signal

____ 4. The ideal differential amplifier has a common-mode voltage gain of _____.

   a. zero
   b. 100
   c. 100,000
   d. infinity

____ 5. The common-mode rejection ratio is the ratio between the ____.

   a. signal and the noise
   b. ideal voltage gain and the actual voltage gain
   c. differential voltage gain and the common-mode voltage gain
   d. none of the above

____ 6. What is a major advantage of a differential amplifier?

   a. the ability to reject common-mode signals
   b. a high ratio of active to passive
   c. the low number of transistors used in construction
   d. the use of dual voltage sources

15.2 Instrumentation Amplifiers

____ 7. How many op-amps does it take to construct an instrumentation amplifier?

   a. 1
   b. 2
   c. 3
   d. 4

_____ 8. What is an advantage of an instrumentation amplifier over a simple differential amplifier?

    a. It has a higher input impedance.

    b. It has a lower output impedance.

    c. The voltage gain can be easily adjusted.

    d. Both a and c.

15.3 Operational Transconductance Amplifiers (OTAs)

_____ 9. The schematic symbol for an operational transconductance amplifier (OTA) is _____.

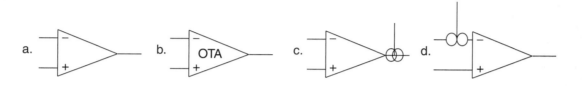

_____ 10. The gain parameter of an OTA is a ratio of output current to input voltage, and the gain is measured in _____.

    a. volts

    b. amps

    c. ohms

    d. transconductance

_____ 11. The voltage gain of an OTA can be calculated using the formula _____.

    a. $A_v = gm \times R_L$

    b. $A_v = R_f/R_i$

    c. $A_v = (R_f/R_i) + 1$

    d. $A_v = 2R_f/R_i$

_____ 12. If an OTA is used as a nonlinear mixer and an audio signal is mixed with an RF signal, the output will be _____ signal.

    a. a square-wave

    b. a triangular wave

    c. an amplitude modulated (AM)

    d. a frequency modulated (FM)

15.4 Optoisolators

_____ 13. Optoisolators are integrated packages that incorporate _____, a photodiode, and a transistor enclosed in a single package.

    a. an op-amp

    b. an OTA

    c. a LED

    d. a JFET

____ 14. The _____ is the ratio between the current flowing through the collector of the transistor and the current flowing through the LED.

    a. current transfer ratio

    b. optional gain

    c. beta of the optoisolator

    d. transconductance

15.5 Voltage Controlled Oscillators (VCOs)

____ 15. A voltage controlled oscillator (VCO) is a circuit in which the output frequency is a function of _____.

    a. the power supply voltage

    b. the alpha voltage

    c. the input voltage

    d. the output voltage

____ 16. The _____ is a general purpose VCO IC.

    a. LM318

    b. LM324

    c. LM565

    d. LM566

____ 17. If a VCO is designed to operate in the RF range and the input of the VCO is connected to an audio signal, the output will be _____ signal.

    a. a square-wave

    b. a triangular wave

    c. an amplitude modulated (AM)

    d. a frequency modulated (FM)

15.6 Phase-Locked Loop (PLL)

____ 18. The phase-locked loop (PLL) consists of three subcircuits: a phase detector, a low-pass filter, and _____.

    a. a voltage controlled oscillator

    b. a differential amplifier

    c. an OTA

    d. a modulator

____ 19. The phase detector in a PLL is followed by a low-pass filter. The low-pass filter passes the _____ and rejects all other frequencies.

    a. input signal

    b. feedback signal

    c. sum of the input and feedback signals

    d. difference frequency of the input and feedback signals

_____ 20. A PLL circuit will obtain lock if the frequency of the input signal enters the _____.
- a. lock range
- b. capture range
- c. bandwidth of the PLL
- d. none of the above

_____ 21. Once locked, a PLL circuit will maintain lock if the frequency of the input signal does not go beyond the _____.
- a. lock range
- b. capture range
- c. bandwidth of the PLL
- d. none of the above

_____ 22. The _____ is a general purpose PLL IC.
- a. LM318
- b. LM324
- c. LM565
- d. LM566

15.7 PLL Applications

_____ 23. PLLs are used in many applications. Which of the following applications does not use a PLL?
- a. FM demodulating
- b. frequency multiplying
- c. frequency synthesis
- d. voltage regulation

_____ 24. The function of the FM demodulator is to convert the frequency deviations in the RF input signal into _____ output signal.
- a. a square-wave
- b. a triangular wave
- c. an audio
- d. an AM

_____ 25. A PLL circuit can be used for frequency multiplication by _____.
- a. designing the circuit to lock to a harmonic of the input
- b. connecting a digital counter in the feedback loop
- c. connecting a summer circuit to the output
- d. both a and b

_____ 26. Frequency _____ is a process by which many frequencies can be obtained from one frequency standard.
- a. multiplying
- b. generation
- c. doubling
- d. synthesis

15.8 Troubleshooting ICs

____ 27. The signal at TP_1 in Figure 15–34 measures 1 mV$_{p-p}$. What is wrong with the circuit?
  a. Op-amp #1 is defective.
  b. R_4 is open.
  c. R_3 is shorted.
  d. The circuit is operating properly.

____ 28. The signal at TP_1 in Figure 15–34 measures 4 V$_{p-p}$. What is wrong with the circuit?
  a. Op-amp #1 is defective.
  b. R_4 is open.
  c. R_3 is shorted.
  d. The circuit is operating properly.

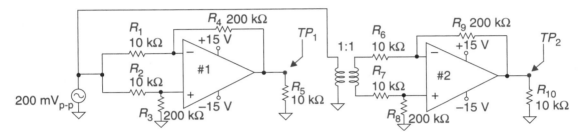

FIGURE 15–34

____ 29. The signal at TP_2 in Figure 15–34 continuously switches between 14 V and −14 V. What is wrong with the circuit?
  a. Op-amp #1 is defective.
  b. R_9 is open.
  c. R_8 is shorted.
  d. The circuit is operating properly.

____ 30. The signal at TP_2 in Figure 15–34 measures 4 V$_{p-p}$. What is wrong with the circuit?
  a. Op-amp #1 is defective.
  b. R_9 is open.
  c. R_8 is shorted.
  d. The circuit is operating properly.

PROBLEMS

1. Calculate the output and draw the output waveforms for the circuit shown in Figure 15–35. (Assume an ideal op-amp.)

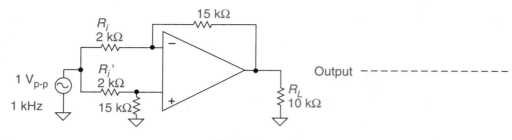

FIGURE 15–35

2. The op-amp in Figure 15–35 has a CMRR equal to 85 dB. Calculate the output and draw
    the output waveforms for the circuit.

3. The op-amp in Figure 15–36 has a CMRR equal to 85 dB. Calculate the output and draw
    the output waveforms for the circuit.

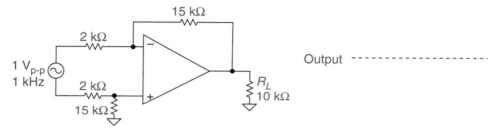

FIGURE 15–36

4. The circuit in Figure 15–37 shows a signal source 500 ft from an amplifier. The 500 ft
    line picks up 100 mV$_{p-p}$ of noise. Calculate the signal output and noise output for the
    circuit.

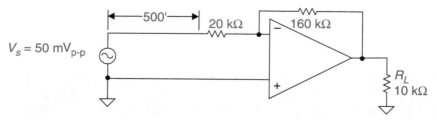

FIGURE 15–37

5. The circuit in Figure 15–38 shows a signal source 500 ft from an amplifier. The 500 ft line picks up 100 mV$_{p-p}$ of common noise. Calculate the signal output and noise output for the circuit. (Assume the circuit has a CMRR = 86 dB.)

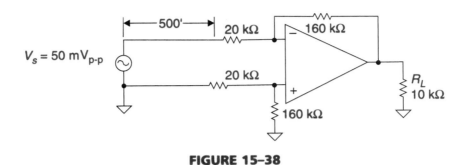

FIGURE 15–38

6. Calculate the output waveforms for both circuits in Figure 15–39. (Assume the op-amp has a *CMRR* = 74 dB.)

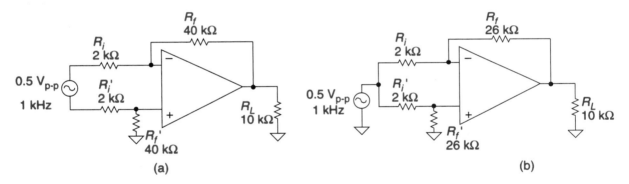

FIGURE 15–39

7. A differential amplifier has a CMRR of 94 dB. If the gain of the amplifier is 1000, the input signal is 8 mV, and the common-mode noise is 35 mV, what is the signal out and the noise out?

8. Calculate the total voltage gain and the circuit input impedance for a circuit like the one in Figure 15–40 using the following components: R_i = 4 kΩ, R_f and R_f' = 40 kΩ, R_i^* and $R_i'^*$ = 5 kΩ, and R_f^* and $R_f'^*$ = 35 kΩ.

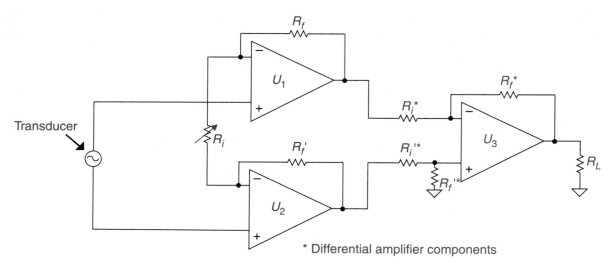

FIGURE 15–40

9. Show the output waveforms for the circuit in Figure 15–41. (All op-amps are using +15 V and –15 V supplies.)

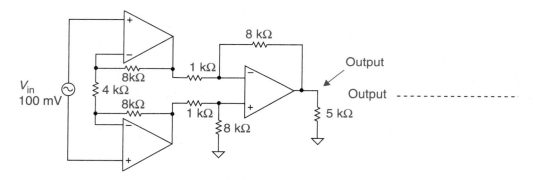

FIGURE 15–41

10. Show the output waveforms for the circuit in Figure 15–42. (All op-amps are using +15 V and –15 V supplies.)

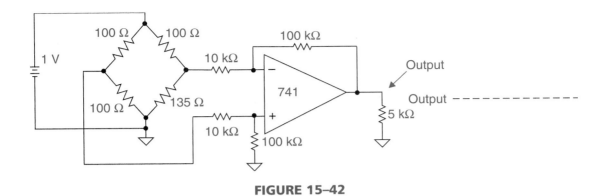

FIGURE 15–42

11. If the OTA constant is 14, calculate the voltage gain for the circuit in Figure 15–43(a).

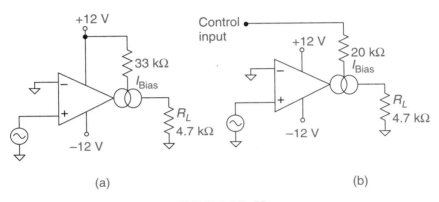

(a)   (b)

FIGURE 15–43

12. Calculate the voltage gain for the circuit in Figure 15–43(b) with a control input of 4 V and with a control input of 7 V. The OTA constant is 12.

13. The optoisolator in Figure 15–44(a) has a current transfer ratio of 200%. Calculate the current through meter M_1 and the current through meter M_2.

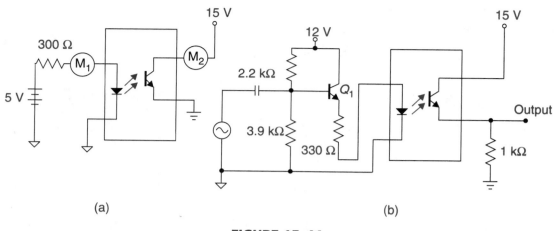

FIGURE 15–44

14. The optoisolator in Figure 15–44(b) has a current transfer ratio of 50%. Calculate the DC voltage at the output with a signal input of zero.

15. The optoisolator in Figure 15–44(b) has a current transfer ratio of 50%. Calculate and draw the signal output if the input signal is a 2 V_{p-p} sine wave.

Output   ------------------------------------.

16. Calculate the output frequency for the circuit in Figure 15–45 for an input voltage of 12 V and for an input voltage of 14 V.

17. Draw the output wave shapes for the signals at output 1 and output 2 for the circuit in Figure 15–45.

    Output 1 — — — — — — — — — —

    Output 2 — — — — — — — — — —

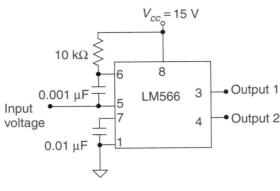

FIGURE 15–45

18. Calculate the free-running frequency, the lock range, and the capture range for the PLL in Figure 15–46.

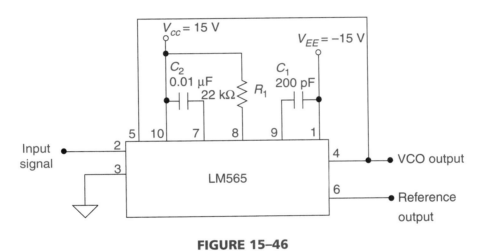

FIGURE 15–46

19. Draw and label a diagram showing how the lock range and the capture range calculated in problem 18 relate to each other.

20. What is the output frequency of the circuit in Figure 15–46 if the frequency of the input signal is 75 kHz?

21. What is the output frequency of the circuit in Figure 15–46 if the frequency of the input signal is 45 kHz?

22. Calculate the output frequency for the frequency synthesis circuit shown in Figure 15–47.

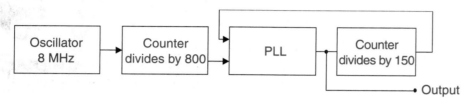

FIGURE 15–47

CHAPTER 16

Power Circuits: Switching and Amplifying

OBJECTIVES

After studying the material in this chapter, you will be able to describe and/or analyze:

- ⭘ the efficiency of electronic circuits,
- ⭘ the advantages of power MOSFETs over power bipolar transistors,
- ⭘ power switching circuits,
- ⭘ the classes of amplifiers,
- ⭘ class-C power amplifiers,
- ⭘ class-B power amplifiers,
- ⭘ integrated power amplifiers,
- ⭘ class-D power amplifiers,
- ⭘ why power devices must utilize heat sinks, and
- ⭘ troubleshooting procedures for power circuits.

16.1 INTRODUCTION TO POWER CIRCUITS

In this chapter, we will examine power circuits that can deliver large quantities of power to electrical loads. We will discuss the efficiency of power circuits, the characteristics of power devices (bipolar transistor, FET, or IC), and the need for heat sinking.

EFFICIENCY

Efficiency is a measurement of how effective a circuit is in delivering signal power to the load. In low power circuitry, the need for efficiency is not as important as in high power circuitry where efficiency becomes extremely important. Efficiency is calculated by dividing the power delivered to the load by the total power delivered to the circuitry. Efficiency is expressed in percentages, so the ratio is multiplied by 100%. The efficiency formula is:

$$Efficiency = \frac{Load\ power}{Total\ power} \times 100\%$$

Figure 16–1 shows block diagrams of an amplifier and a switching circuit. The total power dissipated is equal to the power dissipated by the circuit plus the power dissipated by the load. Examples 16.1 and 16.2 demonstrate this concept.

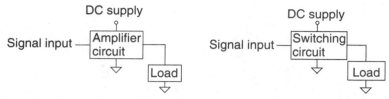

P_{total} = Power loss in amplifier + Load power P_{total} = Power loss in switch + Load power

FIGURE 16–1   Amplifier and switching circuits

EXAMPLE 16.1

Calculate the efficiency of an amplifier system if the total power dissipated by the system is 6 W and the power delivered to the load is 1 W.

$$Efficiency = Load\ power\,/\,Total\ power \times 100\%$$
$$Efficiency = 1\ \text{W}/6\ \text{W} \times 100\% = 16.7\%$$

EXAMPLE 16.2

Calculate the efficiency of a switching system if the total power dissipated by the system is 6 W and the power delivered to the load is 5.8 W.

$$Efficiency = Load\ power\,/\,Total\ power \times 100\%$$
$$Efficiency = 5.8\ \text{W}/6\ \text{W} \times 100\% = 97\%$$

The examples suggest that switching circuits are considerably more efficient than amplifier circuits. It is true that circuits that operate as switches tend to be much more efficient than circuits that operate as amplifiers. Examples 16.3 and 16.4 calculate the efficiencies of a transistor amplifier circuit and a transistor switching circuit.

EXAMPLE 16.3

Calculate the efficiency of the amplifier circuit in Figure 16–2.
Step 1. Calculate the voltage gain of the amplifier.
$$A_v = r_c/r_e = (R_c \parallel R_L)/(R_e + r'_e)$$
$$A_v = 2.1\ \text{k}\Omega/180\ \Omega = 11.7\ (r'_e \text{ is considered negligible.})$$
Step 2. Calculate the RMS signal output voltage.
$$V_{out} = V_{in} \times A_v = 0.5\ \text{V}_{\text{p-p}} \times 11.7 = 5.8\ \text{V}_{\text{p-p}}$$
$$V_{out} = 2.9\ \text{V}_{\text{p}}$$
$$V_{out} = 2\ \text{V}_{\text{RMS}}$$
Step 3. Calculate the signal power delivered to the load.
$$P_{\text{Load}} = (V_{\text{out(RMS)}})^2/R_L = (2\ \text{V}_{\text{RMS}})^2/10\ \text{k}\Omega = 0.4\ \text{mW}$$

EXAMPLE 16.3 continued

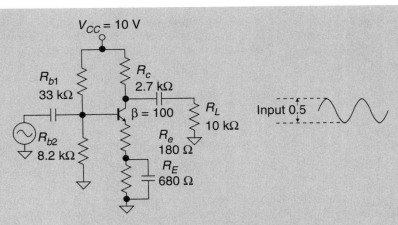

FIGURE 16–2 Transistor amplifier circuit

Step 4. Calculate the quiescent DC power supply current. The total power dissipated is equal to the DC power plus the signal input power. However, the signal input power is normally so small in comparison to the DC power that it can be disregarded.

$$I_{DC} = I_{CQ} + I_{BR1Q} \approx I_{CQ}$$
$$V_B = V_{cc} \times R_{b2}/(R_{b1} + R_{b2}) = 10\text{ V} \times 8.2\text{ k}\Omega/(33\text{ k}\Omega + 8.3\text{ k}\Omega) = 2\text{ V}$$
$$V_E = V_B - 0.7\text{ V} = 2\text{ V} - 0.7\text{ V} = 1.3\text{ V}$$
$$I_{CQ} \approx I_E = V_E/(R_e + R_E) = 1.3\text{ V}/(180\ \Omega + 680\ \Omega) = 1.5\text{ mA}$$
$$P_{\text{total}} \approx P_{DC} \approx V_{cc} \times I_{CQ} = 10\text{ V} \times 1.5\text{ mA} = 15\text{ mW}$$

Step 5. Calculate the efficiency of the amplifier.

$$Efficiency = P_{\text{Load}}/P_{\text{total}} \times 100\%$$
$$Efficiency = 0.4\text{ mW}/15\text{ mW} \times 100\% = 2.7\%$$

EXAMPLE 16.4

Calculate the efficiency of the switching circuit in Figure 16–3 when the transistor is conducting. When the transistor is not conducting, there is no current flow in the circuit; therefore, no power is dissipated in the circuit or the load.

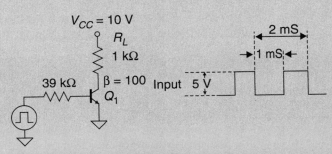

FIGURE 16–3 Transistor switching circuit

EXAMPLE 16.4 continued

> **Step 1.** Calculate the voltage across the load when the transistor is conducting. The voltage across the load is equal to the supply voltage minus the voltage dropped across the transistor. A bipolar transistor in saturation normally drops approximately 0.2 V.
>
> $$V_L = V_{CC} - V_{CE}$$
> $$V_L = 10 \text{ V} - 0.2 \text{ V} = 9.8 \text{ V}$$
>
> **Step 2.** Calculate the load current. The load current will equal the collector current. Since the transistor is operating in the switching mode, the collector current will equal the saturation current (I_{CSat}).
>
> $$I_{CSat} = V_L/R_L$$
> $$I_{CSat} = 9.8 \text{ V}/1 \text{ k}\Omega = 9.8 \text{ mA}$$
>
> **Step 3.** Calculate the power delivered to the load.
>
> $$P_L = I_L \times V_L$$
> $$P_L = 9.8 \text{ mA} \times 9.8 \text{ V} = 96 \text{ mW}$$
>
> **Step 4.** Calculate the power dissipated by the transistor.
>
> $$P_{Q1} = I_C \times V_{CE}$$
> $$P_{Q1} = 9.8 \text{ mA} \times 0.2 \text{ V} = 2 \text{ mW}$$
>
> **Step 5.** Calculate the efficiency of the switching circuit.
>
> $$Efficiency = P_L/(P_L + P_{Q1}) \times 100\%$$
> $$Efficiency = 96 \text{ mW}/(96 \text{ mW} + 2 \text{ mW}) \times 100\% = 98\%$$

Example 16.3 demonstrates that the efficiency of the linear amplifier circuit is extremely low (2.7%). In low power applications, the low efficiency of linear amplifiers is not a major concern. The total power dissipated by the amplifier in Example 16.3 was only 15 mW. The fact that most of the power was dissipated in the circuitry, not in the load, did not matter because 15 mW does not cause excessive circuit heating. However, if a 100 W output is needed, an amplifier with an efficiency of 2.7% is going to cause major problems. With a 2.7% efficiency, the power dissipated by the total circuitry would be 3070 W. The load would receive 100 W, and the rest of the circuitry would have to dissipate 2970 W. The heat generated by 2970 W would surely destroy the circuitry. The solution is to use a more efficient amplifier circuit, which will be covered in later sections of this chapter.

Example 16.4 demonstrates that switching circuits are extremely efficient. This high efficiency of switching circuits has resulted in engineers designing many power applications using switching mode technology. Under high current conditions, the voltage drop across the switching device increases, causing the power dissipated by the circuit to increase. Manufacturers have developed a variety of switching devices to improve circuit performance, so selecting the proper switching device is a major factor in designing good power switching circuits. In the next section, we will examine switching devices.

16.2 POWER MOSFETS VERSUS POWER BIPOLAR TRANSISTORS

The power bipolar transistor has been around for decades, but advances in power MOSFETs over the last decade are making it the device of choice for many power applications. When MOSFETs were first developed, they were capable of handling only low currents; therefore, the first MOSFETs were low power devices. The cross-sectional area of the channel had to be small in order to permit the gate voltage to control channel current flow.

In recent years, a new group of power MOSFETs has been designed. The advancements of power MOSFETs were stimulated by the power supply industry working to improve MOSFETs for use in switching power supplies. These improved MOSFETs are also finding use in the linear power amp field. Power MOSFETs are enhancement MOS devices that utilize a vertical internal construction that permits a wider channel and greater current capability. There are many different designs used to construct power MOSFETs, and each manufacturer names its own designs. The technician will encounter power MOSFETs called VMOS, HEXMOS, TMOS, COOLFETs, and others.

Power MOSFETs have several advantages over bipolar power transistors. Power MOSFETs require less signal power from the driver stage, have the ability to switch off and on more quickly than bipolar transistors, and can be easily connected in parallel to increase their current capacity. The bipolar transistor still has the advantage of a lower voltage drop across the device under high voltage and current conditions.

Power bipolar transistors have a much lower beta than low power bipolar transistors. When troubleshooting, the technician should assume the power bipolar to have a beta of 25 (compared with 100 for low power bipolar transistors). The low beta means the stage driving the power transistor will need considerable current drive to handle the needed base current. For example, a transistor delivering 10 A of load current would require 400 mA of base current. A bipolar Darlington pair can be used to raise the low beta and decrease the needed base current. MOSFETs are voltage driven devices; therefore, the driving stage must deliver the proper voltage to the gate circuitry, but the current requirements are near zero. Power MOSFETs can be driven directly from a CMOS logic circuit or from an open collector TTL logic circuit.

Switching speed is of great importance when circuits are operating at a high frequency. The output waveform of a high speed power MOSFET switching circuit is shown in Figure 16–4(a), and the waveform of a high speed power bipolar transistor is shown in Figure 16–4(b). A device in the off or on state has high efficiency, but its efficiency is greatly reduced when the device is functioning in the linear region. Figure 16–4 shows that the power MOSFET spends much less time in the linear operating region; hence, power MOSFETs tend to be more efficient at high switching speeds.

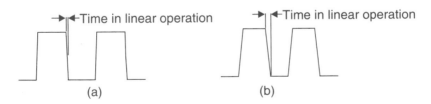

FIGURE 16–4 Switching waveforms

Another advantage of power MOSFETs is the ability to connect them in parallel to increase their current capacity without concern about current hogging. Current hogging is always a concern when bipolars are connected in parallel. The channel in power MOSFETs has a positive coefficient of heat. As the temperature of the MOSFET increases, the resistance of the channel increases. This means power MOSFETs are self-regulating when connected in parallel. If one MOSFET starts to draw more current, its temperature rises, causing its resistance to rise. An increase in resistance causes a decrease in current, bringing the current back in balance with the other parallel MOSFETs. Figure 16–5(a) shows how two MOSFETs can be connected in parallel.

The bipolar transistor has a negative coefficient of heat. As the temperature of the bipolar transistor increases, the resistance between the collector and emitter decreases. If one bipolar

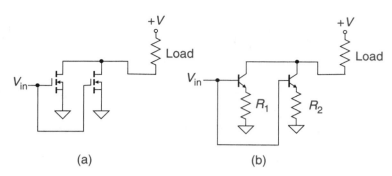

(a)   (b)

FIGURE 16–5   Parallel devices

transistor starts to draw more current, its temperature rises, causing its resistance to fall. A decrease in resistance causes an increase in current. The increased current causes further increases in temperature and further increases in current. The transistor is now in thermal runaway and hogs the current flow. Figure 16–5(b) shows two bipolar transistors connected in parallel. Resistors (R_1 and R_2) can be connected in series with the transistors. The resistors help to maintain a balance in resistance, thus helping to prevent current hogging. In high power circuits, the power lost in the series resistors is considerable, and their ability to eliminate current hogging is not 100% effective. Operating bipolar transistors in parallel is difficult, but operating power MOSFETs in parallel is not difficult.

An advantage the bipolar transistor has over the power MOSFET is a lower voltage drop across the device in the on state under high current conditions. The voltage across a saturated bipolar transistor is approximately 0.2 V. This voltage tends to stay relatively constant, increasing only slightly (may reach over 1 V) with increases in collector current. The voltage across a saturated power MOSFET is equal to $R_{DS(on)} \times I_D$. Power MOSFETs designed with breakdown voltages of less than 200 V have relatively low values for $R_{DS(on)}$. As the breakdown voltage rating of a MOSFET increases, so does the value of $R_{DS(on)}$. If the value of $R_{DS(on)}$ is large, the voltage across the power MOSFET can exceed the voltage across the bipolar transistor. Since the power dissipated in a device is equal to the current through the device times the voltage across the device, it is indeed possible that the bipolar power transistor could be more efficient than the power MOSFET when the device is functioning as a closed switch. This is especially true when the switching circuit is designed to control voltage levels above 200 V.

The insulated gate bipolar transistor (IGBT) was developed in the mid-1980s. The device combines the advantages of the power MOSFET and power bipolar transistor into one component. The input terminal is called a gate and functions like the gate circuit of a normal power MOSFET. The other two terminals are called the emitter and the collector. The device has the high input impedance of a power MOSFET and a low collector-to-emitter voltage drop similar to that of the power bipolar transistor. Figure 16–6 shows the schematic symbol of the IGBT.

The power MOSFET is the device of choice when voltage levels are below 200 V, while the IGBT is becoming the device of choice for high voltage switching (200 V to 1500 V). The power

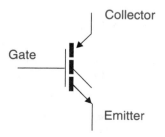

FIGURE 16–6   Insulated gate bipolar transistor

bipolar transistor has been around since the 1950s and is still used in thousands of circuits, but it is gradually becoming obsolete.

16.3 POWER SWITCHING CIRCUITS

When a transistor (bipolar or MOSFET) operates as a switch, it is operating in class-D. Class-D is the most efficient class of operation. When the ideal switching device is off, there is no power dissipated in the device because there is no current flowing through the device. When the ideal switching device is on, there is no power dissipated in the device because the voltage across the device is zero.

Both circuits in Figure 16–7 are designed to control power to a 10 Ω load connected to a 100 V source. When on, the load will draw 10 A of current. The control voltage is a 12 V source. It can be connected directly to the gate of the power MOSFET; however, when a power bipolar transistor is used, a Darlington pair is needed to decrease loading of the control source. As Example 16.5 shows, even with the added buffering circuitry, the bipolar circuit still requires more current drive.

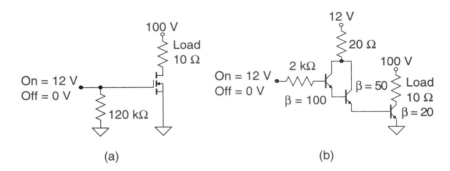

FIGURE 16–7   MOSFET switch versus bipolar switch

EXAMPLE 16.5

Calculate the drive current needed to turn on each switching circuit shown in Figure 16–7.

Step 1. Calculate the drive current needed to turn on the switching circuit in Figure 16–7(a).

$$I = V_{on}/R_G = 12 \text{ V}/120 \text{ k}\Omega = 0.1 \text{ mA}$$

Step 2. Calculate the drive current needed to turn on the switching circuit in Figure 16–7(b). The voltage on the base of the first transistor will be three junction drops above ground (2.1 V).

$$I = (V_{on} - 2.1 \text{ V})/R_B = (12 \text{ V} - 2.1 \text{ V})/2 \text{ k}\Omega = 4.95 \text{ mA}$$

Power bipolar transistors require considerable current drive and cannot be driven directly from standard logic circuits. Power MOSFETs, however, can be directly controlled by logic circuits. Figure 16–8(a) shows a CMOS inverter driving a power MOSFET directly. The voltage at the output of the inverter will switch from 0 V to 12 V. The 0 V level is below the threshold voltage ($V_{GS(th)}$); therefore, the MOSFET will be in the off condition. The 12 V level is considerably above the threshold voltage and will drive the MOSFET into saturation.

Logic circuits using TTL with totem pole outputs have a logic one output of less than 5 V. Driving a power MOSFET directly from TTL is not recommended because MOSFETs may not be driven into saturation. An open collector TTL gate can be used to drive the MOSFET by connecting the pull-up resistor to a voltage high enough to ensure saturation, as shown in Figure 16–8(b).

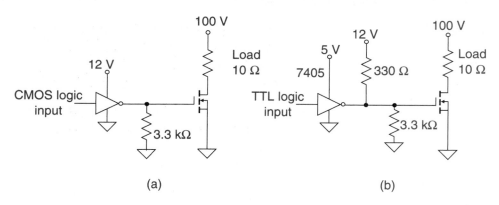

FIGURE 16–8   Logic circuit driving a power MOSFET

To simplify switching circuits, manufacturers have developed a wide variety of switching ICs. Figure 16–9 shows a National Semiconductor LM1921, which is a 1 A industrial switch popular in automotive applications. The IC is designed to permit the switch to operate on the high side of the load. This permits one side of the load to be connected to ground and the other side of the load to be connected to a positive supply voltage through the switch. The IC uses a PNP power bipolar transistor that is controlled by a comparator and an NPN transistor. If the control input (on/off) exceeds 1.2 V, the positive output of the comparator will turn on the NPN transistor, thus supplying the needed base current to turn on the PNP output transistor. If the control input is less than 1.2 V, the low output of the comparator will cut off the NPN transistor, which in turn cuts off the output transistor. The circuit has built-in overvoltage and overtemperature protection. The LM1921 requires only 30 μA of control current, making it compatible with TTL and CMOS logic circuits.

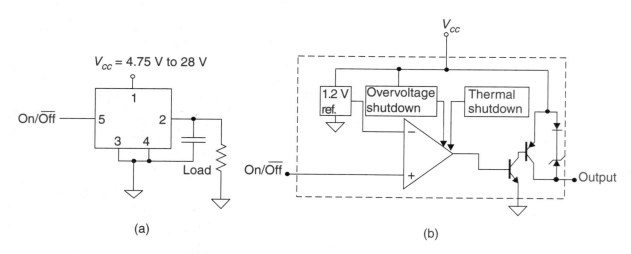

FIGURE 16–9   Industrial switch (LM1921)

16.4 CLASSES OF AMPLIFIERS

Amplifier circuits are classified according to what percentage of the input signal is amplified by the circuit. Figure 16–10 shows three amplifier circuits. Each circuit has the same sine wave input signal; however, the output current flow is different for each circuit. The class-A amplifier circuit has output current flowing for the total input cycle or 360°. The class-B amplifier circuit has output current flowing for one-half of the input cycle or 180°. The class-C amplifier circuit has output current flowing for less than one-half of the input cycle or less than 180°.

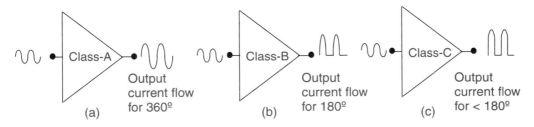

Class-A
Output
current flow
for 360°
(a)

Class-B
Output
current flow
for 180°
(b)

Class-C
Output
current flow
for < 180°
(c)

FIGURE 16–10 Classes of amplifier operations

Class-A amplifiers produce an output signal that is a true reproduction of the input signal, but the efficiency of the class-A amplifier is low. Class-B amplifiers give a true reproduction of only one-half of the input signal, but the efficiency of the class-B amplifier is considerably higher than that of the class-A amplifier. The class-C amplifier produces only short current pulses at the output, so the output is greatly distorted; however, the efficiency of the class-C amplifier is extremely high.

CLASS-A OPERATION

Figure 16–11 shows a class-A common-collector amplifier (emitter follower). This circuit does provide power gain, but it has serious shortcomings for systems where the load requires more than a few tenths of a watt. The main limitation is in the efficiency of class-A operation. Example 16.6 shows that the maximum efficiency that can be achieved using a class-A amplifier with a resistive load is 25%.

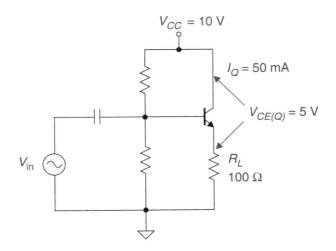

$V_{CC} = 10$ V

$I_Q = 50$ mA

$V_{CE(Q)} = 5$ V

V_{in}

R_L
100 Ω

FIGURE 16–11 Class-A amplifier at quiescence

EXAMPLE 16.6

For the circuit in Figure 16–11, calculate the supply input power, the maximum signal output power, and the maximum efficiency.

Step 1. The input power to an amplifier is equal to the power delivered by the DC power supply. In a class-A amplifier, since the current will vary symmetrically around the Q-point, the average current is equal to the quiescent current flow.

$$P_{\text{in}} = P_{DC} = I_{c(Q)} \times V_{cc} = 50 \text{ mA} \times 10 \text{ V} = 500 \text{ mW}$$

Step 2. Calculate the maximum signal output power. Maximum output power will occur when the voltage output swing across the load is at maximum. Since the Q-point is in the center of the operating range, the voltage can swing +5 V to cutoff and –5 V to saturation. The maximum output voltage swing is 10 $V_{\text{p-p}}$. If the maximum output voltage swing is known, the maximum signal output power can be calculated.

$$V_{\text{out(RMS)}} = V_{\text{out(peak)}} \times 0.707 = 5 \text{ V}_{\text{p}} \times 0.707 = 3.5 \text{ V}_{\text{(RMS)}}$$
$$P_{\text{out}} = V_{\text{out(RMS)}}^2/R_L = 3.5 \text{ V}^2/100 \text{ } \Omega = 125 \text{ mW}$$

Step 3. Calculate the maximum efficiency of the class-A amplifier.

$$Efficiency = P_{\text{out}}/P_{\text{in}} \times 100\%$$
$$Efficiency = 125 \text{ mW}/500 \text{ mW} \times 100\% = 25\%$$

If the output voltage swing is less than full range (which is usually the case), the efficiency will be even lower. Low efficiency is not a real problem when the power is in milliwatts; however, when the power requirements go higher, the efficiency must be improved. The reason that low efficiency cannot be tolerated at higher power ratings is because the power not consumed by the load is consumed by the circuit in the form of heat. If a circuit has an efficiency of 25% and has an input power of 100 W, it will deliver 25 W to the load, and 75 W will be consumed by the circuitry in the form of heat. This heat energy will soon destroy the circuit. The only solution is to not use class-A amplifiers for loads that require more than a few tenths of a watt.

CLASS-B OPERATION

If the bias of a transistor amplifier circuit is set at cutoff, as shown in Figure 16–12, the amplifier is biased class-B. Because the transistor is biased at cutoff, there is zero collector current at

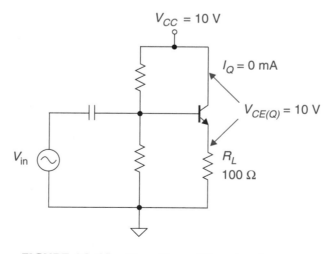

FIGURE 16–12   Class-B amplifier at quiescence

quiescence and zero power consumption with no signal input. When the input signal swings positive on the base of the NPN transistor, the base current begins to flow, causing collector current. If the input swings negative, the NPN transistor is driven further into cutoff, and no output current flows. Output current flows for only one alternation (180°) of the input cycle. The advantage of class-B operation is that at quiescence there is no current flow; therefore, there is no power consumption, thus making it much more efficient than class-A operation. Example 16.7 shows that the maximum efficiency that can be achieved using a class-B amplifier is 78%.

EXAMPLE 16.7

For the circuit in Figure 16–12, calculate the supply input power, the maximum signal output power, and the maximum efficiency.

Step 1. Calculate the peak voltage across the load. At quiescence, 0 V is dropped across the load. If the input signal drives the transistor into saturation, V_{CC} will be dropped across the load.

$$V_{L(\text{peak})} = V_{CC} = 10 \text{ V}$$

Step 2. Calculate the average DC current through the load. The waveshape of the current flowing through the load is a half-wave rectifier. The average current in a half-wave rectifier is equal to the peak current times 0.318.

$$I_{L(\text{peak})} = V_{L(\text{peak})}/R_L = 10 \text{ V}_\text{p}/100 \text{ } \Omega = 100 \text{ mA}$$
$$I_{L(\text{avg})} = I_{L(\text{peak})} \times 0.318 = 100 \text{ mA} \times 0.318 = 31.8 \text{ mA}$$

Step 3. Calculate the input power. The input power is equal to the power delivered by the DC power supply, which is equal to the DC power supply voltage times the average DC current flow.

$$P_{\text{in}} = I_{L(\text{avg})} \times V_{DC} = 31.8 \text{ mA} \times 10 \text{ V} = 318 \text{ mW}$$

Step 4. Calculate the RMS signal voltage across the load. For this step, assume the signal is a sine wave with a peak voltage equal to V_{CC}. The peak voltage is present only during the positive alternation. This fact will be considered in step 5 of our calculations.

$$V_{(\text{RMS})} = V_p \times 0.707 = 10 \text{ V}_p \times 0.707 = 7.07 \text{ V}_{(\text{RMS})}$$

Step 5. Calculate the output signal power. The power is calculated using the normal power formula, then the result is divided by 2 to correct for the fact that power is delivered to the load only during one alternation.

$$P_o = (V_{(\text{RMS})}^2/R_L)/2 = (7.07 \text{ V}^2/100 \text{ } \Omega)/2 = 250 \text{ mW}$$

Step 6. Calculate the maximum efficiency of the class-B amplifier.

$$\text{Efficiency} = P_{\text{out}}/P_{\text{in}} \times 100\%$$
$$\text{Efficiency} = 250 \text{ mW}/318 \text{ mW} \times 100\% = 78.6\%$$

CLASS-C OPERATION

In class-C operation, the active device is biased below cutoff, and output current flows for less than 180° of the input. Normally, the bias is adjusted so output current flows for approximately 90° of each input cycle.

16.5   CLASS-C POWER AMPLIFIERS

The output in class-C operation is only a short duration current pulse, as shown in Figure 16–13(a). The input signal is not truly reproduced at the output.

The uses of class-C amplifiers are limited. Class-C amplifiers are popular in RF (radio frequency) circuits that use resonant tuned circuits as loads. The resonant circuit reinstates the sinusoidal wave shape of the output signal as shown in Figure 16–13(b). The high power current pulses supply energy to the resonant circuit, causing oscillations. The flywheel action of the resonant circuit maintains a sinusoidal output signal. Class-C amplifiers can have efficiencies near 90%. This high efficiency makes them the ideal choice in RF amplifiers used in radio transmitters designed to output kilowatts of signal power.

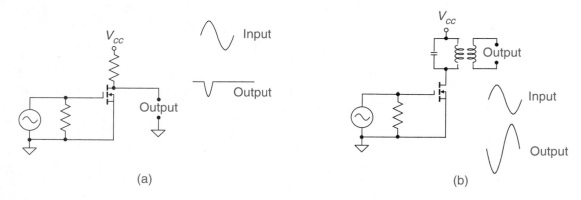

(a)   (b)

FIGURE 16–13   Class-C amplifier

16.6   CLASS-B POWER AMPLIFIERS

MOSFET COMPLEMENTARY PUSH-PULL POWER AMPLIFIERS

In order to utilize the efficiency of class-B operation and still have the output a true reproduction of the input, two power MOSFETs can be used in a complementary push-pull configuration as shown in Figure 16–14. On the positive alternation, the N-channel MOSFET conducts, and on the negative alternation, the P-channel MOSFET conducts. With zero input signal, both MOSFETs are cutoff. The input signal voltage must exceed the threshold voltage ($V_{GS(th)}$) before current will flow through the load; therefore, crossover distortion will exist in the output as shown in Figure 16–14.

Crossover distortion can be eliminated by biasing the circuit as shown in Figure 16–15. The circuit uses voltage divider biasing. The potentiometers are adjusted to permit the power MOSFET to operate slightly above the threshold voltage. This causes a few milliamps of current flow through the MOSFET at quiescence. This is called class-AB operation. When you consider that the peak current through the load (speaker) will exceed 1 A, a quiescent current of a few milliamps is still close to class-B operation. Example 16.8 calculates circuit values using circuit components.

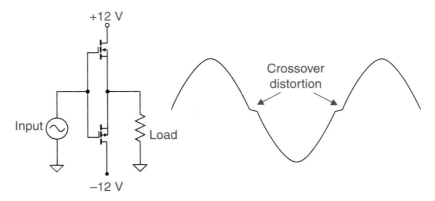

FIGURE 16–14 MOSFET power amplifier with crossover distortion

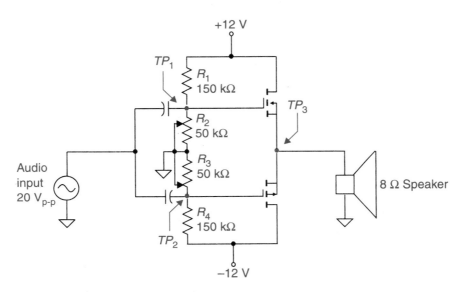

FIGURE 16–15 MOSFET power amplifier with biasing circuitry

EXAMPLE 16.8

Determine the DC voltage and signal values at TP_1, TP_2, and TP_3 and calculate input impedance, power out, and the power gain for the circuit in Figure 16–15. Assume the potentiometers are set for 40 kΩ.

Step 1. Calculate the DC voltages. The voltages at TP_1 and TP_2 can be calculated using the voltage divider formula. Power amplifiers are designed for maximum signal swing; therefore, the voltage at TP_3 will be set to halfway between the positive and the negative supply voltages. Since the voltages are of equal magnitude but opposite polarity, the voltage at TP_3 will be zero volts.

$$V_{TP1} = +V \times R_2/(R_1 + R_2) = 12 \text{ V} \times 40 \text{ k}\Omega/(150 \text{ k}\Omega + 40 \text{ k}\Omega) = 2.5 \text{ V}$$
$$V_{TP2} = -V \times R_3/(R_4 + R_3) = -12 \text{ V} \times 40 \text{ k}\Omega/(150 \text{ k}\Omega + 40 \text{ k}\Omega) = -2.5 \text{ V}$$
$$V_{TP3} = 0 \text{ V}$$

Step 2. Determine the signal levels at TP_1, TP_2, and TP_3. Since the circuit is a source follower circuit with unity voltage gain, all three test points will have the same signal level (20 V$_{p-p}$).

EXAMPLE 16.8 continued

Step 3. Calculate the input impedance. The input impedance is a function of the bias circuitry. In this case, all four of the resistors in the gate circuitry appear in parallel to the input signal.

$$z_{in} = R_1 \parallel R_2 \parallel R_3 \parallel R_4 = 150 \text{ k}\Omega \parallel 40 \text{ k}\Omega \parallel 40 \text{ k}\Omega \parallel 150 \text{ k}\Omega = 16.8 \text{ k}\Omega$$

Step 4. Calculate the output power. The output signal power is equal to the RMS voltage across the load squared and then divided by the resistance to the load.

$$V_{out(p\text{-}p)} = 20 \text{ V}_{p\text{-}p}$$

$$V_{out(RMS)} = V_{out(P)} = 10 \text{ V}_p \times 0.707 = 7.07 \text{ V}_{RMS}$$

$$P_L = (V_{RMS})^2/R_L = 7.07 \text{ V}^2/8 \ \Omega = 6.25 \text{ W}$$

Step 5. Calculate the power gain. The power gain is equal to the current gain times the voltage gain. Since the voltage gain of the circuit is one, the power gain equals the current gain.

$$I_{in} = V_{in}/z_{in} = 7.07 \text{ V}_{(RMS)}/16.8 \text{ k}\Omega = 0.45 \text{ mA}_{(RMS)}$$

$$I_{out} = V_{out}/R_L = 7.07 \text{ V}_{(RMS)}/8 \ \Omega = 0.88 \text{ A}_{(RMS)}$$

$$A_i = I_{out}/I_{in} = 0.88 \text{ A}_{(RMS)}/0.45 \text{ mA}_{(RMS)} = 1956$$

$$A_p = A_i \times A_v = 1956 \times 1 = 1956$$

A disadvantage of the circuit in Figure 16–15 is that it uses dual power supplies. The circuit in Figure 16–16 overcomes this disadvantage by using a single supply and a large capacitor. During the positive alternation, current flows through Q_1, C_1, and R_L, causing the left side of C_1 to charge positive. During the negative alternation, Q_2 is turned on, and C_1 discharges through Q_2 and the speaker. This causes current to flow through the load (speaker) in the opposite direction.

On the next positive alternation, the capacitor will charge, and it will discharge on the next negative alternation. In this type of circuit, the capacitor must be large enough to supply energy to the load during the negative alternation; these capacitors often range in the thousands of microfarads. Circuits designed to deliver large amounts of power to the load normally use dual

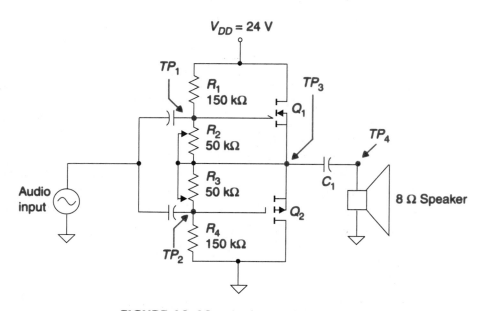

FIGURE 16–16   Single supply push-pull

power supplies, thus eliminating the need for a large output capacitor. Example 16.9 shows how DC voltage readings can be determined.

EXAMPLE 16.9

Determine the DC at the test points in the circuit shown in Figure 16–16. Assume the potentiometers are set to 40 kΩ.

Step 1. Calculate the voltages at TP_1 and TP_2.

$$V_{TP1} = +V_{DD} \times ((R_2 + R_3 + R_4)/(R_1 + R_2 + R_3 + R_4)) = 24 \text{ V} \times (230 \text{ k}\Omega/(380 \text{ k}\Omega) = 14.5 \text{ V}$$
$$V_{TP2} = +V_{DD} \times (R_4/(R_1 + R_2 + R_3 + R_4)) = 24 \text{ V} \times (150 \text{ k}\Omega/(380 \text{ k}\Omega) = 9.5 \text{ V}$$

Step 2. Determine the voltage at TP_3. The voltage at TP_3 will be halfway between the positive and negative supply voltages. The positive supply voltage is 24 V, and the negative supply voltage is 0 V.

$$V_{TP3} = 12 \text{ V}$$

Step 3. Determine the voltage at TP_4. The capacitor blocks the DC voltage from reaching the load.

$$V_{TP4} = 0 \text{ V}$$

BIPOLAR COMPLEMENTARY PUSH-PULL POWER AMPLIFIERS

Figure 16–17 shows a bipolar complementary push-pull amplifier. The support circuitry for the bipolar power amplifier is considerably more complex than that of the MOSFET power amplifier. Transistor Q_1 is the driver transistor used to provide the power transistors with the signal voltage necessary to meet the power output requirements. In addition, the driver buffers the previous stage from the low input impedance of the power transistors. The driver transistor (Q_1) operates as a class-A common-emitter amplifier. Darlington pairs are used in the output circuit. This increases the input impedance of the power amp stage and helps reduce the loading on the driver.

The driver stage (Q_1) uses voltage feedback biasing, which permits maximum signal swing to be delivered to the output transistors. There are two negative feedback paths: one for bias feedback and another for signal feedback. The feedback voltage is taken from the output. The two-stage voltage feedback helps stabilize the total power amplifier. By breaking the negative feedback into two

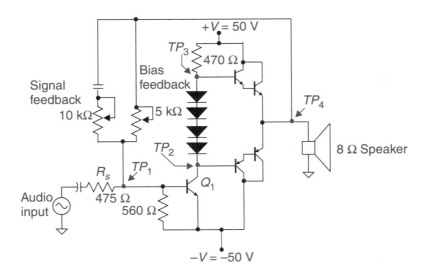

FIGURE 16–17 Bipolar power amplifier

circuits, each can be adjusted independently for maximum system performance. The input impedance of Q_1 will be only a few ohms because of its high emitter current. Resistor R_s is connected in series with the input impedance of Q_1 to raise the input impedance of the circuit.

Four diodes must be used to separate the bases of the Darlington pairs because of the four junction voltage drops. Diodes are normally mounted on the same heat sink as power transistors and provide compensation for changes in junction drops caused by heat. As the temperature increases, the junction voltage drop of the power transistors decreases. The increase in temperature also causes the voltage drop across the diodes to decrease. The reduction in diode voltage decreases the voltage between the bases of the Darlington pairs and helps to maintain the quiescent point. The signal impedance of the diodes is near zero, so the signal levels on the bases of both Darlington pairs will be approximately the same. Example 16.10 shows how DC voltages and signal levels can be determined.

EXAMPLE 16.10

Assume the driver stage in Figure 16–17 has an input impedance of 25 Ω (not considering R_s) and a voltage gain of 200. Calculate the DC voltage at the test points, the signal levels at the test points, and the signal power output with a signal input of 8 Vp-p.

Step 1. Calculate the voltage at the test points. An easy way to estimate DC is to assume that the voltage at the emitters of the output transistors is halfway between the supply voltages. This is a good assumption, since the engineer will design the power amplifier for maximum signal swing. The other DC voltages can be determined by assuming a forward-biased junction drop of 0.7 V.

$$V_{TP4} = 0 \text{ V}$$
$$V_{TP3} = 1.4 \text{ V (Two junction drops higher than } TP_4.)$$
$$V_{TP2} = -1.4 \text{ V (Two junction drops lower than } TP_4.)$$
$$V_{TP1} = -49.3 \text{ V (One junction drop higher than } -50 \text{ V.)}$$

Step 2. Calculate the signal at TP_1. Since the input impedance is known, the voltage divider formula can be used to calculate the signal level at TP_1.

$$v_{TP1} = V_{in} \times (z_{in}/(R_s + z_{in})) = 8 \text{ V}_{p-p} \times (25 \ \Omega/(475 \ \Omega + 25 \ \Omega)) = 0.4 \text{ V}_{p-p}$$

Step 3. Calculate the signal at TP_2. Since the gain of the divider stage is known, the signal at TP_2 can be easily calculated.

$$v_{TP2} = A_v \times v_{TP1} = 200 \times 0.4 \text{ V}_{p-p} = 80 \text{ V}_{p-p}$$

Step 4. Calculate the signals at TP_3 and TP_4. Since the diodes are forward biased with over 100 mA of bias current, their dynamic resistances will be near zero; therefore, the signal voltage at TP_3 will be the same as the signal voltage at TP_2. The Darlington pairs are functioning as common-collector amplifiers with unity gain so the signal voltage on the emitter will be the same as the signal voltage on the base.

$$v_{TP2} = v_{TP3} = v_{TPi} = 80 \text{ V}_{p-p}$$

Step 5. Calculate the signal output power. The signal can be calculated by dividing the square of the RMS signal voltage output by the resistance of the load.

$$V_{out(peak)} = V_{out(p-p)}/2 = 80 \text{ V}_{p-p}/2 = 40 \text{ V}_p$$
$$V_{out(RMS)} = V_{out(peak)} \times 0.707 = 40 \text{ V}_p \times 0.707 = 28.28 \text{ V}_{(RMS)}$$
$$P_{out} = (V_{out(RMS)})^2/R_L = (28.28 \text{ V}_{(RMS)})^2/8 \ \Omega = 100 \text{ W}$$

IC DRIVERS AND DISCRETE POWER TRANSISTORS

The difficulty of designing support circuitry for a bipolar power transistor can be simplified by using an IC driver circuit. Figure 16–18(a) shows the pin-outs of National Semiconductor's LM391 audio power driver designed to drive external bipolar power transistors. Figure 16–18(b) shows a simplified circuit using the LM391.

The LM391 circuit is similar to a noninverting operational amplifier. The signal in is applied to the noninverting input, and the input impedance is equal to the value of R_{in}. The feedback loop determines the gain [$A_v = (R_f/R_i) + 1$], which is normally set between 20 and 200 for the driver. The outputs at pins 5 and 8 drive the power transistors. The signal level is the same at pins 5 and 8; however, the needed bias separation is provided by the internal circuitry of the LM391. Pin 9 provides the IC with a feedback that senses the signal being delivered to the load for internal control.

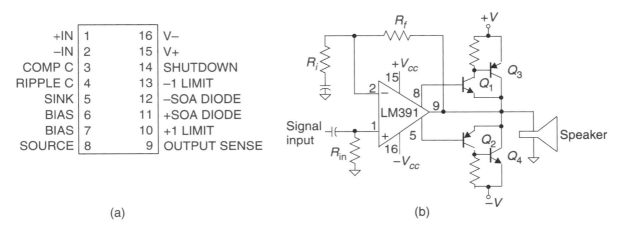

+IN	1	16	V–
–IN	2	15	V+
COMP C	3	14	SHUTDOWN
RIPPLE C	4	13	–1 LIMIT
SINK	5	12	–SOA DIODE
BIAS	6	11	+SOA DIODE
BIAS	7	10	+1 LIMIT
SOURCE	8	9	OUTPUT SENSE

(a)   (b)

FIGURE 16–18 LM391 audio power driver

Figure 16–19 shows a functional power amplifier circuit using an LM391. Several necessary components have been added. Power amplifiers have power gain in the millions. Positive feedback at high frequencies will cause oscillations if precautions are not taken to ensure stability. The following components provide high frequency filtering and help prevent oscillations: R_1, R_2, L_1,

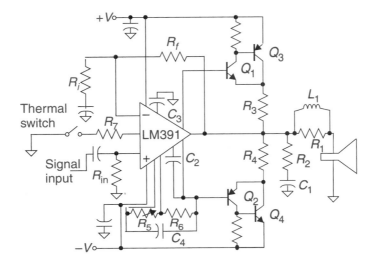

FIGURE 16–19 Functional power amplifier using the LM391

C_1, C_2, and C_3. Resistors R_3 and R_4 are emitter degeneration resistors and give thermal stability to the output stage quiescent current. Since output current flows through these resistors, their values are typically small (in the range of a few tenths of an ohm).

Resistors R_5, R_6, and capacitor C_4 make up the bias circuitry for setting the bias current in the output stage. Resistor R_5 is a potentiometer, so the bias current in the output transistor can be adjusted. Resistor R_7 connects the shutdown control to an external thermal switch. If the switch senses an overheat condition, it will close, thus causing the system to shut down.

16.7   INTEGRATED POWER AMPLIFIERS

National Semiconductor's LM380 power audio amplifier is an excellent example of an IC power amp. The IC can deliver 2.5 W of power to a 4 Ω, 8 Ω, or 16 Ω load. Figure 16–20 shows the pin-outs of the IC and a simple circuit using the LM380 to drive an 8 Ω speaker. The LM380 has an input impedance of 150 kΩ and a fixed voltage gain of 50, making it an ideal output circuit for many consumer applications.

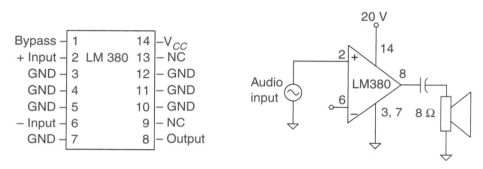

FIGURE 16–20   LM380 power audio amplifier

The LM380 has differential inputs; however, internal biasing permits the signal source to be connected to one input, and the other input can be left floating for simple audio amplifier applications. The output pin will automatically be biased at one-half of V_{cc}. The multiple ground pins are provided to permit external heat sinking. Often the circuit board design will contain a large copper area connected to the ground pins of the LM380. The copper acts as a heat sink.

Figure 16–21 shows a practical audio amplifier using an LM380. The 2.7 resistor and the 0.1 µF capacitor connected to the output pin perform high frequency filtering and suppress high frequency oscillations. The volume control sets the magnitude of the input signal by setting up a voltage divider between the 75 kΩ resistor and the 25 kΩ potentiometer. The tone control provides a variable low impedance path to ground for the higher audio frequencies.

One way to increase the power output capability of an IC power amplifier is to use two identical ICs and operate them in a differential bridge amplifier configuration. Figure 16–22 shows two LM380s connected in a bridge amplifier capable of delivering 5 W to the load. One LM380 is capable of delivering only 2.5 W to a load. The input signal is connected to the noninverting input of one amplifier and the inverting input of the other amplifier. The load is connected between the outputs of the two amplifiers. As the output of one amplifier swings positive, the output of the other amplifier swings negative, so the load receives twice the peak-to-peak signal. A 1 MΩ potentiometer is connected between pin 1 of both ICs and then adjusted so the quiescent voltage across the speaker is 0 V DC.

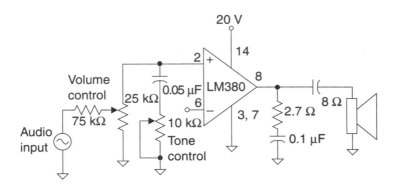

FIGURE 16–21   Practical audio amplifier using an LM380

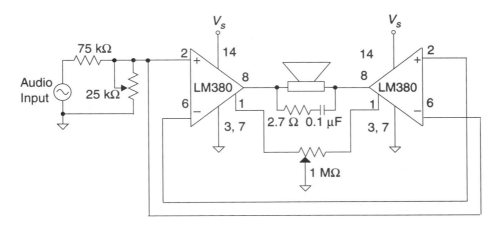

FIGURE 16–22   IC bridge power amplifier

16.8   CLASS-D POWER AMPLIFIERS

Class-B push-pull amplifiers have a theoretical efficiency of 78%, but in practice, efficiencies in the 60% to 70% range are more realistic. In the past, if you wanted a true reproduction of the input signal, class-B push-pull amplifiers were the most efficient circuits available. The advances in switching technology have made class-D power amplifiers possible, which can reach efficiency levels in the area of 90%.

Figure 16–23 shows a block diagram of a class-D power amplifier. Understanding the function of the pulse width modulator (PWM) is the key to understanding the circuit. A pulse width modulator is a rectangular wave oscillator circuit operating at a fixed frequency. The control input (TP_1) to the PWM controls the duty cycle of the output signal of the PWM. The output of the PWM controls the switching of the power MOSFET. When the output of the PWM is high, the MOSFET is on, and 120 V is present at TP_2. When the output of the PWM is low, the MOSFET is off, and 0 V is present at TP_2. The switching signal is connected to a low-pass filter consisting of the components shown inside the dotted line in Figure 16–23.

The fixed frequency of the PWM is in the range of 250 kHz to 1 MHz. The audio input signal is in the range of 20 Hz to 20 kHz. The waveforms in Figure 16–23 show that as the input signal at TP_1 swings positive, the duty cycle of the PWM increases (active state high), and as the signal swings negative, the duty cycle decreases. The low-pass filter passes the average

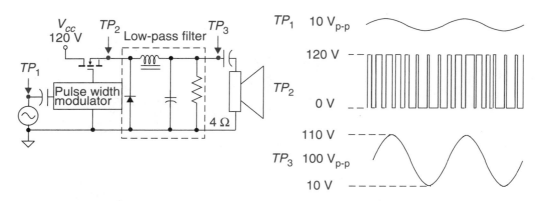

FIGURE 16–23   Class-D switching power amplifier

voltage level present at TP_2. For duty cycles near 100%, the output at TP_3 will approach 120 V, and for duty cycles near 0%, the output at TP_3 will approach 0 V.

The 10 $V_{p\text{-}p}$ audio sine wave input in Figure 16–23 causes the duty cycle to vary from 8% to 92%. The variations in the duty cycle at TP_2 are converted back to a sine wave that varies between 10 V and 110 V at TP_3. The output capacitor blocks the DC component but permits the 100 $V_{p\text{-}p}$ audio signal to reach the speaker. Since the power MOSFET is operating class-D, the circuit has a theoretical efficiency of 100%, but in reality efficiencies of 90% are obtainable.

16.9   HEAT SINKING POWER DEVICES

Devices used in power circuits must be designed to dissipate considerable power. Normally, the larger the package, the greater the power the device is able to handle. Figure 16–24 shows a few different power device packages. Devices that are designed to handle higher power dissipation have a flat metal surface that is designed to be connected to a heat sink. Heat sinks are metal structures connected to the case of power devices to help transfer the heat to the surrounding air and are effective in keeping power systems operating within safe temperature ranges. The power dissipation ratings given in Figure 16–24 are based on the assumption that the proper heat sink is used with each device.

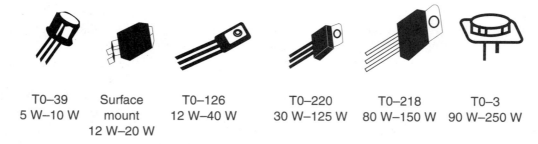

TO–39	Surface	TO–126	TO–220	TO–218	TO–3
5 W–10 W	mount	12 W–40 W	30 W–125 W	80 W–150 W	90 W–250 W
	12 W–20 W				

FIGURE 16–24   Power device packages

It is important for power circuits to be designed to be extremely efficient; however, no circuit is ever 100% efficient. Not all of the power consumed will be delivered to the load; some will be dissipated in the components of the circuitry. The power dissipated in the circuit components is given off in the form of heat. Power ICs, power transistors (bipolar and MOSFET), and rectifier diodes can generate heat levels that are self-destructive. Engineers must design power systems so the heat generated by the power device is quickly transferred away. The temperature of the device should not rise above a safe operating temperature.

Figure 16–25 shows the three heat transfers that must take place to remove the heat from the device. The heat generated in the silicon wafer is transferred to the case of the device, the case transfers the heat to the heat sink, and the heat sink transfers the heat to the surrounding air. These transfers are in series, and the effectiveness of the total system depends on each transfer functioning properly.

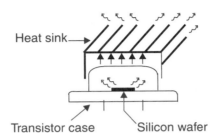

FIGURE 16–25 Heat transfers

The heat transferred from the silicon wafer to the case is a function of the device, and the technician has no control over improving this component of the heat transfer system. The heat transferred from the case to the heat sink is a function of the contact area of the heat sink with the case and the quality of the connection. The technician can improve this component of the heat transfer system by making sure the heat sink is properly installed. When installing a heat sink, a light coat of silicon gel should be applied between the case and the heat sink. The silicon gel improves the heat transfer by filling in any microscopic holes in the surface of the case or heat sink.

The cases of power devices are often electrically the same point as one terminal of the device. For example, the metal case of most bipolar transistors is normally the collector. Sometimes it is necessary to electrically isolate the case of the device from the heat sink. A thin mylar insulation is used. The mylar insulation is designed to transfer heat but still provide electrical insulation. The heat sink needs to be tightly connected to the metal case of the device. If the heat sink is electrically insulated from the case of the device, be sure to use the correct hardware to provide electrical insulation, and always use an ohms meter to validate that insulation is provided.

The surface area of a heat sink is the main factor that determines the ability of a heat sink to transfer heat to air. Most heat sinks are designed with fins that greatly increase their surface area and permit more heat to be transferred. The physical placement of the heat sink can greatly affect its heat transferring capability. Warm air rises, so the heat sink needs to be placed so the rising air can more easily move through the fins. Cooling fans are sometimes used to force air through the fins of the heat sink. The technician needs to make sure these fans are operating properly. If air filters are used, they need to be cleaned regularly so they do not inhibit air flow. High power systems sometimes use liquid cooling systems. Specially designed heat sinks permit liquid to flow through them, thus removing the heat from the system.

Figure 16–26 shows a thermal derating curve for a 10 W power device. The importance of maintaining the case temperature of the device as cool as possible becomes apparent when reviewing the curve. At room temperature (25°C), the device can safely dissipate 10 W. At 100°C, the device is limited to a safe power dissipation of 4 W. Power devices can maintain low operating temperatures only when proper heat sinking systems are used.

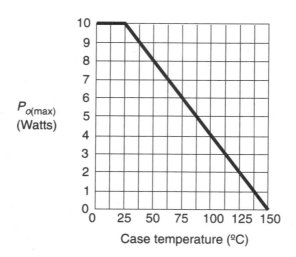

FIGURE 16–26   Thermal derating curve

16.10 TROUBLESHOOTING POWER CIRCUITS

The ideal switch permits all of the power to be delivered to the load and, therefore, is 100% efficient. Of course, ideal switching devices do not exist, but switching circuits are extremely efficient when compared to linear circuits. The high efficiency of the switching circuit depends on the switching device functioning near the ideal. When troubleshooting switching circuits, the technician measures the voltage across the switching device in the on condition. If the voltage is above rated values, the switching device is defective or the control circuitry is not providing the correct signal to drive the switching device into saturation. In the off condition, the voltage across the switching device will equal the supply. If not, the switching device is defective or the control circuitry is not providing the proper signal to cut off the device. Switching circuits may function correctly under static conditions, but under dynamic switching conditions, they may have slow switching times. The technician needs to use an oscilloscope to check switching voltage levels and switching times.

Power amplifiers have exceedingly high gains and are prone to break into oscillations. Oscillations occur because of positive feedback. Positive feedback can occur on the DC power bus if high frequency filter capacitors are defective. Routing cables carrying high powered signals near the input circuitry also can cause positive feedback. Measuring the frequency of the unwanted oscillations is sometimes helpful in isolating the cause of the oscillations.

The technician should begin troubleshooting class-B power amplifiers with the signal input set to zero. Class-B power amplifiers should draw only a small amount of current with no input signal, and all components should be at approximately room temperature. If a component is hot, it indicates a problem with the bias circuitry or a bad power device. The signal output power of an amplifier is directly proportional to the magnitude of the output signal. For this reason, engineers design push-pull power amplifiers with maximum signal swings. Knowing this fact helps the technician quickly determine voltage readings in the circuit. The output power can be determined by measuring the peak-to-peak value of the sine wave voltage at the output and converting the peak-to-peak voltage to RMS voltage. Now, the formula $P = V^2/R_L$ can be used to solve for output signal power.

Remember, in most cases, heat sinking is not an option but a requirement. The heat sink must be properly installed before a final test can be performed on a unit. Also remember that some heat sinks are electrically isolated. Be sure to check for proper isolation before applying power.

Power Amplifiers

1. Calculate the DC voltage at the test points for the circuit shown in Figure 16–27 with 3.3 kΩ resistors in the gate circuitry of the FETs. Then calculate the DC voltages at the test points with 10 kΩ resistors replacing the 3.3 kΩ resistors.

   3.3 kΩ resistors:

   $TP_1 =$ _____ $TP_2 =$ _____ $TP_3 =$ _____ $TP_4 =$ _____

   10 kΩ resistors:

   $TP_1 =$ _____ $TP_2 =$ _____ $TP_3 =$ _____ $TP_4 =$ _____

2. Simulate the circuit in Figure 16–27 with the 3.3 kΩ resistors in the gate circuitry of the FETs and measure and record the DC voltages at the test points. Then replace the 3.3 kΩ resistors and measure and record the DC voltages at the test points. (Note: Set the input signal to 0 V.)

   3.3 kΩ resistors:

   $TP_1 =$ _____ $TP_2 =$ _____ $TP_3 =$ _____ $TP_4 =$ _____

   10 kΩ resistors:

   $TP_1 =$ _____ $TP_2 =$ _____ $TP_3 =$ _____ $TP_4 =$ _____

3. With an input signal of 300 mV peak, calculate the waveforms at the test points for the circuit in Figure 16–27 with the 3.3 kΩ resistors in the gate circuitry of the FETs. Then replace the 3.3 kΩ resistors with 10 kΩ resistors and calculate the waveforms at the test points. (Assume $V_{GS\text{(th)}} = 3.5$ V for each FET.)

   3.3 kΩ resistors:

   TP_1 - - - - - - - - - - TP_2 - - - - - - - - - -

   TP_3 - - - - - - - - - - TP_4 - - - - - - - - - -

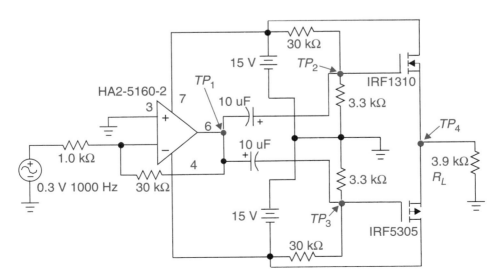

FIGURE 16–27

10 kΩ resistors:

TP_1 - - - - - - - - - - TP_2 - - - - - - - - - -

TP_3 - - - - - - - - - - TP_4 - - - - - - - - - -

4. Simulate the circuit in Figure 16–27 and measure and record the waveforms at the test points with the 3.3 kΩ resistors in the gate circuitry of the FETs. Then replace the 3.3 kΩ resistors with 10 kΩ resistors and measure and record the waveforms at the test points. (Note: Input signal is set to 300 mV peak.)

3.3 kΩ resistors:

TP_1 - - - - - - - - - - TP_2 - - - - - - - - - -

TP_3 - - - - - - - - - - TP_4 - - - - - - - - - -

10 kΩ resistors:

TP_1 - - - - - - - - - - TP_2 - - - - - - - - - -

TP_3 - - - - - - - - - - TP_4 - - - - - - - - - -

5. With 10 kΩ resistors in the gate circuitry of the circuit in Figure 16–27, calculate the signal power input, the signal power output, and the power gain. (Note: Input signal is set to 300 mV peak.)

P_{in} = _____ P_{out} = _____ A_p = _____

6. With the 10 kΩ resistors in the gate circuitry, simulate the circuit in Figure 16–27 and measure the signal power input and the signal power output using the wattmeter. Calculate the power gain using measured values. (Note: Input signal is set to 300 mV peak.)

P_{in} = _____ P_{out} = _____ A_p = _____

Power Amplifiers

I. Objective

To analyze and experimentally verify the operation of a power amplifier circuit.

II. Test Equipment

(1) +/–12 V power supply with a minimum rating of 500 mA

(1) VOM or DMM   (1) Oscilloscope

(1) Function generator

III. Components

Resistors: (1) 1 kΩ, (2) 4.7 kΩ, (3) 100 kΩ, (2) 50 kΩ potentiometers

Capacitors: (2) 0.01 μF, (2) 10 μF

Power MOSFETs: (1) IRF630 N-ch, (1) IRF9630 P-ch

IC: (1) LM741

Speaker: (8 Ω) or 8.2 Ω/5 W resistor

Optional: microphone

IV. Procedure

1. Calculate the DC voltage at the test points for the circuit shown in Figure 16–28 with both potentiometers set to minimum ohms and with both potentiometers set to 25 kΩ.

    Minimum ohms setting:

    $TP_1 =$ _____ $TP_2 =$ _____ $TP_3 =$ _____ $TP_4 =$ _____

    25 kΩ setting:

    $TP_1 =$ _____ $TP_2 =$ _____ $TP_3 =$ _____ $TP_4 =$ _____

2. With an input signal of 150 mV$_{p-p}$, calculate the peak-to-peak signal at the test points for the circuit shown in Figure 16–28 with both potentiometers set to minimum ohms and with both potentiometers set to 25 kΩ. (Assume $V_{GS(th)} = 2.7$ V for each MOSFET.)

    Minimum ohms setting:

    $TP_1 =$ _____ $TP_2 =$ _____ $TP_3 =$ _____ $TP_4 =$ _____

    25 kΩ setting:

    $TP_1 =$ _____ $TP_2 =$ _____ $TP_3 =$ _____ $TP_4 =$ _____

3. The circuit in Figure 16–28 has a 150 mV$_{p-p}$ input signal. Draw the waveform at TP_4 with both potentiometers set to minimum ohms and with both potentiometers set to 25 kΩ. (Assume $V_{GS(th)} = 2.7$ V for each MOSFET.)

    Minimum ohms setting:- - - - - - - - - - - - - - - - - -

    25 kΩ setting:- - - - - - - - - - - - - - - - - -

4. The circuit in Figure 16–28 has a 150 mV$_{p-p}$ input signal and both potentiometers are set to the correct levels to eliminate crossover distortion. Calculate the power delivered to the load.

    $P_{out} =$ _____

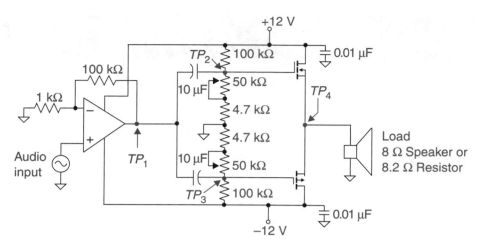

FIGURE 16–28

5. With a signal input of 150 mV$_{\text{p-p}}$, the current from each power supply measures 250 mA. Calculate the efficiency of the circuit in Figure 16–28.

    Efficiency = _____

6. Construct the circuit in Figure 16–28. The power MOSFETs do not need to use heat sinks because they will be operating at a fraction of their power rating (60 W).

7. With no signal input and the potentiometer set to minimum ohms, measure and record the DC voltages at the test points.

    TP_1 = _____ TP_2 = _____ TP_3 = _____ TP_4 = _____

8. Set the input signal to a 1 kHz sine wave with an amplitude of 150 mV$_{\text{p-p}}$. Measure and draw the output waveform.

- -

9. With no input signal, monitor the current flow from 12 V and −12 V power supplies. Adjust the potentiometers until 25 mA flows from each supply.

10. With an input signal of 150 mV$_{\text{p-p}}$, measure and record the current flow from each supply. The currents should be approximately equal. Calculate and record the average of the two currents.

    $+I$ = _____ $−I$ = _____ I_{avg} = _____

11. Measure the bias voltages at TP_2 and TP_3.

    TP_2 = _____ TP_3 = _____

12. Use the average current to calculate the input power.

    P_{in} = _____

13. With an input signal of 150 mV$_{\text{p-p}}$, measure and draw the output waveform. Was the crossover distortion eliminated?

- -

14. Using the measured signal value in step 13, calculate the signal output power.

    P_{out} = _____

15.   Calculate the efficiency of the circuit.

Efficiency = _____

V. Points to Discuss

1.   What is the input impedance of the circuit in Figure 16–28?

2.   If the input impedance of the circuit in Figure 16–28 is 1 MΩ, what is the current gain of the circuit?

3.   If the input impedance of the circuit in Figure 16–28 is 1 MΩ, what is the power gain of the circuit?

4.   Explain why the DC supply current goes up when a signal is applied to the input of the circuit in Figure 16–28.

5.   Determine the maximum signal power the circuit in Figure 16–28 is capable of delivering to the load.

6.   If the load is changed to 4 Ω, would the circuit be capable of greater signal power output? Explain.

QUESTIONS

16.1   Introduction to Power Circuits

____ 1. Three important things to considered when analyzing power circuits is the efficiency of the circuit, the characteristics of the active device, and _____.
   a. the color of the circuit board
   b. the need for dual circuitry
   c. the need for heat sinking
   d. the need to use silver runs

____ 2. _____ is a measurement of how effective a circuit is in delivering signal power to the load.
   a. Power ratio
   b. Power factor
   c. Quality factor
   d. Efficiency

____ 3. _____ circuits are extremely efficient.
   a. Switching
   b. Linear
   c. Class-A
   d. Class-B

16.2   Power MOSFETs Versus Power Bipolar Transistors

____ 4. The advancements of power MOSFETs was stimulated by the _____.
   a. power supply industry working to improve switching power supplies
   b. power supply industry working to improve linear power supplies
   c. computer industry working to improve memory circuitry
   d. computer industry working to improve communication circuitry

____ 5. Power FETs are _____ devices that utilize a vertical internal construction, permitting a wider channel and greater current capability.
   a. switching JFET
   b. enhancement MOS
   c. depletion MOS
   d. linear FET

____ 6. Which of the following is not an advantage of power MOSFETs over bipolar power transistors?
   a. Power MOSFETs require less signal power from the driver stage.
   b. Power MOSFETs have the ability to switch off and on more quickly.
   c. Power MOSFETs can be easily connected in parallel.
   d. Power MOSFETs have a lower voltage drop across the device under high voltage and current conditions.

_____ 7. When troubleshooting, the technician should assume the power bipolar transistor to have a beta of _____.

　　a.　0

　　b.　25

　　c.　100

　　d.　500

16.3 Power Switching Circuits

_____ 8. When a transistor operates as a switch, it is operating in _____.

　　a.　class-A

　　b.　class-B

　　c.　class-C

　　d.　class-D

_____ 9. When the ideal switching device is on, there is no power dissipated in the device because _____.

　　a.　the current through the device is zero

　　b.　the voltage across the device is zero

　　c.　the voltage across the load is zero

　　d.　the voltage across the load equals the voltage across the device

16.4 Classes of Amplifiers

_____ 10. The ____ amplifier circuit has output current flowing for the total input cycle or 360º.

　　a.　class-A

　　b.　class-B

　　c.　class-C

　　d.　class-D

_____ 11. The ____ amplifier circuit has output current flowing for half the input cycle or 180º.

　　a.　class-A

　　b.　class-B

　　c.　class-C

　　d.　class-D

_____ 12. The maximum efficiency that can be achieved using a class-A amplifier is _____.

　　a.　25%

　　b.　50%

　　c.　78%

　　d.　90%

_____ 13. The maximum efficiency that can be achieved using a class-B amplifier is _____.

　　a.　25%

　　b.　50%

　　c.　78%

　　d.　90%

16.5 Class-C Power Amplifiers

____ 14. Class-C amplifiers are popular in _____ circuits.
   a. audio frequency
   b. video frequency
   c. low frequency
   d. radio frequency

____ 16. Class-C amplifiers output high power current pulses that supply energy to the resonant circuit, causing oscillations. The _____ of the resonant circuit maintains a sinusoidal output signal.
   a. popping effect
   b. low impedance
   c. high impedance
   d. flywheel action

16.6 Class-B Power Amplifiers

____ 16. In order to utilize the efficiency of class-B operation and still have the output a true reproduction of the input, two power devices can be used in a _____.
   a. complementary push-pull configuration
   b. complementary pull-pull configuration
   c. complementary push-push configuration
   d. all of the above

____ 17. Crossover distortion can be eliminated by biasing the circuit _____.
   a. class-AB
   b. class-B
   c. class-C
   d. class-D

____ 18. Circuits designed to deliver large amounts of power to the load normally use dual power supplies, thus _____.
   a. eliminating the need for a large output capacitor
   b. enabling class-B operation
   c. enabling class-AB operation
   d. eliminating power losses in the power devices

16.7 Integrated Power Amplifiers

____ 19. The LM380 can deliver 2.5 W of power to a _____ load.
   a. 4 Ω
   b. 8 Ω
   c. 16 Ω
   d. all of the above

____ 20. Two identical power ICs can be operated in a differential bridge amplifier configuration to increase the _____ of the ICs.
   a. frequency response
   b. power output capability
   c. input impedance
   d. none of the above

16.8 Class-D Power Amplifiers

____ 21. The control input to the PWM controls the _____ of the output signal of the PWM.
   a. period
   b. frequency
   c. amplitude
   d. duty cycle

____ 22. In a class-D audio power amplifier, the output of the switch stage is connected _____.
   a. to a PWM
   b. to a low-pass filter
   c. to a high-pass filter
   d. directly to the load

16.9 Heat Sinking Power Devices

____ 23. Heat sinks are _____ structures connected to the case of power devices to help transfer the heat to the surrounding air.
   a. plastic
   b. semiconductor
   c. metal
   d. silicon

____ 24. The power dissipated in the circuit components is normally given off in the form of _____.
   a. light
   b. movement
   c. heat
   d. none of the above

____ 25. Three heat transfers must take place to remove the heat from the device. The heat generated in the silicon wafer is transferred to the case of the device, the case transfers the heat to the heat sink, and the heat sink transfers the heat to the _____.
   a. surrounding air
   b. surrounding components
   c. conductor connecting the device to the circuit
   d. heat control box

____ 26. _____ improves the heat transfer by filling in any microscopic holes in the surface of the case or heat sink.
  a. Heat sink glue
  b. Silicon gel
  c. Solder
  d. Plastic transfer washers

16.10 Troubleshooting Power Circuits

____ 27. If the heat sink is electrically insulated from the case of the power device, be sure to use the correct hardware to provide electrical insulation, and use a meter to validate that _____.
  a. insulation is provided before applying power
  b. insulation is provided after applying power
  c. the case of the power device is at zero volts
  d. the cases of all power transistors are common

____ 28. A system is overheating. What is the probable cause?
  a. Heat sinks were turned 90º during installation.
  b. A cooling fan is defective.
  c. An air filter is dirty.
  d. All of the above.

____ 29. The voltage across the switching device in the on condition should measure _____.
  a. zero volts
  b. a low voltage
  c. a voltage equal to the supply voltage
  d. a changing voltage level between ground and the supply voltage

____ 30. The voltage across the switching device in the off condition should measure _____.
  a. zero volts
  b. a low voltage
  c. a voltage equal to the supply voltage
  d. a changing voltage level between ground and the supply voltage

____ 31. A class-B power amplifier with no input signal has one of its output power transistors radiating excessive heat. What is wrong with the circuit?
  a. The bias circuit is causing excessive current flow through the hot transistor.
  b. The heat sink is not installed properly.
  c. The hot transistor is open.
  d. None of the above.

____ 32. The power output is low, but the voltage amplitude of the signal across the load is correct. What is wrong with the circuit?
  a. The bias circuitry is defective.
  b. The power supply voltage is low.
  c. The power supply voltage is high.
  d. The resistance of the load has increased.

PROBLEMS

1. Calculate the efficiency of an amplifier system if the total power dissipated by the system is 12 W and the power delivered to the load is 7.8 W.

2. Calculate the efficiency of the circuit in Figure 16–29.

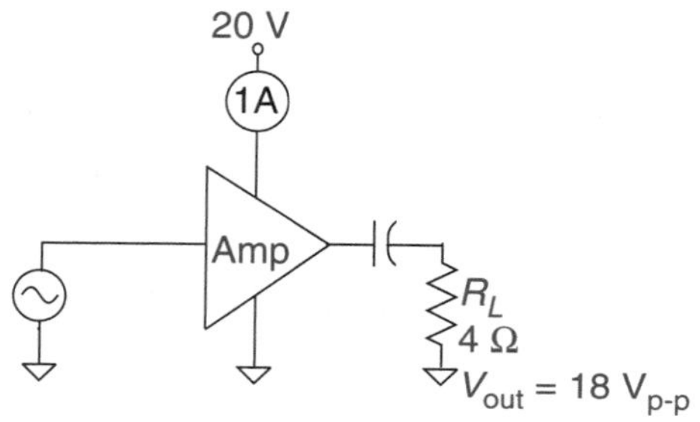

FIGURE 16–29

3. Calculate the efficiency of the amplifier circuit in Figure 16–30.

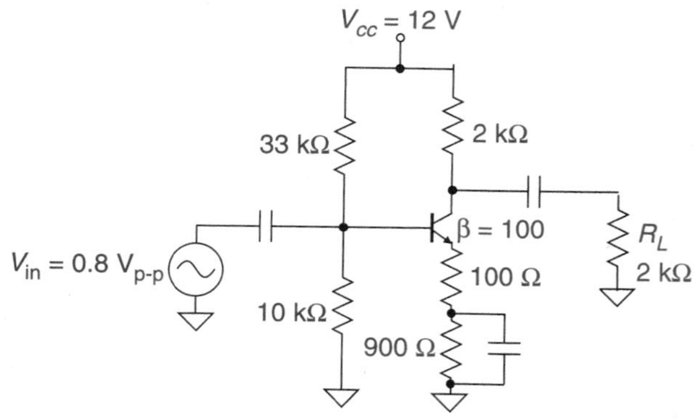

FIGURE 16–30

4. Calculate the efficiency of the switching circuit in Figure 16–31 when the transistor is conducting.

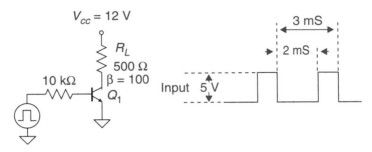

FIGURE 16–31

5. Calculate the drive current needed to turn on each switching circuit shown in Figure 16–32.

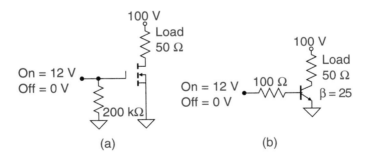

FIGURE 16–32

6. Determine the DC voltage and signal values at TP_1, TP_2, and TP_3 and calculate input impedance, power out, and the power gain for the circuit in Figure 16–33.

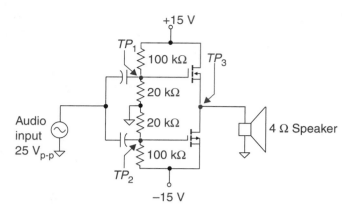

FIGURE 16–33

7. Determine the DC voltage and signal values at TP_1, TP_2, TP_3 and TP_4 and calculate input impedance, power out, and the power gain for the circuit in Figure 16–34.

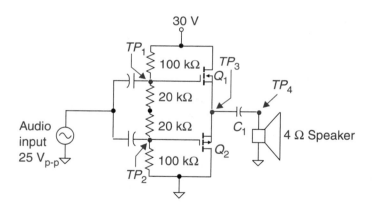

FIGURE 16–34

8. Assume the efficiency of the circuit in Figure 16–34 is 60% with a 25 V_{p-p} input. Calculate the current flow from the 30 V power supply.

9. Assume the driver stage in Figure 16–35 has an input impedance of 20 Ω (not considering R_s), an audio input of 10 V$_{\text{p-p}}$, and a voltage gain of 150. Calculate the DC voltage and signal levels at the test points.

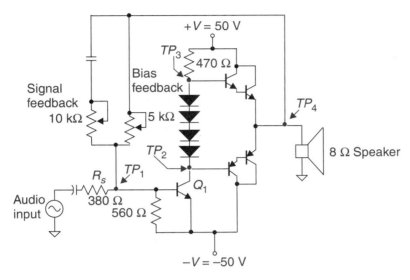

FIGURE 16–35

10. The circuit in Figure 16–35 has a signal input of 10 V$_{\text{p-p}}$. What is the signal power output?

11. What is the signal power gain of the circuit in Figure 16–35?

12. Without increasing the signal output voltage, how can the signal output power be doubled for the circuit in Figure 16–35?

13. What is the efficiency of the circuit in Figure 16–35, with a signal output of 75 V$_{\text{p-p}}$, if the DC supply current is 3 A?

14. It is desired to supply 60 W to an 8 Ω load. If the voltage is sinusoidal, what will be the peak-to-peak voltage required?

16. In the problem above, the power supply is inputting 95 W. What is the efficiency?

16. A practical push-pull amplifier is operated at 62% efficiency. The load dissipates 40 W. What is the approximate power dissipated by the power MOSFETs?

17. What is the maximum signal power that an automobile radio can output to a 4 Ω speaker if the DC input is 12 V and the radio does not use a DC to DC converter to step up the supply voltage?

18. A system operating from 13.4 V is fused at 5 A and has an efficiency of 72%. What is the approximate output power? (Typical fuses operate at two-thirds of their fuse rating.)

19. To test an 8 Ω speaker at 60 Hz, it is connected directly to the wall outlet. How much power must it be able to handle to survive?

20. A system that plugs into 120 V AC and drives a public address speaker system has an input fuse with a 15 A rating. How much power is the system designed to take in? How much power would you estimate could be supplied to the speakers by this system?

CHAPTER 17

Thyristors

OBJECTIVES

After studying the material in this chapter, you will be able to describe and/or analyze:

○ the theory of SCRs,

○ SCR circuits,

○ the theory of triacs,

○ triac circuits,

○ other thyristor circuits, and

○ troubleshooting techniques used with thyristor circuits.

17.1 INTRODUCTION TO THYRISTORS

Thyristors are four-layer semiconductor devices that can be used to design circuits that are capable of delivering large currents to high power loads while being controlled by low power circuitry. Recall that bipolar transistors and FETs can be turned on or off or can operate in the linear region between saturation and cutoff. This is not the case with thyristors. They have only two states: off or on. This chapter will cover seven of the most popular thyristors: SCR, triac, GTO, SCS, diac, PUT, and the Shockley diode. The unijunction transistor (UJT) will also be covered. The UJT is not a true thyristor, but it does function in two modes (on/off) and is often used in circuits with thyristors.

17.2 SILICON-CONTROLLED RECTIFIERS (SCRs)

The silicon-controlled rectifier (SCR) is the most popular unidirectional thyristor. Figure 17–1 shows the symbol for an SCR. It is a three-terminal device having an anode, cathode, and gate. In order for current to flow through the SCR, the anode voltage must be more positive than the cathode, and the proper trigger signal must be applied to the gate.

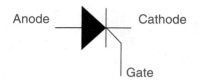

FIGURE 17–1   SCR symbol

Figure 17–2(a) shows an SCR acting as an open switch even when forward biased. This is because no gate trigger is applied. Figure 17–2(b) shows the SCR acting as a closed switch when forward biased and the proper gate trigger is applied.

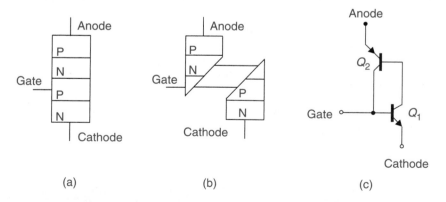

FIGURE 17–2   SCR operation

Figure 17–3(a) shows the four layers of an SCR and the connection points of the three terminals. Figure 17–3(b) shows the same four layers, but the device is separated into two parts to help you understand the inner workings of an SCR.

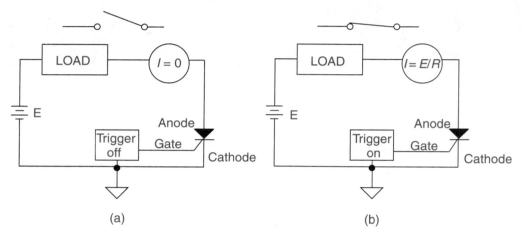

FIGURE 17–3   SCR equivalent

Figure 17–3(c) is an equivalent circuit for an SCR. The lower part is equivalent to an NPN transistor (Q_1), and the upper part is equivalent to a PNP transistor (Q_2). The base of Q_1 is connected to the collector of Q_2, and the base of Q_2 is connected to the collector of Q_1.

OPERATION OF THE SCR IN FORWARD BIAS

Figure 17–4 shows the anode of the SCR equivalent circuit connected to a positive supply voltage through a load resistor. The cathode is connected to ground. Like a normal diode, when the anode is positive with respect to the cathode, the SCR is forward biased.

The gate is connected to the trigger circuitry. If the gate trigger circuit does not output a positive pulse, the emitter base junction of Q_1 is turned off, and the collector current of Q_1 is equal to zero. This means the base current of Q_2 is also zero. With the base current in both transistors equal to zero, both transistors are cut off, and no current flows through the load.

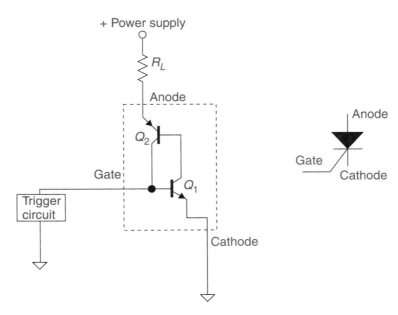

FIGURE 17–4   SCR with load

Now let us see what happens when the trigger circuit outputs a positive pulse to the base of Q_1. This positive pulse causes Q_1 to have base current, which in turn causes collector current in Q_1. The collector current in Q_1 causes base current in Q_2, which in turn causes base current in Q_1. Because each transistor has current gain between the base and collector, a small input pulse to the base of Q_1 is amplified into a larger base signal by the time it has been around the positive feedback loop. This quickly forces both transistors into saturation. With both transistors in saturation, the voltage between the anode and the cathode of the SCR drops to near zero, and the current is limited only by the size of the load resistor.

Let us assume values in the circuit to make this point clear. Assume the trigger circuit puts out a positive pulse that causes 1 µA of base current in Q_1. A beta of 100 will cause a collector current of 100 µA. This 100 µA will be the base current of Q_2, and a beta of 100 for Q_2 will cause a collector current of 10 mA. This 10 mA is now the base current for Q_1. One more time around the loop and the base current at Q_1 will equal 100 A. In other words, once the trigger circuit turns Q_1 on with a positive pulse, a feedback loop is set up that quickly drives both transistors into saturation. The only thing to limit current flow is the value of the external components. An important thing to note is that once the base of Q_1 is turned on by the trigger circuit, it is held on by the feedback loop so the trigger can be removed and both transistors still conduct.

SUMMARY OF THE SCR IN FORWARD BIAS

If the anode of an SCR is positive with respect to the cathode, no current will flow through the device until a positive pulse is received on the gate. This gate pulse initiates conduction, and the SCR quickly goes into saturation. Once the gate signal turns on the SCR, it loses control and cannot be used to turn the SCR off or to control the amount of conduction. Once the SCR is turned on in forward bias, it takes on the characteristics of a normal diode in forward bias. Once turned on, the anode current must be reduced to near zero before the SCR will stop conducting.

SCR IN REVERSE BIAS

If the anode of the SCR is negative with respect to the cathode, the SCR is reverse biased. This keeps the base/emitter junction of Q_2 in Figure 17–4 reverse biased. Even if a positive pulse is received on the gate input, the SCR will not turn on. Therefore, in reverse bias, there is no current flow through the SCR, and it acts like a normal reverse-biased diode.

SCR CURRENT VERSUS VOLTAGE CHARACTERISTIC CURVE

Figure 17–5 shows a graph of the current versus voltage characteristic curve for an SCR. We will examine this curve and discuss the essential points. Look at the reverse-biased part of the curve. As with a normal diode, only a small amount of leakage current flows when the SCR is reverse biased. However, if the reverse-biased voltage is continuously increased, a point will be reached when the SCR goes into avalanche conduction. If this reverse breakdown voltage (V_{rb}) is exceeded, the SCR will be destroyed.

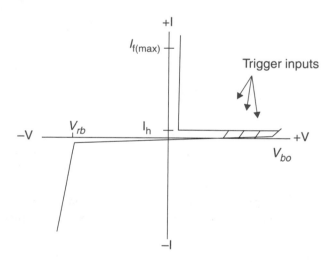

FIGURE 17–5   SCR characteristic curve

Now look at the forward-biased part of the curve. The curve illustrates that even though the SCR is forward biased, it still blocks current flow in the forward-biased direction until a trigger is received on the gate. Once a trigger is received, the voltage from the anode to cathode drops to approximately 1 V, and the current increases rapidly, being limited only by external components. This forward-biased current must be limited to less than the maximum current ($I_{f(\max)}$) the SCR is rated for in the forward direction or the SCR will be destroyed. Once in the forward-biased conduction mode, the SCR will continue to conduct until the anode current drops below the hold current. If the anode current drops below the hold current, the SCR will return to the forward blocking mode until another trigger is received on the gate.

If no trigger is received and the forward-biased voltage is continuously increased, a point will be reached where enough leakage current is created to cause the SCR to break down into conduction. This voltage is called *breakover voltage* (V_{bo}). Once the breakover voltage is reached, the SCR goes into the normal forward conduction mode.

By summarizing the important points of interest on the SCR current-versus-voltage curve, we see that there are four important values. These will be noted in data sheets. Reverse breakdown voltage (V_{rb}) and maximum forward current ($I_{f(max)}$), if exceeded, will destroy the SCR. It is necessary to know reverse breakdown voltage (V_{rb}) and maximum forward current ($I_{f(max)}$) to understand how an SCR will respond when operating on the forward-biased part of the curve. Two more important values given in technical literature, but not shown on the curve, are the gate trigger current (I_{gt}) and gate trigger voltage (V_{gt}). Gate trigger current is the amount of gate current needed to turn the SCR on, and gate trigger voltage is the amount of voltage needed between the gate and cathode to cause the gate current to flow.

SCR SUMMARY

The SCR is a rectifier device that in forward bias will function in one of two modes: off or on. The SCR is turned on by a positive pulse on the gate with respect to the cathode and, once on, can be turned off only by reducing the anode current below the value of the hold current.

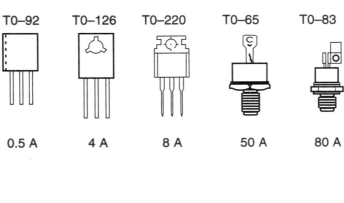

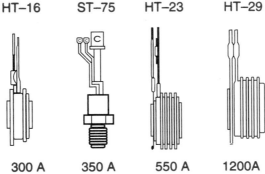

FIGURE 17–6 SCR packages

SCRs come in all shapes and sizes. Small SCRs look physically like transistors in T0-5 or T0-220 type packages. Small SCRs can handle currents up to approximately 10 A. Larger SCRs come in less familiar packages that are designed to handle large power and amp ratings ranging up to 2000 A. Figure 17–6 shows the different types of packages and the approximate amp rating for a variety of SCRs.

17.3   TRIACS

A triac is a three-terminal device having a gate terminal and two main terminals designated as MT_1 and MT_2. Figure 17–7(a) shows the symbol for a triac. A triac is similar to an SCR in that it is designed to be an electronic switch that can be turned on by applying a control pulse to the gate terminal. A triac is different from an SCR because it can allow current to flow in both directions.

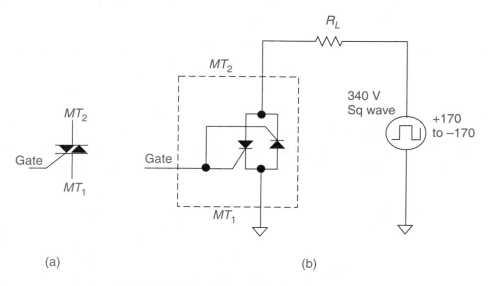

(a)   (b)

FIGURE 17–7   Triac symbol and equivalent

A triac can be thought of as two SCRs connected back-to-back with the gate terminals tied together, as shown in Figure 17–7(b). When MT_2 is positive with respect to MT_1, a gate pulse will turn on the left SCR. If MT_2 is negative with respect to MT_1, a gate pulse will turn on the right SCR. Once the trigger pulse turns on the triac in either direction, it loses control and cannot cut off conduction or control the amount of conduction.

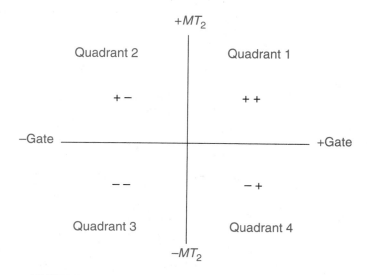

FIGURE 17–8   Four quadrants for triac triggering

Figure 17–8 shows the possible combinations of MT_2 and gate voltages needed to cause a triac to conduct. All voltages are with respect to MT_1. There are four possibilities, which are shown in Figure 17–8 as four quadrants. Quadrant 1 represents the condition where MT_2 is positive and the gate signal is positive. Quadrant 2 represents the condition where MT_2 is positive and the gate signal is negative. Quadrant 3 represents the condition where MT_2 is negative and the gate signal is negative. Quadrant 4 represents the condition where MT_2 is negative and the gate signal is positive. All triacs operate in quadrants 1 and 3, and many triacs can operate in all four quadrants.

TRIAC CURRENT-VERSUS-VOLTAGE CHARACTERISTIC CURVE

Figure 17–9 shows a graph of the current-versus-voltage characteristic curve for a triac. We will examine this curve and discuss the essential points. The right side of the graph looks the same as the right side of the graph for an SCR. A triac blocks current flow until a trigger is received on the gate, then the voltage across the triac drops to approximately 1 V, and the current is limited only by external components. If the voltage between MT_2 and MT_1 is continuously increased and no trigger is applied to the gate, a breakover voltage (V_{bo}) will be reached where the triac will break into conduction without a trigger pulse.

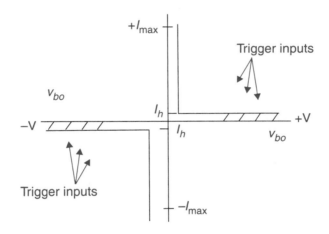

FIGURE 17–9   Triac characteristic curve

The left side of the graph is identical to the right side of the graph with the voltages reversed. The current is blocked from flowing until a trigger pulse is received or until the breakover voltage is exceeded. Note that once triggered, the current will continue to flow as long as it is not reduced below the hold current. This is true for current in both directions. If the current is reduced below the hold current, the triac will return to the off state.

17.4   GATE-TURNOFF SCR (GTO)

Figure 17–10 shows the symbol for a gate-turnoff SCR (GTO). This device is also referred to as a gate-control switch (GCS). This device is similar to an SCR, but the GTO can be turned off with a signal to the gate terminal. As with the SCR, the GTO will not permit current flow when forward biased until a positive trigger is received on the gate terminal. However, unlike the SCR, the GTO can be turned off by applying a negative pulse to the gate terminal. The magnitude of this negative pulse must be ten to twenty times larger than the positive pulse needed to turn on the GTO.

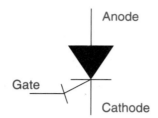

FIGURE 17–10   GTO symbols

17.5   SILICON-CONTROLLED SWITCH (SCS)

Figure 17–11 shows a silicon-controlled switch (SCS). This device has four terminals: anode, cathode, and two gates. The cathode gate is like a normal SCR gate and turns the SCS on when the device is forward biased. A signal to the anode gate of the SCS stops conduction. After conduction has ceased, a positive pulse to the cathode gate is required to restart conduction. The SCS is limited to low power circuits, and the anode current is limited to a few tenths of an amp.

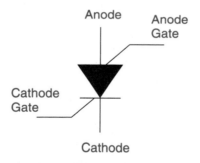

FIGURE 17–11   SCS symbols

17.6   SHOCKLEY DIODE

Figure 17–12(a) shows the symbol for the Shockley diode. The Shockley diode is a two-terminal device having an anode and a cathode. In the reverse-biased direction, it functions like a normal diode, and no current flows. In the forward-biased direction, the current is blocked from flowing until the breakover voltage is reached. Once the breakover voltage is reached, the voltage across the Shockley diode drops, and the current is limited by external components. Once in the conduction mode, conduction will continue unless the anode current is reduced below the hold current.

By examining the current-versus-voltage graph for the Shockley diode in Figure 17–12(b), it can be seen that the Shockley diode resembles an SCR without a gate terminal. It has, however, been designed to have a low breakover voltage.

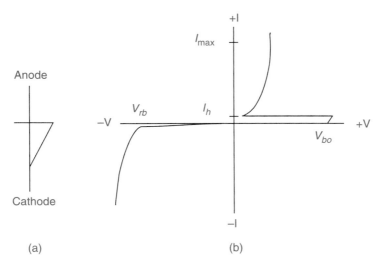

FIGURE 17–12   Shockley diode

17.7   DIACS

Figure 17–13(a) shows the symbol for a diac. The diac is a bidirectional, two-terminal device. Figure 17–13(b) shows the current-versus-voltage graph. The diac blocks current flow in both directions until a breakover voltage is reached. Once the diac breaks over into conduction, it will continue to conduct until the current is reduced below the hold current. Figure 17–13(b) shows that the breakover voltage is symmetrical. Not all diacs have this symmetrical breakover voltage.

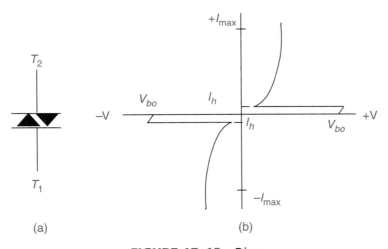

FIGURE 17–13   Diac

17.8   UNIJUNCTION TRANSISTORS (UJTs)

The unijunction transistor (UJT) is not a thyristor because it is not made of four layers of semiconductor material. However, it does operate in two states and is often used to control an SCR. The UJT is a three-terminal device having two bases (B_1 and B_2) and an emitter terminal.

The symbol for the UJT is shown in Figure 17–14(a). As the name suggests, the UJT is made with one PN junction. Figure 17–14(b) shows a drawing of the construction of the UJT. It consists of a channel made with lightly doped N-type material with bases B_1 and B_2 connected at opposite ends and a pocket of heavily doped P-type material approximately midway in the channel. The emitter terminal is connected to the P-type material.

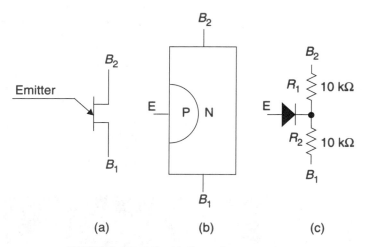

(a)   (b)   (c)

FIGURE 17–14   Unijunction transistor

Because the N-type channel is lightly doped, there is considerable resistance between B_1 and B_2. This resistance is evenly spread along the channel, with approximately one-half below the emitter and the other half above the emitter. Figure 17–14(c) is an equivalent circuit of the UJT. It shows two 10 kΩ resistors connected between B_1 and B_2 and the emitter connected through a diode to the junction of the two resistors.

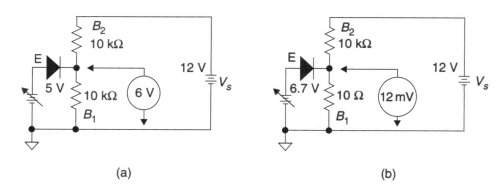

(a)   (b)

FIGURE 17–15   UJT equivalent circuits

The two resistors form a voltage divider that divides the supply voltage as shown in Figure 17–15(a). The voltage at the junction of the two resistors is equal to one-half the supply voltage, or 6 V for this example. Because the emitter voltage is less than 6.7 V, the emitter is reverse biased and does not affect the circuit. If the emitter voltage is increased beyond 6.7 V, the emitter will become forward biased. Recall that the N-type material is lightly doped and the P-type is heavily doped. Because of this, the N-type material between the emitter and B_1 is flooded with current carriers that are injected from the P-type emitter. This large increase in current carriers causes the resistance of the channel between B_1 and the emitter to reduce drastically from approximately 10 kΩ to 10 Ω, as shown in Figure 17–15(b). The channel resistance between B_1 and the emitter will remain low as long as the emitter is forward biased and continues to inject current carriers. If the emitter is reverse biased, the resistance of the channel between B_1 and the emitter will return to its high resistance state. The actual resistance values of the upper and lower sections will depend on the type of UJT being used.

The voltage needed to forward bias the emitter is called the *standoff voltage* and is dependent upon the ratio of the resistance of the two sections of the N-type channel and the supply voltage. The resistance of the sections of the UJT are unknown; however, the manufacturer provides the standoff ratio (eta) on data sheets. If the standoff ratio and the power supply voltages (V_s) are known, the standoff voltage can be calculated by eta times V_s. The typical range of standoff ratio is from 0.5 to 0.8.

UJT RELAXATION OSCILLATOR

Figure 17–16(a) shows a circuit for a UJT relaxation oscillator. When voltage is applied, capacitor C_1 starts to charge through R_1. All the current flowing through R_1 flows into C_1 because the emitter junction of the UJT is reverse biased and there is no current in the emitter leg. When the voltage across capacitor C_1 reaches the standoff voltage, the emitter terminal becomes forward biased, and the resistance between the emitter and B_1 drops to nearly zero. This connects resistor R_2 (100 Ω) across capacitor C_1, causing C_1 to discharge through resistor R_2. Once the voltage across C_1 drops to nearly zero, the emitter will stop conducting. At this point, C_1 starts to charge again through R_1, and the cycle starts over.

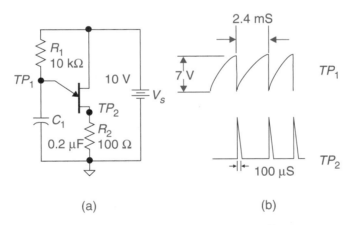

(a)   (b)

FIGURE 17–16   UJT relaxation oscillator

Figure 17–16(b) shows the voltage waveforms across C_1 and R_2. The waveform across the capacitor (TP_1) shows the capacitor charging up through R_1 on the rising part of the waveform and discharging through R_2 on the falling part of the waveform. The voltage waveform across R_2 (TP_2) is a function of the discharge current of C_1 flowing through R_2. If R_2 is small in comparison to R_1, the discharge time will be short, causing a narrow voltage pulse across R_2.

The period of the waveform can be calculated by knowing the standoff ratio and the values of R_1 and C_1. It is not necessary to consider the value of R_2 in the calculation as long as it is small in comparison to R_1; the discharge time can be disregarded.

Figure 17–17 shows the universal time constant curve. The equation for the charging curve is $Y = 1 - e^{-x}$, where Y is the percent of charge, e is the natural base, and X is the time in time constants. The standoff ratio of the UJT sets the percentage of charge the capacitor will reach before turning on the UJT. For example, if the standoff ratio for the UJT used in the circuit in

Figure 17–16 is 0.7, the voltage across the capacitor will reach 70% of the supply voltage before turning on the UJT. Example 17.1 shows how the period of the waveform can be calculated.

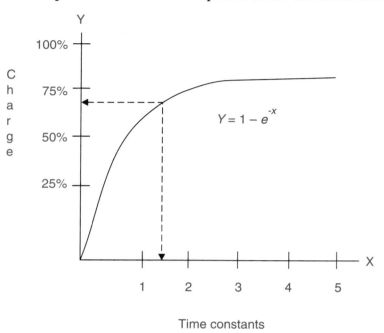

FIGURE 17–17   Universal time constant curve

EXAMPLE 17.1

Calculate the period and frequency of the waveform produced by the circuit in Figure 17–16 if the UJT has a standoff ratio of 0.7.

Step 1. The 0.7 standoff ratio is equal to Y in the equation $Y = 1 - e^{-x}$. All values are known except X. Solve the equation for X.

$$Y = 1 - e^{-x}$$
$$e^{-x} = 1 - Y$$
$$LN\, e^{-x} = LN\, (1 - Y)$$
$$-X = LN\, (1 - Y)$$
$$Y = 0.7$$
$$-X = LN\, (1 - 0.7)$$
$$-X = -1.2$$
$$X = 1.2$$

Step 2. X is time expressed in time constants. The next step is to find the RC time constant for the charging circuit.

$$\tau = R_1 \times C_1$$
$$\tau = 10\ k\Omega \times 0.2\ \mu F$$
$$\tau = 2\ mS$$

Step 3. Calculate the charging period. Since the discharge time is short, it will be considered negligible, and the charging period will be considered to equal the waveform period.

$$P = \tau \times X$$
$$P = 2\ mS \times 1.2 = 2.4\ mS$$

EXAMPLE 17.1 continued

> **Step 4.** Calculate the frequency.
> $$F = 1/p = 1/2.4 \text{ mS} = 417 \text{ Hz}$$

17.9 PROGRAMMABLE UNIJUNCTION TRANSISTOR (PUT)

The programmable unijunction transistor (PUT) is a thyristor that functions like the UJT, but its standoff ratio can be set by two external resistors; hence, the name *programmable*. Figure 17–18 shows the symbol for the PUT. It has three terminals: anode, cathode, and gate. In order for the PUT to conduct, the anode has to be made more positive than both the cathode and the gate. The cathode is usually tied to ground, and the gate is connected to the junction of two resistors, which forms a voltage divider and sets the standoff ratio.

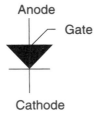

FIGURE 17–18   Programmable UJT symbol

Figure 17–19(a) shows a PUT used in a relaxation oscillator. Resistors R_2 and R_3 form a voltage divider that sets the voltage on the gate. The anode is connected to the RC circuit, which is made up of resistor R_1 and capacitor C_1. Capacitor C_1 charges through resistor R_1 until the voltage at the top of C_1 exceeds the voltage on the gate set by the voltage divider of R_2 and R_3. At this point, the PUT turns on and connects R_4 across C_1. Once turned on, the PUT acts like a closed switch, and capacitor C_1 discharges through the small resistance of R_4. When the voltage across C_1 drops to near zero, the current through the PUT drops below the hold current, and the PUT cuts off. At this point, C_1 again starts to charge through R_1, and the cycle starts over. Figure 17–19(b) shows the waveforms at the two test points.

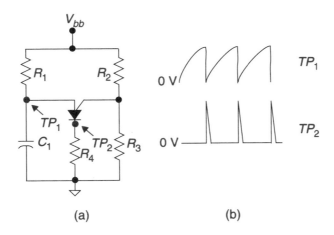

(a)   (b)

FIGURE 17–19   Relaxation oscillator

17.10 SCR PHASE-CONTROL CIRCUITS

So far we have discussed the various types of thyristors and how each device functions. We will now look at how these devices along with other components can be used to control power loads. The loads thyristors control are normally industrial loads such as lights, heaters, and motors. The SCR phase-control circuit discussed in this section is an example of a power control circuit.

An SCR phase-control circuit controls the power delivered to the load by controlling the percentage of time that current will flow through the load during each cycle. Figure 17–20 is a circuit that demonstrates the use of an SCR to control the power to a load. If the SCR is triggered at the beginning of each positive alternation, the load will receive current flow for the full duration of the positive alternation, and the average current will equal 0.637 of peak current. However, the trigger circuit that provides the gate pulse can delay the gate trigger, causing the SCR to turn on later in the cycle and delivering current to the load for only part of the positive alternation. This will reduce the average current, thus reducing the power to the load.

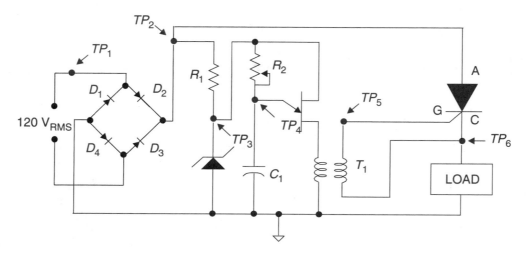

FIGURE 17–20   SCR power control

Figure 17–21 shows the waveforms at the various test points in Figure 17–20. By analyzing the waveforms and the schematic, a good understanding of the circuitry can be obtained. The input waveform at TP_1 is a sinusoidal signal with 170 V_p. The rectifier converts this signal to a pulsating DC at TP_2. The signal at TP_2 is applied across the series combination of the SCR and the load. As with all series circuits, the sum of the instantaneous voltage drops must add up to the source. So, in this case, the sum of the voltage across the SCR and the load equals the voltage at TP_2. The SCR acts like a switch; it is either on, having zero volts across it, or off, having all the input voltage across it. The signal delivered to the load is dependent on whether the SCR is on or off, and this is dependent on the trigger control circuitry.

TP_3 shows the voltage developed across the zener diode in series with the dropping resistor, R_1. On each positive alternation, the zener does not conduct until the zener voltage is reached (12 V). The zener diode and resistor R_1 form a 12 V power source that resets to zero at the end of each alternation as the voltage at TP_2 drops to zero. This power source is used to supply power to the UJT circuitry.

The UJT and associated components form a UJT relaxation oscillator. With 12 V applied to the oscillator, C_1 begins to charge toward 12 V; however, when the voltage at TP_4 reaches the voltage needed to turn on the UJT, the capacitor discharges through the primary of the

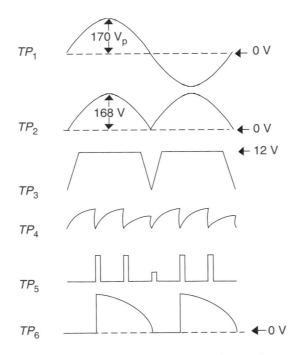

FIGURE 17–21 SCR power control waveforms

transformer, causing a pulse (at TP_5) in the secondary of the transformer that triggers the SCR. As the voltage at TP_4 falls near zero, the UJT cuts off, causing the capacitor to start its charging cycle again. The time needed for the capacitor to charge is controlled by the setting of rheostat R_2. If R_2 is set to a low resistance value, the relaxation oscillator can complete several cycles during the positive alternation. Figure 17–21 shows the waveforms at TP_4 and TP_5. The first pulse at TP_5 turns on the SCR, and the following pulses have no effect during the alternation.

TP_6 shows the voltage across the load. You can see in Figure 17–21 that there is no voltage across the load until the first pulse from the control circuit turns on the SCR. After the SCR is turned on, the input voltage is across the load until it swings to zero. The reduced input voltage at TP_2 will cause current through the SCR to drop below the hold current, and the SCR will stop conducting. The SCR will stay off until it receives the next gate trigger in the next positive alternation.

The purpose of the pulse transformer is to provide DC isolation between the gate of the SCR and the control circuit. This is necessary because the SCR cathode is not at ground potential. It is possible to reverse the position of the load and the SCR and reference the cathode of the SCR to ground. If this is done, the pulse transformer could be replaced with a resistor. However, this would mean that neither side of the load would be at ground reference, which is usually not desirable for safety reasons.

17.11 TRIAC PHASE-CONTROL CIRCUITS

Figure 17–22(a) shows a phase control circuit using a triac. The triac controls the current through the load when triggered by the control circuitry. The triac turns off at the end of each alternation as the circuit current passes through zero. The control circuitry is comprised of a diac ($V_{bo} = 20$ V), capacitor C_1, and rheostat R_1.

When the triac is not conducting, capacitor C_1 is in series with rheostat R_1 and the load. When the voltage across the capacitor reaches the breakover voltage of the diac, the diac conducts, causing the capacitor to deliver a trigger pulse to the triac. The triac then starts conducting and continues to conduct until the triac current decreases below the hold current. The triac current will drop below the hold current as the input voltage approaches zero crossover. This causes the triac to stop conducting until a trigger pulse is received on the next alternation. Since the triac and diac are both bidirectional devices, this circuit works on both positive and negative alternations. Figure 17–22(b) shows the circuit waveforms at the test points.

The amount of power delivered to the load is determined by how long a delay there is before the triac fires. This delay is a function of the time required for the voltage across the capacitor to reach the breakover voltage of the diac. The time needed for the voltage across the capacitor to reach the breakover voltage of the diac is a function of two things: one is the voltage divider set up by the resistance R_1 and the capacitive reactance of C_1, and the second is the phase shift caused by the RC circuit of R_1 and C_1.

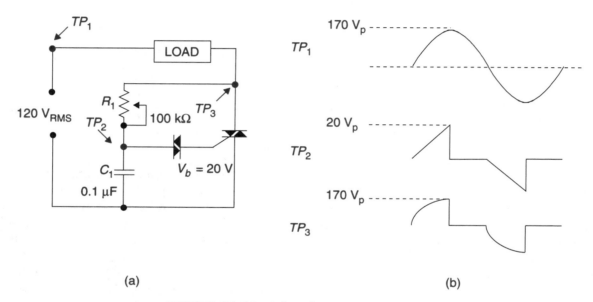

(a)   (b)

FIGURE 17–22   Triac phase control circuit

Before the breakover voltage of the diac is reached, both the diac and the triac are off, leaving an RC circuit made up of R_1 and C_1 (the load resistance is negligible). Let us examine this circuit for two settings of R_1. Figures 17–23 and 17–24 show phasor and waveform diagrams for each setting of R_1. In all cases, the vector sum of the voltages across R_1 and C_1 must equal the generator voltage, which is shown as the reference at zero degrees.

In Figure 17–23, R_1 is set to a large resistance value in comparison to the reactance of C_1. In this case, the voltage developed across C_1 is small, as shown in the diagram. This means the 20 V needed for breakover will not be reached until the voltage across the capacitor is near its peak. Next, notice that the voltage across C_1 is lagging the generator voltage by nearly 90°. This adds additional delay because the voltage across C_1 will not cross into the positive alternation until the generator voltage is approaching 90° in the positive alternation. With this setting of R_1, the breakover voltage needed for the diac is reached late in the positive alternation of the generator, and the overall delay will approach 180° as shown in Figure 17–23.

In Figure 17–24, the resistance value of R_1 is set much smaller than the reactance of C_1. In this case, the voltage developed across C_1 is much larger, as shown in the diagram. This means the 20 V needed for breakover will be reached very quickly in the positive alternation. Next, notice that the voltage across C_1 is only lagging the generator voltage by a few degrees. The fact that the voltage across the capacitor is much larger and is lagging the generator voltage by only

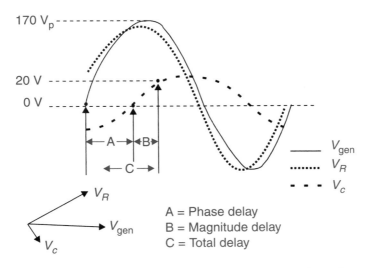

FIGURE 17–23 Diagram for a large resistance value of R_1

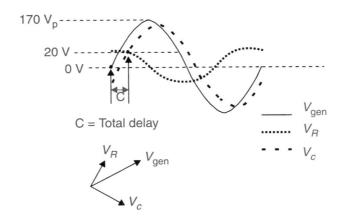

FIGURE 17–24 Diagram for a small resistance value of R_1

a few degrees means that the voltage needed for breakover of the diac will occur earlier in the positive alternation. With this setting of R_1, breakover voltage for the diac is reached in the first few degrees of the positive alternation of the generator as shown in Figure 17–24.

If R_1 has sufficient range, the delay of the triac firing can be controlled from a few degrees to nearly 180º in each alternation. The waveforms shown in Figure 17–22(b) are for a value of R_1 that will cause a 90º delay. The waveforms show the bidirectional characteristics of the triac and diac that make the circuit operation on the negative alternation the same as for the positive alternation.

17.12 MOS-GATED THYRISTORS

Under high current conditions, thyristors have a lower on-state voltage drop than bipolars, IGBTs, or MOSFETs. For this reason, thyristors are preferred for controlling loads that draw hundreds of amps. The drive current needed to turn on large thyristors can be high, thus requiring complex circuitry. The semiconductor industry is developing a new class of thyristors called *MOS-gated thyristors*. These thyristors have MOSFET gate circuitry that permits high power thyristors to be switched on with a voltage level. Since a large gate current pulse is not required, the driving circuitry can be simplified.

There is strong motivation to make improvements in GTO type devices. A MOSFET input GTO type device makes it possible to turn on and turn off high power loads with only a low power voltage pulse. The emitter switched thyristor (EST) is a new device that holds great promise in being an on/off switch that can be controlled with a low power input.

17.13 TROUBLESHOOTING THYRISTOR CIRCUITS

There are a variety of thyristors on the market, but they all have one thing in common: they are on/off devices. SCRs and triacs are turned on by a trigger pulse and are turned off by dropping the current below the hold current. If an SCR or triac is not switching on, a technician can momentarily apply a gate pulse by connecting the gate to a voltage level through a current limiting resistor. If the device turns on, the control circuitry is defective. If it does not, the SCR or triac is defective.

If the SCR or triac is functioning as a continuous short, the technician can power down the system. This will cause the current through the device to go to zero. With the power removed, the gate circuit is disabled. This can normally be accomplished by connecting a jumper between the gate and the cathode on an SCR or between the gate and MT_1 on a triac. Make sure this action does not damage the gate circuitry. With the gate disabled, return power to the system. The device should not conduct. If it does, the device (SCR or triac) needs to be replaced. If the device does not conduct, the drive circuitry is defective and needs to be repaired.

SCR Phase Control

1. Calculate and draw the waveforms at the test points for the circuit shown in Figure 17–25 with the potentiometer set for 4 kΩ, 20 kΩ, and 36 kΩ.

	Pot. = 4 Ω	Pot. = 20 kΩ	Pot. = 36 kΩ
TP_1	- - - - - -	- - - - - -	- - - - - -
TP_2	- - - - - -	- - - - - -	- - - - - -
TP_3	- - - - - -	- - - - - -	- - - - - -
TP_4	- - - - - -	- - - - - -	- - - - - -

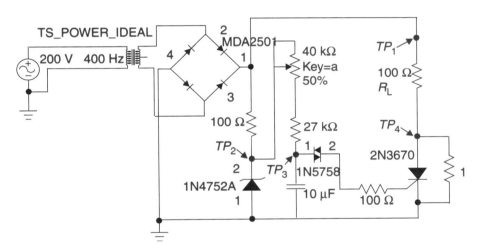

FIGURE 17–25

2. Simulate the circuit in Figure 17–25 and draw the waveforms at the test points for the potentiometer settings of 4 kΩ, 20 kΩ, and 36 kΩ.

	Pot. = 4 Ω	Pot. = 20 kΩ	Pot. = 36 kΩ
TP_1	- - - - - -	- - - - - -	- - - - - -
TP_2	- - - - - -	- - - - - -	- - - - - -
TP_3	- - - - - -	- - - - - -	- - - - - -
TP_4	- - - - - -	- - - - - -	- - - - - -

SCR Phase Control

Caution! This experiment uses line voltage that can cause severe electrical shock.

I. Objectives

1. To experimentally determine the basic operation of an SCR.
2. To construct and analyze an SCR phase-control circuit.

II. Test Equipment

DMM or VOM   5 V power supply
20 V power supply   Oscilloscope
Isolation transformer

III. Components

Resistors: (1) 100 Ω, (1) 330 Ω, (1) 390 Ω/1 W, (1) 1 kΩ, (2) 10 kΩ, (1) 12 kΩ/1 W, (1) 100 kΩ potentiometer
Capacitor: (1) 1 µF
Diodes: (1) 12 V zener, (1) bridge rectifier
SCR: (1) C106B1 or equivalent
UJT: (1) ECG6410 or equivalent
Light bulb: (1) 100 W maximum

IV. Procedure

1. Construct the circuit in Figure 17–26(a).
2. Measure and record the voltage across and the current through the SCR.

    V_{SCR} = _____ I_{SCR} = _____

3. Reverse the supply voltage of the circuit in Figure 17–26(a). Measure and record the voltage across and the current through the SCR.

    V_{SCR} = _____ I_{SCR} = _____

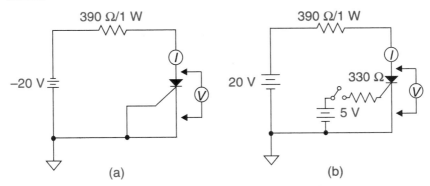

FIGURE 17–26

4. Construct the circuit in Figure 17–26(b).

5. With the switch open, measure and record the voltage across and the current through the SCR.

    V_{SCR} = _____ I_{SCR} = _____

6. Close the switch and measure and record the voltage across and the current through the SCR.

    V_{SCR} = _____ I_{SCR} = _____

7. Analyze the circuit in Figure 17–27 and determine and record the waveshapes for test points TP_1 through TP_5 for the following settings of the potentiometer. Assume the standoff ratio equals 0.63.

	Pot. = 0 Ω	Pot. = 41.5 kΩ	Pot. = 100 kΩ
TP_1	- - - - - -	- - - - - -	- - - - - -
TP_2	- - - - - -	- - - - - -	- - - - - -
TP_3	- - - - - -	- - - - - -	- - - - - -
TP_4	- - - - - -	- - - - - -	- - - - - -
TP_5	- - - - - -	- - - - - -	- - - - - -

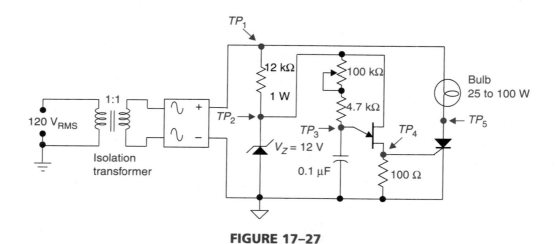

FIGURE 17–27

8. Construct the circuit in Figure 17–27 and measure and record the waveshapes for test points TP_1 through TP_5 for the following load conditions: full power to load, half power to load, and no power to load.

	Full power	Half power	No power
TP_1	- - - - - -	- - - - - -	- - - - - -
TP_2	- - - - - -	- - - - - -	- - - - - -
TP_3	- - - - - -	- - - - - -	- - - - - -
TP_4	- - - - - -	- - - - - -	- - - - - -
TP_5	- - - - - -	- - - - - -	- - - - - -

V. Points to Discuss

1. What is the function of the 330 Ω resistor in the circuit in Figure 17–26(b)?

2. Why is it a good idea to use an isolation transformer when constructing the circuit in Figure 17–27?

3. Explain the function of the zener diode in Figure 17–27.

4. Explain the function of the UJT in Figure 17–27.

5. What would be the waveshape of the voltage across the load added to the voltage across the SCR under all load conditions?

Triac Phase Control

1. Calculate and draw the waveforms at the test points for the circuit shown in Figure 17–28 with the potentiometer set for 10 kΩ, 50 kΩ, and 90 kΩ.

	Pot. = 10 Ω	Pot. = 50 kΩ	Pot. = 90 kΩ
TP_1	- - - - - -	- - - - - -	- - - - - -
TP_2	- - - - - -	- - - - - -	- - - - - -
TP_3	- - - - - -	- - - - - -	- - - - - -

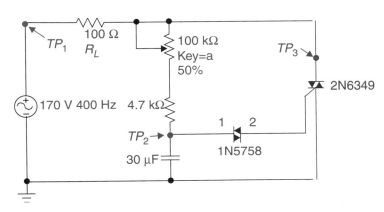

FIGURE 17–28

2. Simulate the circuit shown in Figure 17–28 and draw the waveforms at the test points for the potentiometer settings of 10 kΩ, 50 kΩ, and 90 kΩ.

	Pot. = 10 Ω	Pot. = 50 kΩ	Pot. = 90 kΩ
TP_1	- - - - - -	- - - - - -	- - - - - -
TP_2	- - - - - -	- - - - - -	- - - - - -
TP_3	- - - - - -	- - - - - -	- - - - - -

Triac Phase Control

Caution! This experiment uses line voltage that can cause severe electrical shock.

I. Objective

To construct and analyze a triac phase-control circuit.

II. Test Equipment

60 Hz isolation transformer   Oscilloscope
DMM or VOM

III. Components

Resistors: (1) 4.7 kΩ, (1) 100 kΩ potentiometer
Capacitor: (1) 0.2 µF
Light bulb: (1) 25 W to 100 W
Diac: (1) ECG6408 or equivalent
Triac: (1) ECG5608 or equivalent

IV. Procedure

1. Analyze the circuit in Figure 17–29 and calculate and record the waveshapes for test points TP_1 through TP_3 for the following load conditions: no power to load, half power to load, and full power to load.

	Full power	Half power	No power
TP_1	- - - - - -	- - - - - -	- - - - - -
TP_2	- - - - - -	- - - - - -	- - - - - -
TP_3	- - - - - -	- - - - - -	- - - - - -

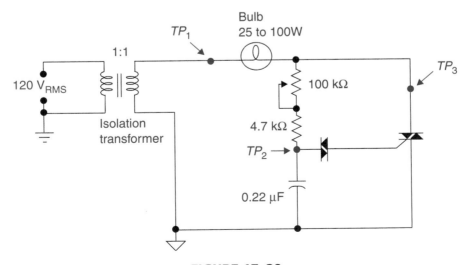

FIGURE 17–29

2. Construct the circuit in Figure 17–29 and measure and record the waveshapes for test points TP_1 through TP_3 for the following load conditions: no power to load, half power to load, and full power to load.

	Full power	Half power	No power
TP_1	- - - - - -	- - - - - -	- - - - - -
TP_2	- - - - - -	- - - - - -	- - - - - -
TP_3	- - - - - -	- - - - - -	- - - - - -

3. In Figure 17–29, move the ground of the oscilloscope to TP_3 and measure the signal across the load with one channel and the signal across the triac with the other channel. Use the invert function and add function of the oscilloscope. Demonstrate that the signal across the load plus the signal across the triac always equal the applied voltage for all power settings.

V. Points to Discuss

1. Explain the function of the diac in Figure 17–29.

2. What is the waveshape of the voltage across the load added to the voltage across the triac under all load conditions? Explain.

3. What is the approximate power dissipated by the triac for the following load conditions: no power to load, half power to load, and full power to load? Explain.

QUESTIONS

17.1   Introduction to Thyristors

____ 1. Thyristors are constructed of how many layers of semiconductor material?
   a. two
   b. three
   c. four
   d. five

____ 2. Thyristors have _____ state(s) of operation.
   a. one
   b. two
   c. three
   d. four

17.2   Silicon-Controlled Rectifiers (SCRs)

____ 3. What is the schematic symbol for an SCR?

   a.   b.   c.   d.   e.   f.   g.   h.

____ 4. What are the three terminals of an SCR called?
   a.   anode, cathode, and base
   b.   anode, cathode, and gate
   c.   anode, emitter, and gate
   d.   collector, emitter, and gate

____ 5. The gate terminal of an SCR is used to _____.
   a.   turn an SCR off
   b.   turn an SCR on
   c.   control the current flow through an SCR
   d.   control the voltage amplitude on the cathode

____ 6. In order for current to flow through the SCR, _____.
   a.   the anode voltage must be more positive than the cathode
   b.   the anode voltage must be more negative than the cathode
   c.   the proper trigger signal must be applied to the gate
   d.   both a and c

17.3 Triacs

_____ 7.   What is the schematic symbol for a triac?

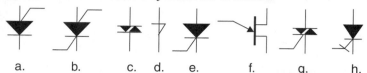

        a.   b.   c.   d.   e.   f.   g.   h.

_____ 8.   What are the three terminals of a triac called?

   a.   anode, cathode, and base

   b.   anode, cathode, and gate

   c. B_1, B_2, and emitter

   d. MT_1, MT_2, and gate

_____ 9.   A triac is different from an SCR because _____.

   a.   it can handle large current flow

   b.   it can handle large voltages

   c.   the gate signal can turn off the current flow

   d.   it can allow current to flow in both directions

_____ 10. All triacs operate in _____, and many triacs can operate in all four quadrants.

   a.   quadrant 1 and quadrant 2

   b.   quadrant 1 and quadrant 3

   c.   quadrant 2 and quadrant 4

   d.   quadrant 3 and quadrant 4

_____ 11. Once triggered, the current through a triac will continue to flow as long as _____.

   a.   a gate signal is present

   b. MT_2 is positive with respect to MT_1

   c. MT_2 is positive with respect to the gate

   d.   the current is not reduced below the hold current

17.4 Gate-Turnoff SCR (GTO)

_____ 12. What is the schematic symbol for a GTO?

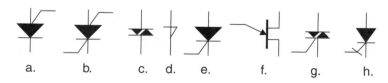

        a.   b.   c.   d.   e.   f.   g.   h.

_____ 13. GTO is also referred to as a _____.

   a.   gate control switch (GCS)

   b.   gated triac

   c.   gated SCR (GSCR)

   d.   gate turn-on SCR

17.5 Silicon-Controlled Switch (SCS)

____ 14. What is the schematic symbol for an SCS?

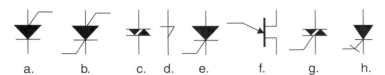

    a.   b.   c.   d.   e.   f.   g.   h.

____ 15. The silicon-controlled switch has four terminals: anode, cathode, _____.
   a. gate, and source
   b. emitter, and collector
   c. emitter, and source
   d. and two gates

17.6 Shockley Diode

____ 16. What is the schematic symbol for a Shockley diode?

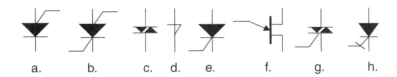

    a.   b.   c.   d.   e.   f.   g.   h.

____ 17. When forward biased, the Shockley diode blocks current flow until the _____.
   a. gate signal is reversed
   b. forward conduction current is reached
   c. trigger voltage is reached
   d. breakover voltage is reached

17.7 Diacs

____ 18. What is the schematic symbol for a diac?

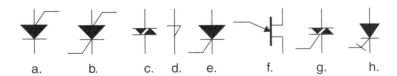

    a.   b.   c.   d.   e.   f.   g.   h.

____ 19. The diac blocks current flow in both directions until a breakover voltage is reached.
         Once the diac breaks over into conduction, it will continue to conduct until _____.
   a. the gate signal goes negative
   b. MT_2 is positive with respect to MT_1
   c. MT_2 is positive with respect to the gate
   d. the current is reduced below the hold current

17.8 Unijunction Transistors (UJTs)

____ 20. What is the schematic symbol for a UJT ?

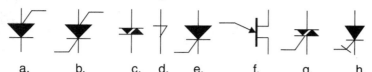

     a.   b.   c.   d.   e.   f.   g.   h.

____ 21. The unijunction transistor is a three-terminal device having _____.
   a. an emitter, base, and collector
   b. a source, gate, and drain
   c. a cathode, gate, and anode
   d. two bases (B_1 and B_2) and an emitter terminal

____ 22. The voltage needed to forward bias the UJT is called the _____.
   a. standoff voltage
   b. breakover voltage
   c. forward bias voltage
   d. UJT voltage

17.9 Programmable Unijunction Transistor (PUT)

____ 23. What is the schematic symbol for a PUT?

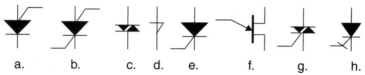

     a.   b.   c.   d.   e.   f.   g.   h.

____ 24. The PUT has three terminals: _____.
   a. emitter, base, and collector
   b. source, gate, and drain
   c. cathode, gate, and anode
   d. two bases (B_1 and B_2) and an emitter terminal

17.10 SCR Phase-Control Circuits

____ 25. An SCR phase-control circuit controls the power delivered to the load by controlling the _____.
   a. time that current will flow through the load each cycle
   b. resistance in series with the load
   c. resistance in parallel with the load
   d. resistance of the load

____ 26. The sum of the instantaneous voltage drops across the load and the SCR in a phase-control circuit must add up to _____.
   a. zero volts
   b. 120 V AC
   c. the source
   d. none of the above

____ 27. An SCR used in a phase-control circuit may receive more than one gate pulse per alternation. The first pulse turns on the SCR, and the second pulse _____.
   a. turns on the load
   b. turns off the SCR
   c. increases the current flow through the SCR
   d. has no effect

17.11 Triac Phase-Control Circuits

____ 28. The triac turns off at the end of each alternation because _____.
   a. the voltage reaches 170 V_p
   b. the current reaches zero
   c. there is an increase in current flow through the triac
   d. none of the above

____ 29. The triac and diac are both _____, permitting the phase control circuit to work on both positive and negative alternations.
   a. voltage control devices
   b. current control devices
   c. bidirectional devices
   d. bipolar devices

17.12 MOS-Gated Thyristors

____ 30. MOS-gated thyristors have MOSFET gate circuitry that permits high power thyristors to be switched on with a _____.
   a. current pulse
   b. light pulse
   c. resistive change
   d. voltage pulse

____ 31. A MOSFET input GTO type device makes it possible _____ high power loads with only a low power voltage pulse.
   a. to turn on
   b. to turn off
   c. to turn on and turn off
   d. to control the amplitude of current to

17.13 Troubleshooting Thyristor Circuits

____ 32. If an SCR or triac is not switching on, a technician can momentarily apply a gate pulse by connecting the gate to _____.
   a. a voltage level through a current limiting resistor
   b. ground
   c. the cathode or MT_1
   d. the anode or MT_2

___ 33. The gate circuitry of an SCR or triac can normally be disabled by connecting a jumper between the gate and _____.

    a. a current limiting resistor

    b. ground

    c. the cathode or MT_1

    d. the anode or MT_2

___ 34. You check the signal at TP_4 in the circuit in Figure 17–30 and there is a positive pulse occurring at 30° in each alternation, but the bulb is not lit. What is the most probable component failure?

    a. The zener diode is defective.

    b. The UJT is defective.

    c. The SCR is shorted between the cathode and anode.

    d. The gate of the SCR is open.

___ 35. The bulb will not light in the circuit in Figure 17–30. You check the signal at TP_4 and there are no positive pulses occurring. TP_3 is at a DC voltage equal to the zener diode voltage. What is the most probable component failure?

    a. The zener diode is defective.

    b. The UJT is defective.

    c. The SCR is shorted between the cathode and anode.

    d. The gate of the SCR is open.

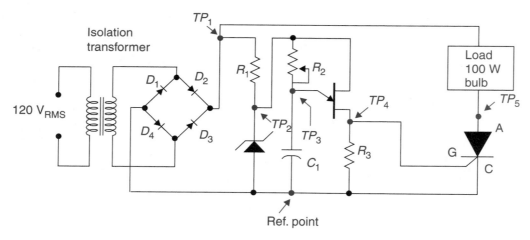

FIGURE 17–30

___ 36. The bulb in the circuit in Figure 17–31 is on full brightness and cannot be controlled by the 100 kΩ potentiometer. A check at TP_2 tells you that the voltage across the 0.2 µF capacitor can be controlled by the setting of the potentiometer. What is the most probable component failure?

    a. The diac is open.

    b. The 10 kΩ resistor is open.

    c. The triac is shorted between MT_1 and MT_2.

    d. The gate of the triac is open.

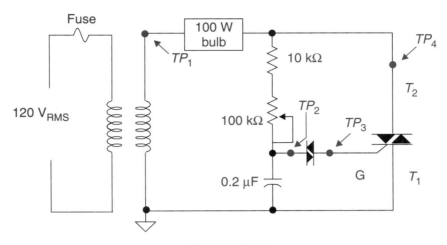

FIGURE 17–31

____ 37. The fuse in the circuit in Figure 17–31 opens. The bulb is removed and a second fuse opens when power is applied. What is the most probable component failure?

    a. One of the windings of the transformer is open.

    b. One of the windings of the transformer is shorted.

    c. The triac is shorted between MT_1 and MT_2.

    d. The 0.2 µF capacitor is shorted.

PROBLEMS

1. An SCR is connected to a 150 V DC source through a 10 Ω load. What will be (a) the current through the load and (b) the power dissipated in the load after the SCR is triggered?

2. Draw the current-versus-voltage characteristic curve for each device in Figure 17–32.

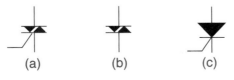

      (a)   (b)   (c)

FIGURE 17–32

3. When an SCR is conducting 8 A of current, estimate the power dissipated in the SCR.

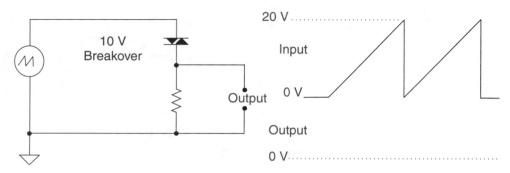

FIGURE 17–33

4. For the circuit in Figure 17–33, draw a diagram showing the output waveform. Show the voltage magnitude on the vertical axis and the timing on the horizontal axis.

5. For the circuit in Figure 17–34, draw a diagram showing the output waveform. Show the voltage magnitude on the vertical axis and the timing on the horizontal axis.

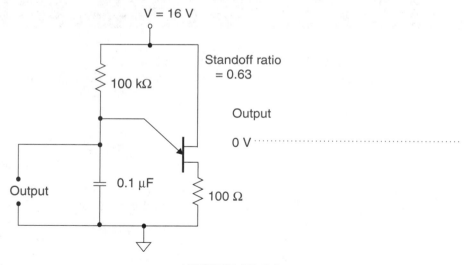

FIGURE 17–34

6. For the circuit in Figure 17–35, draw a diagram showing the output waveform. Show the voltage magnitude on the vertical axis and the timing on the horizontal axis.

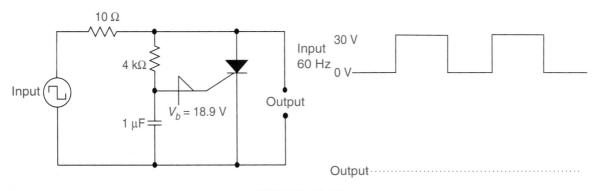

FIGURE 17–35

7. Calculate the frequency of the UJT oscillator shown in Figure 17–36. (Assume the standoff ratio for the UJT is 0.63).

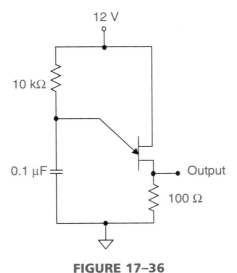

FIGURE 17–36

8. Select resistor and capacitor values to modify the circuit in Figure 17–36 so that the output frequency will be 2 kHz.

9. Draw and label the output waveform for the circuit shown in Figure 17–37 for maximum power to the load.

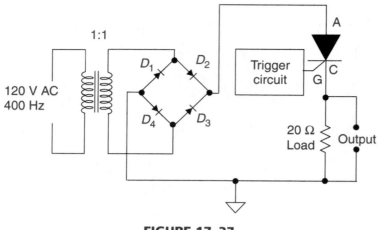

FIGURE 17–37

10. Calculate the maximum output power for the circuit in Figure 17–37.

11. Estimate the power dissipated by the SCR when the circuit in Figure 17–37 is set to deliver maximum output power.

12. Draw and label the output waveform for the circuit shown in Figure 17–37 for half power to the load.

13. Draw and label the output waveform for the circuit shown in Figure 17–38 for minimum power to the load.

14. Estimate the power dissipated by the triac when the circuit in Figure 17–38 is set to deliver minimum output power.

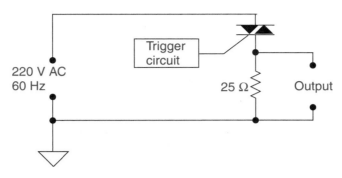

FIGURE 17–38

15. Draw and label the output waveform for the circuit shown in Figure 17–38 for half power to the load.

CHAPTER 18

Power Supplies

OBJECTIVES

After studying the material in this chapter, you will be able to describe and/or analyze:

○ a block diagram of a power supply system,
○ load regulation and line regulation,
○ the difference between linear and switching power supplies,
○ the function of a transformer in a linear power supply,
○ rectifier and filter circuits in a linear power supply,
○ linear regulators,
○ IC linear regulators,
○ the buck switching regulator circuit,
○ the boost switching regulator circuit,
○ the flyback switching regulator circuit,
○ off-line switching regulators,
○ IC switching regulators, and
○ troubleshooting procedures for power supplies.

18.1 INTRODUCTION TO POWER SUPPLIES

Electronics equipment generally requires DC power for proper operation, but electrical power distribution systems are designed to deliver AC power. A power supply, however, may input AC voltage from the distribution system and convert it to the desired DC voltage level needed to operate electronics equipment. For example, a 120 V AC line can be converted to 5 V DC, which is the voltage level needed to power a computer system. This type of power supply is called an AC-DC converter. Sometimes it is necessary to convert one DC voltage level to a second DC voltage level. For example, the −48 V used by the telephone industry may need to be converted to the +28V needed to power a cellular phone relay station. This type of power supply is called a DC-DC converter.

Figure 18–1 is a block diagram of an AC-DC converter power supply. The power supply contains a rectifier, filter, and regulator. The waveforms are shown at the input and output of each section of the power supply. The rectifier changes the AC input voltage to pulsating DC voltage. The filter section removes most of the ripple component and provides an unregulated DC (raw DC) input voltage to the regulator section. The regulator section is designed to deliver a constant voltage to the load under varying circuit conditions.

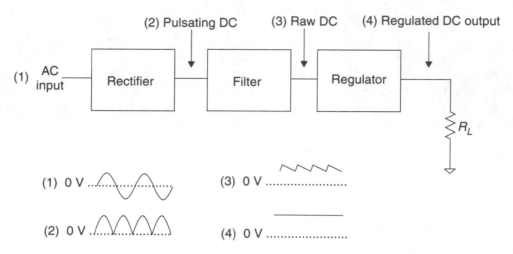

FIGURE 18–1   Block diagram of a power supply

Figure 18–2 shows a power supply in a single block. The two factors that can cause the voltage across the load to vary are fluctuations in input voltage and changes in load current requirements. If the input voltage increases, the output voltage increases, and if the input voltage decreases, the output voltage decreases. It is the function of the power supply circuitry to correct for input voltage fluctuations and maintain a constant voltage across the load. Changes in load current caused by load resistance variations can also change the output voltage of the power supply. A good power supply circuit will maintain a constant voltage across the load with changes in load current.

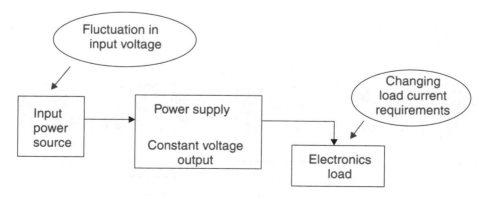

FIGURE 18–2   Power supply fundamentals

POWER SUPPLY RATINGS

Power supplies are rated for a wattage output at a given voltage. For example, a power supply could be rated for 240 W at 12 V. This power supply would be able to deliver 20 A of current to the load and maintain a constant 12 V output.

Regulation is a measurement of how well a power supply can maintain a constant voltage across the load under circuit variations. There are two types of regulations to be considered: load regulation and line regulation.

Load regulation is a measurement of how well the power supply maintains a constant voltage across the load with changes in load current. It is expressed as a percentage of the rated output. The formula for calculating load regulation is:

$$Load\ regulation = ((V_{NL} - V_{FL})/V_{FL}) \times 100\%$$

V_{NL} equals the voltage output with no current flowing through the load. V_{FL} equals the voltage output with maximum current flowing through the load. Some power supplies are not designed to operate with zero load current. For these power supplies, the output voltage is measured with minimum rated load current and maximum rated load current. Example 18.1 shows how the load regulation of a power supply can be calculated.

EXAMPLE 18.1

A power supply is rated for a 10 V output. The output voltage measures 10 V with no load and measures 9.8 V under full load. Calculate the load regulation.

$$Load\ regulation = ((V_{NL} - V_{FL})/V_{FL}) \times 100\%$$
$$Load\ regulation = ((10\ V - 9.8\ V)/9.8\ V) \times 100\% = 2\%$$

Traditionally, the output voltage with no load equals the rated voltage, and the output voltage decreases slightly as the current increases to full load current. However, many manufacturers today design their power supplies so the rated output voltage occurs at midrange load current. The load regulation value given is considered to be a plus/minus value from the rated output voltage. For example, a power supply is rated for 10 V and has a no load output voltage of 10.1 V and a full load output voltage of 9.9 V. If the traditional formula is used to calculate load regulation, the result would be 2%. However, the manufacturer states the power supply has a load regulation of 1%, since the output is always within 1% of the rated output voltage. The percent of regulation stated on data sheets should always be considered a plus/minus value unless otherwise stated.

EXAMPLE 18.2

Calculate the range of output voltage for a power supply with a rated output of 12 V and load regulation of 1.5%. If the power supply is rated for 360 W, what is the maximum load current the supply can deliver?

Step 1. Calculate the permitted change in output voltage. The percent of regulation is considered a plus and minus change. A 1.5% regulation would mean it can fluctuate 1.5% above the rated value and 1.5% below the rated value.

$$Change\ in\ voltage = 12\ V \times 1.5\% = 0.18\ V$$

Step 2. Calculate the output voltage range.

$$V_{out(max)} = 12\ V + 0.18\ V = 12.18\ V$$
$$V_{out(min)} = 12\ V - 0.18\ V = 11.82\ V$$

Step 3. Calculate the maximum load current.

$$I_{L(max)} = P_{max}/V_{out(rated)}$$
$$I_{L(max)} = 360\ W/12\ V$$
$$I_{L(max)} = 30\ A$$

Line regulation is a measurement of how well the power supply maintains a constant output voltage with changes in input voltage. Power supply systems are designed to operate over a given range of input voltages. Line regulation is given as a percentage of rated output for the given range of input voltages. It is also expressed as a percentage of change in output voltage per volt of change in input voltage. The input voltage is an AC voltage for an AC-DC converter and is a DC voltage for a DC-DC converter.

$$Line\ regulation = (\Delta V_{out}/V_{out(rated)}) \times 100\%\ (result\ in\ \%)$$
$$Line\ regulation = ((\Delta V_{out}/V_{out(rated)}) \times 100\%)/\Delta V_{input}\ (result\ in\ \%/V)$$

ΔV_{out} is the change in the output voltage caused by the change in the input voltage (ΔV_{input}) from the maximum rated input to the minimum rated input. Example 18.3 shows how line regulation is calculated.

EXAMPLE 18.3

> A power supply with a rated output of 12 V is designed to operate with a line input voltage of 120 V AC plus or minus 10%. What is the line regulation if the output measures 11.8 V with an input of 108 V AC and measures 12.1 V with an input of 132 V AC?
>
> $$Line\ regulation = (\Delta V_{out}/V_{out(rated)}) \times 100\%$$
> $$Line\ regulation = (0.3\ V/12\ V) \times 100\% = 2.5\%$$
>
> Or:
>
> $$Line\ regulation = ((\Delta V_{out}/V_{out(rated)}) \times 100\%)/\Delta V_{input}$$
> $$Line\ regulation = ((0.3\ V/12\ V) \times 100\%)/24\ V = 0.1\%/V$$

When line regulation is given as a percent (%), the range of change in input voltage needs to be known in order to make the percentage value meaningful. When line regulation is given as a percent per volt (%/V), the technician can calculate the change in output voltage caused by a one volt change in input voltage. In Example 18.3, the line regulation was 0.1%/V, so the output voltage will change 12 mV (12 V × 0.1%) for each volt of change in input voltage.

18.2 LINEAR VERSUS SWITCHING POWER SUPPLIES

There are two main categories of power supplies: linear power supplies and switching power supplies. They differ based on the type of regulator circuit used. Figure 18–3(a) shows a block diagram of a linear regulator circuit. The unregulated DC voltage drives a series circuit consisting of a variable resistance and a load. The control circuit consists of a voltage reference, a voltage divider, and a comparator amplifier. The idea is simple: the unregulated DC voltage input is always higher than the regulated output voltage, and the voltage difference is dropped across the series variable resistance. If the output voltage decreases, the control circuitry reduces the resistance of the series variable resistance, causing the output to increase. If the output voltage increases, the control circuitry increases the resistance of the series variable resistance, causing the output voltage to decrease. By controlling the voltage dropped across the resistance in series with the load, the circuitry maintains a constant output voltage across the load.

Figure 18–3(b) shows a block diagram of a switching regulator. The unregulated DC voltage drives a series circuit consisting of a switch, a filter, and the load. The control circuit consists of a voltage reference, a voltage divider, and a pulse width modulator (PWM). Figure 18–4 shows that the output voltage of the switch is equal to the unregulated DC when the switch is closed

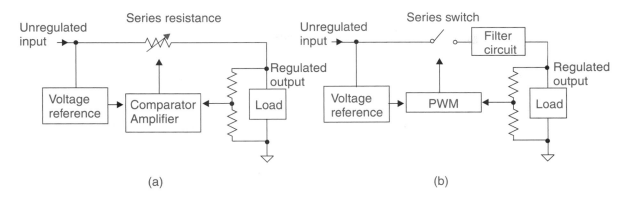

FIGURE 18–3 Linear regulation versus switching regulation

and equal to zero volts when the switch is open. The output of the filter is a DC voltage that equals the average of the voltage present at the output of the switch. During the "on time" of the switch, the output of the filter ramps up, and during the "off time" of the switch, the output of the filter ramps down. The fluctuation in the output voltage of the filter is only a few millivolts, so from a practical standpoint, the load receives steady-state DC voltage.

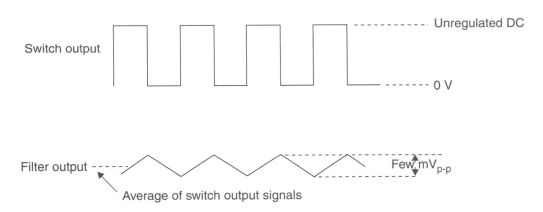

FIGURE 18–4 Switching waveforms

If the output voltage decreases, the control circuitry increases the "on time" of the switch. This causes the output to increase. If the output voltage increases, the control circuitry decreases the "on time" of the switch, causing the output to decrease. By controlling the switch timing, the circuitry maintains a constant output voltage across the load. The switching frequency is constant, normally between 50 kHz to 500 kHz. The switch on/off timing is controlled by the pulse width modulator (PWM), which controls the duty cycle of the switching signal.

A transistor (bipolar or MOSFET) will function as the variable resistance in linear regulators or as the switch in switching regulators. The transistor in linear regulators will operate in the linear area at all times. Considerable power will be dissipated by the transistor in linear operation, since the transistor has a voltage across it and a current flowing through it. The transistor in the switching regulator is in one of two states: on or off. If the transistor is on, there is current through the transistor but no voltage across the transistor and thus no power dissipation. If the transistor is off, there will be voltage across the transistor but no current through the transistor and thus no power dissipation. This results in switching regulators having greater efficiency than linear regulators. Example 18.4 calculates the theoretical efficiency of a linear regulator and a switching regulator.

EXAMPLE 18.4

Calculate the efficiency of the two regulator circuits shown in Figure 18–5.

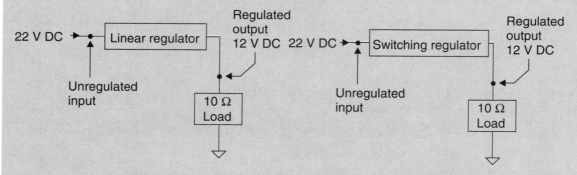

FIGURE 18–5   Linear regulator and switching regulator

Step 1. Calculate the load current for both circuits. Both circuits are delivering 12 V to a 10 Ω load, so both circuits will have the same load current.

$$I_L = V_{out}/R_L$$

$$I_L = 12 \text{ V}/10 \text{ Ω} = 1.2 \text{ A}$$

Step 2. Calculate the output power for both circuits.

$$P_{out} = I_L \times V_{out}$$

$$P_{out} = 1.2 \text{ A} \times 12 \text{ V} = 14.4 \text{ W}$$

Step 3. Calculate the voltage across the linear regulator.

$$V_{reg} = V_{in} - V_{out}$$

$$V_{reg} = 22 \text{ V} - 12 \text{ V} = 10 \text{ V}$$

Step 4. Calculate the power dissipated in the linear regulator. Since the linear regulator is in series with the load, the current through the regulator is equal to the load current.

$$I_{reg} = I_L$$

$$P_{reg} = V_{reg} \times I_{reg}$$

$$P_{reg} = 10 \text{ V} \times 1.2 \text{ A} = 12 \text{ W}$$

Step 5. Calculate the power dissipated in the switching regulator. Since the switching regulator operates as an on/off switch, theoretically there is zero power dissipated.

$$P_{reg} = 0 \text{ W}$$

Step 6. Calculate the efficiency of the linear regulator.

$$P_{total} = P_{out} + P_{reg}$$

$$P_{total} = 14.4 \text{ W} + 12 \text{ W} = 26.4\text{W}$$

$$Efficiency = (P_{out}/P_{total}) \times 100\%$$

$$Efficiency = (14.4 \text{ W}/26.4 \text{ W}) \times 100\%$$

$$Efficiency = 55\%$$

Step 7. Calculate the efficiency of the switching regulator.

$$P_{total} = P_{out} + P_{reg}$$

$$P_{total} = 14.4 \text{ W} + 0 \text{ W} = 14.4 \text{ W}$$

EXAMPLE 18.4 continued

$$Efficiency = (P_{\text{out}}/P_{\text{total}}) \times 100\%$$
$$Efficiency = (14.4 \text{ W}/14.4 \text{ W}) \times 100\%$$
$$Efficiency = 100\%$$

Example 18.4 makes it clear that switching regulators are inherently more efficient than linear regulators. Switching regulators are not, however, 100% efficient. Some power is dissipated during rise and fall times because for that short period of time, the switching transistor is in the linear operating area. Power is also dissipated when the transistor is in the "on" state because there will be a small voltage drop across the "on" (saturated) transistor. Even though there are some losses, switching power supplies have efficiencies in the area of 80% to 90%, whereas linear power supply efficiencies are around 30% to 60%.

Another advantage of the switching regulator is its ability to operate with a wide range of DC input voltages. If a switching power supply has a 25 V DC output, the signal out of the transistor switch must have an average of 25 V. Figure 18–6 shows two signals that have a 25 V average. If the unregulated DC is 50 V, a 50% duty cycle will generate a 25 V average. If the unregulated DC is 100 V, a 25% duty cycle will generate a 25 V average. The efficiency of the switching regulator is not greatly affected by changes in duty cycle; hence, the efficiency is not greatly affected by changes in the DC input voltage.

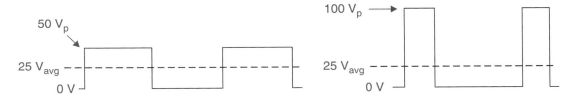

FIGURE 18–6 Duty cycle versus output voltage

The efficiency of linear regulators is greatly affected by changes in the DC input voltage. The larger the difference between the unregulated input voltage and the regulated output voltage, the less efficient the linear regulator. The load current flows through the linear regulator transistor, so a larger voltage drop across the transistor means increased power dissipation in the regulator circuitry without increased power being delivered to the load.

The switching regulator has high efficiency and the ability to operate over a wide range of input voltages, can step down or step up the output voltage, and can even change the polarity of the output voltage. The linear regulator has the advantage of simpler circuitry that does not require an inductor. This means the linear regulator can be totally integrated. Power supplies with power ratings less than 50 W often use linear regulators. Power supplies with power ratings above 50 W normally use switching regulators.

Switching regulators have become popular within the last fifteen years. The technician who is working on old equipment, however, may encounter some high power linear regulators.

18.3 LINEAR POWER SUPPLIES

TRANSFORMER SECTION

Figure 18–7 shows a schematic of a linear power supply. The power supply is designed to deliver 30 V DC to a load drawing a current between 0 mA to 600 mA. Let us examine the circuit

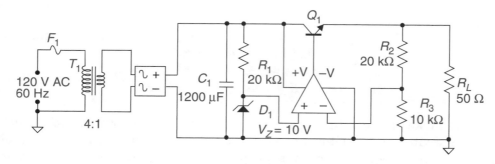

FIGURE 18–7   Linear power supply circuit

operation. The 120 V AC is connected to the primary of a 4:1 step-down transformer through a fuse. The transformer serves two useful functions: first, it permits the AC voltage to be stepped up or stepped down, and second, it provides an isolated AC output at the secondary. The AC signal on the primary (120 V_{RMS}) swings from positive 170 V_{peak} to negative 170 V_{peak} with reference to earth ground (AC common). The AC signal across the secondary (30 V_{RMS}) is isolated and swings from positive 42.4 V_{peak} to negative 42.4 V_{peak}, but the voltage is not in reference to earth ground. The isolated voltage on the secondary of the transformer will permit a ground reference to be connected to the output of the power supply without shorting a diode in the bridge rectifier.

Figure 18–8(a) shows a bridge rectifier connected to the output of a 1:1 transformer. The transformer isolates the input of the bridge from ground, permitting the output of the bridge to be referenced to ground. Figure 18–8(b) shows the input of the bridge connected directly to the AC line. AC common is earth ground. If the output of the bridge is referenced to ground, serious problems result. Effectively, such grounding shorts out diode D_3. During the negative half cycle of the AC input, there is a path for current flow directly through diode D_4. Destruction of diode D_4 is inevitable.

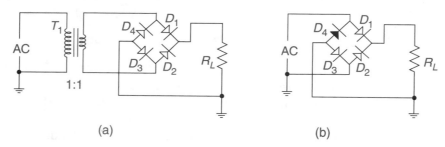

FIGURE 18–8   Bridge rectifier grounding problem

RECTIFIER SECTION

The secondary of the transformer connects to a bridge rectifier circuit that will full-wave rectify the 42.4 V_{peak} present at the secondary of the transformer. Figure 18–9 shows the pulsating DC

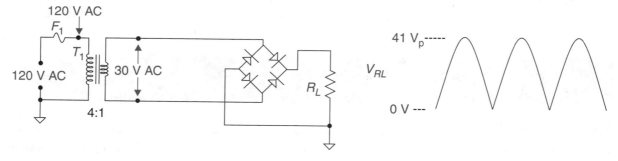

FIGURE 18–9   Rectifier circuit

voltage that would be present at the output of the rectifier if the rectifier were connected directly to a load without a filter circuit. The two diode voltage drops in the bridge reduce the peak output of the rectifier to 41 V.

FILTER SECTION

The function of the filter is to change the pulsating DC to raw DC by filtering out the ripple frequency. Capacitor C_1 (Figure 18–7) connects across the output of the rectifier and performs the needed filtering. The filter capacitor charges when the rectifier diodes are conducting and discharges when the diodes are not conducting. The magnitude of the ripple voltage ($V_{R(p-p)}$) will depend on the load current (I_L), the time the rectifier diodes are off (T_{off}), and the size of the filter capacitor. The magnitude of the ripple can be approximated using the following formula:

$$V_{R(p-p)} = I_L \times T_{off}/C$$

The numerator in the formula is equal to the total charge the capacitor must return to the circuit in each ripple cycle. If the input to the power supply is 60 Hz, the ripple will be 120 Hz. The period of 120 Hz ripple frequency is 8.3 mS. Figure 18–10(a) shows the relationship between the conduction and nonconduction time of the rectifier diodes. A typical power supply will have the diodes off 85% of the ripple period. For a 120 Hz ripple, the off time of the diodes is approximately 7 mS. Figure 18–10(b) shows the unregulated DC output of the filter section. During the positive slope of the waveform, the diodes are conducting, and during the negative slope, the diodes are off. It is not necessary for the filter section to remove all of the ripple component because the filter section will be followed by a regulator that will correct for any variations in the raw DC.

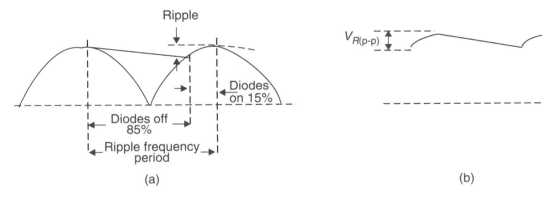

FIGURE 18–10　Rectifier diodes conduction period

VOLTAGE REGULATORS

Figure 18–11 shows the linear regulator section of the circuitry. The input to the regulator section is an unregulated DC voltage that is several volts higher than the desired regulated DC output. The regulated circuitry has been divided into four blocks: sense circuit, comparator circuit, reference voltage circuit, and series pass transistor. The regulated output is monitored by the sense circuit, which feeds back a portion of the output voltage to the comparator circuit. The comparator circuit compares the signal from the sense circuit to a signal from a reference voltage circuit. If there are any changes in the output voltage, the comparator circuit will send a correction signal to the series pass transistor, causing it to adjust its voltage drop to correct the output voltage.

The voltage reference circuit, comprised of a zener diode (D_1) and a resistor (R_1), provides a stable reference voltage of 10 V even with changes in the input voltage. The sense circuit consists

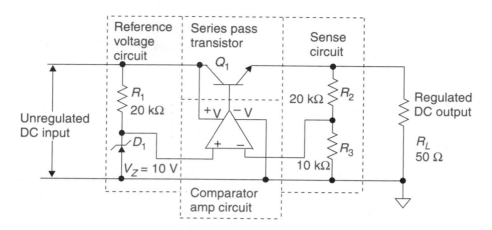

FIGURE 18–11   Series regulator block diagram

of a voltage divider across the output made up by resistors R_2 and R_3. The op-amp is used as a differential amplifier and compares the reference voltage (10 V) to the sense voltage, which equals one-third of the output voltage. The output of the op-amp provides a variable drive to the series pass transistor to maintain a constant output in spite of input voltage changes or output current changes. If the sense voltage is high, the drive to the pass transistor is reduced, thus reducing the current to the load until the output voltage drops to the point that the sense voltage is equal to the reference voltage. If the sense voltage is too low, additional drive is provided to the transistor until output current increases to the point that the sense voltage is equal to the reference voltage. In either case, the result is a regulated output voltage.

If the output of the op-amp is not in saturation, the voltage between the inputs must be zero. The output of the op-amp will not be in saturation if the regulator is operating within design limits. The key to calculating the output voltage is to realize that the circuit will automatically adjust itself to maintain the feedback voltage at the input of the op-amp equal to the reference voltage. Example 18.5 shows how the voltages and waveforms for a power supply circuit are determined.

EXAMPLE 18.5

Calculate and draw the waveforms at the test points shown in Figure 18–12. All test points are measured with respect to ground except TP_2, which is a differential voltage measured across the secondary of the transformer.

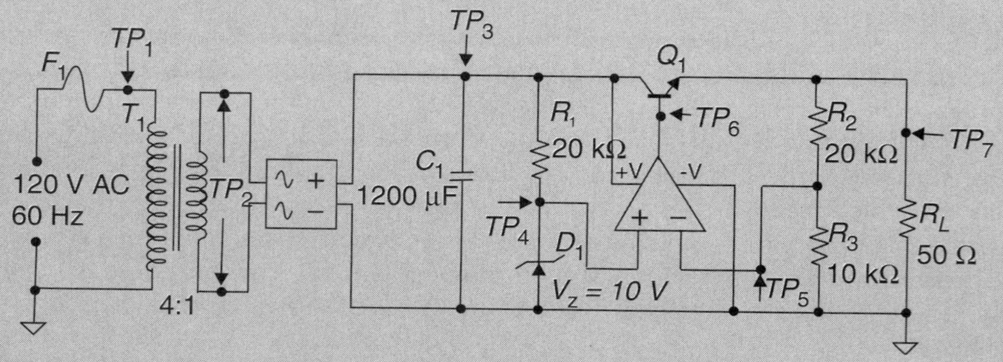

FIGURE 18–12   Linear power supply

EXAMPLE 18.5 continued

Step 1. Calculate the two DC input voltages to the comparator circuit (TP_4 and TP_5). The reference voltage is set by the value of the zener diode. The input from the sense circuit is assumed to be equal to the zener voltage, since the op-amp is not operating in saturation.

$$TP_4 = V_z = 10 \text{ V}$$
$$TP_5 = TP_4 = 10 \text{ V}$$

Step 2. Calculate the DC output voltage (TP_7). The voltage across R_3 was calculated in step 1. Knowing the voltage across R_3 and realizing the current drawn by the op-amp inputs is negligible, the voltage across R_2 can be easily calculated. The sum of the voltage drops across R_2 and R_3 equals the output voltage.

$$I_{R3} = V_{R3}/R_3 = 10 \text{ V}/10 \text{ k}\Omega = 1 \text{ mA}$$
$$I_{R2} = I_{R3} = 1 \text{ mA (Negligible current drawn by op-amp.)}$$
$$V_{R2} = I_{R2} \times R_2 = 1 \text{ mA} \times 20 \text{ k}\Omega = 20 \text{ V}$$
$$V_{\text{out}} = V_{TP7} = V_{R2} + V_{R3} = 20 \text{ V} + 10 \text{ V} = 30 \text{ V}$$

Step 3. Calculate the load current.

$$I_L = V_{\text{out}}/R_L$$
$$I_L = 30 \text{ V}/50 \text{ }\Omega = 0.6 \text{ A}$$

Step 4. Calculate the voltage at TP_6. The emitter-base junction of the NPN pass transistor is forward biased; therefore, the base will be 0.7 V higher than the emitter.

$$V_E = 30 \text{ V}$$
$$V_B = V_E + V_{BE}$$
$$V_B = 30 \text{ V} + 0.7 \text{ V} = 30.7 \text{ V}$$

Step 5. Calculate the peak-to-peak voltage across the secondary of the transformer (TP_2).

$$V_{s(\text{RMS})} = V_{p(\text{RMS})}/a \text{ } (a \text{ equals the transformer turns ratio.})$$
$$V_{s(\text{RMS})} = 120 \text{ V}/4 = 30 \text{ V}_{\text{RMS}}$$
$$V_{s(\text{peak})} = V_{s(\text{RMS})}/0.707$$
$$V_{s(\text{peak})} = 30 \text{ V}_{\text{RMS}}/0.707 = 42.4 \text{ V}$$
$$V_{s(\text{p-p})} = 84.8 \text{ V}$$

Step 6. Calculate the peak value of the raw DC voltage at TP_3. The peak DC voltage will equal the peak voltage at the secondary of the transformer minus two diode drops.

$$V_{TP3} = V_{s(\text{peak})} - 1.4 \text{ V}$$
$$V_{TP3(\text{peak})} = 42.4 \text{ V}_{\text{peak}} - 1.4 \text{ V} = 41 \text{ V}$$
$$V_{TP3(\text{peak})} = 41 \text{ V}$$

Step 7. Calculate the peak-to-peak ripple voltage at TP_3. The magnitude of the ripple voltage ($V_{Rp\text{-}p}$) will depend on the load current (I_L), the time the rectifier diodes are off (T_{off}), and the size of the filter capacitor. The filter capacitor and load current are known. The off time of the diodes can be calculated, since the input frequency is known.

$$F_{\text{ripple}} = 2 \times F_{\text{input}} \text{ (for a full-wave rectification.)}$$
$$F_{\text{ripple}} = 2 \times 60 \text{ Hz} = 120 \text{ Hz}$$
$$P_{\text{ripple}} = 1/F_{\text{ripple}}$$

EXAMPLE 18.5 continued

$$P_{ripple} = 1/120 \text{ Hz} = 8.3 \text{ mS}$$
$$T_{off} = P_{ripple} \times 85\%$$
$$T_{off} = 8.3 \text{ mS} \times 85\%$$
$$T_{off} = 7 \text{ mS}$$
$$V_{R(p\text{-}p)} = I_L \times T_{off}/C$$
$$V_{R(p\text{-}p)} = 0.6 \text{ A} \times 7 \text{ mS}/1200 \text{ μF}$$
$$V_{R(p\text{-}p)} = 3.5 \text{ V}_{p\text{-}p}$$

Step 8. Calculate the minimum voltage at TP_3. The minimum voltage at TP_3 will be equal to the maximum peak voltage minus the peak-to-peak value of the ripple voltage.

$$TP_{3(min)} = TP_{3(max)} - V_{R(p\text{-}p)}$$
$$TP_{3(min)} = 41 \text{ V} - 3.5 \text{ V} = 37.5 \text{ V}$$

Step 9. Draw the waveforms at all the test points for the circuit in Figure 18–12. Figure 18–13 shows the waveforms.

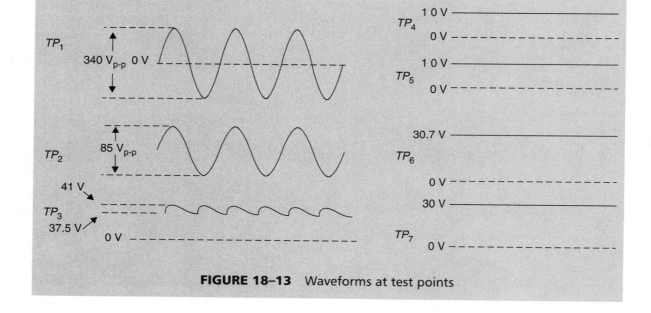

FIGURE 18–13 Waveforms at test points

18.4   IC LINEAR REGULATORS

Integrated linear regulators can provide the entire function of voltage regulation using a single IC. Figure 18–14 shows a three-terminal fixed regulator. The unregulated DC is applied to the input terminal, the circuit ground is connected to the ground pin, and the regulated DC is available at the output pin. Small capacitors (0.1 μF to 1 μF) are normally connected across the input and output to prevent high frequency oscillations.

Fixed voltage linear IC regulators are available in a variation of voltages, ranging from −24 V to +24 V. The current handling capacity of these ICs ranges from 0.1 A to 3 A. The input voltage can vary over a wide range, typically from two volts above the regulated output to 35 V. The closer

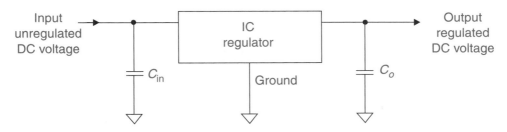

FIGURE 18–14 Integrated regulator

the unregulated voltage is to the regulated voltage, the greater the efficiency of the circuit. Example 18.6 shows how easily a power supply circuit can be designed using a fixed voltage linear IC regulator.

EXAMPLE 18.6

Design a 12 V DC power supply to deliver up to 1 A of current to the load. The input to the supply is 60 Hz at 120 V AC.

Step 1. Draw the schematic diagram of a power supply circuit. The circuit is shown in Figure 18–15. National Semiconductor makes the 78XX series of positive fixed regulator ICs. The XX in the part number is replaced with the voltage of the regulator. The negative fixed regulator uses the part number 79XX. Our design will use the 7812, which is a positive 12 V regulator rated for 1.5 A of output current when using the proper heat sink.

Step 2. Determine the value of bypass capacitors C_2 and C_3. The data sheet on the 78XX series of regulators states that for best stability, the input bypass capacitor (C_2) should be physically close to the input of the regulator and should have a value of 0.22 µF. The input bypass capacitor is needed even if the filter capacitor is physically close to the regulator. The large electrolytic capacitor filter will have high internal inductance and not function as a high frequency bypass; therefore, the large filter capacitor is bypassed by a small capacitor with a good high frequency response. The output bypass capacitor (C_3) improves the transient response of the regulator, and the data sheet recommends a value of 0.1 µF.

Step 3. Calculate a filter capacitor (C_1) that will limit the ripple voltage to 3.5 V_{p-p}. The formula $V_{R(p-p)} = I_L \times T_{off}/C$ can be solved for the value of C. Since the power supply has an input frequency of 60 Hz, the ripple frequency of the full-wave ripple will be 120 Hz. The off-time of the diodes for a 120 Hz ripple was calculated to be 7 mS (Example 18.4).

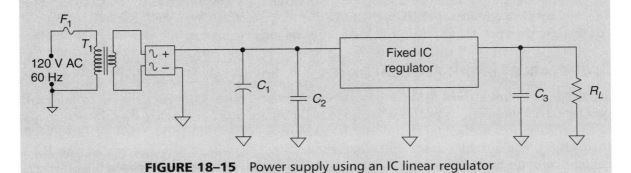

FIGURE 18–15 Power supply using an IC linear regulator

EXAMPLE 18.6 continued

$$C_1 = I_L \times T_{\text{off}}/V_{R(\text{p-p})}$$
$$C_1 = 1 \text{ A} \times 7 \text{ mS}/3.5 \text{ V}_{R(\text{p-p})}$$
$$C_1 = 2000 \text{ }\mu\text{F (Next higher standard size is 2200 }\mu\text{F.)}$$
$$C_1 = 2200 \text{ }\mu\text{F}$$

Step 4. Determine the maximum peak of unregulated DC voltage. The data sheet states the input voltage needed to maintain regulation is 14.6 V. This means the minimum value of the unregulated DC should never drop below 14.6 V. If the ripple voltage is 3.5 $V_{\text{p-p}}$, the unregulated DC will vary between 18.1 V and 14.6 V.

$$V_{\text{unreg(max)}} = 18.1 \text{ V}$$

Step 5. Calculate the secondary voltage needed. Two diodes conduct in the full-wave bridge rectifier on each alternation; therefore, the peak of the secondary voltage must be two diode drops higher than the peak of the unregulated DC.

$$V_{\text{sec(peak)}} = V_{\text{unreg(max)}} + 1.4 \text{ V}$$
$$V_{\text{sec(peak)}} = 18.1 \text{ V} + 1.4 \text{ V} = 19.5 \text{ V}$$
$$V_{\text{sec(RMS)}} = 0.707 \times V_{\text{sec(peak)}}$$
$$V_{\text{sec(RMS)}} = 0.707 \times 19.5 \text{ V}_{\text{p}} = 13.8 \text{ V}_{\text{RMS}}$$

Step 6. Select the transformer. The line voltage is rated for 120 V AC. However, let us design our circuit so it will function properly if the line voltage is lowered by 10% to 108 V AC. If the input to the primary of the transformer is 108 V AC and the output of the secondary needs to be 13.8 V AC, the transformer needs a turns ratio of 8:1 (108 V/13.8 V). The average current flowing in the secondary of the transformer will equal the DC load current. The power supply is designed to deliver 1 A of load current, so the secondary winding of the transformer needs to be rated for 1 A. The diodes in the bridge rectifier will also need a 1 A rating.

Step 7. Calculate the primary current and the rating of the fuse. The current through a fuse is normally two-thirds of the rated value of the fuse. Once the primary current is calculated, the fuse rating can be calculated.

$$I_P = I_S/a$$
$$I_P = 1 \text{ A}/8 = 125 \text{ mA}$$
$$I_{\text{fuse rating}} = I_P/0.67$$
$$I_{\text{fuse rating}} = 125 \text{ mA}/0.67 = 187 \text{ mA}$$

The power supply would use the next higher value fuse—in this case, a 250 mA fuse. Figure 18–16 shows the circuit with all the calculated values and the waveforms that will be present at the test points when the input is 120 V AC and the load current is 1 A.

ADJUSTABLE LINEAR REGULATOR ICs

Figure 18–17 shows the LM317 adjustable regulator. The LM317 is designed to maintain a constant 1.2 V between its output pin and its ADJ pin (across R_1). The current flowing in or out of the ADJ pin is negligible; therefore, the current flowing through resistor R_1 will also flow through R_2. The output voltage will equal the sum of the voltage drops across R_1 and R_2. The current through the two resistors is set by the value of R_1, but the resistance value of R_2 can be

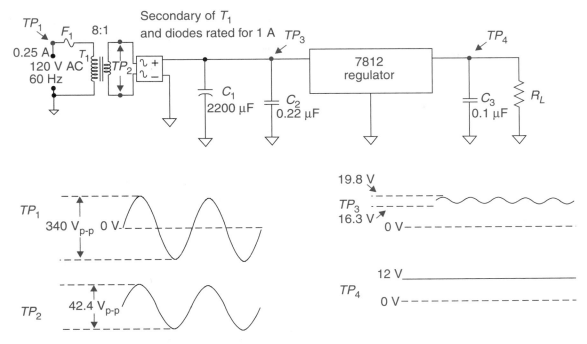

FIGURE 18–16   Power supply and waveforms

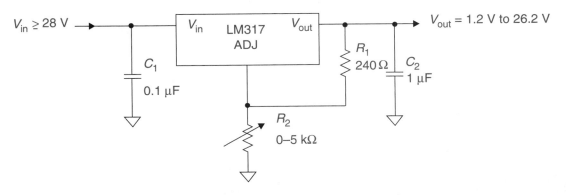

FIGURE 18–17   Adjustable IC regulator

adjusted to control the output voltage. The unregulated DC input must be higher than the regulated output voltage. As Example 18.7 demonstrates, if the value of the adjustable resistor is known, the output voltage can be easily calculated.

EXAMPLE 18.7

If the potentiometer in Figure 18–17 is adjusted to 3 kΩ, calculate the output voltage.

Step 1. Calculate the current through R_1. The voltage across the resistor connected between the output pin and the ADJ pin will always be 1.2 V for the LM317 regulator.

$$I_{R1} = 1.2 \text{ V}/R_1$$
$$I_{R1} = 1.2 \text{ V}/240 \text{ }\Omega$$
$$I_{R1} = 5 \text{ mA}$$

EXAMPLE 18.7 continued

> **Step 2.** Calculate the voltage across the potentiometer. Since there is no current flowing in or out of the ADJ pin of the IC, the current flowing through R_1 will also be flowing through R_2.
>
> $$V_{R2} = R_2 \times I_{R2}$$
> $$V_{R2} = 3 \text{ k}\Omega \times 5 \text{ mA}$$
> $$V_{R2} = 15 \text{ V}$$
>
> **Step 3.** Calculate the output voltage.
>
> $$V_{\text{out}} = V_{R1} + V_{R2}$$
> $$V_{\text{out}} = 1.2 \text{ V} + 15 \text{ V}$$
> $$V_{\text{out}} = 16.2 \text{ V}$$

Since all linear IC regulators are designed to maintain a constant voltage between the output pin and the common pin, any three-terminal IC regulator can be configured as in Figure 18–17 to obtain an adjustable output. The regulation is adversely affected, however, by current flowing in the common leg of the regulator. Because of this, manufacturers have designed adjustable regulators like the LM317 with near zero current flow in the common leg (ADJ).

Most IC regulators have built-in protection from overcurrent conditions and overheating. Regulator ICs are rated for a maximum output current. If circuit conditions cause the output current to reach maximum, then current limiting will kick in and hold the output current constant. If the temperature exceeds a safe operating range, a built-in circuit will sense an overheating condition and shut down the regulator until the temperature drops within the safe range.

Fixed linear IC regulators are extremely popular for applications when power requirements are under 25 W. They can be used as DC-to-DC step-down converters. Often fixed linear IC regulators are mounted directly on each card of an electronic system, thus distributing the task of regulation.

18.5   SWITCHING REGULATORS

INTRODUCTION

Linear regulators have one major disadvantage: a large part of the power is dissipated by the regulator rather than delivered to the load. A switching regulator offers improvements in system efficiency by dissipating only small amounts of energy in the regulator. By turning the pass transistor on and off instead of regulating it in a linear fashion, efficiency is improved.

BASIC SWITCHING REGULATOR OPERATION

The basic components that comprise the switching regulator circuit are a switching device, an inductor, and a diode. These components can be arranged into three configurations called the buck, boost, and flyback. The buck configuration is a step-down regulator, and the unregulated input voltage is always larger than the regulated output voltage. The boost configuration is a step-up regulator, and the unregulated input voltage is always smaller than the regulated output voltage. The flyback configuration permits the output to be smaller or larger than the unregulated input voltage; however, the output is always opposite in polarity.

Figure 18–18 depicts a buck regulator circuit. The switching device is shown as a simple switch but is actually a switching transistor (bipolar or MOSFET). Let us examine the waveforms in Figure 18–18 to gain an understanding of switching operations. When the switch is turned on, the inductor opposes current change, so the current through the switch (I_s) and inductor (I_L) ramps up slowly, rather than in a step increase. The increasing current stores energy in the inductor. When the switch is turned off, the current flowing through the switch goes to zero; however, the energy stored in the inductor causes current to continue to flow through the inductor. The diode provides a path for current flow during the off time of the switch. The current through the inductor (I_L) and diode (I_D) ramps down during the off time of the switch as the inductor returns its stored energy to the circuit.

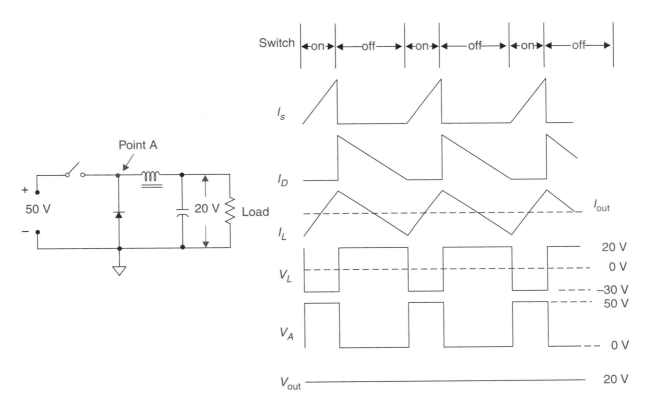

FIGURE 18–18 Buck regulator circuit

The voltage across an inductor equals the size of the inductor (L) times the change in current (ΔI) divided by the change in time (Δt) ($V_L = L \times \Delta I/\Delta t$). Since the up and down ramps of the current are at a constant rate, the voltage across the inductor will switch between two constant voltage levels. In our example, the voltage across the inductor (V_L) is switching between −30 V and 20 V. The voltage at the output of the switch (point A), switches between zero and the unregulated DC. In our example, the unregulated DC is 50 V, so the voltage at point A (V_A) switches between 50 V and 0 V. The output voltage will be the difference between the voltage at point A and the voltage across the inductor. In our example, the output voltage (V_{out}) would be a constant 20 V DC, as shown in Figure 18–18. The load current is equal to the average current through the inductor. Actually, there will be a small ripple voltage present on the output, but the capacitor across the output reduces this ripple to only a few millivolts peak-to-peak, which is acceptable in most applications.

Figure 18–19 shows the boost regulator circuit. The current flowing through the switch, inductor, and diode is the same as for the buck regulator. The difference is the way the voltage across the inductor adds to the unregulated DC to produce the output voltage. When the switch is turned on, the current flowing in the inductor stores energy in the inductor. The inductor produces a voltage that opposes the input voltage. When the switch is turned off, the inductor returns its energy to the circuit by producing a voltage that adds to the input voltage. The input voltage and the inductor add together to forward bias the diode. The current flowing through the diode charges the capacitor. In our example, the input is 50 V and the voltage across the inductor is 30 V, so the output voltage is 80 V.

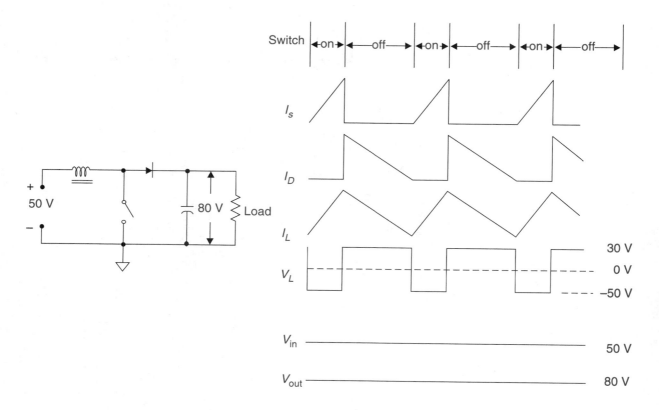

FIGURE 18–19   Boost regulator circuit

Figure 18–20 shows the flyback regulator circuit. The current flowing through the switch, inductor, and diode is the same as for the buck regulator. The difference is the way the input voltage and the voltage across the inductor interact. When the switch is turned on, the positive input voltage is present on the cathode of the diode, reverse biasing the diode. The input voltage across the inductor causes inductor current, which stores energy in the inductor. When the switch is turned off, the collapsing field around the inductor produces a negative voltage that forward biases the diode. The current through the diode delivers energy to the load and stores energy in the capacitor. The capacitor will maintain the negative voltage across the load during the time no diode current flows.

All three switching regulators previously discussed are operating in continuous mode. In continuous mode operation, the current through the inductor ramps up and ramps down, but there is always current flowing through the inductor. All three configurations (buck, boost, and flyback) can be operated in the discontinuous mode. In the discontinuous mode, the current through the inductor

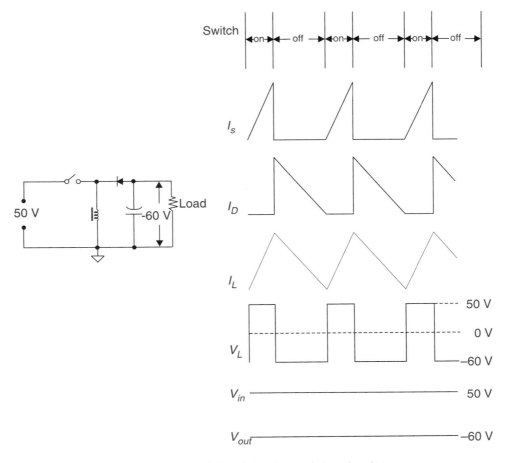

FIGURE 18–20   Flyback regulator circuit

ramps up and ramps down. When all the energy has been removed from the inductor, there is an idle time (t_i) when no current flows through the inductor. Figure 18–21 shows the current waveforms for discontinuous mode operation. Our discussion will be limited to continuous mode operation; however, the general concepts can be applied to discontinuous mode operation.

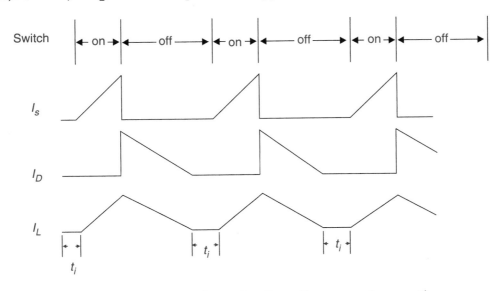

FIGURE 18–21   Waveforms for discontinuous mode operation

CONTROLLING THE OUTPUT VOLTAGE

Duty cycle is equal to switch "on time" divided by total time, multiplied by 100% [$(T_{on}/T_{total}) \times 100\%$]. The formulas below illustrate that the output voltages of buck, boost, or flyback regulators are dependent on the duty cycle (D) of the switching device. In all three configurations, the longer the duty cycle, the higher the output voltage.

Buck	Boost	Flyback
$V_{out} = V_{in}D$	$V_{out} = V_{in}/(1 - D)$	$V_{out} = V_{in}D/(1 - D)$

Figure 18–22 shows a diagram of a buck switching regulator with control circuitry. The pass transistor is cycled on and off at a fixed frequency. Within the period defined by the fixed frequency, the pulse width modulator controls the "on time" (duty cycle) of the transistor as necessary to provide the proper output voltage. The pulse width modulator is controlled by a comparator that compares a reference voltage to a feedback signal from the output. The conduction time of the pass transistor is controlled to keep the load voltage constant under varying load requirements.

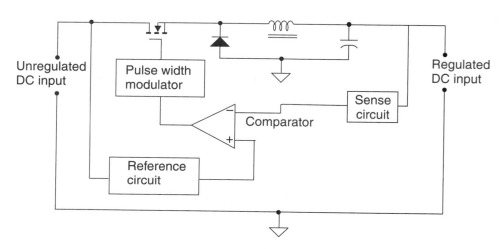

FIGURE 18–22   A basic switching regulator

Figure 18–23 shows the switching of the pass transistor under varying load requirements. Under minimum load conditions (low current drawn by the load), the pass transistor conducts for a short time each cycle. However, under maximum load conditions (high current drawn by the load), the pass transistor conducts for a longer time each cycle. The energy delivered to the load is a function of the time the pass transistor is conducting.

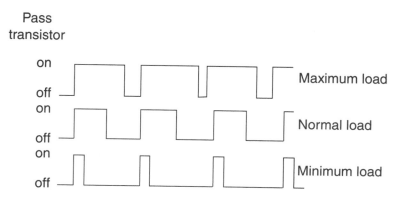

FIGURE 18–23   Output of the pulse width regulator

Suitable frequencies for switching the pass transistor are above 20 kHz (supersonic) and as high as 1 MHz. Energy storage requirements decrease as frequency increases, allowing the inductors and capacitors to be smaller. High frequencies are limited by power losses in the switching device caused by high speed switching.

When the pass transistor is off, no power is dissipated in the pass transistor, and when on, limited power is dissipated. This is because the voltage dropped across the transistor is small when the transistor is on (in saturation). Since very little power is dissipated in the pass transistor, the switching regulator is extremely efficient.

OFF-LINE SWITCHING POWER SUPPLIES

Since the magnitude of the unregulated DC voltage input is critical in linear regulators, power supplies using linear regulation require an input transformer to step the AC voltage up or down so the unregulated DC voltage will be at the proper amplitude. The amplitude of the unregulated DC is not as critical for switching regulators; therefore, a power supply using switching regulation can operate off-line without an input transformer as shown in Figure 18–24.

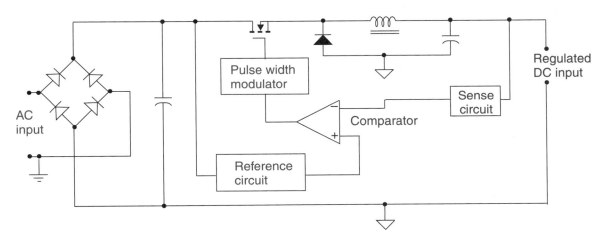

FIGURE 18–24   Off-line switching power supply

The AC line connects directly to the bridge rectifier circuit. The output of the bridge charges a filter (hold-up) capacitor to a DC value. If the AC line is 120 V$_{RMS}$, the bridge rectifier will charge the hold-up capacitor to approximately 170 V DC. This will serve as the unregulated DC input to the buck switching regulator. There are two shortcomings of the circuit shown in Figure 18–24. One, if the regulated DC output voltage is small compared to the unregulated DC voltage, the duty cycle of the switching transistor is going to be extremely short, and the peak current through the diodes will be excessive. Second, one side of the AC line is referenced to earth ground. The regulated output must remain isolated from earth ground or a diode in the bridge rectifier will short out, and the bridge will be destroyed.

A high frequency transformer can be added to the circuit in Figure 18–24 that will provide isolation and increase the duty cycle of the switching circuit. Figure 18–25 shows a buck switching regulator using a transformer. The switching transistor is in series with the primary of the transformer. When the switch turns on, a positive voltage pulse is delivered to the output inductor. The current in the inductor ramps up, and energy is stored. During the off time, the current through the inductor ramps down, and the energy is returned to the circuit. The sense circuit monitors the regulated output voltage and couples a signal back to the comparator through an optoisolator. The comparator compares the feedback signal to the output of the voltage

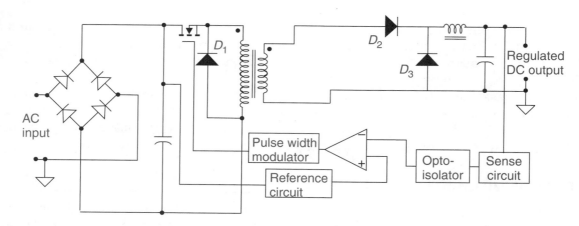

FIGURE 18–25   Transformer coupled buck switching regulator

reference circuit. The correction signal out of the comparator is the control signal to the pulse width modulator. The pulse width modulator adjusts the on/off time of the switching transistor as necessary to maintain a constant output voltage. Diodes D_1 and D_2 block the negative kick across the transformer from reaching the output circuitry.

The transformer permits the voltage on the secondary to be stepped down, which permits the circuit to operate with a longer duty cycle and reduces the peak current. The transformer along with the optoisolator provides isolation, permitting the regulated output voltage to be referenced to earth ground. The transformer also can be used for phase inversion if an output of opposite polarity is desired. The transformer provides the advantage of the power transformers used in linear power supplies; however, one big advantage of transformers used in switching regulators is their high frequency of operation. The physical size and cost of these high frequency transformers are just a fraction of the 60 Hz power transformers used in linear supplies.

Figure 18–26 is a flyback switching power supply using a high frequency switching transformer. This circuit is similar to the buck regulator; however, the transformer has the additional function of serving as the inductor for the flyback regulator. When the switching transistor is turned on, current flows through the primary, storing energy in the transformer. The primary and secondary of the transformer have opposite polarities, so during the "on time," the diode in the secondary is reverse biased, and the voltage across the load is maintained by the output filter capacitor. When the switching transistor is turned off, the polarity of the secondary is reversed,

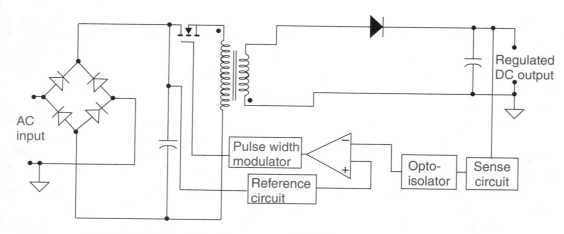

FIGURE 18–26   Transformer coupled flyback switching regulator

and the energy stored in the transformer is returned to the circuit, causing current to flow through the diode, delivering current to the load, and recharging the output capacitor.

18.6 IC SWITCHING REGULATORS

There are numerous integrated circuits available to support switching power supply designs. The National Semiconductor LM78S40 was designed for flexibility and minimum external part count. The LM78S40 contains all the building blocks needed to build a switching regulator in an uncommitted arrangement so it can be utilized in buck, boost, or flyback configurations. Figure 18–27 is a block diagram of the LM78S40 showing the building blocks. The blocks are two transistors used for switching, a 1.3 V reference, a diode, and a pulse width modulator consisting of an oscillator, comparator, and a logic circuit. There is an extra op-amp available on the IC that is usually not used in switching regulator designs but is available if needed.

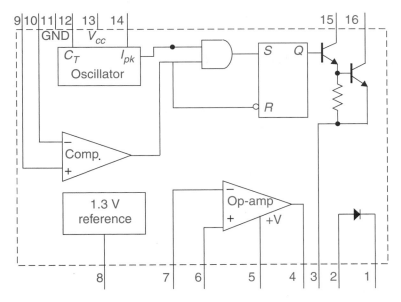

FIGURE 18–27   LM78S40 switching regulator IC

Figure 18–28 shows the LM78S40 configured as a buck regulator. The components in the shaded area are part of the LM78S40. Studying the control circuitry of the LM78S40 will give you an understanding of pulse width modulation. The voltage divider set up by the 1.3 kΩ resistor and the 15 kΩ potentiometer is adjusted so 1.3 V is fed back to the inverting input of the comparator when 10 V is present at the output. The 1.3 V reference voltage is connected to the noninverting input of the comparator.

Figure 18–29 shows a timing diagram for the control signals. The 0.01 µF timing capacitor sets the clock frequency to approximately 25 kHz. The clock signal drives one input of an AND gate and also drives an inverter. The output of the inverter drives the reset input of the SR flip-flop. The clock going low will cause the SR flip-flop set input (S) to go low and the reset input (R) to go high. These inputs will force the flip-flop output (Q) low, which will turn off the switching transistors. The output voltage will start dropping, causing the feedback signal to the comparator to drop.

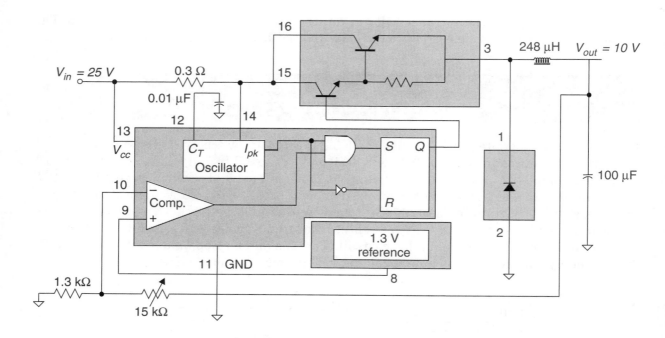

FIGURE 18–28 Buck regulator

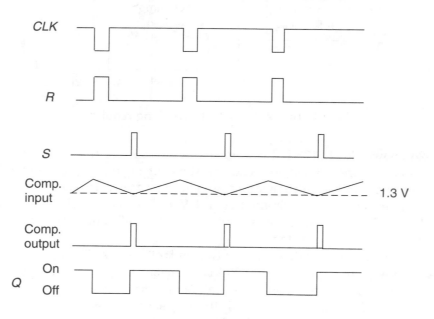

FIGURE 18–29 Buck regulator waveforms

When the feedback signal drops below the 1.3 V reference, the output of the comparator will switch high. The high signal from the comparator is ANDed with the high clock state, forcing the set input (S) of the flip-flop high, which causes the output of the flip-flop (Q) to go high. The high output of the flip-flop turns on the transistors, which causes the output voltage to rise until the flip-flop is reset by the next low state of the clock.

Figure 18–30 shows the timing diagram for an increased load current. An increase in load current causes the output voltage and feedback signals to drop more quickly. The comparator switches earlier in the cycle, which sets the flip-flop and turns on the transistors earlier. The transistors still turn off on the falling edge of the clock signal. The end result is that the transistors are on for a larger part of the cycle. By examining Figure 18–30, it can be seen that an increase in load current causes an increase in duty cycle. The opposite is also true; a decrease in load current causes a decrease in the duty cycle.

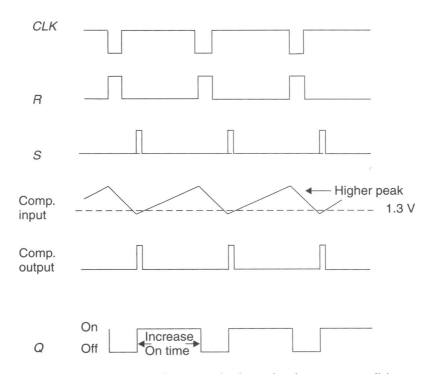

FIGURE 18–30 Waveforms under large load current conditions

The LM78S40 uses a 0.3 Ω resistor to monitor the peak current flow. If the voltage across the resistor exceeds 0.3 V, the clock output switches to a low state, which turns off the switching transistor and stops current flow. The current in Figure 18–28 is limited to a maximum peak of 1 A (0.3 V/0.3 Ω). Current limiting protects the IC from overcurrent conditions that could damage the LM78S40.

Figure 18–31 shows the LM78S40 configured as a boost regulator. When the transistor turns on, the right side of the inductor is connected to ground, and current flow increases through the inductor, storing energy. When the transistor turns off, the voltage across the inductor reverses and adds to the input voltage. This forward biases the diode, permits current to flow to the load, and charges the output filter capacitor. Control and current limiting circuitry function the same in boost and buck configurations.

Figure 18–32 shows the LM78S40 configured for an inverting flyback regulator. An external PNP transistor and diode are needed for this application. The built-in diode cannot be used

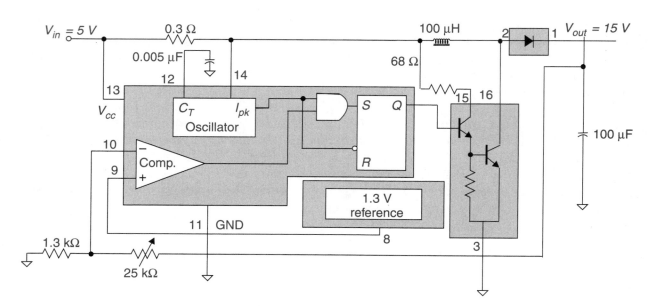

FIGURE 18–31   Boost regulator

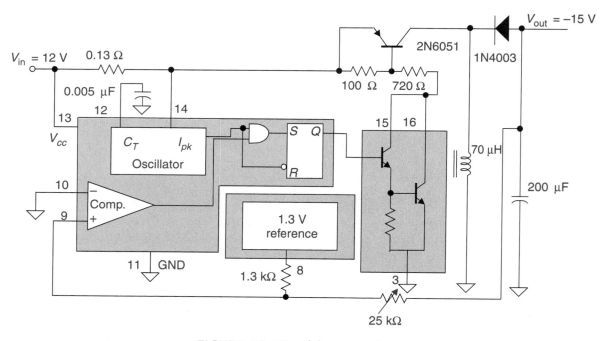

FIGURE 18–32   Flyback regulator

because the substrate of the IC is referenced to ground, and driving the cathode of the internal diode below ground will forward bias the substrate to the cathode junction. When the internal transistor turns on, the base of the external PNP transistor is driven negative, causing it to turn on. Current flows through the inductor, storing energy in the inductor. During this period of time, the top of the inductor is positive, and the diode is reverse biased. When the control circuitry cuts off the transistors, the voltage across the inductor reverses, and the top of the inductor becomes negative. The negative voltage across the inductor forward biases the diode, thus permitting current to the load, and charges the capacitor.

Since the output voltage is negative, the feedback circuitry is modified so the comparator will switch on a feedback signal of −1.3 V. The 1.3 kΩ resistor is connected to the 1.3 V positive reference. The voltage at the junction of the 1.3 kΩ resistor and the 25 kΩ potentiometer will equal zero when the feedback signal equals −1.3 V. The noninverting input of the comparator is connected to this point, and the inverting input is referenced to ground. This arrangement causes switching when the feedback signal crosses the −1.3 V level. The peak current limit is set to 2.3 A by the 0.13 Ω resistor (0.3 V/0.13 Ω).

Energy storage units (capacitors and inductors) in switching power supplies operate at high frequencies (20 kHz to 1 MHz) compared to linear power supplies that normally operate at 60 Hz. Therefore, they do not have to store as much energy during each cycle, allowing smaller value components to be used. At higher frequencies, the iron content in transformers and inductors can be reduced for the same reason. Since the transistor switch has less power to dissipate compared to the pass transistor of the linear power supply, it, too, can be smaller. With less power being dissipated, heat sinks also can be smaller. Hence, high efficiency and high operating frequencies permit switching power supplies to be more compact than linear power supplies.

18.7 TROUBLESHOOTING POWER SUPPLIES

The one thing all electronic systems have in common is that they all have a power source. If the system does not operate directly from batteries, the system is using a power supply circuit. The technician must have an understanding of power supply circuitry, since it is a part of most electronic systems. In fact, the first thing the technician normally checks is the DC voltage readings at the output of the power supply circuitry. If the readings are correct, the technician can proceed to troubleshoot the rest of the system. If the readings are incorrect, the power supply circuitry must be repaired.

Most power supplies have built-in protection circuitry to prevent over current, over voltage, or over temperature conditions. It is possible the power supply is operating normally but another section of the system has failed and is loading down the power supply. If the power supply can be disconnected from the system and connected to a dummy load, it can be checked for correct operation under proper loading conditions.

The technician should always do a good physical inspection of the power supply circuitry. Since all the energy consumed by the system is supplied through the power supply circuitry, there is sufficient energy available to do visual damage to the components.

A measurement of the raw DC voltage driving the regulator is a good place for the technician to start when troubleshooting power supplies. This reading permits the technician to determine whether the failure is in the rectifier and filter sections or in the regulator section. By measuring the peak-to-peak ripple voltage of the raw DC, an estimation of the load current can be determined. Low ripple voltage magnitude means low load current, and high ripple voltage magnitude means high load current.

Linear Regulated Power Supply

1. Calculate and record the DC voltages at the test points for the circuit in Figure 18–33.

   $TP_1 =$ _____ $TP_2 =$ _____

2. Calculate and record the load current for the circuit shown in Figure 18–33 with R_L set to 100 Ω and with R_L set to 15 Ω.

   $R_L = 100$ Ω $I_L =$ _____
   $R_L = 15$ Ω $I_L =$ _____

3. Calculate and record the peak-to-peak ripple voltage at test point TP_1 for the circuit shown in Figure 18–33 with R_L set to 100 Ω and with R_L set to 15 Ω.

   $R_L = 100$ Ω $TP_1 =$ _____
   $R_L = 15$ Ω $TP_1 =$ _____

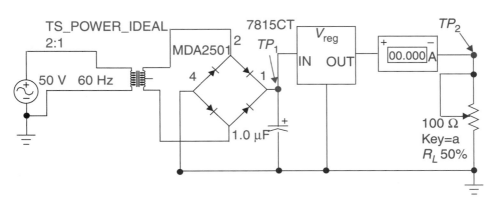

FIGURE 18–33

4. Simulate the circuit in Figure 18–33 and measure the DC voltages at the test points with R_L set to 100 Ω and with R_L set to 15 Ω.

   $R_L = 100$ Ω $TP_1 =$ _____ $TP_2 =$ _____
   $R_L = 15$ Ω $TP_1 =$ _____ $TP_2 =$ _____

5. Using the values measured in step 4, calculate the load regulation of the power supply circuit in Figure 18–33. Measure the no-load output voltage with a load current of 0.15 A. Measure the full-load output voltage with a load current of 1 A.

   $Reg_{(Load)} =$ _____

6. Simulate and record the peak-to-peak ripple voltage at test point TP_1 for the circuit shown in Figure 18–33 with R_L set to 100 Ω and with R_L set to 15 Ω.

   $R_L = 100$ Ω $TP_1 =$ _____
   $R_L = 15$ Ω $TP_1 =$ _____

Linear Regulated Power Supply

Caution! This experiment uses line voltage that can cause severe electrical shock.

I. Objective

To construct and analyze a regulated power supply and measure the load and line regulation.

II. Test Equipment

Voltmeter   Oscilloscope   Variac

III. Components

Resistors: (1) 100 Ω/2 W, (1) 220 Ω, (1) 15 Ω/10 W, (1) 2.5 kΩ potentiometer

Capacitors: (2) 0.1 μF, (1) 4700 μF

Transformer: (1) 120 V primary, 12.6 V center-tapped secondary rated at 1 A

ICs: (1) LM317, (1) 7812

(1) Heat sink

(1) Bridge rectifier

IV. Procedure

1. Construct the circuit in Figure 18–34 with no load ($R_L = \infty$) connected. (Heat sink 7812.)

2. Measure and record the DC and ripple values at TP_1 with no load.
   $V_{TP1(DC)} =$ _17.1 V_ $V_{TP1(ripple)} =$ _16.8mVp-p_

3. Measure and record the DC and ripple values at TP_2 with no load.
   $V_{TP2(DC)} =$ _12.1 V_ $V_{TP2(ripple)} =$ _17.6mVp-p_

4. Measure and record the DC and ripple values at TP_1 with full load ($I_L = 800$ mA and $R_L = 15\,\Omega/10\,W$). 100Ω
   $V_{TP1(DC)} =$ _17.1 V_ $V_{TP1(ripple)} =$ _12.5mVp-p_

5. Measure and record the DC and ripple values at TP_2 with full load ($I_L = 800$ mA and $R_L = 15\,\Omega/10\,W$). 100Ω
   $V_{TP2(DC)} =$ _12.2 V_ $V_{TP2(ripple)} =$ _4.80mVp-p_

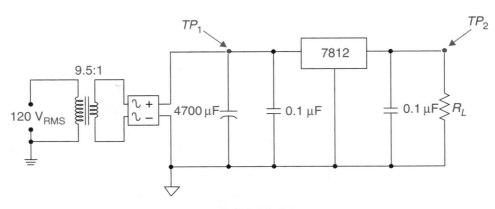

FIGURE 18–34

6. Calculate the load regulation for the power supply. $(V_{NL} - V_{FL})/V_{FL} \times 100\%$
    $Reg_{(Load)} =$ _82.6%_ $(12.2V - 12.1V)/12.1V \times 100\% = 82.6\%$

7. Calculate the efficiency of the regulator stage under full load operation.
    $Efficiency =$ _71%_ $(12.2/17.1) \times 100 = 71\%$

8. Connect the primary of the transformer to a variac and measure the DC voltage at TP_2
    with the variac adjusted to 132 V_{RMS}. ($R_L =$ ~~15 Ω~~ 100 /10 W.)
    $V_{TP2(DC)} =$ _12.3V_

9. Adjust the variac to 108 V_{RMS} and measure the DC voltage at TP_2. ($R_L =$ ~~15 Ω~~ 100 /10 W.)
    $V_{TP2(DC)} =$ _11.9V_

10. Calculate the line regulation for the power supply. $(\Delta V_{out}/V_{out(rated)}) \times 100\%$
    $Reg_{line} =$ _96.7%_ $(11.9V/12.3V) \times 100\%$
    $= 96.7\%$

11. Calculate and record the minimum and maximum output voltages for the circuit in Figure
    18–35.
    $V_{out(min)} =$ _12.3V_ $V_{out(max)} =$ _13.9V_

12. Construct the circuit in Figure 18–35. (Heat sink 317.)

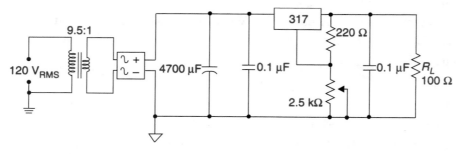

FIGURE 18–35

13. Adjust the potentiometer through the full range and record the minimum and maximum
    output voltages.
    $V_{out(min)} =$ _____ $V_{out(max)} =$ _____

V. Points to Discuss

1. Why is the ripple voltage larger at full load? Explain.

2. Under full load conditions (800 mA), what is the power dissipated by the regulator IC?
    Explain.

3. Under full load conditions (800 mA), what is the power dissipated by the bridge rectifier?
    Explain.

4. If the transformer in Figure 18–34 is replaced with a transformer with a 9:1 turns ratio, would the efficiency of the power supply improve? Explain.

5. Calculate the efficiency of the circuit in Figure 18–35 for a minimum output voltage and a maximum output voltage.

   $EffV_{out(min)}$ = _____ $EffV_{out(max)}$ = _____

6. What needs to be modified in the circuit in Figure 18–35 to permit the output voltage range to equal 1.25 V to 20 V?

Switching Regulators

1. Calculate and draw the waveforms at the test points for the circuit shown in Figure 18–36 for an input waveform with a duty cycle of 25%, 50%, and 75%.

   Duty Cycle 25%

   TP_1 ——————————————
   TP_2 ——————————————

   Duty Cycle 50%

   TP_1 ——————————————
   TP_2 ——————————————

   Duty Cycle 75%

   TP_1 ——————————————
   TP_2 ——————————————

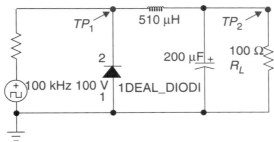

FIGURE 18–36

2. Simulate, measure, and draw the waveforms at the test points for the circuit shown in Figure 18–36 for an input waveform with a duty cycle of 25%, 50%, and 75%. (Note: Be sure to let the output voltage reach steady-state before making measurements.)

   Duty Cycle 25%

   TP_1 ——————————————
   TP_2 ——————————————

   Duty Cycle 50%

   TP_1 ——————————————
   TP_2 ——————————————

   Duty Cycle 75%

   TP_1 ——————————————
   TP_2 ——————————————

Switching Regulators

I. Objective

To construct and analyze a step-down switching regulator, a step-up switching regulator, and an inverting (flyback) switching regulator.

II. Test Equipment

Voltmeter   Oscilloscope

III. Components

Resistors: (1) 0.33 Ω or three 1 Ω in parallel, (1) 51 Ω/2 W, (1) 100 Ω, (1) 150 Ω/2 W, (1) 470 Ω, (1) 510 Ω/2 W, (1) 12 kΩ, (1) 100 kΩ (1) 100 kΩ potentiometer

Capacitors: (1) 0.01 μF, (1) 0.1 μF, (1) 470 μF

Inductor: (1) 500 μH

Transistors: (1) ECG232 PNP Darlington pair

IC: (1) LM78S40

IV. Procedure

1. Construct the circuit in Figure 18–37 with no load ($R_L = \infty$) connected and adjust the potentiometer for an output voltage of 10 V.

2. Measure and record the DC and ripple values at the output with no load.

   $V_{\text{out(DC)}} =$ _____ $V_{\text{out(ripple)}} =$ _____

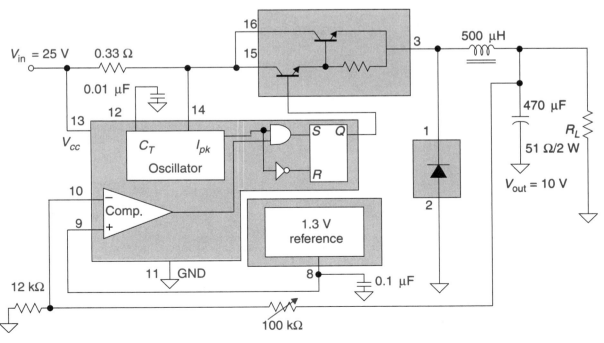

FIGURE 18–37

3. Measure and record the DC and ripple values at the output with full load ($I_L = 200$ mA and $R_L = 51\ \Omega/2$ W).

   $V_{out(DC)} =$ _____ $V_{out(ripple)} =$ _____

4. Calculate the load regulation for the power supply.

   $Reg_L =$ _____

5. Measure and record the input current.

   $I_{in} =$ _____

6. Calculate the efficiency of the regulator stage under full load operation.

   $Efficiency =$ _____

7. Construct the circuit in Figure 18–38 with no load ($R_L = \infty$) connected and adjust the potentiometer for an output voltage of 25 V.

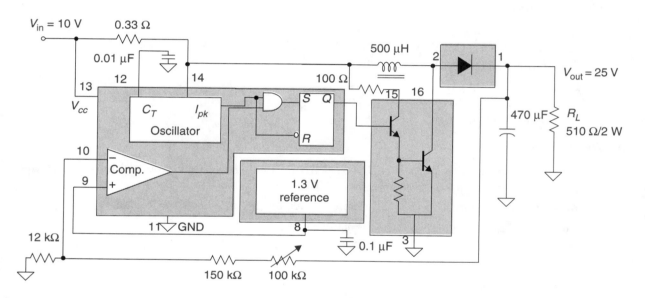

FIGURE 18–38

8. Measure and record the DC and ripple values at the output with no load.

   $V_{out(DC)} =$ _____ $V_{out(ripple)} =$ _____

9. Measure and record the DC and ripple values at the output with full load ($I_L = 50$ mA and $R_L = 510\ \Omega/2$ W).

   $V_{out(DC)} =$ _____ $V_{out(ripple)} =$ _____

10. Calculate the load regulation for the power supply.

    $Reg_L =$ _____

11. Measure and record the input current.

    $I_{in} =$ _____

12. Calculate the efficiency of the regulator stage under full load operation.

    $Efficiency =$ _____

13. Construct the circuit in Figure 18–39 with no load ($R_L = \infty$) connected and adjust the potentiometer for an output voltage of −15 V.

14. Measure and record the DC and ripple values at the output with no load.

    $V_{out(DC)} =$ _____ $V_{out(ripple)} =$ _____

15. Measure and record the DC and ripple values at the output with full load ($I_L = 100$ mA and $R_L = 150\ \Omega/2$ W).

    $V_{out(DC)} =$ _____ $V_{out(ripple)} =$ _____

16. Calculate the load regulation for the power supply.

    $Reg_L =$ _____

17. Measure and record the input current.

    $I_{in} =$ _____

18. Calculate the efficiency of the regulator stage under full load operation.

    $Efficiency =$ _____

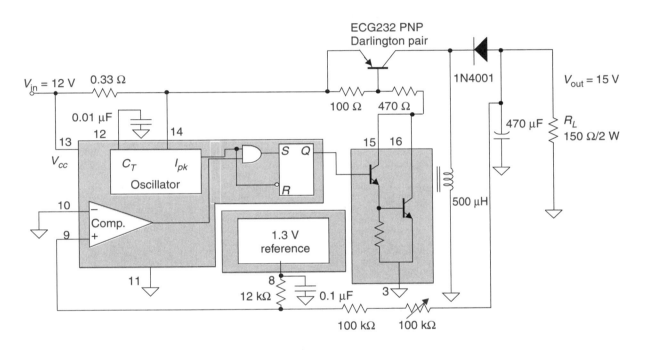

FIGURE 18-39

V. Points to Discuss

1. Compare the efficiency of the step-down switching regulator in Figure 18–37 to the maximum efficiency that could be obtained from a linear regulator with a 25 V input and a 10 V output.

2. The circuit in Figure 18–38 is a step-up switching regulator. Why is it possible to have a step-up regulator circuit using switching technology, but not linear technology?

3. The circuit in Figure 18–39 is a flyback switching regulator that gives an output that is opposite in polarity from the DC input voltage. Can linear technology be used to invert the polarity of the raw DC input? Explain.

QUESTIONS

18.1   Introduction to Power Supplies

____ 1. An AC-DC converter power supply contains all of the following except a _____.
    a. rectifier circuit
    b. filter circuit
    c. sample-and-hold circuit
    d. regulator circuit

____ 2. The ideal voltage regulator maintains a constant DC output voltage regardless of changes in _____.
    a. its input voltage
    b. its output voltage demand
    c. its load current demand
    d. both a and c

____ 3. Under full load conditions, _____.
    a. the input voltage is at its maximum value
    b. the load resistance is at a minimum value
    c. no load resistance is present
    d. the load current is at a minimum value

____ 4. _____ is a measurement of how well the power supply maintains a constant voltage across the load with changes in load current.
    a. Voltage control
    b. Load voltage control
    c. Load regulation
    d. Line regulation

____ 5. What is the load regulation of a power supply with a no load voltage of 16.5 V and a full load voltage of 15 V?
    a. 1.5%
    b. 9.1%
    c. 10%
    d. 90.9%
    e. 0%
    f. none of the above

____ 6. _____ is a measurement of how well the power supply maintains a constant output voltage with changes in input voltage.
    a. Voltage control
    b. Load voltage control
    c. Load regulation
    d. Line regulation

18.2 Linear Versus Switching Power Supplies

____ 7. A _____ maintains a constant output voltage by controlling the voltage dropped across the resistance in series with the load.

    a. controller regulator

    b. linear regulator

    c. switching regulator

    d. pulse regulator

____ 8. A _____ maintains a constant output voltage by controlling the on/off time of a switch in series with the load.

    a. controller regulator

    b. linear regulator

    c. switching regulator

    d. pulse regulator

____ 9. How will the output voltage differ from the input voltage of a linear regulator?

    a. It will be higher, but with much less ripple.

    b. It will be higher, but with about the same amount of ripple.

    c. It will be lower, but with much less ripple.

    d. It will be lower, but with about the same amount of ripple.

____ 10. The pass transistor in linear regulators will _____.

    a. operate in the linear area at all times

    b. be in cutoff at all times

    c. be in saturation at all times

    d. switch between cutoff and saturation

____ 11. Switching regulators have _____ than linear regulators.

    a. longer life

    b. simpler circuitry

    c. a higher cost in all cases

    d. greater efficiency

____ 12. Linear regulators have _____ than switching regulators.

    a. longer life

    b. simpler circuitry

    c. a higher cost in all cases

    d. greater efficiency

18.3 Linear Power Supplies

____ 13. The transformer serves two useful functions in a linear power supply: first, it permits the _____, and second, it provides an isolated AC output at the secondary.

    a. AC voltage to be stepped up

    b. AC voltage to be stepped down

    c. AC voltage to be stepped up or stepped down

    d. none of the above

_____ 14. The rectifier circuit converts _____.
   a.   AC input voltage to pulsating DC voltage
   b.   AC input voltage to filtered raw DC voltage
   c.   AC input voltage to regulated DC voltage
   d.   none of the above
_____ 15. What is the frequency of the ripple signal at the output of a full-wave rectifier?
   a.   one-fourth the input
   b.   the same as the input
   c.   1000 Hz
   d.   one-half the input
   e.   twice the input
   f.   none of the above
_____ 16. What is the voltage at TP_1 for the circuit shown in Figure 18–40?
   a.   0 V
   b.   12 V
   c.   0.7 V
   d.   20 V
   e.   8 V
   f.   none of the above

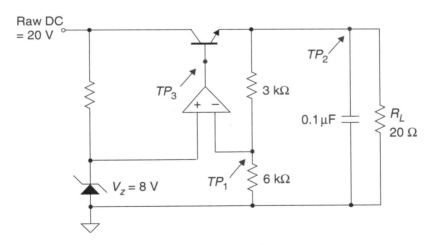

FIGURE 18–40

_____ 17. What is the voltage at TP_2 for the circuit shown in Figure 18–40?
   a.   0 V
   b.   12 V
   c.   0.7 V
   d.   20 V
   e.   8 V
   f.   none of the above

_____ 18.   What is the voltage at TP_3 for the circuit shown in Figure 18–40?
   a.   0 V
   b.   12 V
   c.   0.7 V
   d.   20 V
   e.   8 V
   f.   none of the above

18.4 IC Linear Regulators

_____ 19.   When using an IC linear regulator, the closer the unregulated voltage is to the regulated output voltage, the _____.
   a.   lower the output current
   b.   lower the output voltage
   c.   greater the efficiency of the circuit
   d.   lower the efficiency of the circuit

_____ 20.   The LM317 is _____.
   a.   a 17 V fixed linear regulator
   b.   a 31.7 V fixed linear regulator
   c.   an adjustable linear regulator
   d.   none of the above

_____ 21.   What is the waveform at TP_2 for the circuit shown in Figure 18–41?

(a)   0 V ———————————

(b)   0.7 V
      0 V

(c)   4 V
      0 V

(d)   0 V   8 V$_{\text{p-p}}$

(e)   8 V ———————————

(f)   none of the above

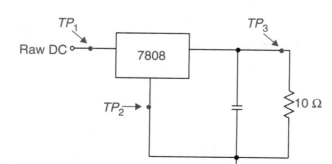

FIGURE 18–41

_____ 22. What is the waveform at TP_3 for the circuit shown in Figure 18–41?

(a)   0 V _____

(b)   0.7 V
       0 V

(c)   4 V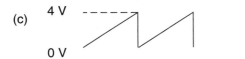
       0 V

(d)   0 V $8\ V_{p\text{-}p}$

(e)   8 V _____

(f)   none of the above

_____ 23. What is the power dissipated in the IC regulator in Figure 18–41?
    a.   0 W
    b.   8 W
    c.   9.6 W
    d.   3.2 W
    e.   6.4 W
    f.   none of the above

_____ 24. A series regulator circuit, like the one shown in Figure 18–41, dissipates 10 W, and the load dissipates 18 W. What is the efficiency of the circuit?
    a.   28 W
    b.   36%
    c.   10 W
    d.   18 W
    e.   64%
    f.   none of the above

18.5 Switching Regulators

_____ 25. The three configurations of switching regulator circuits are called the _____.
    a.   buck, boost, and flyback
    b.   buck, boost, and fixed
    c.   buck, doubler, and flyback
    d.   pass, stop, and inverter

_____ 26. When a switching regulator is operating in continuous mode operation, current always flows through the _____; however, in the discontinuous mode there is an idle time when no current flows through the _____.
    a.   load, load
    b.   inductor, load
    c.   inductor, inductor
    d.   capacitor, load

_____ 27. The output voltage of a switching regulator is dependent upon the _____.
  a. frequency of the switching
  b. duty cycle of the switching
  c. resistance of the load
  d. amplitude of the raw DC

_____ 28. An off-line switching power supply _____.
  a. is designed to function when the AC goes off-line
  b. is designed to operate portable equipment
  c. has the AC input line connected directly to the load
  d. has the AC input line connected directly to the rectifier

_____ 29. What is the waveform at TP_1 for the circuit shown in Figure 18–42?

  (a)   0 V _____

  (b)   0.7 V
         0 V

  (c)   20 V

         0 V

  (d)   0 V   20 V$_{p-p}$

  (e)   12 V _____

  (f)   none of the above

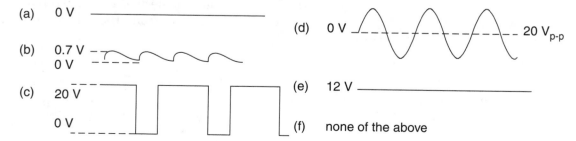

FIGURE 18–42

_____ 30. What is the waveform at TP_2 for the circuit shown in Figure 18–42?

  (a)   0 V _____

  (b)   0.7 V
         0 V

  (c)   20 V

         0 V

  (d)   0 V   20 V$_{p-p}$

  (e)   12 V _____

  (f)   none of the above

18.6 IC Switching Regulators

___ 31. The LM78S40 contains the building blocks needed to build a _____ mode switching regulator.

    a. buck

    b. boost

    c. flyback

    d. all of the above

___ 32. A linear regulator can be totally integrated, but a switching regulator cannot be totally integrated because _____.

    a. the switching transistor must be larger than the series pass transistor

    b. an inductor is needed in the output circuitry to store energy

    c. switching regulators dissipate too much power to be totally integrated

    d. it is impossible to integrate a PWM

18.7 Troubleshooting Power Supplies

___ 33. The one thing all electronic systems have in common is that they all have a _____.

    a. battery

    b. linear power supply

    c. switching power supply

    d. power source

___ 34. When troubleshooting an electronic system, the first thing the technician normally checks is the _____.

    a. bias voltages on the output power transistors

    b. DC voltage readings

    c. signal voltage gains of all amplifiers

    d. frequency responses of all circuitry

___ 35. A measurement of the raw DC voltage driving the regulator indicates there is a larger than normal ripple voltage present. This is a good indication that the load is _____.

    a. open

    b. drawing less than normal current

    c. drawing greater than normal current

    d. resistance

PROBLEMS

1. A power supply is rated for a 15 V output. The output voltage measures 15 V with no load and measures 14.6 V under full load. Calculate the load regulation.

2. A power supply is rated for a 15 V output. The output voltage measures 15.2 V with no load and measures 14.8 V under full load. Calculate the load regulation.

3. Explain how the manufacturers may express the values calculated in problems 1 and 2.

4. Calculate the maximum range of output voltage for a power supply with a rated output of 5 V and a load regulation of +/– 0.5%. If the power supply is rated for 200 W, what is the maximum load current the supply can deliver?

5. A power supply with a rated output of 25 V is designed to operate with a line input voltage of 120 V AC, plus or minus 10%. What is the line regulation if the output measures 24.8 V with an input of 108 V AC and measures 25.3 V with an input of 132 V AC? (Express the answer in % and in %/V.)

6. Calculate the efficiency of the regulator circuit shown in Figure 18–43.

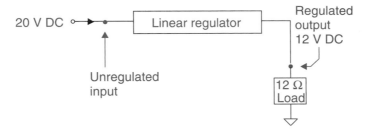

FIGURE 18–43

7. Calculate the ideal efficiency of the regulator circuit shown in Figure 18–44.

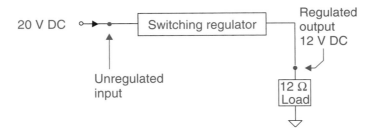

FIGURE 18–44

8. Calculate and draw the waveforms at the test points shown in Figure 18–45. All test points are measured with respect to ground except TP_2, which is a differential voltage measured across the secondary of the transformer.

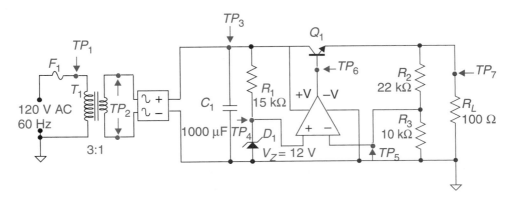

FIGURE 18–45

9. What is the approximate power dissipated in resistor R_1 shown in Figure 18–45?

10. Calculate the current flow through F_1 in Figure 18–45.

11. Calculate the efficiency of the regulator circuit shown in Figure 18–45.

12. Calculate and draw the waveforms at the test points shown in Figure 18–46. All test points are measured with respect to ground except TP_2, which is a differential voltage measured across the secondary of the transformer.

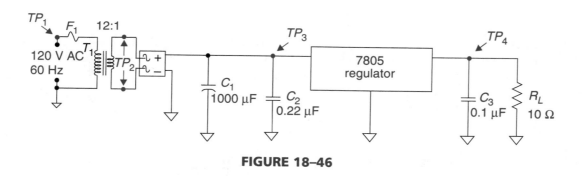

FIGURE 18–46

13. Calculate the current flow through F_1 in Figure 18–46.

14. Calculate the efficiency of the regulator circuit shown in Figure 18–46.

15. Design a 15 V DC power supply, using a linear regulator IC, to deliver up to 800 mA of current to the load. The input to the supply is 60 Hz at 120 V AC.

16. If the potentiometer in Figure 18–47 is adjusted to 10 kΩ, calculate the output voltage.

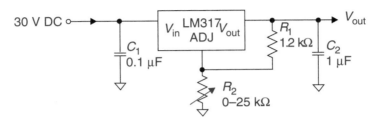

FIGURE 18–47

17. What is the adjustable range of output voltage for the circuit in Figure 18–47?

18. Sketch a power supply to provide a variable output from 1.2 V to 20 V, regulated by an LM317 IC regulator.

19. Draw the waveforms for the signals at TP_1 through TP_3 for the circuit in Figure 18–48. Assume the variable pulse width modulator is operating at 50 kHz and a 45% duty cycle. (Show amplitude and timing.)

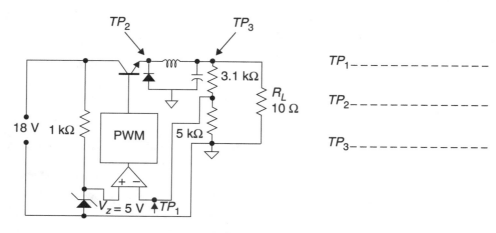

FIGURE 18–48

20. If the load resistor in Figure 18–48 is changed to 8 Ω, what effect would this have on the duty cycle of the variable pulse width modulator?

21. If a power supply filter circuit is to provide 8 A, with 4 $V_{p\text{-}p}$ of ripple, calculate the size of the filtering capacitor. (Assume the system is operating from 60 Hz.)

22. A filter capacitor chosen for the raw DC portion of a power supply limits ripple to 4 $V_{p\text{-}p}$ under full load conditions. How much ripple will there be when the supply is operating at half of its rated load current?

23. To develop 5 V at 3 A starting with 24 V DC, what is the maximum efficiency if switching technology is not used? Estimate the efficiency that can be obtained if switching technology is used.

CHAPTER 19

Data Conversion

OBJECTIVES

After studying the material in this chapter, you will be able to describe and/or analyze:

○ the differences between analog and digital signals,

○ the resolution of conversion systems,

○ how a DAC can be utilized to make a digital controlled amplifier,

○ a digital-to-analog conversion circuit,

○ a flash analog-to-digital converter,

○ a successive approximation analog-to-digital converter,

○ a sample-and-hold circuit, and

○ troubleshooting procedures for conversion systems.

19.1 INTRODUCTION TO DATA CONVERSION SYSTEMS

In this text you have been studying analog signals and circuits. An analog signal has a given range with an infinite number of points within that range. An analog signal with a range of 0 V to 10 V could have infinite possibilities. For example, it could be 5.9 V, or 5.99 V, or 5.999 V, or any point within the range. The world is generally analog in nature. For example, the temperatures in central Florida may range from 20°F to 100°F, with infinite possibilities between the two extremes.

In the last twenty-five years, the digital computer has become the brain of electronic controls. Figure 19–1 shows a system that is controlled electronically. Electronic sensors monitor the system under control. The sensors output analog signals that are conditioned by analog circuitry. The conditioned analog signals are converted to digital signals. The digital signals are processed by the digital computer and then converted back to analog signals, which are amplified to drive actuators that control the system. The digital computer will not be covered in this text, but the circuitry used to convert analog-to-digital and digital-to-analog will be covered in this chapter.

Digital signals, like analog signals, have a range; however, unlike analog signals, digital signals have a definite number of possibilities within the range. Another difference is the number of

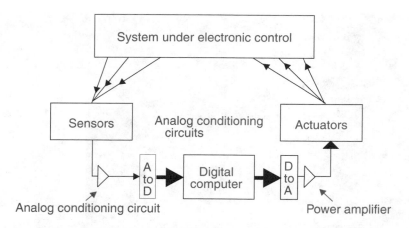

FIGURE 19–1   Electronic control system

conductors needed to support the signal transfer. Several conductive paths are needed to transfer a parallel digital signal, whereas an analog signal can be transferred over a single path.

19.2   RELATIONSHIP BETWEEN ANALOG AND DIGITAL SIGNALS

The graph in Figure 19–2 shows the relationship between an analog signal with a range of 0 to 10 and a 4-bit digital signal with a range of 0000 to 1111 base two. The graph can be used to convert an analog value to a digital value, or to convert a digital value to an analog value. Examples 19.1 and 19.2 demonstrate conversion using the graph.

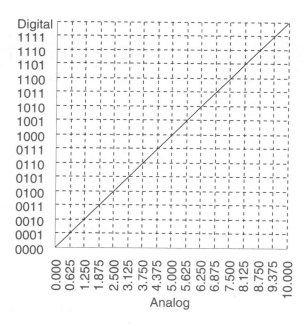

FIGURE 19–2   Digital versus analog graph

EXAMPLE 19.1

Convert an analog signal of 6.1 to its digital equivalent.

Step 1. Start the analog-to-digital conversion by finding the value 6.1 on the horizontal axis (x axis). (See Figure 19–3.)

Step 2. Move up vertically from the point located in step 1 until the diagonal line is intersected.

Step 3. Move left horizontally from the intersection of the diagonal line until the vertical axis (y axis) is intersected.

Step 4. Read the digital value from the vertical axis. In this case, the value is 1001. Read the digital value at the intersection point or the next value below the intersection point. Always round down because the electronic converter will always round down.

$$6.1 = 1001$$

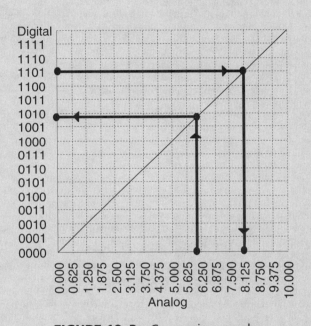

FIGURE 19–3 Conversion graph

EXAMPLE 19.2

Convert a digital signal of 1101 to its analog equivalent.

Step 1. Start the digital-to-analog conversion by finding the value 1101 on the vertical axis (y axis). (See Figure 19–3.)

Step 2. Move right horizontally from the point located in step 1 until the diagonal line is intersected.

Step 3. Move down vertically from the intersection of the diagonal line until the horizontal axis (x axis) is intersected.

Step 4. Read the analog value from the horizontal axis. In this case, the value is 8.125.

$$1101 = 8.125$$

Figure 19–4 shows a block diagram of an analog-to-digital converter (ADC) and a digital-to-analog converter (DAC) with the inputs and outputs determined (Examples 19.1 and 19.2). If the analog range and the digital range of a converter are known, the output for a given input can be calculated using a ratio. Examples 19.3 and 19.4 show this procedure.

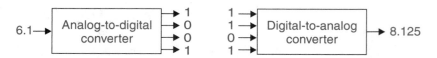

FIGURE 19–4   Analog-to-digital and digital-to-analog converters

EXAMPLE 19.3

If the ADC in Figure 19–4 has an input of 4.2 V, what is the digital output? The analog input range of the converter is 0 to 10, and the digital output range is 0 to 16.

Step 1. Set up and solve the ratio for the unknown value of the converter.

$$4.2 \text{ V}/10 \text{ V} = X/16$$

$$X = 6.72$$

Step 2. Round the number down to the highest digital possibility.

$$X = 6.72 \text{ rounded down } X = 6$$

Step 3. Convert base ten to base two using a digital word size that equals the word size of the converter.

$$X = 6_{10} = 0110_2$$

EXAMPLE 19.4

If the DAC in Figure 19–4 has an input of 1100, what is the analog output voltage? The digital input range is 0 to 16, and the analog output range is 0 to 10.

Step 1. Convert the base two input to a base ten number.

$$1100_2 = 12_{10}$$

Step 2. Set up and solve the ratio for the output voltage of the converter.

$$12/16 = V_{\text{out}}/10 \text{ V}$$

$$V_{\text{out}} = 7.5 \text{ V}$$

19.3   RESOLUTION OF CONVERSION SYSTEMS

One of the main factors that determines the quality of a conversion system is its resolution. The resolution of a converter is equal to the number of bits in the digital word of the converter. Figure 19–5 shows two conversion systems. Both systems have an input signal of 10 $V_{\text{p-p}}$ riding on a 5 V reference. The input signals are converted to digital signals, and then converted back to analog signals. One system has a resolution of four bits, and the other system has a resolution of five bits. Notice the output of the 5-bit system is a closer reproduction of the input. Resolution is a measurement of how close a digital signal can represent the analog signal.

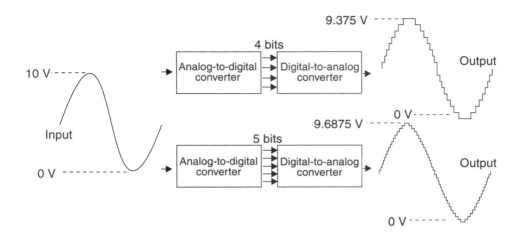

FIGURE 19–5 Resolution comparison

The number of bits determines the number of digital combinations that will represent the analog signal. The number of combinations can be calculated by raising 2 to the number of bits in the digital word (2^n where n is the number of bits). The 5-bit conversion system has 32 (2^5) unique digital words to represent the analog range. The 4-bit conversion system only has 16 (2^4) unique digital words to represent the analog range. The smallest analog increment a conversion system can detect or output is equal to the value of the least significant bit (LSB). The value of LSB is calculated by dividing the analog range of the converter by the maximum number of digital combinations supported by the converter. Example 19.5 shows the importance of resolution and how to calculate the LSB.

EXAMPLE 19.5

Calculate the smallest analog change that can be detected or output by both systems in Figure 19–5.

Step 1. Calculate the number of combinations in the digital word of each system. The number of possible combinations in a digital word is found by raising 2 to the number of bits in the word.

The system with 4-bit resolution equals: $2^4 = 16$

The system with 5-bit resolution equals: $2^5 = 32$

Step 2. Calculate the value of the LSB for each system.

The system with 4-bit resolution equals: 10 V/16 = 0.625 V

The system with 5-bit resolution equals: 10 V/32 = 0.3125 V

The system with a 5-bit resolution can detect a change in the analog input of 0.3125 V, but the 4-bit system is only able to detect a change of 0.625 V. A system with an 8-bit resolution and 0 V to 10 V input range can detect a change of 39 mV, and a system with a 10-bit resolution can detect a change of approximately 10 mV in the analog input signal. The size of the analog range also has a bearing on the smallest detectable analog increment (LSB). For example, a 10-bit system with an analog range of 0 V to 5 V would be able to detect an analog step of 4.9 mV (5 V/1024).

19.4   DIGITAL-TO-ANALOG CONVERSION

Figure 19–6 shows a diagram of a digital-to-analog converter. The digital inputs to the DAC control analog switches, which connect a –8 V reference or ground to the inputs of a D-to-A binary ladder network. If the digital input is a logic one, the reference voltage is connected. If the digital input is a logic zero, the ground is connected. The binary ladder is a network of resistors containing only two values. The actual ohm values of the resistors are not important as long as the ratio is 2:1. For example, if all the 1 R resistors equal 1 kΩ, then all the 2 R resistors must equal 2 kΩ. The currents developed by each leg of the ladder network combine and flow through resistor R_f, generating the output voltage. The binary ladder network can be extended to accommodate any length of digital input.

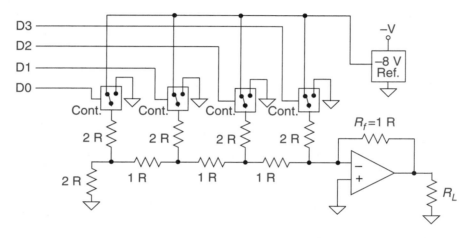

FIGURE 19–6   Binary ladder circuit

If the circuit in Figure 19–6 has a digital input of 1000_2, the switching circuits will connect the D3 input of the ladder to the reference voltage, and D2, D1, and D0 will be connected to ground. Figure 19–7(a) shows the condition that will exist when the circuit has an input of 1000_2 and the 1 R value equals 1 kΩ. The key to analyzing the circuit is understanding points labeled A, B, C, and D in Figure 19–7(a). The resistance to ground, looking left from any of these points, is always equal to 2 R, or 2 kΩ in this case.

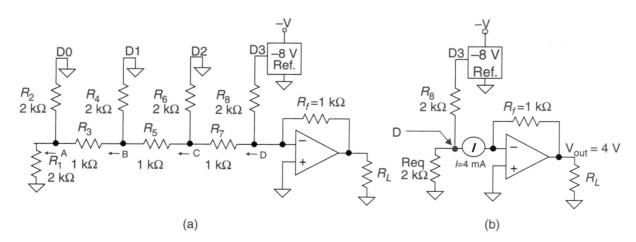

(a)   (b)

FIGURE 19–7   Ladder network with digital input of 1000_2

When looking left from point A, there is only one resistor (R_1), so the resistance to ground is 2 kΩ (2 R). Looking left from point B, resistors R_1 and R_2 are in parallel with an equivalent resistance of 1 kΩ. This parallel combination is in series with R_3 for a total resistance of 2 kΩ to ground looking left from point B. If the resistors to the left of point B are replaced with their equivalent resistance of 2 kΩ and the resistance to the left of point C is calculated, it will also turn out to equal 2 kΩ. The fact is, no matter how large the ladder network, the resistance from the junction points of the 1 R resistors to ground is always 2 R.

Figure 19–7(b) is a simplified circuit for a digital input of 1000_2. The resistors to the left of point D have been replaced with a single 2 kΩ equivalent resistance. The inverting input of the op-amp is at virtual ground; therefore, the current flowing through the current meter will equal the reference voltage divided by R_8 or 4 mA (–8 V/2 kΩ). The 4 mA flows through the feedback resistor (R_f), causing the output to equal 4 V. Note that the negative reference voltage causes the output voltage to be positive.

Figure 19–8(a) shows a condition that will exist when the circuit has an input of 0100_2. Figure 19–8(b) is a simplified circuit for a digital input of 0100_2. The resistors to the left of point C have been replaced with a single 2 kΩ equivalent resistance. Resistor R_7 connects to virtual ground of the op-amp; therefore, the current flowing through R_7 also flows through R_f and generates the output voltage. The current through R_7 calculates to 2 mA as shown in Example 19.6. The 2 mA flowing through R_f forces the output to 2 V.

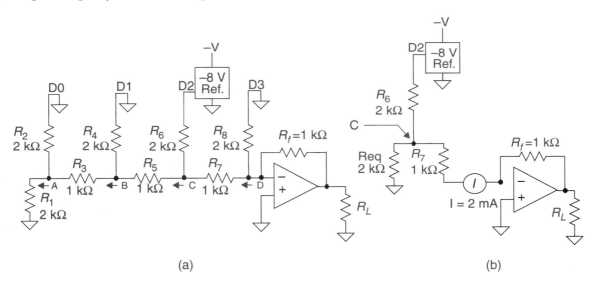

(a)   (b)

FIGURE 19–8 Ladder network with digital input of 0100_2

EXAMPLE 19.6

Step 1. Calculate the parallel equivalent $(R_{P(eq)})$ of R_{eq} and R_7.
$$R_{P(eq)} = R_{eq} \parallel R_7 = 2\ \text{k}\Omega \parallel 1\ \text{k}\Omega = 667\ \Omega$$

Step 2. Calculate the voltage across the parallel equivalent.
$$V_{RP(eq)} = V_{Ref} \times (R_{P(eq)}/(R_{P(eq)} + R_6))$$
$$V_{RP(eq)} = -8\ \text{V} \times (667\ \Omega/(667\ \Omega + 2\ \text{k}\Omega))$$
$$V_{RP(eq)} = -2\ \text{V}$$

Step 3. Calculate the current flow through R_7.
$$I_{R7} = V_{RPeq}/R7$$
$$I_{R7} = -2\ \text{V}/1\ \text{k}\Omega = -2\ \text{mA (The negative sign indicates direction of current flow.)}$$

EXAMPLE 19.6 continued

> **Step 4.** Calculate the output voltage (V_{out}).
> $$I_{Rf} = I_{R7} = 2 \text{ mA}$$
> $$V_{out} = I_{Rf} \times R_f = 2 \text{ mA} \times 1 \text{ k}\Omega = 2 \text{ V}$$

The important thing to note is that the most significant bit (D3) causes an output of 4 V, and bit D2 causes an output of 2 V. If this pattern continues, D1 will cause an output of 1 V, and D0 will cause an output of 0.5 V. If more than one bit is high, the voltage caused by each bit will add at the output. For example, an input of 1010_2 would drive the output to 5 V.

19.5   INTEGRATED DAC

Figure 19–9 shows a digital-to-analog converter circuit using a DAC0808. The DAC0808 contains a ladder network and needed switching circuitry to make a complete DAC. The accuracy of the output is dependent on the reference current supplied to pin 14 ($V_{\text{Ref}(+)}$) of the DAC0808. The circuit in Figure 19–9 uses an LH0070 precision 10 V voltage reference IC and a precision 5.000 kΩ resistor to obtain a reference current of precisely 2 mA. In applications where accuracy is not of great concern, the 2 mA current reference could be set by connecting a 2.5 kΩ resistor between pin 14 and the 5 V supply. The output at pin 4 is a current that ranges from 0 mA for an input of $0000\ 0000_2$ to 1.992 mA for an input of $1111\ 1111_2$. The current at pin 4 is converted to a voltage by using an op-amp as shown in Figure 19–9. The direction of current at pin 4 forces the output of the op-amp positive. The 5 kΩ feedback resistor sets the output voltage range from 0 V to 9.96 V.

The circuit in Figure 19–10 is a modified DAC circuit that has an output voltage swing from –5 V to 5 V. The output op-amp sums two currents: the current from pin 4 of the DAC and the current flowing through resistor R_3. The current through R_3 is a constant 1 mA (5 V/5 kΩ). With a digital input of $0000\ 0000_2$, the current at pin 4 of the DAC equals zero. Therefore, with a digital input of $0000\ 0000_2$, 1 mA flows through the feedback resistor (R_4), which drives the output to –5 V. As the digital signal increases, the current flow at pin 4 increases. This current flow through

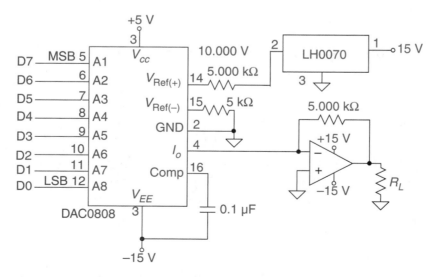

FIGURE 19–9   DAC circuit using a DAC0808 IC

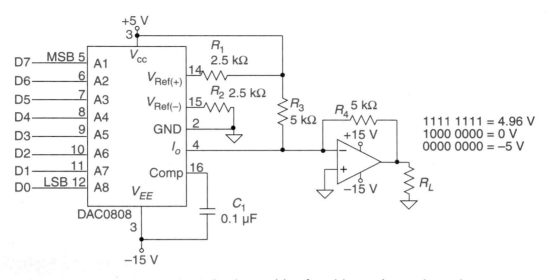

FIGURE 19–10 DAC circuit capable of positive and negative swing

the feedback resistor cancels the current caused by R_3. At a digital input of 1000 0000$_2$, the current flow at pin 4 is 1 mA, which totally cancels the current from R_3, causing 0 mA to flow through the feedback resistor and causing zero output voltage. At a digital input of 1111 1111$_2$, the current flow at pin 4 is 2 mA (actually 1.96 mA), causing a difference current of 1 mA through the feedback resistor, thus forcing the output to +5 V.

19.6 DIGITAL CONTROLLED AMPLIFIER

Figure 19–11 shows how a DAC can be used to design a digital controlled analog amplifier. Pin 14 of the DAC0808 acts as a virtual ground. The current flowing into pin 14 in Figure 19–11(a)

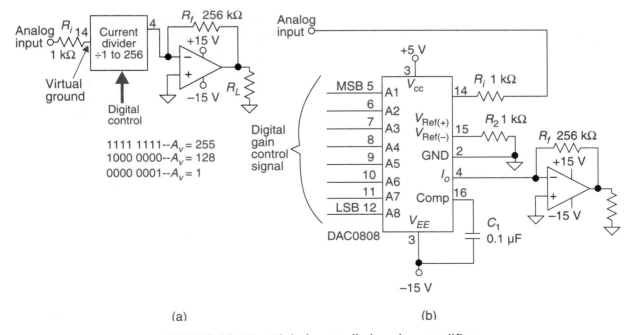

(a)   (b)

FIGURE 19–11 Digital controlled analog amplifier

is a function of the analog input voltage divided by resistor R_i. If the digital input is $1111\ 1111_2$, the current out of pin 4 equals the input current at pin 14 (actually 99.6%), and the gain is a function of R_f/R_i. Changing the digital control signal changes the current at pin 4 to equal a fractional part of the input current. For example, if the digital control signal is set to $1000\ 0000_2$, the current at pin 4 will equal one-half of the input current, and the voltage gain of the system will be reduced by one-half. The voltage gain of the circuit can be adjusted from 1 to 255 using an 8-bit digital input ranging from $0000\ 0001$ to $1111\ 1111_2$. A digital input of $0000\ 0000$ would always result in an output of zero volts. Figure 19–11(b) shows the complete circuit using the DAC0808. The analog input must be riding on a DC reference because pin 14 of the IC must remain positive at all times.

Figure 19–12 shows how a DAC can be used as an attenuator. The ratio of R_f to R_i is set to 1:1. The input signal is 10 V_{p-p} riding on a 6 V DC reference. If the digital control signal is $1111\ 1111_2$, the output signal is only slightly attenuated to a 9.96 V_{p-p} signal. However, when the digital input is set to $0000\ 0001_2$, the output signal is attenuated by –48 dB, and only 39 mV_{p-p} is present at the output of the op-amp.

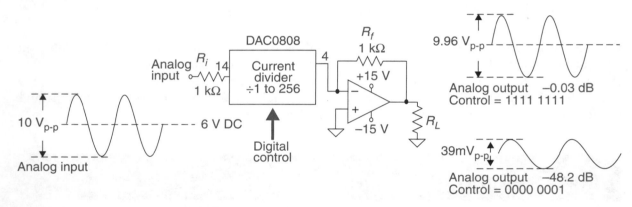

FIGURE 19–12   Digital controlled attenuator

19.7   ANALOG-TO-DIGITAL CONVERSION

There are two popular types of analog-to-digital converters: flash converters and successive approximation converters. As the name would indicate, the flash converter is quicker than the successive approximation converter, but it is also more expensive and consumes more power.

FLASH CONVERSION

Figure 19–13 shows a 2-bit flash converter. A reference voltage is applied to a voltage divider circuit, which divides the reference voltage into voltage levels that supply a switching reference for each comparator. The noninverting inputs of all comparators are connected to the analog input. Comparators are designed to output digital levels; negative saturation is equal to a logic zero, and positive saturation is equal to a logic one. The output of the comparator is converted to binary code by an encoder circuit. The binary code out of the encoder is loaded into an output latch. The enable signal permits the output of the converter to be sampled by digital circuitry.

An analog input between 0 V and 1 V will cause the circuit in Figure 19–13 to have a digital output of 00_2; an analog input between 1 V and 2 V will cause the circuit to have a digital output

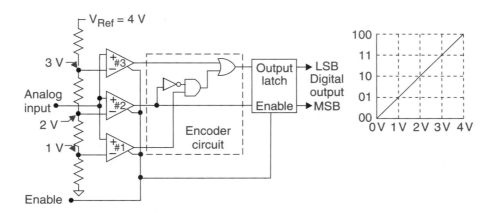

FIGURE 19-13 Flash converter

of 01_2; an analog input between 2 V and 3 V will cause the circuit to have a digital output of 10_2; and an analog input between 3 V and 4 V will cause the circuit to have a digital output of 11_2.

The biggest disadvantage of the flash converter is the complexity of the circuitry. The number of comparators needed depends on the number of output bits and is equal to 2 raised to the n power minus 1 $(2^n - 1)$, where n is the number of output bits. For example, an 8-bit flash converter requires 255 comparators followed by a complex encoder circuit. The large amount of circuitry dissipates considerable power. For example, National ADC0881 dissipates 600 mW compared to a range of 20 to 50 mW for successive approximation converters.

The advantage of the flash converter is speed; the only delay between the input and output is the switching time of the comparators and the propagation time through the decoder circuitry. The conversion time of a flash converter is measured in nanoseconds. National ADC0881, for example, has a conversion time of 50 nS, which is equal to 20,000,000 conversions per second. Successive approximation converters have conversion times in the range of 30 to 100 µS. Because of the speed of the flash converter, digital data is available on command. When the enable signal is activated, the digital data appears on the output data lines as shown in Figure 19–14.

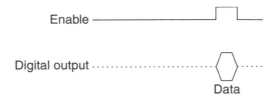

FIGURE 19-14 Flash converter timing diagram

SUCCESSIVE APPROXIMATION CONVERSION

A block diagram of a successive approximation converter and its timing diagram are shown in Figure 19–15. In our example, consider all signals to be active high. When the chip select and the start conversion pulse are activated together, this starts the conversion and causes the conversion complete signal to go inactive. When the conversion complete signal goes active, digital data is available. The digital data can be read by activating the chip select signal and the read signal together.

Figure 19–16 shows how 3.6 V is converted to a 4-bit digital equivalent using successive approximation. The conversion starts with the most significant bit (MSB), since it divides the total range in half. In our example, if D3 equals zero, the voltage is less than 2.5 V, and if D3 equals one, the voltage is over 2.5 V. With an analog input of 3.6 V, D3 will equal one. With D3

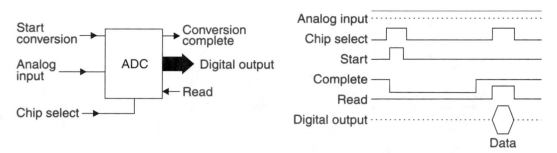

FIGURE 19–15   Successive approximation converter timing diagram

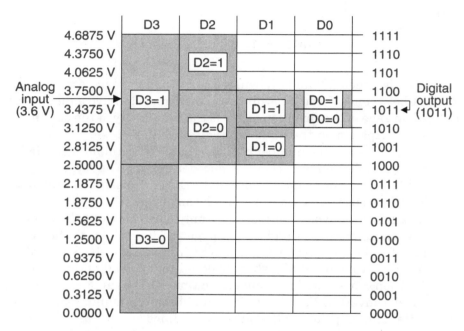

FIGURE 19–16   Successive approximation conversion

equal to one, the range is narrowed to 2.5000 V to 5.0000 V. The condition of D2 will divide the remaining range in half. Since the voltage (3.6 V) is between 2.5000 V and 3.7500 V, D2 will equal zero. This process of cutting the range in half continues until the value of the least significant bit (LSB) is determined. Review Figure 19–16. Notice how the digital approximation of the analog signal gets closer as we move toward the LSB. The analog voltage of 3.6 V is represented by the digital word 1011. Actually, the digital word 1011 represents any analog value from 3.4375 V to 3.7000 V. The longer the digital output word, the better the resolution; however, since the condition of each bit is determined in sequence from MSB to LSB, the longer the digital word, the longer the conversion time.

Figure 19–17 is a detailed block diagram of a 4-bit successive approximation converter. When the chip select (CS) signal and the start signal go high together, a pulse is generated that resets all the flip-flops and counter. The falling edge of this pulse starts the clock circuitry, which outputs eight clock cycles to drive the counter and the successive approximation register (SAR) circuitry. The output of the counter drives a decoder in the SAR.

Each flip-flop requires two clock cycles to be set to the proper level. Let us follow the sequence for the MSB (D3). At the start of the conversion, all flip-flops are reset, and the counter contains count 00, thus driving the decoder output Y0 high. On the falling edge of the first clock, Q3 will be set. The SAR now has an output of 1000, which drives the DAC. On the rising edge of the first

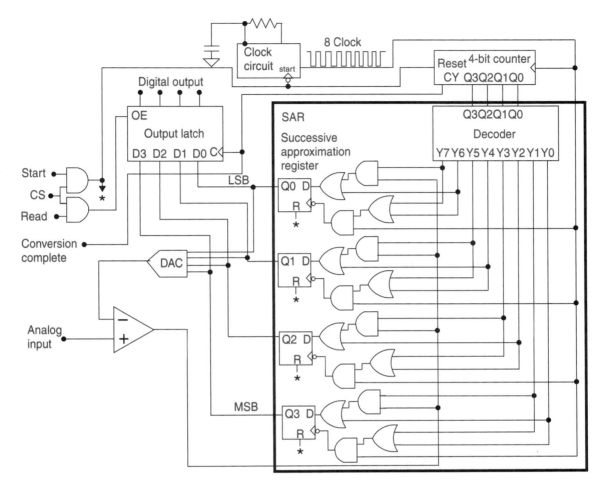

FIGURE 19–17   Four-bit successive approximation converter

clock, the counter is incremented to count 01, and the Y1 output of the decoder is driven high. The output of the DAC is compared to the analog input. The comparator output is a logic one if the analog signal is greater than the DAC output and a logic zero if the analog signal is less than the DAC output. On the falling edge of the second clock, the output of the comparator will be clocked into Q3. The rising edge of the second clock will increment the counter to count 02. The value in Q3 will remain for the rest of the conversion cycle, since the MSB flip-flop is enabled only on count 00 and count 01.

During the next six clock cycles, the remaining flip-flops will be set to the correct value that represents the analog input. On the rising edge of clock eight, the carry signal (CY) out of the counter will latch the digital word into the output register and drive the conversion complete signal active. The digital data can now be read from the output register by activating the chip select signal and the read signal together.

19.8   INTEGRATED ADC

Figure 19–18 shows an analog-to-digital converter circuit using an ADC0804. The ADC0804 is a successive approximation converter designed to interface to microprocessor systems. The circuit in Figure 19–18 has the ADC0804 configured so it automatically converts the analog input

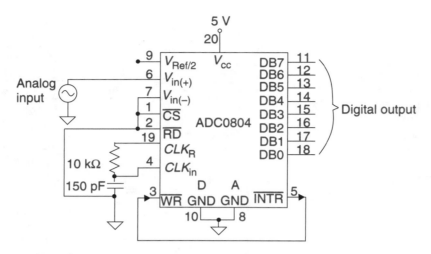

FIGURE 19–18   ADC0804 successive approximation converter

to an 8-bit digital word. The chip select (/CS) and the read (/RD) pins are tied active low, so the IC is selected and the digital word is present at the output at all times. The /INTR pin goes active low when the conversion is completed. When the /WR (write) pin is driven low, conversion starts. If the /INTR pin is connected back to the /WR pin, the ADC0804 will start a new conversion each time a conversion is completed.

The conversion time of the ADC0804 is controlled by the RC network formed by the 10 kΩ resistor and the 150 pF capacitor. The maximum conversion speed is 100 µS. If pin 9 ($V_{Ref}/2$) is left open, the analog voltage range is 0 V to 5 V. The analog range can be adjusted downward by connecting a voltage level to pin 9. The analog voltage range is equal to twice the voltage level connected to pin 9. For example, if 2 V is connected to pin 9, the voltage range would be 0 V to 4 V. The ADC0804 uses differential analog inputs. The circuit in Figure 19–18 has $V_{in(-)}$ connected to ground, permitting the ADC to operate with a single-ended input. With pin 7 ($V_{in(-)}$) not connected to ground, the ADC0804 can handle a differential analog signal input between pins 6 ($V_{in(+)}$) and 7 ($V_{in(-)}$).

19.9   SAMPLE-AND-HOLD CIRCUIT

It is important the analog signal into an ADC does not change during the conversion process. This is not a major problem with flash converters, since the analog signal is sampled one time and conversion is performed on the sample; however, with successive approximation converters, a changing analog input can cause erroneous results, since the input analog signal is compared several times during the conversion process. The answer is to use a sample-and-hold circuit preceding the ADC.

Figure 19–19 shows a sample-and-hold circuit consisting of two op-amp buffers, an analog switch, and a hold capacitor. The analog signal is connected to an input buffer, and the output of the input buffer is connected to a MOSFET switch. When the switch is turned on by the control input, the hold capacitor charges to the voltage level of the input signal. When the control signal turns off the switch, the voltage charge on the hold capacitor is retained and is coupled to the output through the output buffer. Since the output buffer has a high input impedance, the hold capacitor will remain charged. If the control input goes active, the hold capacitor quickly charges or discharges to the magnitude of the analog input, as shown by the waveform in Figure 19–19.

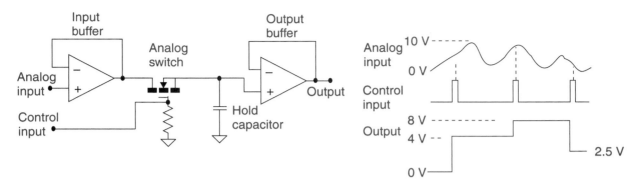

FIGURE 19–19 Sample-and-hold circuit

There are many IC sample-and-hold circuits available. One of these ICs may precede a successive approximation converter. At the start of conversion, the sample-and-hold circuit is enabled and an analog value is captured that will hold the input of the ADC constant through the conversion process. There are a number of ADCs on the market that have built-in sample-and-hold circuitry.

19.10 TROUBLESHOOTING CONVERSION SYSTEMS

Troubleshooting ADC and DAC circuits requires the technician to have a knowledge of analog electronics, digital electronics, and some programming skills. Most often converters are used in conjunction with microprocessors or microcomputers. The computer can be programmed to assist in troubleshooting. For example, let us assume an 8-bit DAC is connected to a computer system. Since the DAC is 8 bits, there will be a possibility of 256 digital input words. The computer can be programmed to continuously count through all possible input words (00H, 01H, 02H, . . ., FFH, 00H). If an oscilloscope is connected to the output of the DAC while the program is being executed, a sawtooth waveform with 256 discreet steps will be generated, as shown in Figure 19–20. If there is a problem with the DAC, the output sawtooth will be distorted.

The computer also can be used to troubleshoot an ADC circuit. A program can be written that continuously inputs the ADC and then outputs the results to a display. The analog input can be adjusted across the range while the display is monitored. If the ADC is functioning properly, all digital combinations will be displayed as the analog signal is adjusted from minimum to maximum.

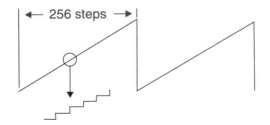

FIGURE 19–20 Output of DAC with input counting through all combinations

ADC and DAC

1. Calculate and record the digital output and the analog output for the circuit in Figure 19–21 for the input voltages shown in Table 19–1.

Analog Input	Calculated Digital Output	Simulated Digital Output	Calculated Analog Output	Simulated Analog Output
0 V				
1 V				
2 V				
3 V				
4 V				
5 V				

TABLE 19–1

2. Simulate the circuit in Figure 19–21 and adjust the potentiometer for the input voltages shown in Table 19–1. Then measure and record the digital output and the analog output at each level in Table 19–1.

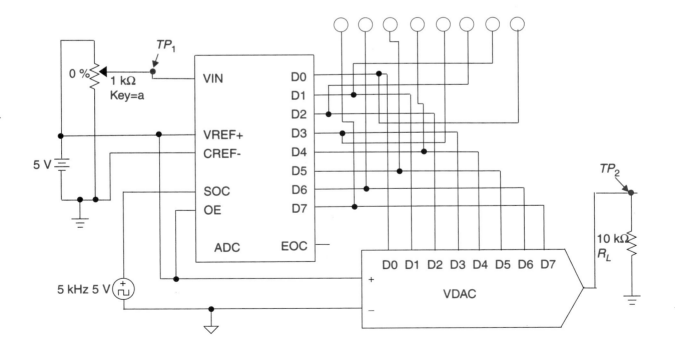

FIGURE 19–21

3. Replace the potentiometer with a sine wave signal source of 200 Hz set to 5 $V_{p\text{-}p}$ riding on a 2.5 V DC reference. Calculate and draw the waveforms you would expect to see at TP_1 and TP_2 for the circuit in Figure 19–21.

TP_1 ————————————

TP_2 ————————————

4. Simulate and record the waveforms you measure at TP_1 and TP_2 for the circuit in Figure 19–21.

TP_1 ————————————

TP_2 ————————————

5. Place a low pass RC filter (1 kΩ and 0.1 μF) with a cutoff frequency of 1600 Hz between the output of the DAC and R_L. Simulate and record the waveforms you measure at TP_1 and TP_2 for the circuit in Figure 19–21.

TP_1 ————————————

TP_2 ————————————

ADC and DAC

I. Objective

To experimentally verify the output of an ADC and a DAC.

II. Test Equipment

5 V power supply	–12 V power supply
Voltmeter	Dual-trace oscilloscope
Function generator	

III. Components

(1) ADC0804, (1) DAC0808, (1) 74LS240, (1) LM741

Resistors: (8) 330 Ω, (3) 2.4 kΩ, (2)10 kΩ, (1) 10 kΩ potentiometer

Capacitors: (1) 0.1 μF, (1) 150 pF

(8) LEDs

IV. Procedure

1. The analog input range of the ADC circuit in Figure 19–22 is 0 V to 5 V. Calculate the digital output for the following analog inputs:

Analog Input	Digital Output
0 V	_____
1 V	_____
2.5 V	_____
4 V	_____
5 V	_____

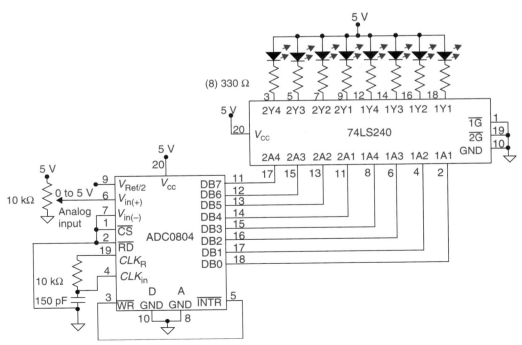

FIGURE 19–22

2. Construct the circuit in Figure 19–22.

3. Adjust the analog input to each of the following values and record the digital word present on the LED at each setting.

Analog Input	Digital Output
0 V	_____
1 V	_____
2.5 V	_____
4 V	_____
5 V	_____

4. Calculate the digital output and the analog output of the circuit in Figure 19–23 for the analog values given below:

Analog Input	Digital Output	Analog Output
0 V	_____	_____
1.25 V	_____	_____
2.5 V	_____	_____
3.75 V	_____	_____
5 V	_____	_____

5. Construct the circuit in Figure 19–23.

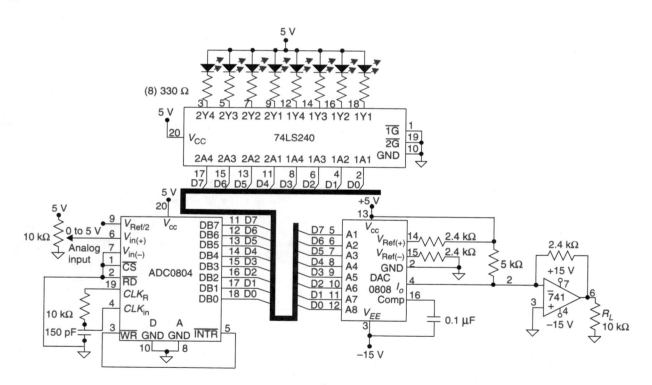

FIGURE 19–23

6. Adjust the analog input to each of the following values and record the digital word present on the LED and the analog output at each setting.

Analog Input	Digital Output	Analog Output
0 V	_____	_____
1.25 V	_____	_____
2.5 V	_____	_____
3.75 V	_____	_____
5 V	_____	_____

7. The conversion time can be determined by connecting an oscilloscope to the jumper between pins 3 and 5 (/INTR-/WR) and measuring the period between pulses. What is the conversion time? _____

8. Adjust the function generator to a 4 V_{p-p} sine wave riding on a 2.5 V DC offset at a frequency of 500 Hz. Remove the 0 V to 5 V DC source (10 kΩ potentiometer) and connect the function generator to the analog input.

9. Using a dual-trace oscilloscope, compare the analog input to the analog output. Draw the waveform below.

    V_{in} -

    V_{out} -

10. Increase the frequency of the function generator to 2 kHz. Compare the analog input to the analog output. Draw the waveform below.

    V_{in} -

    V_{out} -

V. Points to Discuss

1. Discuss why the digital output of the ADC seems to change immediately when the analog input changes even though the ADC0804 is a relatively slow successive approximation converter.

2. How could the conversion time of the ADC be changed?

3. Discuss why the analog output signal at 500 Hz appears closer to the ideal sine wave than the analog output signal at 2 kHz.

QUESTIONS

19.1 Introduction to Data Conversion Systems

____ 1. An analog signal has a range from 0 V to 4 V. What is the total number of analog possibilities within this range?

   a. 3

   b. 4

   c. 40

   d. infinite

____ 2. In the last twenty-five years, the _____ has become the brain of electronic controls.

   a. op-amp

   b. analog computer

   c. digital computer

   d. electronic sensor

____ 3. Digital signals have _____ possibilities within the range.

   a. two

   b. eight

   c. a definite number of

   d. an infinite number of

19.2 Relationship Between Analog and Digital Signals

____ 4. If the analog and digital ranges of a converter are known, the output for a given input can be calculated _____.

   a. by raising the product of the ranges to the second power

   b. by using the formula $V_{out} = DAC \times ADC$

   c. by using a ratio

   d. none of the above

____ 5. When converting an analog level to a digital level, if the result is not precisely one of the digital possibilities, then _____.

   a. the result is always rounded down

   b. the result is always rounded up

   c. a new digital possibility is created

   d. the conversion is not possible

19.3 Resolution of Conversion Systems

____ 6. Resolution is _____.

   a. only a concern when working with ADC

   b. only a concern when working with DAC

   c. a function of the speed of the conversion system

   d. a function of the number of bits in the digital word

_____ 7. The number of bits in a digital word determines the number of digital combinations that will be represented. The number of combinations can be calculated by _____.
   a. raising 2 to the number of bits in the analog word
   b. raising 2 to the number of bits in the digital word
   c. multiplying 2 times the number of bits in the digital word
   d. determining the range of the analog input

_____ 8. The smallest analog increment an ADC can detect is calculated by _____.
   a. dividing the analog range by the total number of digital combinations
   b. dividing V_{cc} by the total number of digital combinations
   c. dividing V_{cc} by the number of bits in the digital word
   d. dividing the analog range by the number of bits in the digital word

19.4 Digital-to-Analog Conversion

_____ 9. The _____ is a network of resistors containing only two values.
   a. binary divider
   b. binary ladder
   c. analog divider
   d. analog ladder

_____ 10. What is the resistance from point A to ground in Figure 19–24?
   a. 2 kΩ
   b. 4 kΩ
   c. 10 kΩ
   d. 22 kΩ

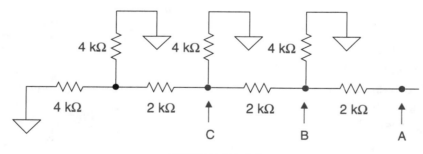

FIGURE 19–24

_____ 11. If a logic one on the most significant bit (D8) of a DAC causes an output of 1 V, a logic one on the next most significant bit (D7) will cause an output of _____ .
   a. 0.25 V
   b. 0.5 V
   c. 1 V
   d. 2 V

19.5 Integrated DAC

_____ 12. The DAC0808 contains a _____ network and needed switching circuitry to make a complete DAC.

a. binary divider

b. binary ladder

c. analog divider

d. analog ladder

_____ 13. The DAC0808 IC is a digital-to-analog converter that outputs a _____ that is proportional to the digital input signal.

a. current value

b. voltage value

c. resistance change

d. log value

19.6 Digital Controlled Amplifier

_____ 14. The digital controlled amplifier has an analog input and analog output. It also has _____ that controls the voltage gain of the amplifier.

a. an analog input

b. a log input

c. a digital input

d. none of the above

_____ 15. By changing the resistor value in the feedback loop of the op-amp, the digital controlled amplifier can be converted to _____.

a. an analog controlled amplifier

b. a digital controlled attenuator

c. a digital controlled resistor

d. none of the above

19.7 Analog-to-Digital Conversion

_____ 16. The two popular types of analog-to-digital converters discussed in this chapter were the _____.

a. flash converter and successive approximation converter

b. flash converter and analog converter

c. linear converter and successive approximation converter

d. linear converter and analog converter

_____ 17. The biggest disadvantage of the _____ is the complexity of the circuitry.

a. analog converter

b. flash converter

c. linear converter

d. successive approximation converter

____ 18. The longer the digital output word, the better the resolution; however, since the successive approximation converter determines each bit in sequence from MSB to LSB, _____.

     a. the converter is limited to an 8-bit word

     b. the converter is limited to a 10-bit word

     c. the longer the digital word, the shorter the conversion time

     d. the longer the digital word, the longer the conversion time

19.8 Integrated ADC

____ 19. The ADC0804 is _____ designed to interface to microprocessor systems.

     a. an analog to analog converter

     b. a flash converter

     c. a linear converter

     d. a successive approximation converter

____ 20. The conversion time of the ADC0804 is controlled by the _____ .

     a. analog network

     b. ladder network

     c. RC network

     d. crystal oscillator

19.9 Sample-and-Hold Circuit

____ 21. A sample-and-hold circuit consists of _____.

     a. an op-amp buffer, an analog switch, and a hold inductor

     b. an op-amp buffer, an analog switch, and a hold capacitor

     c. two op-amp buffers, an analog switch, and a variable resistor

     d. two op-amp buffers, an analog switch, and a hold capacitor

____ 22. A sample-and-hold circuit _____ a successive approximation converter and ensures that the analog signal remains constant through the conversion process.

     a. follows

     b. precedes

     c. is in parallel with

     d. is connected between the flash converter and

19.10 Troubleshooting Conversion Systems

____ 23. Troubleshooting ADC and DAC circuits requires the technician to have some knowledge of _____.

     a. analog electronics

     b. digital electronics

     c. programming

     d. all the above

____ 24. A system outputs a continuous ascending digital count to a DAC. The output of the DAC will be a _____.

  a. sine wave

  b. square wave

  c. sawtooth wave

  d. triangular wave

____ 25. A system continuously inputs the data from a 6-bit ADC and then outputs the digital results to a display. What will be displayed if the ADC has an analog range of 0 V to 8 V and the analog input is 5.2 V?

  a. 00 1001

  b. 01 1011

  c. 10 1001

  d. 10 1010

PROBLEMS

1. An 8-bit DAC has an output voltage range of 0 V to 5 V. What are the analog outputs for the following inputs?

Input	Output
0011 1000	_____
0100 0111	_____
1000 0100	_____
1101 0011	_____

2. A 6-bit DAC has an output voltage range of 0 V to 4 V. What are the analog outputs for the following inputs?

Input	Output
00 1011	_____
00 0111	_____
10 0100	_____
11 0011	_____

3. An ADC has an input voltage range of 0 V to 5 V and a resolution of 8 bits. What are the digital outputs for the following input voltages?

Input	Output
0.7 V	_____
2.3 V	_____
3.1 V	_____
4.6 V	_____

4. An ADC has an input voltage range of 0 V to 8 V and a resolution of 6 bits. What are the digital outputs for the following input voltages?

Input	Output
0.3 V	_____
1.2 V	_____
5.2 V	_____
7.1 V	_____

5. Determine the number of analog steps possible at the output of a DAC with a resolution of 9 bits and an output analog range of 0 V to 8 V.

6. Determine the value of the LSB for a DAC with a resolution of 10 bits and an output analog range of 0 V to 5 V.

7. Calculate the smallest analog change that can be detected by an ADC with a resolution of 10 bits and an analog input range of 0 V to 8 V.

8. Calculate the voltage gain of the circuit in Figure 19–25.

9. Draw and label the output for the circuit in Figure 19–25.

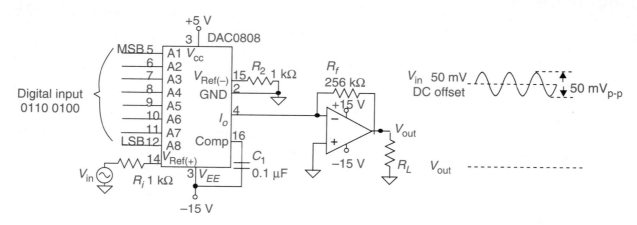

FIGURE 19–25

CHAPTER 20

Optoelectronics

OBJECTIVES

After studying the material in this chapter, you will be able to describe and/or analyze:

- ○ the cathode ray tube,
- ○ liquid-crystal displays,
- ○ light sensing devices,
- ○ photoactive devices,
- ○ optoisolators and optical sensors,
- ○ lasers,
- ○ laser diodes, and
- ○ fiber optics.

20.1 INTRODUCTION TO OPTOELECTRONICS

The area of study that deals with the use of light in electronic circuitry is called *optoelectronics*. The hardware in this field can be divided into two general areas: those devices that input an electrical signal and output light used as a light source or display, and those devices that input light and output an electrical signal. In the first category, we will study CRTs, LCDs, and LEDs. These devices are used in electronic systems as displays. In the second category, we will study photoresistors, photodiodes, phototransistors, photoSCRs, and phototriacs. These devices are transducers that enable electronic circuitry to sense light and can be considered the eyes of electronics.

20.2 CATHODE RAY TUBES (CRTs)

INTRODUCTION

Thirty-five years ago when transistors were replacing tubes in all types of electronic circuitry, I had a college teacher who believed the CRT was an example of one tube that would never be replaced.

743

Well, never is a long time, and maybe in another ten years the CRT will be history, but today it is still one of the most popular display devices used.

BASIC ELECTRON TUBE THEORY

The CRT is a type of electron vacuum tube, and for this reason, we will briefly study basic tube theory. Figure 20–1 shows the symbol for an electron vacuum tube. This tube is comprised of three elements and a heater. All of these are enclosed in a glass case in which the air has been pumped out to form a vacuum.

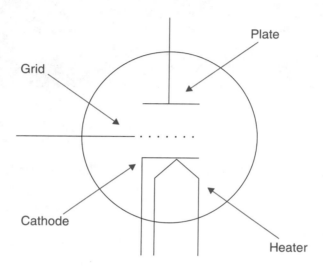

FIGURE 20–1   Electron vacuum tube

   A current is sent through the heater that causes it to become hot and heat the cathode. The cathode is made from a material that, when heated, emits electrons in the form of a space cloud, much like steam boiling out of a hot pan of water. If the plate is made positive with respect to the cathode, the electrons in the space cloud will travel toward the plate. The grid is made of screen-like material so the electrons can pass through it and reach the plate. In other words, there is a current path through the tube from the cathode to the plate. If the grid is made negative with respect to the cathode, it will reduce the number of electrons going to the plate. If the negative voltage on the grid is increased, a point will be reached that cuts off the flow of electrons from the cathode to the plate. If the grid is made less negative with respect to the cathode, the number of electrons flowing to the plate increases. The grid is the control terminal and controls the magnitude of current flow between the cathode and the plate.

CRT THEORY

Figure 20–2 shows the inner construction of a CRT. The heater, cathode, and grids make up a section known as the electron gun. The purpose of the electron gun is to produce a stream of electrons that can be aimed at a phosphor coated screen. When the electrons hit the screen, they cause the phosphor to emit visible light. The heater and the cathode perform the same function as they did in the basic vacuum tube: to provide a cloud of electrons. The function of the control grid is to control the magnitude of the electron beam, which controls the intensity of the emitted light. The accelerating grids and the focusing grid are used in conjunction with each other to speed up the electron beam and to provide electrical optics that focus the electron beam to hit one small point on the phosphor. Control of these grids is accomplished by setting the voltage levels on each grid.

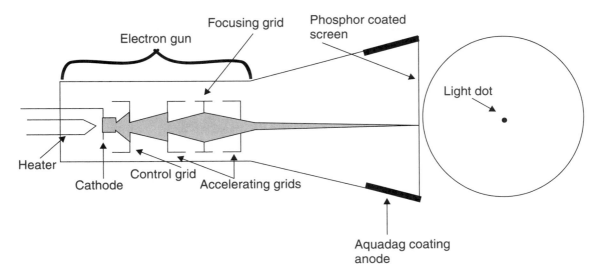

FIGURE 20–2 Inner construction of a CRT

Once the electrons leave the electron gun, they are further accelerated toward the phosphor coated screen by the anode (plate). The anode consists of an aquadag coating that is deposited inside the glass enclosure. The anode is at a high voltage in the range of 1000 V to 25,000 V. At this point in our discussion, we have a single dot on the front of the CRT, and we can control the intensity with the voltages on the control grid. We can also control the sharpness or focus by the voltage on the focusing grid. This dot will become our pencil point for writing on the CRT screen.

DEFLECTION (MOVING THE LIGHT DOT)

Figure 20–3 is the same as Figure 20–2 except vertical deflection plates have been added. If a positive potential is placed on one plate and a negative potential is placed on the other, the electron beam will be attracted by the positive plate and repelled by the negative plate. This action bends the electron beam, which will move the light dot on the screen vertically. By placing horizontal deflection plates at right angles to the vertical deflection plates, the light dot can be moved horizontally. By using a combination of both, the light dot can be moved anywhere on the

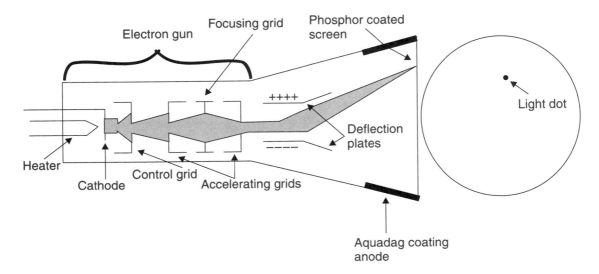

FIGURE 20–3 CRT with vertical deflection plates

face of the screen. If the light dot is moved back and forth between two points fast enough, the screen will appear to have a line drawn between the two points. There are two reasons the moving dot appears as a solid line: the phosphor continues to glow for a short time after being hit by the electron beam, and the human eye cannot detect movement faster than one-fiftieth of a second. Actually, any image can be drawn on the screen, but for the image to look stationary on the screen, it is necessary that the beam retrace the image continuously.

The deflection used in Figure 20–3 is called *electrostatic deflection*. The plates are placed inside the tube, and the electrical potential on the plates causes deflection. Another method of obtaining deflection is to use *electromagnetic deflection*. In electromagnetic deflection, deflection coils are placed on the outside of the tube around the tube neck. There are four coils (two vertical and two horizontal), and they usually come in a single unit called a *deflection yoke*. By sending a current through the coils, magnetic fields are formed, and these fields cause the electron beam to be deflected. The advantage of electromagnetic deflection is that stronger deflection fields can be produced, and in large CRTs, this is necessary so that the beam can be deflected to the edge of the screen. Electrostatic deflection, however, is faster and more linear. Electromagnetic deflection is normally used in equipment like TV sets and computer monitors where large screens are used, and electrostatic deflection is normally used in equipment like oscilloscopes where screens are small and where high-speed, linear response is needed.

An oscilloscope uses two input voltages to control the location of the electron beam. The horizontal input voltage is normally generated by the time base circuitry and is a sawtooth waveform that causes the beam to sweep across the screen from left to right. The vertical input voltage is connected to the signal the technician wants to display. This causes the beam to move vertically, as it sweeps horizontally, drawing the waveform on the screen.

Television sets and computer monitors use scanning circuitry to move the beam on the screen. The scanning circuitry generates two sawtooth waveforms. The high frequency sawtooth is connected to the horizontal deflection coils, and the low frequency sawtooth is connected to the vertical deflection coils. Figure 20–4 shows the waveforms and the resulting pattern of the beam on the screen of the CRT. The horizontal signal drives the beam from left to right at a constant rate. Once the beam reaches the right side of the screen, it is quickly driven back to the left side of the screen by the falling edge of the sawtooth signal. The lower frequency vertical sawtooth signal causes the beam to move gradually from the top of the screen to the bottom. Notice, in Figure 20–4, that the beam moves across the screen five times during the time it takes the beam to move from top to bottom. TV sets and computer monitors have a large number of horizontal sweeps during one vertical sweep. For example, a TV used in the USA has 262.5 horizontal sweeps for each vertical sweep, and a VGA computer monitor has 480 horizontal sweeps for each vertical sweep.

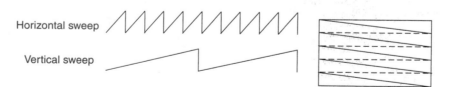

Horizontal sweep

Vertical sweep

FIGURE 20–4   CRT scan pattern

The scanning action causes the beam to continuously paint the full screen. The information to be displayed on the screen is applied to the control grid of the CRT and controls the flow of electrons in the beam. If the signal on the control grid turns off the beam, the screen will be dark. If the signal on the control grid turns the beam on to maximum, the screen will be bright. By controlling the voltage on the control grid, the brightness of the screen can be controlled. As the beam sweeps from left to right, a video signal is applied to the control grid that illuminates the desired display. On the retrace (right to left movement), a blanking signal is applied to the control grid that cuts off the beam.

COLOR CRT

The type of phosphor coating the screen determines the color of the light emitted from the CRT. Popular monocolor CRTs are white, green, or amber, but other colors are possible. CRTs can be designed to emit any color. These color CRTs use a method of on screen color mixing to obtain their array of colors. A color CRT has three electron guns: a red, green, and blue gun. Electrically, all three guns are the same, but the electron beam of each gun hits a different color of phosphor. Figure 20–5 shows the front of a color CRT. The screen is actually coated with dots of phosphors, not a solid sheet. The dots are made by three different types of phosphors: one type emits red light, the second emits blue, and the third emits green.

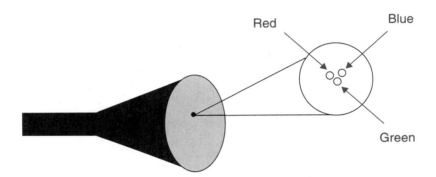

FIGURE 20–5   Front screen of a color CRT

The dots are so physically close together that the human eye cannot distinguish the individual dots. However, as shown in the blowup of Figure 20–5, adjacent dots are made of different color emitting phosphors. Each of the three electron guns is aimed to hit and illuminate only one color dot. If the guns for the blue and green dots were turned off, the only dot to be illuminated would be the red dot. If the beam of the red gun was deflected so it scanned the whole screen, the whole screen would be red. If the red and green guns were turned off and the blue gun on, the screen would be blue. Likewise, if the red and blue were off and the green gun on, the screen would be green. By controlling the intensity of the electron beam coming from each gun, different amounts of red, blue, and green can be obtained. Because of the closeness of the dots, you see only one color from the three dots. This color is a mixture of red, blue, and green, and the actual color depends on the intensity of each gun. By adjusting the intensity of the three primary colors, any color can be produced on the screen.

SUMMARY OF CRTs

The cathode ray tube (CRT) is one of the most popular displays. Any image can be formed on the screen by causing the electron beam to illuminate a pattern of phosphor dots. The electron beam is produced by an electron gun made up of a heater, cathode, control grid, focus grid, and acceleration grids. The beam becomes the pencil for writing on the screen of the CRT. The movement of the beam on the screen is accomplished by the deflection circuitry using either electrostatic or electromagnetic deflection. By using a screen that has adjacent dots of red, blue, and green phosphors and three electron guns, a color CRT can be made that produces any color.

20.3   LIQUID-CRYSTAL DISPLAYS (LCDs)

The name *liquid-crystal* comes from nematic liquid-crystals. Nematic liquid-crystal molecules are long and rod-shaped and align themselves in a symmetrical crystalline structure. Light can travel through the rod-shaped molecules. Light waves are polarized, and as the light travels through the liquid-crystal material, its polarization can be changed.

Figure 20–6 shows the components that comprise liquid-crystal displays (LCD). Two pieces of glass form a sealed unit that sandwiches a thin layer of nematic liquid-crystals. The front glass has a vertical polarized coating, and the rear glass has a horizontal polarized coating. The back of the unit is coated with a reflective coating. Vertical polarized light enters the front of the unit by passing through the vertical polarizer. The light then travels through the liquid-crystal material. The orientation of the molecules in the liquid-crystal twist the light 90º so the light enters the horizontal polarizer with the correct polarization to pass through and strike the rear reflector. The light is reflected back through the unit. In this state, the unit is acting as a transparent window with a mirror coating on the rear glass. The light that enters the unit is reflected back.

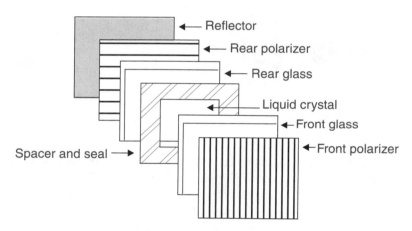

FIGURE 20–6   LCD construction

A seven-segment display can be constructed by adding transparent electrodes. The transparent electrodes are deposited on the inner surface of the glass. These electrodes are in the pattern of the desired display (in this case, a seven-segment display as shown in Figure 20–7). Each segment on the front glass is connected to an individual lead, so each segment can be controlled. All the segments on the rear glass are connected together and return to ground.

With no electrical signal applied, the window is transparent. Light travels through the window, hits the back reflective coating, and is reflected back through the window. The result is that the display appears as a uniform light color. If an electrical signal is applied to a segment, an electrical field is formed between the front and rear electrodes. This electrical field will cause the nematic molecules under the selected segment to align. Once the molecules are aligned, the nematic liquid crystal will not twist the light 90º; therefore, the light reaching the rear horizontal polarizer will be the wrong polarity to pass through to the reflector. Since the light is no longer reflected, the selected segment will appear black.

The LCD unit we have been discussing is classified as a reflective mode LCD unit. It depends on external light and works well in high intensity light environments. On the other hand, it is not good for displays in low level light conditions. Transmissive mode LCD units are available

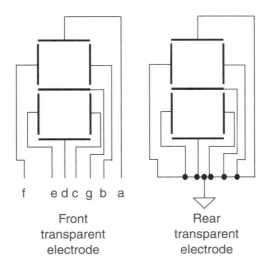

FIGURE 20–7 Transparent electrodes

to work in low light environments. These units are designed with backlighting. A lighting source behind the display is designed to provide uniform lighting. This light is transmitted through the LCD window, and the activated segment blocks the light to form characters on the display.

Flat-panel displays are designed using transmissive mode. Figure 20–8 shows dot segments in a matrix. Each segment is a pixel (picture element), and its location can be defined by a row and a column. For example, the white pixel in Figure 20–8 is in row 5 and column 8. Passive-matrix displays use scanning circuitry to update the display one row at a time. Data is sent to the column drivers. The first row is selected, and all the pixels in row one respond. New data is sent to the column drivers, and the second row is selected. When the last row is updated, the scanning starts over.

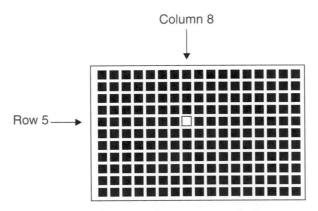

FIGURE 20–8 LCD flat-panel display

The main disadvantage of passive-matrix displays is slow response time. New data will not be displayed until the scanning circuitry activates the row where the new data resides. Because LCD response time is slow, the scan rate is slow, and it can take several hundred milliseconds for data to be updated. Computers using passive-matrix displays cannot display fast moving graphics. Passive-matrix displays require relatively few transistor drivers. For example, a VGA monitor has 640 columns and 480 rows. This would require 640 transistor drivers for the columns and 480 transistor drivers for the rows, for a total of 1120 transistor drivers.

Active-matrix displays do not use scanning circuitry. Instead, there is a transistor driver for each pixel. This means when new data is available, it immediately updates data on the screen. There are a large number of transistor drivers required for an active-matrix display. A 640 × 480 VGA monitor, for example, would require 307,200 transistor drivers. To make matters worse, for color displays, each pixel has three segments (red, green, and blue). This means a color VGA monitor requires 921,600 transistor drivers. Manufacturers have integrated the driver circuitry into large integrated circuits that actually form the back plate of the display panel. The disadvantage of the active-matrix display is high cost. An active-matrix display can add several hundred dollars to the cost of a notebook computer.

The LCD electrically looks like a small value capacitance in parallel with a large resistance. Recall that a capacitor is two conductive plates separated by a dielectric. The two electrodes form the conductive plates, and the liquid crystal is the dielectric. The resistance of the liquid crystal material is in the hundreds of megaohms. Figure 20–9 shows the electrical equivalent circuit of an LCD.

The drive voltage needed for LCDs is a small AC signal in the range of 3 to 7 V AC. Larger voltages will break down the thin liquid crystal layer. The drive voltage can be either a sine wave or a square wave, but it must be symmetrical with respect to ground. A DC drive voltage cannot be used because it would result in electrolysis, which would cause the electrodes to become coated and destroy the display. If a square wave is used, it must be symmetrical with respect to ground and have a fifty percent duty cycle or a damaging DC offset voltage will result. The frequency of the drive signal should be in the range of 30 to 50 Hz. Lower frequencies will cause the display to flicker, and higher frequencies will lower the input impedance of the display.

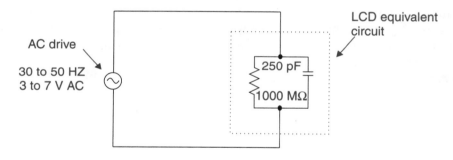

FIGURE 20–9   LCD equivalent circuit

Because of the high impedance of the LCD, the current drawn is in the microamp range, and the power consumption is in microwatts. Low power consumption is one of the major advantages of LCDs and has made them the best choice for portable systems.

20.4 LEDs

LEDs were discussed in Chapter 3 and will not be covered in detail in this section. However, recall that LEDs are specially designed diodes that emit visible light when forward biased. Electrically, they function like normal diodes except they have a higher forward voltage drop (usually about 2 V). LEDs are often connected together to form display units. These units have the advantage of being low in cost and easy to connect, and they provide a readable display even in total darkness. The disadvantages of LEDs are their high power consumption and a tendency not to be visible in high intensity light.

20.5 LIGHT SENSING DEVICES

INTRODUCTION

Light sensing devices are transducers that take in light energy and convert it to electrical energy, which is a quantity that can be detected by electronic circuitry. In general, light sensing devices can be divided into two groups: photocell and photoactive devices. In the first group are the photoconductive cells and photovoltaic cells. In the second group, we will study the operation of the photodiode, which is the basis of all other devices in this group. Then the phototransistor, photoFET, photoSCR, and phototriac will be covered.

PHOTOCONDUCTIVE CELL (PHOTORESISTOR)

Photoconductive cells are two-terminal devices that vary their resistances with exposure to light. The resistance of the device is inversely proportional to light intensity. Because these devices vary their resistances with light, they are often called *photoresistors*.

Photoconductive cells are made from either cadmium selenide (CdSe) or cadmium sulfide (CdS). A thin layer of these substances is deposited on a substrate of glass or ceramic material, and connections are made to lead pins at each end. A window is placed over the material, which is then encapsulated in a plastic case. Figure 20–10(a) shows the physical layout, and Figure 20–10(b) shows the symbol of the photoconductive cell.

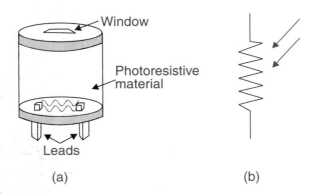

FIGURE 20–10 Photoconductive cell

Figure 20–11(a) shows the response curve for the human eye, a photoconductive cell made of cadmium sulfide (CdS), and one made of cadmium selenide (CdSe). You can see that the response curve for the cadmium sulfide (CdS) is closer to that for the human eye than is the response curve for cadmium selenide (CdSe). The response of the cadmium selenide (CdSe) to longer wavelength light (lower frequency) moves it closer to infrared.

Construction and the material used will determine the ratio of dark-to-light resistance of photoconductive cells. The range is from 100:1 to 10,000:1. Advantages of the photoconductive cell include its extreme sensitivity to light, low cost, and simple support circuitry. Figure 20–11(b) shows a simple voltage divider circuit where one of the resistors is a photoconductive cell. Because of the large ratio between dark-to-light resistance, this works well to produce a signal that is dependent on light.

Photoconductive cells can withstand relatively large voltages, with typical ratings in the range of 100 V to 300 V. Power ratings, however, are small, with typical maximum ranges of 30 mW to

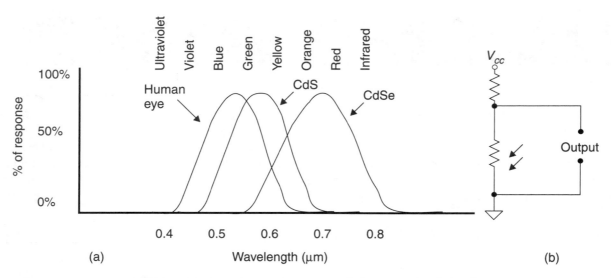

FIGURE 20–11   Response curves of photoconductive cells

300 mW. The resistance of a photoconductive cell is bidirectional, and it can be used in AC as well as in DC circuits.

The disadvantages of photoconductive cells are their slow response time, poor temperature stability, narrow response curve, and an effect called *light history*. Photoconductive cells respond to light changes in milliseconds. The cadmium sulfide photoconductive cell has a response time in the range of 100 mS, while the cadmium selenide is somewhat quicker in the range of 10 mS. Poor temperature stability means that photoconductive cells are sensitive to heat as well as light, so precaution must be taken to keep the ambient temperature of the device relatively constant. Light history effects can be compared to residual magnetism in magnets. Once exposed to light, if the device is returned to total darkness, its resistance will increase, but it will not return to its original value of resistance. In other words, previous illumination has an effect on the resistance of a photoconductive cell.

PHOTOVOLTAIC CELLS (SOLAR CELLS)

Photovoltaic cells are two-terminal devices that produce a voltage between the terminals when exposed to light. The voltage produced is directly proportional to the light intensity. Figure 20–12(a) shows the construction of a silicon photovoltaic cell. The cell is comprised of a PN junction with two metal electrodes with leads attached. The top electrode is formed into a ring shape so a window can be placed in the center for light to enter the silicon material. Light rays are theoretically made up of tiny particles called *photons*. Figure 20–12(b) shows a blowup of a portion of the PN junction. A photon entering the silicon material penetrates the P-type material and hits an atom in the depletion region. The energy of the photon is transferred to the atom and causes an electron in the valence ring to break out of its covalent bond. This causes a free electron/hole pair. The electron moves into the N-type material, adding a negative charge, and the hole drifts toward the P-type material, adding a positive charge. This process is repeated millions of times and causes the P-type material to become approximately 0.5 V positive with respect to the N-type material. A larger voltage difference is not possible because the energy transferred by the photon to the atom is not large enough to force the electron to enter the negatively charged N-type material when the voltage difference exceeds approximately 0.5 V.

If an external load is applied across the photovoltaic cell, current will flow. Actually, a silicon photovoltaic cell under load conditions will only have a terminal voltage of approximately 0.4 V. The current capability of a photovoltaic cell is directly proportional to the area of the cell. On a

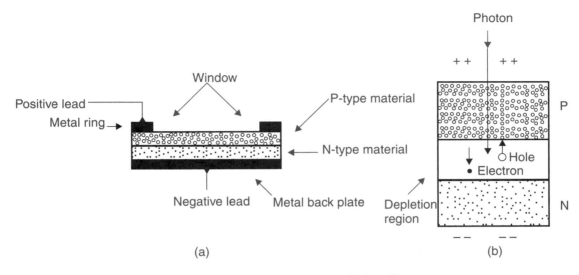

FIGURE 20–12 Photovoltaic cell

clear day when the sun is directly overhead, a silicon photovoltaic cell with an area of 0.5 cm^2 can deliver 25 mA at 0.4 V, and one with an area of 75 cm^2 can deliver 2 A at 0.4 V.

In order to obtain usable power levels, it is necessary to connect cells in series and parallel. As with other voltage sources, connecting cells in series increases the output voltage, and connecting them in parallel increases the output current capabilities. Silicon cells are the most popular photovoltaic cells because they have the greatest power output for a given area. Silicon cells respond best to infrared light rather than light in the visible spectrum, and selenium photocells are used when it is desirable to have a spectral response closer to that of the human eye.

20.6 PHOTOACTIVE DEVICES

All photoactive devices are based on the same principles. In your study of basic semiconductors, you learned that the magnitude of minority current carriers was mainly a function of heat energy. The heat energy causes a breakdown of covalent bonding and causes an electron/hole pair. In N-type material, the holes become minority carriers, and in P-type material, the electrons become minority carriers. If a device is made so that the semiconductor material can be exposed to light, the light energy will have the same effect as heat and increase the number of minority carriers in the material. By controlling the number of minority carriers, the conduction of a PN junction can be controlled in the reverse-biased mode. This fact is put to use in all photoactive devices.

PHOTODIODES

A window permits light energy to strike the depletion region of a PN junction, causing an increase in minority carriers. The increase in minority carriers causes the resistance of the reverse-biased junction to drop. A photodiode is essentially a light sensitive resistor: the greater the light intensity, the lower the reverse-biased resistance. The photodiode has the fastest response to changes in light intensity of all photo devices and is the building block for all photoactive devices. Figure 20–13 shows the symbol for the photodiode.

FIGURE 20–13   Photodiode symbol

PHOTOTRANSISTORS

Figure 20–14(a) shows an equivalent circuit for a phototransistor. Light activates a photo-diode, causing base current in the transistor, which causes the transistor to conduct. Phototransistors are three-terminal devices with a collector, an emitter, and a base. Some phototransistors connect only the collector and the emitter to external leads, while others bring all three elements out to leads. Figure 20–14(b) shows the symbol for a three-terminal phototransistor, and Figure 20–14(c) shows the symbol for a two-terminal phototransistor. In the three-terminal phototransistor, the base terminal is used for external biasing.

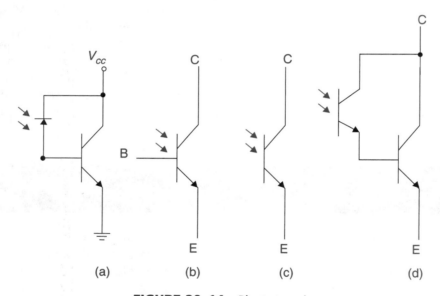

(a)   (b)   (c)   (d)

FIGURE 20–14   Phototransistors

A phototransistor has greater output current capability and greater sensitivity to light than a photodiode because of the amplification of the transistor. However, the construction of the base region is very thin, which causes the phototransistor to have a relatively high capacitance. This high capacitance causes the response time of the phototransistor to be slower than that of the photodiode. Higher current output phototransistors are available using a Darlington pair as the output device [Figure 20–14(d)].

PHOTOFETs

Manufacturers can produce a photoFET by forming a photodiode between the gate and the channel as shown in Figure 20–15(a). Under dark conditions, the photodiode has no effect, and

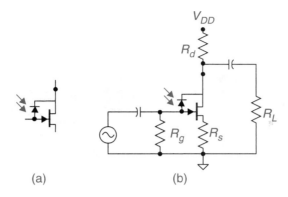

FIGURE 20–15 PhotoFET circuit

the device functions like a normal JFET. Under light conditions, the photodiode conducts, causing current to flow in the gate leg.

The circuit in Figure 20–15(b) is a light sensitive amplifier designed using a photoFET. Under dark conditions, the photodiode is an open, thus, the circuit functions like a normal self-biased JFET amplifier. The gate terminal is at 0 V, and the source terminal is at a positive voltage determined by the size of R_s and the current flowing through the FET. Let us assume that the source is at positive 2 V; the bias voltage, therefore, is negative 2 V. Now expose the circuit to light. When the photodiode is exposed to light, leakage current flows. This leakage current flows through R_g, causing the top of R_g to become positive (assume 0.5 V). The positive 0.5 V on the gate reduces the bias voltage between the gate and the source to –1.5 V. A drop in bias voltage will cause the gm to increase, which will increase the voltage gain of the circuit. The brighter the light, the higher the voltage gain of the circuit.

LIGHT ACTIVATED SCRs AND TRIACS

Figure 20–16(a) shows the symbol for the light activated SCR (LASCR), which is the same as that used for the normal SCR except arrows show that the device is activated by light. When light activates the gate, the SCR turns on. Once on, the SCR will stay on until the anode current drops below the hold current. Figure 20–16(b) shows the symbol for a light activated triac. It is

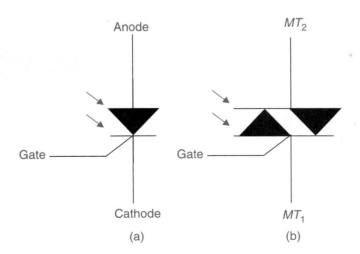

FIGURE 20–16 Light activated SCR and triac

the same as the normal triac symbol except the arrows have been added to indicate that it is a light activated device.

20.7 OPTOISOLATORS AND OPTICAL SENSORS

Figure 20–17 shows several types of optoisolators. An optoisolator is a device that contains an optotransmitter and an optosensor in a single package. The input circuitry of the optoisolator is an LED. The output can be several types of devices. Figure 20–17(a) is an optoisolator with a transistor output. The optoisolator in Figure 20–17(b) is also a transistor output circuit, but two LEDs are used in the input circuitry, permitting the circuit to be activated by current flow in either direction.

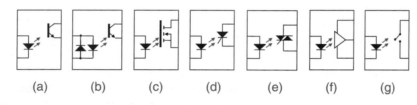

(a)   (b)   (c)   (d)   (e)   (f)   (g)

FIGURE 20–17 Optoisolators

Figure 20–17(c)–(e) shows the output device as a MOSFET, an SCR, and a triac, respectively. The output can also be a logic signal as shown in Figure 20–17(f) or it can control an analog switch as shown in Figure 20–17(g). When the output is a switch, the device is normally referred to as a *solid state relay*.

Figure 20–18 shows several types of optical sensors. The sensors contain an optical transmitter (LED) and an optical sensing device, normally an optotransistor. Sensors A through D are the slot type of optical sensors. The slot optical sensor has a slot separating the LED and the optotransistor. The light output of the infrared LED causes the optotransistor to conduct. If an object enters the slot, the light beam will be interrupted, causing the optotransistor to cut off. Sensors F through E are the reflective type of optical sensors. An LED transmits infrared light away from the sensors. If an object passes in front of the sensor, the light energy will be reflected back to the sensor and actuate an optotransistor. The object must be within range of the reflective optical sensor to be detected.

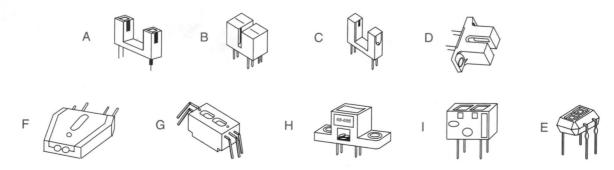

FIGURE 20–18 Optical sensors

20.8 LASERS

Laser is the acronym for light amplification by stimulated emission of radiation. In simplest terms, a laser can be defined as a high energy light beam, but the laser has certain characteristics that separate it from other light sources.

BASIC LIGHT THEORY

Light signals are sinusoidal in nature and are measured in terms of frequency, period, and wavelength. The visible spectrum starts at 390,000,000,000,000 Hz for infrared and ranges up to 770,000,000,000,000 Hz for ultraviolet. Because these frequencies are so high, it is easier to express them in wavelengths. The relationship between wavelength and frequency is *wavelength = velocity of light / frequency*. The velocity of light is equal to 300,000,000 meters/second and is a constant. This means infrared has a wavelength of 0.77×10^{-6} meters, and ultraviolet has a wavelength of 0.39×10^{-6} meters. It is usually more convenient to express light wavelength in angstrom (10^{-10}). Figure 20–19 shows the visible spectrum with light wavelength expressed in angstroms.

In Figure 20–19, you can see that different colors are determined by the wavelength of the light energy. White light is a mixture of light energy from across the visible spectrum. Let us examine an incandescent lamp as a source of light energy and see how light is radiated. Figure 20–20 shows that radiated light energy is omnidirectional and that the amount of energy hitting a plane is inversely related to the distance the plane is located from the light source.

The reason the light energy reduces with distance is because the light rays are not parallel and become further apart as they move away from the light source. Rays of light are comprised of photons that vary in a sinusoidal pattern and determine the frequency and the color of the light. For an incandescent light source, there are frequencies representing all the colors of the rainbow, but we perceive the light as one shade dependent on the magnitude of each frequency present. Light rays can be formed into a directional beam by using reflectors; however, this beam of light would not be a laser.

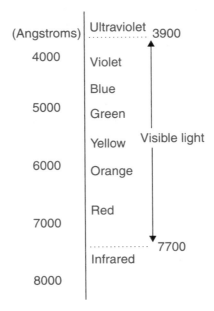

FIGURE 20–19 Visible spectrum

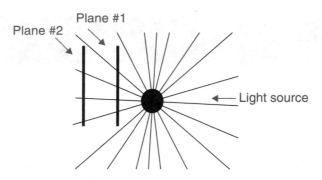

FIGURE 20–20   Omnidirectional light energy

LASER THEORY

There are three factors that distinguish a beam of light from a laser. A laser beam must have light that is monochromatic, collimated, and coherent. Monochromatic means one color—or stated another way, all the light in a laser must be of one frequency. If a light beam is collimated, all of the rays of light are parallel to each other. Figure 20–21(a) shows a light source with rays separating as they move away from the source. The light source is directed into a collimating lens. The light on the right side of the lens is collimated. Note that a plane would receive the same density of light regardless of the distance from the source.

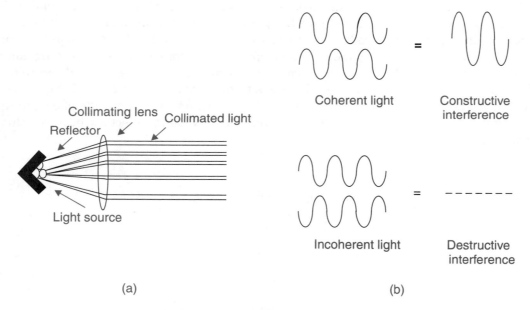

(a)   (b)

FIGURE 20–21   Collimated light and coherent light

If a light beam is monochromatic (one color), it is comprised of one frequency of light. This does not mean, however, that all the rays are in phase with each other. From studying electrical signals, we know that if signals that are in phase are combined, they add, but if they are 180º out of phase, they cancel. The same is true for light signals. If two light rays are in phase, they add; this is called *constructive interference*. On the other hand, if they are out-of-phase, they subtract. Maximum cancellation occurs when the two rays are 180º out-of-phase. In order to create a laser, it is necessary to obtain light signals in phase in order for constructive interference to take place. Light that has all of its rays in phase is called *coherent light*. Figure 20–21(b) illustrates the concepts of coherent light and constructive interference. A perfect laser beam will be totally monochromatic (one frequency), coherent (in phase), and collimated (parallel rays).

20.9 LASER DIODES

INTRODUCTION

One type of device that can generate a laser beam is an injection laser diode. The laser diode is similar to an LED because it is forward biased and a current flows, causing electrons to combine with holes. As the electrons fall into the holes, they release energy in the form of photons or light energy. However, the light produced by LEDs is low energy and incoherent light. The laser diode is designed using gallium arsenide (GaAs) and aluminum gallium arsenide (AlGaAs) so that a single frequency coherent light beam is generated when a large forward-biased current is passed through the device.

STIMULATED EMISSION

At low current values, photons are radiated at random, and the light produced is incoherent. As the current is increased, the atoms are brought to a higher energy level. When these high energy atoms are hit by photons, stimulated emission occurs. In stimulated emission, when a photon hits an atom, it will cause two photons to be released. When this process occurs repeatedly, it results in light amplification. Another important thing to note about stimulated emission is that the two new photons radiated have identical wavelengths, phase, and direction as the photon causing their release.

OPTICAL RESONANCE

The physical construction of the laser diode is shown in Figure 20–22. The length of the laser diode and the two reflective ends form an optical resonance cavity. One end is 100% reflective,

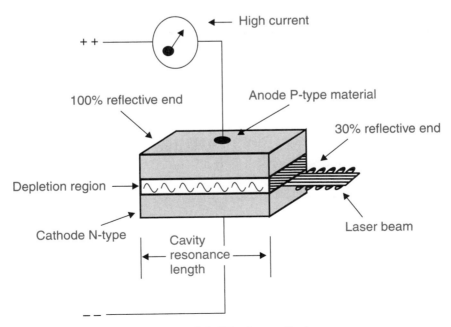

FIGURE 20–22 Laser diode

and all of the photons hitting this end are reflected back into the cavity. The other end is 30% to 40% reflective, and 30% to 40% of the photons are reflected back into the cavity. Because the length of the cavity is a function of the wavelength of the light being transmitted, a standing wave is set up in the cavity. This standing wave is reinforced by the reflected photons.

The laser beam leaves the laser diode at the partially reflected end. Since only 30% to 40% of the photons are reflected back, 60% to 70% exit the diode and form the laser beam. The light in the laser beam is all one wavelength (monochromatic), and it is coherent (in phase). It is also fairly collimated (parallel rays), although not as collimated as beams produced by other types of lasers. If it is necessary, the light beam of the laser diode can be collimated by attaching a collimating lens system to the output. Large amounts of energy leave the laser diode in the form of light, and this energy must be provided by an energy pump of some type. In the case of the laser diode, it is the high forward-bias current that keeps the atoms at high energy levels and allows the laser action to be a continuous process.

20.10 FIBER OPTICS

INTRODUCTION

Humans have often used light beams as a way to communicate, but the medium for this communication has been the atmosphere, which has two shortcomings. First, the atmosphere is constantly changing (rain, snow, fog, pollution), and these changes affect light transmission. Second, light beams in the atmosphere are line of sight, so communication in the atmosphere is limited between points that are in sight of each other. In order to use light successfully as a means of transferring information, it is necessary to use light conductors, which we call *fiber optic cables*.

The question you may be asking is why use light at all? Why not stay with RF transmission and electrical signals over copper conductors? The fact is that light transmission has several advantages over electrical transmissions. Two of the more important advantages are that light signals can support a much wider bandwidth, which means much more information can be transmitted, and that light signals are not affected by electromagnetic interference.

LIGHT TRANSMISSION IN FIBER OPTIC CABLES

Fiber optic cables are made from glass or plastic fibers, and the basic idea is that light will enter one end and travel down these mediums to the other end. In order for this to work, the light must be contained in the fiber and not permitted to exit the sides of the fiber rod. This is illustrated in Figure 20–23, which shows a light ray being reflected from side to side as it moves down a fiber rod. The fiber rod has a higher density than the air surrounding the rod. Light is reflected from the sides of the rod because light rays will not exit a dense material to a less dense material unless the angle of intersection is less than the critical angle.

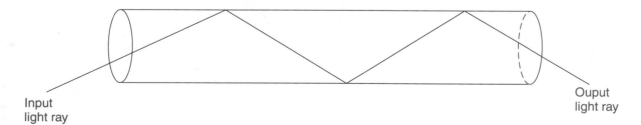

Input
light ray

Ouput
light ray

FIGURE 20–23   Light ray traveling down a fiber rod

The critical angle ($\emptyset$) is measured with reference to a line perpendicular to the junction of two materials, as shown in Figure 20–24. Three light rays are shown in Figure 20–24. Light ray #1 (at a low angle) is slightly bent but leaves the more dense material and enters the less dense material. Light ray #2 is at the critical angle and is reflected back into the more dense material. Light ray #3 is at an angle greater than the critical angle and is also reflected back into the more dense material.

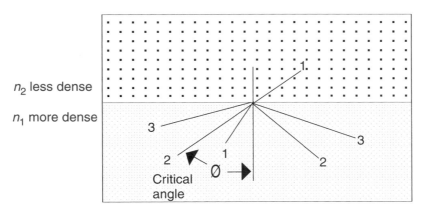

FIGURE 20–24 Critical angle needed for reflection

Refraction of light rays is caused by the fact that light travels at different speeds in different materials. The more dense the material, the slower the light will travel.

Optically transparent materials have a constant called the index of refraction (n). The index of refraction is the ratio of the speed of light in a vacuum to the speed of light in the material. For example, the speed of light in a vacuum is 300,000,000 meter/sec, and the speed of light in glass is 187,500,000 meter/sec, so the index of refraction for glass is equal to 1.6.

$$n = \frac{\text{Speed of light in vacuum}}{\text{Speed of light in glass}} = \frac{300,000,000 \text{ m/s}}{187,500,000 \text{ m/s}} = 1.6$$

If the index of refraction is known for the two materials in Figure 20–24, the critical angle can be calculated. The formula for calculating the critical angle is:

$$S_{\text{in}} \emptyset = \frac{n_2}{n_1}$$

Where:

$\emptyset$ is the critical angle

n_1 is the more dense material

n_2 is the less dense material

Light rays can travel down optical fibers as long as they intersect the side of the fiber at an angle equal to or greater than the critical angle so they will be reflected back and not exit the fiber. The critical angle is a function of both the material the fiber is made from and the material surrounding the fiber material.

EXAMPLE 20.1

Calculate the critical angle for a glass rod surrounded by air. The index of refraction of glass is 1.6, and the index of air is approximately 1.

EXAMPLE 20.1 continued

Solution:

$$S_{in} \varnothing = \frac{n_2}{n_1} = \frac{1}{1.6} = .625$$

$$\varnothing = 38.7°$$

CONSTRUCTION OF FIBER OPTIC CABLES

The fiber core of a fiber optic cable is made of glass or plastic. Glass is more expensive and generally harder to work with, but it supports light transmission with less loss per meter than plastic. While the glass or plastic fiber could be uncovered and air used as the second material surrounding the core to determine the critical angle, most often a covering called *cladding* is added, as shown in Figure 20–25. The cladding is made of a material less dense than the core and becomes the second material when calculating the critical angle.

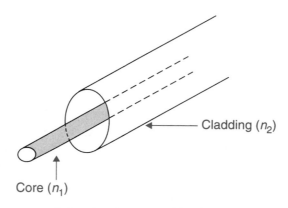

Core (n_1)

Cladding (n_2)

FIGURE 20–25   Cable cladding

TYPES OF FIBER OPTIC CABLES

We can classify fiber optic cables into three groups by using the method in which the light propagates down the fiber: multimode step-index, single-mode step-index, and multimode graded-index. The mode refers to the path the light will travel through the material. The index, as we have previously discussed, is a constant that indicates the speed at which light will travel through the fiber material.

Figure 20–26(a) shows a multimode step-index fiber. The core is made of glass or plastic of a single index value, and the cladding material also has a single index value. There is a single step between the two index values, and the difference in index (Δn) values is relatively large. This means the critical angle is small, and there will be many paths available for light to travel down the fiber. The light rays that intersect the junction of n_1 and n_2 with an angle greater than the critical angle will be reflected back.

Figure 20–26(a) shows three light rays traveling three paths or modes. Since they all are traveling in a core that has a constant index, they are all traveling at the same speed. Because of the different paths, they arrive at the output at different times. Ray #1 travels the shortest distance and arrives at the output first. On the other hand, ray #3 travels the farthest and is the last to arrive at the output. This spreading of arrival times at the output causes a distortion called *modal dispersion*. The effect of modal dispersion is shown in the waveforms of Figure 20–27(a). The figure shows a

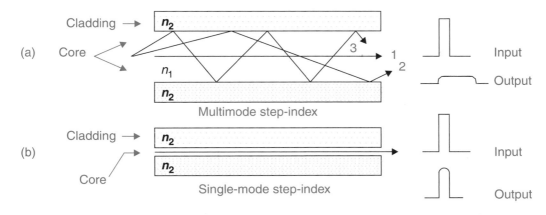

FIGURE 20–26 Step-index fiber

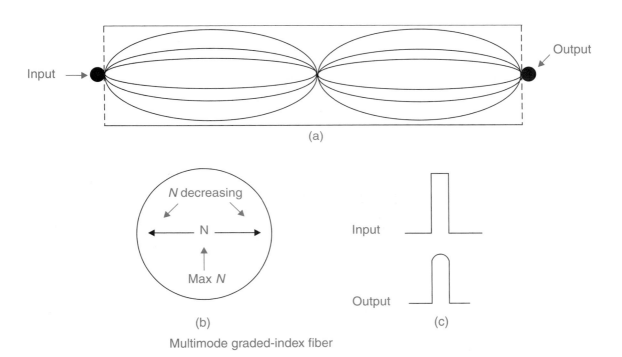

FIGURE 20–27 Graded-index fiber

narrow pulse of light energy at the input and the output energy attenuated and dispersed over a wide period of time. Modal dispersion limits the data rate that the fiber optic cable can handle. As the data rate goes up, the input pulses become narrower and closer together. Because of modal dispersion, the output pulses start to run together at higher data rates.

To overcome modal dispersion, the single-mode step-index fiber was developed. It still uses a core of single index and a cladding of a different index, but two things have been changed. First, the difference in the index (Δn) between the core and the cladding is small. This causes a large critical angle (near 90º) which means the light rays must travel close to parallel with the junction of n_1 and n_2 or they will exit the core. Second, the size of the core has been greatly reduced so the distance from side to side is small, and all paths through the core are approximately the same distance. The net effect of these two changes is that all light rays traveling through the core travel the same distance, and modal dispersion is greatly reduced. The input and output waveforms in Figure 20–26(b) for the single-mode step-index fiber show the reduction in output distortion.

The single-mode step-index fiber can handle higher data rates than the multimode step-index fiber (higher bandwidth). In order to make the cable single-mode, the size of the core has to be reduced to approximately 15 microns. This causes single-mode step-index fiber cables to be higher in cost and harder to work with. In addition, the small size makes it difficult to inject light energy into the cable, and only laser transmitter devices can be used.

Another approach to overcoming modal dispersion is to use multimode graded-index fibers as shown in Figure 20–27. Figure 20–27(b) shows a fiber core that is made with a center having a high index of refraction, with the index gradually decreasing with distances away from the center. Recall that the higher the index, the slower the light rays travel. The light rays traveling down the middle will travel the slowest, and those near the edge will travel the fastest. Figure 20–27(a) shows light rays traveling down the core. Near the edge they must travel farther, but because they travel at greater velocity, they reach the output at the same time as those traveling near the center. Since light rays that enter together exit together, modal dispersion is prevented.

Notice that a distinct junction did not exist between two materials of different indexes, but instead, the index of the core gradually decreased. The light rays are maintained in the core material not by reflection from a single junction but by a continuous bending of the rays as they move away from the center of the core. This bending action is called *continuous refraction*. You can see by examining the waveforms in Figure 20–27(c) that multimode graded-index fibers have a performance between multimode step-index fibers and single-mode step-index fibers. They are also in the middle when it comes to price and workability, and they have found widespread use.

FIBER OPTIC DATA COMMUNICATION LINKS

A fiber optic data communication link can be configured using commercially available components. An example of such a system is shown in Figure 20–28. It is comprised of a transmitter, an optic cable, and a receiver. The transmitter consists of an LED in an enclosure that is designed to connect easily to the optic cable. The cable is a single-step plastic fiber. An in-line connector is shown that is designed to splice two cable links. The receiver consists of a photodetector and a DC amplifier in an enclosure that is designed to connect to the optic cable. These components will form a low cost fiber optic link that can transfer data at a rate up to 5 million pulses per second over a distance of 12 meters.

Let us use the system in Figure 20–28 and consider what limits the length of the optic link. Let us assume the link is being used for digital transmission, and the output of the receiver will

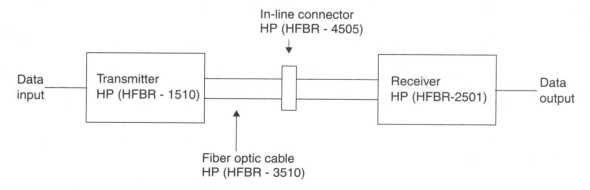

FIGURE 20–28   Fiber optic data communication link

be a logic 1 (+5 V) or a logic 0 (0 V), depending on the information sent by the transmitter. In order for the receiver to respond, it must have a minimum input power level from the transmitter. The transmitter is designed to output a power level that exceeds the power input needed by the receiver. However, losses in the line between the transmitter and receiver reduce the power until, at some point, the input to the receiver is too low. The output power of the transmitter is rated in dBm, which is a dB measurement using 1 mW as a reference. Example 20.2 demonstrates how line loss limits the length of the optic link.

EXAMPLE 20.2

Using worst case conditions from the technical data, calculate the maximum length of the optic link in Figure 20–28.

Solution:

From the data sheet:

Transmitter output power	–14.1 dBm (min)
Receiver input power	–21.6 dBm (min)
Cable attenuation	–0.43 dB/meter (max)
Connector attenuation	–2.8 dB (max)

Total line losses permitted = Tx min. output – Rec. min. input

–14.1 dB – (–21.6 dB) = 7.5 dB

7.5 dB – 2.8 dB = 4.7 dB (one connector)

4.7 dB/0.43 dB/m = 10.9 meter

A distance of only 10.9 meters may seem like a short distance, but this is a low cost system intended for in-house applications such as high voltage isolation, remote photo interrupters, and industrial controls. Higher priced components that handle data of up to 40 million pulses per second for distances of over 1200 meters are available from Hewlett Packard and other companies. The optic cables used in the more expensive systems typically have an attenuation of –5 dB/km or –0.005 dB/m.

Optical Sensors and Optoisolators

1. A piece of paper is in the slot sensor and there is nothing above the reflective sensor. Calculate and record the DC voltage at the following test points shown in Figure 20–29.

    $TP_1 = $ _____ $TP_2 = $ _____ $TP_3 = $ _____ $TP_4 = $ _____ $TP_5 = $ _____

    $TP_6 = $ _____ $TP_7 = $ _____ $TP_8 = $ _____ $TP_9 = $ _____ $TP_{10} = $ _____

2. A piece of paper is in the slot sensor and there is nothing above the reflective sensor. Determine and draw the waveforms at the following test points shown in Figure 20–29.

    TP_5 ——————————————————

    TP_6 ——————————————————

3. With nothing in the slot sensor and a white piece of paper 3 mm above the reflective sensor, calculate and record the DC voltages at the following test points shown in Figure 20–29. (Assume the reflective sensor has an output current of 60 µA.)

    $TP_1 = $ _____ $TP_2 = $ _____ $TP_3 = $ _____ $TP_4 = $ _____ $TP_5 = $ _____

    $TP_6 = $ _____ $TP_7 = $ _____ $TP_8 = $ _____ $TP_9 = $ _____ $TP_{10} = $ _____

4. With nothing in the slot sensor and a white piece of paper 3 mm above the reflective sensor, determine and draw the waveforms at the following test points shown in Figure 20–29. (Assume the reflective sensor has an output current of 60 µA.)

    TP_5 ——————————————————

    TP_6 ——————————————————

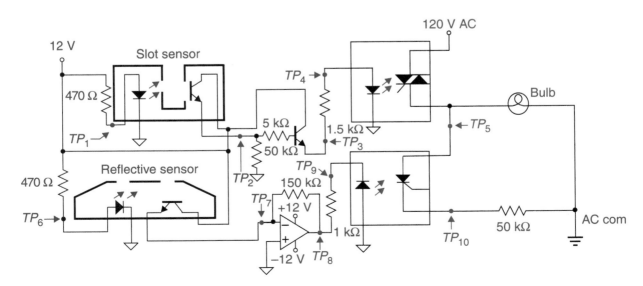

FIGURE 20–29

5. With a piece of paper in the slot sensor and a white piece of paper 3 mm above the reflective sensor, determine and draw the waveforms at the following test points shown in Figure 20–29.

TP_5 ——————————————————-

TP_6 ——————————————————-

6. Calculate the forward-biased current flow through all LEDs in Figure 20–29.

$I_{F(slot\ sensor)} =$ _____ $I_{F(reflective\ sensor)} =$ _____
$I_{F(triac)} =$ _____ $I_{F(SCR)} =$ _____

LAB 20.1

Optical Sensors and Optoisolators

Caution! This experiment uses line voltage that can cause severe electrical shock. Use an isolated AC source.

I. Objective

To analyze a circuit that uses optical sensors and optoisolators.

II. Test Equipment

±12 V power supplies   Oscilloscope   Voltmeter

III. Components

Resistors: (2) 330 Ω, (2) 1 kΩ, (1) 4.7 kΩ, (2) 100 kΩ, (1) 220 kΩ

Transistor: (1) 2N3905

IC: (1) LM741

Optical sensors: (1) Slot type EE-SX1042, (1) Reflective type EE-SY148

Optoisolators: (1) Triac type MOC3020, (1) SCR type 4N40IS

IV. Procedure

1. A piece of paper is in the slot sensor and there is nothing above the reflective sensor. Calculate and record the DC voltages at the following test points shown in Figure 20–30.

   $TP_1 =$ _____ $TP_2 =$ _____ $TP_3 =$ _____ $TP_4 =$ _____ $TP_5 =$ _____

   $TP_6 =$ _____ $TP_7 =$ _____ $TP_8 =$ _____ $TP_9 =$ _____ $TP_{10} =$ _____

2. A piece of paper is in the slot sensor and there is nothing above the reflective sensor. Determine and draw the waveforms at the following test points shown in Figure 20–30.

   TP_5 ————————————————

   TP_{10} ————————————————

3. With nothing in the slot sensor and a white piece of paper 3 mm above the reflective sensor, calculate and record the DC voltages at the following test points shown in Figure 20–30. (Assume the reflective sensor has an output current of 60 µA.)

   $TP_1 =$ _____ $TP_2 =$ _____ $TP_3 =$ _____ $TP_4 =$ _____ $TP_5 =$ _____

   $TP_6 =$ _____ $TP_7 =$ _____ $TP_8 =$ _____ $TP_9 =$ _____ $TP_{10} =$ _____

4. With nothing in the slot sensor and a white piece of paper 3 mm above the reflective sensor, determine and draw the waveforms at the following test points shown in Figure 20–30. (Assume the reflective sensor has an output current of 60 µA.)

   TP_5 ————————————————

   TP_{10} ————————————————

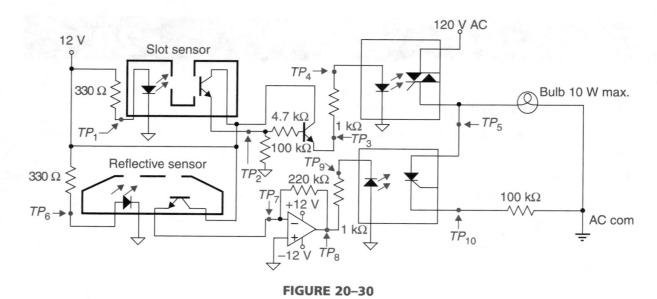

FIGURE 20–30

5. Construct the circuit in Figure 20–30.

6. Place a piece of paper in the slot sensor and make sure nothing is above the reflective sensor. Measure and record the DC voltages at the following test points shown in Figure 20–30.

    $TP_1 = $ _____ $TP_2 = $ _____ $TP_3 = $ _____ $TP_4 = $ _____ $TP_5 = $ _____
    $TP_6 = $ _____ $TP_7 = $ _____ $TP_8 = $ _____ $TP_9 = $ _____ $TP_{10} = $ _____

7. Place a piece of paper in the slot sensor and make sure nothing is above the reflective sensor. Measure and draw the waveforms at the following test points shown in Figure 20–30.

    TP_5 ——————————————————-

    TP_{10} ——————————————————-

8. The paper is removed from the slot sensor and a white piece of paper is placed 3 mm above the reflective sensor. Measure and record the DC voltages at the following test points shown in Figure 20–30.

    $TP_1 = $ _____ $TP_2 = $ _____ $TP_3 = $ _____ $TP_4 = $ _____ $TP_5 = $ _____
    $TP_6 = $ _____ $TP_7 = $ _____ $TP_8 = $ _____ $TP_9 = $ _____ $TP_{10} = $ _____

9. The paper is removed from the slot sensor and a white piece of paper is placed 3 mm above the reflective sensor. Measure and draw the waveforms at the following test points shown in Figure 20–30.

    TP_5 ——————————————————-

    TP_{10} ——————————————————-

10. Place a piece of paper in the slot sensor and place a white piece of paper 3 mm above the reflective sensor. Measure and draw the waveforms at the following test points shown in Figure 20–30.

TP_5 ——————————————-

TP_{10} —————————————-

V. Points to Discuss

1. What is the approximate power dissipated in each of the 330 Ω resistors in Figure 20–30?

2. Assume the ±12 V is supplied by batteries. A lightning strike induces 1000 V onto the 120 V AC line. Will this voltage spike cause any damage to the control circuit in Figure 20–30? Explain.

3. One of the sensors is sensitive to the color of the paper, but the other sensor is not color sensitive. Explain why.

4. The slot sensor not only controls the signal to the bulb but also controls the signal to the 100 kΩ output resistor connected to the SCR. How could the circuit be modified to permit the reflective sensor to have full control of the signal across the 100 kΩ resistor?

QUESTIONS

20.1 Introduction to Optoelectronics

____ 1. The area of study that deals with the use of _____ in electronic circuitry is called optoelectronics.
   a. optolenses
   b. light
   c. optotrons
   d. LLDs

____ 2. CRTs, LCDs, and LEDs are devices used in electronic systems as _____.
   a. input devices
   b. light sensors
   c. coupling devices
   d. displays

20.2 Cathode Ray Tubes (CRTs)

____ 3. The CRT is _____.
   a. designed from a large sheet of semiconductor material
   b. a type of electron vacuum tube
   c. a glass tube structure designed with thousands of LEDs
   d. the newest type of display device

____ 4. The grid of a vacuum tube controls the magnitude of current flowing between the _____.
   a. source and drain
   b. emitter and collector
   c. cathode and plate
   d. emitter and plate

____ 5. The purpose of the electron gun in a CRT is to _____.
   a. produce a beam of electrons
   b. capture loose ions
   c. prevent overheating of the tube
   d. none of the above

____ 6. The CRT system, which uses electromagnetic deflection, normally has a _____ placed around the neck of the tube.
   a. deflection yoke
   b. deflection plate
   c. heat sink
   d. head support

____ 7. The vertical input of an oscilloscope is connected to the _____.
   a. vertical time-base circuitry
   b. vertical oscillator

    c. scan generator circuitry

    d. signal the technician wants to display

____ 8. The electron beam continuously paints the full screen when a CRT is used in a scanning type display. The information to be displayed on the screen is applied to the control _____ of the CRT.

    a. gate

    b. grid

    c. plate

    d. source

20.3 Liquid-Crystal Displays (LCDs)

____ 9. Nematic liquid-crystal molecules are _____ shaped.

    a. square

    b. triangular

    c. ball

    d. rod

____ 10. With no electrical signal applied to an LCD, the liquid-crystal display functions as _____.

    a. a transparent window

    b. a light emitting window

    c. an opaque window

    d. a dark window

____ 11. LCDs use _____ electrodes that are deposited on the inner surface of the glass. These electrodes are in the pattern of the desired display.

    a. gold

    b. silver

    c. copper

    d. transparent

____ 12. Transmissive mode LCD units are available to work in low light environments. These units are designed with _____.

    a. backlighting

    b. endlighting

    c. frontlighting

    d. sidelighting

____ 13. The main disadvantage of active-matrix displays is _____.

    a. they do not work in low light environments

    b. their slow response time

    c. the large number of transistor drivers required

    d. they cannot display fast moving graphics

____ 14. The LCD electrically looks like a _____.

    a. large value capacitance in parallel with a small resistance

    b. large value capacitance in series with a large resistance

    c. small value capacitance in series with a large resistance

    d. small value capacitance in parallel with a large resistance

20.4 LEDs

____ 15. LEDs are specially designed diodes that emit visible light when _____.

  a. forward biased

  b. reverse biased

  c. the reverse breakdown voltage is exceeded

  d. the current flowing through the diode exceeds 500 mA

____ 16. Electrically, LEDs function like normal diodes except they have a forward voltage drop of approximately _____.

  a. 0.3 V

  b. 0.7 V

  c. 2 V

  d. 5 V

20.5 Light Sensing Devices

____ 17. Photoconductive cells are two-terminal devices that vary their _____ with exposure to light.

  a. output voltage

  b. output current

  c. resistances

  d. all of the above

____ 18. A photoconductive cell made of _____ has a response curve close to that of the human eye.

  a. cadmium selenide

  b. cadmium sulfide

  c. silicon sulfide

  d. silicon selenide

____ 19. Photovoltaic cells are two-terminal devices that vary their _____ with exposure to light.

  a. output voltages

  b. output currents

  c. resistances

  d. all of the above

____ 20. In order to obtain usable power levels from photovoltaic cells, it is necessary to connect cells in series and parallel. Connecting photovoltaic cells in parallel _____.

  a. increases their output voltages

  b. increases their output voltage capabilities

  c. increases their output currents

  d. increases their output current capabilities

20.6 Photoactive Devices

____ 21. All photoactive devices are based on the principle that the magnitude of minority current carriers is mainly a function of _____.

    a. forward-biased voltage

    b. reverse-biased voltage

    c. the energy applied to their PN junction

    d. none of the above

____ 22. The _____ has the fastest response to changes in light intensity of all photo devices and is the building block for all photoactive devices.

    a. photoresistor

    b. photodiode

    c. phototransistor

    d. photogate

____ 23. What is the minimum number of leads a phototransistor can be connected with and still operate?

    a. 1

    b. 2

    c. 3

    d. 4

____ 24. Manufacturers can produce a _____ by forming a photodiode between the gate and the channel.

    a. phototriac

    b. phototransistor

    c. photoFET

    d. photoSCR

20.7 Optoisolators and Optical Sensors

____ 25. An optoisolator is a device containing _____ in a single package.

    a. a photodiode and an optosensor

    b. a phototransistor and an optosensor

    c. an optotransmitter and an LED

    d. an optotransmitter and an optosensor

____ 26. The slot optical sensor has a slot separating _____.

    a. a photodiode and an optosensor

    b. a phototransistor and an optosensor

    c. an optotransmitter and an LED

    d. an LED and an optotransistor

20.8 Lasers

____ 27. A laser beam must have light that is _____.

    a. monochromatic

    b. collimated

    c.   coherent

    d.   all of the above

____ 28.   If a light beam is collimated, all of the rays of light are _____.

    a.   parallel to each other

    b.   the same frequency

    c.   in phase

    d.   a single color

20.9 Laser Diodes

____ 29.   Large amounts of energy leave the laser diode in the form of light, and this energy must be provided by an energy source of some type. In the case of the laser diode, it is the _____ that provides energy and allows the laser action to be a continuous process.

    a.   high reverse-bias voltage

    b.   high forward-bias current

    c.   high reverse-bias current

    d.   none of the above

20.10 Fiber Optics

____ 30.   One important advantage a fiber optic cable has over a copper conductor is its _____.

    a.   low cost

    b.   ease of repair

    c.   ability to handle large current flow

    d.   wide bandwidth

____ 31.   The _____ is the ratio of the speed of light in a vacuum to the speed of light in the material of interest.

    a.   critical angle

    b.   index of refraction

    c.   light ratio

    d.   perpendicular ratio

____ 32.   The _____ is measured with reference to a line perpendicular to the junction of two optically transparent materials.

    a.   critical angle

    b.   index of refraction

    c.   light ratio

    d.   perpendicular ratio

____ 33.   The core material of a fiber optic cable is surrounded by a second less dense material called _____.

    a.   casing

    b.   the fiber barrier

    c.   the surrounding shield

    d.   cladding

_____ 34. A type of fiber optic cable is _____.
   a.   multimode step-index
   b.   single-mode step-index
   c.   multimode graded-index
   d.   all of the above

PROBLEMS

1. The CRT in Figure 20–31 is using scanned deflection circuitry. The video signal is being applied to the grid of the CRT. Show the video signal for the two horizontal sweeps in Figure 20–31. (Assume the beam is off for the dark bars.)

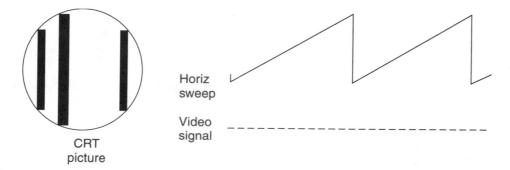

FIGURE 20–31

2. A portable computer has a screen resolution of 320 × 200. If the system is using a passive-matrix display, how many transistor drivers will be required? If the system is using an active-matrix display, how many transistor drivers will be required?

3. Sketch a circuit using a photoconductive cell that outputs a TTL logic one when illuminated. Assume the photoconductive cell has a ratio of dark-to-light resistance of 1000:1, and its dark resistance is 1 mΩ.

4. Sketch a circuit using photovoltaic cells that will provide 2.4 V with a load current of 200 mA. (Assume each is rated for 0.4 V at 50 mA.)

5. Sketch a circuit using a phototransistor in which the output voltage will decrease when illuminated.

6. The optoisolator in Figure 20–32 has a current transfer ratio of 80%. What resistance of R_1 will cause the output voltage to equal 5 V? (The current transfer ratio was discussed in Chapter 15.)

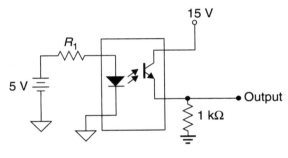

FIGURE 20–32

7. A light activated SCR (LASCR) is connected in a circuit as shown in Figure 20–33(a). When exposed to light, it starts conducting. Explain what happens when the circuit is returned to a dark environment.

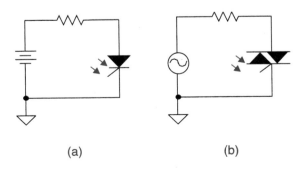

(a)   (b)

FIGURE 20–33

8. A light activated triac is connected in a circuit as shown in Figure 20–33(b). When exposed to light, it starts conducting. Explain what happens when the circuit is returned to a dark environment.

9. Compute the distance light travels in 1 nS in a vacuum.

10. A fiber optic cable with a core index of refraction of 1.5 and a cladding index of refraction of 1.1 has a critical angle of _____.

11. Calculate the critical angle for a fiber rod surrounded by air. The index of refraction of the fiber rod is 1.5.

12. Express 18 dBm in watts.

13. An optical transmitter is rated for a minimum output power of –4 dBm, and an optical receiver has a minimum sensitivity of –18 dBm. If a fiber optic cable with a maximum attenuation of –0.15 dB/meter is used, what is the maximum distance the receiver can be from the transmitter?

CHAPTER 21

Transducers and Actuators

OBJECTIVES

After studying the material in this chapter, you will be able to describe and/or analyze:

- ○ temperature transducers,
- ○ displacement transducers,
- ○ pressure transducers,
- ○ flow transducers,
- ○ sensor transducer signal conditioning,
- ○ solenoids,
- ○ relays,
- ○ DC motors,
- ○ AC motors,
- ○ stepper motors, and
- ○ speakers.

21.1 INTRODUCTION TO TRANSDUCERS AND ACTUATORS

Devices that convert one form of energy to another form of energy are called transducers. Transducers can be grouped into two broad areas: sensor and actuator transducers. Sensor transducers have a nonelectrical input (temperature, pressure, flow, light, and so on) and an electrical output. These sensors give an electronics circuit the ability to sense what is happening in the world. Sensor transducers can be compared to the five senses of the human body.

Actuators take the output of an electronics system and convert it to a form of energy that is useful for running and controlling automated systems (mechanical movement, sound, light, and so on). Actuator transducers could be considered the arms, legs, and mouth of the electronics system. Many of the actuator devices we will discuss make use of magnetism, and it is assumed you have a knowledge of basic magnetic principles.

Figure 21–1 shows a system that is controlled electronically. In order to have electronic control, it is necessary to have sensors that sense the status of the system and feed this information to the electronic control system. The electronic control system processes the data from the sensors and outputs control signals to the actuators. If the necessary control is complex, the electronic control system may be a digital computer system. If a digital computer system is used, the analog signals from the transducers must be converted to digital using an ADC. After processing the output of the digital system, it must be converted back to analog using a DAC to drive the actuators. The study of digital processing is covered in microcomputer textbooks, but this chapter will cover some of the most popular sensors and actuators.

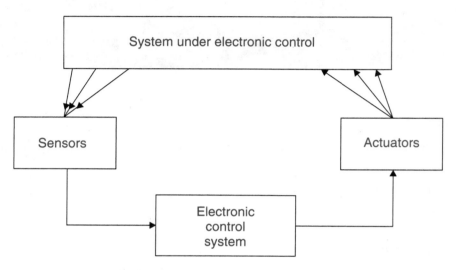

FIGURE 21–1   Electronic control system

21.2   TEMPERATURE SENSORS

INTRODUCTION

Temperature is a quantity that humans have been interested in measuring for hundreds of years. Galileo was credited with inventing the thermometer in 1592. As industry became more and more complex, the need to know the precise temperature of materials and equipment became more important. We will discuss five types of temperature transducers: thermometer, thermocouple, RTD, thermistor, and the IC sensor.

Temperatures can be expressed using three scales: Fahrenheit, Celsius, or Kelvin. The freezing point of water on the Fahrenheit scale is 32°F, and water boils at 212°F; the freezing point of water on the Celsius scale is 0°C, and water boils at 100°C; and the freezing point of water on the Kelvin scale is 273°K, and water boils at 373°K. The Kelvin scale is also called the absolute scale, and zero degrees Kelvin is theoretically the lowest temperature possible. To convert Celsius to Kelvin, simply add 273° to the Celsius value. To convert Celsius to Fahrenheit, multiply the Celsius temperature times 1.8 and add 32°. Example 21.1 demonstrates these conversions.

EXAMPLE 21.1

To convert 22º Celsius to the Kelvin and Fahrenheit equivalents:
Step 1. Convert Celsius to Kelvin.

$$Degrees\ Kelvin = Degrees\ Celsius + 273º$$
$$Degrees\ Kelvin = 22º + 273º = 295ºK$$

Step 2. Convert Celsius to Fahrenheit.

$$Degrees\ Fahrenheit = (Degrees\ Celsius \times 1.8) + 32º$$
$$Degrees\ Fahrenheit = (22º \times 1.8) + 32º = 71.6ºF$$

THERMOMETERS

The temperature transducer that is the most familiar is the glass thermometer. It consists of a glass tube that contains mercury in a bulb at the bottom of the tube. An increase in temperature causes the liquid mercury to expand. The height of the liquid in the tube is calibrated in degrees of temperature: Fahrenheit or Celsius. Although the thermometer is not electrical in nature, it is a simple linear way to measure temperature. Two disadvantages of glass thermometers are that they are fragile and slow to respond to temperature changes. Since mercury is a good conductor, a simple temperature control switch can be designed as shown in Figure 21–2. One contact point is connected to the pool of mercury while the other contact point is placed up the tube. When the mercury expands, it makes contact with the second contact point, and the switch closes.

FIGURE 21–2   Mercury switch

THERMOCOUPLES

Thermocouples are electrical transducers that produce a voltage proportional to temperature. Thermocouples function on the basis that different metals have different thermionic characteristics. Thomas Seebeck discovered in 1821 that if two wires of dissimilar metals were connected, as in Figure 21–3, and the junction was heated, a voltage was produced. This voltage is called the *Seebeck voltage* and is directly related to the temperature of the thermocouple.

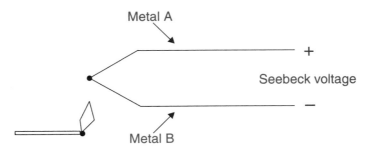

FIGURE 21–3   Seebeck voltage

Figure 21–4 shows a graph of temperature versus Seebeck voltage for the most popular types of thermocouples. Thermocouples have reasonable linearity over wide temperature ranges. The chromel-constant thermocouple, for example, is good to about 900°C, has reasonable linearity, and will provide approximately 65 millivolts at 900°C. Other alloys operate up to 2200°C (4000°F). The higher the maximum possible temperature, the less the sensitivity of the thermocouple. Thermionically more active metals that generate relatively high voltage outputs will melt down at high temperatures.

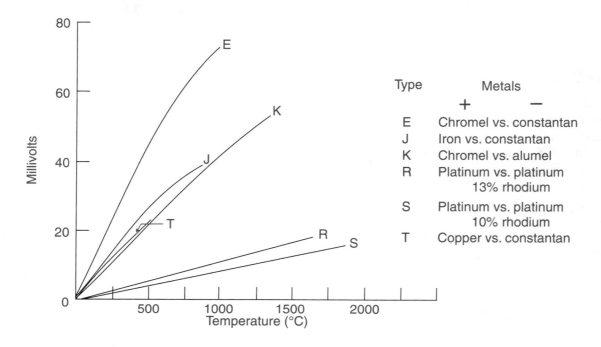

FIGURE 21–4   Thermocouple voltage versus temperature curve

In Figure 21–4, the letters identifying the curves are used as standard industrial identifications. For example, if a technician is using a J-type thermocouple, it is constructed by connecting a wire made of iron to a wire made of constantan (an alloy of copper and nickel).

Unfortunately, Seebeck voltage cannot be read directly with a voltmeter because connecting the copper leads of the voltmeter forms additional thermocouples. Figure 21–5 shows a solution to the problem. The two thermocouples formed by the copper leads are held at a constant temperature of 0°C; thus, the voltage across these thermocouples will be zero, and the meter will read the 40 mV produced by the K-type thermocouple. In the early days of instrumentation, the

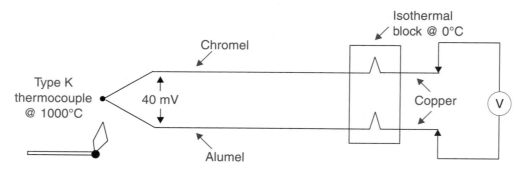

FIGURE 21–5   Isothermal block connection

isothermal block containing the copper junctions was submerged in an ice bath to hold its temperature at 0ºC. The junctions were known as *cold junctions* after this process, a name that is still used today.

Today an electronics compensation circuit can be added to correct for changes in the voltage drops across the cold junctions. Figure 21–6 shows the compensation circuit using ICs available from Linear Technology. The cold junctions are permitted to change temperature with ambient temperature. A cold junction compensator IC (LT1025) is connected in series with the thermo-couple and the cold junctions. The voltage produced by the cold junctions will vary with changes in ambient temperature; however, the output of the LT1025 also changes with the surrounding temperature. The LT1025 is designed to output a voltage of opposite polarity that tracks the magnitude of voltage generated by the cold junctions. The voltage (V_c) produced by the LT1025 cancels the voltage (V_{cj}) generated by the cold junctions so the only voltage present at the input of the amplifier is the Seebeck voltage (V_s) produced by the thermocouple. The amplifier stage provides buffering and voltage gain. The LT1025 is designed with multiple output pins. By selecting the correct output, the LT1025 can provide compensation to the six popular types of thermocouples.

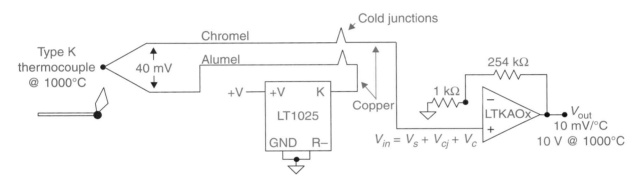

FIGURE 21–6   Cold junction compensation circuit

RESISTANCE TEMPERATURE DETECTORS (RTDs)

Resistance temperature detectors (RTDs) are a class of devices that depend on the positive temperature coefficients of metals. RTDs are available in different materials, with platinum being the most frequently used. A standard platinum RTD has a resistance of 100 Ω at 0ºC. The resistance increases quite linearly to 331 Ω at 600ºC. This means the temperature coefficient of the RTD is 0.385 Ω per degrees Celsius.

To determine temperature with an RTD, it is necessary to determine the value of resistance first and then translate that resistance value into the correct temperature. Typically, the resistance value is converted to a voltage that can be easily displayed or recorded. Figure 21–7 shows a resistance bridge network with three resistors and an RTD. From your study of basic

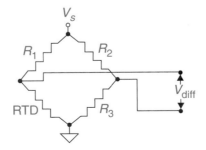

FIGURE 21–7   Bridge circuit connected to an RTD

electronics, you know that if the resistance of the RTD changes and the other resistors remain constant, there will be a change in the differential voltage across the bridge. The change in voltage is proportional to the temperature change. Example 21.2 shows how the differential output voltage can be calculated for a given RTD at a known temperature.

EXAMPLE 21.2

A standard platinum RTD having a resistance of 100 Ω at 0°C and a temperature coefficient of 0.385 Ω per degrees Celsius is connected in a bridge circuit as shown in Figure 21–7. If $R_1, R_2,$ and R_3 all equal 100 Ω, what will be the differential output voltage if the RTD is at a temperature of 24°C and V_s equals 1 V?

Step 1. Calculate the resistance of the RTD. First, calculate the change in temperature from the reference temperature of 0°C. Next, calculate the change in resistance by multiplying the change in temperature times the temperature coefficient (α) of the RTD. Finally, calculate the current resistance by adding the change in temperature to the reference temperature.

$$\Delta T = T - 0°C = 24°C - 0°C = 24°C$$
$$\Delta R = \Delta T \times \alpha = 24°C \times 0.385\ \Omega = 9.24\ \Omega$$
$$R = R_{\text{ref}} + \Delta R = 100\ \Omega + 9.24\ \Omega = 109.24\ \Omega$$

Step 2. Calculate the voltage across the RTD and the voltage across R_3.

$$V_{\text{RTD}} = V_s \times R_{\text{RTD}}/(R_1 + R_{\text{RTD}}) = 1\ \text{V} \times 109.24\ \Omega/(100\ \Omega + 109.24\ \Omega) = 0.522\ \text{V}$$
$$V_3 = V_s \times R_3/(R_2 + R_3) = 10\ \text{V} \times 100\ \Omega/(100\ \Omega + 100\ \Omega) = 0.500\ \text{V}$$

Step 3. Calculate the difference output voltage.

$$V_{\text{diff}} = V_{\text{RTD}} - V_3 = 5.22\ \text{V} - 5.00 = 0.022\ \text{V}$$

Often it is necessary to locate the RTD some distance from the other three bridge resistors. Because the change in resistance of an RTD is small, the lead resistance to and from the RTD becomes a factor that introduces errors into the system. For example, 100 feet of number 22 wire has 1.614 Ω of resistance at 20°C. In the circuit in Figure 21–8(a), the resistance of two wire segments is added to the resistance of the RTD, which upsets the balance of the bridge. Resistance could be added to R_1, but this would not solve the problem because the wire segments will vary in resistance with temperature. The lead resistance problem can be overcome by connecting the RTD to the bridge circuit using three wire segments, as shown in Figure 21–8(b). Wire segment "a" is in series with resistor R_1, and wire segment "b" is in series with the RTD. If the wire segments are made of the same material and are the same length, they will not affect the balance

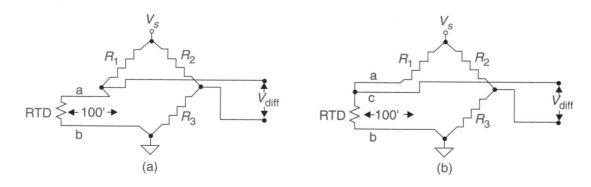

FIGURE 21–8   Three-wire bridge connection

of the bridge. It is important that the output of the bridge be connected to a high impedance so there is minimum current in wire segment "c."

RTDs are available in two general types: wire and metal film. When an RTD is made from wire, it is necessary to wind the wire into a helix to maximize resistance. Support of the helix must be in such a manner as to minimize mechanical strain because of expansion occurring during heating. Wire RTDs tend to be fragile and expensive. The metal film is the other type of RTD. A platinum film is deposited onto a small flat ceramic substrate and then etched with a laser and sealed. Metal film RTDs are less stable than wire RTDs but are lower in cost and smaller.

RTDs are passive devices and do not generate internal potential differences but must depend on external supplies. Precautions must be taken to prevent self-heating of the transducers by energy from the external source. The precautions involve keeping the voltage and current applied to the RTD as small as possible. Larger RTDs can handle more energy without self-heating but have slow response time to temperature changes and may cause thermal shunting. Thermal shunting occurs when the temperature being measured is changed by the insertion of the transducer.

RTDs are the most stable, linear, and accurate temperature transducers, which makes them a favorite for research and laboratory work. However, they are fragile, expensive, and require relatively complex support circuitry, which limits their field applications.

THERMISTORS

Thermistors are thermally sensitive resistors manufactured from semiconductor materials that normally have a negative temperature coefficient (NTC). As the temperature of the thermistor increases, its resistance decreases. There are a few thermistors with a positive temperature coefficient (PTC), and these will be discussed later. Thermistors are the most sensitive of all the temperature transducers, with a typical change in resistance of a few percent per degrees Celsius. For example, a thermistor could have a temperature coefficient of –5%/°C, meaning a one degree change would decrease the resistance by 5%. Thermistors are also the most nonlinear of all the temperature transducers. The temperature coefficient is normally given for a temperature of 25°C, and because of nonlinearity, it will vary over the operating range of the thermistor. A temperature versus resistance curve and the symbol for the thermistor are shown in Figure 21–9.

Example 21.3 shows how a thermistor can be used to detect a small change in temperature. The circuit can function over a wide range of temperatures, but because of the nonlinearity of

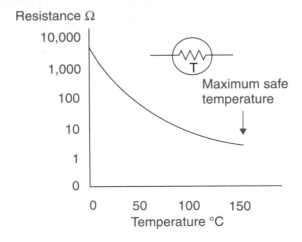

FIGURE 21–9 Thermistor symbol and temperature versus resistance curve

thermistors, the math needed to calculate the resistance of the thermistor with wide temperature variations is beyond the scope of this text. If available, the technician can easily read the resistance of the thermistor at any temperature from a temperature versus resistance graph. Once the resistance is known, the output voltage can be easily calculated.

EXAMPLE 21.3

Calculate the output voltage of the circuit in Figure 21–10 at the temperatures of 24ºC, 25ºC, and 27ºC. The thermistor has a temperature coefficient of –5%/ºC and resistance of 2 kΩ at 25ºC.

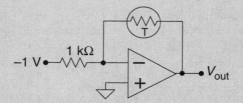

FIGURE 21–10   Thermistor circuit configuration

Step 1. Calculate the resistance of the thermistor at each temperature. This procedure is only an approximation and should be used only for a few degrees variation.
$$R = (\Delta T \times \alpha \times R_{ref}) + R_{ref}$$
At 23ºC: $R = (-2°C \times -5\% \times 2\ k\Omega) + 2\ k\Omega = 2.2\ k\Omega$
At 25ºC: $R = (0°C \times -5\% \times 2\ k\Omega) + 2\ k\Omega = 2\ k\Omega$
At 28ºC: $R = (3°C \times -5\% \times 2\ k\Omega) + 2\ k\Omega = 1.7\ k\Omega$

Step 2. Calculate the current through the thermistor.
$$I = V_{in}/R_i = 1\ V/1\ k\Omega = 1\ mA$$

Step 3. Calculate the output voltage at each temperature.
$$V_{out} = I \times R_f$$
At 23ºC: $V_{out} = 1\ mA \times 2.2\ k\Omega = 2.2\ V$
At 25ºC: $V_{out} = 1\ mA \times 2\ k\Omega = 2\ V$
At 28ºC: $V_{out} = 1\ mA \times 1.7\ k\Omega = 1.7\ V$

A distinct advantage of thermistors over RTDs is their high resistance. A typical resistance for thermistors at 25ºC is several thousand ohms, where the RTD only has a resistance of approximately 100 Ω. Lead resistance is not a problem with thermistors because of their high resistance. Another advantage of thermistors is that they can be made very small, which means they do not cause thermal shunting and will respond quickly to temperature changes. In addition to being nonlinear, thermistors also have a limit in temperature range, with a typical maximum rating of 200ºC. Being semiconductor devices, they are susceptible to permanent damage if the maximum temperature is exceeded. Like RTDs, thermistors are susceptible to self-heating.

Thermistors with a positive temperature coefficient (PTC) actually decrease slightly in resistance with increases in temperatures until the transition temperature is reached. At the transition temperature, the PTC thermistor switches quickly to a high resistance value. Figure 21–11 shows the resistance versus temperature curve for the PTC thermistor. PTC thermistors are available with transition temperatures in the range of 30ºC to 150ºC.

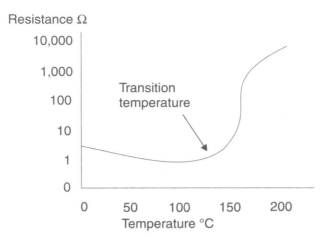

FIGURE 21–11   Resistance versus temperature curve for a PTC thermistor

INTEGRATED CIRCUIT TEMPERATURE TRANSDUCERS

IC temperature sensors are integrated circuits that are designed to give a linear output voltage that is proportional to changes in temperature. Their output is the most linear of all the temperature transducers discussed. Fahrenheit, Celsius, or Kelvin IC sensors normally output 10 mV per degree. Figure 21–12 shows three circuits using National Semiconductor IC temperature sensors. The LM34 Fahrenheit sensor and the LM35 Celsius sensor require no external components and can operate with any supply input between 5 V and 30 V. The temperature range of each can be extended in the negative direction by additional components. The LM335 is a special zener diode designed to drop a voltage equal to the Kelvin temperature times 10 mV. The supply voltage and R_1 must be selected to maintain a current through the device between 0.4 mA to 5 mA. Resistor R_2 provides a calibrated adjustment and is adjusted for the correct output voltage at a known temperature.

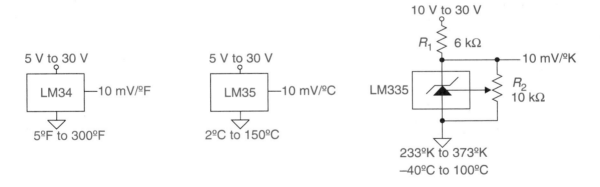

FIGURE 21–12   IC temperature sensors

SUMMARY OF TEMPERATURE TRANSDUCERS

Table 21–1 summarizes the advantages and disadvantages of the four types of temperature transducers discussed. Most of the items in the table are self-explanatory. However, we will define four terms used in the table to clarify their meanings with regard to transducers.

1. *Accuracy:* Closeness of a reading or indication of a measurement device to the actual value of the quantity being measured.
2. *Linearity:* The ability of an instrument to generate a straight line response.

3. *Sensitivity:* The minimum change in input signal to which an instrument can respond.

4. *Stability:* The quality of an instrument or sensor to maintain a consistent output when a constant input is applied.

TABLE 21–1   Comparison of temperature transducers

	IC Sensor	Thermistor	RTD	Thermocouple
Advantages	1. Most linear 2. Inexpensive 3. Highest output	1. Fast 2. High output 3. Large ΔR	1. Most stable 2. Most accurate 3. Linear	1. Widest temperature range 2. Rugged 3. Inexpensive
Disadvantages	1. Limited temperature range 2. Self-heating 3. Power supply required 4. Fragile	1. Most nonlinear 2. Limited temperature range 3. Self-heating 4. Fragile	1. Expensive 2. Small ΔR 3. Self-heating	1. Nonlinear 2. Low output 3. Least stable 4. Least sensitive

One point to note is that the IC sensor is listed for the highest output, but its output is typically in millivolts or microamps, so all of these transducers require analog signal conditioning before the signal can be processed.

21.3   DISPLACEMENT SENSORS

LINEAR POTENTIOMETER AS A DISPLACEMENT SENSOR

Displacement is a measurement of the distance an object moves. The SI (Standard International) unit for displacement is the meter. As shown in Figure 21–13, a potentiometer can be used as a transducer to convert mechanical displacement into an electrical signal. The advantages of this

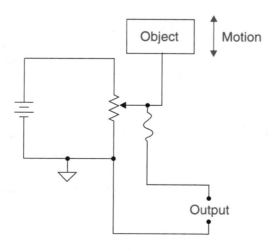

FIGURE 21–13   Potentiometer displacement transducer

transducer include its simplicity, its high output level, and the resistance of the potentiometer can be designed so the output voltage will be linear and proportional to the displacement. The disadvantages are that repeated use can cause mechanical wear, it has limited resolution, and it tends to have high electronic noise.

LINEAR VARIABLE DIFFERENTIAL TRANSFORMER (LVDT)

The linear variable differential transformer, or LVDT, is a popular displacement transducer. It is comprised of a transformer with a movable core, a primary, and two secondaries as shown in Figure 21–14. If the core is centered between the two secondaries, as shown in Figure 21–14(a), the outputs of both secondaries will be equal.

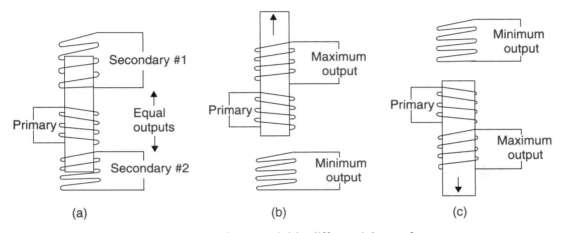

FIGURE 21–14 Linear variable differential transformer

However, if the core is moved upward, as shown in Figure 21–14(b), the upper secondary will have greater magnetic coupling with the primary and have a greater voltage output than the lower secondary. In Figure 21–14(c), the lower secondary has the larger voltage output because the core has moved down and allows greater magnetic coupling between the primary and the lower secondary.

Of course, the input and the output of an LVDT must be an AC signal. However, by adding some conditioning circuitry to the output of the LVDT (Figure 21–15), a linear DC output voltage can be obtained that is proportional to the displacement of the core.

The two secondaries are connected to form a center tap secondary. The center tap is connected to ground reference. The ends of the secondary are connected to positive half-wave rectifiers (D_1 and D_2). The output of the rectifiers are connected to opposite ends of the output filter capacitor

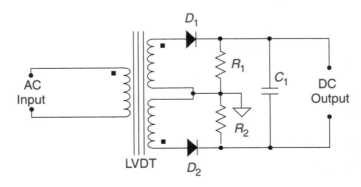

FIGURE 21–15 LVDT circuit with DC output

C_1. If the core is in the center position, the secondary voltages will be balanced, and outputs of the rectifiers will be equal; therefore, the difference voltage across the output capacitor will be zero. If the core is off-center, one of the rectified voltages will be greater, causing an output voltage to be developed across the capacitor. LVDTs are sensitive and can detect displacements as small as 0.002 mm.

STRAIN GAGES

The strain gage is the most universal device for converting mechanical displacements into electrical quantities and is a fundamental part of many mechanical to electrical transducers. The strain gage works on the principal that a material under mechanical stress will change its physical shape.

At a fixed temperature, solid objects have a constant volume. If a conductor is stretched in length, its cross-sectional area must decrease in order for the volume to remain constant. If the length is decreased by a compression force, the cross-sectional area must increase for the volume to remain constant. These changes in physical shape cause changes in resistance.

The basic strain gage is shown in Figure 21–16. A wire or foil conducting material is mounted on a thin paper or plastic backing. The conductor material is laid out in the pattern shown so the wire length is sufficient to exhibit a usable resistance, and so the strain gage will be sensitive only in one direction (unidirectional). In Figure 21–16, if tensile stress or compression stress is applied along the horizontal axis, the small conductors will lengthen or shorten respectively. If, however, the stress is applied in the direction of the vertical axis, the conductors will unfold, and the overall length of the conductors will remain the same. The tabs are for the connection of external wires.

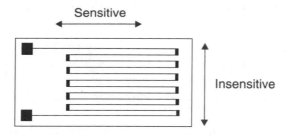

FIGURE 21–16   Single-element strain gage

The strain gage is bonded with glue to the object to be measured. The backing is stretched slightly before being glued to the object under test. As the test object is elongated, the gage is further stretched, causing the conductors to decrease their cross-sectional area and increase in length, thereby increasing the resistance of the strain gage. If the test object is compressed, the conductors will shorten, and their cross-sectional area will increase, thereby causing a reduction in resistance. The change in resistance is linear and proportional to the mechanical displacement.

Standard metal strain gages are available with normal resistances of 120 Ω, 350 Ω, 600 Ω, and 700 Ω. The first two, 120 Ω and 350 Ω, are the most popular. Sizes ranging from 0.6 mm to 150 mm are commercially available. The ability of a strain gage to convert a change in mechanical displacement to a change in electrical resistance is measured by the gage factor (GF). Gage factor is equal to the fractional change in gage resistance to the fractional change in length. The gage factor is a measurement of the gage sensitivity. A typical number for GF is 2, meaning a one percent change in length would provide a two percent change in resistance.

If you have a bathroom scale with a digital display, then you use a strain gage every morning you step on the scale. Figure 21–17(a) shows how strain gages can be used to convert weight to an electrical signal. The scale has a steel beam with supports on each end. The load (your weight) is applied to the center of the beam. Two strain gages are attached to the beam: one on the top side and one on the bottom side. When the load is applied to the beam, it bends, resulting in gage 1 being compressed and gage 2 being expanded. The resistance of gage 1 will decrease, and the resistance of gage 2 will increase.

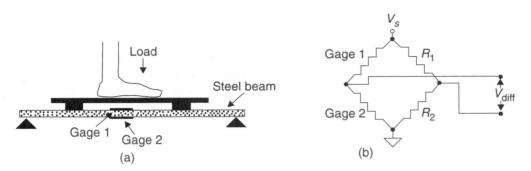

FIGURE 21–17 Bathroom scale using strain gages

One problem with strain gages is that the changes in resistance caused by mechanical displacement is small. The design of the scale helps overcome this problem by using two gages properly placed so when the resistance of one goes up the resistance of the other goes down. Figure 21–17(b) shows the strain gages of the scale connected into a bridge circuit. Since two gages are used, the output voltage of the bridge will be doubled. Example 21.4 will show how the output voltage of the bridge can be calculated for known changes in the length of the strain gages.

EXAMPLE 21.4

Calculate the output voltage of the bridge circuit in Figure 21–17 if the scale is weighing a 160 lb person. The scale is using 350 Ω strain gages with a GF of 2. When a 160 lb person stands on the scale, the gages change their lengths by 1.5%. Resistor R_1 equals R_2, and the supply voltage is a 9 V battery.

Step 1. Calculate the resistance of the strain gages. Multiplying the percent of change in length times the gage factor (GF) will give the percent of change in resistance.

$$\%\Delta R = \%\Delta L \times GF = 1.5\% \times 2 = 3\%$$
$$\Delta R = R \times \%\Delta R = 350\ \Omega \times 3\% = 10.5\ \Omega$$
$$R_{gage1} = R - \Delta R = 350\ \Omega - 10.5\ \Omega = 339.5\ \Omega$$
$$R_{gage2} = R + \Delta R = 350\ \Omega + 10.5\ \Omega = 360.5\ \Omega$$

Step 2. Calculate the voltage across gage 2 and the voltage across R_2. The voltage across the gage can be found by using the voltage divider formula. The voltage across R_2 will equal one-half the supply, since R_1 equals R_2.

$$V_{gage2} = V_s \times R_{gage2}/(R_{gage1} + R_{gage2}) = 9\ V \times 360.5\ \Omega/(339.5\ \Omega + 360.5\ \Omega) = 4.635\ V$$
$$V_{R2} = V_s \times 0.5 = 9\ V \times 0.5 = 4.5\ V$$

Step 3. Calculate the difference output voltage.

$$V_{diff} = V_{gage2} - V_2 = 4.635\ V - 4.5\ V = 0.122\ V$$

Semiconductor strain gages have become increasingly popular. Like the metal strain gage, semiconductor strain gages change their resistance under changing mechanical displacement. However, in semiconductor strain gages, the change in resistance is due to a phenomenon called the *piezoresistive effect*. This phenomenon causes a change in resistance due mainly to changes in the crystal structure that alter the electron and hole mobility. Semiconductor strain gages are more sensitive, having gage factors in the range of 50 to 200, compared to a typical gage factor of 2 for metal strain gages. Even though semiconductor strain gages are much more sensitive, their use is limited because they are highly nonlinear.

21.4   PRESSURE TRANSDUCERS

INTRODUCTION

Pressure transducers are used to measure pressure. Pressure measurement, like voltage measurement, must have a reference value. There are three types of pressure measurements, depending on the reference used: absolute, gage, and differential. Absolute pressure is the pressure relative to zero. To measure absolute pressure, one side of the transducer must be at zero pressure, or exposed to a vacuum. Gage pressure measurements are the most common where it is desired to know the pressure in a container relative to ambient pressure. One side of the transducer is exposed to the pressure to be measured, while the other side is vented to the surrounding atmospheric pressure. Differential pressure measurements are made with one side of the transducer exposed to one pressure and the other side of the transducer exposed to a second pressure.

The SI unit for pressure is Newtons per square meter. This unit has been given the name *pascal* (Pa). The English unit for pressure is the pound per square inch, usually written psi.

DIAPHRAGM PRESSURE TRANSDUCER

Figure 21–18 shows a diaphragm pressure transducer. It consists of two chambers separated by a diaphragm. Figure 21–18(a) shows that the diaphragm is not deflected if the pressure is the same in both chambers. In Figure 21–18(b), the pressure in the left chamber is greater than the pressure in the right; therefore, there is deflection of the diaphragm. If a strain gage is bonded to the diaphragm, the deflection will cause the output of the gage to vary in resistance.

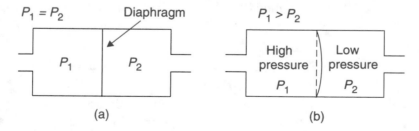

FIGURE 21–18   Diaphragm pressure transducer

There are usually four strain gages arranged so that two are at right angles to the other two. Resistance will increase in two of the gages when strain occurs, and decrease in the other two, depending on their orientation on the diaphragm. If a bridge circuit is formed, as shown in Figure 21–19, the output voltage will represent the pressure.

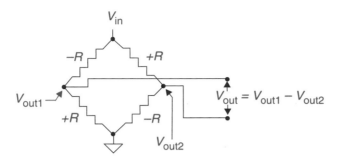

FIGURE 21–19   Bridge circuit

If all four resistors in the bridge circuit have the same value of resistance with no differential in input pressure, there is no output voltage. The output voltage that does exist with zero input pressure is called the *null offset voltage*. External resistors may be used to eliminate the null offset voltage at a cost of reduced sensitivity.

Pressure transducers using semiconductor strain gages are more sensitive than ones with metal strain gages. However, pressure transducers using metal strain gages are more linear and can be used in environments that the semiconductor devices cannot tolerate.

21.5   FLOW TRANSDUCERS

INTRODUCTION

The control of the flow of liquids or gases is essential for many industrial processes. In order to control the flow, it is first necessary to have a means of measuring the flow. The flow transducer converts the flow rate into an electrical response that is proportional to the flow. The two general classes of flow meters are the differential pressure type and the positive displacement type.

DIFFERENTIAL PRESSURE FLOW TRANSDUCERS

The laws of physics state that the volume of fluid flowing per unit time through a pipe is the same at all points along the pipe. If the pipe has a cross-sectional area that is varied, as shown in Figure 21–20, the velocity of the fluid must change to maintain a constant volume of fluid moving through the pipe.

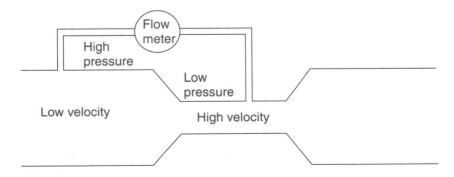

FIGURE 21–20   Differential pressure flow transducer

The Bernoulli principle states that an increase in velocity causes a decrease in pressure. Figure 21–20 shows how this principle is used to measure the flow of a liquid. A differential pressure sensor is connected between the large and small sections of the pipe. Since the flow is proportional to the difference in pressure between the two points, the display of the meter can be calibrated in units of flow (for example, gallons per minute).

POSITIVE DISPLACEMENT FLOW TRANSDUCERS

Positive displacement flow transducers use the principle of mechanical movement caused by fluid flow. In the example in Figure 21–21, a turbine is installed in the pipe, and the fluid flow causes the turbine to rotate. The higher the flow, the faster the rotation. The shaft of the turbine is connected to a tachometer. The output of the tachometer is a frequency dependent upon the speed of rotation. The frequency of the output signal, therefore, is proportional to the flow of the liquid in the pipe.

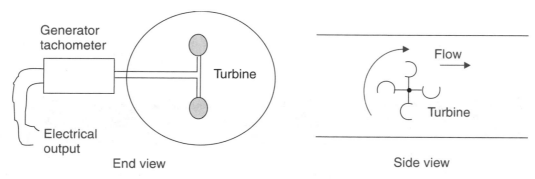

FIGURE 21–21   Positive displacement flow transducer

21.6 ACCELERATION SENSORS

LVDT ACCELERATION SENSORS

Acceleration can be measured by using two laws of physics: Newton's law and Hooke's law. Newton's law states that force (F) is equal to mass (m) times acceleration (a), and Hooke's law states that force (F) exerted by a spring is equal to the spring constant (k) times the distance the spring is extended (s). If the forces developed by Newton's and Hooke's laws are equal, then a formula for acceleration can be developed as follows:

$F = ma$ (Newton's law)

$F = ks$ (Hooke's law)

If the two forces are equal, then:

$ma = ks$

$a = ks/m$

Figure 21–22 shows how these principles can be put to practical application to measure acceleration using an LVDT. Figure 21–22(a) shows the acceleration sensor at rest (or at a constant speed). The core of the LVDT is supported to allow easy side-to-side motion, being restrained only by the spring. In Figure 21–22(b), as the sensor is accelerated at a constant rate, the spring extends to create enough force to overcome the mass of the core of the LVDT. The distance of the spring extension (s) can be sensed electrically by the output of the LVDT.

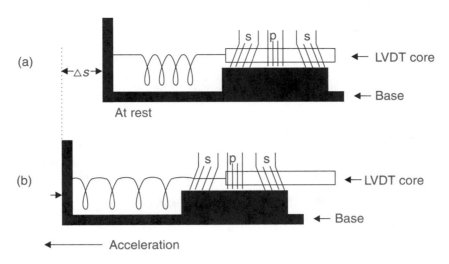

(a)

At rest

(b)

Acceleration

FIGURE 21–22   LVDT Acceleration transducer

If the spring constant (k) and mass of the core (m) are known, the acceleration can be calculated ($a = ks/m$). Since the mass of the core and the spring constant do not change, the output of the LVDT is directly proportional to acceleration.

PIEZOELECTRIC ACCELERATION TRANSDUCERS

Figure 21–23 shows a piezoelectric acceleration transducer. Again, the spring constant and the mass are known, and the displacement is measured. When acceleration occurs, the piezoelectric crystal is stressed and outputs a voltage that is proportional to the displacement.

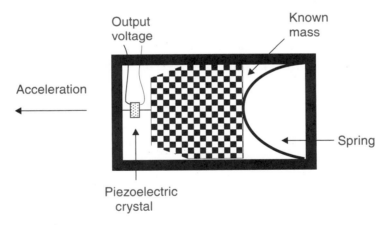

FIGURE 21–23   Piezoelectric accelerator transducer

SUMMARY OF ACCELERATION TRANSDUCERS

All acceleration transducers have a natural resonance because they use a spring-mass system. If a mass is connected to a spring and the spring is extended by moving the mass, when the mass is released, it will not return immediately to its place of rest but will overshoot and oscillate about the resting point before finally coming to rest. The frequency of these oscillations is called *natural resonance frequency*. In applications, the natural resonance of the acceleration transducer must be avoided. This can be accomplished by making sure the mechanical frequency of acceleration and deceleration are well below the natural resonance of the accelerometer selected.

LVDT transducers have an excellent linearity and a natural resonance in the range of 80 Hz. Piezoelectric transducers are not as linear as LVDTs, but they have a natural resonance in the 5 kHz range. Because of this, LVDT acceleration transducers are used in steady-state acceleration testing where the acceleration is nonperiodic. Piezoelectric acceleration transducers, however, find applications in vibration testing where the acceleration is periodic.

21.7   MAGNETIC SENSORS

VARIABLE RELUCTANCE AND INDUCTION SENSORS

Recall from your study of magnetism that if the magnetic field around a coil of wire changes, a voltage will be generated across the coil. The variable reluctance and the induction sensor operate on this principle. Figure 21–24(a) shows the concept of a variable reluctance sensor where a coil is placed around a permanent magnet. The flux in the magnet is constant; hence, there is no output voltage. If a ferromagnetic material is moved nearer the air gap, the flux lines will pass through the material, reducing the reluctance of the magnetic circuit and causing the flux to increase. The change in magnetic flux causes an output voltage across the coil.

Figure 21–24(b) shows an induction sensor where a coil is placed around a ferromagnetic core. When a permanent magnet is moved near the coil, a changing magnetic field will be produced in the ferromagnetic core. This magnetic field will cause a voltage to be generated across the coil. An important thing to note about both of these sensors is that there is an output voltage produced only when a changing magnetic field is present. The changing magnetic field is a function of the relative movement between the permanent magnet and the ferromagnetic material.

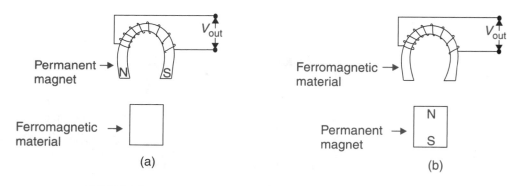

FIGURE 21–24   Variable reluctance and induction sensors

Figure 21–25 shows how a tachometer could be designed using a variable reluctance sensor. A gear is monitored by the variable reluctance sensor. Each time a tooth of the gear passes the sensor, a pulse is generated that drives the input (pin 1) of LM2907. The LM2907 is a National Semiconductor IC designed for frequency to voltage conversion. The voltage present at the output (pin 4) is a function of the supply voltage (V_s), capacitor C_1, resistor R_1, and the input frequency (pin 1). All the variables are constant except the frequency; therefore, the output voltage is a function of the input frequency. The output voltage of the LM2907 is connected to an op-amp that drives a 1 mA meter movement in its feedback loop. The face of the meter movement is dimensioned in rpm's. Example 21.5 shows how to calculate the range of the tachometer using the information given in Figure 21–25.

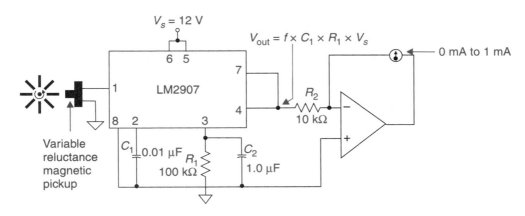

FIGURE 21–25 Tachometer circuit

EXAMPLE 21.5

Calculate the minimum and maximum range of the tachometer in Figure 21–25.

Step 1. Calculate the rpm's when the current flowing through the meter is 0 mA. In order for 0 mA to flow through the meter, the voltage at the output of the LM2907 would have to equal zero volts. This is only possible if the input frequency is zero, which means the gear is not turning (zero rpm's).

$$0 \text{ mA} = 0 \text{ rpm}$$

Step 2. Calculate the output voltage of an LM2907 needed to cause 1 mA to flow through the meter. If 1 mA flows through the meter, then 1 mA also flows through R_2. Since one end of R_2 is connected to virtual ground, the voltage across R_2 equals the output of the LM2907.

$$V_{R2} = I_M \times R_2 = 1 \text{ mA} \times 10 \text{ k}\Omega = 10 \text{ V}$$

Step 3. Calculate the input frequency required to cause the output of the LM2907 to equal 12 V. Solve the formula given in Figure 21–25 for frequency.

$$f = V_{out}/(C_1 \times R_1 \times V_s) = 10 \text{ V}/(0.01 \text{ μF} \times 100 \text{ k}\Omega \times 12 \text{ V}) = 833 \text{ Hz}$$

Step 4. Convert the input frequency to rpm's. Since there are eight teeth on the gear, one revolution will generate eight pluses. The revolutions per second can be calculated by dividing the frequency by the number of teeth on the gear. Multiplying the revolutions per second times sixty will give the revolutions per minute.

$$Revolutions/sec \ (RPS) = f/\# \ of \ teeth = 833 \text{ Hz}/8 = 104$$
$$RPM = RPS \times 60 = 104 \times 60 = 6240 \text{ rpm}$$

HALL-EFFECT SENSOR

The Hall effect occurs when a current carrying conductor is placed into a magnetic field. A voltage will be generated perpendicular to the current flow. Figure 21–26 illustrates the principle of the Hall effect. A constant current is flowing through a block of semiconductor material. If this block is in a magnetic field, a voltage called the Hall voltage will be produced. The magnitude of the Hall voltage is directly proportional to the strength of the magnetic field.

The small voltage generated by the Hall effect must be amplified. Manufacturers have designed single-chip Hall-effect sensor ICs that incorporate the circuitry to convert the small Hall-effect signal into useful voltage levels. Some sensors have an analog output that is proportional to the

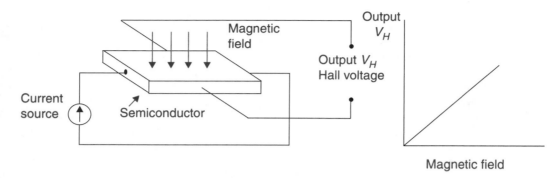

FIGURE 21–26   Hall-effect principle

strength of the magnetic field. Others are designed with a digital output that switches logic levels when the strength of the magnetic field exceeds a certain level. Figure 21–27 shows the internal circuitry of an analog Hall-effect sensor. The internal regulator permits the IC to operate over a wide range of supply voltages, typically 5 V to 18 V. The output of the Hall element is amplified by a differential amplifier that drives the output transistor. The emitter of the output transistor connects to the external load, forming a common collector. As the graph shows, the output of the sensor varies linearly as long as the magnetic input does not drive the amplifier into saturation.

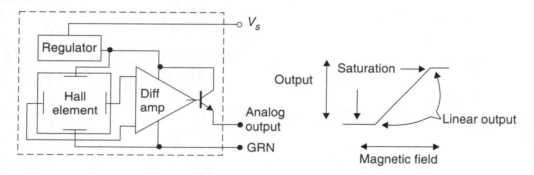

FIGURE 21–27   Analog Hall-effect sensor

Figure 21–28 shows the circuitry of a digital output Hall-effect sensor. The circuitry is similar to the analog output sensor except the differential amplifier drives a Schmitt trigger circuit instead of an output transistor. The Schmitt trigger has a built-in hysteresis that causes the device to have two trip points. As the magnetic field increases, the device will switch from a low to a high logic level when the upper trip point (UTP) is reached. On a decreasing magnetic field,

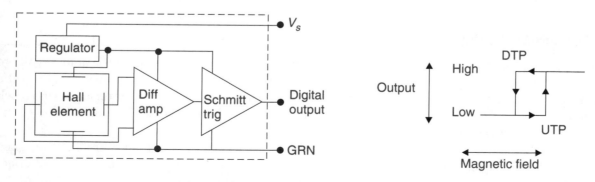

FIGURE 21–28   Digital Hall-effect sensor

the device will not switch from a high to a low logic level until the downward trip point (DTP) is reached. Output logic levels are normally TTL compatible.

21.8 SENSOR SIGNAL CONDITIONING AND CALIBRATION

The output of a sensor is normally a small electrical signal, often in a noisy environment. In order to make this output useful, analog amplification is usually used before further processing can be accomplished. If the output of the sensor is a changing resistance, it normally is converted to a changing voltage by using a bridge network. The instrumentation amplifier is an excellent amplifier stage to follow the sensor. The instrumentation amplifier has high impedance and differential inputs. This high input impedance is needed so the voltage out of the sensor or bridge circuit will not be loaded by the amplifier. The differential input, with a high common-mode rejection, is helpful because the signal is small, and if line noise is not canceled, the signal will be lost in the noise.

Figure 21–29 shows a Linear Technology (LT1101) integrated instrumentation amplifier connected to a bridge circuit. The differential output voltage of the bridge is connected to the IC, which has a gain of 100. The sensor in Figure 21–29 is located 100 feet from the amplifier. To help reduce noise, a shielded cable is connected from the bridge to the amplifier. It is important to note that the shield of the cable is connected to ground reference only on one end. If both ends of the shield are connected to ground, a conductive loop is formed that acts like the secondary of a transformer and inductively couples noise into the system.

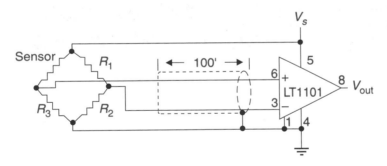

FIGURE 21–29   Integrated instrumentation amplifier

Calibration is a process of checking an instrument for the correct output when measuring a known standard. If the instrument is not reading correctly, the circuitry is adjusted until the instrument reads the known standard accuracy. If adjustments are not possible, a correction table can be prepared. The correction table will permit personnel to convert the reading of the instrument to the correct value. Since we base our decisions on the input from instruments, it is imperative that instrumentation is operating correctly. The only way to ensure that instruments are operating properly is to periodically perform calibration tests on all instrumentation. One problem that is often encountered in calibration is finding a standard for comparison. The setup in Figure 21–30 is an example of one way to calibrate a pressure sensor.

When the pump moves hydraulic fluid to the chamber on the left, the platform holding the weight will rise. When it rotates freely, the pressure in the hydraulic chamber is exactly equal to the total weight of the platform and its contents divided by the cross-sectional area of the post supporting the platform. The gage reading corresponding to this pressure is noted. Additional

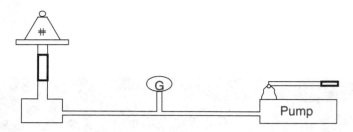

FIGURE 21–30   Calibrating a pressure sensor

weight is added, and the process repeats for as many steps as desired. The cross-sectional area of the post is typically a convenient quantity such as one square inch. Pressure readings are then pounds per square inch (psi), with the pounds being the weight of the platform and its contents. Table 21–2 shows the data that can be obtained by running a calibration test using the setup shown in Figure 21–30.

TABLE 21–2

Weight	Meter reading	Correction
10 lb	10.1 psi	−0.1 psi
20 lb	19.6 psi	+0.4 psi
30 lb	29.5	+0.5 psi
40 lb	40.3	−0.3 psi

21.9   SOLENOIDS

A solenoid is an electromagnetic device that when actuated causes a linear movement. Figure 21–31 shows a solenoid consisting of a coil, movable core, and spring. When switch S_1 is closed, the current flows through the coil, causing a magnetic field that attracts the core, which causes the push rod to extend. If switch S_1 is open, the spring will return the core to its original position. When the rod extends, the solenoid is a push type, but if the rod reacts inward when power is applied, the solenoid is a pull type.

Solenoids are available in a variety of operating voltages, with 12 V and 24 V DC and 120 V AC being the most popular. Data sheets give the resistances of the coils so the power needed to drive the solenoid can be calculated. The mechanical output is rated in lift and stroke. Lift is the load

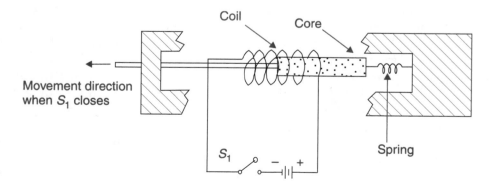

FIGURE 21–31   Solenoid

weight in ounces, and stroke is the distance the load can be moved in inches. Data sheets give a minimum and maximum length of stroke rating; for example, a solenoid could have a minimum rating of 100 oz at 0.12" and a maximum rating of 20 oz at 1". This means a weight of 100 oz will move a minimum of 0.12" when the solenoid is energized or a weight of 20 oz will move 1" when the solenoid is energized. The less the weight, the greater the distance of movement. The typical stroke of most solenoids is less than one inch. The mechanical output is a function of the electrical energy applied and the efficiency of the solenoid. Solenoids capable of handling large mechanical loads require considerable electrical power.

21.10 RELAYS

ELECTROMAGNETIC RELAYS

A relay is an electromagnetic device that permits a low power input signal to control a large output current. The control signal energizes a coil that causes electrical contact points to open or close. An open contact functions like an open switch, and a closed contact functions like a closed switch. A normally open (NO) contact is opened when the relay coil is de-energized, and a normally closed (NC) contact is closed when the coil is de-energized.

Figure 21–32 shows a relay circuit. When switch S_1 is opened, the magnetic field is removed, and the relay is de-energized. When switch S_1 is closed, the stationary core and coil form an electromagnet that attracts the moveable relay arm. This causes the normally open contact to close and the normally closed contact to open. The relay in our example performs the function of a single-pole double-throw (SPDT) switch. Relays can have many poles, all of which are energized by a single coil.

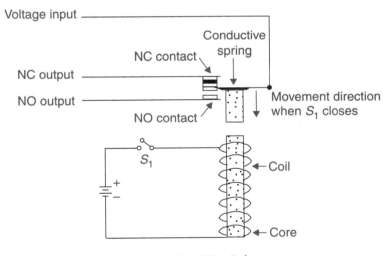

FIGURE 21–32 Relay

Relays are available in a variety of coil operating voltages and contact current ratings. The most popular operating voltages are 12 V and 24 V DC and 120 V AC. Data sheets will give the voltage rating of the coil and the resistance of the coil so the needed drive current from the control circuitry can be calculated. Data sheets will also give the current rating of the contacts. The current through the contacts can be AC or DC and can range from a few amps up to several hundred amps.

Figure 21–33 shows a relay circuit designed to control two 500 W AC loads using a 12 V DC as the control input. Figure 21–33 shows two schematic formats. The schematic on the left shows the relay package as one unit, and the schematic on the right breaks the relay apart, giving a logical layout. The relay coil is labeled (CR_1), and all contacts associated with that coil are given the same label.

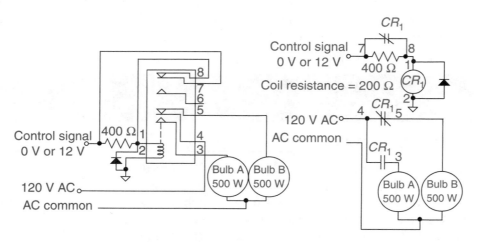

FIGURE 21–33   Relay circuit

The diode connected across the relay coil is reverse biased for the control signal; therefore, it has no effect when the relay is energized. However, when the relay is de-energized, the current through the coil drops quickly. This fast changing current generates a high induced voltage across the coil that could damage components in the control circuitry. The diode is forward biased for the induced voltage and quickly absorbs the energy stored in the coil while maintaining the induced voltage at a safe level.

When the control voltage is 0 V, the relay coil is de-energized, and AC current flows through the normally closed relay contact, thus lighting bulb B. When the control voltage switches to 12 V, the relay is energized, and the normally open contact closes and the normally closed contact opens. This causes bulb B to turn off and bulb A to turn on. It also removes the short circuit across the 400 Ω resistor and effectively places the 400 Ω resistor in series with the relay coil. This increases the resistance, thus reducing the current flowing through the coil. The relay requires a high current, called the *pull-in current*, to energize the relay; however, once energized, the coil current can be reduced to a value called the *hold current*, which is sufficient to maintain the energized state. A reduction in current reduces the load on the control circuitry and saves power. Example 21.6 will demonstrate this concept.

EXAMPLE 21.6

Calculate the pull-in current, hold current, wattage consumed by the coil during pull-in time, and wattage consumed by the coil after the relay is energized for the circuit in Figure 21–33.

Step 1. Calculate the pull-in current.

$$I_{\text{pull-in}} = V_{\text{control}}/R_{\text{coil}} = 12 \text{ V}/200 = 60 \text{ mA}$$

Step 2. Calculate the hold current.

$$I_{\text{hold}} = V_{\text{control}}/(R_s + R_{\text{coil}}) = 12 \text{ V}/(400 \ \Omega + 200 \ \Omega) = 20 \text{ mA}$$

EXAMPLE 21.6 continued

> **Step 3.** Calculate the power required from the control circuitry during pull-in.
> $$P = I_{\text{pull-in}} \times V_{\text{control}} = 60 \text{ mA} \times 12 \text{ V} = 720 \text{ mW}$$
> **Step 4.** Calculate the power required from the control circuitry after pull-in.
> $$P = I_{\text{hold}} \times V_{\text{control}} = 20 \text{ mA} \times 12 \text{ V} = 240 \text{ mW}$$

The control power needed to keep this relay circuit energized is only 240 mW. That is quite impressive when you consider the relay is delivering 500 W to the load. Most of us consider relays to be old technology, but they are "good" technology, and new systems are designed every day that make use of electromagnetic relays.

SOLID STATE RELAYS

Solid state relays are not electromagnetic devices but are complete electronic control circuits that are sealed into single packages. Many use optocouplers on the input to isolate the low power control circuitry from the high power circuitry that is being controlled. The output circuitry normally consists of an SCR or a triac. Because of their high input impedance and good isolation, solid state relays often can be driven directly by a microprocessor controller. Solid state relays have some disadvantages. An electromagnetic relay rated for 25 A will cost approximately $5, while a solid state relay rated for 25 A will set you back about $20. Electromagnetic relays are available with multipoles, whereas solid state relays perform the function of a single-pole single-throw (SPST) switch. Most solid state relays are designed for AC output current, and they depend on zero crossover voltage to terminate conduction. There are DC output current devices available, but they are expensive. Solid state relays have relatively low output current ratings, normally below 100 A.

21.11 MOTORS

INTRODUCTION

Motors are actuators that convert electrical energy into rotating mechanical force. In the following sections, we will discuss DC and AC motors. First, let us examine the input and output ratings and the efficiency of motors in general.

The electrical input of a motor is rated in watts. The output of an electrical motor is rated in horsepower. One horsepower is equal to 746 watts (1 hp = 746 W). If you know the input voltage and current, you can calculate the input power. By knowing the input and output power, efficiency can be calculated. When motors are operating from an AC source, they tend to be an inductive load, so for accurate input power calculations for AC motors, the formula $P = I \times V \times PF$ should be used. Power factor (PF) is equal to the cosine of the phase angle (the angle between input voltage and input current): $PF = \cos \varnothing$. Example 21.7 shows how the efficiency of a motor can be calculated.

EXAMPLE 21.7

> Calculate the efficiency of a 2 hp electric motor operating at 240 V AC at 8 A with a power factor of 0.9.

EXAMPLE 21.7 continued

> **Step 1.** Calculate the input power.
> $$P_{in} = I \times V \times PF = 8 \text{ A} \times 240 \text{ V} \times 0.9 = 1728 \text{ W}$$
> **Step 2.** Convert the output to watts.
> $$P_{out} = 2 \text{ hp} \times 746 \text{ W} = 1492 \text{ W}$$
> **Step 3.** Calculate the efficiency of the motor.
> $$Efficiency = (P_{out}/P_{in}) \times 100\% = (1492 \text{ W}/1728 \text{ W}) \times 100\% = 86.3\%$$

DC MOTORS

Figure 21–34 shows the construction of a DC motor. The magnetic poles are stationary and may be permanent magnets (as shown) or electrical magnets. The armature coil is suspended on bearings (not shown), so it is free to rotate in the magnetic field created by the magnetic poles. Power is applied to the armature coil by a commutator and brush arrangement. The brushes make electrical contact with the rotating commutator. The commutator provides the reversal of polarity needed to permit armature rotation.

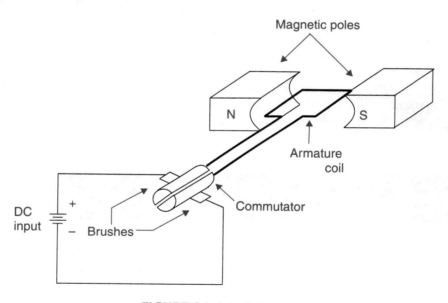

FIGURE 21–34   DC motor

When voltage is applied to the armature coil through the brushes and commutator, current flows in the coil. Current flowing through the armature coil causes a magnetic field around the coil. Figure 21–35 shows how the magnetic field of the armature coil interacts with the magnetic field of the magnetic pole pieces to cause motor action. Figure 21–35(a) is a front view of the two ends of the armature coil. The end labeled "a" is shown as a cross (+) to represent electron current flowing into the coil. The other end is shown as a dot (•) to represent electron current flowing out of the coil. The circle-arrow around the coil ends represents the magnetic flux lines caused by armature current, and the parallel lines represent the magnetic flux lines of the magnetic pole pieces.

If you examine the end of the coil labeled "a" in Figure 21–35(a), you will notice that the magnetic flux lines caused by the armature coil interact with the magnetic flux lines of the magnetic poles. The armature coil flux lines below end "a" add to the magnetic flux lines that are caused by the pole pieces, thus making the magnet field stronger below end "a" of the armature. Above end "a"

FIGURE 21–35 Interaction of magnetic fields in a DC motor

the magnetic flux lines cancel, making the magnetic field weaker. From magnetic theory we know there will be a force created in the direction of the weaker field, or in this case, a force up as shown. If you examine the opposite end of the coil in Figure 21–35(a), you will notice that the force created by the interacting magnetic fields is in the down direction. This is because the current is flowing out of the coil. Since the armature is free to rotate, it will start rotating in a counterclockwise direction.

Follow end "a" as it rotates 90° as shown in Figure 21–35(b). Notice that the electron current is still flowing into "a," and the force created by the interactions of the magnetic fields continues to push the armature in a counterclockwise direction. Figure 21–35(c) shows end "a" rotated 90° more. The arrow still indicates a counterclockwise rotation; however, notice that the electron current is now flowing out of end "a." This current reversal is caused by the commutator and is necessary to keep the interaction of the magnetic fields in the proper direction to support rotation.

Figure 21–35(d) shows a 90° rotation; the forces are now in the same direction. Another 90° rotation will bring the armature back to the position shown in Figure 21–35(a). Notice that there has been another current reversal between Figures 21–35(d) and 21–35(a).

For a two-pole motor, there is a current reversal every 180° caused by the commutator. Figure 21–36 shows that the commutator is split into two segments. For 180° of rotation, brush "1" is in contact with segment "a" of the commutator, and brush "2" is in contact with segment "b" of the commutator, causing section "a" to be positive. For the next 180° of rotation, brushes are in contact with opposite segments, causing segment "b" to be positive. There is always one segment of the commutator for each pole of the motor; so in a four-pole motor, there would be a current reversal every 90°. This is necessary to keep the interaction of the magnetic fields supporting rotation.

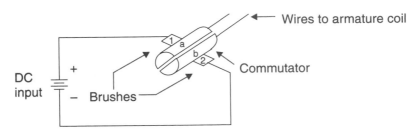

FIGURE 21–36 Commutator and brushes

Normally, only fractional horsepower motors use permanent magnetic poles. Larger DC motors use electromagnetic poles and have two coils: a low resistance armature coil and a high resistance field coil.

The speed of a DC motor can be controlled by controlling the current through the armature coil or the current through the field coil. The speed is directly proportional to armature coil current and indirectly proportional to field coil current. This can be understood by discussing the equilibrium that a motor reaches once it obtains constant speed. Figure 21–37 shows a DC motor where the field coil and the armature coil have separate power sources. When switch S_1 is closed, power is applied to the field, causing a magnetic field to be created in the pole pieces. The current through the field is limited to 0.1 A by the resistance of the field coil.

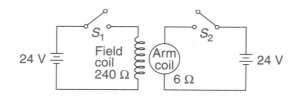

FIGURE 21–37   Separately excited DC motor

If switch S_2 is closed, current will flow in the armature coil. If the armature is mechanically locked so that it cannot rotate, the only thing limiting armature current will be the armature coil resistance, and the current will be 4 A. However, if the armature is free to rotate, it will start spinning. The spinning of the armature coil in the magnetic field of the pole pieces causes generator action. This generator action produces a countervoltage, or counterelectromotive force (CEMF), that opposes the DC armature supply and limits armature current. The faster the armature rotates, the greater the countervoltage and the less the armature current. The armature current can be calculated using the formula $I_A = (V_{in} - V_{CEMF})/R_A$. At some speed, depending on the motor characteristics and the mechanical load, an equilibrium will be reached. For our example in Figure 21–37, let us say the equilibrium is reached when the armature current drops to 0.5 A (0.5 A = (24 V – 21 V)/6 Ω). As long as the mechanical load is constant, the motor will try to obtain the equilibrium of approximately 0.5 A of armature current.

Now, if we increase the armature supply voltage by 4 V to 28 V, the armature current will increase and the motor will speed up, causing the CEMF to increase. When the CEMF increases to approximately 25 V, the motor will again reach its equilibrium (0.5 A = (28 V – 25 V)/6 Ω) at a new speed. Actually, the armature current at its new equilibrium will be slightly higher than 0.5 A. A decrease in armature voltage will require the motor to reduce its CEMF by lowering its speed, and a new equilibrium will be reached slightly below 0.5 A of armature current.

Let us examine what happens when the armature voltage is held constant and the field is varied. A reduction in field current causes a decrease in the flux density of the stationary magnetic field. This means that the armature must spin faster to maintain a constant CEMF and to limit armature current, so a decrease in field current causes an increase in speed. If the field current is increased, the armature can produce the needed CEMF at a lower speed, and the speed will decrease.

It should be apparent that CEMF is the main factor limiting current in the armature coil. If the mechanical load is increased, the rotor (armature) speed will slow down, causing the CEMF to decrease. The reduced CEMF causes the armature current to increase, which increases the torque, thus causing the motor to increase speed. The motor speed reaches a new equilibrium at a speed slightly lower than the original speed. If a motor is mechanically locked, there will be no CEMF, and therefore, excessive armature current will destroy the motor.

The direction of rotation can be reversed by reversing the current through the field coil or the current through the armature coil. If either one of the interactive magnetic fields is reversed, the rotation of the motor will be reversed. It should be noted that if the direction of current is changed in both the field and the armature, the direction of rotation will remain the same.

Figure 21–38 shows three types of DC motors based on the way the fields and coils are connected. The type of connection is a determining factor of how the motor will operate under mechanical loads. The shunt motor in Figure 21–38(a) is good at maintaining constant speeds under changing mechanical loads. The series motor in Figure 21–38(b) has good starting torque. Figure 21–38(c) shows a compound motor where the field coil has been divided into a series section and a shunt section. Compound motors can be designed that incorporate the advantages of both series and shunt motors.

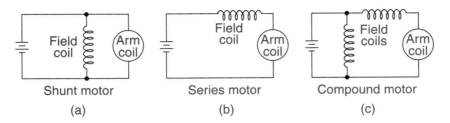

FIGURE 21–38   Types of DC motors

AC SYNCHRONOUS MOTORS

In DC motors, pole pieces supply a stationary magnetic field, and the magnetic field in the rotating armature is switched to make rotation possible. In AC motors, the rotor has a constant magnetic field, and rotation is made possible by electrically revolving the magnetic field in the stator.

Figure 21–39 shows a three-phase AC synchronous motor. The rotor is a magnet with a fixed magnetic field. It is normally an electromagnet, and connection to the rotor coil is made through brushes and slip rings. The stator has three pairs of poles. Each pair is fed by a different phase from three-phase power. Note in Figure 21–39 that phase A (∅A) is 120º out of phase with phase B (∅B), and phase B is 120º out of phase with phase C (∅C). Assume that when the sine wave is at positive peak, the pole (A, B, or C) being fed by that phase is at maximum magnetic north, and the prime pole of the pair (A', B', or C') is at maximum magnetic south. By analyzing the three phase currents flowing through the field coils, you can see that the magnetic field revolves counterclockwise around the stator. The speed at which the magnetic field revolves is called the *synchronous speed* and can be calculated in rpm's by using the formula:

$$Synchronous\,speed = \frac{120 \times Frequency}{Number\,of\,poles}$$

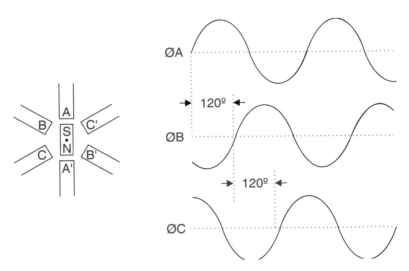

FIGURE 21–39   Synchronous motor

If the rotor can be rotated at near synchronous speed, the rotor will lock with the revolving field, and the speed of the rotor will be the synchronous speed. Some large synchronous motors are not self-starting. They must be started as an induction motor (covered in the next section) or have a starting motor connected to the same shaft to get the rotor spinning near synchronous speed. Once near synchronous speed, voltage is applied to the rotor, locking it to synchronous speed.

Synchronous motors have been popular because their speed can be precisely controlled by the frequency of the input signal to the rotating field. However, because of advances in electronics, DC motors also can be precisely controlled using a closed-loop system. One advantage of synchronous motors over other types of motors is that their power factor can be controlled by the excitement of the rotor. This permits a synchronous motor to be used to cancel out the inductive load caused by other motors.

AC INDUCTION MOTORS

The three-phase induction motor uses the same principle of field rotation as the synchronous motor, but the way the rotor receives energy is different. The induction motor depends on inductive coupling between the stator and the rotor. The stator acts like the primary of a transformer, and the rotor acts like a shorted secondary of the transformer, as shown in Figure 21–40(a).

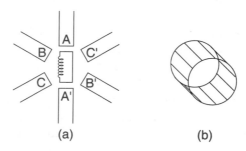

(a)   (b)

FIGURE 21–40   Induction motor

As the field rotates, it cuts the winding of the rotor coil and induces a voltage into the rotor. Since the rotor is shorted, the induced voltage causes a large current, which causes a magnetic field to be developed around the rotor. The squirrel-cage is the most popular type of induction motor. It uses a rotor made of large gage copper conductors connected to shorting rings as shown in Figure 21–40(b). This arrangement provides a low resistance secondary capable of handling large current flow.

Once the rotor starts revolving, the relative motion between the rotating field (primary) and rotor (secondary) decreases. This means less voltage will be induced into the rotor and less current flows through the squirrel-cage winding. An equilibrium speed is obtained below the synchronous speed. The rotor speed always has to be slower than the synchronous speed. This relative motion of the stator magnetic field will cause enough energy to transfer to the rotor to support rotation. The difference in synchronous speed and rotor speed can be expressed as percentage of slip and calculated by the following formula.

$$Percent\ slip = \frac{Synchronous\ speed - Rotor\ speed}{Synchronous\ speed} \times 100\%$$

Induction motors are popular when precise speed control is not needed. The squirrel-cage type is the most popular because it requires no brushes. There are wire-wound induction motors that do use brushes and slip rings to connect external resistors in series with the rotor coils. This gives a means of speed control and also permits higher starting torque.

SINGLE-PHASE AC MOTORS

The AC motors we have discussed to this point have operated with three-phase power. Operating an AC motor from single-phase sources requires additional circuitry in order for the motor to be self-starting. The problem with single-phase sources is the field will not revolve around the stator but instead switches back and forth between the two pole pieces. This problem is overcome by adding a starting field winding that operates from current that has been phase shifted with respect to the current through the main running winding.

Figure 21–41 shows two methods of shifting the phase of the current through the starting winding. Figure 21–41(a) shows a high resistance starting winding. Since the starting winding is high in resistance and low in inductance, the current through the winding tends to be only a few degrees out of phase with the input voltage, as shown in Figure 21–41(c). The current through the running winding is approximately 90° out of phase with the input voltage because the winding has low resistance and high inductance. These out-of-phase currents through the windings permit the motor to be self-starting. Once the motor starts, the starting winding is normally removed from the circuit with a centrifugal switch, as shown in Figure 21–41.

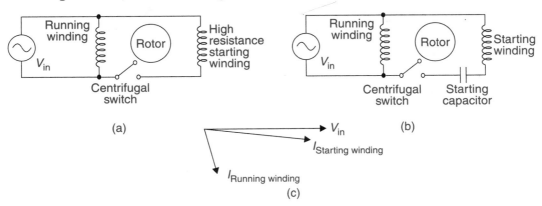

FIGURE 21–41   Starting a single-phase motor

Figure 21–41(b) shows that the same effects can be obtained by using a capacitor in series with a low resistance starting winding. In this case, the capacitive reactance will cancel the inductive reactance, causing the current through the starting winding to be in phase with the input voltage.

The direction of rotation of a single-phase motor can be reversed by reversing the direction of current flowing through the starting winding. The single-phase motor is very popular in the home and in light industry because only single-phase power is available. However, three-phase AC motors supply torque more evenly throughout the rotation and tend to be more efficient.

BRUSHLESS DC MOTORS

The advancements in Hall-effect sensors and integrated control circuitry have led to the widespread use of brushless DC motors. A brushless DC motor operates similarly to an AC motor; the rotor is the field (permanent magnet), and the stator is the armature. The electronic control circuitry switches power MOSFETs on and off in proper sequence so the current flows through the stator windings, generating a rotating magnetic field. The rotor field is locked with the moving stator field so it rotates at the speed of field rotation.

Figure 21–42 shows a simplified schematic of a brushless DC motor and control circuit. A pair of power MOSFETs connect to each stator coil. If the P-channel MOSFET is on, the coil is connected to the DC source, and if the N-channel MOSFET is on, the coil is connected to ground. Therefore, it is possible to reverse the magnetic field in each coil by reversing the current flow through each coil. The switching of the MOSFETs must be synchronized with the physical location of the rotor. The physical location of the rotor is detected by Hall-effect sensors. Motorola's MC33035 is an integrated controller that monitors the Hall-effect sensors and outputs the needed drive signals to the power MOSFETs. The MC33035 controls the power delivered to the motor by using pulse width modulation (PWM).

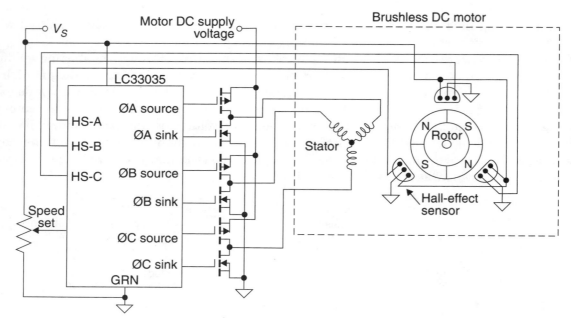

FIGURE 21–42 Brushless DC motor

One obvious advantage of the brushless DC motor is there are no brushes. Brushes wear out and also generate electrical noise. When a normal DC motor is used in closed-loop control systems, some type of tachometer sensor must be added to the system. The brushless DC motor has the advantage of already having built-in Hall-effect sensors that can be used in closed-loop control systems. Motorola's MC33039 IC, for example, is designed to work in conjunction with the MC33035 to provide a complete closed-loop motor control system.

STEPPER MOTORS

Stepper motors have become popular in automation systems because they can be used to provide repetitive mechanical movement without the complexity of a closed-loop control system. The shaft of a stepper motor moves in discrete steps. Figure 21–43 shows a stepper motor in operation. In the example, the stator has four poles, and the rotor is designed with six teeth. Figure 21–43(a) shows the darkened tooth under the energized pole, which is also darkened. Figure 21–43(b) shows the first pole is de-energized and a second pole 90° counterclockwise is energized. The magnetic field of the energized pole attracts the closest tooth, which causes the rotor to move 30° clockwise, as shown by the darkened tooth. As you follow the drawings across, through Figure 21–43(e), you will see that each time the electrical field moves 90°, it causes a 30° step in the placement of the rotor. By adding additional poles and teeth, it is possible to reduce the step size to less than 1°. Most popular step sizes are 1.8°, 3.6°, 7.5°, and 15°.

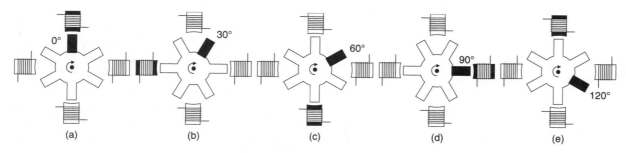

FIGURE 21–43 Stepper motors

Four-phase stepper motors usually are connected to driving circuits with six wires. The color codes shown in Figure 21–44 are commonly used. Some stepper motors combine the black and white wires into one black/white wire. Driving circuits consist of transistor switches controlled by inputs from logic circuitry. A logic one on the input side of an inverter will turn on the transistor switch. If the switches are continuously sequenced in proper order, the shaft will rotate. The speed of rotation is controlled by the speed at which the transistors are switched through the sequence of steps.

The tables in Figure 21–44 show the sequence needed to rotate the shaft clockwise. If the four-step sequence shown in the first table is followed, the motor will advance a full step for each new digital input. For example, a stepper motor with a step angle rating of 15° per step would advance 15°. The same circuit could be used, but the digital input sequence could be modified to the eight-step sequence shown in the second table. The motor would now half step. For example, a stepper with a step angle rating of 15° per step would advance 7.5°. Half stepping does require additional current drive, since at times current is flowing through two windings. The direction of rotation can be reversed by reversing the switching sequence.

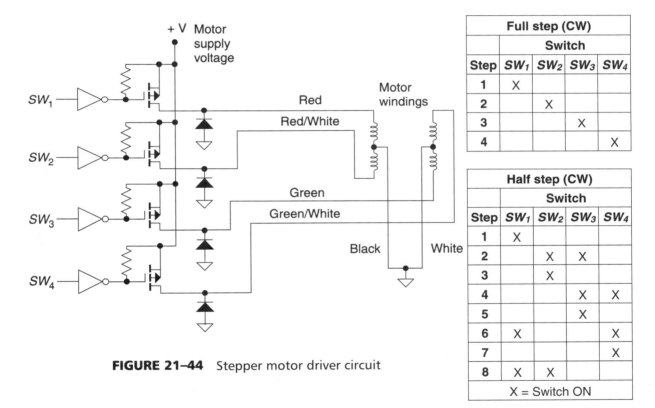

FIGURE 21–44 Stepper motor driver circuit

Full step (CW)				
	Switch			
Step	SW₁	SW₂	SW₃	SW₄
1	X			
2		X		
3			X	
4				X

Half step (CW)				
	Switch			
Step	SW₁	SW₂	SW₃	SW₄
1	X			
2		X	X	
3		X		
4			X	X
5			X	
6	X			X
7				X
8	X	X		
X = Switch ON				

Four-phase stepper motors contain four motor windings. The current flows through each winding in one direction (unipolar). Two-phase steeper motors only have two motor windings. The drive circuit must use a complementary push-pull that permits current to flow both ways through the windings. The most important stepper motor characteristics are listed and defined below.

> *Detent torque:* The torque needed to move the rotor a full step when no power is applied to the system.
>
> *Holding torque:* The torque needed to move the rotor a full step when power is applied to the system.
>
> *Number of phases:* Number of motor windings, normally two or four windings.

Phase resistance: The resistance of the motor winding. The current load can be approximated by dividing the voltage by the phase resistance.

Step angle: The degrees of movement for full step operation.

V_{DC}: The DC operating voltage of the motor.

21.12   SPEAKERS

Speakers are electromagnetic devices that input an electrical signal and output sound. As shown in Figure 21–45, the speaker is comprised of four basic parts: a permanent magnet, a frame, a cone, and a voice coil. The cone is a diaphragm made from flexible material (often paper) that is suspended from a frame so it is free to move. Attached to the base of the cone is a voice coil. A permanent magnet is attached to the base of the frame.

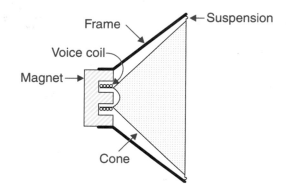

FIGURE 21–45   Speaker

If an electrical signal is applied to the voice coil, a magnetic field will be created around the voice coil. The magnetic field of the voice coil interacts with the field of the permanent magnet, causing motor action that moves the cone. The amount and rate of movement of the cone depends on the magnitude and frequency of the signal applied to the voice coil. The cone movement modulates the air, thus causing sound.

Speakers are designed to operate in the audio frequency range (20 Hz to 20 kHz). A full-range speaker is one that attempts to cover the total frequency range. However, it is impossible to design a single speaker that will work well across the total range. For this reason, better quality speakers are designed to respond to only a portion of the audio range. Speakers can be subdivided based on their frequency response. Woofer speakers are designed to respond to low range frequencies, midrange speakers are designed to respond to midrange frequencies, and tweeter speakers are designed to respond to high range frequencies. Good speakers do not guarantee a quality speaker system. In order for a speaker to function properly, it must be built into a well-designed speaker cabinet.

Fan Control System

1. If the temperature is 20°C, determine the signal at the following test points for the circuit in Figure 21–46. If the signal is a DC voltage, record the voltage. If the signal is a waveform, draw and label the waveform. (Assume the supply voltage of the motor is 3 V, which causes the motor to run at 4000 rpm.)

$TP_1 =$ _____ $TP_2 =$ _____ $TP_3 =$ _____

$TP_4 =$ _____ $TP_5 =$ _____ $TP_6 =$ _____

$TP_7 =$ _____ $TP_8 =$ _____ $TP_9 =$ _____

2. At what temperature does the motor switch on? _____

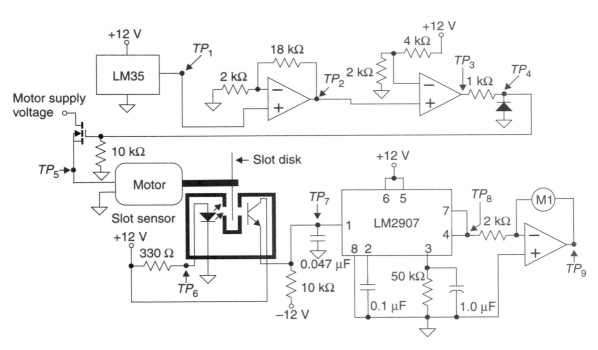

FIGURE 21–46

3. If the temperature is 45°C, determine the signal at the following test points for the circuit in Figure 21–46. If the signal is a DC voltage, record the voltage. If the signal is a waveform, draw and label the waveform. (Assume the supply voltage of the motor is 3 V, which causes the motor to run at 4000 rpm.)

$TP_1 =$ _____ $TP_2 =$ _____ $TP_3 =$ _____

$TP_4 =$ _____ $TP_5 =$ _____ $TP_6 =$ _____

$TP_7 =$ _____ $TP_8 =$ _____ $TP_9 =$ _____

4. Meter M_1 has a full-scale current reading of 3 mA. If the meter is reading half scale, at what speed is the motor running? _____

Fan Control System

I. Objective

To analyze and construct a fan control system.

II. Test Equipment

±12 V power supplies	Adjustable power supply
Oscilloscope	Current meter (10 mA)

III. Components

Resistors: (1) 470 Ω, (3) 1 kΩ, (1) 2 kΩ, (1) 2.7 kΩ, (2) 10 kΩ, (1) 100 kΩ
Capacitors: (1) 0.047 µF, (1) 0.1 µF, (1) 1 µF
ICs: (1) LM2907, (1) LM324
Sensors: (1) slot type EE-SX1042, (1) LM35 IC temperature sensor
Transistor: (1) Power MOSFET N-ch IRF630
Motor: (1) small DC motor

IV. Procedure

1. If the temperature is 25°C, determine the signal at the following test points for the circuit in Figure 21–47. If the signal is a DC voltage, record the voltage. If the signal is a waveform, draw and label the waveform. (Assume the supply voltage of the motor is 3 V, which causes the motor to run at 3000 rpm.)

   $TP_1 =$ _____ $TP_2 =$ _____ $TP_3 =$ _____

   $TP_4 =$ _____ $TP_5 =$ _____ $TP_6 =$ _____

   $TP_7 =$ _____ $TP_8 =$ _____ $TP_9 =$ _____

2. If the temperature is 36°C, determine the signal at the following test points for the circuit in Figure 21–47. If the signal is a DC voltage, record the voltage. If the signal is a waveform, draw and label the waveform. (Assume the supply voltage of the motor is 3 V, which causes the motor to run at 3000 rpm.)

   $TP_1 =$ _____ $TP_2 =$ _____ $TP_3 =$ _____

   $TP_4 =$ _____ $TP_5 =$ _____ $TP_6 =$ _____

   $TP_7 =$ _____ $TP_8 =$ _____ $TP_9 =$ _____

3. Meter M_1 has a full-scale current reading of 10 mA. If the meter is reading half scale, at what speed is the motor running? _____

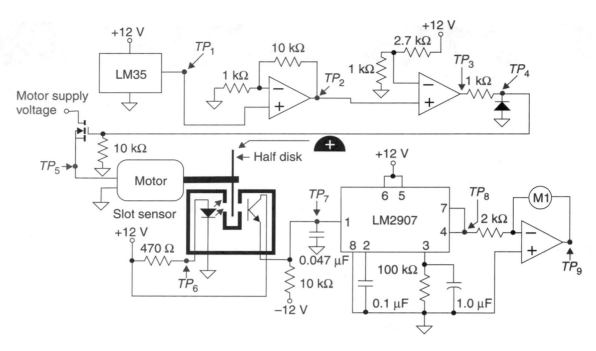

FIGURE 21–47

4. From cardboard, construct a half disk with a small hole for the motor shaft as shown in Figure 21–47. Press the disk onto the shaft of the motor. Construct the circuit in Figure 21–47.

5. Temporarily connect the gate of the MOSFET to 12 V DC. This should cause the motor to run. Note the direction of rotation. If not clockwise, reverse the direction of rotation by changing the motor connections. Adjust the power supply of the motor until the tachometer indicates the motor is running at 3000 rpm. Remove the temporary jumper.

6. With the temperature sensor at room temperature (approximately 21°C), measure the signal at the following test points for the circuit in Figure 21–47. If the signal is a DC voltage, record the voltage. If the signal is a waveform, draw and label the waveform.

$TP_1 = $ _____ $TP_2 = $ _____ $TP_3 = $ _____

$TP_4 = $ _____ $TP_5 = $ _____ $TP_6 = $ _____

$TP_7 = $ _____ $TP_8 = $ _____ $TP_9 = $ _____

7. With the temperature sensor at body temperature (approximately 36°C), measure the signal at the following test points for the circuit in Figure 21–47. If the signal is a DC voltage, record the voltage. If the signal is a waveform, draw and label the waveform. (Holding the temperature sensor tightly between your fingers should raise the temperature of the sensor to near body temperature.)

$TP_1 = $ _____ $TP_2 = $ _____ $TP_3 = $ _____

$TP_4 = $ _____ $TP_5 = $ _____ $TP_6 = $ _____

$TP_7 = $ _____ $TP_8 = $ _____ $TP_9 = $ _____

V. Points to Discuss

1. Is it important to have the supply voltage for the LM35 exactly at 12 V? Explain.

2. The circuit in Figure 21–47 is designed to switch the motor on at a certain temperature. Explain how the circuit could be modified so the motor turns on at 40ºC.

3. Explain the effect armature voltage has on the speed of rotation.

4. Explain the effect armature voltage has on the direction of rotation.

QUESTIONS

21.1 Introduction to Transducers and Actuators

____ 1. Devices that convert one form of energy to another form of energy are called ____.
   a. converters
   b. energy converters
   c. transformers
   d. transducers

____ 2. _____ can be compared to the five senses of the human body.
   a. Converters
   b. DACs
   c. Sensors
   d. Actuators

____ 3. _____ could be considered the arms, legs, and mouth of the electronic system.
   a. Converters
   b. DACs
   c. Sensors
   d. Actuators

21.2 Temperature Sensors

____ 4. Temperatures can be expressed using the _____ scale.
   a. Fahrenheit
   b. Celsius
   c. Kelvin
   d. all the above

____ 5. _____ is an electrical transducer that produces a voltage proportional to the temperature and functions on the basis that different metals have different thermionic characteristics.
   a. An IC temperature sensor
   b. A thermistor
   c. An RTD
   d. A thermocouple

____ 6. The voltage produced by a thermocouple is called the _____.
   a. hot junction voltage
   b. cold junction voltage
   c. Seebeck voltage
   d. Hooke voltage

____ 7. _____ depends on the positive temperature coefficients of metals and often is made from platinum.
   a. An IC temperature sensor
   b. A thermistor

    c. An RTD

    d. A thermocouple

____ 8. Often it is necessary to use three wires to connect _____ to a bridge, but normally a two-wire connection works well when connecting _____.

    a. an IC temperature sensor, an RTD

    b. a thermistor, an RTD

    c. an RTD, a thermistor

    d. a thermocouple, an IC temperature sensor

____ 9. _____ is a thermally sensitive resistor manufactured from a semiconductor material that has a negative temperature coefficient.

    a. An IC temperature sensor

    b. A thermistor

    c. An RTD

    d. A thermocouple

____ 10. _____ are the most nonlinear of all the temperature transducers.

    a. IC temperature sensors

    b. Thermistors

    c. RTDs

    d. Thermocouples

____ 11. _____ are the most linear of all the temperature transducers.

    a. IC temperature sensors

    b. Thermistors

    c. RTDs

    d. Thermocouples

____ 12. The _____ temperature transducer has the widest temperature range.

    a. IC sensor

    b. thermistor

    c. RTD

    d. thermocouple

21.3 Displacement Sensors

____ 13. The SI (Standard International) unit for displacement is the _____.

    a. inch

    b. foot

    c. yard

    d. meter

____ 14. The _____ is comprised of a transformer with a movable core, a primary, and two secondaries.

    a. strain gage

    b. accelerometer

    c. LVDT

    d. transformer displacement sensor

____ 15. The strain gage works on the principle that a solid material under mechanical stress will change its _____.

    a. temperature

    b. volume

    c. molecular makeup

    d. physical shape but maintain a constant volume

____ 16. The ability of a strain gage to convert a change in mechanical displacement to a change in electrical resistance is measured by the _____.

    a. gage factor

    b. gage ratio displacement factor

    c. displacement factor

    d. displacement ratio

21.4 Pressure Transducers

____ 17. The three types of pressure measurements are _____.

    a. exact, reflective, and referenced

    b. absolute, exact, and differential

    c. absolute, gage, and differential

    d. absolute, gage, and reflective

____ 18. A diaphragm pressure transducer consists of two chambers separated by a diaphragm with _____ connected to the diaphragm.

    a. a strain gage

    b. a pressure gage

    c. an RTD

    d. an optotransistor

21.5 Flow Transducers

____ 19. A differential pressure flow transducer is based on the _____ principle that states that an increase in velocity causes a decrease in pressure.

    a. Bernoulli

    b. Shook

    c. Haas

    d. Russell

____ 20. A _____ flow transducer consists of a turbine installed in a pipe that is connected to a tachometer.

    a. direct

    b. turbine-tach

    c. differential pressure

    d. positive displacement

21.6 Acceleration Sensors

____ 21.   Acceleration can be measured by using two laws of physics: _____.
   a.   Newton's law and Haas's law
   b.   Russell's law and Hooke's law
   c.   Russell's law and Shook's law
   d.   Newton's law and Hooke's law

____ 22.   The _____ are two types of acceleration transducers.
   a.   LVDT and the piezoelectric
   b.   magnetic and the piezoelectric
   c.   RTD and the magnetic
   d.   LVDT and the RTD

21.7 Magnetic Sensors

____ 23.   The _____ magnetic sensors operate on the principle of dectecting the change in reluctance of a magnetic circuit using a coil.
   a.   differential and the Hall-effect
   b.   Hall-effect and the induction
   c.   variable reluctance and the induction
   d.   differential and the displacement

____ 24.   A constant current is flowing through a block of semiconductor material. The block is placed into a magnetic field, and a voltage is generated perpendicular to the current flow. The voltage produced is directly proportional to the strength of the magnetic field and is called the _____.
   a.   magnetic voltage
   b.   Hall voltage
   c.   magnetic displacement
   d.   piezomagnetic output

21.8 Sensor Signal Conditioning and Calibration

____ 25.   If the output of the sensor is a changing resistance, it normally is converted to a changing voltage by using _____.
   a.   a resistance to voltage converter
   b.   an op-amp
   c.   a bridge network
   d.   an RC network

____ 26.   _____ amplifier has high impedance and differential inputs and is an excellent amplifier stage to follow the sensor.
   a.   A sensor
   b.   An instrumentation
   c.   A transducer
   d.   A Hall

____ 27. Calibration is a process of checking an instrument for the correct output when measuring _____.

    a. an input that causes a full-scale reading

    b. an input that causes a midscale reading

    c. a known standard

    d. both a and b

21.9 Solenoids

____ 28. A solenoid is an electromagnetic device that when actuated causes a _____.

    a. rotation movement

    b. linear movement

    c. switch to open or close

    d. valve to open or close

____ 29. The mechanical output of a selenoid is rated in _____.

    a. rpm

    b. psi

    c. horsepower

    d. lift and stroke

s

21.10 Relays

____ 30. The diode is often connected across the coil of a relay to prevent _____.

    a. high induced voltage across the coil from damaging the control circuitry

    b. large current from flowing through the relay coil

    c. the relay from energizing on opposite polarity voltage

    d. the coil from breaking into high frequency oscillation

____ 31. The relay requires a high current, called the _____ current, to energize the relay; however, once energized, the coil current can be reduced to a value called the hold current.

    a. turn-on

    b. pull-in

    c. start

    d. energizing

____ 32. _____ relays are not electromagnetic devices but are complete electronic control circuits that are sealed into single packages.

    a. Switch control

    b. Auto-state

    c. Electronic switch

    d. Solid state

21.11 Motors

____ 33. One horsepower is equal to _____.
    a. 15.5 W
    b. 100 W
    c. 746 W
    d. 1276 W

____ 34. _____ use brushes to make electrical contact with the rotating commutator. The commutator provides the reversal of polarity needed to permit armature rotation.
    a. Synchronous motors
    b. Stepper motors
    c. Induction motors
    d. DC motors

____ 35. The speed of a DC motor can be controlled by controlling the _____.
    a. current through the armature coil
    b. current through the field coil
    c. polarity of the voltage across the armature coil
    d. both a and b

____ 36. The main factor limiting current through the armature coil is _____.
    a. the resistance of the armature
    b. the inductance of the armature
    c. the size of the wire in the field windings
    d. CEMF

____ 37. The rotor of the _____ has a constant magnetic field, and the rotor rotates at the speed of the electrically revolving magnetic field in the stator.
    a. synchronous motor
    b. stepper motor
    c. induction motor
    d. DC motor

____ 38. The _____ depends on inductive coupling between the stator and the rotor. The stator acts like the primary of a transformer, and the rotor acts like a shorted secondary of the transformer.
    a. synchronous motor
    b. stepper motor
    c. induction motor
    d. DC motor

____ 39. Brushless DC motors use _____ to detect the position of the rotor.
    a. slip rings
    b. optical sensors
    c. Hall-effect sensors
    d. RTD sensors

____ 40. Stepper motors have become popular in automation systems because _____.

    a. of their low cost

    b. they have the greatest power output of all motors

    c. they can operate directly with a DC or AC power source

    d. they can be used to provide repetitive mechanical movement

21.12 Speakers

____ 41. A speaker is comprised of four basic parts: _____.

    a. a permanent magnet, a frame, a cone, and a voice coil

    b. a permanent magnet, a frame, a cone, and a voice capacitor

    c. a permanent magnet, a frame, a voice box, and a voice coil

    d. a permanent magnet, a frame, a voice box, and a voice capacitor

____ 42. A _____ is a type of speaker designed to respond to a given audio range.

    a. woofer

    b. midrange

    c. tweeter

    d. all of the above

PROBLEMS

1. Convert 18° Celsius to the Kelvin and Fahrenheit equivalents.

2. Convert 78° Fahrenheit to the Kelvin and Celsius equivalents.

3. Determine (from the graph in Figure 21–4) the output voltage of the following thermocouples at 500°C: iron/constantan, chromel/alumel, copper/constantan, and chromel/constantan.

4. Common thermocouple materials are iron/constantan, chromel/alumel, copper/constantan, and chromel/constantan. From the graph in Figure 21–4, determine which thermocouple has the highest sensitivity.

5. A standard platinum RTD having a resistance of 100 Ω at 0ºC and a temperature coefficient of 0.385 Ω per degrees Celsius is connected in a bridge circuit as shown in Figure 21–48. If $R_1, R_2,$ and R_3 all equal 100 Ω, what will be the differential output voltage if the RTD is at a temperature of 34ºC and V_s equals 2 V?

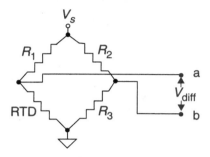

FIGURE 21–48

6. Calculate the output voltage of the circuit in Figure 21–49 at the temperatures of 23ºC, 25ºC, and 27ºC. The thermistor has a temperature coefficient of –4%/ºC and a resistance of 3 kΩ at 25ºC.

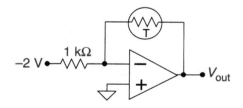

FIGURE 21–49

7. A 10 mm strain gage with a GF of 2 elongates to 11 mm under mechanical stress. If it is connected in the circuit in Figure 21–50 as R_4 and has an unstressed resistance of 350 Ω, what will be the output voltage of the circuit under stress? ($R_1, R_2,$ and R_3 = 350 Ω.)

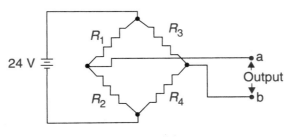

FIGURE 21–50

8. Two 10 mm strain gages with a GF of 2 elongate to 11 mm under mechanical stress. If they are connected in the circuit in Figure 21–50 as R_1 and R_4 and have an unstressed resistance of 350 Ω, what will be the output voltage of the circuit under stress? (R_2 and R_3 = 350 Ω.)

9. Calculate the output voltage of the bridge circuit in Figure 21–51 if the scale is weighing a 190 lb person. The scale is using 350 Ω strain gages with a GF of 3. When a 190 lb person stands on the scale, the gages change their lengths by 2%. Resistor R_1 equals R_2, and the supply voltage is a 9 V battery.

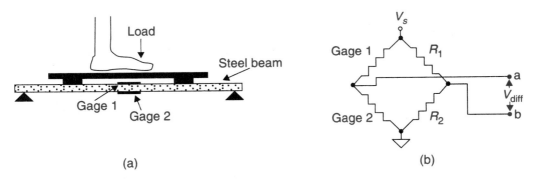

(a)   (b)

FIGURE 21–51

10. Calculate the minimum and maximum range of the tachometer in Figure 21–52.

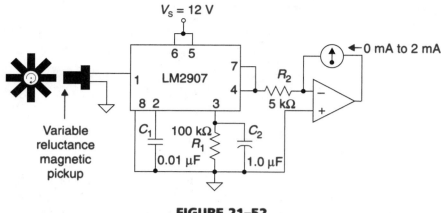

FIGURE 21–52

11. Calculate the efficiency of a 1.6 hp electric motor operating at 240 V AC at 7 A with a power factor of 0.9.

12. A 0.5 hp DC motor is operating from 48 V and drawing 10 A. What is its efficiency?

13. A 24 V DC motor is drawing 15 A and has an efficiency of 85%. Calculate the output horsepower.

14. A 10 hp AC motor is operating from 220 V_{RMS} and drawing 40 A with a PF of 0.9. What is its efficiency?

15. Calculate the pull-in current, hold current, wattage consumed by the coil during pull-in time, and wattage consumed by the coil after the relay is energized for the circuit in Figure 21–53.

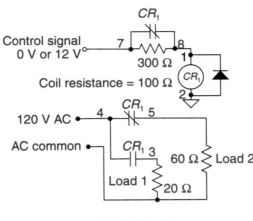

FIGURE 21–53

16. A six-pole synchronous motor is operating from 60 Hz. What is the rpm of the rotor?

17. A six-pole induction motor is operating from 60 Hz under a load, which causes a 5% slip. What is the rpm of the rotor?

APPENDIX A

Components List

INTRODUCTION

Electronic components are available from numerous mail-order houses. The author has ordered from the following companies and has found their services to be satisfactory.

Digi-Key Corp.
701 Brooks Ave. South
Thief River Falls, MN 56701-0677
1-800-344-4539
www.digikey.com

Jameco Electronic Components
1355 Shoreway Rd.
Belmont, CA 94002-4100
1-800-831-4242
www.jameco.com

JDR Microdevices
1850 South 10th St.
San Jose, CA 95112-4108
1-800-538-5000
www.jdr.com

COMPONENTS LIST

The parts listed below have a Digi-Key part number.

BULBS

Qty.	Description	Part #	Qty.	Description	Part #
(1)	12 V @ 80 mA	CM7382-ND	(1)	5 V @ 60 mA	CM683-ND
(1)	120 V AC (15 W to 100 W)				

CAPACITORS

Qty.	Description	Part #	Qty.	Description	Part #
(1)	100 pF	P4800-ND	(1)	150 pF	P4802-ND
(2)	0.001 µF	P4812-ND	(2)	0.01 µF	P4824-ND
(2)	0.022 µF	EF1223-ND	(1)	0.047 µF	EF1474-ND
(2)	0.1 µF	E1104-ND	(1)	0.22 µF	E1224-ND
(1)	1 µF	EF1105-ND	(5)	2.2 µF	EF1225-ND
(4)	10 µF	P5979-ND	(1)	100 µF	P6362-ND
(1)	4700 µF	P6369-ND			

CRYSTALS

Qty.	Description	Part #
(1)	4MHz	X405-ND

DIODES

Qty.	Description	Part #	Qty.	Description	Part #
(4)	1N4004	1N4004CT-ND	(4)	1N4148	1N4148CT-ND
(1)	bridge rectifier	DF04M-ND	(1)	12 V zener	1N5242BCT-ND

HEAT SINK

Qty.	Description	Part #
(2)	Type for TO-220 package	HS106-ND

INDUCTORS

Qty.	Description	Part #	Qty.	Description	Part #
(1)	100 µH	DN7425-ND	(1)	500 µH	M5256-ND

ICs

Qty.	Description	Part #	Qty.	Description	Part #
(1)	74LS240	DM74LS240N-ND	(1)	ADC0804	ADC0804LCN-ND
(1)	DAC0808	DAC0808LCN-ND	(2)	LM311	LM311N-ND
(1)	LM317	LM317T-ND	(1)	LM318	LM318N-ND
(1)	LM324	LM324N-ND	(1)	LM340 (78xx)	LM340-AT-12-ND
(2)	LM555	LM555CN-ND	(1)	LM565	LM565CN-ND
(1)	LM566	LM566CN-ND	(2)	LM741	LM741CN-ND
(1)	LM2907	LM2907N-8-ND	(1)	LM3080	LM3080AN-ND
(1)	LM78S40	LM78S40CN-ND			

LEDs

Qty.	Description	Part #	Qty.	Description	Part #
(8)	Red	P308-ND	(1)	7-seg. MAN72A	P510-ND

OPTOISOLATORS

Qty.	Description	Part #	Qty.	Description	Part #
(1)	MOC3020 triac	MOC3020QT-ND	(1)	4N40IS SCR	4N40IS-ND

POTENTIOMETERS

Qty.	Description	Part #	Qty.	Description	Part #
(1)	1 k	CT2202-ND	(1)	2.5 k	CT2203-ND
(1)	5 k	CT2204-ND	(1)	10 k	CT2205-ND
(2)	50 k	CT2207-ND	(1)	100 k	CT2208-ND

RESISTORS

1/4 W (Part # = Value + QBK-ND. Example: 100 Ω = 100QBK-ND.)

Qty.	Description	Qty.	Description	Qty.	Description
(3)	1 Ω	(1)	4.7 Ω	(1)	47 Ω
(2)	100 Ω	(1)	150 Ω	(1)	180 Ω
(1)	200 Ω	(1)	220 Ω	(8)	330 Ω
(1)	470 Ω	(1)	620 Ω	(1)	680 Ω
(1)	820 Ω	(5)	1 k Ω	(2)	1.2 kΩ
(1)	1.5 kΩ	(1)	1.8 kΩ	(1)	2 kΩ
(2)	2.2 kΩ	(3)	2.5 kΩ	(1)	3.3 kΩ
(2)	3.9 kΩ	(2)	4.7 kΩ	(4)	5.6 kΩ
(3)	8.2 kΩ	(4)	10 kΩ	(1)	12 kΩ
(1)	15 kΩ	(1)	18 kΩ	(2)	22 kΩ
(1)	27 kΩ	(2)	33 kΩ	(2)	39 kΩ
(1)	67 kΩ	(2)	82 kΩ	(1)	91 kΩ
(3)	100 kΩ	(1)	220 kΩ	(1)	330 kΩ
(1)	1 M				

High Wattage

Qty.	Description	Part #	Qty.	Description	Part #
(1)	8.2 Ω/5 W	8.2-5-ND	(1)	15 Ω/10 W	15-10-ND
(1)	51 Ω/2 W	51W-2-ND	(1)	100 Ω/2 W	100W-2-ND
(1)	150 Ω/2 W	150W-2-ND	(1)	390 Ω/1 W	390W-1-ND
(1)	510 Ω/2 W	510W-2-ND	(1)	12 kΩ/1 W	12KW-1-ND

SENSORS

Qty.	Description	Part #	Qty.	Description	Part #
(1)	EE-SX1042 slot	OR518-ND	(1)	EE-SY148 reflective	OR516-ND
(1)	IC temp.	LM35DZ-ND			

THYRISTORS

Qty.	Description	Part #	Qty.	Description	Part #
(1)	C106B1 SCR	T106B1-ND	(1)	ECG5608 triac	Q4004F31-ND
(1)	ECG6408 diac	HT-32-ND			

TRANSISTORS

Qty.	Description	Part #	Qty.	Description	Part #
(1)	2N7000 E-MOSFET	2N7000 DIC-ND			
(1)	N-ch E-MOSFET	IRF630-ND	(1)	P-ch E-MOSFET	IRF9630-ND

The parts listed below have an Electronix Express part number.

TRANSFORMERS

Qty.	Description	Part #	Qty.	Description	Part #
(1)	1:1 audio	16A600-600	(1)	12.6 V power	16P112-1

TRANSISTORS

Qty.	Description	Part #	Qty.	Description	Part #
(2)	NPN 2N3904	11 2N3904	(1)	PNP 2N3906	11 2N3906
(1)	NPN Darlington	11 2N6037	(1)	PNP Darlington	11 2N6034
(1)	JFET MPF102	11 MPF102	(1)	D-MOSFET	11 MFE201
(1)	UJT ECG6410	11 2N4870		40673	

OPTIONAL COMPONENTS

Qty.	Description	Part #	Qty.	Description	Part #
(1)	Microphone	26MPCY	(1)	Speaker	26PPC

The part listed below has a Jameco Electronic Components part number.

MOTOR

Qty.	Description	Part #
(1)	DC motor	105849

APPENDIX B

Components Data

Diodes

Type	V_{RRM}	I_o	I_{FSM}	V_F	I_R	trr
1N4004	400 V	1 A	30 A	1.1 V	5 mA	...
1N4148	75 V	150 m	450 mA	1 V	25 nA	400 µS
Bridge	400V	1 A	45 A	1.1 V	...	...
LED	4 V	25 mA	...	2 V	...	...
LED 7-seg.	6 V	30 mA	...	2 V	...	...

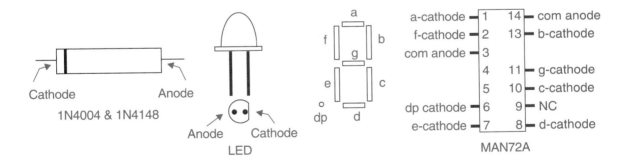

1N4004 & 1N4148

LED

MAN72A

Zener

Type	Voltage	Power
Zener 1N5242	12 V	1/2 W

1N5242

Bipolar Transistors

Type	P_T	I_C	Breakdown V_{CB} V_{CE} V_{EB}			Beta (H_{FE})
2N3904 NPN	1.2 W	800 mA	75 V	40 V	6 V	100–300
2N3906 PNP	1.2 W	1 A	80 V	80 V	5 V	50–250
2N6034 PNP	40 W	4 A	80 V	80 V	5 V	750 min
2N6037 NPN	40 W	4 A	80 V	80 V	5 V	750–2000

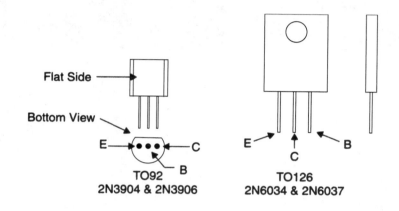

TO92
2N3904 & 2N3906

TO126
2N6034 & 2N6037

UJT

Type	P_T	I_E	V_{B2B1}	n
UJT ECG6410	300 mW	50 mA max	35 V max	0.70–0.85

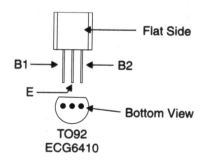

TO92
ECG6410

FETs

Type	P_T	I_D	$V_{GS(off)}$ or $V_{G(th)}$	gm_0	$r_{D(on)}$	Breakdown V_{GS} V_{DS}	
40673 N-ch D-MOSFET	330 mW	50 mA	−2 V	12,000 µS	...	6 V	20 V
MPF102 N-ch JFET	310 mW	50 mA	−4 V max	7000 µS	...	25 V	...
2N7000 N-ch E-MOSFET	400 mW	200 mA	3 V max	...	5 Ω max	...	60 V
IRF630 N-ch MOSFET	74 W	9 A	...	...	0.4 Ω	...	200 V
IRF9630 P-ch MOSFET	74 W	4 A	...	...	0.8 Ω	...	200 V

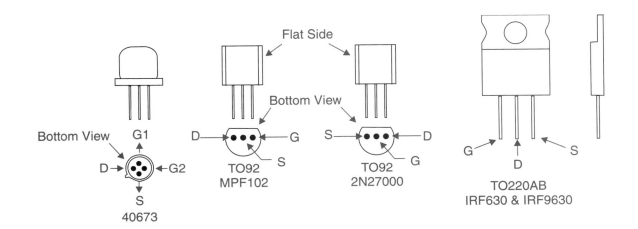

Bottom View

G1

D ← → G2

S

40673

Flat Side

Bottom View

D ● ● ● G

S

TO92
MPF102

S ● ● ● D

G

TO92
2N27000

G → ↓ ← S

D

TO220AB
IRF630 & IRF9630

Thyristors

Type	$I_{T(RMS)}$	$V_{E(blocking)}$	I_G	V_G	V_{on}
SCR C106B1	4 A	200 V	200 µA	2 V max	2.2 V max
Triac ECG5608	4 A	200 V	10 mA	2 V max	1.6 V max

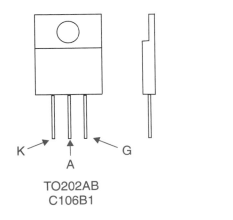

K → ↑ ← G
A

TO202AB
C106B1

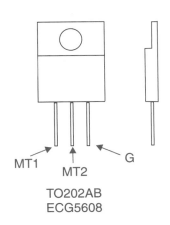

MT1 → ↑ ← G
MT2

TO202AB
ECG5608

Diac

Type	V_{BO}	I_{max}
Diac ECG6408	27 V min to 37 V max	2 A

SCR Optoisolator

Part #	Isolation Voltage	Trigger Current (I_F)	Blocking Voltage	Output Current (I_{out})
4N40IS	660 V	14 mA	400 V	300 mA

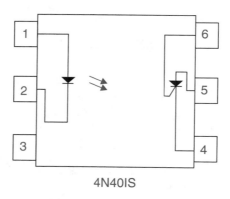

4N40IS

Triac Optoisolator

Part #	Isolation Voltage	Trigger Current (I_F)	Blocking Voltage	Output Current (I_{out})
MOC3020	5300 V	30 mA	400 V	100 mA

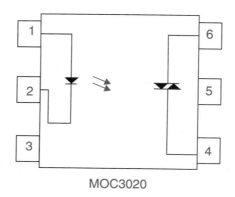

MOC3020

ADC0801/ADC0802/ADC0803/ADC0804/ADC0805
8-Bit µP Compatible A/D Converters

General Description

The ADC0801, ADC0802, ADC0803, ADC0804 and ADC0805 are CMOS 8-bit successive approximation A/D converters that use a differential potentiometric ladder—similar to the 256R products. These converters are designed to allow operation with the NSC800 and INS8080A derivative control bus with TRI-STATE® output latches directly driving the data bus. These A/Ds appear like memory locations or I/O ports to the microprocessor and no interfacing logic is needed.

Differential analog voltage inputs allow increasing the common-mode rejection and offsetting the analog zero input voltage value. In addition, the voltage reference input can be adjusted to allow encoding any smaller analog voltage span to the full 8 bits of resolution.

Features

- Compatible with 8080 µP derivatives—no interfacing logic needed - access time - 135 ns
- Easy interface to all microprocessors, or operates "stand alone"

- Differential analog voltage inputs
- Logic inputs and outputs meet both MOS and TTL voltage level specifications
- Works with 2.5V (LM336) voltage reference
- On-chip clock generator
- 0V to 5V analog input voltage range with single 5V supply
- No zero adjust required
- 0.3" standard width 20-pin DIP package
- 20-pin molded chip carrier or small outline package
- Operates ratiometrically or with 5 V_{DC}, 2.5 V_{DC}, or analog span adjusted voltage reference

Key Specifications

- Resolution   8 bits
- Total error   ± ¼ LSB, ± ½ LSB and ± 1 LSB
- Conversion time   100 µs

Typical Applications

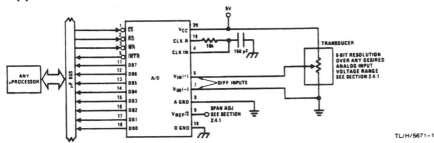

TL/H/5671–1

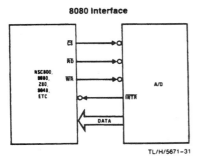

8080 Interface

TL/H/5671–31

Part Number	Full-Scale Adjusted	$V_{REF}/2 = 2.500$ V_{DC} (No Adjustments)	$V_{REF}/2 =$ No Connection (No Adjustments)
ADC0801	± ¼ LSB		
ADC0802		± ½ LSB	
ADC0803	± ½ LSB		
ADC0804		± 1 LSB	
ADC0805			± 1 LSB

Error Specification (Includes Full-Scale, Zero Error, and Non-Linearity)

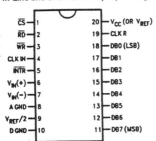

ADC080X Dual-In-Line and Small Outline (SO) Packages

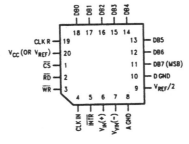

ADC080X Molded Chip Carrier (PCC) Package

Semiconductor

DAC0808/DAC0807/DAC0806 8-Bit D/A Converters

General Description

The DAC0808 series is an 8-bit monolithic digital-to-analog converter (DAC) featuring a full scale output current settling time of 150 ns while dissipating only 33 mW with ±5V supplies. No reference current (I_{REF}) trimming is required for most applications since the full scale output current is typically ±1 LSB of 255 I_{REF}/ 256. Relative accuracies of better than ±0.19% assure 8-bit monotonicity and linearity while zero level output current of less than 4 μA provides 8-bit zero accuracy for $I_{REF} \geq 2$ mA. The power supply currents of the DAC0808 series are independent of bit codes, and exhibits essentially constant device characteristics over the entire supply voltage range.

The DAC0808 will interface directly with popular TTL, DTL or CMOS logic levels, and is a direct replacement for the MC1508/MC1408. For higher speed applications, see DAC0800 data sheet.

Features

- Relative accuracy: ±0.19% error maximum (DAC0808)
- Full scale current match: ±1 LSB typ
- 7 and 6-bit accuracy available (DAC0807, DAC0806)
- Fast settling time: 150 ns typ
- Noninverting digital inputs are TTL and CMOS compatible
- High speed multiplying input slew rate: 8 mA/μs
- Power supply voltage range: ±4.5V to ±18V
- Low power consumption: 33 mW @ ±5V

Block and Connection Diagrams

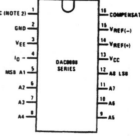

Dual-In-Line Package

Order Number
**DAC0808, DAC0807,
or DAC0806**
See NS Package
Number J16A,
M16A or N16A

TL/H/5687–1

Small-Outline Package

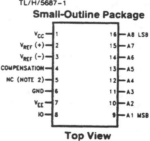

Top View

Typical Application

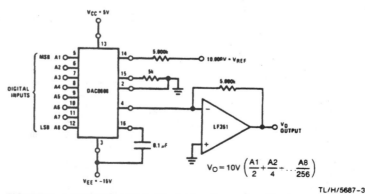

$$V_O = 10V \left(\frac{A1}{2} + \frac{A2}{4} + \cdots \frac{A8}{256} \right)$$

TL/H/5687–3

FIGURE 1. +10V Output Digital to Analog Converter (Note 7)

National Semiconductor

DM54ALS240A/DM74ALS240A/DM74ALS241A
Octal TRI-STATE® Bus Driver

General Description

These octal TRI-STATE bus drivers are designed to provide the designer with flexibility in implementing a bus interface with memory, microprocessor, or communication systems. The output TRI-STATE gating control is organized into two separate groups of four buffers. The ALS240A control inputs symmetrically enable the respective outputs when set logic low, while the ALS241A has complementary enable gating. The TRI-STATE circuitry contains a feature that maintains the buffer outputs in TRI-STATE (high impedance state) during power supply ramp-up or ramp-down. This eliminates bus glitching problems that arise during power-up and power-down.

Features

- Advanced low power oxide-isolated ion-implanted Schottky TTL process
- Functional and pin compatible with the DM54/74LS counterpart
- Improved switching performance with less power dissipation compared with the DM54/74LS counterpart
- Switching response specified into 500Ω and 50 pF load
- Switching response specifications guaranteed over full temperature and V_{CC} supply range
- PNP input design reduces input loading
- Low level drive current:
  54ALS = 12 mA, 74ALS = 24 mA

Connection Diagram

Dual-In-Line Package

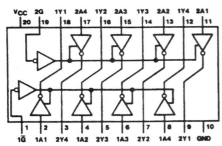

Top View

TL/F/6210–1

Order Number DM54ALS240AJ, DM74ALS240AWM, DM74ALS240AN or DM74ALS240ASJ
See NS Package Number J20A, M20B, M20D or N20A

Dual-In-Line Package

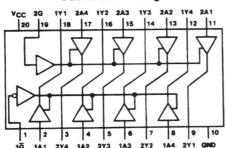

Top View

TL/F/6210–2

Order Number DM74ALS241AWM or DM74ALS241AN
See NS Package Number M20B or N20A

Function Tables

'ALS240A

Input		Output
$\overline{G}$	A	Y
L	L	H
L	H	L
H	X	Z

'ALS241A

Input		Output
2G	2A	Y
H	L	L
H	H	H
L	X	Z

'ALS241A

Input		Output
1G	1A	Y
L	L	L
L	H	H
H	X	Z

H = High Level Logic State
L = Low Level Logic State
X = Don't Care (Either Low or High Level Logic State)
Z = High Impedance (Off) State

OMRON

EE-SX1042

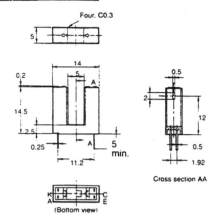

Four. C0.3

Cross section AA

(Bottom view)

Specifications

■ ABSOLUTE MAXIMUM RATINGS ($T_A = 25°C$)

	Item	Symbol	Rated value
Emitter	Forward current	I_F	50 mA*
	Reverse voltage	V_R	4 V
Detector	Collector-emitter voltage	V_{CEO}	30 V
	Collector current	I_C	20 mA
	Collector dissipation	P_C	100 mW*
Ambient temperature	Operating	Topr	−25° to 85°C**
	Storage	Tstg	−30° to 100°C

*Refer to Engineering Data if the ambient temperature is not within the normal room temperature range.
**The operating temperature of the EE-SX1041/1070 is −25° to 95°C.

■ CHARACTERISTICS ($T_A = 25°C$)

	Item	Symbol	EE-SX1018/1025/1041/1042/1070/1071	
			Value	**Condition**
Emitter	Forward voltage	V_F	1.2 V typ. 1.5 V max.	I_F = 30 mA
	Reverse current	I_R	0.01 μA typ. 10 μA max.	V_R = 4 V
	Peak emission wavelength	$\lambda p(L)$	940 nm typ.	I_F = 20 mA
Detector	Dark current	I_D	2 nA typ. 200 nA max.	V_{CE} = 10 V 0/x
	Peak spectral sensitivity wavelength	$\lambda p(P)$	850 nm typ.	V_{CE} = 10 V
Combination	Light current (collector current)	I_L	0.5 mA min. 14 mA max.	I_F = 20 mA V_{CE} = 10 V
	Collector-emitter saturated voltage	V_{CE} (sat)	0.1 V typ. 0.4 V max.	I_F = 20 mA I_L = 0.1 mA
	Rising time (see note)	tr	4 μs typ.	V_{CC} = 5 V R_L = 100 Ω I_L = 5 mA
	Falling time (see note)	tf	4 μs typ.	

EE-SX1041/1042/1070/1071

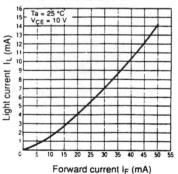

Note: The following illustrations show the rising time, tr, and the falling time, tf.

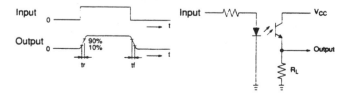

OMRON

PHOTOMICROSENSOR | EE-SY148

Reflective Photomicrosensor with Fixed Sensing Distance

- ■ Low-profiled design facilitates stacked installation of sensor units
- ■ Phototransistor output configuration
- ■ Low cost

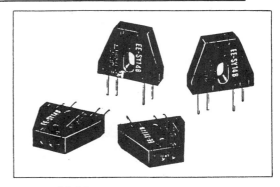

Ordering Information

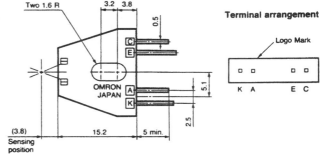

Terminal arrangement

Sensing method	Reflective type
Sensing distance	3 mm(white paper with a reflection factor of 90%)
Output configuration	Phototransistor
Terminal type	PC board
No.of terminals	4
Model	EE-SY148

Specifications

Absolute Maximum Ratings (Ta = 25°C)

	Item	Symbol	Value
Input	Forward current*	I_F	50 mA
	Pulse forward current**	I_{FP}	1 A
	Reverse voltage	V_R	3 V
Output	Collector-emitter voltage	V_{CEO}	30 V
	Collector current	I_C	20 mA
	Collector power dissipation*	P_C	50 mW
Operating temperature***		Topr	−40° to 80°C
Storage temperature		Tstg	−40° to 80°C

* Refer to "Temperature rating diagram" when ambient temperature is not at 25°C.
** The data indicated are those measured with a pulse width of 10 μs at a pulse recurrence of 100 Hz.
*** Without freezing or condensation.

Electrical Characteristics (Ta = 25°C)

	Item	Symbol	Limits	Test conditions
Input	Forward voltage	V_F	1.7 V max.	I_F = 40 mA
	Reverse current	I_R	100 μA max.	V_R = 3 V
Output	Dark current	Id	100 nA max.	V_{CE} = 10 V, 0 ℓx
Input/ output (coupled)	Light current* (Collector current)	I_L	10 μA min. 60 μA typ.	I_F = 40 mA V_{CE} = 5 V
	Leakage current**	I_{LEAK}	1 μA max.	I_F = 40 mA V_{CE} = 5 V
	Rise time (See Note.)	tr	10 μs typ.	V_{CC} = 20 V R_L = 1 kΩ
	Fall time (See Note.)	tf	8 μs typ.	I_L = 1 mA

* Measured with white paper having a reflection factor of 90% at a sensing distance of 3.8 mm.
** Measured with non-reflective object at 0 ℓx.

Input/output characteristics

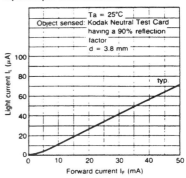

Note:
Refer to the following timing diagram for tr and tf.

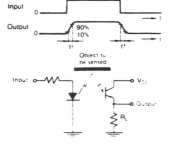

LM35/LM35A/LM35C/LM35CA/LM35D
Precision Centigrade Temperature Sensors

General Description

The LM35 series are precision integrated-circuit temperature sensors, whose output voltage is linearly proportional to the Celsius (Centigrade) temperature. The LM35 thus has an advantage over linear temperature sensors calibrated in ° Kelvin, as the user is not required to subtract a large constant voltage from its output to obtain convenient Centigrade scaling. The LM35 does not require any external calibration or trimming to provide typical accuracies of ±¼°C at room temperature and ±¾°C over a full −55 to +150°C temperature range. Low cost is assured by trimming and calibration at the wafer level. The LM35's low output impedance, linear output, and precise inherent calibration make interfacing to readout or control circuitry especially easy. It can be used with single power supplies, or with plus and minus supplies. As it draws only 60 µA from its supply, it has very low self-heating, less than 0.1°C in still air. The LM35 is rated to operate over a −55° to +150°C temperature range, while the LM35C is rated for a −40° to +110°C range (−10° with improved accuracy). The LM35 series is available packaged in hermetic TO-46 transistor packages, while the LM35C, LM35CA, and LM35D are also available in the plastic TO-92 transistor package. The LM35D is also available in an 8-lead surface mount small outline package and a plastic TO-202 package.

Features

- Calibrated directly in ° Celsius (Centigrade)
- Linear + 10.0 mV/°C scale factor
- 0.5°C accuracy guaranteeable (at +25°C)
- Rated for full −55° to +150°C range
- Suitable for remote applications
- Low cost due to wafer-level trimming
- Operates from 4 to 30 volts
- Less than 60 µA current drain
- Low self-heating, 0.08°C in still air
- Nonlinearity only ±¼°C typical
- Low impedance output, 0.1 Ω for 1 mA load

Connection Diagrams

TO-46
Metal Can Package*

BOTTOM VIEW

TL/H/5516–1

*Case is connected to negative pin (GND)

Order Number LM35H, LM35AH, LM35CH, LM35CAH or LM35DH
See NS Package Number H03H

TO-92
Plastic Package

+Vs Vout GND

BOTTOM VIEW

TL/H/5516–2

Order Number LM35CZ, LM35CAZ or LM35DZ
See NS Package Number Z03A

SO-8
Small Outline Molded Package

Vout	1	8	+Vs
N.C.	2	7	N.C.
N.C.	3	6	N.C.
GND	4	5	N.C.

TL/H/5516–21

Top View
N.C. = No Connection

Order Number LM35DM
See NS Package Number M08A

TO-202
Plastic Package

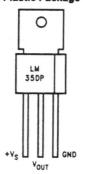

+Vs   GND
       Vout

TL/H/5516–24

Order Number LM35DP
See NS Package Number P03A

Typical Applications

+Vs
(4V TO 20V)

OUTPUT
0 mV +10.0 mV/°C

TL/H/5516–3

FIGURE 1. Basic Centigrade
Temperature
Sensor (+2°C to +150°C)

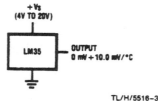

TL/H/5516–4

Choose R_1 = $-V_S$/50 µA

V_{OUT} = +1,500 mV at +150°C
     = +250 mV at +25°C
     = −550 mV at −55°C

FIGURE 2. Full-Range Centigrade
Temperature Sensor

LM111/LM211/LM311 Voltage Comparator

General Description

The LM111, LM211 and LM311 are voltage comparators that have input currents nearly a thousand times lower than devices like the LM106 or LM710. They are also designed to operate over a wider range of supply voltages: from standard ±15V op amp supplies down to the single 5V supply used for IC logic. Their output is compatible with RTL, DTL and TTL as well as MOS circuits. Further, they can drive lamps or relays, switching voltages up to 50V at currents as high as 50 mA.

Both the inputs and the outputs of the LM111, LM211 or the LM311 can be isolated from system ground, and the output can drive loads referred to ground, the positive supply or the negative supply. Offset balancing and strobe capability are provided and outputs can be wire OR'ed. Although slower than the LM106 and LM710 (200 ns response time vs

40 ns) the devices are also much less prone to spurious oscillations. The LM111 has the same pin configuration as the LM106 and LM710.

The LM211 is identical to the LM111, except that its performance is specified over a −25°C to +85°C temperature range instead of −55°C to +125°C. The LM311 has a temperature range of 0°C to +70°C.

Features

- Operates from single 5V supply
- Input current: 150 nA max. over temperature
- Offset current: 20 nA max. over temperature
- Differential input voltage range: ±30V
- Power consumption: 135 mW at ±15V

Typical Applications**

Offset Balancing

Strobing

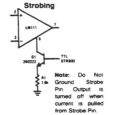

Note: Do Not Ground Strobe Pin. Output is turned off when current is pulled from Strobe Pin.

**Note: Pin connections shown on schematic diagram and typical applications are for H08 metal can package.

Increasing Input Stage Current*

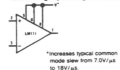

*Increases typical common mode slew from 7.0V/μs to 18V/μs.

Detector for Magnetic Transducer

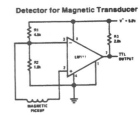

Digital Transmission Isolator

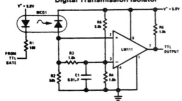

Relay Driver with Strobe

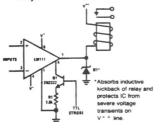

*Absorbs inductive kickback of relay and protects IC from severe voltage transients on V⁻ + line.

Note: Do Not Ground Strobe Pin.

Strobing off Both Input* and Output Stage

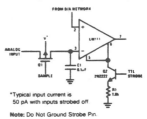

*Typical input current is 50 pA with inputs strobed off.

Note: Do Not Ground Strobe Pin.

Connection Diagrams*

Metal Can Package

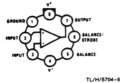

TL/H/5704–6

Top View

Note: Pin 4 connected to case

**Order Number LM111H, LM111H/883*, LM211H or LM311H
See NS Package Number H08C**

Dual-In-Line Package

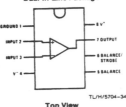

TL/H/5704–34

Top View

**Order Number LM111J-8, LM111J-8/883*, LM211J-8, LM211M, LM311J-8, LM311M or LM311N
See NS Package Number J08A, M08A or N08E**

Dual-In-Line Package

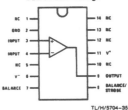

TL/H/5704–35

Top View

**Order Number LM111J, LM111J/883*, LM211J, LM311J or LM311N-14
See NS Package Number J14A or N14A**

LM117A/LM117/LM317A/LM317
3-Terminal Adjustable Regulator

General Description

The LM117 series of adjustable 3-terminal positive voltage regulators is capable of supplying in excess of 1.5A over a 1.2V to 37V output range. They are exceptionally easy to use and require only two external resistors to set the output voltage. Further, both line and load regulation are better than standard fixed regulators. Also, the LM117 is packaged in standard transistor packages which are easily mounted and handled.

In addition to higher performance than fixed regulators, the LM117 series offers full overload protection available only in IC's. Included on the chip are current limit, thermal overload protection and safe area protection. All overload protection circuitry remains fully functional even if the adjustment terminal is disconnected.

Normally, no capacitors are needed unless the device is situated more than 6 inches from the input filter capacitors in which case an input bypass is needed. An optional output capacitor can be added to improve transient response. The adjustment terminal can be bypassed to achieve very high ripple rejection ratios which are difficult to achieve with standard 3-terminal regulators.

Besides replacing fixed regulators, the LM117 is useful in a wide variety of other applications. Since the regulator is "floating" and sees only the input-to-output differential voltage, supplies of several hundred volts can be regulated as long as the maximum input to output differential is not exceeded, i.e., avoid short-circuiting the output.

Also, it makes an especially simple adjustable switching regulator, a programmable output regulator, or by connecting a fixed resistor between the adjustment pin and output, the LM117 can be used as a precision current regulator. Supplies with electronic shutdown can be achieved by clamping

the adjustment terminal to ground which programs the output to 1.2V where most loads draw little current.

For applications requiring greater output current, see LM150 series (3A) and LM138 series (5A) data sheets. For the negative complement, see LM137 series data sheet.

LM117 Series Packages and Power Capability

Part Number Suffix	Package	Rated Power Dissipation	Design Load Current
K	TO-3	20W	1.5A
H	TO-39	2W	0.5A
T	TO-220	20W	1.5A
MP	TO-202	2W	0.5A
E	LCC	2W	0.5A

Features

■ Guaranteed 1% output voltage tolerance (LM117A, LM317A)
■ Guaranteed max. 0.01%/V line regulation (LM117A, LM317A)
■ Guaranteed max. 0.3% load regulation (LM117A, LM117)
■ Guaranteed 1.5A output current
■ Adjustable output down to 1.2V
■ Current limit constant with temperature
■ P+ Product Enhancement tested
■ 80 dB ripple rejection
■ Output is short-circuit protected

Typical Applications

1.2V–25V Adjustable Regulator

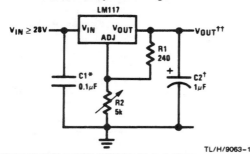

TL/H/9063–1

Full output current not available at high input-output voltages

*Needed if device is more than 6 inches from filter capacitors.

†Optional—improves transient response. Output capacitors in the range of 1 μF to 1000 μF of aluminum or tantalum electrolytic are commonly used to provide improved output impedance and rejection of transients.

$$^{\dagger\dagger}V_{OUT} = 1.25V \left(1 + \frac{R2}{R1}\right) + I_{ADJ}(R2)$$

Digitally Selected Outputs

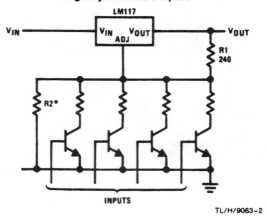

INPUTS

TL/H/9063–2

*Sets maximum V_{OUT}

LM118/LM218/LM318
Operational Amplifiers

General Description

The LM118 series are precision high speed operational amplifiers designed for applications requiring wide bandwidth and high slew rate. They feature a factor of ten increase in speed over general purpose devices without sacrificing DC performance.

The LM118 series has internal unity gain frequency compensation. This considerably simplifies its application since no external components are necessary for operation. However, unlike most internally compensated amplifiers, external frequency compensation may be added for optimum performance. For inverting applications, feedforward compensation will boost the slew rate to over 150V/μs and almost double the bandwidth. Overcompensation can be used with the amplifier for greater stability when maximum bandwidth is not needed. Further, a single capacitor can be added to reduce the 0.1% settling time to under 1 μs.

The high speed and fast settling time of these op amps make them useful in A/D converters, oscillators, active filters, sample and hold circuits, or general purpose amplifiers. These devices are easy to apply and offer an order of magnitude better AC performance than industry standards such as the LM709.

The LM218 is identical to the LM118 except that the LM218 has its performance specified over a −25°C to +85°C temperature range. The LM318 is specified from 0°C to +70°C.

Features

- 15 MHz small signal bandwidth
- Guaranteed 50V/μs slew rate
- Maximum bias current of 250 nA
- Operates from supplies of ±5V to ±20V
- Internal frequency compensation
- Input and output overload protected
- Pin compatible with general purpose op amps

Connection Diagrams

Dual-In-Line Package

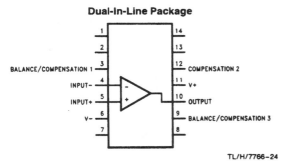

TL/H/7766-24

Top View

Order Number LM118J/883*
See NS Package Number J14A

Dual-In-Line Package

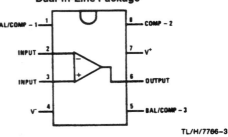

TL/H/7766-3

Top View
Order Number LM118J-8, LM118J-8/883*
LM318J-8, LM318M or LM318N
See NS Package Number J08A, M08A or N08B

Metal Can Package**

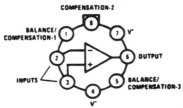

TL/H/7766-2

Top View
**Pin connections shown on schematic diagram
and typical applications are for TO-5 package.

Order Number LM118H, LM118H/883*
LM218H or LM318H
See NS Package Number H08C

*Available per JM38510/10107.

LM124/LM224/LM324, LM2902
Low Power Quad Operational Amplifiers

General Description

The LM124 series consists of four independent, high gain, internally frequency compensated operational amplifiers which were designed specifically to operate from a single power supply over a wide range of voltages. Operation from split power supplies is also possible and the low power supply current drain is independent of the magnitude of the power supply voltage.

Application areas include transducer amplifiers, DC gain blocks and all the conventional op amp circuits which now can be more easily implemented in single power supply systems. For example, the LM124 series can be directly operated off of the standard $+5$ V_{DC} power supply voltage which is used in digital systems and will easily provide the required interface electronics without requiring the additional ±15 V_{DC} power supplies.

Unique Characteristics

- In the linear mode the input common-mode voltage range includes ground and the output voltage can also swing to ground, even though operated from only a single power supply voltage
- The unity gain cross frequency is temperature compensated
- The input bias current is also temperature compensated

Advantages

- Eliminates need for dual supplies
- Four internally compensated op amps in a single package
- Allows directly sensing near GND and V_{OUT} also goes to GND
- Compatible with all forms of logic
- Power drain suitable for battery operation

Features

- Internally frequency compensated for unity gain
- Large DC voltage gain   100 dB
- Wide bandwidth (unity gain)   1 MHz
  (temperature compensated)
- Wide power supply range:
  Single supply   3 V_{DC} to 32 V_{DC}
  or dual supplies ±1.5 V_{DC} to ±16 V_{DC}
- Very low supply current drain (700 μA)—essentially independent of supply voltage
- Low input biasing current   45 nA_{DC}
  (temperature compensated)
- Low input offset voltage   2 mV_{DC}
  and offset current   5 nA_{DC}
- Input common-mode voltage range includes ground
- Differential input voltage range equal to the power supply voltage
- Large output voltage swing   0 V_{DC} to $V^+ - 1.5$ V_{DC}

Connection Diagram

Dual-In-Line Package

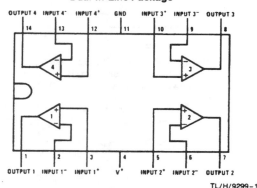

TL/H/9299–1

Top View

Order Number LM124J, LM124AJ, LM124J/883**, LM124AJ/883*, LM224J, LM224AJ, LM324J, LM324AJ, LM324M, LM324AM, LM2902M, LM324N, LM324AN or LM2902N
See NS Package Number J14A, M14A or N14A

*LM124A available per JM38510/11006
**LM124 available per JM38510/11005

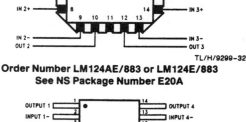

TL/H/9299–32

Order Number LM124AE/883 or LM124E/883
See NS Package Number E20A

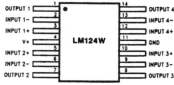

TL/H/9299–33

Order Number LM124AW/883 or LM124W/883
See NS Package Number W14B

LM140A/LM140/LM340A/LM340/LM7800/LM7800C
Series 3-Terminal Positive Regulators

General Description

The LM140A/LM140/LM340A/LM340/LM7800/LM7800C monolithic 3-terminal positive voltage regulators employ internal current-limiting, thermal shutdown and safe-area compensation, making them essentially indestructible. If adequate heat sinking is provided, they can deliver over 1.0A output current. They are intended as fixed voltage regulators in a wide range of applications including local (on-card) regulation for elimination of noise and distribution problems associated with single-point regulation. In addition to use as fixed voltage regulators, these devices can be used with external components to obtain adjustable output voltages and currents.

Considerable effort was expended to make the entire series of regulators easy to use and minimize the number of external components. It is not necessary to bypass the output, although this does improve transient response. Input bypassing is needed only if the regulator is located far from the filter capacitor of the power supply.

The entire series of regulators is available in the steel TO-3 power package. The LM340A/LM340/LM7800/LM7800C series is also available in the TO-220 plastic power package.

Features

- Complete specifications at 1A load
- Output voltage tolerances of ±2% at $T_j = 25°C$ and ±4% over the temperature range (LM140A/LM340A)
- Line regulation of 0.01% of V_{OUT}/V of ΔV_{IN} at 1A load (LM140A/LM340A)
- Load regulation of 0.3% of V_{OUT}/A (LM140A/LM340A)
- Internal thermal overload protection
- Internal short-circuit current limit
- Output transistor safe area protection
- P+ Product Enhancement tested

Device	Output Voltages	Packages
LM140A/LM140	5, 12, 15	TO-3 (K)
LM340A/LM340	5, 12, 15	TO-3 (K), TO-220 (T)
LM7800	8, 18, 24	TO-3 (K), TO-220 (T)
LM7800C	5, 6, 8, 12, 15, 18, 24	TO-3 (K), TO-220 (T)

Typical Applications

Fixed Output Regulator

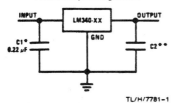

TL/H/7781-1

*Required if the regulator is located far from the power supply filter.

**Although no output capacitor is needed for stability, it does help transient response. (If needed, use 0.1 µF, ceramic disc).

Adjustable Output Regulator

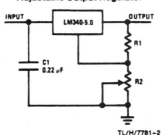

TL/H/7781-2

$V_{OUT} = 5V + (5V/R1 + I_Q) R2$ $5V/R1 > 3 I_Q$, load regulation $(L_r) \approx [(R1 + R2)/R1] (L_r$ of LM340-5).

Current Regulator

TL/H/7781-3

$$I_{OUT} = \frac{V2\text{-}3}{R1} + I_Q$$

$\Delta I_Q = 1.3$ mA over line and load changes.

Connection Diagrams and Ordering Information

TO-3 Metal Can Package (K and KC)

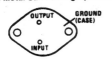

TL/H/7781-11

Bottom View

Steel Package Order Numbers:

LM140AK-5.0	LM140AK-12	LM140AK-15
LM140K-5.0	LM140K-12	LM140K-15
LM140AK-5.0/883	LM140AK-12/883	LM140AK-15/883
LM140K-5.0/883	LM140K-12/883	LM140K-15/883
LM340AK-5.0	LM340AK-12	LM340AK-15
LM340K-5.0	LM340K-12	LM340K-15
LM7806CK	LM7808CK	LM7808K
LM7818CK	LM7818K	LM7824CK
	LM7824K	

See Package Number K02A

TO-220 Power Package (T)

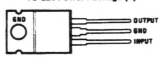

Top View

Plastic Package Order Numbers:

LM340AT-5.0	LM340T-5.0
LM340AT-12	LM340T-12
LM340AT-15	LM340T-15
LM7805CT	LM7812CT
LM7815CT	LM7806CT
LM7808CT	LM7818CT
LM7824CT	

See Package Number T03B

TO-39 Metal Can Package (H)

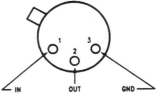

Top View

Metal Can Order Numbers†:

LM140H-5.0/883	LM140H-6.0/883
LM140H-8.0/883	LM140H-12/883
LM140H-15/883	LM140H-24/883

See Package Number H03A

LM555/LM555C Timer

General Description

The LM555 is a highly stable device for generating accurate time delays or oscillation. Additional terminals are provided for triggering or resetting if desired. In the time delay mode of operation, the time is precisely controlled by one external resistor and capacitor. For astable operation as an oscillator, the free running frequency and duty cycle are accurately controlled with two external resistors and one capacitor. The circuit may be triggered and reset on falling waveforms, and the output circuit can source or sink up to 200 mA or drive TTL circuits.

Features

- Direct replacement for SE555/NE555
- Timing from microseconds through hours
- Operates in both astable and monostable modes

- Adjustable duty cycle
- Output can source or sink 200 mA
- Output and supply TTL compatible
- Temperature stability better than 0.005% per °C
- Normally on and normally off output

Applications

- Precision timing
- Pulse generation
- Sequential timing
- Time delay generation
- Pulse width modulation
- Pulse position modulation
- Linear ramp generator

Applications Information

MONOSTABLE OPERATION

In this mode of operation, the timer functions as a one-shot (*Figure 1*). The external capacitor is initially held discharged by a transistor inside the timer. Upon application of a negative trigger pulse of less than $1/3\ V_{CC}$ to pin 2, the flip-flop is set which both releases the short circuit across the capacitor and drives the output high.

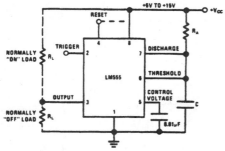

TL/H/7851-5

FIGURE 1. Monostable

The voltage across the capacitor then increases exponentially for a period of $t = 1.1\ R_A\ C$, at the end of which time the voltage equals $2/3\ V_{CC}$. The comparator then resets the flip-flop which in turn discharges the capacitor and drives the output to its low state. *Figure 2* shows the waveforms generated in this mode of operation. Since the charge and the threshold level of the comparator are both directly proportional to supply voltage, the timing internal is independent of supply.

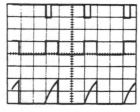

$V_{CC} = 5V$

$TIME = 0.1\ ms/DIV.$

$R_A = 9.1\ k\Omega$

$C = 0.01\ \mu F$

Top Trace: Input 5V/Div.

Middle Trace: Output 5V/Div.

Bottom Trace: Capacitor Voltage 2V/Div.

TL/H/7851-6

FIGURE 2. Monostable Waveforms

During the timing cycle when the output is high, the further application of a trigger pulse will not effect the circuit so long as the trigger input is returned high at least 10 μs before the end of the timing interval. However the circuit can be reset during this time by the application of a negative pulse to the reset terminal (pin 4). The output will then remain in the low state until a trigger pulse is again applied.

When the reset function is not in use, it is recommended that it be connected to V_{CC} to avoid any possibility of false triggering.

Figure 3 is a nomograph for easy determination of R, C values for various time delays.

NOTE: In monostable operation, the trigger should be driven high before the end of timing cycle.

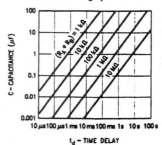

TL/H/7851-7

FIGURE 3. Time Delay

ASTABLE OPERATION

If the circuit is connected as shown in *Figure 4* (pins 2 and 6 connected) it will trigger itself and free run as a multivibrator. The external capacitor charges through $R_A + R_B$ and discharges through R_B. Thus the duty cycle may be precisely set by the ratio of these two resistors.

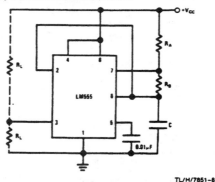

TL/H/7851-8

FIGURE 4. Astable

In this mode of operation, the capacitor charges and discharges between $1/3\ V_{CC}$ and $2/3\ V_{CC}$. As in the triggered mode, the charge and discharge times, and therefore the frequency are independent of the supply voltage.

Applications Information (Continued)

Figure 5 shows the waveforms generated in this mode of operation.

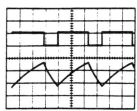

TL/H/7851-9

V_CC = 5V   Top Trace: Output 5V/Div.
TIME = 20 μs/DIV.   Bottom Trace: Capacitor Voltage 1V/Div.
R_A = 3.9 kΩ
R_B = 3 kΩ
C = 0.01 μF

FIGURE 5. Astable Waveforms

The charge time (output high) is given by:
$$t_1 = 0.693 (R_A + R_B) C$$
And the discharge time (output low) by:
$$t_2 = 0.693 (R_B) C$$
Thus the total period is:
$$T = t_1 + t_2 = 0.693 (R_A + 2R_B) C$$
The frequency of oscillation is:
$$f = \frac{1}{T} = \frac{1.44}{(R_A + 2 R_B) C}$$

Figure 6 may be used for quick determination of these RC values.

The duty cycle is:
$$D = \frac{R_B}{R_A + 2R_B}$$

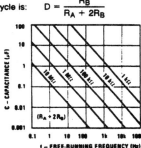

TL/H/7851-10

FIGURE 6. Free Running Frequency

FREQUENCY DIVIDER

The monostable circuit of *Figure 1* can be used as a frequency divider by adjusting the length of the timing cycle. *Figure 7* shows the waveforms generated in a divide by three circuit.

Connection Diagrams

Metal Can Package

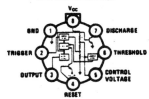

TL/H/7851-2

Top View

Order Number LM555H or LM555CH
See NS Package Number H08C

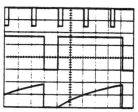

TL/H/7851-11

V_CC = 5V   Top Trace: Input 4V/Div.
TIME = 20 μs/DIV.   Middle Trace: Output 2V/Div.
R_A = 9.1 kΩ   Bottom Trace: Capacitor 2V/Div.
C = 0.01 μF

FIGURE 7. Frequency Divider

PULSE WIDTH MODULATOR

When the timer is connected in the monostable mode and triggered with a continuous pulse train, the output pulse width can be modulated by a signal applied to pin 5. *Figure 8* shows the circuit, and in *Figure 9* are some waveform examples.

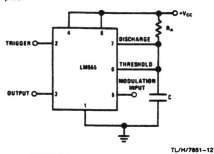

TL/H/7851-12

FIGURE 8. Pulse Width Modulator

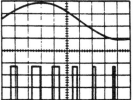

TL/H/7851-13

V_CC = 5V   Top Trace: Modulation 1V/Div.
TIME = 0.2 ms/DIV.   Bottom Trace: Output Voltage 2V/Div.
R_A = 9.1 kΩ
C = 0.01 μF

FIGURE 9. Pulse Width Modulator

PULSE POSITION MODULATOR

This application uses the timer connected for astable operation, as in *Figure 10*, with a modulating signal again applied to the control voltage terminal. The pulse position varies with the modulating signal, since the threshold voltage and hence the time delay is varied. *Figure 11* shows the waveforms generated for a triangle wave modulation signal.

Dual-In-Line and Small Outline Packages

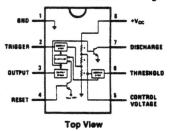

Top View

Order Number LM555J, LM555CJ,
LM555CM or LM555CN
See NS Package Number J08A, M08A or N08E

National
Semiconductor

LM565/LM565C Phase Locked Loop

General Description

The LM565 and LM565C are general purpose phase locked loops containing a stable, highly linear voltage controlled oscillator for low distortion FM demodulation, and a double balanced phase detector with good carrier suppression. The VCO frequency is set with an external resistor and capacitor, and a tuning range of 10:1 can be obtained with the same capacitor. The characteristics of the closed loop system—bandwidth, response speed, capture and pull in range—may be adjusted over a wide range with an external resistor and capacitor. The loop may be broken between the VCO and the phase detector for insertion of a digital frequency divider to obtain frequency multiplication.

The LM565H is specified for operation over the −55°C to +125°C military temperature range. The LM565CN is specified for operation over the 0°C to +70°C temperature range.

Features

- 200 ppm/°C frequency stability of the VCO
- Power supply range of ±5 to ±12 volts with 100 ppm/% typical
- 0.2% linearity of demodulated output
- Linear triangle wave with in phase zero crossings available
- TTL and DTL compatible phase detector input and square wave output
- Adjustable hold in range from ±1% to > ±60%

Applications

- Data and tape synchronization
- Modems
- FSK demodulation
- FM demodulation
- Frequency synthesizer
- Tone decoding
- Frequency multiplication and division
- SCA demodulators
- Telemetry receivers
- Signal regeneration
- Coherent demodulators

Connection Diagrams

Metal Can Package

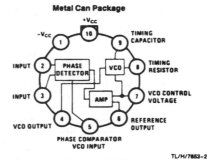

TL/H/7853–2

Order Number LM565H
See NS Package Number H10C

Dual-In-Line Package

TL/H/7853–3

Order Number LM565CN
See NS Package Number N14A

LM566C Voltage Controlled Oscillator

General Description

The LM566CN is a general purpose voltage controlled oscillator which may be used to generate square and triangular waves, the frequency of which is a very linear function of a control voltage. The frequency is also a function of an external resistor and capacitor.

The LM566CN is specified for operation over the 0°C to +70°C temperature range.

Features

- Wide supply voltage range: 10V to 24V
- Very linear modulation characteristics
- High temperature stability
- Excellent supply voltage rejection
- 10 to 1 frequency range with fixed capacitor
- Frequency programmable by means of current, voltage, resistor or capacitor

Applications

- FM modulation
- Signal generation
- Function generation
- Frequency shift keying
- Tone generation

Connection Diagram

Dual-In-Line Package

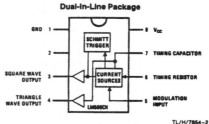

TL/H/7854–2

Order Number LM566CN
See NS Package Number N08E

Typical Application

1 kHz and 10 kHz TTL Compatible
Voltage Controlled Oscillator

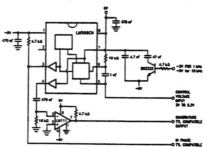

LM741 Operational Amplifier

General Description

The LM741 series are general purpose operational amplifiers which feature improved performance over industry standards like the LM709. They are direct, plug-in replacements for the 709C, LM201, MC1439 and 748 in most applications.

The amplifiers offer many features which make their application nearly foolproof: overload protection on the input and output, no latch-up when the common mode range is exceeded, as well as freedom from oscillations.

The LM741C/LM741E are identical to the LM741/LM741A except that the LM741C/LM741E have their performance guaranteed over a 0°C to +70°C temperature range, instead of −55°C to +125°C.

Connection Diagrams

Metal Can Package

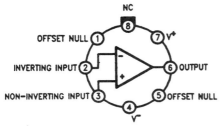

TL/H/9341–2

**Order Number LM741H, LM741H/883*, LM741AH/883
LM741CH or LM741EH
See NS Package Number H08C**

Ceramic Dual-In-Line Package

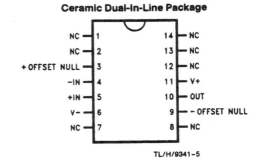

TL/H/9341–5

Order Number LM741J-14/883*, LM741AJ-14/883
See NS Package Number J14A**

*also available per JM38510/10101
**also available per JM38510/10102

Dual-In-Line or S.O. Package

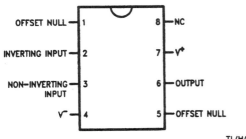

TL/H/9341–3

**Order Number LM741J, LM741J/883, LM741CJ,
LM741CM, LM741CN or LM741EN
See NS Package Number J08A, M08A or N08E**

Ceramic Flatpak

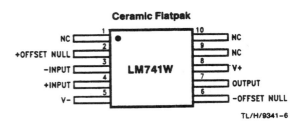

TL/H/9341–6

**Order Number LM741W/883
See NS Package Number W10A**

LM2907/LM2917 Frequency to Voltage Converter

General Description

The LM2907, LM2917 series are monolithic frequency to voltage converters with a high gain op amp/comparator designed to operate a relay, lamp, or other load when the input frequency reaches or exceeds a selected rate. The tachometer uses a charge pump technique and offers frequency doubling for low ripple, full input protection in two versions (LM2907-8, LM2917-8) and its output swings to ground for a zero frequency input.

Advantages

- Output swings to ground for zero frequency input
- Easy to use; $V_{OUT} = f_{IN} \times V_{CC} \times R1 \times C1$
- Only one RC network provides frequency doubling
- Zener regulator on chip allows accurate and stable frequency to voltage or current conversion (LM2917)

Features

- Ground referenced tachometer input interfaces directly with variable reluctance magnetic pickups
- Op amp/comparator has floating transistor output
- 50 mA sink or source to operate relays, solenoids, meters, or LEDs

- Frequency doubling for low ripple
- Tachometer has built-in hysteresis with either differential input or ground referenced input
- Built-in zener on LM2917
- ±0.3% linearity typical
- Ground referenced tachometer is fully protected from damage due to swings above V_{CC} and below ground

Applications

- Over/under speed sensing
- Frequency to voltage conversion (tachometer)
- Speedometers
- Breaker point dwell meters
- Hand-held tachometer
- Speed governors
- Cruise control
- Automotive door lock control
- Clutch control
- Horn control
- Touch or sound switches

Block and Connection Diagrams Dual-In-Line and Small Outline Packages, Top Views

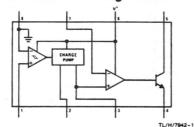

TL/H/7942-1

Order Number LM2907N-8
See NS Package Number N08E

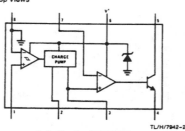

TL/H/7942-2

Order Number LM2917N-8
See NS Package Number N08E

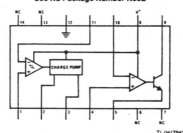

TL/H/7942-3

Order Number LM2907N
See NS Package Number N14A

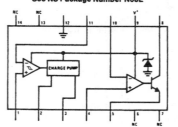

TL/H/7942-4

Order Number LM2917M or LM2917N
See NS Package Number M14A or N14A

National
Semiconductor

LM3080
Operational Transconductance Amplifier

General Description

The LM3080 is a programmable transconductance block intended to fulfill a wide variety of variable gain applications. The LM3080 has differential inputs and high impedance push-pull outputs. The device has high input impedance and its transconductance (g_m) is directly proportional to the amplifier bias current (I_{ABC}).

High slew rate together with programmable gain make the LM3080 an ideal choice for variable gain applications such as sample and hold, multiplexing, filtering, and multiplying.

The LM3080N and LM3080AN are guaranteed from 0°C to +70°C.

Features

- Slew rate (unity gain compensated): 50 V/μs
- Fully adjustable gain: 0 to $g_m \cdot R_L$ limit
- Extended g_m linearity: 3 decades
- Flexible supply voltage range: ±2V to ±18V
- Adjustable power consumption

Dual-In-Line Package

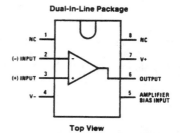

Top View
Order Number LM3080AN, LM3080M or LM3080N
See NS Package Number M08A or N08E

National
Semiconductor

LM78S40
Universal Switching Regulator Subsystem

General Description

The LM78S40 is a monolithic regulator subsystem consisting of all the active building blocks necessary for switching regulator systems. The device consists of a temperature compensated voltage reference, a duty-cycle controllable oscillator with an active current limit circuit, an error amplifier, high current, high voltage output switch, a power diode and an uncommitted operational amplifier. The device can drive external NPN or PNP transistors when currents in excess of 1.5A or voltages in excess of 40V are required. The device can be used for step-down, step-up or inverting switching regulators as well as for series pass regulators. It features wide supply voltage range, low standby power dissipation, high efficiency and low drift. It is useful for any stand-alone, low part count switching system and works extremely well in battery operated systems.

Features

- Step-up, step-down or inverting switching regulators
- Output adjustable from 1.25V to 40V
- Peak currents to 1.5A without external transistors
- Operation from 2.5V to 40V input
- Low standby current drain
- 80 dB line and load regulation
- High gain, high current, independent op amp
- Pulse width modulation with no double pulsing

Block and Connection Diagrams

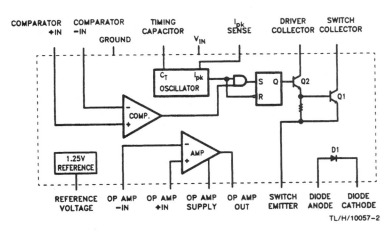

TL/H/10057–2

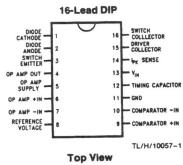

16-Lead DIP

Top View

TL/H/10057–1

Ordering Information

Part Number	NS Package	Temperature Range
LM78S40J LM78S40J/883	J16A Ceramic DIP J16A Ceramic DIP	−55°C to +125°C
LM78S40N	N16E Molded DIP	−40°C to +125°C
LM78S40CJ LM78S40CN	J16A Ceramic DIP N16E Molded DIP	0°C to +70°C

CHAPTER 1

QUESTIONS

1. a   3. a   5. c   7. d   9. a   11. d   13. d   15. a

CHAPTER 2

QUESTIONS

1. c   3. a   5. a   7. b   9. a   11. a   13. b   15. d   17. c.   19. d

PROBLEMS

1. a. 0.7 V   b. 18 V
3. a. 0 W   b. 2.1 W
5.
    0 V
  −250 V

7.
  100 V
7 V of 120 Hz ripple

9. 0.6 V DC
11.
  50 V
   0 V

13. 7 W

15. a. 1.4%   b. 38.9%

17.

100 V

4 kHz ripple

19.

10 V

−4.7 V

21.

−2.3 V

−4 V

23.

−2.3 V

−30.3 V

CHAPTER 3

QUESTIONS

1. b   3. b   5. c   7. c   9. b   11. d   13. d   15. c   17. b   19. b
21. b   23. a   25. e   27. g

PROBLEMS

1.

5 V ─────────────

V_{out} 0 V -

3. 80 mA
5. 7.5 V
7. (a) 63 mW   (b) 136 mW
9. 70 mW
11. 0 A
13. 2.52 MHz
15. 1.2 kΩ
17. 26 mA
19.

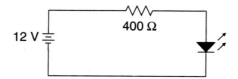

12 V   400 Ω

CHAPTER 4

QUESTIONS

1. a   3. b   5. c   7. b   9. c   11. a   13. d   15. b   17. c   19. a
21. c   23. c   25. c   27. d   29. c   31. b   33. d   35. d   37. b   39. d

PROBLEMS

1. $I_B = 10\ \mu A$ $I_E = 0.81$ mA $I_C = 0.8$ mA $V_{CE} = 5.2$ V
3. $\beta = 200$
5. $V_c = 20$ V
7. $\beta = 9.3$
9. $R_B = 37.5$ kΩ
11. $R_B = 510$ kΩ
13. $V_C = 0.2$ V
15. $\beta = 89$
17. $A_v = 200$
19. $V_c = 6.4$ V $V_B = 0.7$ V $V_E = 0$ V $V_{CE} = 6.4$ V
    $z_{in} = 2$ kΩ $z_{out} = 6$ kΩ $V_{out} = 4.58$ V $A_v = 229$

CHAPTER 5

QUESTIONS

1. d   3. d   5. c   7. d   9. c   11. a   13. d   15. c   17. b   19. b
21. d   23. b   25. d   27. c   29. c   31. a   33. b

PROBLEMS

1. $V_c = 10.25$ V $V_B = 3.8$ V $V_E = 3.1$ V $A_v = 1.5$
   $V_{CE} = 7.15$ V $z_{in} = 5.8$ kΩ $z_{out} = 2$ kΩ $V_{out} = 30$ mV$_{p\text{-}p}$
3. $V_c = 10.25$ V $V_B = 3.8$ V $V_E = 3.1$ V $A_v = 12$
   $V_{CE} = 7.15$ V $z_{in} = 3.9$ kΩ $z_{out} = 2$ kΩ $V_{out} = 240$ mV$_{p\text{-}p}$
5. $C_1 = 4\ \mu F$ $C_2 = 5.3\ \mu F$ $C_3 = 159\ \mu F$
7. Stage 1:
   $V_B = 4.8$ V $V_E = 4.1$ V $V_C = 8.8$ V
   $A_v = 6.1$ $z_{in} = 3.8$ kΩ $z_{out} = 1$ kΩ
   Stage 2:
   $V_B = 3.2$ V $V_E = 2.5$ V $V_C = 8$ V
   $A_v = 12.7$ $z_{in} = 1$ kΩ $z_{out} = 390$ Ω
   Total Circuit:
   $A_v = 77$ $z_{in} = 3.8$ k $z_{out} = 390$ Ω
9. Stage 1:
   $V_B = 0$ V $V_E = -0.7$ V $V_C = 3.8$ V
   $A_v = 1.7$ $z_{in} = 360$ kΩ $z_{out} = 6$ kΩ

9. (continued)
Stage 2:

$V_B = 3.8$ V	$V_E = 4.5$ V	$V_C = 0$ V
$A_v = 4.6$	$z_{in} = 740$ kΩ	$z_{out} = 7.4$ kΩ

Total Circuit:

$z_{in} = 360$ k	$z_{out} = 7.4$ kΩ	$A_v = 7.8$

CHAPTER 6

QUESTIONS

1. d   3. a   5. c   7. d   9. c   11. d   13. c   15. a   17. c   19. c
21. a   23. d   25. b   27. a

PROBLEMS

1. $V_B = 6.7$ V $V_E = 6$ V $V_C = 12$ V
    $V_{CE} = 6$ V $I_E = 60$ mA
3. $A_i = 7.8$ $A_p = 7.8$
5. $V_{BQ1} = 9.4$ V $V_{EQ1} = 8.7$ V $V_{CQ1} = 18$ V $V_{BQ2} = 8.7$ V
    $V_{EQ2} = 8$ V $V_{CQ2} = 18$ V $I_{EQ2} = 320$ mA
7. $A_i = 40$ $A_p = 40$
9. a. 3 μW   b. 6.2 mW
11. $A_v = 162$ $z_{in} = 25$ Ω $z_{out} = 6.8$ kΩ
    Max. output swing ≈ 12.8 V_{p-p}
13. $A_{v(single-ended)} = 132$ $A_{v(diff)} = 264$

CHAPTER 7

QUESTIONS

1. a   3. c   5. c   7. a   9. b   11. a   13. b   15. c   17. a   19. b
21. d   23. d   25. a   27. b   29. d   31. d   33. c   35. d

PROBLEMS

1. $gm = 2.4$ mS
3. $z_{in} = 560$ k $z_{out} = 3.3$ kΩ $A_v = 5.8$ $V_{out} = 116$ mV
5. $V_G = 0$ V $V_s = 2$ V $V_D = 6$ V $z_{in} = 2$ MΩ
    $z_{out} = 3$ kΩ $A_v = 5.5$ $V_{out} = 27.5$ mV
7. $z_{in} = 500$ kΩ $z_{out} = 2$ kΩ $A_v = 4.2$ $V_{out} = 84$ mV
9. $z_{in} = 2$ M $z_{out} = 4.7$ kΩ $A_v = 7$ $V_{out} = 210$ mV$_{p-p}$
11. $z_{in} = 300$ k $z_{out} = 3$ kΩ $A_v = 6$ $V_{out} = 300$ mV$_{p-p}$
13. $V_G = 0$ V $V_s = -2$ V $V_D = -11.8$ V $V_{DS} = -9.8$ V
    $z_{in} = 220$ kΩ $z_{out} = 5.6$ kΩ $A_v = 10.8$ $V_{out} = 1.08$ V$_{p-p}$

15.

```
   ┌─┐ ┌───┐                Control signal
 ──┘ └─┘   └──
   ∿∿∿╲───∿∿╱                100 mV_{p-p}
```

CHAPTER 8

QUESTIONS

1. b   3. d   5. c   7. c   9. b   11. c   13. c   15. d   17. a   19. a
21. a   23. d   25. d   27. d   29. c

PROBLEMS

1. $V_{GS(off)} = -3$ V $I_{DSS} = 7.5$ mA
3. $z_{in} = 220$ k $z_{out} = 1.2$ k $A_v = 2.6$ $V_{out} = 130$ mV
5. $V_G = 0$ V $V_S = 0$ V $V_D = -4.5$ V $V_{DS} = -4.5$ V
7. $V_{GS(th)} = 2$ V
9. $z_{in} = 27$ kΩ $z_{out} = 3$ k $A_v = 6.5$ $V_{out} = 260$ mV
11. FET on: $V_{out} = 4$ mV$_{p-p}$ FET off: $V_{out} = 333$ mV$_{p-p}$

CHAPTER 9

QUESTIONS

1. c   3. d   5. b   7. d   9. a   11. c   13. b   15. c   17. b   19. b
21. d   23. b   25. d

PROBLEMS

1. $z_{in} =$ Very high $A_v = 1$ $V_{out} = 6$ V$_{p-p}$

Output   6 V$_{p-p}$

3. $z_{in} =$ Very high $A_v = 4.4$ $V_{out} = 440$ mV$_{p-p}$

Output   440 mV$_{p-p}$

5. $R_f = 99$ kΩ
7. $z_{in} = 2$ k $A_v = 6$ $V_{out} = 600$ mV$_{p-p}$

Output   600 mV$_{p-p}$

9. $R_f = 100$ kΩ

11. a. z_{in} = Very high $A_v = 5$ $V_{out} = 5$ V$_{p-p}$
    b. z_{in} = 3 k $A_v = 4$ $V_{out} = 4$ V$_{p-p}$

CHAPTER 10

QUESTIONS

1. c   3. b   5. a   7. b   9. d   11. a   13. c   15. a   17. a   19. b
21. c   23. b   25. d   27. d   29. d   31. d

PROBLEMS

1. Input bias current = 90 nA
3. Input offset voltage = 2.5 mV
5. 4 V$_{p-p}$
7. 1.7 V$_{p-p}$
9. a. $A_v = 20$ b. $A_v = 10$ c. $A_v = 1.8$
11. 4 MHz
13. 48.1 dB
15. $A_v = 121$ $A_v = 41.6$ dB $V_{out} = 1.21$ V
    $V_{out(1dec)} = 0.121$ V $V_{out(2dec)} = 0.0121$ V

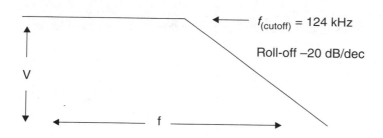

17.

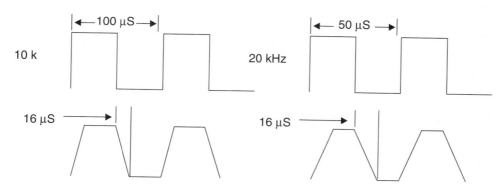

19. 25 kHz
21.

23.

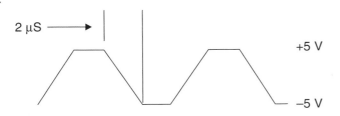

25.

2 µS ⟶

+5 V

−5 V

27.

429 mV$_p$

CHAPTER 11

QUESTIONS

1. a   3. c   5. d   7. d   9. a   11. a   13. d   15. b   17. a   19. d
21. c   23. b   25. b   27. a   29. b   31. c   33. e   35. d   37. b

PROBLEMS

1. $R_1 = 100$ kΩ $R_2 = 300$ kΩ $R_3 = 1$ M
3. a. 49 µV   b. 2 V
5. −3 V
7.

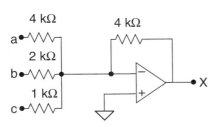

9. 0 V
11.

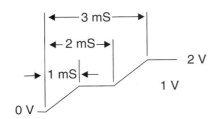

13. 7 mS

15.

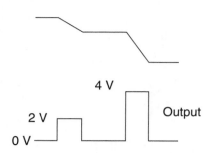

17.

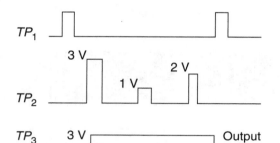

19.

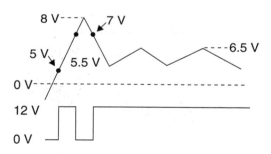

21.

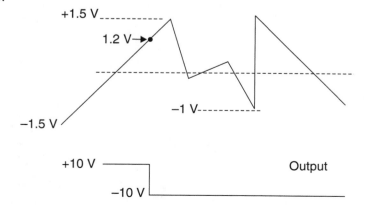

23.

CHAPTER 12

QUESTIONS

1. b   3. a   5. d   7. d   9. c   11. d   13. b   15. a   17. b   19. b
21. c   23. d   25. a   27. c   29. d   31. a   33. a   35. a   37. c   39. d
41. d

PROBLEMS

1. $V_R = 5.6$ V $V_{xc} = 5.6$ V
3. $V_{out(cutoff)} = 7.07$ V $V_{out(1dec)} = 1$ V $V_{out(2dec)} = 0.1$ V
5. $f_{(cutoff)} = 1$ kHz   Roll-off = 20 dB/dec

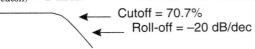

Cutoff = 70.7%
Roll-off = −20 dB/dec

7. $f_{(cutoff)} = 3$ kHz   Roll-off = 20 dB/dec

Cutoff = 70.7% ⟶
Roll-off = −20 dB/dec ⟶

9. $f_{(cutoff)} = 1.5$ kHz   Roll-off = 60 dB/dec or 18 dB/oct
11. $f_o = 100$ kHz $BW = 60$ kHz $Q = 1.7$ Roll-off = 12 dB/oct
13. 2.8 V
15. f
17. b
19. d
21. $f_o = 3.56$ kHz $Q = 3.16$ $BW = 1.13$ kHz
    $f_1 = 3$ kHz $f_2 = 4.12$ kHz
23. $V_B = 2.7$ V $V_E = 2$ V $V_C = 15$
    $V_{CE} = 13$ V $f_o = 563$ kHz
    $BW = 6$ kHz $f_1 = 560$ kHz $f_2 = 566$ kHz

CHAPTER 13

QUESTIONS

1. b   3. b   5. c   7. c   9. c   11. c   13. c   15. a   17. d   19. b
21. c   23. a   25. a   27. d   29. b

PROBLEMS

1.

TP_1 4 V_{p-p}, 796 Hz

1. (continued)

TP_2 4 V$_{p-p}$, 796 Hz

TP_3 12 V$_{p-p}$, 796 Hz

3.

TP_1 14 V$_{p-p}$

TP_2 4.7 V$_{p-p}$

TP_3 4.7 V$_{p-p}$

5. $R_x = 58$ kΩ
7. 1592 Hz
9. 6
11. $V_B = 2.7$ V $V_E = 2$ V $V_c = 10.4$ V $V_{CE} = 8.4$ V
13. 411 kHz
15. 20

CHAPTER 14

QUESTIONS

1. d   3. c   5. b   7. c   9. a   11. d   13. c   15. a   17. b   19. d
21. a   23. d   25. d   27. a

PROBLEMS

1.

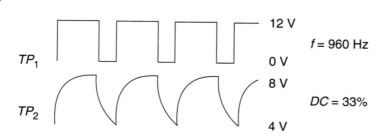

12 V

$f = 960$ Hz

TP_1 0 V

8 V

$DC = 33\%$

TP_2

4 V

3. $R_a = 36 \text{ k}\Omega$ $R_b = 18 \text{ k}\Omega$

5.

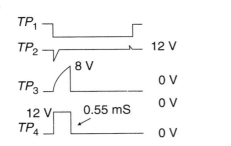

7.

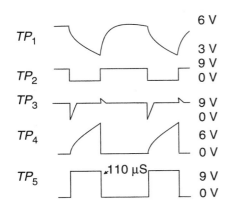

9.

11. Reduce the power supply voltages to +/−11 V or use a zener diode to limit the output swing.

13. 2 MHz

15.

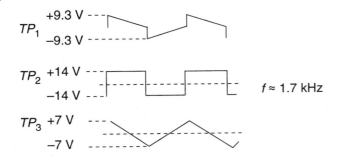

CHAPTER 15

QUESTIONS

1. c   3. b   5. c   7. c   9. c   11. a   13. c   15. c   17. d   19. d
21. a   23. d   25. d   27. d   29. b

PROBLEMS

1. 0 V

3.

7.5 V$_{p-p}$

5. $V_{out(signal)} = 400$ mV$_{p-p}$ $V_{out(noise)} = 40$ µV$_{p-p}$
7. $V_{out(signal)} = 8$ V$_{p-p}$ $V_{out(noise)} = 0.7$ mV$_{p-p}$
9.

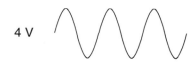

4 V

11. $A_v = 46$ $gm = 9.9$ mS
13. $M_1 = 10$ mA $M_2 = 20$ mA
15.

9.1 V

6.1 V

17.

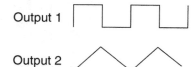

Output 1

Output 2

19.

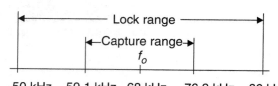

Lock range

Capture range

f_o

50 kHz   59.1 kHz   68 kHz   76.9 kHz   86 kHz

21. 68 kHz

CHAPTER 16

QUESTIONS

1. c   3. a   5. b   7. b   9. b   11. b   13. c   15. d   17. a   19. d
21. d   23. c   25. a   27. a   29. b   31. a

PROBLEMS

1. 65%
3. 16%
5. a. 60 µA   b. 113 mA
7. $TP_1 = 17.5$ V DC, 25 V$_{p-p}$ $TP_2 = 12.5$ V DC, 25 V$_{p-p}$
   $TP_3 = 15$ V DC, 25 V$_{p-p}$ $TP_4 = 0$ V DC, 25 V$_{p-p}$
   $z_{in} = 9$ k $P_{out} = 19.5$ W $A_p = 2,241$

9. $TP_1 = -49.3$ V DC, 0.5 V$_{p\text{-}p}$ $TP_2 = -1.4$ V DC, 75 V$_{p\text{-}p}$
     $TP_3 = 1.4$ V DC, 75 V$_{p\text{-}p}$ $TP_4 = 0$ V DC, 75 V$_{p\text{-}p}$

11. 2,820

13. 29%

15. 63%

17. 4.5 W

19. 1800 W

CHAPTER 17

QUESTIONS

1. c   3. e   5. b   7. g   9. d   11. d   13. a   15. d   17. d   19. d
21. d   23. a   25. a   27. d   29. c   31. c   33. c   35. b   37. b

PROBLEMS

1. a. = 15 A   b. ≈ 2,250 W

3. ≈ 8 W

5.

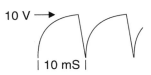

7. 1 kHz

9.

11. 6 W

13. ------------------------------- 0 V

15.

CHAPTER 18

QUESTIONS

1. c   3. b   5. c   7. b   9. c   11. d   13. c   15. e   17. b   19. c
21. a   23. d   25. a   27. b   29. c   31. d   33. d   35. c

PROBLEMS

1. 2.7%
3. Many times manufacturers state load regulation as a plus/minus value with respect to rated output. The answer to problem #1 would equal +/−2.7%, but since the rated value in problem #2 is at midrange, the load regulation can be stated as +/−1.35%.
5. a. 2%   b. 0.083%/V
7. 100%
9. 129 mW
11. 70%
13. 42 mA
15.

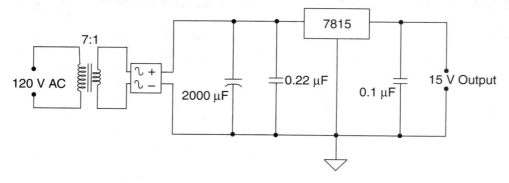

17. 1.2 V to 26.2 V
19.

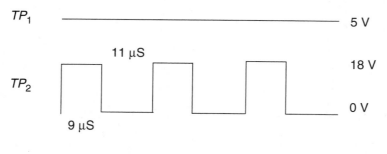

21. 14,000 μF
23. a. 21%   b. 90%

CHAPTER 19

QUESTIONS

1. d   3. c   5. a   7. b   9. b   11. b   13. a   15. b   17. b   19. d
21. d   23. d   25. c

PROBLEMS

1. a. 10.9 V   b. 1.37 V   c. 2.58 V   d. 4.12 V
3. a. 0010 0011   b. 0111 0101   c. 1001 1110   d. 1110 1011
5. 512
7. 7.8 mV
9.

CHAPTER 20

QUESTIONS

1. b   3. b   5. a   7. d   9. d   11. d   13. c   15. a   17. c   19. a
21. c   23. b   25. d   27. d   29. b   31. b   33. d

PROBLEMS

1.

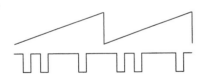

3.

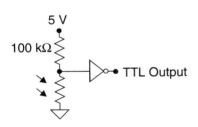

5.

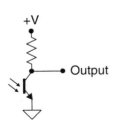

7. The circuit will continue to conduct.
9. 0.3 M
11. 42°
13. 93 M

CHAPTER 21

QUESTIONS

1. d   3. d   5. d   7. c   9. b   11. a   13. d   15. d   17. c   19. a
21. d   23. c   25. c   27. c   29. d   31. b   33. c   35. d   37. a   39. c
41. a

PROBLEMS

1. a. 291° K   b. 64.4°F
3. a. 25 mV   b. 20 mV   c. 21 mV   d. 42 mV
5. 61 mV
7. $V_{ab} = -1.1$ V
9. $V_{ab} = 0.27$ V
11. $I_{pull-in} = 120$ mA $I_{hold} = 30$ mA $P_{pull-in} = 1.44$ W $P_{hold} = 0.036$ W
13. 78%
15. 94%
17. 1140 rpm

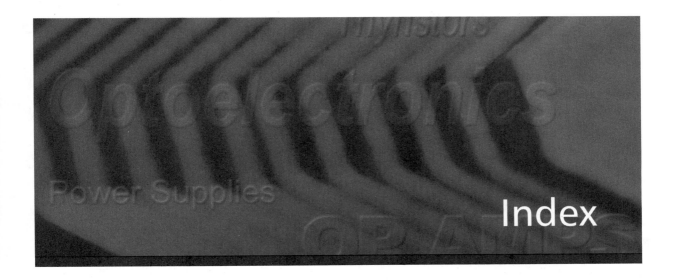

Index